james halliday

australia & new zealand

wine companion

2000 EDITION

james halliday

australia & new zealand

wine companion

2000 EDITION

GRUB STREET • LONDON

This edition published in 1999 by
Grub Street
The Basement
10 Chivalry Road
London
SW11 1HT
Email: food@grubstreet.co.uk

A CIP record for this book is available from the British Library

ISBN 1 902304 38 1

Printed in Australia

contents

how to use this book

The *Wine Companion* is arranged with wineries in alphabetical order, and the entries should be self-explanatory, but here I will briefly take you through the information for each entry.

winery entries

cape mentelle ★★★★★
Off Wallcliffe Road, Margaret River, WA 6285 **region** Margaret River
ph (08) 9757 3266 **fax** (08) 9757 3233 **open** 7 days 10–4.30
winemaker John Durham **prod.** 50 000 **est.** 1970
prod. range ($20–44 R) Chardonnay, Semillon Sauvignon Blanc, Cabernet Sauvignon, Cabernet Merlot, Shiraz, Zinfandel, Trinders Cabernet Merlot.
summary Notwithstanding majority ownership by Veuve Clicquot, David Hohnen remains very much in command of one of Australia's foremost medium-sized wineries. Exceptional marketing skills and wine of the highest quality, with the back-up of New Zealand's Cloudy Bay, are a potent combination. The Chardonnay and Semillon Sauvignon Blanc are among Australia's best, the potent Shiraz usually superb, and the berry/spicy Zinfandel makes one wonder why this grape is not as widespread in Australia as it is in California.

winery name Cape Mentelle

Although it might seem that stating the winery name is straightforward, this is not necessarily so. To avoid confusion, wherever possible I use the name that appears most prominently on the wine label and do not refer to any associated trading name.

ratings ★★★★★

The winery star system may be interpreted as follows:

★★★★★ Outstanding winery regularly producing exemplary wines.
★★★★☆ Extremely good; virtually on a par with a five-star winery.
★★★★ Consistently produces high-quality wines.
★★★☆ A solid, reliable producer of good wine.
★★★ Typically good, but may have a few lesser wines.
★★☆ Adequate.
★★ Hard to recommend.

If the ratings seem generous, so be it. The fact is that Australia is blessed with a marvellous climate for growing grapes, a high degree of technological skill, and a remarkable degree of enthusiasm and dedication on the part of its winemakers. Across the price spectrum, Australian wines stand tall in the markets of the world. I see no reason, therefore, to shrink from recognising excellence. NR = not rated, either because the winery is new or because I have not tasted enough of its wines.

address Off Wallcliffe Road, Margaret River, WA 6285
ph (08) 9757 3266 **fax** (08) 9757 3233

The details are usually those of the winery and cellar door but in a few instances may simply be of the winery; this occurs when the wine is made at another winery under contract and is sold only through retail.

region Margaret River

The mapping of Australia into Zones and Regions with legally defined boundaries is now well underway. This edition sees radical changes (and additions) to the regional names and boundaries. Wherever possible the official 'Geographic Indication' name has been adopted, and where the registration process is incomplete, I have used the most likely name. Occasionally you will see 'Warehouse' as the region. This means the wine is made from purchased grapes in someone else's winery. In other words, it does not have a vineyard or winery home in the ordinary way.

cellar door sales hours **open** 7 days 10–4.30

Although a winery might be listed as not open or only open on weekends, some may in fact be prepared to open by appointment. Many will, some won't; a telephone call will establish whether it is possible or not. Also, virtually every winery that is shown as being open only for weekends is in fact open for public holidays as well. Once again, a telephone call will confirm this.

winemaker John Durham

In the large companies the winemaker is simply the head of a team; there may be many executive winemakers actually responsible for specific wines.

prod. 50 000

This figure given (representing the number of cases produced each year) is merely an indication of the size of the operation. Some wineries (principally but not exclusively the large companies) regard this information as confidential; in that event, NFP (not for publication) will appear. NA = information was not available.

year of establishment **est.** 1970

A more or less self-explanatory item, but keep in mind that some makers consider the year in which they purchased the land to be the year of establishment, others the year in which they first planted grapes, others the year they first made wine, others the year they first offered wine for sale, and so on. There may also be minor complications where there has been a change of ownership or a break in production.

price range and prod. range ($20–44 R) Chardonnay, Semillon Sauvignon Blanc, Cabernet Sauvignon, Cabernet Merlot, Shiraz, Zinfandel, Trinders Cabernet Merlot.

The **price range** given covers the least expensive through to the most expensive wines usually made by the winery in question (where the information was available). Hence there may be a significant spread. That spread, however, may not fully cover fluctuations that occur in retail pricing, particularly with the larger companies. Erratic and often savage discounting remains a feature of the wine industry, and prices must therefore be seen as approximate.

The Australian winery prices are for purchase in Australia, in Australian dollars; those for New Zealand are for purchase in New Zealand, in New Zealand dollars.

I have indicated whether the price is cellar door (CD), mailing list (ML) or retail (R). By and large, the choice has been determined by which of the three methods of sale is most important to the winery. The price of Australian and New Zealand wines in other countries is affected by a number of factors, including excise and customs duty, distribution mark-up and currency fluctuations. Contact the winery for details.

product range Particularly with the larger companies, it is not possible to give a complete list of the wines. The saving grace is that these days most of the wines are simply identified on their label by their varietal composition.

summary Notwithstanding majority ownership by Veuve Clicquot, David Hohnen remains very much in command of one of Australia's foremost medium-sized wineries. Exceptional marketing skills and wine of the highest quality, with the back-up of New Zealand's Cloudy Bay, are a potent combination. The Chardonnay and Semillon Sauvignon Blanc are among Australia's best, the potent Shiraz usually superb, and the berry/spicy Zinfandel makes one wonder why this grape is not as widespread in Australia as it is in California.

My summary of the winery. Little needs to be said, except that I have tried to vary the subjects I discuss in this part of the winery entry.

❦ The vine leaf symbol indicates wineries that are new entries in this year's listing.

wine entries and tasting notes

Cape Mentelle Cabernet Sauvignon
The wine which started the Cape Mentelle juggernaut, with the '82 and '83 vintages winning the Jimmy Watson Trophy in successive years. Both style and quality wandered somewhat in the second half of the 1980s, but has steadied (and improved greatly) in the 1990s as David Hohnen has come to terms with the now fully mature vineyard (and it with him).

🍷🍷🍷🍷🍸 **1995** Medium to full red-purple; a bouquet with uncommon depth and complexity, and a range of secondary earthy/cedary/berry characters already starting to appear. A wine with similarly good structure and depth to the palate although the flavours are tending more towards the savoury end of the spectrum than the opulently fruity. **rating:** 90

best drinking 2000–2010 **best vintages** '76, '78, '82, '83, '86, '90, '91, '93, '94, '95 **drink with** Loin of lamb • $43.20

wine name Cape Mentelle Cabernet Sauvignon

In most instances, the wine's name will be prefaced by the name of the winery.

ratings 🍷🍷🍷🍷🍸

Two ratings are given for each wine; the ratings apply to the vintage reviewed, and may vary from one year to the next.

Points scale	Glass symbol	
98–100	–	Perfection which exists only as an idea.
94–97	🍷🍷🍷🍷🍷	As close to perfection as the real world will allow.
90–93	🍷🍷🍷🍷🍸	Excellent wine full of character; of gold medal standard.
85–89	🍷🍷🍷🍷	Very good wine; clear varietal definition/style; silver verging on gold medal standard.
80–84	🍷🍷🍷🍸	Good fault-free, flavoursome; high bronze to silver medal standard.

You will see that nearly all of the wines reviewed in this book rate 84 points (3½ glasses) or better. This is not wanton generosity on my part. It simply reflects the fact that the 1000 or so wines selected for specific review are the tip of more than 5000 tasting notes accumulated over the past year. In other words, the wines described are among Australia's top 20 per cent. NR = not rated.

background The wine which started the Cape Mentelle juggernaut, with the '82 and '83 vintages winning the Jimmy Watson Trophy in successive years. Both style and quality wandered somewhat in the second half of the 1980s, but has steadied (and improved greatly) in the 1990s as David Hohnen has come to terms with the now fully mature vineyard (and it with him).

Like the summary information given in the winery entries, I have tried to vary the approach of my discussions.

🍷🍷🍷🍷🍷 **1995** Medium to full red-purple; a bouquet with uncommon depth and complexity, and a range of secondary earthy/cedary/berry characters already starting to appear. A wine with similarly good structure and depth to the palate although the flavours are tending more towards the savoury end of the spectrum than the opulently fruity. **rating:** 90

The tasting note opens with the vintage of the wine tasted. With the exception of a very occasional classic wine, this tasting note will have been made within the 12 months prior to publication. Even that is a long time, and during the life of this book the wine will almost certainly change. More than this, remember that tasting is a highly subjective and imperfect art. NV = non-vintage.

best drinking 2000–2010

I will usually give a range of years or a more specific comment (such as 'quick-developing style'), but whatever my best drinking recommendation, always consider it with extreme caution and as an approximate guide at best. When to drink a given wine is an intensely personal decision, which only you can make.

best vintages '76, '78, '82, '83, '86, '90, '91, '93, '94, '95

Self-explanatory information, but a note of caution: wines do change in the bottle, and it may be that were I to taste all of the best vintages listed, I would demote some and elevate some not mentioned.

drink with Loin of lamb

Again, merely a suggestion – a subliminal guide to the style of wine.

price • $43.20

This is a recommended retail price only. NA = information not available.

Abbreviation: mlf = malolactic fermentation

australian wineries and wines

wine regions of australia

key to regions

1 Lower Hunter Valley
2 Upper Hunter Valley
3 Hastings River
4 Mudgee
5 Orange
6 Cowra
7 Murray Darling and Swan Hill
8 Riverina
9 Pericoota
10 Hilltops
11 Canberra District
12 Tumbarumba
13 Shoalhaven
14 Far South West Victoria
15 Grampians
16 Pyrenees
17 Ballarat
18 Bendigo
19 Goulburn Valley
20 Central Victorian High Country
21 Glenrowan
22 Rutherglen
23 King Valley
24 Ovens Valley
25 Gippsland
26 Mornington Peninsula
27 Yarra Valley
28 Geelong
29 Sunbury
30 Macedon Ranges
31 Northern Tasmania
32 Southern Tasmania
33 Mount Gambier
34 Robe
35 Coonawarra
36 Wrattonbully
37 Mount Benson
38 Padthaway
39 Langhorne Creek
40 McLaren Vale
41 Adelaide Hills
42 Eden Valley
43 Adelaide Plains
44 Barossa Valley
45 Riverland
46 Clare Valley
47 Port Lincoln
48 Great Southern
49 Pemberton
50 Blackwood Valley
51 Margaret River
52 Geographe
53 South-west Coast
54 Perth Hills
55 Swan District
56 Burnett Valley
57 Granite Belt

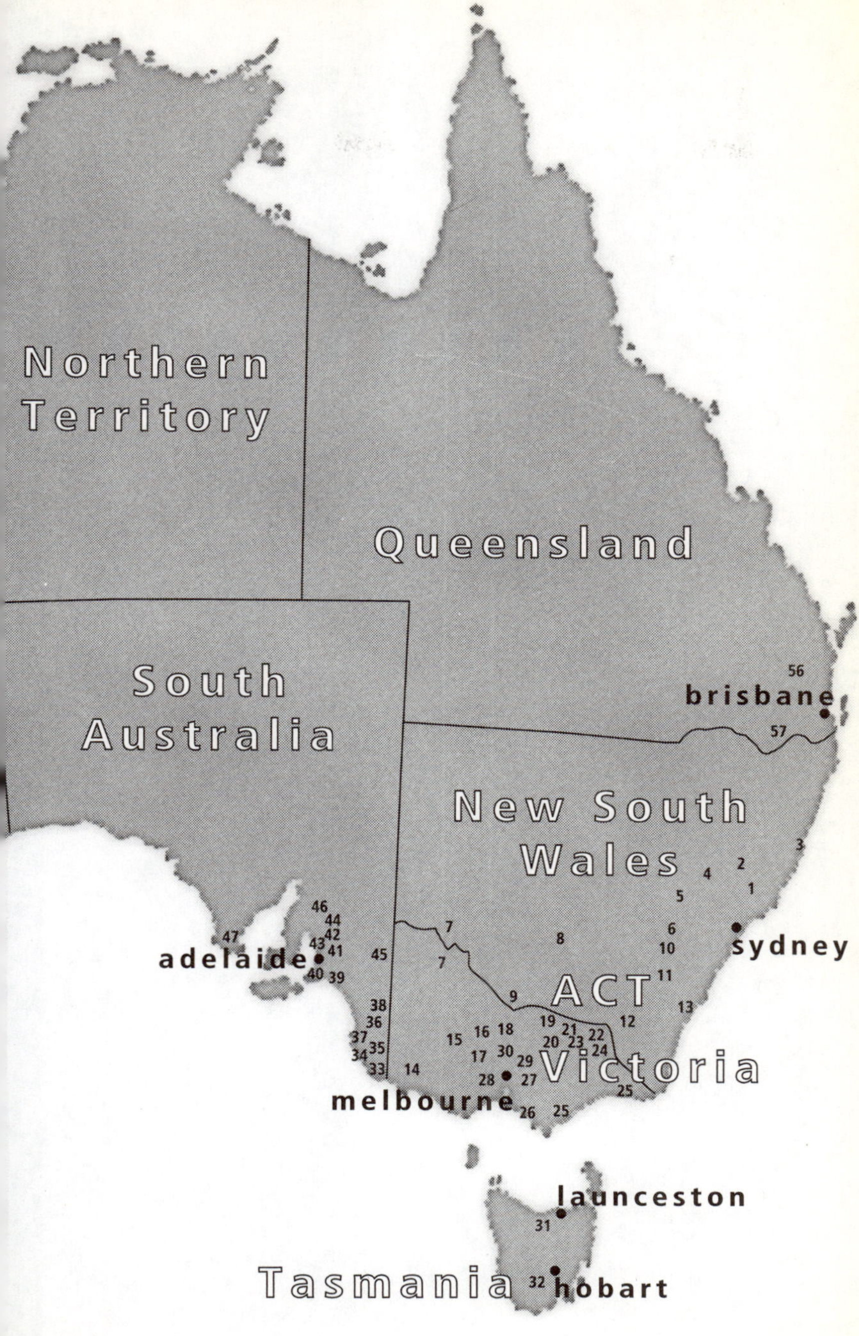
Northern Territory
Queensland
South Australia
New South Wales
ACT
Victoria
Tasmania
brisbane
sydney
adelaide
melbourne
launceston
hobart
1 2 3 4 5 6 7 8 9 10 11 12 13 14 15 16 17 18 19 20 21 22 23 24 25 26 27 28 29 30 31 32 33 34 35 36 37 38 39 40 41 42 43 44 45 46 47 56 57

12 acres ★★☆

Nagambie–Rushworth Road, Bailieston, Vic 3608 **region** Goulburn Valley
ph (03) 5794 2020 **fax** (03) 5794 2020 **open** Thur–Mon 10–6, July weekends only
winemaker Peter Prygodicz, Jana Prygodicz **prod.** 650 **est.** 1994
prod. range ($14–16 CD) Shiraz, Merlot, Cabernet Sauvignon.
summary The charmingly named 12 Acres is a red wine specialist, with Peter and Jana Prygodicz making the wines on site in a tiny winery. The wines could benefit from renewal of the oak in which they are matured, for they are all quite astringent.

abbey vale ★★★★

Wildwood Road, Yallingup, WA 6282 **region** Margaret River
ph (08) 9755 2121 **fax** (08) 9755 2286 **open** 7 days 10–5
winemaker Dorham Mann **prod.** 21 000 **est.** 1986
prod. range ($14–28 CD) Festival White, Semillon, Dry Verdelho, Sauvignon Blanc, Sunburst Verdelho, Chardonnay, Merlot Shiraz, Cabernet Merlot, Cabernet Sauvignon, Reserve Cabernet Sauvignon; Moonshine Ale brewed on the premises.
summary Abbey Vale has gone from strength to strength in recent years, vinifying an ever-increasing proportion of the production from its large 30-hectare vineyard (plus another 60 under contract), and winning a significant number of show awards.

abercorn NR

Cassilis Road, Mudgee, NSW 2850 **region** Mudgee
ph (02) 6373 3106 **fax** (02) 6373 3108 **open** By appointment
winemaker John Baruzzi (Contract) **prod.** NA **est.** 1996
prod. range ($16.50–19.50 CD) Chardonnay, Cabernet Sauvignon.
summary Tim and Connie Stevens acquired the 25-year-old Abercorn Vineyard in 1996 which, while admirably located next door to Huntington Estate, had become somewhat run down. It is being rejuvenated, and the first wines (from 1997) were contract-made by John Baruzzi. The Stevens plan to build their own winery and to be making the wines within the next three years.

ada river ★★★

Main Road, Neerim South, Vic 3831 **region** Gippsland
ph (03) 5628 1221 **fax** (03) 5466 2333 **open** 10–6 weekends and public holidays
winemaker Peter Kelliher, Chris Kelliher **prod.** 1000 **est.** 1983
prod. range ($12–18 CD) From Gippsland-grown grapes Chardonnay and Cabernet Sauvignon; from Yarra Valley grapes Traminer, Chardonnay and Pinot Noir; and Pinot Noir drawn from both regions.
summary The Kelliher family first planted vines on their dairy farm at Neerim South in 1983, extending the original Millstream Vineyard in 1989, and increasing plantings yet further by establishing the nearby Manilla Vineyard in 1994. The family also has the Goondalahg Vineyard at Steels Creek in the Yarra Valley under long-term lease, thus providing two distinct wine ranges. Wine production began in 1991, the first wines going on sale in 1995.

Ada River Yarra Traminer

A delicate wine showing the Yarra Valley's ability to produce fine Gewurztraminer which age well.
🍷🍷🍷🍷 **1997** Light yellow-green; the bouquet is light, but has quite clear lychee and lime aromas. The palate likewise has spice, lychee and lime flavours running through to a clean but soft finish.

rating: 86

best drinking 1999–2003 **best vintages** NA **drink with** Asian prawns • $12

affleck NR

154 Millynn Road off Gundaroo Road, Bungendore, NSW 2621 **region** Canberra District
ph (02) 6236 9276 **fax** (02) 6236 9090 **open** Weekends and public holidays or by appointment
winemaker Ian Hendry **prod.** 200 **est.** 1976
prod. range ($16–18 CD) Chardonnay, Pinot Noir, Semillon, Sweet White, Cabernet Shiraz, Muscat, Ruby Port.
summary The cellar door and mail order price list says that the wines are 'grown, produced and bottled on the estate by Ian and Susie Hendry with much dedicated help from family and friends'. The original 2.5-hectare vineyard was virtually doubled by the planting of an additional 2 hectares of merlot in 1998, and it is hoped to have a new tasting room (offering light lunches) open by the end of 1999.

aldinga bay winery NR

Main South Road, Aldinga, SA 5173 **region** McLaren Vale
ph (08) 8556 3179 **fax** (08) 8556 3350 **open** 7 days 10–5
winemaker Nick Girolamo **prod.** 5000 **est.** 1979
prod. range ($7.80–14.80 CD) Chardonnay, Sauvignon Blanc, Riesling, Cabernet Sauvignon, Shiraz, Cabernet Franc, Merlot, Port.
summary The former Donolga Winery has had a name and image change since Nick Girolamo, the son of founders Don and Olga Girolamo, returned from Roseworthy College with a degree in Oenology. Nick Girolamo has taken over both the winemaking and marketing; prices remain modest, though not as low as they once were, reflecting an increase in the quality and an upgrade in packaging. Aldinga Bay also has some very interesting varietal plantings, 12 in all, including 1 hectare each of petit verdot and sangiovese.

alkoomi ★★★★★

Wingeballup Road, Frankland, WA 6396 **region** Great Southern
ph (08) 9855 2229 **fax** (08) 9855 2284 **open** 7 days 10.30–5
winemaker Michael Staniford, Merv Lange **prod.** 50 000 **est.** 1971
prod. range ($13–50 R) Classic White, Riesling, Chardonnay, Sauvignon Blanc, Late Harvest Riesling, Classic Red, Malbec, Shiraz, Cabernet Sauvignon, Sparkling Alkoomi, Tawny Port.
summary For those who see the wineries of Western Australia as suffering from the tyranny of distance, this most remote of all wineries shows there is no tyranny after all. It is a story of unqualified success due to sheer hard work, and no doubt to Merv and Judy Lange's aversion to borrowing a single dollar from the bank. The substantial production is entirely drawn from the ever-expanding estate vineyards, which by 1999 amounted to 50 hectares. Wine quality across the range is impeccable, always with precisely defined varietal character.

Alkoomi Riesling

As with all of the Alkoomi wines, produced from estate-grown fruit. Yet another example of the symbiotic relationship between Mount Barker and Riesling, for it is usually an excellent wine.

🍷🍷🍷🍷 **1998** Very pale straw–green; the fragrant bouquet has aromas of lime and herb leading into a crisp, clean palate with bracingly fresh acid. Good now, but will become significantly better with time in bottle. **rating:** 87

best drinking 2002–2008 **best vintages** '94, '95, '96, '98 **drink with** Salad • $17

Alkoomi Sauvignon Blanc

Alkoomi has 2.5 hectares of sauvignon blanc, and was one of the first vineyards in the region to experiment with the variety. Some of the early vintages lacked varietal character, but every year since 1995 have displayed great depth of flavour and strong varietal character.

🍷🍷🍷🍷🍷 **1998** 1998 may have been a tough year for sauvignon blanc in South Australia, but posed no problems in the west. There are voluminous passionfruit and gooseberry characters on a highly aromatic but not sweaty bouquet; similar flavours come through on a crisp, long and sustained palate. Alkoomi has this variety down pat. **rating:** 92

best drinking 1999–2000 **best vintages** '95, '96, '97, '98 **drink with** Ginger prawns • $19

Alkoomi Wandoo Semillon Sauvignon Blanc

The current releases from Alkoomi in the Frankland River region of Western Australia underline both the quality of the region and in particular that of the 25-year-old vineyards of Alkoomi (not to mention first class winemaking). A new arrival is the '97 Wandoo Semillon Sauvignon Blanc, with crisp and powerful herb, nettle and more grassy flavours, a long finish, and a subliminal touch of oak.

🍷🍷🍷🍷🍷 **1997** Potent and powerful fruit aromas running through the herbal/grassy spectrum, aided by subtle oak, lead directly into a powerful, crisp and fresh palate, with a long finish, and oak just perceptible in the background. **rating:** 90

best drinking 1998–2003 **best vintages** '97 **drink with** Mussels • $30

Alkoomi Classic White

An anonymous blend, with the varieties not specified, and probably not relevant, although one would suspect the presence of both Sauvignon Blanc and Riesling.

🍷🍷🍷🍷 **1998** Pale, almost white colour; a highly aromatic, scented and striking passionfruit bouquet is followed by a similarly high-flavoured gooseberry, herb and passionfruit palate. Italian white wine drinkers would simply hate this. **rating:** 88

best drinking 1999–2000 **best vintages** '98 **drink with** Smoked chicken salad • $15

Alkoomi Frankland River Chardonnay

Produced from 5 hectares of estate plantings, and barrel-fermented and matured in a mixture of oak, predominantly French Nevers and Vosges from Seguin Moreau.

🍷🍷🍷🍷🍷 **1997** A wine which has come on impressively since first tasted in February 1998, with quite lovely melon, citrus and pear fruit aromas, soft and clean; the palate is as elegant as the bouquet promises, with predominantly citrus and stone fruit flavours, and some melon. Subtle spicy oak throughout. **rating:** 93

best drinking 1998–2002 **best vintages** '85, '88, '90, '92, '94, '97 **drink with** Stir-fried chicken with cashew nuts • $21

Alkoomi Blackbutt

A new super-premium release from Alkoomi, first made in 1994 from a blend of Cabernet Sauvignon, Malbec and Merlot, with the first blend made after the components had already been in barrel for 20 months, thereafter being transferred to 100% new French oak for a further eight months before bottling. Only very small quantities are made.

🍷🍷🍷🍷🍷 **1995** Dense red-purple; the bouquet is rich, complex and concentrated, with luscious dark berry and dark chocolate fruit, followed by an equally rich and luscious palate, charged with blackberry, currant and chocolate fruit. Positive but balanced cedary oak adds lustre to a superb wine. **rating:** 96

best drinking 2000–2015 **best vintages** '94 **drink with** Rare beef • $50

allandale ★★★★

Lovedale Road, Lovedale, NSW 2320 **region** Lower Hunter Valley
ph (02) 4990 4526 **fax** (02) 4990 1714 **open** Mon–Sat 9–5, Sun 10–5
winemaker Bill Sneddon, Steve Langham **prod.** 15 000 **est.** 1978

prod. range ($16–22 R) Hilltops Riesling, Late Picked Hilltops Riesling, Hilltops Semillon, Semillon, Sauvignon Blanc, Chardonnay, Chardonnay Semillon, Lombardo (Pinot Noir Shiraz blend), Matthew Shiraz, Mudgee Cabernet Sauvignon, William Méthode Champenoise.

summary Unostentatious, medium-sized winery which has been under the control of winemaker Bill Sneddon for well over a decade. Has developed something of a reputation as a Chardonnay specialist, but does offer a broad range of wines of good quality, with an increasing number of wines produced from grapes grown in the Hilltops region. The wines are exported to the United Kingdom, US and Switzerland.

Allandale Hunter Valley Semillon

Utilises 2 hectares of estate plantings, part of which are used for Semillon Sauvignon Blanc and Chardonnay Semillon blends. The best component is released as a straight varietal, made in the classic style without oak influence. An interesting comment on the back label of the '97 says 'The vintage conditions were both difficult and challenging. Past history has shown the wetter years often produce the best Semillons.' A true statement.

🍷🍷🍷🍷 **1998** Light straw-green; the bouquet is clean, of moderate intensity with overtones of herb. The palate shows good varietal character, predominantly grassy but with some citrus and stone fruit flavours. **rating:** 86

best drinking 2000–2004 **best vintages** '86, '91, '94, '96, '97 **drink with** Summer salads • $16

Allandale Hunter River Chardonnay

Draws upon 3 hectares of estate plantings. At its best, the wine shows excellent use of a mix of French and American oak. The wine is given extended lees contact and taken through partial malolactic fermentation.

🍷🍷🍷🍷🍷 **1998** Medium to full yellow-green; rich and complex barrel-ferment oak aromas are wound around tangy fruit on the bouquet. Overall, a very good example of uninhibited use of clove/spice oak because the fruit is there to support it. **rating:** 90

best drinking 1999–2000 **best vintages** '91, '94, '96, '98 **drink with** Smoked salmon • $18

allanmere NR

Allandale Road, Allandale via Pokolbin, NSW 2321 **region** Lower Hunter Valley

ph (02) 4930 7387 **fax** (02) 4930 7900 **open** 7 days 9.30–5

winemaker Greg Silkman **prod.** 7000 **est.** 1984

prod. range ($15–20 CD) Gold Label Chardonnay, Semillon, Trinity White (Chardonnay, Semillon, Sauvignon Blanc), Cabernet Sauvignon, Trinity Red (Cabernet blend), Cabernet Shiraz. Durham Chardonnay is top-of-the-range Chardonnay.

summary Now owned by Monarch Winemaking Services. While it has a relatively low profile in conventional retail markets, cellar-door sales are flourishing in response to the ever-increasing tourist traffic in the Hunter Valley. No recent tastings. Exports to Japan, Switzerland, US, Taiwan, UK.

allinda NR

119 Lorimers Lane, Dixons Creek, Vic 3775 **region** Yarra Valley

ph (03) 5965 2450 **fax** (03) 5965 2467 **open** Weekends and public holidays 10–6

winemaker Al Fencaros **prod.** 2000 **est.** 1991

prod. range ($16.50–24 CD) Riesling, Sauvignon Blanc, Chardonnay, Late Harvest Riesling, Cabernets.

summary Winemaker Al Fencaros is a graduate of Bachelor of Wine Science from Charles Sturt University, and was formerly employed by De Bortoli in the Yarra Valley. All of the Allinda wines are produced on site; all except the Shiraz (from Heathcote) are estate-grown from a little over 3 hectares of vineyards. Limited retail distribution in Melbourne and Sydney.

all saints ★★★★☆

All Saints Road, Wahgunyah, Vic 3687 **region** Rutherglen
ph (02) 6033 1922 **fax** (02) 6033 3515 **open** Mon–Sat 9–5.30, Sun 10–5.30
winemaker Peter Brown **prod.** NFP **est.** 1864
prod. range ($9.50–62 CD) Riesling, Chenin Blanc, Muscadelle, Chardonnay, Marsanne, Orange Muscat (a winery specialty), Aleatico, Merlot, Shiraz, Cabernet Sauvignon, Late Harvest Semillon, Late Picked Muscadelle, Sparkling. The real focus is on Classic Release Muscat and Tokay and on Show Reserve Muscat and Tokay.
summary The winery rating principally reflects the Show Reserve fortified wines, but the table wines are more than adequate. An excellent winery restaurant makes this a compulsory and most enjoyable stop for any visitor to the northeast. All Saints and St Leonards are now wholly owned by Peter Brown; the vast majority of the wines are sold through cellar door and by mailing list.

All Saints Marsanne

Produced from estate plantings which are now of considerable age. It follows the Chateau Tahbilk approach, rather than that of Mitchelton, in presenting a stainless steel-fermented, early bottled style.

🍷🍷🍷🍷 **1998** Medium yellow-green; there is above-average fruit concentration to the honeysuckle aromas; the palate likewise shows crystal clear honeysuckle fruit, with a nice grip to the finish. Virtually guaranteed to flower given time in bottle. **rating:** 86

best drinking 2000–2005 **best vintages** NA **drink with** Sweetbreads • $15.50

All Saints Classic Release Tokay

An exceptionally good wine at the price, with an average blend age of eight years. The varietal definition is excellent, the blend a delightful amalgam of old and young material.

🍷🍷🍷🍷🍷 **NV** Light golden-brown; rich and sweet malt and tea-leaf aromas with barely perceptible fortifying spirit. The palate is of medium to full weight, with malty/toffee/caramel/tea-leaf flavours, finishing with well-balanced acidity. **rating:** 94

best drinking 1999–2009 **best vintages** NA **drink with** Try it as an aperitif • $19.80

All Saints Show Reserve Tokay

While some of the best stocks of All Saints were sold to other purchasers before Brown Brothers completed the acquisition of the entire property, certain of the very best material remained. This has in turn formed the base of this show blend, which has an average age of 20 years, and which has already accumulated a large number of trophies and gold medals.

🍷🍷🍷🍷🍷 **NV** Deep golden-brown; wonderfully concentrated and rich caramel, toffee and molasses aromas lead on to a sensuously luscious and complex toffee, tea-leaf and butterscotch-flavoured palate. **rating:** 96

best drinking 1999–2009 **best vintages** NA **drink with** A meal in itself • $62

All Saints Classic Release Muscat

The blend has an average age of ten years, and, even if not showing the same outstanding varietal character as its sister wine, the Liqueur Tokay, is an impressive wine, attesting to the depth of the fortified stocks still held at All Saints. Won two trophies at the 1998 Sydney International Wine Competition, including the Millers Trophy for runner-up to Best Wine of Show. With a price of $19.80 retail, it was placed in the top ten Best Value Wines.

🍷🍷🍷🍷🍷(half) **NV** Medium tawny with just a hint of brown on the rim; soft, raisiny varietal muscat with fractionally earthy spirit. The palate is quite luscious, with good raisined fruit, starting sweet and finishing with cleansing acidity. **rating:** 90

best drinking 1999–2009 **best vintages** NA **drink with** Coffee and chocolates • $19.80

amberley estate ★★★★

Thornton Road, Yallingup, WA 6282 **region** Margaret River
ph (08) 9755 2288 **fax** (08) 9755 2171 **open** 7 days 10–4.30
winemaker Eddie Price, Greg Tilbrook **prod.** 60 000 **est.** 1986
prod. range ($14–32 CD) Semillon Sauvignon Blanc, Classic Margaret River Semillon Chardonnay, Semillon, Chardonnay, Chenin Blanc, Nouveau, Cabernet Merlot, Shiraz.
summary Based its initial growth on the basis of its ultra-commercial, fairly sweet Chenin Blanc which continues to provide the volume for the brand, selling out well prior to the following release. However, the quality of all of the other wines has risen markedly over recent years as the 31 hectares of estate plantings have become fully mature. Production has virtually doubled over the past few years, and exports to the UK and Japan have been established.

Amberley Estate Semillon

Typically spends some months in French oak, and some wild yeast fermentations are incorporated.
🍷🍷🍷🍷🍸 **1998** Medium yellow-green; the bouquet is quite intense and complex, with just a whisker of mineral or matchbox which progressively dissipated as the wine aired. It has excellent mouthfeel, flavour and length, with that extra horsepower which is unique to Margaret River.
rating: 93
best drinking 1999–2004 **best vintages** '95, '98 **drink with** Roast chicken • $21

anderson NR

Lot 12 Chiltern Road, Rutherglen, Vic 3685 **region** Rutherglen
ph (02) 6032 8111 **open** 7 days 10–5
winemaker Howard Anderson **prod.** 1000 **est.** 1993
prod. range ($11–23 CD) Semillon, Chenin Blanc, Doux Blanc, Chardonnay, Soft Cabernet, Shiraz, Cabernet Merlot, Cabernet Sauvignon, Late Harvest Tokay, Pinot Chardonnay, Fortifieds.
summary With a winemaking career spanning 30 years, including a stint at Seppelt Great Western, Howard Anderson and family have started their own winery, ultimately intending to specialise in sparkling wine made entirely on site.

Anderson Riesling

Competently made, no-frills style, probably made from grapes grown in the King Valley.
🍷🍷🍷🍷 **1997** Medium green-yellow; the bouquet is starting to develop secondary characters with an attractive touch of toast, while the wine offers plenty of soft generosity in the mouth making it a good food style. **rating:** 87
best drinking 1999–2002 **best vintages** '97 **drink with** Kassler • $12.50

andrew garrett ★★★★

Kangarilla Road, McLaren Vale, SA 5171 **region** McLaren Vale
ph (08) 8323 8853 **fax** (08) 8323 8271 **open** 7 days 10–4
winemaker Phillip Reschke, Charles Hargreaves **prod.** 20 000 **est.** 1983
prod. range ($15 R) Semillon, Sauvignon Blanc, Chardonnay, Cabernet Merlot, Bold Shiraz, Vintage Pinot Chardonnay.
summary Effectively another brand in the Mildara Blass wine group, with many of the wines now not having a sole McLaren Vale source, but instead drawn from regions across south-eastern Australia. Over the past few years, winemaker Phillip Reschke has produced some excellent wines which provide great value for money, particularly for the smooth, peachy Chardonnay. Exports and distribution via Mildara Blass.

Andrew Garrett Chardonnay

A blend of 55% Padthaway, 30% McLaren Vale and 15% Cowra fruit barrel-fermented in a mix of American and French oak hogsheads. Over the past few vintages has been particularly good, a testimonial to the winemaking skills of Phillip Reschke.

1997 Medium yellow-green; astonishing depth and complexity to the tangy, ripe melon-accented bouquet; the palate likewise has good flavour, dipping slightly in the middle, but picking up again on the finish. Clever oak handling. The winner of two trophies and two gold medals. **rating:** 90

best drinking 1999–2000 **best vintages** '95, '97 **drink with** Loin of pork • $15

Andrew Garrett Bold Shiraz

I'm not too sure about the brand name for this wine, as it recalls a Kaiser Stuhl Bold Red which failed dismally 20 years ago. This is in fact a far better wine than either the name or the price would suggest, and what is more has been given Rolls Royce oak treatment in new French and American oak barrels for 12 months.

1997 Excellent purple-red; bright and fresh cherry and earth varietal fruit on both bouquet and palate continue the style established by this wine in previous vintages. Oak plays only a minor role. **rating:** 84

best drinking 1999–2000 **best vintages** NA **drink with** Barbecued red meat • $15

andrew garrett vineyard estates NR

McLarens on the Lake, Kangarilla Road, McLaren Vale, SA 5171 **region** McLaren Vale

ph (08) 8323 8911 **fax** (08) 8323 9010 **open** 7 days 10–5

winemaker Andrew Garrett **prod.** NA **est.** 1986

prod. range ($9.70–32 CD) There are three premium wine ranges: from the Yarra Valley in Victoria comes the Yarra Glen label, from the Adelaide Hills in South Australia the Springwood Park label, and Marandoo Run Cabernet Sauvignon from Langhorne Creek. In much lower price and quality category come the McLarens on the Lake blended wines of Colombard Semillon Chardonnay, Chardonnay, Cabernet Shiraz, Grenache Shiraz and Brut Cuvée.

summary The irrepressible Andrew Garrett has risen once again after 20 years in the wine industry as 'winemaker, innovator, entrepreneur, marketer and personality' (to use his own words). Andrew Garrett Vineyard Estates is now the umbrella for the Yarra Valley-based Yarra Glen label, the Adelaide Hills-based Springwood Park label and for the far longer established and more humble McLarens on the Lake range. 1997 saw the first releases under the Yarra Glen and Springwood Park labels, with three good wines.

Andrew Garrett Vineyard Estates Yarra Glen Chardonnay

Produced from the 100-hectare Yarra Valley vineyard established by the Andrew Garrett Group in the mid-1990s. The wine is not given any oak treatment but is given some skin contact to give the wine extra weight, and has quite high acidity. Shows considerably more character and strength than most unwooded Chardonnays.

1997 Medium yellow-green; the bouquet is smooth with abundant but soft yellow peach aromas. The palate is fine and unforced, yet quite rich in both flavour and texture. A far better unoaked Chardonnay than most. **rating:** 85

best drinking 1998–2004 **best vintages** NA **drink with** Baby calamari • $18

Andrew Garrett Vineyard Estates Springwood Park Chardonnay

Produced from grapes grown by a number of independent growers in the Adelaide Hills, including the Piccadilly Valley. It is barrel-fermented in French barriques, and matured on yeast lees for six months.

🍷🍷🍷🍷 **1997** Medium yellow-green; with sweet and quite intense peach fruit aromas. The sweet peach flavours continue on the full-blooded palate, which is slightly cloying and betrays its 14.3% alcohol. Well handled oak undoubtedly helps the wine, but it is quite pricey. **rating:** 83

best drinking 1999–2001 **best vintages** NA **drink with** Smoked salmon risotto • $30

Andrew Garrett Vineyard Estates Springwood Park Pinot Noir

The wine is made from grapes provided by contract growers in the Piccadilly Valley subregion of the Adelaide Hills. It is matured in a mix of Troncais, Alliers and Nevers oak casks.

🍷🍷🍷🍷 **1997** Medium red-purple; the bouquet has a mix of plum, earth and forest aromas which provide authentic varietal character. It falls away slightly on the rather earthy and tannic palate, which veers towards medium-bodied dry red, rather than fine Pinot Noir. Nonetheless, a wine of considerable character. **rating:** 83

best drinking 1998–2001 **best vintages** NA **drink with** Confit of duck • $32

andrew harris vineyards ★★★★☆

Sydney Road, Mudgee, NSW 2850 **region** Mudgee

ph (02) 6373 1213 **fax** (02) 6373 1296 **open** Not

winemaker Frank Newman **prod.** 12 000 **est.** 1991

prod. range ($16–45 R) Premium Chardonnay, Shiraz, Premium Cabernet Sauvignon; at the top come Reserve Chardonnay, Reserve Shiraz and Reserve Cabernet Sauvignon; super premium is Shiraz Cabernet Sauvignon The Vision.

summary Andrew and Debbie Harris have lost no time since purchasing a 300-hectare sheep station southeast of Mudgee in 1991. The first 6 hectares of vineyard were planted in that year, and have since been expanded to 90 hectares. A substantial portion of the production is sold to others, but right from the first vintage limited quantities of high-quality wines have been made under the Andrew Harris label (by Simon Gilbert as contract-winemaker) which have deservedly enjoyed considerable show success. Production has more than doubled over the past 18 months, and Frank Newman, with 30 years experience as a winemaker, is now in charge.

Andrew Harris Premium Shiraz

The second tier of the Andrew Harris range, but 100% estate-grown and given good oak.

🍷🍷🍷🍷🍷 **1997** Medium to full red-purple; the clean, moderately intense bouquet offers a neat marriage between sweet fruit and oak, the flavours of cherry and mint, touched with chocolate, coming through clearly on the well-balanced full-flavoured palate. **rating:** 90

best drinking 2001–2006 **best vintages** '97 **drink with** Butterfly leg of lamb • $15

Andrew Harris Reserve Shiraz

Like the Premium Shiraz, estate-grown, but in this instance the pick of the crop. The wine spends 18 months in new oak, but the concentration of the fruit is such that the wine remains in balance.

🍷🍷🍷🍷🍷 **1997** Dark red-purple; the bouquet opens with powerful, sweet dark berry fruit with lots of smoothly integrated oak. The palate, likewise, is concentrated, with ripe dark cherry fruit, well-handled oak and full but soft tannins on the finish. **rating:** 94

best drinking 2002–2012 **best vintages** '97 **drink with** Rich casserole • $22

Andrew Harris Reserve Cabernet Sauvignon

The upper tier of the two Cabernets, which is 100% estate-grown and utilises the best of the 20 hectares of cabernet sauvignon grapes. Neither the '97 Premium nor the '96 Reserve could be described as overoaked, which is an unambiguously good feature of the two. On the other hand, the fruit is less intense than that to be found in the Shiraz.

🍷🍷🍷🍷 **1996** Medium red, with just a touch of purple remaining; the bouquet is sweet, with earthy cabernet varietal character, but not concentrated. The pleasant earthy, chocolatey, berry palate is nicely balanced, but not in the same league as the Shiraz. **rating:** 86

best drinking 2000–2005 **best vintages** NA **drink with** Osso bucco • $22

angove's ★★★☆

Bookmark Avenue, Renmark, SA 5341 **region** Riverland
ph (08) 8595 1311 **fax** (08) 8595 1583 **open** Mon–Fri 9–5
winemaker Garry Wall **prod.** 1 million **est.** 1886
prod. range ($3.15–45 CD) Classic Reserve Riesling, Colombard, Sauvignon Blanc, Chardonnay, Shiraz, Mondiale Shiraz Cabernet, Cabernet Sauvignon; Mondiale (White), Mondiale Shiraz Cabernet; Sarnia Farm Chardonnay, Cabernet Sauvignon; Floreate; Cheaper Butterfly Ridge varietals and Misty Vineyards generics; also specialist Brandy producer; Fortifieds.
summary Exemplifies the economies of scale achievable in the Australian Riverland without compromising potential quality. Very good technology provides wines which are never poor and which can sometimes exceed their theoretical station in life. The white varietals are best. Angove's expansion into Padthaway has resulted in estate-grown premium wines at the top of the range.

Angove's Sarnia Farm Cabernet Sauvignon

A new venture for Angove's which has hitherto produced almost all of its wines from its Nanya Vineyard in the Riverland, making occasional Limited Release/Winemaker Selection with grapes purchased from various premium areas. 1993 was the inaugural vintage. As with the Chardonnay, gained the name 'Sarnia Farm' with the second release, and really impresses with its most recent release. It may well be that a maturing vineyard and/or better viticulture is playing a role.

🍷🍷🍷🍷 **1996** Medium to full red-purple; an attractive bouquet with sweet cassis and cedar aromas is followed by a palate which, while only of medium weight, is quite luscious, with sweet cassis/berry fruit supported by sweet oak. Soft tannin finish. **rating:** 87

best drinking 1999–2004 **best vintages** '93, '96 **drink with** Beef Provençale • $13.15

antcliff's chase NR

RMB 4510, Caveat via Seymour, Vic 3660 **region** Central Victorian High Country
ph (03) 5790 4333 **fax** (03) 5790 4333 **open** Weekends 10–5
winemaker Chris Bennett, Ian Leamon **prod.** 800 **est.** 1982
prod. range ($14–30 CD) Riesling, Chardonnay, Pinot Noir, Cabernet Franc.
summary A small family enterprise which commenced planting the vineyards at an elevation of 600 metres in the Strathbogie Ranges in 1982, but which has only recently commenced wine production from the 4-hectare vineyard. As the scarecrow label indicates, birds are a major problem for remote vineyards such as this.

apsley gorge vineyard ★★★★☆

Rosedale Road, Bicheno, Tas 7215 **region** Southern Tasmania
ph (03) 6375 1221 **fax** (03) 6375 1589 **open** By appointment
winemaker Andrew Hood (Contract) **prod.** 1500 **est.** 1988
prod. range ($22 ML) Chardonnay, Pinot Noir.

summary While nominally situated at Bicheno on the east coast, Apsley Gorge is in fact some distance inland, taking its name from a mountain pass. Clearly, it shares with the other east coast wineries the capacity to produce Chardonnay and Pinot Noir of excellent quality, with skilled winemaking by Andrew Hood doing the rest. Retail distribution through Sutherland Cellars, Melbourne.

aquila estate ★★★

85 Carabooda Road, Carabooda, WA 6033 **region** Swan District
ph (08) 9561 5415 **fax** (08) 9561 5415 **open** Not
winemaker Andrew Spencer Wright **prod.** 19 500 **est.** 1993
prod. range ($9.99–18.99 R) Semillon, Sauvignon Blanc, Boyup Brook Chardonnay, Reflections (white blend), Cabernet Sauvignon, Flame (red blend).
summary As Aquila Estate has matured, so have its grape sources centred on the Margaret River (principally) and Boyup Brook. The white wines are quite attractive, particularly the Margaret River Sauvignon Blanc and Boyup Brook Chardonnay.

arlewood estate NR

Harmans Road South, Willyabrup, WA 6284 **region** Margaret River
ph (08) 9755 6267 **fax** (08) 9755 6267 **open** Fri–Mon 11–4, 7 days during school holidays
winemaker Jurg Muggli (Contract) **prod.** 1800 **est.** 1988
prod. range ($14–27 CD) Semillon, Sauvignon Blanc Semillon, Margaret River Classic, Liaison (sweet), Cabernet Sauvignon Reserve, Port.
summary Liz and John Wojturski have expanded their initial plantings of 2.5 hectares to over 9 hectares, and now have limited distribution through agents in Melbourne, Sydney, country New South Wales and Perth.

Arlewood Estate Semillon

With a vineyard as well sited as that of Arlewood, it makes sense to let the fruit quality do the talking, as it does here.

🍷🍷🍷🍷🍸 **1998** Light yellow-green; the bouquet is clean, with that grassy citrussy fruit which so typifies Margaret River Semillon, with good bite and tang. The well-balanced and stylish palate ripples along with zesty varietal fruit into a lingering slightly minerally, finish. **rating:** 90

best drinking 1999–2002 **best vintages** '98 **drink with** Richly-sauced fish • $18.50

armstrong vineyards NR

Lot 1 Military Road, Armstrong, Vic 3381 **region** Grampians
ph (08) 8277 6073 **fax** (08) 8277 6035 **open** Not
winemaker Tony Royal **prod.** 1000 **est.** 1989
prod. range ($39 R) Shiraz.
summary Armstrong Vineyards is the brain- or love-child of Tony Royal, former Seppelt Great Western winemaker who now runs the Australian business of Seguin Moreau, the largest of the French coopers. Armstrong Vineyards has 6.5 hectares of shiraz, the first two hectares planted in 1989, the remainder in 1995–6. Low yields (4.5 to 5.5 tonnes per hectare) mean the wine will always be produced in limited quantities, with a likely maximum of 500 dozen per year.

Armstrong Vineyards Shiraz

The 1996 Shiraz is the initial release from Armstrong, with the '97 following in August 1999. The wine was made with a four-day pre-fermentation cold-soak, and is then fermented in small open fermenters which are hand-plunged. Fermentation is completed in French oak (70% new, 30% one-year-old, Seguin Moreau, of course) where it spends a little under two years.

ⲨⲨⲨⲨ **1997** Medium red-purple; the strong, minty aromas are very typical of Central Victoria, and similarly pronounced minty fruit dominates the palate. The structure is good, with subtle oak and fine tannins; it all depends on how much mint you like in your wine. **rating:** 86

best drinking 2001–2007 **best vintages** '96 **drink with** Yearling steak • $39

arranmore vineyard NR

Rangeview Road, Carey Gully, SA 5144 **region** Adelaide Hills
ph (08) 8390 3034 **fax** (08) 8390 3034 **open** By appointment
winemaker John Venus **prod.** 120 **est.** 1998
prod. range ($11.50–21 ML) Sauvignon Blanc, Chardonnay, Pinot Noir.
summary One of the tiny operations which are appearing all over the beautiful Adelaide Hills. The 2-hectare vineyard is planted to pinot noir, chardonnay and sauvignon blanc, and the wines are basically sold through word of mouth and mail order.

arrowfield ★★★

Denman Road, Jerry's Plains, NSW 2330 **region** Upper Hunter Valley
ph (02) 6576 4041 **fax** (02) 6576 4144 **open** 7 days 10–5
winemaker Don Buchanan **prod.** 100 000 **est.** 1968
prod. range ($13–21 R) Top-of-the-range Show Reserve Chardonnay, Semillon, Shiraz, Cabernet Sauvignon; Cowra Chardonnay, Merlot, Late Harvest Gewurztraminer; Arrowfield varietals Chardonnay, Semillon Chardonnay, Sauvignon Blanc, Traminer Riesling, Late Harvest Riesling, Shiraz, Cabernet Merlot; Sparkling and Fortifieds; also Simon Whitlam range of Semillon, Semillon Chardonnay, Shiraz, Cabernet Sauvignon.
summary After largely dropping the Arrowfield name in favour of Mountarrow and a plethora of other brands, this Japanese-owned company has come full circle, once again marketing the wines solely under the Arrowfield label. Its principal grape sources are Cowra and the Upper Hunter, but it does venture further afield from time to time. Tragically, during the 1999 vintage, the Arrowfield winery (and some of its wine stocks) were severely damaged in a $10 million fire.

arthurs creek estate ★★★★★

Strathewen Road, Arthurs Creek, Vic 3099 **region** Yarra Valley
ph (03) 9827 6629 **fax** (03) 9824 0252 **open** Not
winemaker Mitchelton (Contract), Gary Baldwin (Consultant) **prod.** 1500 **est.** 1976
prod. range ($23.65–25.35 R) Chardonnay, Cabernet Sauvignon.
summary A latter-day folly of leading Melbourne QC, S E K Hulme, who planted 1.5 hectares each of semillon, chardonnay, and cabernet sauvignon at Arthurs Creek in the mid-1970s, and commenced to have wine made by various people for 15 years before deciding to sell any of it. A ruthless weeding-out process followed, with only the best of the older vintages offered. The Cabernets from the 1990s are absolutely outstanding, deeply fruited and marvellously structured.

ashbrook estate ★★★★★

Harmans Road South, Willyabrup, WA 6284 **region** Margaret River
ph (08) 9755 6262 **fax** (08) 9755 6290 **open** 7 days 11–5
winemaker Tony Devitt, Brian Devitt **prod.** 7500 **est.** 1975
prod. range ($15–23 CD) Gold Label Riesling, Black Label Riesling, Semillon, Sauvignon Blanc, Chardonnay, Verdelho, Cabernet Sauvignon.
summary A fastidious maker of consistently outstanding estate-grown table wines but which shuns publicity and the wine show system alike, and is less well known than it deserves to be, selling much of its wine through the cellar door and by an understandably very loyal mailing list clientele. All of the white wines are of the highest quality, year in, year out. Small quantities of the wines now find their way to Japan, Singapore, Hong Kong, US and Taiwan.

Ashbrook Estate Semillon

All of the Ashbrook Estate wines are estate-grown, the Semillon from 2.6 hectares of vines. Fermented in stainless steel, and not given any time in oak, the wine shows the herbaceous style of Margaret River Semillon to best advantage.

🍷🍷🍷🍷 **1998** Light to medium yellow-green; the aromas of the bouquet are almost juicy, so strong is the citrus-accented fruit; the palate, too, is bright, fresh and elegant with juicy lime flavours neatly balanced by crisp acidity. **rating:** 88

best drinking 1999–2005 **best vintages** '87, '92, '93, '94, '95, '97 **drink with** Blanquette of veal • $15

---

asher NR

360 Goldsworthy Road, Lovely Banks, Geelong, Vic 3231 **region** Geelong
ph (03) 5276 1365 **open** Sat, public holidays 10–5, Sun 12–5
winemaker Brian Moten **prod.** Minuscule **est.** 1975
prod. range ($10 CD) Sauvignon Blanc, Cabernet Sauvignon, Malbec.
summary A tiny, semi-home-winemaking operation situated at the picturesquely named town of Lovely Banks on the outskirts of Geelong.

ashton hills ★★★★☆

Tregarthen Road, Ashton, SA 5137 **region** Adelaide Hills
ph (08) 8390 1243 **fax** (08) 8390 1243 **open** Weekends 11–5.30
winemaker Stephen George **prod.** 1750 **est.** 1982
prod. range ($16–30 R) Chardonnay, Riesling, Salmon Brut, Blanc de Blancs, Pinot Noir, Obliqua (Cabernet Merlot).
summary Stephen George wears three winemaker hats: one for Ashton Hills, drawing upon a 3.5-hectare estate vineyard high in the Adelaide Hills; one for Galah Wines and one for Wendouree. It would be hard to imagine three wineries producing more diverse styles, with the elegance and finesse of Ashton Hills at one end of the spectrum, the awesome power of Wendouree at the other. The Riesling has moved into the highest echelon. Export markets have been developed in the UK and US.

Ashton Hills Riesling

A wine of unusually consistent style, and which invariably ages slowly. The natural acidity is high, and the fruit sometimes tight and inexpressive when the wine is young, slowly opening up as it evolves in the bottle. These elegant wines are ideal to complement food.

🍷🍷🍷🍷🍷 **1998** Light green-yellow; a very pure manifestation of Riesling, with lovely, crisp lime and mineral aromas. The palate, likewise, has crystal-clear varietal fruit, lime-accented, with great length. Will develop beautifully. **rating:** 94

best drinking 2001–2011 **best vintages** '89, '90, '91, '93, '94, '96, '97, '98 **drink with** Fresh asparagus • $16

Ashton Hills Chardonnay

Like all the Ashton Hills wines, relatively slow-developing, progressively building complexity as it ages.

🍷🍷🍷🍷🍷(half) **1997** Medium to full yellow-green; while the fruit aromas of melon and some citrus are relatively discreet, the overall impression of the bouquet is quite intense, filled out by a touch of subtle oak. The palate is tight, with melon/citrus fruit and a long finish; evolving slowly, and will continue to do so. **rating:** 90

best drinking 2000–2007 **best vintages** '95, '97 **drink with** Roast veal • $22

Ashton Hills Reserve Pinot Noir

The vines at the Ashton Hills vineyard are now over 15 years old, producing better grapes year by year. Stephen George likewise has become ever more skilled in dealing with pinot noir.

🍷🍷🍷🍷🍸 **1997** Dense red-purple; a complex bouquet with ripe plum and more foresty characters is followed by a generous, ripe multiflavoured palate, typical of the Adelaide Hills, and developing well. **rating:** 92

best drinking 2000–2005 **best vintages** NA **drink with** Smoked duck • $30

Ashton Hills Obliqua Cabernet Merlot

The Ashton Hills vineyard is in a distinctly cool part of the Adelaide Hills, and the Bordeaux varieties – particularly cabernet sauvignon – struggle to achieve the flavour ripeness Australians are used to. Some would say vive la difference.

🍷🍷🍷🍷 **1998** Medium to full red-purple; there are some slight green canopy aromas, but also cedar, cigar and berry to compensate. The palate is powerful and long, sustained by tannins on the finish. For those who like austerity. **rating:** 86

best drinking 2000–2006 **best vintages** NA **drink with** Marinated beef • $26

ashwood grove NR

Wood Wood, Swan Hill, Vic 3585 **region** Murray Darling and Swan Hill

ph (03) 5030 5291 **fax** (03) 5030 5605 **open** By appointment

winemaker Andrew Peace **prod.** 50 000 **est.** 1995

prod. range ($9.95–15.95 R) Sauvignon Blanc, Chardonnay, Colombard Chardonnay, Mourvedre Shiraz, Grenache Shiraz, Shiraz, Cabernet, Merlot.

summary The Peace family has been a major Swan Hill grape grower since 1980, and moved into winemaking with the opening of a $3 million winery in 1997. The modestly priced wines are aimed at supermarket-type outlets in Australia and, in particular, at the export market in the major destinations for Australian wine.

ashworths hill NR

Ashworths Road, Lancefield, Vic 3435 **region** Macedon

ph (03) 5429 1689 **fax** (03) 5429 1689 **open** 7 days 10–6

winemaker Anne Manning **prod.** 100 **est.** 1982

prod. range ($12–20 CD) Macedon Ranges Cabernet Sauvignon is the flagship; Victorian Riesling and Chardonnay also available.

summary Peg and Ken Reaburn offer light refreshments throughout the day, the property offering scenic views of the Macedon Ranges.

auldstone ★★★☆

Booths Road, Taminick via Glenrowan, Vic 3675 **region** Glenrowan

ph (03) 5766 2237 **open** Thur–Sun 9–5

winemaker Michael Reid **prod.** 2500 **est.** 1987

prod. range ($10–21 CD) Riesling, Traminer Riesling, Chardonnay, Late Picked Riesling, Shiraz, Cabernet, Herceynia Tawny Port, Boweya Muscat, Sparkling Shiraz.

summary Michael and Nancy Reid have restored a century-old stone winery and have replanted the largely abandoned 24-hectare vineyard around it. Gourmet lunches are available on weekends.

austin's barrabool ★★★★

50 Lemins Road, Waurn Ponds, Vic 3221 **region** Geelong

ph (03) 5241 8114 **fax** (03) 5241 8122 **open** By appointment

winemaker John Ellis (Contract), Pamela Austin **prod.** 2500 **est.** 1982

prod. range ($15–25 CD) Riesling, Chardonnay, Reserve Shiraz, Cabernet Sauvignon.

summary Pamela and Richard Austin have quietly built their business from a tiny base, but which is now poised for much bigger things. The vineyard has been extended to 16.4 hectares, and instead of selling part of the production to others, they intend to significantly increase wine production under their own label – it has already doubled between 1998 and 1999. Wine quality, too, has risen sharply, with high-quality wines being skilfully made by John Ellis at Hanging Rock.

Austin's Barrabool Riesling

A wine which distinguished itself (on my score sheet, at least) at the 1998 *Winewise* Small Makers Competition.

🍷🍷🍷🍷 **1997** Medium yellow-green; the bouquet is light but crisp, with fine lime and toast aromas, the palate with interesting lime/lemon tingle flavours; good weight, length and balance. Retasted December 1998 and developing surely but slowly. **rating:** 87

best drinking 1999–2003 **best vintages** NA **drink with** Marinated scallops • $15

Austin's Barrabool Chardonnay

The wine is made from the Waurn Ponds Vineyard of Austin's, from vines which are now 15 years old. Skilled winemaking is very evident in the wine.

🍷🍷🍷🍷🍷 **1997** Light to medium yellow-green; the bouquet is clean, and quite fresh, with pronounced nutty/creamy malolactic fermentation influences. The same characters show on the palate, which has very nice feel and structure, once again with emphatic mlf characters, yet these do not strip the wine of fruit flavour and are not coarse. **rating:** 90

best drinking 1999–2004 **best vintages** '97 **drink with** Pan-fried scallops • $22

Austin's Barrabool Reserve Shiraz

Estate-grown, and the only wine in the roster to attract the Reserve status on the label. It comes from both the older Waurn Ponds Vineyard and the newer, north-facing slopes of the Southerns Creek Vineyard.

🍷🍷🍷🍷🍷 **1997** Great purple-red colour, full but clear. The bouquet is concentrated, with a marvellous array of blackberry, liquorice, forest and earthy fruit supported by subtle oak. There is more of the same attractive varietal fruit on the concentrated blackberry-flavoured palate. Fine tannins and good length mark a wine of distinction, with further improvement in front of it. **rating:** 93

best drinking 2001–2010 **best vintages** '97 **drink with** Fillet of kangaroo • $39

avalon vineyard ★★☆

RMB 9556 Whitfield Road, Wangaratta, Vic 3678 **region** King Valley
ph (03) 5729 3629 **fax** (03) 5729 3635 **open** 7 days 10–5
winemaker Doug Groom **prod.** 1000 **est.** 1981

prod. range ($9–18 CD) Chardonnay, Sauvignon Blanc Semillon, Cabernet Sauvignon, Pinot Noir, Pinot Noir Méthode Champenoise.

summary Avalon Vineyard is situated in the King Valley, 4 kilometres north of Whitfield. Much of the production from the 10-hectare vineyard is sold to other makers, with limited quantities made by Doug Groom, a graduate of Roseworthy, and one of the owners of the property.

avalon wines NR

1605 Bailey Road, Glen Forrest, WA 6071 **region** Perth Hills
ph (08) 9298 8049 **open** By appointment
winemaker Lyndon Crockett **prod.** 100 **est.** 1986

prod. range ($10–14 CD) Chardonnay, Semillon, Cabernet Sauvignon.

summary One of the newer wineries in the Perth Hills, drawing upon three-quarters of a hectare each of chardonnay, semillon and cabernet sauvignon.

bacchus estate NR

381 Milbrodale Road, Broke, NSW 2330 **region** Lower Hunter Valley
ph (02) 6579 1069 **fax** (02) 6579 1069 **open** By appointment
winemaker Andrew Margan (Contract) **prod.** 1000 **est.** 1993
prod. range ($14–20 CD) Chardonnay, Unwooded Chardonnay, Shiraz.
summary Another of the wineries springing up like mushrooms after spring rain in the Hunter Valley, in this instance at Broke. A little over 10 hectares of vineyard have been established, with a further 12 hectares to be planted in the near future. As the new plantings mature, production will increase significantly.

baileys of glenrowan ★★★★★

Cnr Taminick Gap Road and Upper Taminick Road, Glenrowan, Vic 3675 **region** Glenrowan
ph (03) 5766 2392 **fax** (03) 5766 2596 **open** Mon–Fri 9–5, weekends 10–5
winemaker Allan Hart **prod.** 5000 **est.** 1870
prod. range ($15–50 CD) Classic Chardonnay, Riesling, Touriga, Shiraz, 1920's Block Shiraz are the principal wines; Warby Range, Founders and Winemaker's Selection Tokay and Muscat; Phantom's Lake Chardonnay and Shiraz were introduced in 1997 in fancy bottles.
summary Now part of the sprawling Mildara Blass empire, inherited via the Rothbury takeover. Has made some excellent Shiraz in recent years, but its greatest strength lies in its fortified wines. It is for these wines that the winery rating is given. Exports and distribution via Mildara Blass.

Baileys 1920's Block Shiraz

Baileys was founded in 1870 by Varley Bailey, and in 1920 – to celebrate the 50th anniversary – a special block of shiraz was planted. It is from these vines that the wine is made, first produced in 1991 and immediately proclaiming its class. Only 900 cases of the wine are produced each year.

🍷🍷🍷🍷🍷 **1997** Medium to full red-purple; very ripe black cherry fruit is accompanied by some vanilla oak on the bouquet. The sweet, ripe and concentrated palate finishes with soft, lingering tannins; the oak is in balance. **rating** 90

best drinking 2002–2007 **best vintages** '91, '92, '93, '96, '97 **drink with** Rare rump steak, venison • $24

Baileys Founder Liqueur Tokay

The midpoint of the Tokays produced by Baileys, coming between Warby Range at the bottom end and Winemaker's Selection at the top end. In the totally distinctive, rich and sweet Baileys' style, with more accent on complexity and less on primary fruit than that of Morris; one of the great classics.

🍷🍷🍷🍷 **NV** Medium golden brown; full, complex sweeter style of Tokay with clean spirit. The palate is rich, full and textured with flavours of butterscotch and sweet biscuit, and a chewy texture, but finishing long and clean. **rating** 89

best drinking 1999–2000 **best vintages** NA **drink with** Winter aperitif; summer after dinner • $18

Baileys Winemaker's Selection Old Liqueur Tokay

Made in very limited quantities, and these days released on strict allocation as the popularity of these very old wines has deservedly grown. Made from muscadelle (as are all northeastern Victorian Tokays) and is preferred by many winemakers to Muscat because of its greater elegance.

🍷🍷🍷🍷🍷 **NV** Deep brown; the bouquet shows obvious barrel-aged rancio characters, very complex but still retaining good varietal character. The wine has outstanding structure in the mouth, with complexity apparent immediately the wine is tasted. The flavours run in the cold tea/butterscotch/brandysnap spectrum, finishing with cleansing acidity. **rating:** 94

best drinking 1999–2000 **best vintages** NA **drink with** After coffee; alternative to Cognac • $50

Baileys Founder Liqueur Muscat

Made from Brown Frontignac, otherwise known as Brown Muscat, and the best known of the fortified wines of northeastern Victoria. As one would expect, in the rich Baileys mould. One of four Muscats produced, starting at the bottom with Warby Range, then Founder Liqueur, then Gold Label and ultimately Winemaker's Selection.

🍷🍷🍷🍷 **NV** Medium red-brown; an arresting bouquet with hints of spice to the sweet, ripe, complex fruit. The same unusual spicy/cinnamon aspects are apparent on the rich and complex palate. **rating:** 88

best drinking 1999–2000 **best vintages** NA **drink with** After coffee; alternative to Cognac • $18

Baileys Winemaker's Selection Old Liqueur Muscat

For many, Muscat is the greatest expression of northeastern Victorian fortified wines. This particular wine shows the essential blending skills, with a balance of very old and younger material, simultaneously providing complexity and freshness.

🍷🍷🍷🍷🍷 **NV** Dark brown tending to olive on the rim; complex, faintly earthy/spirity aromas, with an almost nutty edge to the raisined muscat fruit. A tremendously powerful and concentrated wine in the mouth with excellent balance and a wonderful finish. Iron fist in a velvet glove. **rating:** 94

best drinking 1999–2009 **best vintages** NA **drink with** After coffee; alternative to Cognac • $50

baldivis estate ★★☆

Lot 165 River Road, Baldivis, WA 6171 **region** South West Coast

ph (08) 9525 2066 **fax** (08) 9525 2411 **open** Mon–Fri 10–4, weekends, holidays 11–5

winemaker Jane Gilham **prod.** 6000 **est.** 1982

prod. range ($9.95–28.95 CD) Wooded and Unwooded Chardonnay, Classic White, Late Picked Semillon, Blue Rock Pinot Noir Cabernet, Cabernet Merlot, Cabernet Sauvignon Reserve.

summary Part of a very large mixed horticultural enterprise on the Tuart Sands of the coastal plain. There is ample viticultural and winemaking expertise; although the wines are pleasant, soft and light-bodied, they tend to lack concentration.

Baldivis Estate Chardonnay

The '97 Balidivis Estate wines were all competently made, and had good fruit. Unfortunately, the same could not be said for the '98 vintage wines.

🍷🍷🍷🍷 **1997** Light green-yellow; the bouquet is fresh, with light citrus and melon fruit and just a hint of oak. The palate follows on logically, with zesty, lively melon and grapefruit flavours and a nicely handled touch of spicy oak. **rating:** 86

best drinking 1998–1999 **best vintages** NA **drink with** Western Australian crayfish • $22

Baldivis Estate Nobile

An interesting wine which specifies neither variety nor region on the label, and which seems as if it may have been made in the manner used to produce the Porphyrys of bygone days. This involved the addition of sweet concentrate to already fermented wine; the alcohol of 11 degrees lends some indirect support to this assumption. Regardless, it is an attractive wine.

🍷🍷🍷🍷🍷 **1997** Deep gold-yellow; the bouquet is rich and sweet, the palate with similarly rich honey, peach and honeycomb flavours supported by subtle oak. **rating** 90

best drinking 1998–2000 **best vintages** NA **drink with** Fruit tart • $25

bald mountain ★★★☆

Hickling Lane, Wallangarra, Qld 4383 **region** Granite Belt
ph (07) 4684 3186 **fax** (07) 4684 3433 **open** 7 days 10–5
winemaker Simon Gilbert (Contract) **prod.** 5000 **est.** 1985
prod. range ($11–19 CD) Classic Queenslander (in fact 100% Sauvignon Blanc), Chardonnay, Late Harvest Sauvignon Blanc, Shiraz, Shiraz Cabernet.
summary Denis Parsons is a self-taught but exceptionally competent vigneron who has turned Bald Mountain into the viticultural showpiece of the Granite Belt. In various regional and national shows since 1988, Bald Mountain has won almost 70 show awards, placing it at the forefront of the Granite Belt wineries. The two Sauvignon Blanc-based wines, Classic Queenslander and the occasional non-vintage Late Harvest Sauvignon Blanc, are interesting alternatives to the mainstream wines. Future production will also see grapes coming from new vineyards near Tenterfield just across the border in New South Wales.

balgownie estate ★★★★

Hermitage Road, Maiden Gully, Vic 3551 **region** Bendigo
ph (03) 5449 6222 **fax** (03) 5449 6506 **open** Mon–Sat 10.30–5
winemaker Lindsay Ross **prod.** 12 000 **est.** 1969
prod. range ($20 R) Estate-produced Shiraz and Cabernet Sauvignon; Premier Cuvée (second, non-estate label), Chardonnay and Cabernet Shiraz.
summary It is ironical that, just at the time when Balgownie Estate was returning to some of its former glory with its Shiraz and Cabernet Sauvignon (the 1996 and 1997 vintages are excellent) it should be offered for sale by Mildara Blass, and will no doubt be in different ownership by the time this book appears in print. Exports and distribution via Mildara Blass.

Balgownie Estate Shiraz

Produced from fully-mature estate vineyards established in the early 1970s, and with a proud history dating back to that time. After an uncertain period in the second half of the 1980s and early '90s, has returned to form in no uncertain fashion.

🍷🍷🍷🍷🍷 **1997** Dense red-purple; the bouquet is rich and concentrated, with dark berry, prune and chocolate aromas. The palate is lusciously rich and concentrated, with lingering tannins and well-integrated oak. **rating:** 93

best drinking 2007–2017 **best vintages** '91, '93, '96, '97 **drink with** Beef in red wine • $20

Balgownie Estate Cabernet Sauvignon

The best known of the Balgownie wines in the height of its glory days of the mid-1970s, partly because there were so few small-winemaker Cabernet Sauvignons around at the time. After a period in the wilderness, has bounced back to top form.

🍷🍷🍷🍷🍷 **1997** Dense red-purple; brooding blackberry and prune fruit aromas with some vanilla oak lead into a powerful, rich palate with essencey fruit, sweet tannins and vanilla oak. A milkshake style, but very good. **rating:** 93

best drinking 2002–2012 **best vintages** '75, '76, '80, '94, '96, '97 **drink with** Beef casserole • $20

ballandean estate ★★★

Sundown Road, Ballandean, Qld 4382 **region** Granite Belt
ph (07) 4684 1226 **fax** (07) 4684 1288 **open** 7 days 9–5
winemaker Mark Ravenscroft **prod.** 9000 **est.** 1970

prod. range ($8.50–28 CD) Riesling, Semillon, Semillon Sauvignon Blanc, Black Label Sauvignon Blanc, Sylvaner Late Harvest, White Pearl (semi sweet white), Lambrusco, Granite Range Shiraz, Black Label Shiraz, Merlot, Port, Sparkling.

summary The senior winery of the Granite Belt, and by far the largest. The white wines are of diverse but interesting styles, the red wines smooth and usually well made. The estate specialty Sylvaner Late Harvest is a particularly interesting wine of good flavour. No recent tastings.

balnarring ★★★☆

62 Bittern-Dromana Road, Balnarring, Vic 3926 **region** Mornington Peninsula

ph (03) 5983 5258 **open** 7 days 10–4

winemaker Bruce Paul, Stan Paul **prod.** 1200 **est.** 1982

prod. range ($11–15 ML) Chardonnay, Riesling, Gewurztraminer, Pinot Noir, Merlot, Cabernet Merlot.

summary Over the years, the wines of Balnarring have been made at various wineries under contract, but have shown a consistent vineyard style, with the red wines in particular possessing exceptional colour and depth of flavour. Winemaking is now carried out by owners Bruce and Stan Paul.

balnaves of coonawarra ★★★★

Riddoch Highway, Coonawarra, SA 5263 **region** Coonawarra

ph (08) 8737 2946 **fax** (08) 8737 2945 **open** Mon–Fri 9–5, weekends 10–5

winemaker Peter Bissell **prod.** 10 000 **est.** 1975

prod. range ($18–28 R) Chardonnay, Sparkling Cabernet, The Blend (Merlot Cabernet Franc), Shiraz, Cabernet Merlot, Cabernet Sauvignon.

summary Former Hungerford Hill vineyard manager and now viticultural consultant-cum-grape grower Doug Balnaves established his vineyard in 1975, but did not launch into winemaking until 1990, with colleague Ralph Fowler as contract-maker in the early years. A striking new 300-tonne winery was built and was in operation for the 1996 vintage, with former Wynns Coonawarra Estate assistant winemaker Peter Bissell in charge. The expected leap in quality has indeed materialised with the 1996 and subsequent vintages. The wines are sold through Majestic Wines in the UK, and are also exported to Japan.

Balnaves Chardonnay

Produced from a careful selection from part of the estate vineyards. The clarified juice is barrel-fermented in new Seguin Moreau French oak hogsheads followed by lees stirring for two months, followed by a further two months oak maturation before relatively early bottling.

🍷🍷🍷🍷 **1998** Light to medium yellow-green; slightly sawdusty oak over light melon and nectarine fruit on the bouquet is a wobbly start, but the palate is well-weighted and balanced, with melon and fig fruit, and the oak better integrated. **rating:** 86

best drinking 2000–2002 **best vintages** '92, '93, '94, '96 **drink with** Robe crayfish • $18

Balnaves Shiraz

Only produced when seasonal conditions permit; not made, for example, in either 1994 or 1995. Selected from estate-grown grapes from the highest and stoniest part of the Balnaves shiraz plantings. The fermentation in a vinimatic is completed in new and second-use American oak barrels.

🍷🍷🍷🍷🍷(half) **1997** Medium to full purple-red; the bouquet is quite fragrant, with gentle spice and red cherry fruit supported by good oak. A nicely flavoured and balanced wine, with vanilla oak making a positive contribution, and finishing with soft tannins. **rating:** 90

best drinking 2000–2010 **best vintages** '93, '96, '97 **drink with** Cotechino sausages • $18

Balnaves Cabernet Merlot

A blend of 75% Cabernet Sauvignon and 25% Merlot, matured in used French and American hogsheads for 12 months. Until 1991, the wine was fairly light in style, suggesting high yields. If the '96 sets the pattern for the future, all will be well.

YYYYY **1996** Medium red-purple; a strongly herbaceous bouquet with minty/leafy/berry fruit, and virtually identical flavours on the palate. Has balance and nice tannins. Top gold 1998 National Wine Show. **rating:** 92

best drinking 2002–2006 **best vintages** '90, '91, '92, '96 **drink with** Calf's liver Italian style • $23

Balnaves Cabernet Sauvignon

Drawn from a little over 16 hectares of estate plantings, the majority of the grapes are sold to other leading Coonawarra winemakers. As with the Cabernet Merlot, vintages since 1991 have been on the light side, and one also has to wonder whether there is an element of over-cropping. The '96 won the trophy for Best Varietal Cabernet in Show at the 1998 Royal Sydney Wine Show.

YYYYY **1996** Medium to full red-purple; the best of the Cabernet family wines produced at Balnaves in 1996, with sweet cassis berry fruit and subtle oak. There is plenty of concentration to the cassis/red berry fruit on the palate, supported by fine, lingering tannins and the same subtly sweet oak as the bouquet. Retasted March 1999 with similar points and rating. **rating:** 94

best drinking 2001–2005 **best vintages** '90, '91, '95, '96 **drink with** Veal chops • $28

bannockburn vineyards ★★★★★

Midland Highway, Bannockburn, Vic 3331 **region** Geelong
ph (03) 5281 1363 **fax** (03) 5281 1349 **open** Not
winemaker Gary Farr **prod.** 8000 **est.** 1974
prod. range ($18–105 R) Riesling, Sauvignon Blanc, Chardonnay, Pinot Noir, Saignee (Rosé), Serré, Shiraz, Cabernet Merlot, Cabernet Sauvignon.
summary With the qualified exception of the Cabernet Merlot, which can be a little leafy and gamey, produces outstanding wines across the range, all with individuality, style, great complexity and depth of flavour. The low-yielding estate vineyards play their role, but so does the French-influenced winemaking of Gary Farr. Export markets have been established in the UK, US, Hong Kong, New Zealand and Switzerland.

Bannockburn Chardonnay

As with all the Bannockburn wines, 100% estate-grown from plantings made in 1974, 1981 and 1987 which are typically low-yielding and produce fully ripe grapes with very concentrated flavour. Made with what I describe as traditional French techniques, with roughly settled juice, barrel-fermented, natural yeast and natural malolactic fermentation, resulting in wines of great complexity.

YYYYY **1996** Medium to full yellow-green; the bouquet is extremely complex, with pronounced high-toast barrel-ferment oak aromas, but on the palate intense melon and fig fruit comes up to balance that oak. Lots and lots happening here. **rating:** 95

best drinking 1999–2004 **best vintages** '88, '90, '91, '92, '94, '96 **drink with** Rich white meat dishes • $45

banrock station NR

Holmes Road, off Sturt Hwy, Kingston-on-Murray, SA 5331 **region** Riverland
ph (08) 8583 0299 **fax** (08) 8583 0288 **open** Mon–Sat 9–5, Sun 10–4
winemaker Glenn James **prod.** NFP **est.** 1994

prod. range ($6.30–14.95 R) Semillon Chardonnay, Unwooded Chardonnay, Shiraz Cabernet, Shiraz, Oak-matured Port (available at the visitors' centre only). The Semillon Chardonnay and Shiraz Cabernet are packaged both in 750 ml bottles and 2-litre casks; also Sparkling Chardonnay.

summary The $1 million visitors centre at Banrock Station was opened in October 1998. Owned by BRL Hardy, the Banrock Station property covers over 1700 hectares, with 230 hectares of vineyard and the remainder being a major wildlife and wetland preservation area. The Unwooded Chardonnay of Banrock Station has been of consistently good quality, and excellent value at its price.

Banrock Station Unwooded Chardonnay

Right from day one, the Banrock Chardonnay has exhibited above-average flavour (and quality) for a Riverland wine. What is more, it is an unwooded Chardonnay which succeeds where so many fail.

🍷🍷🍷🍷 **1998** Light to medium yellow-green; as ever, nice citrus, melon and peach fruit comes through strongly on the bouquet, matched by well-balanced flavour and length on the palate. Gold medal 1998 Cowra Wine Show. **rating:** 86

best drinking 1999–2000 **best vintages** NA **drink with** Takeaway • $10.90

Banrock Station Shiraz

As one should expect with a wine at this price, the oak influence seems to be provided by the use of oak chips, but it avoids the evil of over use.

🍷🍷🍷🍷(half) **1997** Medium red-purple; the bouquet is firm and fresh with a mix of spice and earth varietal fruit. The berry and leaf fruit flavours of the palate are cleverly augmented by the use of American oak chips; no phenolic bitterness. **rating:** 84

best drinking 1999–2000 **best vintages** NA **drink with** Pizza • $10.90

barak estate NR

Barak Road, Moorooduc, Vic 3933 **region** Mornington Peninsula

ph (03) 5978 8439 **fax** (03) 5978 8439 **open** Weekends and public holidays 11–5

winemaker James Williamson **prod.** 450 **est.** 1996

prod. range ($14–17 CD) Chardonnay, Shiraz, Cabernet Sauvignon, Cherry Port.

summary When James Williamson decided to plant vines on his 4-hectare Moorooduc property and establish a micro-winery, he already knew it was far cheaper to buy wine by the bottle than to make it. Undeterred, he ventured into grape growing and winemaking, picking the first grapes in 1993, and opening Barak Estate in 1996. Old telegraph poles, railway sleepers, old palings and timber shingles have all been used in the construction of the picturesque winery. Four vintages of Chardonnay ('95–'98) were available in early 1999.

barambah ridge NR

79 Goschnicks Road, Redgate via Murgon, Qld 4605 **region** Other Wineries of Qld

ph (07) 4168 4766 **fax** (07) 4168 4770 **open** 7 days 10–5

winemaker Bruce Humphery-Smith **prod.** 10 000 **est.** 1997

prod. range ($10.50–16.50 CD) Semillon, Ridge Semillon, Ridge White, Chardonnay, French Oaked Chardonnay, Ridge Red, Reserve Shiraz, Cabernet Sauvignon.

summary Barambah Ridge is owned by South Burnett Wines, an unlisted public company, and is a major new entrant on the Queensland wine scene. A winery has been constructed for the 1998 vintage, with an anticipated crush of 150 tonnes. The 1997 wines were made by the omnipresent Bruce Humphery-Smith, winning an array of medals at the annual Sheraton/*Courier-Mail* Brisbane Queensland Wine Awards, including the trophy and gold medal for Best White Wine (with the 1997 unwooded Chardonnay).

baratto's NR

Farm 678, Hanwood, NSW 2680 **region** Riverina/Griffith
ph (02) 6963 0171 **fax** (02) 6963 0171 **open** 7 days 10–5
winemaker Peter Baratto **prod.** NA **est.** 1975
prod. range Semillon, Chardonnay, Dry White, Barbera, Shiraz, Cabernet Sauvignon.
summary Baratto's is in many ways a throwback to the old days. Peter Baratto has 15 hectares of vineyards, and sells the wine in bulk or in 10 and 20-litre casks from the cellar door at old-time prices, from as little as $3 per litre.

barletta bros NR

Polish Hill River Estate, Sevenhill, SA 7453 **region** Clare Valley
ph (08) 8342 3395 **fax** (08) 8344 2180 **open** By appointment
winemaker Neil Pike (Contract) **prod.** 2000 **est.** 1993
prod. range ($14–24 CD) Bros Hand Picked Riesling, Shiraz, Dry Grown Grenache Shiraz.
summary There have been many twists since 1993 when Mario, Ben and Julio Barletta started an 'own brand' for their then-retail business, Walkerville Cellars. They no longer own Walkerville Cellars, and the winemaking business is now the sole business of Mario, who has an active involvement with contract-winemaker Neil Pike. The Barletta tanks and barrels reside at Pikes, but Mario Barletta intends to ultimately establish his own independent vineyard and winery and operation.

barossa ridge wine estate NR

Light Pass Road, Tanunda, SA 5352 **region** Barossa Valley
ph (08) 8563 2811 **fax** (08) 8563 2811 **open** By appointment
winemaker Marco Litterini **prod.** 1000 **est.** 1987
prod. range ($19–22 CD) Valley of Vines Merlot Cabernet Franc Cabernet Sauvignon Petit Verdot, Old Creek Shiraz, Mardia's Vineyard Cabernet Franc, Bamboo Creek Merlot.
summary A grape grower turned winemaker, with a small list of interesting red varietals, including the Valley of Vines blend of Merlot, Cabernet Franc, Cabernet Sauvignon and Petit Verdot (what a mouthful), the only such wine produced in the Barossa Valley. Production has doubled, and The National Wine Merchant been appointed as distributor through all States.

barossa settlers ★★★

Trial Hill Road, Lyndoch, SA 5351 **region** Barossa Valley
ph (08) 8524 4017 **fax** (08) 8524 4519 **open** Mon–Sat 10–4, Sun 11–4
winemaker Howard Haese **prod.** 500 **est.** 1983
prod. range ($13.50–26.50 CD) Gully Winds Riesling, Semillon, Woolshed Flat Chardonnay, Megan's White (Semillon Chardonnay), Festive Champagne, Millstowe Royale (Sparkling Red), Late Harvest Riesling, Old Home Block (Light Red), Hoffnungsthal Settlement Shiraz, Rostock Red (Shiraz), Cabernet Sauvignon, Port, Sherry.
summary A superbly located cellar door is the only outlet (other than mail order) for the wines from this excellent vineyard owned by the Haese family. Production has slowed in recent years, with the grapes from the 31-hectare vineyard being sold to others, picking up pace again in 1997.

Barossa Settlers Gully Winds Riesling

Made from the best block of riesling on the substantial vineyard. First tasted in 1997, and retasted March 1999. Has matured very well; the tasting note is from the March 1999 tasting.

🍷🍷🍷🍷🍸 **1997** Glowing yellow-green; most attractive bottle-developed toasty aromas followed by lime, honey and toast on the palate. Quite delicious, but has developed rapidly and will probably be at its best sooner rather than later. **rating:** 91

best drinking 2000–2002 **best vintages** '97 **drink with** Sweetbreads • $15.50

Barossa Settlers Hoffnungsthal Shiraz

Produced from century-old shiraz vines, although the wine is not as intense or rich as one might have expected.

🍷🍷🍷🍷 **1996** Medium red; the bouquet is clean, of light to medium intensity, with touches of tobacco and cedar. The palate is pleasant, well-balanced, with gently ripe, sweet fruit matched by soft oak. **rating:** 85

best drinking 1999–2006 **best vintages** NA **drink with** Ragout of veal • $26.50

barossa valley estate ★★★☆

Heaslip Road, Angle Vale, SA 5117 **region** Adelaide Plains

ph (08) 8284 7000 **fax** (08) 8284 7219 **open** Mon–Fri 9–5, Sat 10–5

winemaker Natasha Mooney **prod.** 60 000 **est.** 1984

prod. range ($10–50 R) Spires Chardonnay Semillon, Shiraz Cabernet Sauvignon; Moculta Semillon, Chardonnay, Shiraz, Cabernet Merlot; Ebenezer Chardonnay, Sparkling Pinot Noir, Shiraz, Cabernet Sauvignon Merlot; and the premium E & E Sparkling Shiraz and Black Pepper Shiraz.

summary Barossa Valley Estate is now part-owned by BRL Hardy, marking the end of a period during which it was one of the last significant co-operative-owned wineries in Australia. Across the board, the wines are full-flavoured and honest. E & E Black Pepper Shiraz is an upmarket label with a strong reputation and following, the Ebenezer range likewise. Over-enthusiastic use of American oak (particularly with the red wines) has been the Achilles heel in the past. The wines are distributed in Australia and UK by BRL Hardy, and by independent distributors in North America.

Barossa Valley Estate Moculta Semillon

Made entirely from Barossa semillon, unwooded, and which won a gold medal at the 1998 Royal Melbourne Wine Show.

🍷🍷🍷½ **1998** Bright yellow-green; the bouquet is big, solid and ripe, with just a hint of toast. A big, chunky full-on style, typical of the Barossa, with the weight and density of Chardonnay rather than Semillon, and hence is an ideal substitute for those looking for an alternative to wooded Chardonnay. **rating:** 84

best drinking 1999–2000 **best vintages** NA **drink with** Salmon caviar • $13.50

Barossa Valley Estate Ebenezer Chardonnay

A light-year in front of previous Ebenezer Chardonnays and, indeed, close to a benchmark for the Barossa Valley. While the wine will in all probability develop very quickly, it has far more finesse than one normally expects or encounters.

🍷🍷🍷🍷½ **1998** Medium to full yellow-green; tangy melon and citrus fruit is married with well-integrated clove spice oak. The same characters appear on the palate, which is neither overblown nor overoaked, and which has a long, fine finish. **rating:** 92

best drinking 1999–2001 **best vintages** '98 **drink with** Roast chicken • $22

Barossa Valley Estate Sparkling Shiraz

In my view, Sparkling Shiraz demands considerable time both on lees prior to disgorgement and thereafter on cork. Unless and until they are around ten years of age, I find them clumsy and difficult to match with food.

🍷🍷🍷🍷 **1995** Dark, dense red-purple; the bouquet is rich, with dark berry fruits and little or no oak evident. The palate has a strong fruit base, just a hint of oak (no more is acceptable) and with undoubted potential. If given another five years, will doubtless merit much higher points. **rating:** 86

best drinking 2005–2015 **best vintages** NA **drink with** Wild duck • $37.50

Barossa Valley Estate E & E Black Pepper Shiraz

The flagship of Barossa Valley Estate, made from grapes grown on a number of low-yielding vineyards from the northeastern corner of the unirrigated vineyards, planted on ironstone-laced soil over a limestone base, yields as little as 1 tonne per acre, and the grapes are hand-picked when very ripe. Notwithstanding the implication of the name, pepper and spice are not aromas or flavours to be found in this luscious, traditional Barossa style.

🍷🍷🍷🍷 **1995** Full red-purple; the bouquet is rich, ripe and concentrated with ripe fruit, chocolate and vanilla aromas, and the palate follows track, with masses of dark blackberry fruits and lots of vanilla oak. Like raw oysters, I am more than happy to leave it to others. **rating:** 89

best drinking 2000–2010 **best vintages** '90, '91, '94 **drink with** Rich red meat dishes • $42

Barossa Valley Estate Spires Shiraz Cabernet

The Spires range, initially comprising a Chardonnay Semillon blend together with this wine, was launched in February 1999. It is made from grapes sourced solely from within the Barossa Valley and is exceptionally well priced.

🍷🍷🍷🍷 **1998** Medium to full red-purple; quite solid, sweet fruit with a slight gamey touch on the bouquet leads into a palate with abundant, sweetly ripe fruit and just a hint of oak. **rating:** 85

best drinking 1999–2000 **best vintages** NA **drink with** Pizza • $9

barratt ★★★★★

PO Box 204, Summertown, SA 5141 **region** Adelaide Hills

ph (08) 8390 1788 **fax** (08) 8390 1788 **open** By appointment

winemaker Jeffrey Grosset (Contract) **prod.** 1000 **est.** 1993

prod. range ($25–35 ML) Chardonnay, Pinot Noir.

summary Lindsay and Carolyn Barratt purchased the Uley Vineyard, situated at the northern end of the Piccadilly Valley at a height of 500 metres, from the late Ian Wilson in August 1990. Part of the production from the 5.2-hectare vineyard is sold to other makers, with a small proportion being contract-made by Jeffrey Grosset. Both wines are complex and of high quality. Limited quantities are sold in the UK, US and Singapore.

Barratt Chardonnay

Made from hand-harvested grapes selected from particular blocks within the Uley Vineyard, and matured in a mix of new and one-year-old French oak barriques for approximately ten months. Forty per cent of the wine undergoes malolactic fermentation.

🍷🍷🍷🍷🍷(half) **1997** Medium yellow-green; the bouquet offers complex and quite concentrated tangy melon fruit clearly speaking of its cool-grown origins. An elegant wine which builds flavour progressively into a long, powerful fruit-driven finish. **rating:** 91

best drinking 2001–2007 **best vintages** '98 **drink with** Trout mousse • $25

barretts wines NR

Portland-Nelson Highway, Portland, Vic 3305 **region** Far South West Victoria

ph (03) 5526 5251 **open** 7 days 11–5

winemaker Rod Barrett **prod.** 1000 **est.** 1983

prod. range ($12–15 CD) Riesling, Traminer, Late Harvest Riesling, Pinot Noir, Cabernet Sauvignon, Port.

summary The second (and newer) winery in the Portland region. The initial releases were made at Best's, but since 1992 all wines have been made on the property by Rod Barrett.

Barretts Pinot Noir

It is not surprising that Barretts should come up with a wine such as this, for Seppelt produces some exceptional Pinot Noir from its nearby Drumborg Vineyard.

▼▼▼▼▽ **1997** Medium red-purple; the aromas are of plum and violets, with attractive sweet oak in support. A well-made and balanced wine, with attractive, sweet, silky fruit and a clean, long finish augmented by well-handled oak. **rating:** 93

best drinking 1999–2003 **best vintages** '97 **drink with** Peking duck • $15

barrington estate ★★★☆

Yarraman Road, Wybong, NSW 2333 **region** Upper Hunter Valley
ph (02) 6547 8118 **fax** (02) 6547 8039 **open** By appointment
winemaker Simon Gilbert (Contract) **prod.** 20 000 **est.** 1967
prod. range ($22.50 R) Yarraman Road is the top label with Black Clay Chardonnay and Sandy Slopes Cabernet Shiraz; Barrington Estate is cheaper label.
summary Yarraman Road/Barrington Estate is the reincarnation of Penfolds Wybong Estate, into which Penfolds poured millions of dollars between 1960 and 1978, then selling the winery and surrounding vineyards to Rosemount Estate. Rosemount removed most of the unproductive vineyards, and used the winery for red wine production until 1992, then converting it to pure storage area. In 1994 Gary and Karen Blom purchased the property from Rosemount after they returned from the United States where Australian-born entrepreneur Gary Blom had a highly successful career; their main investment is the IMAX Theatre in Darling Harbour, but they intend to spend $3–4 million in redeveloping Yarraman Road.

barwang vineyard ★★★★

Postal c/o McWilliam's Wines, Doug McWilliam Road, Yenda, NSW 2681 **region** Hilltops
ph (02) 9722 1200 **fax** (02) 9707 4408 **open** Not
winemaker Jim Brayne **prod.** NA **est.** 1975
prod. range ($15.95–19.95 R) Chardonnay, Semillon, Shiraz, Cabernet Sauvignon.
summary Peter Robertson pioneered viticulture in the Young region when he planted his first vines in 1975 as part of a diversification programme for his 400-hectare grazing property. When McWilliam's acquired Barwang in 1989, the vineyard amounted to 13 hectares; today the plantings exceed 100 hectares. Wine quality has been exemplary from the word go, always elegant, restrained and deliberately understated, repaying extended cellaring. Interestingly, the name has been adopted for a large volume, relatively low-priced range of wines for distribution in the US.

Barwang Semillon

A new addition to the Barwang range, although the vineyard does have some 20-year-old semillon vines. The wine is barrel-fermented in new French oak, and matured in a mix of new and used oak, and given lees contact. Notwithstanding this oak treatment, the wine is driven by its fruit.

▼▼▼▼ **1997** Light to medium yellow-green; the bouquet is potent and lively, with lemon/citrus aromas and just a hint of spicy oak. The palate has a wholly admirable piercing delicacy; it will be fascinating to watch the development of the wine. **rating:** 86

best drinking 1998–2004 **best vintages** NA **drink with** Crab cakes • $15.95

Barwang Shiraz

Overall, the Shiraz has been the best performer in a classy stable. Each vintage has been extremely good, with very clear varietal character, the only problem being a slight hardness in the mouth in the lesser years.

▼▼▼▼ **1997** Dark red, with touches of purple; solid dark berry and bitter chocolate fruit aromas are supported by subtly sweet oak; a mouthfilling wine, with dark berry and earth fruit flavours, finishing with slightly dry tannins. **rating:** 87

best drinking 2002–2007 **best vintages** '90, '91, '92, '93, '94, '97 **drink with** Kraft Australian parmesan • $19.95

Barwang Cabernet Sauvignon

The continental climate of the Barwang Vineyard, marked by cold nights and warm summer days, but with the growing season finishing much later than it does in either the Hunter Valley or Mudgee, produces red wines of considerable flavour, power and extract, exemplified by this wine.

🍷🍷🍷🍷 **1997** Medium purple-red; sweet fruit and slightly dusty, vanilla oak on the bouquet is followed by a sweet, cassis berry-flavoured palate, rounded off with a slice of vanilla oak and soft tannins. **rating:** 86

best drinking 2001–2006 **best vintages** '89, '91, '92, '93, '96 **drink with** Beef Wellington • $19.95

basedow ★★★☆

161–165 Murray Street, Tanunda, SA 5352 **region** Barossa Valley

ph (08) 8563 3666 **fax** (08) 8563 3597 **open** Mon–Fri 10–5, weekends and public holidays 11–5

winemaker Craig Stansborough **prod.** 100 000 **est.** 1896

prod. range ($9.95–60 CD) Eden Valley Riesling, Barossa Valley Semillon (White Burgundy), Sauvignon Blanc Semillon, Barossa Chardonnay, Unwooded Chardonnay, Late Harvest, Oscar's Heritage, Grenache, Barossa Shiraz, Mistella, Old Tawny Port; Museum Release Watervale Riesling, Barossa Carbernet Sauvignon.

summary An old and proud label, particularly well known for its oak-matured Semillon (called White Burgundy on the Australian market), which underwent a number of changes of ownership during the 1990s. Overall, a reliable producer of solidly flavoured wines. Exports to the UK, US, Canada, Hong Kong, Netherlands, NZ, Singapore, Switzerland and Thailand.

Basedow Semillon

Has been selected in the Top 100 of the Sydney International Wine Competition for three years in a row, although constant name changes (Oscar's Traditional White Burgundy, Semillon, etc coming and going) haven't really helped the cause.

🍷🍷🍷🍷 **1997** Medium to full yellow; a traditional, full-blown Barossa Semillon style, with considerable oak input to both the bouquet and palate, yet not compromising the citrus and herbal fruit flavours on the palate. Presently has nice mouthfeel, but, as ever, this wine will be at its best in the year following vintage, and should under no circumstances be cellared. **rating:** 85

best drinking 1998–1999 **best vintages** NA **drink with** Tortellini • $11.45

basket range wines NR

c/o PO Basket Range, SA 5138 **region** Adelaide Hills

ph (08) 9390 1515 **open** Not

winemaker Phillip Broderick **prod.** 500 **est.** 1980

prod. range ($18 ML) A single Bordeaux-blend of Cabernet Sauvignon, Cabernet Franc, Merlot, Malbec drawn from 3 hectares of estate plantings.

summary A tiny operation known to very few, run by civil and Aboriginal rights lawyer Phillip Broderick, a most engaging man with a disarmingly laid-back manner.

bass phillip ★★★★★

Tosch's Road, Leongatha South, Vic 3953 **region** Gippsland

ph (03) 5664 3341 **fax** (03) 5664 3209 **open** 7 days 11–6 summer and autumn

winemaker Phillip Jones **prod.** 700 **est.** 1979

prod. range ($24–90 ML) Tiny quantities of Pinot Noir in three categories: standard, Premium and an occasional barrel of Reserve. A hatful of Chardonnay also made; plus Pinot Rosé and Gamay.

summary Phillip Jones has retired from the Melbourne rat-race to hand-craft tiny quantities of superlative Pinot Noir which, at its best, has no equal in Australia. Painstaking site selection, ultra-close vine spacing and the very, very cool climate of South Gippsland are the keys to the magic of Bass Phillip and its eerily Burgundian Pinots. The quality (and longevity) of the '96 and '97 wines have added to the reputation of the brand.

batista NR

Franklin Road, Middlesex, WA 6258 **region** Pemberton
ph (08) 9772 3530 **fax** (08) 9772 3530 **open** By appointment
winemaker Bob Peruch, Brendan Smith (Contract) **prod.** 600 **est.** 1993
prod. range ($26 CD) Pinot Noir
summary Batista is in fact the baptismal name of owner Bob Peruch, a Pinot Noir devotee whose father planted 1 hectare of vines back in the 1950s, although these have since gone. Between 1993 and 1996 Bob Peruch has planted 1.5 hectares of pinot noir, a hectare each of shiraz and the cabernet family, and half a hectare of chardonnay destined for sparkling wine. He is assisted by the Pannell family in his winemaking, and has received rave reviews for the 1997 Pinot Noir. Randall Pollard of Bannockburn Cellars in Victoria says 'I really cannot think of a better Western Australian Pinot, and it has to be one of the best value Pinots in the world'. Coming from Pollard, who really knows a great deal about Pinot Noir, that is praise indeed.

baxter stokes wines NR

65 Memorial Avenue, Baskerville, WA 6065 **region** Swan District
ph (08) 9296 4831 **fax** (08) 9296 4831 **open** 9.30–5 weekends and public holidays
winemaker Greg Stokes **prod.** 750 **est.** 1988
prod. range ($10–14 CD) Chardonnay, Verdelho, Shiraz Pinot Noir, Shiraz Cabernet Sauvignon.
summary A weekend and holiday operation for Greg and Lucy Stokes, with the production sold by mail order and through cellar door.

beckett's flat NR

Beckett Road, Metricup, WA 6280 **region** Margaret River
ph (08) 9755 7402 **fax** (08) 9755 7402 **open** 7 days 10–6
winemaker Belizar Ilic, Bernard Abbott **prod.** 2000 **est.** 1992
prod. range ($15–19 CD) Sauvignon Blanc Semillon, Oak Matured Semillon Sauvignon Blanc, Autumn Harvest Semillon, Cabernet Merlot Shiraz, Cabernet Sauvignon.
summary Bill and Noni Ilic opened Beckett's Flat in September 1997. Situated just off the Bussell Highway, at Metricup, midway between Busselton and the Margaret River, it draws upon 8 hectares of estate vineyards, first planted in 1992. As from 1998 the wines have been made at the on-site winery. Accommodation is available.

Beckett's Flat Oak Matured Sauvignon Blanc Semillon

Produced from 3 hectares of estate sauvignon blanc and 0.8 hectares of semillon. Winner of a silver medal at the Australian Small Makers Wine Show, and a good example of the style.

🍷🍷🍷🍷 **1998** Light to medium green-yellow; obvious clove spice oak is neatly set against fresh fruit on the bouquet. The crisp and lively palate has good length; here fruit is the dominant part, the oak more subtle. **rating:** 85

best drinking 1999–2001 **best vintages** NA **drink with** Fresh prawns • $17

Beckett's Flat Cabernet

Produced from 3.5 hectares of cabernet sauvignon and 0.8 hectares of merlot. A gold medal winner at the Australian Small Makers Wine Show.

ΨΨΨΨ **1997** Medium to full red-purple; the bouquet is clean, of moderate intensity, with some slightly earthy/gamey notes common to the Margaret River. The palate is smooth, with blackberry and mint fruit, subtle oak pleasant tannins. **rating:** 87

best drinking 2000–2005 **best vintages** NA **drink with** Ox kidney • $19

belbourie ★★☆

Branxton Road, Rothbury, NSW 2330 **region** Lower Hunter Valley

ph (02) 4938 1556 **open** Weekends, holidays 10–sunset

winemaker Bob Davies **prod.** 2000 **est.** 1963

prod. range ($15–16 CD) Barramundi Chardonnay, Belah Semillon Chardonnay, Hermitage.

summary A winery with a rich, and at times highly controversial, history of wine and winemaking, but these days tending more to the conventional. It has always sought to encourage cellar-door and mailing list sales, focusing on monthly wine and food events, and has a loyal clientele.

bellingham vineyard ★★★

Pipers Brook, Tas 7254 **region** Northern Tasmania

ph (03) 6382 7149 **open** By appointment

winemaker Greg O'Keefe (Contract) **prod.** 700 **est.** 1984

prod. range ($5.50–8 ML) Riesling, Chardonnay, Pinot Noir, Cabernet Sauvignon.

summary Dallas Targett sells most of the grapes from his 13-hectare vineyard to Greg O'Keefe; a small part has been made for the Bellingham label.

Bellingham Pinot Noir

An intriguing wine which stood up well in the strong class of the 1999 Tasmanian Wines Show, showing excellent handling in the winery, but suggesting some element of unripeness in the vineyard.

ΨΨΨΨ **1997** Medium purple-red; both the bouquet and palate have green, stemmy characters running through a wine which otherwise has exemplary pinot noir flavour and structure with hints of black cherry, briar and cedar. Certainly makes for a complex wine. **rating:** 89

best drinking 1999–2003 **best vintages** '97 **drink with** Duck risotto • NA

belubula valley vineyards NR

Golden Gully, Mandurama, NSW 2798 **region** Orange

ph (02) 6367 5236 **fax** (02) 9362 4726 **open** Not

winemaker David Somervaille **prod.** 1000 **est.** 1986

prod. range Cabernet Sauvignon.

summary Belubula Valley is a foundation member of the Central Highlands Grapegrowers Association, centred on Orange; the vineyard is located on the Belubula River, near Carcoar, and the small amounts of wine made to date have not yet been commercially released. David Somervaille, incidentally, was Chairman of partners of the national law firm Blake Dawson Waldron.

beresford wines ★★★

49 Fraser Avenue, Happy Valley, SA 5159 **region** McLaren Vale

ph (08) 8322 3611 **fax** (08) 8322 3610 **open** Mon–Fri 9–5, weekends 11–5

winemaker Robert Dundon, John Davey **prod.** 158 000 **est.** 1985

prod. range ($8–18 CD) Beacon Hill Semillon Chardonnay, Shiraz Cabernet, Tawny Port; Highwood Sauvignon Blanc, Chardonnay, Shiraz; the Saints range of St Yvette Chardonnay, St Helene Cabernet Shiraz, St Martine Sparkling Brut; followed by the Belleville range of Riesling, Chardonnay, Semillon Sauvignon Blanc, Shiraz and Cabernet Merlot.

summary The Beresford brand sits at the top of a range of labels primarily and successfully aimed at export markets in the UK, US, Hong Kong and China. The accent is on price, and the wines do not aspire to great complexity. Quality, however, seems to have improved in the latter part of the 1990s.

berri estates ★★

Sturt Highway, Glossop, SA 5344 **region** Riverland
ph (08) 8582 0300 **fax** (08) 8583 2224 **open** Mon–Sat 9–5
winemaker Paul Kasselbaum, Peter Hensel, Graham Buller **prod.** NFP **est.** 1916

prod. range Light Fruity Lexia, Fruity Gordo Moselle, Chablis, Claret, Rosé, White Lambrusco, all in cask form.

summary Strictly a producer of cask and bulk wine with no pretensions to grandeur, and with a substantial part of the production exported in bulk. Part of the BRL Hardy Group.

best's wines ★★★★☆

1 kilometre off Western Highway, Great Western, Vic 3377 **region** Grampians
ph (03) 5356 2250 **fax** (03) 5356 2430 **open** Fri–Sat 10–5, Sun if holiday
winemaker Viv Thomson, Michael Unwin **prod.** 30 000 **est.** 1866

prod. range ($8–46 R) Great Western Chardonnay, Great Western Gewurztraminer, Victoria Riesling, Victoria Chenin Blanc, Great Western Pinot Noir, Great Western Dolcetto, Great Western Merlot, Great Western Shiraz, Bin O Shiraz, Thomson Family Shiraz, Great Western Cabernet Sauvignon, together with a large range of fortified wines sourced from St Andrews at Lake Boga. Some of these wines are available only at cellar door.

summary An historic winery, owning some priceless vineyards planted as long ago as 1867 (other plantings are, of course, much more recent) which has consistently produced elegant, supple wines which deserve far greater recognition than they in fact receive. The Shiraz is a classic; the Thomson Family Shiraz magnificent. Exports to the UK, Canada, Holland, Belgium and Switzerland.

Best's Thomson Family Shiraz

An outstanding wine first released in late 1994 to commemorate the Centenary of Best's Great Western Vineyards. What might have been a one-off event is now an annual one (vintage permitting) under the Thomson Family label. The '95 was a superb successor to the Centenary vintage, made entirely from vines planted adjacent to the winery in 1867 by Henry Best, and matured in small French oak. The '96 is in the same class as the '95.

🍷🍷🍷🍷🍷 **1996** Very good red-purple; dark cherry fruit, followed by evident but not aggressive oak. There is a similar interplay on the palate, with sweet oak complementing the dark cherry/berry fruit and soft, lingering tannins on the finish. Altogether stylish. **rating:** 94

best drinking 2001–2011 **best vintages** '95, '96 **drink with** Stir-fried beef • $46

bethany wines ★★★★

Bethany Road, Bethany via Tanunda, SA 5352 **region** Barossa Valley
ph (08) 8563 2086 **fax** (08) 8563 0046 **open** Mon–Sat 10–5, Sun 1–5
winemaker Geoff Schrapel, Robert Schrapel **prod.** 25 000 **est.** 1977

prod. range ($12.50–65.80 CD) Riesling (Reserve Dry, Special Select Late Harvest), Chardonnay, Wood Aged Semillon, Steinbruch, Cabernet Merlot, Shiraz, Family Reserve Shiraz, Grenache Pressings, Old Quarry Barossa Tawny Port, Old Quarry Barossa Fronti (White Port).

summary The Schrapel family has been growing grapes in the Barossa Valley for over 140 years, but the winery has only been in operation since 1977. Nestling high on a hillside in the site of an old quarry, it is run by Geoff and Rob Schrapel, who produce a range of consistently well-made and attractively packaged wines. They have 36 hectares of vineyards in the Barossa Valley, 8 in the Eden Valley and (recently and interestingly) 2 hectares of chardonnay and 2 hectares of cabernet sauvignon on Kangaroo Island. The wines enjoy national distribution in Australia, and are exported to the UK, New Zealand, Europe, Taiwan and the US.

Bethany Barrel Fermented Semillon

Produced from semillon vines which are now 80 years old. Unlike the majority of oaked Barossa Semillons, is not excessively heavy or phenolic.

🍷🍷🍷🍷 **1998** Light green-yellow; the bouquet is clean, with restrained oak influence. The palate offers the same subtle approach to oak, with ample fresh honey and lime marmalade fruit flavours; good length to the finish. **rating:** 89

best drinking 1999–2002 **best vintages** NA **drink with** Salmon pizza • $15.90

Bethany Wines GR3 Reserve Shiraz

A new deluxe, Reserve bottling for Bethany, which has always made good Shiraz. Replete with ultra-heavy imported bottle.

🍷🍷🍷🍷🍷 **1995** Medium red-purple; the bouquet is clean, with fragrant cherry/cherry pip fruit and nicely handled oak. The palate has plenty of sweet cherry fruit on the mid-palate, complemented by soft vanillin oak. **rating:** 90

best drinking 2000–2005 **best vintages** NA **drink with** Rack of lamb • $65.70

beyond broke vineyard NR

Cobcroft Road, Broke, NSW 2330 **region** Lower Hunter Valley

ph (02) 6026 2043 **fax** (02) 6026 2043 **open** Tastings available at Broke Village Store 10–4

winemaker Pete Howland (Contract) **prod.** 4000 **est.** 1996

prod. range ($12.50–18.30 R) Semillon, Verdelho, Chardonnay, Unwooded Chardonnay, Sparkling Semillon, Shiraz.

summary Beyond Broke Vineyard is the reincarnation of a former Lindemans vineyard purchased by Bob and Terry Kennedy in 1996. In a more than slightly ironical twist, the 1997 Beyond Broke Semillon won two trophies at the Hunter Valley Wine Show of that year, the first for the Best Current Vintage Semillon and the second, the Henry John Lindeman Memorial Trophy for the Best Current Vintage Dry White Wine. Subsequent shows have been less spectacularly kind, but there is nothing surprising in that, and its turn will come again when vintage conditions permit.

bianchet ★★★

Lot 3 Victoria Road, Lilydale, Vic 3140 **region** Yarra Valley

ph (03) 9739 1779 **fax** (03) 9739 1277 **open** 7 days 10–6

winemaker Keith Salter, Martin Williams (Consultant) **prod.** 2000 **est.** 1976

prod. range ($15–23 CD) Traminer, Semillon, Chardonnay, Verduzzo, Verduzzo Gold, Pinot Noir, Shiraz, Cabernet Sauvignon, Merlot.

summary After a protracted period, Bianchet has been purchased by a small Melbourne-based syndicate, with Keith Salter (one of the syndicate members) and consultant Martin Williams taking on winemaking responsibilities. The wines are still basically sold through cellar door, although there is limited export to Japan. One of the most unusual wines from the winery is Verduzzo Gold, a late-harvest, sweet white wine made from the Italian grape variety.

Bianchet Verduzzo

Verduzzo is a rare Italian variety which receives only the briefest of mentions in Jancis Robinson's bible, *Vines, Grapes and Wines*, and giving no real clue about its essential varietal character. Bianchet has produced the wine for many years now, producing both a normal and late-harvest (Verduzzo Gold) version, and I am not sure I am any the wiser. However both versions of the wine are pleasant.

🍷🍷🍷🍷 **1996** Medium to full yellow-green; the bouquet is clean and quite rich with honeyed aromas, though beyond that not particularly distinctive. The palate likewise offers soft, honeyed fruit supported by just the barest touch of oak. **rating:** 86

best drinking 1999–2003 **best vintages** NA **drink with** Pasta with smoked salmon • $19

Bianchet Chardonnay

Produced from 20-year-old, dry-grown estate vines, and one of the first releases after Bianchet changed hands. A truly promising start for the new owners, typical of the concentration produced by the small berries and small bunches of the '97 vintage.

🍷🍷🍷🍷🍷 **1997** Medium yellow-green; the bouquet is sweet with nectarine and fig fruit aromas, the palate elegant, harmonious and supple. Fruit-driven, and not particularly complex, but with lovely pure Chardonnay varietal character. **rating:** 90

best drinking 1999–2004 **best vintages** '97 **drink with** Veal saltimbocca • $19

bimbadgen estate NR

Lot 21 McDonalds Road, Pokolbin, NSW 2321 **region** Lower Hunter Valley

ph (02) 4998 7585 **fax** (02) 4998 7732 **open** Weekends, holidays 10–5

winemaker Kees Van De Scheur, Thomas Jung **prod.** 50 000 **est.** 1968

prod. range ($10–25 R) Semillon, Chardonnay, Shiraz, Cabernet Shiraz; Grand Ridge Estate is the second label.

summary Established as McPherson Wines, then successively Tamalee, then Sobels, then Parker Wines and now Bimbadgen, this substantial winery has had what might politely be termed a turbulent history. It has the great advantage of having 45 hectares of estate plantings, mostly with now relatively old vines.

bindi wine growers ★★★★★

343 Melton Road, Gisborne, Vic 3437 **region** Macedon

ph (03) 5428 2564 **fax** (03) 5428 2564 **open** Not

winemaker Michael Dhillon, Stuart Anderson **prod.** 900 **est.** 1988

prod. range ($30–45 ML) Chardonnay, Original Vineyard Pinot Noir, Block 5 Pinot Noir.

summary A relatively new arrival in the Macedon region, which has gone from strength to strength. The Chardonnay is top shelf, the Pinot Noir as remarkable (albeit in a very different idiom) as Bass Phillip, Giaconda or any of the other tiny-production, icon wines. Notwithstanding the tiny production, the wines are exported (in small quantities, of course) to the UK, Italy, Singapore and the US.

Bindi Chardonnay

Bindi produces complex rich and notably long-lived Chardonnay. A four-vintage mini vertical tasting in January 1998 produced top points for the still-youthful, complex and powerful '91 vintage (92 points), followed by the citrussy '94, with an exceptionally long, fine palate (87 points), then a curiously garlicky '95, which I did not like at all (73 points), and the very different '96, most deeply coloured of all of the wines, and showing what appears to be some botrytis influence. Sourced from 2 hectares of estate plantings. Not surprisingly, the '97 is near the top of the tree, and I fancy will rate even higher points when it reaches its peak some years down the track.

▼▼▼▼▽ **1997** Medium yellow-green; the bouquet is discreet and understated, reminiscent of Giaconda, with a mix of cashew, mineral and melon aromas. The palate is tight, with melon/citrus fruit, minerally acidity, and subtle oak. Will live for years. **rating:** 90

best drinking 2002–2007 **best vintages** '91, '94, '96, '97 **drink with** Corn-fed chicken • $30

Bindi Block 5 Pinot Noir

A new departure for Bindi, which in 1997 yielded only 1 tonne per acre, less even than the Original Vineyard which produced 1.3 tonnes per acre – this on close-planted vines. I suspect I am meant to have preferred Block 5; although it is a close call, the points indicate my vote goes the other way.

▼▼▼▼▼ **1997** Dark red-purple; marginally less deep that the Original Vineyard. The bouquet, likewise, is less forthcoming, but this is a relative – and perhaps unfair – judgment, there is an exceptional array of forest, plum, cedar and earth aromas. The palate is concentrated and complex and, like the Original Vineyard, has an almost indefinite cellaring future. **rating:** 96

best drinking 2000–2010 **best vintages** '97 **drink with** Rich game • $45

Bindi Original Vineyard Pinot Noir

I cannot help but wonder what Rick Kinzbrunner (of Giaconda) would achieve if given the Bindi pinot noir grapes to vinify. In saying this I am not suggesting any shortcoming on the part of Michael Dhillon or Stuart Anderson; on the contrary. It is simply that I think the quality of the Bindi grapes, in all except the wettest, coldest years, is utterly exceptional, and in years like 1997, to die for if you are a passionate Pinot Noir maker. Since 1997 there have been two Bindi Pinot Noirs: Original Vineyard and Block 5.

▼▼▼▼▼ **1997** Very deep, dark red–purple; the bouquet is immensely rich and ripe, bursting at the seams with damson plum and sweet tobacco aromas. The palate is gloriously ripe and expressive, with damson plum, spice of all kinds, and fine tannins. **rating:** 97

best drinking 2000–2010 **best vintages** '93, '94, '96, '97 **drink with** Squab • $30

birdwood estate ★★★☆

Mannum Road, Birdwood, SA 5234 **region** Adelaide Hills

ph (08) 8263 0986 **fax** (08) 8263 0986 **open** Not

winemaker Oli Cucchiarelli **prod.** 700 **est.** 1990

prod. range ($11–20 ML) Chardonnay, Riesling, Merlot, Cabernet Sauvignon.

summary Birdwood Estate draws upon 5 hectares of estate vineyards progressively established since 1990. The quality of the white wines, and in particular the Chardonnay, has generally been good. The tiny production is principally sold through retail in Adelaide, with limited distribution in Sydney and Melbourne.

Birdwood Estate Adelaide Hills Riesling

Estate-grown, and in the manner of the Adelaide Hills Rieslings, offering excellent varietal definition and equally certain longevity.

▼▼▼▼▽ **1997** Light yellow-green; the bouquet is powerful and concentrated with pure varietal lime aromas, the palate possessing plenty of lime pastille mid-palate flavour followed by a long finish. **rating:** 90

best drinking 2000–2007 **best vintages** '97 **drink with** Asparagus and prosciutto • $14

birnam wood wines NR

Turanville Road, Scone, NSW 2337 **region** Upper Hunter Valley

ph (02) 6545 3286 **fax** (02) 6545 3431 **open** 7 days 10–4

winemaker Simon Gilbert (Contract) **prod.** 4000 **est.** 1994

prod. range ($14.95–18.95 CD) Semillon, Semillon Sauvignon Blanc, Chardonnay, Premium Reserve Chardonnay.

summary Former Sydney car dealer Mike Eagan and wife Min moved to Scone to establish a horse stud; the vineyard came later (in 1994) but is now a major part of the business, with over 30 hectares of vines. Most of the grapes are sold; part only is vinified for Birnam Wood.

black george NR

Black Georges Road, Manjimup, WA 6258 **region** Pemberton
ph (08) 9772 3569 **fax** (08) 9772 3102 **open** 7 days 10.30–4.45
winemaker Dr Shelley E Wilson **prod.** 4000 **est.** 1991
prod. range ($17.50–32 CD) Unwooded Chardonnay, The Captains Reserve Chardonnay, Late Picked Verdelho, Classic White, Pinot Noir, Cabernet Franc Merlot.
summary A relatively recent arrival on the scene, with particular aspirations to make high-quality Pinot Noir. As with so much of the Pemberton region, it remains to be seen whether the combination of soil and climate will permit this. Distributors have been appointed in New South Wales, Victoria and Queensland, with direct exports to the UK.

blackjack vineyards ★★★☆

Cnr Blackjack Road and Calder Highway, Harcourt, Vic 3453 **region** Bendigo
ph (03) 5474 2355 **fax** (03) 5474 2355 **open** Weekends and public holidays 11–5
winemaker Ian McKenzie, Ken Pollock **prod.** 2000 **est.** 1987
prod. range ($22 CD) Shiraz, Cabernet Merlot.
summary Established by the McKenzie and Pollock families on the site of an old apple and pear orchard in the Harcourt Valley. Best known for some very good Shirazes. Ian McKenzie, incidentally, is not to be confused with Ian McKenzie of Seppelt Great Western.

BlackJack Shiraz

Produced from the hillside vineyards of BlackJack, and aged in American and French oak barriques for 18 months prior to bottling. The name 'BlackJack' derives from an American sailor who jumped ship during the gold-rush days of the 1850s and 1860s, and who earned for himself this nickname.

🍷🍷🍷🍷 **1997** Medium purple-red; the impact of new oak is obvious if slightly raw; clean positive fruit will sustain the bouquet until the components come together. The palate is clean, with cherry fruit and again slightly unintegrated oak. Has good development potential, and continues the form shown by the '96 vintage. **rating:** 86

best drinking 2002–2007 **best vintages** '93, '96, '97 **drink with** Barbecued T-bone • $22

blackwood crest wines ★★☆

RMB 404A Boyup Brook, WA 6244 **region** Other Wineries of WA
ph (08) 9767 3029 **fax** (08) 9767 3029 **open** 7 days 10–6
winemaker Max Fairbrass **prod.** 2000 **est.** 1976
prod. range ($12–18 CD) Riesling, Blackwood Classic (Sauvignon Blanc Chardonnay), White Cascade (Semillon Riesling), Shiraz, Cabernet Sauvignon.
summary A remote and small winery which has produced one or two striking red wines full of flavour and character; worth watching. The '98 Classic White, '97 Sauvignon Blanc and '97 Shiraz are pleasant, well-crafted wines.

blanche barkly wines NR

Rheola Road, Kingower, Vic 3517 **region** Bendigo
ph (03) 5443 3664 **open** Weekends, public holidays 10–5
winemaker David Reimers **prod.** NFP **est.** 1972
prod. range ($10–16.50 CD) Shiraz, Cabernet Sauvignon.
summary Sporadic but small production and variable quality seem to be the order of the day; the potential has always been there. No recent tastings.

blaxlands wines NR

Broke Road, Pokolbin, NSW 2320 **region** Lower Hunter Valley
ph (02) 4998 7550 **fax** (02) 4998 7802 **open** 7 days 10.30–4.30
winemaker Trevor Drayton (Contract) **prod.** 400 **est.** 1976
prod. range ($17–18 CD) Chardonnay, Chardonnay Semillon, Shiraz.
summary Chris Barnes is an industry veteran who has run Blaxlands Restaurant and Wine Centre in Pokolbin for almost 20 years. He is also the owner of 1 hectare each of chardonnay and semillon, the wines from which are included in the comprehensive range of Hunter Valley wines available from the Wine Centre (and the restaurant).

bleasdale vineyards ★★★☆

Wellington Road, Langhorne Creek, SA 5255 **region** Langhorne Creek
ph (08) 8537 3001 **fax** (08) 8537 3224 **open** Mon–Sat 9–5, Sun 11–5
winemaker Michael Potts **prod.** 150 000 **est.** 1850
prod. range ($9–24 CD) Langhorne Crossing White and Dry Red; Chardonnay, Sandhill Verdelho, Malbec, Mulberry Tree Cabernet Sauvignon, Bremerview Shiraz, Frank Potts Cabernet Malbec Merlot, Sparkling, Fortified.
summary One of the most historic wineries in Australia drawing upon vineyards that are flooded every winter by diversion of the Bremer River, which provides moisture throughout the dry, cool, growing season. The wines offer excellent value for money, all showing that particular softness which is the hallmark of the Langhorne Creek region. Production has soared, and export markets established in the UK, US, Canada and Switzerland.

blewitt springs winery ★★★☆

12 Victoria Avenue, Woodcroft, SA 5162 **region** McLaren Vale
ph (08) 8322 0210 **fax** (08) 8322 0210 **open** Not
winemaker Brett Howard **prod.** 4000 **est.** 1987
prod. range ($10.60–13.70 R) Riesling, Chardonnay, Semillon, Shiraz, Cabernet Sauvignon.
summary When it first arrived on the scene, attracted much attention and praise for its voluptuous Chardonnays, crammed full of peachy, buttery fruit and vanillin American oak. Oak also plays a major role in the Semillon and the red wines; a lighter touch might please some critics.

bloodwood estate ★★★★

4 Griffin Road, Orange, NSW 2800 **region** Orange
ph (02) 6362 5631 **fax** (02) 6361 1173 **open** By appointment
winemaker Stephen Doyle, Jon Reynolds **prod.** 3500 **est.** 1983
prod. range ($10–25 ML) Riesling, Rosé of Malbec, Schubert, Chardonnay, Ice Riesling, Maurice (Bordeaux-blend), Cabernet, Merlot Noir, Chirac (Pinot Chardonnay).
summary Rhonda and Stephen Doyle are two of the pioneers of the burgeoning Orange district. The wines are sold mainly through cellar door and an energetically and informatively-run mailing list; the principal retail outlet is Ian Cook's Fiveways Cellar, Paddington, Sydney. Bloodwood has done best with elegant but intense Chardonnay and the intermittent releases of super-late harvest Ice Riesling.

Bloodwood Estate Riesling

Usually made as a dry style, but occasionally (as in 1994) produces a spectacular late-harvest style – the '94 was picked on 5 July in the wake of April frosts and winter snows, and was justifiably called 'Ice Riesling'. However, the normal treatment is as a dry wine.

▼▼▼▼ 1998 Light green-yellow; the bouquet is quite fragrant and zesty, with tight lime juice fruit aromas, the palate similarly intense with mid-palate sweetness and lively lime juice flavours. **rating:** 88

best drinking 1999–2005 **best vintages** '88, '90, '92, '94, '95, '98 **drink with** Grilled scallops • $14

Bloodwood Estate Rosé Of Malbec

Bloodwood has always used its malbec to produce a Rosé, with consistent and convincing results. Malbec tends to crop heavily, and while it has high colour potential, tends to be hollow on the mid to back palate when made as a varietal red wine. On the other hand, the juicy flavours are very well suited to Rosé.

▼▼▼▽ 1998 Light, vivid fuschia colour. The bouquet is lively, with a mix of floral and more woody aromas, the palate clean and fresh, with red cherry flavours and a nicely balanced, pleasantly dry finish. **rating:** 84

best drinking 1998–1999 **best vintages** NA **drink with** Mediterranean eggplant • $10

Bloodwood Estate Maurice

Initially released in 1997, the Maurice is a blend of the Bordeaux varieties – principally Merlot, Cabernet Sauvignon and Cabernet Franc, with a little Malbec. Interestingly, the Doyles say they 'went for the most delicate (maybe even fragile?) fruit across the vineyard, and this is the result'. The oak is 25% new Seguin Moreau, the balance two and three-year-old hogsheads. Incidentally, the wine is named in honour of Maurice O'Shea.

▼▼▼▽ 1997 Medium red-purple; a curious but far from unattractive bouquet of ribena-like fruit blended with cedar and tobacco. On the palate, abundant cinnamon spice, spearmint and damson plum flavours appear. The wine is by no means a heavyweight, and the tannins are fine and soft. **rating:** 83

best drinking 1999–2003 **best vintages** NA **drink with** Veal chops • $25

Bloodwood Estate Merlot Noir

A most peculiar name for a wine which is in fact not a blend of Merlot and Pinot Noir, but is simply a Merlot, and Bloodwood's first. It gives solid support to the suitability of the Orange region for Merlot.

▼▼▼▼ 1996 Medium to full red-purple; the bouquet is clean, with a mix of red berry and leaf aromas, but positively varietal. The wine has powerful mint, leaf and berry flavours, with lingering but soft tannins. Had it spent time in new French oak, could have been an outstanding wine. **rating:** 86

best drinking 1999–2004 **best vintages** NA **drink with** Braised lamb shanks • $20

Bloodwood Estate Cabernet

A wine which attests to the ability of the Orange region to produce late-ripening red wine styles as handsomely as it does Chardonnay.

▼▼▼▼ 1996 Light to medium red-purple; the bouquet is clean and relatively light, with leafy/minty notes giving way to more sweet fragrant characters. These are picked up in no uncertain fashion with the lush cassis berry fruit on the mid-palate, which closes with slightly green tannins. **rating:** 85

best drinking 2000–2004 **best vintages** NA **drink with** Rib of veal • $18

blue pyrenees estate ★★★

Vinoca Road, Avoca, Vic 3467 **region** Pyrenees

ph (03) 5465 3202 **fax** (03) 5465 3529 **open** Mon–Fri 10–4.30, weekends and public holidays 10–5

winemaker Kim Hart, Greg Dedman **prod.** 80 000 **est.** 1963

prod. range ($9–29 R) Brut, Midnight Cuvee, Chardonnay, Estate Red; Fiddlers Creek Chardonnay, Sauvignon Blanc, Semillon, Pinot Noir, Cabernet Shiraz, Brut. Leydens Vale has recently been introduced as a mid-priced range between the Estate wines and Fiddlers Creek.
summary Notwithstanding its distinguished French ownership, the perseverance of former winemaker Vincent Gere, a very well-equipped winery, and lavish marketing expenditure, the former Chateau Remy has struggled. The 1996 renaming of the winery as Blue Pyrenees is a sign of that struggle, and also of a progressive shift in production towards still table wine, a sensible move. The second label, multi-region-sourced Fiddlers Creek range makes steady progress and offers real value for money at around $12. Not surprisingly, the wines are exported throughout Asia (frequently through Remy subsidiaries), the UK and US.

boneo plains NR

RMB 1400 Browns Road, South Rosebud, Vic 3939 **region** Mornington Peninsula
ph (03) 5988 6208 **fax** (03) 5988 6208 **open** By appointment
winemaker R D Tallarida **prod.** 2500 **est.** 1988
prod. range ($10–18 CD) Chardonnay, Cabernet Sauvignon; Roch Unwooded Chardonnay, Roch Rosé.
summary A 9-hectare vineyard and winery established by the Tallarida family, well known as manufacturers and suppliers of winemaking equipment to the industry. The Chardonnay is the best of the wines so far released.

bonneyview NR

Sturt Highway, Barmera, SA 5345 **region** Riverland
ph (08) 8588 2279 **open** 7 days 9–5.30
winemaker Robert Minns **prod.** 5000 **est.** 1975
prod. range ($6–25 CD) Riesling, Chardonnay, Frontignan Blanc, Shiraz Petit Verdot, Cabernet Petit Verdot, Cabernet Blend, Fortifieds.
summary The smallest Riverland winery selling exclusively cellar door, with an ex-Kent cricketer and Oxford University graduate as its owner/winemaker. The Shiraz Petit Verdot (unique to Bonneyview) and Cabernet Petit Verdot add a particular dimension of interest to the wine portfolio.

booth's taminick cellars NR

Taminick via Glenrowan, Vic 3675 **region** Glenrowan
ph (03) 5766 2282 **fax** (03) 5766 2151 **open** Mon–Sat 9–5, Sun 10–5
winemaker Cliff Booth, Peter Booth **prod.** 4000 **est.** 1904
prod. range ($6.50–12 CD) Trebbiano, Chardonnay, Late Harvest Trebbiano, Shiraz, Cabernet Merlot, Cabernet Sauvignon, Ports, Muscat.
summary Ultra-conservative producer of massively flavoured and concentrated red wines, usually with more than a few rough edges which time may or may not smooth over.

boston bay wines ★★★☆

Lincoln Highway, Port Lincoln, SA 5606 **region** Other Wineries of SA
ph (08) 8684 3600 **fax** (08) 8684 3600 **open** Weekends, school/public holidays 11.30–4.30
winemaker David O'Leary (Contract) **prod.** 2600 **est.** 1984
prod. range ($10–17 CD) Riesling, Spätlese Riesling, Chardonnay, Cabernet Sauvignon, Merlot, Baudin's Blend (Magnum), Mistelle.
summary A strongly tourist-oriented operation which has extended the viticultural map in South Australia. It is situated at the same latitude as Adelaide, overlooking the Spencer Gulf at the southern tip of the Eyre Peninsula. Say proprietors Graham and Mary Ford, 'It is the only vineyard in the world to offer frequent sightings of whales at play in the waters at its foot'.

botobolar ★★★☆

89 Botobolar Road, Mudgee, NSW 2850 **region** Mudgee
ph (02) 6373 3840 **fax** (02) 6373 3789 **open** Mon–Sat 10–5, Sun 10–3
winemaker Kevin Karstrom **prod.** 5000 **est.** 1971
prod. range ($15–18 CD) Chardonnay, Marsanne, Shiraz, Cabernet Sauvignon; St Gilbert Dry Red and White; Preservative Free White and Red.
summary One of the first organic vineyards in Australia with present owner Kevin Karstrom continuing the practices established by founder Gil Wahlquist. Preservative Free Dry White and Dry Red extend the organic practice of the vineyard to the winery. Shiraz is consistently the best wine to appear under the Botobolar label. Exports to the UK, Denmark, Germany and Japan.

bowen estate ★★★★

Riddoch Highway, Coonawarra, SA 5263 **region** Coonawarra
ph (08) 8737 2229 **fax** (08) 8737 2173 **open** 7 days 10–5
winemaker Doug Bowen **prod.** 11 000 **est.** 1972
prod. range ($18–25 CD) Chardonnay, Shiraz, The Blend, Cabernet Sauvignon, Sanderson Sparkling.
summary One of the best-known names among the smaller Coonawarra wineries with a great track record of red winemaking; Chardonnay and Sanderson Sparkling have joined the band, and the Riesling ended with the '93 vintage. Full-bodied reds at the top end of the ripeness spectrum are the winery trademarks, with a chewy richness uncommon in Coonawarra.

Bowen Estate The Blend

A relatively new direction for Bowen Estate, and intriguingly made (at least on the evidence of the '95) in a very different style from the normal Bowen Estate reds, earlier picked and with less opulently ripe flavour. A Cabernet Sauvignon, Merlot and Cabernet Franc.

🍷🍷🍷🍷 **1996** The colour is starting to show some development, slightly surprising for a '96 vintage wine. The bouquet ranges through leaf, tobacco, cedar and earth, but the palate has much more richness than the bouquet suggests, sweetened by hints of chocolate and vanilla, and supported by soft tannins. **rating:** 86

best drinking 1998–2004 **best vintages** '95 **drink with** Lasagne • $21.95

boynton's of bright ★★★☆

Ovens Valley Highway, Bright, Vic 3747 **region** Ovens Valley
ph (03) 5756 2356 **fax** (03) 5756 2610 **open** 7 days 10–5
winemaker Kel Boynton **prod.** 15 000 **est.** 1987
prod. range ($10–16 CD) Riesling, Semillon, Sauvignon Blanc, Chardonnay, Unoaked Chardonnay, Noble Riesling, Noble Chardonnay, Pinot Noir, Shiraz, Merlot, Cabernet Sauvignon, Port, Vintage Brut.
summary The original 12.5-hectare vineyard expanded almost 16 hectares by 1996 plantings of pinot gris, durif and sauvignon blanc, is situated in the Ovens Valley north of the township of Bright, under the lee of Mount Buffalo. In the early years a substantial part of the crop was sold, but virtually all is now vinified at the winery. Overall, the red wines have always outshone the whites, initially with very strong American oak input, but in more recent years with better fruit/oak balance.

brahams creek NR

Woods Point Road, East Warburton, Vic 3799 **region** Yarra Valley
ph (03) 9560 0016 **fax** (03) 9560 0016 **open** Weekends and public holidays 10–5
winemaker Geoffrey Richardson **prod.** 1000 **est.** 1985

prod. range ($12.50 CD) Chardonnay, Sauvignon Blanc, Pinot Noir, Cabernet Sauvignon, Merlot.

summary Owner Geoffrey Richardson did not start marketing his wines until 1994 and a string of older vintage wines are available for sale at cellar door.

brand's of coonawarra ★★★★

Main Road, Coonawarra, SA 5263 **region** Coonawarra

ph (08) 8736 3260 **fax** (08) 8736 3208 **open** Mon–Fri 9–4.30, weekends 10–4

winemaker Jim Brand, Jim Brayne **prod.** NFP **est.** 1965

prod. range ($13.95–27.95 R) Riesling, Chardonnay, Cabernet Merlot, Shiraz, Merlot, Cabernet Sauvignon.

summary Part of a very substantial investment in Coonawarra by McWilliam's, which first acquired a 50% interest from the founding Brand family, then moved to 100%, and followed this with the purchase of 100 hectares of additional vineyard land. Significantly increased production of the smooth wines for which Brand's is known will follow past the end of the decade.

Brand's Laira Riesling

Just when one makes up one's mind that growing riesling in Coonawarra is a capital offence, or at the very least a terrible waste of rare and finite resource (Coonawarra terra rossa), a wine such as this comes along to confound you.

🍷🍷🍷🍷🍷 **1998** Light straw-green; a very fine, classic mineral and herb bouquet is followed by a tight, steely palate. Very fine, but has flavour and will continue developing for many years. Gold medal 1998 National Wine Show. **rating:** 94

best drinking 1999–2003 **best vintages** '98 **drink with** Caesar salad • $13.95

Brand's Chardonnay

A relatively recent arrival on the scene for Brand's, the style of which has improved greatly since McWilliam's became involved. These wines tend to develop well, usually – but not always – slowly.

🍷🍷🍷🍷 **1997** Quite developed, deep yellow; tangy fruit aromas invest the bouquet with some complexity; the wine has a rich, sweet mid-palate, and shows all the signs of developing very quickly. **rating:** 85

best drinking 1999–2000 **best vintages** '90, '92, '95, '96 **drink with** Pasta with salmon • $16.95

Brand's of Coonawarra Shiraz

As the Coonawarra vineyard resources of McWilliam's (and hence Brand's) have increased, so has the style of the Brand's wines subtly changed, becoming more elegant and (perhaps) understated.

🍷🍷🍷🍷 **1997** Medium to full purple-red; the bouquet is clean, with dark berry, plum and prune aromas; quite powerful fruit, good extract and soft tannins on the finish of the palate all suggest this wine will be a long-term improver. **rating:** 87

best drinking 2002–2010 **best vintages** '90, '91, '96, '97 **drink with** Braised lamb shanks • $17.95

Brand's of Coonawarra Merlot

A new release reflecting the increasing vineyard production at Brand's. Still feeling the way, but with potential.

🍷🍷🍷🍷 **1997** Medium red-purple; distinctly leafy/earthy characters on the bouquet are distinctly varietal, and the palate carries on the theme. The only downside is a touch of bitterness. **rating:** 85

best drinking 2000–2004 **best vintages** NA **drink with** Veal chops • $27.95

Brand's of Coonawarra Cabernet Merlot

A winner of three gold medals, albeit not at the most convincing wine shows, but certainly indicates wine judges are consciously seeking to reward more elegant wines.

ҮҮҮҮ **1996** Light to medium red-purple; the relatively light bouquet and palate offer some elegance, with gently leafy/minty/berry fruit, fine tannins and imperceptible oak. **rating:** 85

best drinking 1999–2004 **best vintages** NA **drink with** Veal cutlets • $18.95

Brand's of Coonawarra Cabernet Sauvignon

First made in 1971, and over the intervening years has produced some wonderful wines; the earlier vintages, in particular, were pacesetters. Since then the field has caught up and indeed passed Brand's, but the viticultural resources are there, and there is no doubting the honesty of the wine. Has been a consistent gold medal winner at national shows over recent vintages.

ҮҮҮҮ **1996** Medium to full red-purple; cassis berry fruit makes the initial impression on the bouquet, followed by a hint of earth and subtle oak. The lively palate runs through mint, leaf and sweet berry flavours, not particularly concentrated but with clear varietal definition. Has collected a couple of regional show gold medals. **rating:** 87

best drinking 2000–2005 **best vintages** '86, '90, '91, '93, '94, '95 **drink with** Yearling steak • $18.95

brangayne of orange ★★★★☆

49 Pinnacle Road, Orange, NSW 2880 **region** Orange

ph (02) 6365 3229 **fax** (02) 6365 3170 **open** By appointment

winemaker Simon Gilbert (Contract) **prod.** 5000 **est.** 1994

prod. range ($15.50–25.50 CD) Sauvignon Blanc, Premium Chardonnay, Reserve Chardonnay, Shiraz, Merlot, The Tristan (Cabernet blend), Cabernet Sauvignon.

summary Orchardists Don and Pamela Hoskins decided to diversify into grape growing in 1994, and have progressively established 25 hectares of high-quality vineyards. With viticultural consultancy advice from Dr Richard Smart, and skilled contract-winemaking by Simon Gilbert, Brangayne has made an extraordinarily auspicious debut, emphatically underlining the potential of the Orange region.

Brangayne of Orange Premium Chardonnay

The first vintage (1997) was a silver medal winner in Class 3 at the 1997 Liquorland National Wine Show, and well-deserving of that award. Skilful winemaking, subtle use of oak and excellent cool-climate fruit are the ingredients.

ҮҮҮҮ **1998** Medium yellow-green; the bouquet is quite intense, with a mix of unusual mineral characters plus more conventional fruit. The palate moves into familiar territory, with sweet melon and nectarine fruit braced by crisp acidity; the oak is barely perceptible at any point. **rating:** 86

best drinking 1999–2002 **best vintages** '97 **drink with** Calamari • $15.50

Brangayne of Orange Reserve Chardonnay

The best chardonnay, picked riper (13.5 degrees versus 13 degrees than the Premium Chardonnay) is barrel-fermented in a mix of Allier and Vosges French oak barriques, before spending five months on lees. Sophisticated winemaking with excellent base material has produced a spectacular first-up wine.

ҮҮҮҮҮ **1998** Medium yellow-green; an unusually elegant and subtle bouquet which shows gentle barrel-ferment and malolactic-ferment aromas which unfold progressively as the wine sits in the glass. The palate is equally refined and elegant, with gentle melon fruit and a hint of citrus. Quality oak is there in the background. **rating:** 94

best drinking 2000–2006 **best vintages** '97, '98 **drink with** Flathead fillets • $20

Brangayne of Orange The Tristan

The first release of a blend of Cabernet Sauvignon, Shiraz and Merlot, and which reinforces the suggestion that the Hoskins are not only skilled viticulturists, but have chosen an exceptionally good vineyard site.

🍷🍷🍷🍷🍷 **1997** Medium to full red-purple; the bouquet is concentrated, with ripe, dense chocolate and cassis fruit; the palate equally shows ripeness and concentration, finishing with sweet tannins. It is to be hoped this is not one of those first-crop wines which are not as good again for another ten years or so. **rating:** 92

best drinking 2002–2007 **best vintages** '97 **drink with** Grain-fed beef • $25.50

bream creek vineyard ★★★★

Marion Bay Road, Bream Creek, Tas 7175 **region** Southern Tasmania

ph (03) 6231 4646 **fax** (03) 6231 4646 **open** At Potters Croft, Dunally tel (03) 6253 5469

winemaker Steve Lubiana (Contract) **prod.** 3000 **est.** 1975

prod. range ($15–17 ML) Riesling, Gewürztraminer, Schonburger, Chardonnay, Pinot Noir, Cabernet Pinot, Cabernet Sauvignon.

summary Until 1990 the Bream Creek fruit was sold to Moorilla Estate, but since that time has been independently owned and managed under the control of Fred Peacock, legendary for the care he bestows on the vines under his direction. Peacock's skills have seen both an increase in production, and also a vast lift in wine quality across the range, headed by the Pinot Noir.

Bream Creek Pinot Noir

Bream Creek has two vineyards, one established many years ago on the east coast of Tasmania, the other more recently in the Tamar Valley. This wine comes entirely from the often temperamental east coast vineyard.

🍷🍷🍷🍷🍷 **1997** Strong red-purple; a veritable cascade of aromas on the bouquet, powerful and complex, running through spice, briar, berry and dark plum. The palate does not disappoint, with rich and powerful fruit (rather than oak) doing the work. The intensity and complexity of the fruit is reminiscent of the best Paringa Estate wines from the Mornington Peninsula. **rating:** 95

best drinking 1999–2003 **best vintages** NA **drink with** Braised duck • $17

bremerton wines ★★★★

Strathalbyn Road, Langhorne Creek, SA 5255 **region** Langhorne Creek

ph (08) 8537 3093 **fax** (08) 8537 3109 **open** 7 days 10–5

winemaker Rebecca Willson **prod.** 15 000 **est.** 1988

prod. range ($11–28 CD) Sauvignon Blanc Semillon, Botrytised Chenin Blanc, Verdelho, Young Vine Shiraz, Old Adam Shiraz, Bremerton Blend (red), Cabernet Sauvignon.

summary The Willsons have been grape growers in the Langhorne Creek region for some considerable time, but their dual business as grape growers and winemakers has expanded significantly over the past few years. Their vineyards have more than doubled to over 100 hectares (predominantly cabernet sauvignon, shiraz and merlot), as has their production of wine under the Bremerton label, no doubt in recognition of the quality of the wines.

Bremerton Shiraz

Produced from the 15.4 hectares of now fully mature estate vines, and having that particular sweetness which the much under-estimated Langhorne Creek region seems to produce with so little effort.

🍷🍷🍷🍷 **1997** Medium red-purple; the bouquet offers a gentle mix of sweet, faintly earthy, fruit and vanilla oak, insufficient preparation for an excellently textured palate with lots of sweet cherry flavour, and a soft, fine tannin finish. **rating:** 86

best drinking 2000–2007 **best vintages** NA **drink with** Butterfly leg of lamb • $15

Bremerton Old Adam Shiraz

The flagship wine for Bremerton, made from the very best grapes coming from the estate vineyards. A fruit-driven style with highly restrained use of oak.

YYYY𝑌 **1997** Medium to full purple-red; dark plum, prune and blackberry fruit aromas introduce a wine with sweet, rich dark berry fruit flavours flooding the palate. Oak and tannins are undoubtedly there, but you barely notice them. **rating:** 91

best drinking 2002–2012 **best vintages** '97 **drink with** Parmesan cheese • $28

Bremerton Cabernet Sauvignon

Draws upon a little over 20 hectares of estate plantings. Matured in a mix of new and two-year-old French oak barriques.

YYYY𝑌 **1997** Medium to full red-purple; the bouquet is full, with attractive dark berry and chocolate fruit. The palate has an equal abundance of sweet, dark berry fruit, and in the Bremerton style of things, soft tannins and subtle oak are there in the background. **rating:** 90

best drinking 2002–2010 **best vintages** '91, '96, '97 **drink with** Braised beef in red wine • $22

brewery hill winery NR

Olivers Road, McLaren Vale, SA 5171 **region** McLaren Vale
ph (08) 8323 7344 **fax** (08) 8323 7355 **open** Mon–Fri 9–5, weekends 10–5
winemaker Various Contract **prod.** 12 000 **est.** 1869
prod. range ($6.20–22 CD) Classic Dry White, Sauvignon Blanc, Riesling, Chardonnay, Botrytis Riesling, Classic Dry Red, Grenache, Shiraz, Shiraz Cabernet, Sparkling and Fortifieds.
summary A change of name and of address for the former St Francis Winery which has moved into the former Manning Park Winery, and is now known as Brewery Hill Winery.

briagolong estate ★★☆

Valencia–Briagolong Road, Briagolong, Vic 3860 **region** Gippsland
ph (03) 5147 2322 **fax** (03) 5147 2400 **open** By appointment
winemaker Gordon McIntosh **prod.** 300 **est.** 1979
prod. range ($35 ML) Chardonnay, Pinot Noir.
summary This is very much a weekend hobby for medical practitioner Gordon McIntosh, who nonetheless tries hard to invest his wines with Burgundian complexity, with mixed success. Dr McIntosh must have established an all-time record with the 15.4% alcohol in the '92 Pinot Noir.

brian barry wines ★★★☆

PO Box 128, Stepney, SA 5069 **region** Clare Valley
ph (08) 8363 6211 **fax** (08) 8362 0498 **open** Not
winemaker Brian Barry **prod.** 10 000 **est.** 1977
prod. range ($14.99–30 R) Jud's Hill Handpicked Riesling, Chardonnay, Handpicked Merlot, Handpicked Cabernet Sauvignon; Gleeson's Ridge Shiraz Merlot and Semillon Chardonnay.
summary Brian Barry is an industry veteran with a wealth of winemaking and show-judging experience. His is nonetheless in reality a vineyard-only operation, with a substantial part of the output sold as grapes to other wineries, and the wines made under contract at various wineries, albeit under Brian Barry's supervision. As one would expect, the quality is reliably good. Retail distribution through all States, and exports to US, Canada, Taiwan, Switzerland and NZ.

briar ridge ★★★☆

Mount View Road, Mount View, NSW 2325 **region** Lower Hunter Valley
ph (02) 4990 3670 **fax** (02) 4990 7802 **open** Mon–Fri 9–5, weekends 9.30–4.30
winemaker Karl Stockhausen, Neil McGuigan **prod.** 16 000 **est.** 1972

prod. range ($17–24 CD) Varietal Range of Methode Champenoise, Early Harvest Semillon, Hand Picked Chardonnay, Verdelho, Botrytis Semillon, Old Vines Shiraz, Cabernet Sauvignon, Tawny Port; Signature Stockhausen Semillon, and Hermitage; and Signature McGuigan Chardonnay and Cabernet Merlot.

summary Semillon and Hermitage, each in various guises, have been the most consistent performers, underlying the suitability of these varieties to the Hunter Valley. The Semillon, in particular, invariably shows intense fruit, and cellars well. Briar Ridge has been a model of stability with the winemaking duo of Neil McGuigan and Karl Stockhausen, and also has the comfort of over 40 hectares of estate vineyards, from which it is able to select the best grapes.

Briar Ridge Stockhausen Hermitage

Named in honour of veteran winemaker Karl Stockhausen, who spent decades with Lindemans in the Hunter Valley, commencing his winemaking career there in 1960. Now a consultant to Briar Ridge, and brings his vast experience into play.

🍷🍷🍷🍷 **1997** Medium to full red; the bouquet is complex, with lots of Hunter Shiraz varietal character supported by well-integrated and balanced oak; quite impressive. The palate seems then to go off on an entirely unexpected direction of its own, with an unusual mix of flavours including slightly unripe raspberries, balanced by more conventional spice and cherry. Lots of oak also present in a Janus-headed wine. **rating:** 87

best drinking 2000–2005 **best vintages** '86, '87, '89, '91, '93, '94, '96 **drink with** Braised lamb shanks • $24

bridgeman downs NR

Barambah Road, Moffatdale via Murgon, Qld 4605 **region** Granite Belt

ph (07) 4168 4784 **fax** (07) 4168 4767 **open** By appointment

winemaker Bruce Humphery-Smith **prod.** NA **est.** NA

prod. range ($11.50–16.50 CD) Cellar White and Red; Chardonnay, Verdelho, Shiraz, Merlot Cabernet.

summary A substantial, albeit new, vineyard with 4 hectares of vines, the major plantings being of verdelho, chardonnay and shiraz, and lesser amounts of merlot and cabernet sauvignon. The perpetual-motion Bruce Humphery-Smith has been retained as consultant winemaker, which should ensure wine quality. The first wines were released from the 1998 vintage, but were not tasted.

bridgewater mill ★★★★

Mount Barker Road, Bridgewater, SA 5155 **region** Adelaide Hills

ph (08) 8339 3422 **fax** (08) 8339 5253 **open** Mon–Fri 9.30–5, weekends 10–5

winemaker Brian Croser **prod.** 20 000 **est.** 1986

prod. range ($15–33 R) Sauvignon Blanc, Chardonnay, Millstone Shiraz, Cabernet Malbec.

summary The second label of Petaluma, which consistently provides wines most makers would love to have as their top label. The fruit sources are diverse, with the majority of the sauvignon blanc and chardonnay coming from Petaluma-owned or managed vineyards, while the Shiraz is made from purchased grapes.

Bridgewater Mill Sauvignon Blanc

A blend of material coming from the Adelaide Hills, Clare Valley and Coonawarra, cold-fermented in stainless steel and relatively early bottled.

🍷🍷🍷🍷 **1998** Light green-yellow; the bouquet is clean, and while not particularly exotic or aromatic, does have nice grassy/herbal aromas. On the palate hints of passionfruit and gooseberry appear which add both to the attraction and balance of a nice wine. **rating:** 87

best drinking 1999–2000 **best vintages** '92, '94, '95 **drink with** Mousseline of scallops • $19.95

Bridgewater Mill Millstone Shiraz

Made from shiraz grown in McLaren Vale and at Langhorne Creek, said to be from low-cropping vines; I am not quite certain what 'low-cropping' means in the context of Langhorne Creek, however. The wine is matured in French oak barriques for one year.

1996 Excellent red-purple; there is an interesting array of aromas on the bouquet, with lively cherry to start with followed by hints of earth and leather. A solid wine on the palate with plum, cherry, mint and earth all adding to the fruit weight. **rating:** 86

best drinking 2001–2006 **best vintages** '90, '91, '92, '94 **drink with** Lamb shanks • $24.50

brindabella hills ★★★★

Woodgrove Close, via Hall, ACT 2618 **region** Canberra District
ph (02) 6230 2583 **fax** (02) 6230 2023 **open** Weekends, public holidays 10–5
winemaker Dr Roger Harris **prod.** 1500 **est.** 1989
prod. range ($16.50–24 CD) Riesling, Sauvignon Blanc Semillon, Chardonnay, Reserve Chardonnay, Shiraz, Cabernet.

summary Distinguished research scientist Dr Roger Harris presides over Brindabella Hills, which increasingly relies on estate-produced grapes, with small plantings of cabernet sauvignon, cabernet franc, merlot, shiraz, chardonnay, sauvignon blanc, semillon and riesling. Wine quality has been consistently impressive. All of the wine is sold direct ex-winery.

Brindabella Hills Reserve Chardonnay

In both 1996 and 1997 the most distinguished of the Brindabella Hills range. The '96 was placed first in the 1998 *Winewise* Small Makers Competition, and the '97 was of similar class and style. The '98 also fared well at the same competition in 1999.

1998 Medium yellow-green; a solidly-constructed wine from start to finish, with good fruit and oak balance and integration. The length of the back palate and finish is the wine's strong suit. **rating:** 87

best drinking 1999–2001 **best vintages** '96, '97 **drink with** Chinese prawns • $24

Brindabella Hills Cabernet

The wine is produced from a blend of Cabernet Sauvignon, Cabernet Franc and Merlot, fermented and hand-plunged in open fermenters, basket pressed, and matured in French and American barriques for two years.

1997 Medium red-purple; high-toned aromas of cassis, briar and leaf are followed by a fresh, intense palate which combines ripeness with elegance; sweet, lingering cassis flavours run through to the very finish. **rating:** 90

best drinking 2000–2004 **best vintages** '90, '91, '93, '95, '96, '97 **drink with** Roast lamb • $20

britannia creek wines NR

75 Britannia Creek Road, Wesburn, Vic 3799 **region** Yarra Valley
ph (03) 5780 1426 **fax** (03) 5780 1426 **open** Weekends 10–6
winemaker Charlie Brydon **prod.** 1200 **est.** 1982
prod. range ($12–18 CD) Sauvignon Blanc, Semillon, Cabernets.

summary The wines (from Britannia Creek Wines) are made under the Britannia Falls label from 4 hectares of estate-grown grapes. A range of vintages (stretching back to 1990) were available from cellar door in 1998, with some interesting, full-flavoured Semillon.

broadview NR

Rowbottoms Road, Granton, Tas 7030 **region** Southern Tasmania
ph (03) 6263 6882 **open** By appointment
winemaker Alain Rousseau (Contract) **prod.** NA **est.** 1996

prod. range Riesling, Chardonnay.

summary David and Kaye O'Neil planted their vineyard in the spring of 1996, picking the first precious 200 kilograms of chardonnay in 1998, producing a fine minerally wine.

broke estate ★★★☆

Wollombi Road, Broke, NSW 2330 **region** Lower Hunter Valley
ph (02) 6579 1065 **fax** (02) 6579 1065 **open** By appointment
winemaker Simon Gilbert (Contract) **prod.** 5000 **est.** 1989

prod. range ($14–48 ML) Limited Edition Chardonnay, Sauvignon Blanc Semillon, First Edition Cabernets, Limited Edition Cabernets.

summary With a high-profile consultant viticulturist (Dr Richard Smart) achieving some spectacular early results, Broke Estate has seldom been far from the headlines. Contrary to what one might expect, the opulent red wines (rather than the whites) have been the most successful.

broken bago vineyards NR

Milligans Road, off Bago Road, Wauchope, NSW 2446 **region** Hastings River
ph (02) 6585 7099 **fax** (02) 6585 7099 **open** 7 days 11-5
winemaker Hill of Hope (Contract) **prod.** 1000 **est.** 1985

prod. range ($10–21.50 CD) Chardonnay, Jazz Classic White, Chambourcin, Sparkling, Tawny Port.

summary Jim and Kay Mobs commenced planting the Broken Bago Vineyards in 1985 with 1 hectare of chardonnay, and have now increased the total plantings to 12 hectares. Contract-winemaking has moved to the Hunter Valley's Hill of Hope winery.

brokenwood ★★★★★

McDonalds Road, Pokolbin, NSW 2321 **region** Lower Hunter Valley
ph (02) 4998 7559 **fax** (02) 4998 7893 **open** 7 days 10–5
winemaker Iain Riggs **prod.** 70 000 **est.** 1970

prod. range ($10–40 R) Semillon, Cricket Pitch Sauvignon Blanc Semillon, ILR Semillon, Cricket Pitch Unwooded Chardonnay, Graveyard Chardonnay, Cricket Pitch Red, Cricket Pitch Cabernet Merlot, Shiraz, Pinot Noir, Cabernet Sauvignon, Rayner Vineyard Shiraz, Graveyard Shiraz.

summary Deservedly fashionable winery producing consistently excellent wines. Cricket Pitch Sauvignon Blanc Semillon has an especially strong following, as has Cabernet Sauvignon; the Graveyard Shiraz is one of the best Hunter reds available today, the unwooded Semillon a modern classic. In 1997 acquired a controlling interest in Seville Estate (Yarra Valley) and has also been involved in the establishment of substantial vineyards in Cowra.

Brokenwood Semillon

A wine much appreciated by the Sydney market, which ensures that each release sells out long before the next becomes available. It is made in traditional style: in other words, without the use of oak, and unforced by techniques such as skin contact. Most is drunk young as a crisp, quasi-Chablis style, but as tastings show, can develop into a Hunter classic. (Specially made Reserve releases are in the maturation pipeline.)

🍷🍷🍷🍷 **1998** Light to medium yellow-green; the bouquet has good weight with some sweeter fruit notes, progressively building in flavour through to the back palate and finish. Excellent acidity and length; very much the result of a benign Hunter vintage. **rating:** 90

best drinking 1999–2003 **best vintages** '85, '86, '89, '92, '94, '95, '96, '97, '98 **drink with** Balmain bugs • $17.50

Brokenwood ILR Semillon

This is part of a Reserve range of Semillons specially made by Iain Riggs for lengthy cellaring prior to release. The 1994 deservedly won a gold medal at the 1999 Sydney Royal Wine Show.

🍷🍷🍷🍷🍷 **1994** Medium to full yellow-green; the bouquet offers classic bottle-developed Hunter Semillon aromas, with a distinct edge of lightly browned toast. The palate is quite delicious, with honey on toast flavours, still a twist of lemon, and a long finish. **rating:** 94

best drinking 2001–2006 **best vintages** NA **drink with** Grilled lobster • $35

Brokenwood Cowra Verdelho

Let there be no doubt about it: this is a nice little earner for Brokenwood, well made as ever, but with no pretensions or claims to greatness.

🍷🍷🍷½ **1998** Light green-yellow; the bouquet is quite scented and rich; altogether quite interesting. On the palate the wine has idenifiable verdelho varietal flavour and character. The first from Brokenwood's Cowra vineyards. **rating:** 83

best drinking 1999–2000 **best vintages** NA **drink with** Takeaway • $17

Brokenwood Rayner Vineyard Shiraz

One of Brokenwood's regional specials, coming from the Rayner Vineyard in McLaren Vale. The wine is, however, made at Brokenwood in the Hunter Valley.

🍷🍷🍷🍷½ **1997** Excellent bright purple-red in colour; the bouquet shows typical McLaren Vale fruit in a rich chocolate, earth and red berry spectrum. A sweet, smooth full-bodied palate follows, with dark berry fruits, soft tannins, subtle oak, and a particularly agreeable finish. **rating:** 90

best drinking 1999–2004 **best vintages** NA **drink with** Marinated beef • $40

Brokenwood Cricket Pitch Red

First made in 1987, taking its name from the fact that the vineyard immediately adjacent to the Brokenwood winery was once the site of a cricket field. The wine in fact has had no greater connection than the name with that vineyard, as it has been sourced from grapes grown in various parts of Australia, typically McLaren Vale, Mount Barker, Cowra, Coonawarra and the Hunter Valley. To add spice to the mix, it is a blend of 48% Merlot, 45% Cabernet Sauvignon and 7% Shiraz. It stands apart from most Brokenwood reds as a smooth early-drinking style.

🍷🍷🍷🍷 **1997** Medium red-purple; the bouquet offers aromas of leaf, mint and earth which are slightly astringent, but the palate is much better, with a mix of almost essencey cherry, chocolate and vanilla oak flavours. **rating:** 85

best drinking 1999–2002 **best vintages** NA **drink with** Pork spare ribs • $18.50

brook eden vineyard NR

Adams Road, Lebrina, Tas 7254 **region** Northern Tasmania

ph (03) 6395 6244 **open** 7 days 10–5

winemaker Jan Bezemer **prod.** 800 **est.** 1988

prod. range ($15–18 CD) Chardonnay, Pinot Noir.

summary Jan and Sheila Bezemer own a 60-hectare Angus beef property which they purchased in 1987, but have diversified with the establishment of 2.5 hectares of vines. Jan Bezemer makes the wine at Delamere, the first vintage being 1993. The vineyard site is beautiful, with viticultural advice from the noted Fred Peacock.

brookland valley ★★★★

Caves Road, Willyabrup, WA 6284 **region** Margaret River

ph (08) 9755 6250 **fax** (08) 9755 6214 **open** Tues–Sun 11–4.30

winemaker Larry Cherubino **prod.** 8000 **est.** 1984

prod. range ($18–30 R) Sauvignon Blanc, Chardonnay, Merlot, Cabernet Merlot.

summary Brookland Valley has an idyllic setting, with its much enlarged Flutes Cafe one of the best winery restaurants in the Margaret River region. In 1997 BRL Hardy acquired a 50% interest in the venture, and has taken responsibility for viticulture and winemaking. The move towards richer and more complex red wines evident before the takeover will no doubt continue in its wake.

Brookland Valley Merlot

A wine which does nothing to answer the eternal question asked in Australia: what should Merlot taste like? For all that, a new direction for Brookland Valley, and a variety which does appear suited to the Margaret River region.

ŸŸŸŸ **1996** Medium red, with just a slightly dull black edge. The bouquet is not promising, although it is certainly unusual, with a mix of dried twigs and berries. The palate redeems the wine, exhibiting authority in a foresty/briary mode but with a dark berry core. **rating:** 86

best drinking 2001–2006 **best vintages** NA **drink with** Devilled steak • $30

brookside vineyard NR

5 Loaring Road, Bickley Valley, WA 6076 **region** Perth Hills

ph (08) 9291 8705 **fax** (08) 9291 5316 **open** Weekends and public holidays 10-5

winemaker Darlington Estate (Contract) **prod.** 450 **est.** 1984

prod. range ($15–17 CD) Chardonnay, Cabernet Sauvignon, Methode Champenoise.

summary Brookside is one of the many doll's house-scale vineyard operations which dot the Perth Hills. It has a quarter of a hectare each of chardonnay and cabernet sauvignon, basically selling the wine through a mailing list. It does, however, offer bed and breakfast accommodation at the house with its attractive views of the Bickley Valley.

brown brothers ★★★★☆

Snow Road, Milawa, Vic 3678 **region** King Valley

ph (03) 5720 5500 **fax** (03) 5720 5511 **open** 7 days 9–5

winemaker Terry Barnett, Wendy Cameron, Matt Fawcett **prod.** 300 000 **est.** 1885

prod. range ($10.95–45.50 R) A kaleidoscopic array of varietal wines, with a cross-hatch of appellations, the broadest being Victorian (e.g. Victorian Shiraz), more specific being King Valley (e.g. NV Brut and Pinot Chardonnay) and Milawa (e.g. Noble Riesling), then the Limited Release, Family Selection (e.g. Very Old Tokay and King Valley Chardonnay) and the Family Reserve ranges.

summary Brown Brothers draws upon a considerable number of vineyards spread throughout a range of site climates, ranging from very warm to very cool, with the climate varying according to altitude. It is also known for the diversity of varieties with which it works, and the wines always represent excellent value for money. Deservedly one of the most successful family wineries in Australia. The wines are exported to over 20 countries spread throughout Europe, UK, Asia and the Far East. Conspicuously, Brown Brothers still remains out of the US market.

Brown Brothers King Valley Riesling

An utterly reliable, full-flavoured Riesling which doesn't aspire to greatness but does offer value for money. Sourced from various growers throughout the King Valley.

ŸŸŸŸ **1997** Medium yellow-green; the bouquet is rich and full, with aromas of lime, honey and toast, the palate similarly smooth and sweet with lime-accented fruit. **rating:** 86

best drinking 1999–2002 **best vintages** NA **drink with** Asparagus salad • $14.60

Brown Brothers Family Reserve Chardonnay

The Family Reserve range sits at the top of the Brown Brothers tree, the white wines held back for at least three years bottle age, the Cabernet-based reds for eight years. The Chardonnay is a vineyard selection from the King Valley, barrel-fermented in a mix of new and used French oak barriques, with 70% being matured in oak for eight months prior to bottling.

🍷🍷🍷🍷 **1995** Medium yellow-green; there is abundant, smooth peachy fruit on both the bouquet and palate, with the appropriate amount of oak in smooth support. **rating:** 87

best drinking 1999–2005 **best vintages** NA **drink with** Veal terrine • $30.55

Brown Brothers King Valley Pinot Chardonnay

A Pinot Noir dominant (typically around 70%) blend with Chardonnay and Pinot Meunier comprising the balance, produced entirely from grapes grown at Brown Brothers Whitlands Vineyard high in the King Valley. Complexity is gained by an extraordinarily broad harvest span extending from early February to late April. The wine undergoes malolactic fermentation before being blended and tiraged, and spends three years on yeast lees prior to disgorgement. Invariably a distinguished wine. The yet to be released 1995 won the Thorp Trophy for Best Sparkling White Wine at the 1998 Royal Sydney Wine Show.

🍷🍷🍷🍷🍷 **1994** Light straw-yellow; there are attractive citrus tinges and a touch of brioche to the bouquet; the palate is well balanced, with the citrus and brioche flavours promised by the bouquet running throughout the entire length, braced with crisp acidity on the finish. **rating:** 92

best drinking 1999–2000 **best vintages** '90, '91, '92, '93, '94 **drink with** Shellfish • $36.45

Brown Brothers Noble Riesling (375 ml)

Traces its origins back to 1934, when the first incidence of botrytis in the riesling at Milawa was noticed by John Brown senior. The wine was made but incorporated into a fortified wine, for there was no market for the style at the time. The first commercial vintage, albeit on a tiny scale, was 1962, the first significant commercial vintage in 1970. But it can legitimately claim to be the first Australian botrytised Riesling commercially released, and the wine has gone from strength to strength, particularly with the 1990s' era decision to bottle and release the wine as a fresher, younger style. It is unlikely Brown Brothers will ever have greater success than they have had with the 1996 vintage which has won seven trophies and 11 gold medals.

🍷🍷🍷🍷🍷 **1996** Deep gold; the bouquet explodes with intense lime and cumquat botrytis fruit; the palate, likewise, is exceptionally rich with honeycomb, lime and cumquat flavours offset by well-balanced acidity. **rating:** 95

best drinking 1999–2005 **best vintages** '92, '94, '96 **drink with** Sticky date pudding • $21.75

Brown Brothers Victorian Shiraz

Sourced from Brown Brothers' estate in Milawa and Mystic Park Vineyards, the balance coming from contract growers in the King Valley. Typically harvested over a long period running from mid-March to late April, which invests the wine with an extra degree of complexity. American oak is used at a level which is in the mainstream of Australian commercial red winemaking.

🍷🍷🍷🍷 **1996** A pleasant, easy-drinking commercial red wine with good colour, soft red berry fruit and vanilla oak aromas, and a smooth, easy, moderately oaky palate. **rating:** 81

best drinking 2000–2004 **best vintages** '91, '92, '93, '94 **drink with** Braised beef Chinese-style • $18.50

Brown Brothers Walpoles Tempranillo

So far as I am aware, the first commercial release of a Tempranillo in Australia. It will, I am certain, be the first of many, for this Spanish grape is both marvellously flexible and capable of producing wines of very high quality in a range of climate and soil. The grapes were picked at the end of March at 13° baumé, and the wine spent 18 months in French and American oak.

🍷🍷🍷🍷 **1996** Medium red-purple; the bouquet has gentle, sweet berry fruit with subtle oak; it is the palate which shows particular promise, with black cherry fruit cradled in a firm structure supported by soft tannins running through from start to finish. **rating:** 86

best drinking 2000–2005 **best vintages** NA **drink with** Barbecued lamb • $15.50

Brown Brothers King Valley Merlot

There are relatively large plantings of merlot in the King Valley. At higher elevations the climate should be ideal; the problems will come in restraining yield – merlot is very nearly as sensitive to over-cropping as is pinot noir.

▼▼▼▽ **1996** A wine which belies its 13.5 degrees alcohol, particularly on the bouquet, which while of medium intensity does not have a great deal of definition or varietal character. The palate shows a bit more, with sweet berry fruit flavours touched by earth; subtle oak and soft tannins are appropriate in the context of the varietal style. **rating:** 80

best drinking 1999–2002 **best vintages** NA **drink with** Lasagne • $18.50

Brown Brothers Family Selection Very Old Tokay

The material used to make the wine averages ten years of age, with the oldest components between 20 and 25 years, the youngest between three and four years.

▼▼▼▼▼ **NV** Medium to full tawny-red; very complex tea-leaf/plum pudding/malty aromas. High-quality Tokay varietal character on the palate, which is lively and intense, with strong tea-leaf/malt/butterscotch flavours, and good balance. **rating:** 94

best drinking 1999–2000 **best vintages** NA **drink with** After coffee • $29.80

browns of padthaway ★★☆

Keith Road, Padthaway, SA 5271 **region** Padthaway

ph (08) 8765 6063 **fax** (08) 8765 6083 **open** At Padthaway Estate

winemaker Contract **prod.** 35 000 **est.** 1993

prod. range ($10–20 R) Classic Diamond, Riesling, Sauvignon Blanc, Non Wooded Chardonnay, Verdelho, T-Trellis Shiraz, Redwood Cabernet Malbec, Myra Family Reserve Cabernet Sauvignon, Sparkling Shiraz.

summary The Brown family has for many years been the largest independent grape grower in Padthaway, a district in which most of the vineyards were established and owned by Wynns, Seppelts, Lindemans and Hardys, respectively. A rapidly expanding range of wines is now appearing under the Browns of Padthaway label, the majority being pleasant but very light in body and flavour.

Browns Of Padthaway Myra Family Reserve Cabernet Sauvignon

A quantum leap for the wines of this producer, offering the concentration and flavour so conspicuously absent in the majority of its other wines.

▼▼▼▼▽ **1996** Medium red, with just a touch of purple; sweet, earthy berry fruit is supported by some vanilla oak on the bouquet of medium intensity. The palate has good concentration of sweet chocolatey berry fruit, finishing with fine but persistent tannins. **rating:** 90

best drinking 2000–2006 **best vintages** '96 **drink with** Roast beef • $20

bullers beverford ★★★

Murray Valley Highway, Beverford, Vic 3590 **region** Swan Hill (Vic)

ph (03) 5037 6305 **fax** (03) 5037 6803 **open** Mon–Sat 9–5

winemaker Richard Buller (Jnr) **prod.** 50 000 **est.** 1952

prod. range ($9.50–15.95 CD) Victoria Chenin Blanc Colombard and Shiraz Grenache Malbec; The Magee Semillon Chardonnay and Cabernet Sauvignon Shiraz; White Label range of Semillon Chenin Blanc, Spätlese Lexia, Rosé, Shiraz, Cabernet Sauvignon; Victoria range of fortifieds including Port, Tokay and Muscat; also Sails Unwooded Chardonnay and Cabernet Shiraz Merlot.

summary Traditional wines which in the final analysis reflect both their Riverland origin and a fairly low-key approach to style in the winery. It is, however, one of the few remaining sources of reasonable quality bulk fortified wine available to the public, provided in 22-litre Valorex barrels at $6.50 per litre. Some recent red wines have impressed.

Bullers Beverford White Label Shiraz

1997 marks the third release of Beverford White Label Shiraz, grown at Bullers Swan Hill vineyard from which the wine takes is name. It is given 12 months barrel maturation in predominantly old oak.

▼▼▼▽ **1997** Medium red-purple; the bouquet is clean, of medium intensity, with red and black cherry together with nuances of tobacco. The palate is fresh and clean, not overripe nor over-extracted, with appropriately soft tannins and imperceptible oak. **rating:** 84

best drinking 1999–2002 **best vintages** NA **drink with** Pasta with meat sauce • $15

---

bullers calliope ★★★★☆

Three Chain Road, Rutherglen, Vic 3685 **region** Rutherglen

ph (02) 6032 9660 **fax** (02) 6032 8005 **open** Mon–Sat 9–5, Sun 10–5

winemaker Andrew Buller **prod.** 5000 **est.** 1921

prod. range ($16–55 CD) Limited Release Shiraz, Mondeuse Shiraz, Grenache Cinsaut; Premium Black Label range and Museum Release range of old and rare material; Rare Liqueur and Liquid Gold Muscat and Tokay.

summary The winery rating is very much influenced by the recent superb releases of Museum fortified wines. Limited releases of Calliope Shiraz and Shiraz Mondeuse can also be exceptionally good.

Bullers Liquid Gold Tokay (375 ml)

Repackaged into the slender, clear glass 'olive oil' type of 375 ml bottle much favoured these days, but done extremely well with an attractive label and capsule design. Externalities to one side, a beautiful expression of young Rutherglen Tokay, the sort of wine one should drink, rather than sip – especially on a cold winter's night.

▼▼▼▼▽ **NV** Glowing golden brown; totally delicious, sweet tea-leaf and caramel varietal aromas leap from the glass, with the flavour precisely tracking the bouquet. Clean spirit, sweet but not the least bit cloying. **rating:** 91

best drinking 1999–2000 **best vintages** NA **drink with** Cake and coffee • $18

Bullers Calliope Rare Liqueur Tokay (375 ml)

Both the Rare Liqueur Muscat and Tokay are of extreme quality and of great age, with the older material in the blend dating back to the 1940s, and picked from Bullers' 1920 Rutherglen dryland vineyard. The base wine is over 20 years old, and only 150 dozen 375 ml bottles are released each year to protect the integrity of the base material. Like the Liquid Gold Tokay, very smartly packaged in a 375 ml bottle.

▼▼▼▼▼ **NV** Deep golden brown; a classic mix of sweet tea-leaf and crème brûlée aromas is followed by an outstanding palate showing the complexity which only age (and first class base material) can bring; some nutty characters join the tea-leaf and crème brûlée of the bouquet. **rating:** 95

best drinking 1999–2009 **best vintages** NA **drink with** Strictly unnecessary, a meal in itself • $55

Bullers Liquid Gold Muscat (375 ml)

The sister wine to the Tokay, again featuring relatively fresh material, and again in the style one can drink in generous quantities. Shares the same attractive packaging.

▼▼▼▼▽ **NV** Orange-brown; classic raisiny/grapey aromas lead into a young Muscat at its very best, with intense raisiny fruit, just a hint of nuttiness, and finishing with clean spirit. **rating:** 90

best drinking 1999–2000 **best vintages** NA **drink with** Walnuts and almonds • $18

Bullers Calliope Rare Liqueur Muscat (375 ml)

Emerged from the shadows in spectacular fashion at the 1994 Sydney International Wine Competition, where it won the trophy for Best Wine of Show. A magnificent wine of great age and complexity. Originally released as Very Old Rutherglen Muscat, but now under the Rare Liqueur label, and in the new package.

ŦŦŦŦŦ **NV** Deep brown with a touch of olive on the rim; full and deep, almost into chocolate, with intense raisined fruit; richly textured, with great structure to the raisined/plum pudding fruit flavours, and obvious rancio age. **rating:** 94

best drinking 1999–2009 **best vintages** NA **drink with** Strictly unnecessary, a meal in itself • $55

---

bungawarra NR

Bents Road, Ballandean, Qld 4382 **region** Granite Belt
ph (07) 4684 1128 **open** 7 days 10.30–4.30
winemaker Bruce Humphery-Smith, Jeff Harden **prod.** 1300 **est.** 1975
prod. range ($10–16 CD) Traminer, Block Six Chardonnay, Foundation Chardonnay, Festival Red, Paragon (Shiraz Cabernet Malbec), Liqueur Muscat.
summary Now owned by Jeff Harden. It draws upon 5 hectares of mature vineyards which over the years have shown themselves capable of producing red wines of considerable character.

burge family winemakers ★★★☆

Barossa Way, Lyndoch, SA 5351 **region** Barossa Valley
ph (08) 8524 4644 **fax** (08) 8524 4444 **open** 7 days 10–5
winemaker Rick Burge **prod.** 3500 **est.** 1928
prod. range ($17.50–40 CD) Olive Hill Riesling, Olive Hill Semillon, Chardonnay, Muscat Blanc Late Harvest, Clochmerle (Grenache), Olive Hill Grenache Shiraz, Draycott Shiraz Grenache, Draycott Shiraz, Draycott Reserve Shiraz, Draycott Cabernet Merlot; Draycott Sparkling Red, Fortifieds.
summary Rick Burge came back to the family winery after a number of years successfully running St Leonards; there was much work to be done, but he has achieved much, using the base of very good fortified wines and markedly improving table wine quality, with Draycott Shiraz (both standard and Reserve) leading the way. The wines are exported to the US, Singapore and Canada.

Burge Family Olive Hill Semillon

Barossa Semillon is not an easy wine to make well, often tending to heaviness and prone to rapid development, particularly when exposed to phenolic, German oak. This wine has been very well handled, 10% being barrel-fermented in new French (Allier) oak. It should, nonetheless, be drunk sooner rather than later.

ŦŦŦŦ **1998** Glowing green-yellow; the bouquet is smooth, with no excess phenolics; the palate is very well balanced, with excellent oak handling; in a rich drink-now style. **rating:** 88

best drinking 1999–2000 **best vintages** NA **drink with** Chicken pasta • $22.80

Burge Family Draycott Shiraz Grenache

A blend of 70% Shiraz and 30% old vine Grenache.

ŦŦŦŦ **1997** Medium to full red-purple; the bouquet shows very ripe grapes with some plum jam characters from the grenache. It is the grenache influence which drives the voluptuously rich, slightly jammy palate. For those who like a boots and all red wine. **rating** 87

best drinking 1999–2004 **best vintages** NA **drink with** Rich casserole • $17.50

Burge Family Olive Hill Shiraz Grenache

A blend of 66% Shiraz, 21% Grenache and 13% Mourvedre, a classic Rhône blend which works wonderfully well in this instance.

🍷🍷🍷🍷🍷 **1998** Dense red-purple; the bouquet has masses of dark plum and prune fruit which swallows up whatever oak is there. The palate is likewise crammed with rich, ripe fruit and welcome tannins to provide balance on the finish. **rating:** 90

best drinking 2000–2010 **best vintages** NA **drink with** Game • $22.80

burnbrae NR

Hill End Road, Erudgere via Mudgee, NSW 2850 **region** Mudgee

ph (02) 6373 3504 **fax** (02) 6373 3601 **open** Wed–Mon 9–5

winemaker Alan Cox **prod.** NFP **est.** 1976

prod. range ($10–18 CD) Sauvignon Blanc, Chardonnay, Pinot Noir, Shiraz, Malbec, Cabernet Sauvignon, Vintage Port, Liqueur Muscat.

summary The founding Mace family sold Burnbrae to Alan Cox in 1996. It continues as an estate-based operation with 23 hectares of vineyards. No recent tastings.

burramurra NR

Barwood Park, Nagambie, Vic 3608 **region** Goulburn Valley

ph (03) 5794 2181 **fax** (03) 5794 2755 **open** Not

winemaker Mitchelton (Contract) **prod.** 800 **est.** 1988

prod. range ($18 R) Cabernet Merlot.

summary Burramurra is the relatively low-profile vineyard operation of the Deputy Premier of Victoria, the Honourable Pat McNamara. Most of the grapes are sold to Mitchelton; a small amount is contract-made for the Burramurra label. Glowing reveiws in the United States have led to brisk export business with that country.

calais estates ★★★

Palmers Lane, Pokolbin, NSW 2321 **region** Lower Hunter Valley

ph (02) 4998 7654 **fax** (02) 4998 7813 **open** Mon–Fri 9–5, weekends 10–5

winemaker Adrian Sheridan **prod.** 11 000 **est.** 1987

prod. range ($12–30 CD) Chenin Blanc, Semillon, Chardonnay, Reserve Chardonnay, Late Harvest Riesling, Sauterne, Shiraz Pinot, Shiraz, Reserve Shiraz, Cabernet Sauvignon.

summary The '97 white wines entered at the 1997 Hunter Valley Wine Show showed raw oak and other problems, but the '96 Shiraz (gold medal and top in its class) was excellent, with rich blackberry/plum pudding fruit and sweet vanilla oak.

cambewarra estate ★★★☆

520 Illaroo Road, Cambewarra NSW 2540 **region** Shoalhaven

ph (02) 4446 0170 **fax** (02) 4446 0170 **open** Weekends and public holidays Wed–Sun 10–5, Mon–Fri by appointment

winemaker Tamburlaine (Contract) **prod.** 3500 **est.** 1991

prod. range ($15–29 CD) Chardonnay (wooded and unwooded), Verdelho, Petit Rouge, Chambourcin, Cabernet Sauvignon, Vintage Port.

summary Geoffrey and Louise Cole founded Cambewarra Estate near the Shoalhaven River on the central southern coast of New South Wales, with contract-winemaking competently carried out (a considerable distance away) at Tamburlaine Winery in the Hunter Valley. Cambewarra continues to produce attractive wines which have had significant success in wine shows.

Cambewarra Estate Chambourcin

Chambourcin is a French-bred hybrid which is highly resistant to mildew and rot, and particularly suited to warmer, wetter growing regions. Cassegrain on the north coast of New South Wales has the largest plantings, but they are scattered through various parts of the State. At Cambewarra, as elsewhere, it produces a strongly coloured wine which is typically best consumed when young. Cambewarra makes two versions: one a lighter style, without oak, the other a fuller, oak-matured wine.

🍷🍷🍷🍷 **1998** Typical glass-staining purple; rich, dark, juicy blackberry and plum aromas with faintly earthy overtones are followed by a juicy berry/minty palate with some tannins on the finish. A very good example of Chambourcin. **rating:** 86

best drinking 1999–2000 **best vintages** '94, '97, '98 **drink with** Italian cuisine • $20

Cambewarra Estate Cabernet Sauvignon

Half a hectare of estate plantings produced the first vintage in 1996. The wine is matured in French oak, and shows excellent varietal character. The '96 won a gold medal at the 1997 Cowra Wine Show, then going on to win the trophy for Best Table Wine in the Small Producers classes at the 1998 Royal Sydney Wine Show, an exceptional achievement for Cambewarra Estate and the Shoalhaven region.

🍷🍷🍷🍷 **1998** Youthful purple-red; the bouquet is likewise youthful with cassis/blackberry fruit and minimal oak. The palate is quite attractive, with cassis and mint fruit followed by a soft finish. Not as callow or unmade as the bouquet suggests. **rating:** 88

best drinking 2001–2005 **best vintages** '96 **drink with** Illabo lamb • $29

campbells ★★★★

Murray Valley Highway, Rutherglen, Vic 3685 **region** Rutherglen
ph (02) 6032 9458 **fax** (02) 6032 9870 **open** Mon–Sat 9–5, Sun 10–5
winemaker Colin Campbell **prod.** 35 000 **est.** 1870
prod. range ($13.80–94 R) Semillon, Riesling, Pedro Ximenez, Trebbiano, Bobbie Burns Chardonnay, Silverburn Dry White and Red, Bobbie Burns Shiraz, The Barkly Durif, Malbec, Shiraz Durif Cabernet, Cabernets, Liquid Gold Tokay, Isabella Tokay, Rutherglen Muscat, Merchant Prince Muscat.

summary A wide range of table and fortified wines of ascending quality and price, which are always honest; as so often happens in this part of the world, the fortified wines are the best, with the extremely elegant Isabella Tokay and Merchant Prince Muscat at the top of the tree. For all that, the table wines are impressive in a full-bodied style; the winery rating is something of a compromise between that for the fortified wines and for the table wines. A feature of the cellar door is an extensive range of back vintage releases of small parcels of wine not available through any other outlet. Significant retail distribution through all States, and exports to the UK and US.

Campbells The Barkly Durif

While Morris Wines has always been known for its Durif, Campbells also has one of the largest vineyard holdings of durif in Rutherglen. The variety was bred by its namesake, Dr Durif, in the Rhône Valley in the 1880s, and introduced into Rutherglen soon thereafter by Hubert De Castella. The small berries (and small bunches) invariably produce a deeply-coloured wine, rich in extract and high in alcohol.

🍷🍷🍷🍷 **1995** Full red-purple, ripe blackberry fruit, tinged with chocolate and mint is immediately obvious on the bouquet. The palate provides an unusual mix of mint, chocolate, cedar and blackberry, rounded off with fine tannins on the finish. The oak contribution is negligible. **rating:** 83

best drinking 2000–2005 **best vintages** NA **drink with** Jugged hare • $35

Campbells Isabella Tokay

One of a pair of super-premium fortified wines produced by Campbells, the other being Campbells Merchant Prince Muscat. The Campbells wine is, and always has been, lighter and fresher than that of the other major producers, with more emphasis thrown on the underlying varietal fruit of the wines. It is a question of style rather than quality; these deluxe wines deserve their price.

🍷🍷🍷🍷🍷 **NV** Light tawny-gold; fragrant grapey, sweet tea-leaf aromas with clean spirit; the palate is luscious with sweet juicy berry and tea-leaf flavours, finishing with good acidity and a very clean aftertaste. **rating:** 93

best drinking 1999–2000 **best vintages** NA **drink with** As fine an aperitif as it is a digestif • $94

Campbells Rutherglen Tokay (375 ml)

Campbells are one of the best makers of young Tokay (average age five years or so) in Rutherglen. The wine has perfectly defined varietal character, and is to be drunk, rather than just sipped.

🍷🍷🍷🍷🍷 **NV** Bright golden brown; lovely young cold tea, tea-leaf varietal aromas flowing without a break into the palate, where the spirit is very harmonious and does not threaten the wonderful tokay flavour. **rating:** 94

best drinking 1999–2000 **best vintages** NA **drink with** As fine an aperitif as it is a digestif • $18.25

Campbells Merchant Prince Muscat

A superbly balanced and constructed Muscat in a distinctly lighter mould than Baileys, Chambers or Morris, the big names of the district. For all that, it has an average age of 25 years, with the oldest component dating back over 60 years. One of those rare Muscats which actually invites a second glass.

🍷🍷🍷🍷🍷 **NV** Light to medium brown; intense but fragrant spice and raisin aromas with clean spirit. The palate is remarkably fresh and light given the age of the wine, with raisin, spice, malt and toffee flavours all intermingling, followed by cleansing acidity. **rating:** 94

best drinking 1998–1999 **best vintages** NA **drink with** Coffee, high-quality biscuits • $94

Campbells Rutherglen Muscat

Again the second tier of Campbells fortifieds, like the Tokay showing largely unmodified varietal character.

🍷🍷🍷🍷🍷 **NV** Light tawny red; very youthful raisiny fruit aromas, with the spirit fractionally jumpy – quite why, I am not sure. High-toned fruit on the palate adds a haunting edge to the flavour, almost floral. Intriguing and delicious. **rating:** 90

best drinking 1999–2000 **best vintages** NA **drink with** Fruit cake • $18.25

candlebark hill ★★★★

Fordes Lane, Kyneton, Vic 3444 **region** Macedon

ph (03) 9836 2712 **fax** (03) 9836 2712 **open** Sunday 10–6 and by appointment

winemaker David Forster, Llew Knight (Consultant) **prod.** 500 **est.** 1987

prod. range ($17–27 CD) Chardonnay, Pinot Noir, Cabernet Merlot, Cabernets, Cabernet Sauvignon.

summary Candlebark Hill has been established by David Forster on the northern end of the Macedon Ranges, enjoying magnificent views over the central Victorian countryside north of the Great Dividing Range. The 3.5-hectare vineyard is planted to pinot noir (1.5 hectares) together with 1 hectare each of chardonnay and the three Bordeaux varieties, complete with half a hectare of shiraz and malbec. The quality of the 1997 vintage wines is exemplary, promising much for the future.

Candlebark Hill Pinot Noir

Produced from 1.5 hectares of estate pinot noir, with plantings dating back to 1987. It spends 15 months in new French oak, yet is in no sense overoaked.

🍷🍷🍷🍷🍷 **1997** Medium red-purple; the bouquet has excellent rich, ripe plummy varietal fruit supported by subtle oak. The promise of the bouquet is fulfilled on the palate with above-average fruit richness, concentration and extract; the oak is evident but not oppressive. **rating:** 91

best drinking 1999–2002 **best vintages** '97 **drink with** Smoked quail • $19.95

Candlebark Hill Cabernet Merlot

Predominantly made from Cabernet Sauvignon and Merlot, with a touch of Cabernet Franc. The wine has achieved an unusually high 14 degrees alcohol.

🍷🍷🍷🍷🍷 **1997** Medium to full red-purple; the bouquet offers sweet, dark chocolate, briar and earth-tinged fruit supported by high-quality oak. The palate is quite excellent, with sweet dark berry fruit and chocolate flavours; ripe, fine tannins run throughout a really good wine. **rating:** 94

best drinking 2000–2007 **best vintages** '97 **drink with** Rack of lamb • $26.95

canobolas-smith ★★★★

Boree Lane, Off Cargo Road, Lidster via Orange, NSW 2800 **region** Orange

ph (02) 6365 6113 **fax** (02) 6365 6113 **open** Weekends, public holidays 11–5

winemaker Murray Smith **prod.** 1800 **est.** 1986

prod. range ($10–22 CD) Aurora, Chardonnay, Highland Chardonnay, Pinot Noir, Highland Red, Cabernets, Alchemy (Cabernet blend).

summary After a tentative start with early experimental vintages, Canobolas-Smith has established itself as one of the leading Orange district wineries with its distinctive blue wrap-around labels. Much of the wine is sold from the cellar door, which is well worth a visit.

cape bouvard NR

Mount John Road, Mandurah, WA 6210 **region** South West Coast

ph (08) 9739 1360 **fax** (08) 9739 1360 **open** 7 days 10–5

winemaker Gary Grierson **prod.** 2000 **est.** 1990

prod. range ($15–20 CD) Chardonnay, Shiraz, Cabernet Sauvignon, Sparkling, Port.

summary Doggerel poet-cum-winemaker Gary Grierson draws upon 1 hectare of estate plantings, but also purchases grapes from other growers for the new Cape Bouvard label. The few wines tasted have been light but inoffensive.

cape clairault ★★★★

Henry Road, Willyabrup, WA 6280 **region** Margaret River

ph (08) 9755 6225 **fax** (08) 9755 6229 **open** 7 days 10–5

winemaker Ian Lewis, Peter Stark **prod.** 8000 **est.** 1976

prod. range ($13–22 CD) Under the Cape Clairault label Sauvignon Blanc, Unwooded Chardonnay, Semillon Sauvignon Blanc, Riesling, Claireau (sweet white), Clairault (Cabernet blend); under the second Cape label, Cape White, Cape Rose, Cape Late Harvest and Cape Red.

summary Ian and Ani Lewis have been joined by two of their sons and, in consequence, have not only decided not to sell the business, but to double its size, with winery capacity being almost doubled from 85 tonnes to 150 tonnes. Notwithstanding increasing production, demand for the wines is so great that Cape Clairault has withdrawn from export to concentrate on the local market. A vineyard specialty is guinea fowl, not to be eaten (I think), but to control grasshopper plagues.

cape jaffa wines NR

Limestone Coast Road, Cape Jaffa, SA 5276 **region** Mount Benson and Robe
ph (08) 8768 5053 **fax** (08) 8768 5040 **open** 7 days 10–5
winemaker Derek Hooper **prod.** 5000 **est.** 1993
prod. range ($16–18 CD) Unwooded Chardonnay (McLaren Vale), Semillon Sauvignon Blanc, Barrel Fermented Chardonnay (Mount Benson and Padthaway), Merlot (McLaren Vale), Shiraz (McLaren Vale) Cabernet Sauvignon (Mount Benson).
summary Cape Jaffa is the first of the Mount Benson wineries to come into production, albeit with most of the initial releases coming from other regions. Ultimately all of the wines will come from the substantial estate plantings of 20 hectares which include the four major Bordeaux red varieties, shiraz, chardonnay, sauvignon blanc and semillon. It is a joint venture between the Hooper and Fowler families, and the winery (built of local paddock rock) has been designed to allow eventual expansion to 1000 tonnes, or 70 000 cases.

capel vale ★★★★

Lot 5 Stirling Estate, Mallokup Road, Capel, WA 6271 **region** Geographe
ph (08) 9727 1986 **fax** (08) 9727 1904 **open** 7 days 10–4
winemaker Rob Bowen, Krister Jonsson **prod.** 90 000 **est.** 1979
prod. range ($10.50–48 R) CV 'Bistro' range of Chenin, Unwooded Chardonnay, Sauvignon Blanc Chardonnay, Pinot Noir, Shiraz, Cabernets Merlot; Capel Vale 'Fine Dining' Riesling, Verdelho, Sauvignon Blanc Semillon, Chardonnay, Merlot, Shiraz, Cabernet Sauvignon; 'Reserve Connoisseur' range of Whispering Hill Riesling, Seven Day Road Sauvignon Blanc, Frederick Chardonnay, Kinaird Shiraz and Howecroft Cabernet Sauvignon Merlot; Tawny Port.
summary Capel Vale continues to expand its viticultural empire, its contract-grape sources and its marketing, the last through the recent introduction of a series of vineyard or similarly named super-premium wines. Against the run of play, as it were, the most successful of these super-premiums are the red wines, for I have long admired the elegance and finesse of the Capel Vale whites. The strong marketing focus the company has always had is driven by its indefatigable owner, Dr Peter Pratten, who has developed export markets throughout Europe, US and Asia.

Capel Vale Howecroft Cabernet Sauvignon Merlot

A blend of 80% Cabernet Sauvignon from the Margaret River and 15% Merlot and 5% Cabernet Franc from Capel. The components were separately fermented, and matured in new French oak for 12 months prior to blending. The wine selected for the Howecroft reserve bottling then spent a further year in barrel before being bottled. Like the Kinnaird Shiraz, presented in a dreadnought-class bottle.

🍷🍷🍷🍷 **1996** Medium red-purple; pungent green leaf/herbal aromas dominate the bouquet; the palate is no less potent, and still basically in the green spectrum. Offsetting those green characters are touches of chocolate and cedar, and the tannins seem fine. **rating:** 88

best drinking 2001–2006 **best vintages** NA **drink with** Rare eye fillet • $48

cape mentelle ★★★★★

Off Wallcliffe Road, Margaret River, WA 6285 **region** Margaret River
ph (08) 9757 3266 **fax** (08) 9757 3233 **open** 7 days 10–4.30
winemaker John Durham **prod.** 50 000 **est.** 1970
prod. range ($20–44 R) Chardonnay, Semillon Sauvignon Blanc, Cabernet Sauvignon, Cabernet Merlot, Shiraz, Zinfandel, Trinders Cabernet Merlot.
summary Notwithstanding majority ownership by Veuve Clicquot, David Hohnen remains very much in command of one of Australia's foremost medium-sized wineries. Exceptional marketing skills and wine of the highest quality, with the back-up of New Zealand's Cloudy Bay, are a potent combination. The Chardonnay and Semillon Sauvignon Blanc are among Australia's

best, the potent Shiraz usually superb, and the berry/spicy Zinfandel makes one wonder why this grape is not as widespread in Australia as it is in California.

Cape Mentelle Semillon Sauvignon Blanc

A wine that has been part of the Cape Mentelle portfolio since 1985, but which in the early years was not particularly exciting. To what extent the Cloudy Bay skills rubbed off is a moot point, but the fact is that in more recent times this has been another faultless wine combining finesse with power.

🍷🍷🍷🍷🍷 **1998** The highly aromatic bouquet shows marvellous fruit expression, predominantly in the gooseberry spectrum, and the palate lives up to the promise of the bouquet. Here the rich fruit is balanced by a touch of mineral and crisp acid on the finish. **rating:** 95

best drinking 1999–2002 **best vintages** '85, '88, '91, '93, '95, '96, '97, '98 **drink with** Fish, Asian cuisine • $21.25

Cape Mentelle Chardonnay

First made in 1988, and immediately established itself as another classic. The 1990 vintage was selected for British Airways First Class; each succeeding year has reached new heights. These are wines of exceptional complexity, Chardonnays made by a red winemaker (but in the best possible way). The 1993 won the George Mackey Award for best wine exported from Australia in 1994; subsequent vintages have all been in the same class.

🍷🍷🍷🍷🍷 **1997** Light to medium yellow-green; an enormously sophisticated wine of the highest quality from start to finish, underlining the capacity of Margaret River to produce truly great Chardonnay. Grapefruit and melon fruit is matched by perfectly balanced and integrated oak on both bouquet and palate, with a long, lingering and intense finish providing a near-perfect rendition of the variety. **rating:** 97

best drinking 2000–2010 **best vintages** '90, '91, '92, '93, '94, '95, '96, '97 **drink with** Tasmanian salmon • $30.60

Cape Mentelle Shiraz

Made its debut in 1981, and over the intervening years has produced some spectacular wines which – to my palate at least – have not infrequently outclassed the Cabernet Sauvignon. The wines typically show wonderful spice, game and liquorice characters reminiscent of the Rhône Valley. Since 1986 a small percentage of Grenache has been included in some years.

🍷🍷🍷🍷🍷 **1997** Very good purple-red; a youthful bouquet with a mix of fresh berry, spice and more earthy aromas; there is an abundance of red cherry/berry fruit on the palate with fairly persistent tannins running through to the finish. So powerful and youthful the 14.5 degrees alcohol does not stand out as it otherwise might. **rating:** 90

best drinking 2002–2012 **best vintages** '86, '88, '90, '91, '93, '94, '96, '97 **drink with** Stir-fried Asian beef • $27.10

Cape Mentelle Zinfandel

Also made its first appearance in 1981, a direct reflection of David Hohnen's early winemaking experiences in California. Despite its exceptional quality, remains the only Zinfandel worth mentioning in Australia; it is most surprising that it has not encouraged others to try.

🍷🍷🍷🍷🍷 **1997** Medium to full red; an exotic array of spices and plummy fruit come rocketing through the bouquet; the palate follows through with blackcurrant, prune and more of those exotic spices. An imperious 15.5 degrees alcohol, yet paradoxically overall has the character of a cool-grown Californian Zinfandel. **rating:** 91

best drinking 1999–2004 **best vintages** '86, '87, '91, '92, '93, '94, '95, '97 **drink with** Rare char-grilled rump steak • $30.60

Cape Mentelle Cabernet Sauvignon

The wine which started the Cape Mentelle juggernaut, with the '82 and '83 vintages winning the Jimmy Watson Trophy in successive years. Both style and quality wandered somewhat in the second half of the 1980s, but has steadied (and improved greatly) in the 1990s as David Hohnen has come to terms with the now fully mature vineyard (and it with him).

🍷🍷🍷🍷🍷 **1995** Medium to full red-purple; a bouquet with uncommon depth and complexity, and a range of secondary earthy/cedary/berry characters already starting to appear. A wine with similarly good structure and depth to the palate although the flavours are tending more towards the savoury end of the spectrum than the opulently fruity. **rating:** 90

best drinking 2000–2010 **best vintages** '76, '78, '82, '83, '86, '90, '91, '93, '94, '95 **drink with** Loin of lamb • $43.20

capercaillie ★★★

Londons Road, Lovedale, NSW 2325 **region** Lower Hunter Valley
ph (02) 4990 2904 **fax** (02) 4991 1886 **open** Mon–Sat 9–5, Sun 10–5
winemaker Alasdair Sutherland **prod.** 5000 **est.** 1995
prod. range ($17–23 CD) Semillon, Unoaked Chardonnay, Chardonnay, Dessert Style Gewurztraminer, Hunter Shoalhaven Chambourcin, Coonawarra Orange Cabernet Sauvignon, C Sparkling Chambourcin.
summary The former Dawson Estate, now run by Hunter Valley veteran Alasdair Sutherland (no relation to Neil Sutherland of Sutherland Estate). The Capercaillie wines, and the Chardonnays in particular, are always extremely full flavoured and generous.

capogreco estate NR

Riverside Avenue, Mildura, Vic 3500 **region** Murray Darling and Swan Hill
ph (03) 5023 3060 **open** Mon–Sat 10–6
winemaker Bruno Capogreco **prod.** NFP **est.** 1976
prod. range ($8–12 CD) Riesling, Moselle, Shiraz-Mataro, Cabernet Sauvignon, Claret, Rosé, Fortifieds.
summary Italian-owned and run, the wines are a blend of Italian and Australian Riverland influences; the herb-infused Rosso Dolce is a particularly good example of its kind.

captain's paddock NR

Booie-Crawford Road, Kingaroy, Queensland 4610 **region** Other Wineries of Qld
ph (07) 4162 4534 **fax** (07) 4162 4502 **open** Weekends 9–4
winemaker Adam Chapman **prod.** 700 **est.** 1995
prod. range ($10–15 CD) Chardonnay, Shiraz, Captain's Red.
summary Don and Judy McCallum planted the first hectare of vineyard in 1995, followed by a further 3 hectares in 1996, focusing on shiraz and chardonnay. It is a family affair; the mud-brick cellar door building was made with bricks crafted by Don McCallum and Judy's screen printing adorns the tables and chairs and printed linen for sale to the public. Their two children are both sculptors, with works on display at the winery. Captain's Paddock is fully licensed, offering either light platters or full dishes incorporating local produce. Meals are served either inside or alfresco in the courtyard, with its views over the Booie Ranges.

carabooda estate NR

297 Carabooda Road, Carabooda, WA 6033 **region** Swan District
ph (08) 9407 5283 **fax** (08) 9407 5283 **open** 7 days 10–6
winemaker Terry Ord **prod.** 2000 **est.** 1989
prod. range ($12.50–18 CD) Sauvignon Blanc, Chenin Blanc, Shiraz, Cabernet Shiraz, Cabernet Sauvignon.

summary 1989 is the year of establishment given by Terry Ord, but it might as well have been 1979 (when he made his first wine) or 1981 (when he and wife Simonne planted their first vines). But it has been a slowly, slowly exercise, with production from the 3 hectares of estate plantings now supplemented by purchased grapes, the first public release not being made until mid-1994. As at March 1999 red wines from the 1994 and 1995 vintages were available ex-cellar door, the whites from 1998. Since that time production has risen significantly.

carbunup estate ★★★

Bussel Highway, Carbunup, WA 6280 **region** Margaret River
ph (08) 9755 1111 **open** 7 days 10–5
winemaker Robert Credaro **prod.** NFP **est.** 1988
prod. range ($12–16 CD) Under the premium Vasse River Wines label: Chardonnay, Semillon, Sauvignon Blanc; under Carbunup Estate label: Verdelho, Shiraz.
summary A relative newcomer, selling part of the grapes produced from the 18 hectares of vineyards, but keeping part for release under the Carbunup Estate and Vasse River labels – strikingly different in design, and giving no clue that they emanate from the same winery. It has had immediate success with its white wines, and in particular its Chardonnay and Semillon.

carindale wines NR

Palmers Lane, Pokolbin, NSW 2321 **region** Lower Hunter Valley
ph (02) 4998 7665 **fax** (02) 4998 7665 **open** Fri–Tues 10–4.30
winemaker Contract **prod.** 1000 **est.** 1996
prod. range ($18–25 CD) Chardonnay, Cabernet Franc, Merlot.
summary Carindale is still in its infancy, drawing upon 2 hectares of chardonnay, 1.2 hectares of cabernet franc and 0.4 hectares of merlot (together with few muscat vines). At this juncture the wines are available only through cellar door and by mail order.

carosa NR

310 Houston Street, Mount Helena, WA 6082 **region** Perth Hills
ph (08) 9572 1603 **fax** (08) 9572 1604 **open** Weekends, holidays 11–5 or by appointment
winemaker James Elson **prod.** 400 **est.** 1984
prod. range ($12–21 CD) Chardonnay, Classic Dry White, Pinot Noir, Cabernet Merlot, Janne Louise Méthode Champenoise, Old Tawny Port, White Port.
summary Very limited production and small-scale winemaking result in wines which can only be described as rustic, but which sell readily enough into the local market. Winemaker Jim Elson has extensive eastern Australia winemaking experience (with Seppelt) so should succeed. Almost all of the wine is sold through cellar door and by mailing list, but there is limited retail distribution through McLaren Vale Cellars, Canberra.

casa fontana NR

4 Cook Street, Lutana, Tas 7009 **region** Southern Tasmania
ph (03) 6272 3180 **open** Not
winemaker Steve Lubiana (Contract) **prod.** 300 **est.** 1994
prod. range Riesling, Chardonnay, Pinot Noir.
summary Mark Fontana, and Japanese wife Shige, planted their first pinot noir in 1994, and over the following two years expanded the vineyard to its present level of 2.6 hectares, 1 hectare each of pinot noir and chardonnay, and 0.6 hectares of riesling. Mark Fontana is a metallurgist with Pasminco, and came into grape growing through his love of fine wine. The '98 Pinot Noir, tasted ex-barrel, was very rich and concentrated, with lovely plum and cherry fruit.

cascabel ★★★☆

Rogers Road, Willunga, SA 5172 **region** McLaren Vale
ph (08) 8557 4434 **fax** (08) 8557 4435 **open** By appointment
winemaker Susana Fernandez, Duncan Ferguson **prod.** 2000 **est.** 1998
prod. range ($14–27 R) Eden Valley Riesling, Sauvignon Blanc, Adelaide Hills Pinot Noir, Shiraz, Grenache Shiraz, Sparkling Shiraz.

summary Cascabel's proprietors, Duncan Ferguson and Susana Fernandez, established Cascabel when they purchased a property at Willunga on the Fleurieu Peninsula and planted it to tempranillo, graciano, monastrel, grenache, shiraz and viognier. The choice of grapes reflected the winemaking experience of the proprietors in Australia, the Rhône Valley, Bordeaux, Italy, Germany and New Zealand – and also Susana Fernandez's birthplace, Spain. Both are fully qualified, and intend to move the production base steadily towards the style of the Rhône Valley, Rioja and other parts of Spain. In the meantime the 4000-case capacity winery which they erected on site prior to the 1998 vintage is being kept busy with grapes sourced from areas throughout South Australia. The initial releases left no doubt that the proprietors know what they are doing, and it will be interesting to watch the development of the wines from the estate plantings.

Cascabel Eden Valley Riesling

Made from grapes purchased from a single vineyard with dry-grown, low-yielding 40-year-old vines. 400 cases made.

🍷🍷🍷🍷 **1998** Light green-yellow; the bouquet has an interesting spectrum of aroma, with sweet spice/nutmeg nuances together with more tropical lime fruit. The palate is clean, fresh, nicely balanced with citrus lime flavours to the fore. Well made. **rating:** 87

best drinking 1999–2003 **best vintages** NA **drink with** Stir-fried prawns • $14

Cascabel Sparkling Shiraz

The base wine came from dry-grown eighty-year-old Clare Valley shiraz vines. It has been aged on lees for four years, and was obviously bought by Cascabel as a cleanskin.

🍷🍷🍷🍷 **NV** Full purple-red; the bouquet has quite rich chocolate and earth aromas, together with some liquorice and leather. Minimal oak influence. The palate is pleasant, not too tannic nor yet oaky nor yet sweet. Whether this relative austerity will appeal to all is a moot point. One hundred and thirty cases made. **rating:** 85

best drinking 1999–2005 **best vintages** NA **drink with** Terrine or paté • $27

casella ★★★

Wakley Road, Yenda, NSW 2681 **region** Riverina
ph (02) 6968 1346 **fax** (02) 6968 1196 **open** Not
winemaker Alan Kennett **prod.** 650 000 **est.** 1969
prod. range ($8.95–14.95 R) Under the newly adopted Carramar Estate label Semillon Sauvignon Blanc, Chardonnay, Unwooded Chardonnay, Botrytis Semillon, Merlot, Shiraz, Shiraz Cabernet, Cabernet Sauvignon.

summary Casella is typical of the new wave sweeping through the Riverina. It draws upon 285 hectares of estate vineyards, selling much of its wine in bulk or as cleanskin bottled wine to other producers, but also marketing a range of varietals under the Carramar Estate label. Casella is part of the Semillon of the Riverina group (promoting dry Semillon), and overall its wines offer good value for money. Predictably, the Botrytis Semillon is the outstanding release. The wines are exported to the US, UK, Japan and New Zealand.

Casella Carramar Estate Botrytis Semillon

Fully botrytised semillon is picked at 23° baumé and cold-fermented at 12 degrees in stainless steel. An excellent example of Riverina Botrytis Semillon, the '97 winning the trophy for Best New South Wales Sweet Wine at the 1998 New South Wales Wine Awards.

🍷🍷🍷🍷🍸 **1997** Medium to full yellow; a very complex bouquet with luscious botrytised fruit and oak. The palate is similarly ripe and luscious, with honeyed peachy/citrus fruit and well-balanced oak. **rating:** 90

best drinking 1999–2004 **best vintages** '97 **drink with** Any rich dessert • $14.95

Carramar Estate Merlot

A good example of modern Riverina winemaking in which yields have been sharply reduced to give colour and flavour undreamt of a few years ago. A strong silver medal performer at the 1999 Sydney Royal Wine Show.

🍷🍷🍷🍷 **1997** Medium red-purple; the bouquet is youthful, with strong earth and berry fruit, rather than oak, leading the way, with more of that very sweet cherry/berry fruit on the palate. **rating:** 89

best drinking 1999–2003 **best vintages** NA **drink with** Osso bucco • $10.95

cassegrain ★★★☆

Hastings River Winery, Fernbank Creek Road, Port Macquarie, NSW 2444 **region** Hastings River

ph (02) 6583 7777 **fax** (02) 6584 0354 **open** 7 days 9–5

winemaker David Barker **prod.** 55 000 **est.** 1980

prod. range ($9.95–29.95 R) Traminer Riesling, Sauvignon Blanc, Semillon, Verdelho, White Pinot, Chardonnay, Rose, Shiraz, Pinot Noir, Cabernet Merlot, Reserve Merlot, Chambourcin, Reserve Chambourcin; Fromenteau Chardonnay; Five Mile Hollow Red and White; Sparkling, Fortified and Dessert Wine.

summary A very substantial operation based in the Hastings Valley on the north coast of New South Wales. In earlier years, it drew fruit from many parts of Australia, but is now entirely supplied by the 162 hectares of estate plantings which offer 14 varieties, including the rare chambourcin, a French-bred cross. Exports to the UK, US, Japan, Switzerland, Germany, Holland, Canada and Asia.

Cassegrain Hastings River Semillon

Produced solely from grapes grown in various vineyards in the Hastings River region, and cold-fermented in a mix of stainless steel tanks and traditional large oak vats. Given one month's lees contact and then racked and bottled mid-year. Traditional Semillon making.

🍷🍷🍷🍷 **1998** Full yellow-green; the bouquet is smooth, quite deep, with some sweet apple fruit evident. The palate is clean, quite full and smooth, again showing a mix of citrus and ripe apple fruit flavours. The wine is helped by a twist of acidity on the finish which prevents it cloying. **rating:** 87

best drinking 1999–2002 **best vintages** '97, '98 **drink with** Oysters • $14.95

castle rock estate ★★★★

Porongurup Road, Porongurup, WA 6324 **region** Great Southern

ph (08) 9853 1035 **fax** (08) 9853 1010 **open** Mon–Fri 10–4, weekends and public holidays 10–5

winemaker Michael Staniford **prod.** 5000 **est.** 1983

prod. range ($14.50–22 CD) Riesling, Late Harvest Riesling, Great Southern White, Chardonnay, Pinot Noir, Cabernet Merlot, Merlot Cabernet Franc, Cabernet Sauvignon, Muscat Liqueur.

summary An exceptionally beautifully sited vineyard and cellar-door sales area with sweeping vistas from the Porongurups, operated by the Diletti family. The standard of viticulture is very high, and the site itself ideally situated (quite apart from its beauty). The Rieslings have always been elegant, and handsomely repaid time in bottle. In the most recent vintages the other wines of Castle Rock have improved considerably, with far greater weight and flavour than hitherto.

Castle Rock Estate Riesling

Consistently the best of the Castle Rock Estate wines, and ages wonderfully well. It is easy to overlook the wine in its youth, when it is typically very toasty, very crisp and on the lean side, but flowers with age. The lesson has been learned again and again, with four- and five-year vertical tastings producing the same result: the wines age superbly, gaining progressively higher points the older they are. In a mini vertical tasting in January 1998 the point scores for the older vintages went as follows: '96 – 88 points; '94 – 90 points; '93 – 92 points.

ΨΨΨΨ **1998** Light yellow-green; the bouquet is clean and crisp with mineral and lime fruit aromas which come on the crisp, delicate minerally palate. Touches of citrus are there too, and the wine will develop well over the next seven to ten years. **rating:** 88

best drinking 2003–2008 **best vintages** '86, '89, '90, '91, '93, '94, '96, '97, '98 **drink with** Seafood salad • $18.50

Castle Rock Estate Great Southern White

One of the cocktail white wines so favoured in the Margaret River and Great Southern regions, made from a blend of three and sometimes four varieties. In fact none are specified on the back label, but it doesn't matter, for the wine fits the bill precisely.

ΨΨΨΨ **1998** Light yellow-green; the bouquet is quite aromatic, with crisp, herbal and some more floral notes. There are corresponding flavours on the palate, quite intense and long. **rating:** 86

best drinking 1999–2001 **best vintages** NA **drink with** Seafood • $14.50

cathcart ridge estate NR

Moyston Road, Cathcart via Ararat, Vic 3377 **region** Grampians

ph (03) 5352 1997 **fax** (03) 5352 1558 **open** 7 days 10–5

winemaker David Farnhill, Simon Clayfield **prod.** 2000 **est.** 1977

prod. range ($9–60 CD) Estate Reserve Chardonnay, Grampians Riesling, Shiraz, Merlot, Cabernet Sauvignon; Rhymney Reef Colombard Chenin Blanc, Sauvignon Blanc, Shiraz, Shiraz Cabernet, Merlot, Cabernet Merlot, Cabernet Sauvignon, Old Tawny Port; and, at the bottom of the price range, Mount Ararat Dry White, Pinot Noir, Shiraz.

summary Now owned and operated by the Farnhill family; it remains to be seen whether the high reputation that Cathcart Ridge enjoyed in the early 1980s can be restored. The 1995 Reserve Merlot underlines the potential of the vineyard, not so much the winery.

Cathcart Ridge Estate Reserve Merlot

A wine which seems to have spent quite a long time in barrel, and which might also have benefited had some of the tannin been removed.

ΨΨΨΨ **1995** Medium red-purple; a fully mature, earthy but quite complex bouquet with faint hints of liquorice and anise is followed by a palate which opens up with sweet fruit but quickly gives way to drying tannins which start to impact from the middle palate onwards. **rating:** 86

best drinking 2000–2005 **best vintages** NA **drink with** Oxtail • $50

catherine vale vineyard ★★★☆

Milbrodale Road, Bulga, NSW 2330 **region** Lower Hunter Valley

ph (02) 6579 1334 **fax** (02) 6579 1334 **open** 7 days 10–5

winemaker John Hordern (Contract) **prod.** 725 **est.** 1994

prod. range ($12.80–14.50 CD) Semillon, Chardonnay.

summary Former schoolteachers Bill and Wendy Lawson have established Catherine Vale as a not-so-idle retirement venture. Both were involved in school athletics and sports programmes, handy training for do-it-yourself viticulturists. Most of the grapes from the 3.5-hectare vineyard are sold to contract-winemaker John Hordern; a small proportion is vinified for the Catherine Vale label.

Catherine Vale Semillon

Produced from 1.75 hectares of estate plantings, made in the traditional simple style, cold-fermented in stainless steel and early bottled.

YYYY **1998** Light to medium green-yellow; the bouquet is clean and crisp, with light grass, herb and mineral aromas. The wine has equally clean, lingering citrus and herb flavours, with bright acidity sustaining the length. Certain to develop well. **rating:** 89

best drinking 2000–2008 **best vintages** '98 **drink with** Vegetarian • $12.80

Catherine Vale Chardonnay

Like the Semillon, drawn from 1.75 hectares of estate plantings, fermented in stainless steel and then transferred to small American oak for maturation.

YYYY **1998** Medium to full yellow-green; abundant toasty barrel-ferment oak characters are followed by peachy, buttery fruit aromatics. The palate has good length and grip to the tangy fruit, together with lots of barrel-ferment oak. Heroic style. **rating:** 88

best drinking 1999–2001 **best vintages** '98 **drink with** Yakitori chicken • $14.50

celtic farm NR

94 Bridge Road, Richmond, Vic 3121 **region** Warehouse
ph (03) 9421 3334 **fax** (03) 9421 3335 **open** Not
winemaker Gerry Taggert **prod.** 2150 **est.** 1997
prod. range ($18–27.50 R) South Block Riesling, The Gridge Pinot Grigio, Raisin Hell Rutherglen Muscat.

summary Yet another Warehouse winery, the brainchild of a marketing and sales team of Mark McKenzie and Gerry Taggert. The proprietors say 'Celtic Farm is produced from classic varieties selected from Australia's premium wine regions and made with a total commitment to quality. While we have a desire to pay homage to our Celtic (drinking) heritage we are also acutely aware that wine should be about enjoyment, fun and not taking yourself too seriously'.

Celtic Farm South Block Riesling

The wine is made from grapes sourced from a number of growers in the Clare Valley; free-run and lightly pressed juice has resulted in an excellent regional style.

YYYYY **1998** Light to medium yellow-green; the bouquet is of medium intensity, with attractive lime blossom fruit aromas and plenty of depth. Attractive though the bouquet is, the wine is even better on the palate, with sustained fruit, great structure and length, and a pleasingly dry finish. **rating:** 93

best drinking 1999–2008 **best vintages** '98 **drink with** Vegetable terrine • $18

Celtic Farm Raisin Hell Rutherglen Muscat

Gerry Taggert and Mark McKenzie must have been in an exceptionally persuasive mode when they managed to acquire this Muscat, for it really is a very good example of moderately aged wine. The base wines for the Solera from which the wine comes were laid down in the early 1970s, but the wine is driven by younger yet high-quality material.

YYYYY **NV** Medium tawny; the aromas are quite lovely, with intense grapey/raisiny fruit, the palate rich and concentrated, with classic raisin, Christmas cake flavours. A powerful argument for the cause of younger rather than older Muscat. **rating:** 90

best drinking 1999–2000 **best vintages** NA **drink with** Appropriate company • $27.50

chain of ponds ★★★★☆

Adelaide Road, Gumeracha, SA 5233 **region** Adelaide Hills
ph (08) 8389 1415 **fax** (08) 8389 1877 **open** 7 days 10.30–4.30
winemaker Caj Amadio (Contract) **prod.** 8000 **est.** 1993
prod. range ($10–29 CD) Riesling, Semillon, Sauvignon Blanc Semillon, Special Release Sauvignon Blanc, Chardonnay, Novello Rosso, Pinot Noir, Ledge Shiraz, Amadeus Cabernet Sauvignon, Diva Pinot Chardonnay.
summary Caj and Genny Amadio are the largest growers in the Adelaide Hills, with 100 hectares of vineyards established on a Scott Henry trellis producing 1000 tonnes of grapes a year, almost all sold to Penfolds, but with a small amount made into wine for sale under the Chain of Ponds label. The first vintage was 1993, and the wines first offered for sale in 1995. The full-flavoured white wines have enjoyed consistent show success.

Chain of Ponds Riesling

Competently made in a direct, no-frills style. The moderately warm vineyards produce a wine slightly more towards Clare Valley in style than the more exotically fruity wines of (say) the Lenswood subdistrict of the Adelaide Hills.

🍷🍷🍷🍷🍷 **1998** Medium yellow-green; the bouquet is classic lime Riesling with a steely core; the palate has plenty of weight, yet is not phenolic, offering flavours of lime and herb, with good acidity sustaining a long finish. **rating:** 93

best drinking 2000–2007 **best vintages** '98 **drink with** Globe artichoke • $15

Chain of Ponds Sauvignon Blanc Semillon

First produced in 1997, and met with immediate success, winning a gold medal at the 1997 Adelaide Hills Wine Show. The '98 is in the same class.

🍷🍷🍷🍷🍷 **1998** The bouquet is quite aromatic, with crisp, herbaceous fruit, the palate lively, tingling and fresh. Two tastings within weeks of each other both provoked the comment that the wine is particularly well balanced. **rating:** 94

best drinking 1999–2000 **best vintages** '98 **drink with** Blue swimmer crab • $20

Chain of Ponds Ledge Shiraz

The wine takes its name from a weathered shale ledge which has given the particular block its surface soil over a well-drained stony base. This type of site is necessary in the Adelaide Hills to guarantee ripening of varieties such as shiraz.

🍷🍷🍷🍷 **1997** Medium to full red-purple; the bouquet is quite intense with savoury, leathery fruit combined with a dusting of vanilla oak. The palate is quite soft and forward, with red berry, vanilla and mocha flavours. The length and finish are, indeed, the best features of a more than useful wine. **rating:** 87

best drinking 2001–2010 **best vintages** NA **drink with** Braised lamb with couscous • $29

Chain of Ponds Amadeus Cabernet Sauvignon

Amadeus may have been the second Christian name of Mozart, but it is also old Latin for Amadio, and hence its adoption for what the Amadios clearly see as their flagship red wine. Like all of the Chain of Ponds wines, estate-grown, and extremely competently made.

🍷🍷🍷🍷 **1997** Medium to full red-purple; the bouquet is strong, with a mix of cedar, cassis, earth and more dusty notes. Once tasted, the wine offers pristine, uncompromising, savoury, dry cabernet fruit with firm tannins. Could be something quite special in five or so years time. **rating:** 86

best drinking 2001–2011 **best vintages** NA **drink with** Rib of beef • $29

chalk hill NR

Field Street, McLaren Vale, SA 5171 **region** McLaren Vale

ph (08) 8556 2121 **fax** (08) 8556 2121 **open** Not

winemaker Contract **prod.** 2500 **est.** 1973

prod. range ($5.90–13 CD) Riesling, Rosé, Shiraz, Cabernet Sauvignon, Port.

summary Chalk Hill is temporarily closed, but has been producing some lovely wine made for its new owners, prominent Yarra Valley grape growers John and Di Harvey (who acquired Chalk Hill in 1996).

Chalk Hill Shiraz

Produced from high-quality McLaren Vale shiraz which has been allowed to ripen to 14.5° baumé (producing 14.5 degrees alcohol) and matured in American oak. The power and density of the fruit happily accommodates that oak.

ŸŸŸŸŸ **1997** Medium to full purple-red; rich and dense ripe black fruit aromas are followed by a palate oozing McLaren Vale chocolate and black cherry fruit from every pore. The wine has gobbled up the oak like a black hole in space. **rating:** 94

best drinking 2002–2017 **best vintages** '97 **drink with** Rich game dishes • $21

chambers rosewood ★★★★☆

Barkley Street, Rutherglen, Vic 3685 **region** Rutherglen

ph (02) 6032 8641 **fax** (02) 6032 8101 **open** Mon–Sat 9–5, Sun 11–5

winemaker Bill Chambers **prod.** 10 000 **est.** 1858

prod. range ($8–75 CD) A range of modestly priced and modestly made table wines, but the real forte of the winery is its Tokay and Muscat, also old Amontillado Sherries and Ports.

summary Bill Chambers is one of the great characters of the industry, but is not given to correspondence or to filling out forms. It remains a matter of record, however, that Bill Chambers makes some of the very greatest, albeit nearly unprocurable, Special Old Liqueur Muscat and Special Old Liqueur Tokay to be found in the northeast of Victoria. No recent tastings; Bill Chambers doesn't believe in wasting wine on wine scribes. Small quantities of the wines find their way to the US and UK.

chapel hill ★★★★

Chapel Hill Road, McLaren Vale, SA 5171 **region** McLaren Vale

ph (08) 8323 8429 **fax** (08) 8323 9245 **open** 7 days 12–5

winemaker Pam Dunsford (Consultant), Angela Meaney **prod.** 45 000 **est.** 1979

prod. range ($13.50–35 CD) Eden Valley Riesling, Unwooded Chardonnay, Reserve Chardonnay, Verdelho, McLaren Vale Shiraz, The Vicar (Cabernet Shiraz), Reserve Cabernet Shiraz, McLaren Vale/Coonawarra Cabernet Sauvignon, Tawny Port.

summary A winery which, in the 1990s, leapt from obscurity to national prominence after a change of ownership, a very large capital injection, and the installation of Pam Dunsford as consultant winemaker. In the early phases of the growth the wines were superb, but continued growth seems to have taken some of the edge off. It is a reasonable expectation that once growth has slowed, quality will return to the very highest level. Exports to the UK, Germany and Switzerland.

Chapel Hill Reserve Chardonnay

Fully deserves the Reserve label. It is sourced from premium vineyards in McLaren Vale, Padthaway and Coonawarra and barrel-fermented in high-quality French oak to produce a wine of great style which is a consistent medal winner in shows.

🍷🍷🍷🍷🍷 **1997** Medium yellow-green; the moderately intense bouquet is clean and smooth, with nicely balanced and integrated oak. A very attractive, unforced style, with elegant fig, melon and citrus fruit, crisp acidity and just a light touch of oak. **rating:** 90

best drinking 1999–2002 **best vintages** '91, '92, '93, '94, '97 **drink with** Slow-cooked fresh salmon • $21

Chapel Hill Shiraz

Yet another outstanding red wine from Chapel Hill, utilising grapes grown at Bakers Gully (halfway up the Adelaide Hills escarpment), at Blewitt Springs, and at Chapel Hill's home vineyard. The wine spent 15 months in American oak, 20% new, the remainder first and second use. As ever, made with a flawless touch.

🍷🍷🍷🍷🍷 **1996** Strong, deep purple-red; rich, concentrated and sweet black cherry fruit is joined by a touch of chocolate on the similarly-flavoured palate. A full wine, very good structure and oak handling. **rating:** 90

best drinking 2001–2006 **best vintages** '91, '94, '95 **drink with** Grilled calf's liver • $26

Chapel Hill The Vicar Cabernet Shiraz

First released with the 1994 vintage, a blend of 85% McLaren Vale and 15% Coonawarra grapes, the varietal mix typically being 55% Cabernet Sauvignon and 45% Shiraz. The wine spent 22 months in a mix of American and French oak. There are no prizes for guessing the inspiration for the name of the wine.

🍷🍷🍷🍷 **1996** Medium to full red-purple; the bouquet is at once complex, soft and seductive with sweet berry fruit and sweet oak in harmony with each other. The palate emphasises the ripeness and sweetness of the berry fruit with that glossy/minty overlay which is so typical of ripe South Australian Shiraz. **rating:** 89

best drinking 2001–2007 **best vintages** '94, '96 **drink with** Entrecôte of beef • $35

Chapel Hill McLaren Vale/Coonawarra Cabernet Sauvignon

A blend of roughly two-thirds McLaren Vale Cabernet Sauvignon and one-third Coonawarra, varying slightly from one vintage to the next, matured in a mix of quality French and American oak. Consistently good over the years, peaking with the '92 vintage, which capped a distinguished show career by winning three trophies at the 1994 National Wine Show in Canberra, including Best Table Wine of Show.

🍷🍷🍷🍷 **1996** Medium to full red-purple; ripe cassis, almost jammy, aromas lead the bouquet and logically introduce lots of cassis, redcurrant and other red fruit flavours on the forepalate, with unexpectedly powerful tannins to close. **rating:** 85

best drinking 2001–2008 **best vintages** '88, '90, '91, '92, '95 **drink with** Marinated beef • $23

chapman's creek vineyard NR

RMS 447 Yelverton Road, Willyabrup, WA 6280 **region** Margaret River
ph (08) 9755 7545 **fax** (08) 9755 7571 **open** 7 days 10.30–4.30
winemaker Various Contract **prod.** 5000 **est.** 1989
prod. range ($13–28 CD) Chenin Blanc, Unoaked Chardonnay, Chardonnay, Merlot, Cabernet Merlot, Tawny Port.

summary Tony Lord is an extremely experienced wine journalist, who for many years was editor and part-owner of *Decanter* magazine of the United Kingdom, one of the leaders in the field. He still writes for Australian magazines and newspapers, and knows the industry extremely well. Notwithstanding this, he has been positively reclusive about Chapman's Creek Vineyard, which he owns. I have unofficially tasted a number of the wines over the past two years, and those I have tasted have been excellent, fully reflecting the outstanding quality of the Willyabrup subregion of the Margaret River.

charles cimicky ★★★☆

Gomersal Road, Lyndoch, SA 5351 **region** Barossa Valley
ph (08) 8524 4025 **fax** (08) 8524 4772 **open** Tues–Sat 10.30–4.30
winemaker Charles Cimicky **prod.** 15 000 **est.** 1972
prod. range ($15–25 CD) Sauvignon Blanc, Chardonnay, Cabernet Franc, Classic Merlot, Cabernet Sauvignon, Signature Shiraz, Old Fireside Tawny Port.
summary These wines are of very good quality, thanks to the lavish (but sophisticated) use of new French oak in tandem with high-quality grapes. The intense, long-flavoured Sauvignon Blanc has been a particularly consistent performer, as has the rich, voluptuous American-oaked Signature Shiraz. Limited retail distribution in South Australia, Victoria, New South Wales and Western Australia, with exports to UK, US, Switzerland, Canada, Malaysia and Hong Kong.

charles melton ★★★★★

Krondorf Road, Tanunda, SA 5352 **region** Barossa Valley
ph (08) 8563 3606 **fax** (08) 8563 3422 **open** 7 days 11–5
winemaker Charlie Melton, Joanne Aherne **prod.** 7500 **est.** 1984
prod. range ($14.90–42.90 R) Rosé of Virginia, Grenache, Shiraz, Nine Popes (Shiraz Grenache Mourvedre), Cabernet Sauvignon, Sparkling Red.
summary Charlie Melton, one of the Barossa Valley's great characters, with wife Virginia by his side, makes some of the most eagerly sought à la mode wines in Australia. Inevitably, the Melton empire grew in response to the insatiable demand, with a doubling of estate vineyards to 13 hectares (and the exclusive management and offtake of a further 10 hectares), and the erection of a new barrel store in 1996. The expanded volume has had no adverse effect on the wonderfully rich, sweet and well-made wines. Exports to the UK, Ireland, Switzerland, France, US and southeast Asia.

Charles Melton Rosé Of Virginia

One of the most interesting Rosés currently made in Australia, produced from grenache, and presumably a partial by-product of Nine Popes juice run-off. It has much more fruit flavour than a standard Rosé, at least being more a cross between a standard Rosé and a Beaujolais style. Tremendous summer drinking.

YYYYY **1998** Vivid purple–red; the floral bouquet has fresh aromas of crushed rose petals and a dusting of spice, the palate with dancing spicy, cherry and crushed rose petal flavours. Finishes refreshingly dry. **rating:** 90

best drinking 1998–2000 **best vintages** '97, '98 **drink with** Light Mediterranean dishes • $14.90

---

charles reuben estate NR

777 Middle Tea Tree Road, Tea Tree, Tas 7017 **region** Southern Tasmania
ph (03) 6268 1702 **fax** (03) 6231 3571 **open** Thur–Sun 10–5
winemaker Tim Krushka **prod.** 350 **est.** 1990
prod. range ($15–22 CD) Semillon Sauvignon Blanc, Chardonnay, Riesling, Pinot Noir, Bordeaux-blend red.
summary Charles Reuben Estate has 1.5 hectares of pinot noir, half a hectare of chardonnay and a few rows of riesling in production. It has also planted 1.2 hectares of the four Bordeaux varieties, headed by cabernet sauvignon with a little cabernet franc, merlot and petit verdot, and 0.6 hectare of sauvignon blanc accompanied by a few rows of semillon. The principal wines will be Pinot Noir, Chardonnay, a Bordeaux-blend red and a Sauvignon Blanc Semillon, although there is an element of trial in the plantings to establish which varieties succeed best on the estate.

charles sturt university winery ★★★☆

Boorooma Street, North Wagga Wagga, NSW 2650 **region** Other Wineries of NSW
ph (02) 6933 2435 **fax** (02) 6933 2107 **open** 7 days 10–4
winemaker Kirsten Munro, James Kelly **prod.** 16 000 **est.** 1977
prod. range ($10–22 R) The precise composition varies from one release to the next, but is divided into two sections: the top-of-the-range Limited Release Series (e.g. Cabernet Sauvignon Shiraz, Cowra Chardonnay, Méthode Champenoise, Cabernet Sauvignon, Botrytis Semillon, Liqueur Port and Liqueur Muscat) and a basic range of lower-priced varietals including Chardonnay, Traminer Riesling, Sauvignon Blanc Semillon and Cabernet Sauvignon Shiraz.
summary Between 1990 and 1996 winemaking at Charles Sturt University was carried out under the direction of Rodney Hooper, who managed to resolve the dual roles of producing commercial wines and teaching students with consummate skill. Neither the '97 nor '98 vintages seemed to be in the same class as those made by Hooper, although they do offer quite good value at their price points. The wines are exported to the US.

charley brothers ★★☆

The Ruins Way, Inneslake, Port Macquarie, NSW 2444 **region** Hastings River
ph (02) 6581 1332 **fax** (02) 6581 0391 **open** Mon–Fri 1–5, weekends 10–5
winemaker Cassegrain (Contract) **prod.** 1500 **est.** 1988
prod. range ($9–14 CD) Semillon, Chardonnay, Semillon Chardonnay, Summer White, Dry Red, Shiraz Cabernet.
summary The property upon which the Charley Brothers vineyards are established has been in the family's ownership since the turn of the century, but in fact had been planted to vines by a Major Innes in the 1840s. After carrying on logging and fruit growing at various times, the Charley family planted vines in 1988 with the encouragement of John Cassegrain. A little over 10.5 hectares of vines have been established. The '98 vintage wines are pleasant, and certainly appropriate to the cellar door.

charlotte plains NR

The Grange, RMB 3180, Dooleys Road, Maryborough, Vic 3465 **region** Bendigo
ph (03) 5361 3137 **open** By appointment
winemaker Roland Kaval **prod.** 80 **est.** 1990
prod. range ($16 ML) Shiraz.
summary Charlotte Plains is a classic example of miniaturism. Production comes from a close-planted vineyard which is only one-third of a hectare, a quarter being shiraz, the remainder sauvignon blanc. The minuscule production is sold solely through the mailing list and by phone, but the '96 Shiraz (the second vintage from Charlotte Plains) was awarded four stars in *Winestate* magazine in mid-1997, and judged as the Equal Best Shiraz for Central Victoria.

chateau doré NR

Mandurang Road, via Bendigo, Vic 3551 **region** Bendigo
ph (03) 5439 5278 **open** Tues–Sun 10–6
winemaker Ivan Gross **prod.** 1000 **est.** 1860
prod. range ($9–14 CD) Riesling, Shiraz, Cabernet Sauvignon, Tawny Port.
summary Has been in the ownership of the Gross family since 1860 with the winery buildings dating back respectively to 1860 and 1893. All wine is sold through cellar door.

chateau dorrien NR

Cnr Seppeltsfield Road. and Barossa Valley Way, Dorrien, SA 5352 **region** Barossa Valley
ph (08) 8562 2850 **fax** (08) 8562 1416 **open** 7 days 10–5
winemaker Fernando Martin **prod.** 2000 **est.** 1983

prod. range ($10–18 CD) Riesling, Semillon Chardonnay, Traminer, Frontignac Traminer, Frontignac Spaetlese, Late Harvest Frontignac, Semillon Chardonnay Sparkling Brut, Prima Vera (light red), Limited Release Grenache, Shiraz, Cabernet Sauvignon, Tawny Port.
summary Unashamedly and successfully directed at the tourist trade.

chateau francois ★★★

Broke Road, Pokolbin, NSW 2321 **region** Lower Hunter Valley
ph (02) 4998 7548 **fax** (02) 4998 7805 **open** Weekends 9–5 or by appointment
winemaker Don Francois **prod.** 700 **est.** 1969
prod. range ($11 ML) Pokolbin Mallee Semillon, Chardonnay, Shiraz Pinot Noir.
summary The retirement hobby of former NSW Director of Fisheries, Don Francois. Soft-flavoured and structured wines which frequently show regional characters, but which are modestly priced and are all sold through the cellar door and mailing list to a loyal following. The tasting room is available for private dinners for 12 to 16 people. Don Francois has sailed through a quadruple-bypass followed by a mild stroke with his sense of humour intact, if not enhanced. A recent newsletter says (inter alia) '... my brush with destiny has changed my grizzly personality and I am now sweetness and light ... Can you believe? Well, almost!' He even promises comfortable tasting facilities.

chateau hornsby NR

Petrick Road, Alice Springs, NT 0870 **region** Alice Springs
ph (08) 8955 5133 **fax** (08) 8955 5133 **open** 7 days 11–4
winemaker Gordon Cook **prod.** 1000 **est.** 1976
prod. range ($12–17 CD) Riesling, Semillon, Chardonnay, Shiraz, Cabernet Sauvignon.
summary Draws in part upon 3 hectares of estate plantings, and in part from grapes and wines purchased from other regions. Very much a tourist-oriented operation, with numerous allied entertainments on offer.

chateau leamon ★★★

5528 Calder Highway, Bendigo, Vic 3550 **region** Bendigo
ph (03) 5447 7995 **fax** (03) 5447 0855 **open** Wed–Mon 10–5
winemaker Ian Leamon **prod.** 1500 **est.** 1973
prod. range ($13–20 CD) Riesling, Semillon, Chardonnay, Shiraz, Reserve Shiraz, Cabernet Merlot.
summary After a period of uncertainty, Chateau Leamon is returning to some of its former glory. Ian Leamon is using both locally grown grapes, but also looking to the Strathbogie Ranges for grapes for other wines, including Pinot Noir. Limited retail distribution in Victoria and Queensland.

Chateau Leamon Reserve Shiraz

First made in 1996, and produced from 26-year-old vines grown at Chateau Leamon's Big Hill Vineyard near Bendigo. It is open fermented, and spends a little over 12 months in 100% new American oak hogsheads.

ŸŸŸŸŸ **1997** Full red-purple; wonderfully rich, sweet and juicy liquorice, black cherry and sweet oak flood the bouquet. The palate is redolent of liquorice and cherry fruit supported by high-toned clove spice American oak. A show-stopper style which has been wonderfully handled. **rating:** 94

best drinking 2002–2012 **best vintages** '97 **drink with** Rich game • $30

chateau pato ★★★★

Thompson's Road, Pokolbin, NSW 2321 **region** Lower Hunter Valley
ph (02) 4998 7634 **open** By appointment
winemaker Nicholas Paterson **prod.** 400 **est.** 1978
prod. range ($14–19 CD) Gewurztraminer, Pinot Noir, Shiraz.
summary Nicholas and Roger Paterson have taken over responsibility for this tiny winery following the death of their father David Paterson during the 1993 vintage. It is a much-loved, if tiny, Hunter landmark, with vintages back to 1993 available at the microscopic cellar door. On all the evidence, David Paterson's inheritance is being handsomely guarded.

chateau tahbilk ★★★★

Goulburn Valley Highway, Tabilk, Vic 3607 **region** Goulburn Valley
ph (03) 5794 2555 **fax** (03) 5794 2360 **open** Mon–Sat 9–5, Sun 11–5
winemaker Alister Purbrick, Neil Larson, Tony Carapetis **prod.** 100 000 **est.** 1860
prod. range ($9.95–99.95 R) Riesling, Chardonnay, Marsanne, Shiraz, Cabernet Sauvignon; 1860 Vines Shiraz is rare flagship, with a Reserve Red released from each vintage.
summary A winery steeped in tradition (with high National Trust classification) which should be visited at least once by every wine-conscious Australian, and which makes wines – particularly red wines – utterly in keeping with that tradition. The essence of that heritage comes in the form of the tiny quantities of Shiraz made entirely from vines planted in 1860. As well as Australian national distribution through Tucker Seabrook, Chateau Tahbilk has agents in every principal wine market, including the UK, European and North American.

Chateau Tahbilk Marsanne

The best known of Chateau Tahbilk's wines, with an illustrious history. Made in a very simple and direct fashion, in total contrast to neighbour Mitchelton, and in particular without the use of oak. Like young traditional Semillon, frequently needs time to come into its own. As a matter of interest, the 40 hectares of marsanne planted at Chateau Tahbilk is the largest single vineyard planting of that variety in the world.

🍷🍷🍷🍷 **1998** Light green-yellow; a clean, crisp lemony bouquet is accompanied by a palate with excellent length and early flavour, all pointing to great development potential. It is very difficult to give appropriate points to a wine barely out of its nappies. **rating:** 89

best drinking 2003–2008 **best vintages** '53, '74, '79, '81, '84, '85, '87, '88, '91, '92, '93, '95, '96, '98 **drink with** Lighter Italian or Asian dishes • $11.95

Chateau Tahbilk Shiraz

A rock of ages, deliberately made by Alister Purbrick in precisely the same way as the wines of ten, 20, 30 and 40 years ago. Normally there is tremendous colour, flavour and extract, with little or no oak influence. The only question about the better vintages is whether they need five, ten or 20 years in the cellar. Quite why the '94, '95 and '96 vintages have been (relatively speaking) weak and somewht dilute I do not know.

🍷🍷🍷🍸 **1996** Medium red, with a touch of purple; the bouquet is clean, but on the light side and slightly washed out. The palate is pleasant, with simple, straightforward cherry fruit and soft tannins; seems to continue a subtle change in style, although it may just have been the vintage speaking. **rating:** 84

best drinking 2000–2005 **best vintages** '61, '62, '65, '68, '71, '74, '76, '78, '81, '84, '85, '86, '91, '92 **drink with** Barbecued T-bone steak • $19.95

Chateau Tahbilk Cabernet Sauvignon

As with all of the Tahbilk wines, estate-grown, produced from a significant percentage of old vines. This is terroir speaking, the hand of the maker being deliberately withdrawn. As from 1995, 10% Cabernet Franc was incorporated in the wine.

🍷🍷🍷🍷 **1996** Medium red; the bouquet shows very ripe, distinctly jammy fruit, the palate similarly jammy and ripe in an almost Spanish mould. **rating:** 84

best drinking 2000–2005 **best vintages** '65, '71, '72, '75, '76, '78, '79, '86, '90, '92 **drink with** Strong mature cheddar, stilton • $19.95

Chateau Tahbilk Special Reserve Cabernet Sauvignon

A wine with a proud history, numbering among its past vintages some of the most memorable full-bodied Australian dry reds made since the Second World War. In earlier decades, oscillated between a shiraz pressings and a cabernet sauvignon base, but has now settled upon cabernet sauvignon. Approximately 500 cases are made each vintage, but this is very much dependent on the year, and in fact none was made in either 1987 or 1989. Since the appearance of the 1991 vintage in late 1998 the name has been changed from Special Bin to Special Reserve and the bin number dropped.

🍷🍷🍷🍷🍷 **1991** Medium to full red, still with touches of purple remaining. The bouquet is clean and rich, with abundant red berry, chocolate and earth aromas. The palate is very powerful, pitting strong tannins against dark chocolatey/berry fruit. It is impossible to tell which force will prevail. **rating:** 90

best drinking 2006–2011 **best vintages** '62, '65, '68, '71, '75, '76, '80, '81, '82, '84, '85 **drink with** Rich red meat dishes • $49.95

chatsfield ★★★★

O'Neil Road, Mount Barker, WA 6324 **region** Great Southern

ph (08) 9851 1704 **fax** (08) 9851 1704 **open** Tues–Sun, public holidays 10–5

winemaker Steve Pester **prod.** 10 000 **est.** 1976

prod. range ($14–17 CD) Mount Barker Riesling, Gewurztraminer, Chardonnay, Cabernet Franc, Shiraz.

summary Irish-born medical practitioner Ken Lynch can be very proud of his achievements at Chatsfield, as can the various contract-winemakers who have taken the high-quality estate-grown material and made such impressive wines, notably the Riesling, vibrant Cabernet Franc (as an unwooded nouveau style) and spicy liquorice Shiraz.

Chatsfield Chardonnay

By no means a flashy wine, made in a reserved style, and unequivocally showing its cool-climate background, often seeming more like a cross between Sauvignon Blanc and Chardonnay than anything else. Will mature well with age.

🍷🍷🍷🍷🍷 **1997** Light to medium green-yellow; the bouquet is crisp and firm with hints of mineral and chalk running through tight lime juice aromas. The elegant palate has that classic balance of fruit and mineral notes, with a long, clean finish. **rating:** 92

best drinking 2001–2011 **best vintages** '87, '89, '90, '93, '97 **drink with** Smoked salmon • $17

Chatsfield Shiraz

Has been the star performer for Chatsfield in most vintages, coming from vineyards which are more than 20 years old.

🍷🍷🍷🍷 **1997** Medium to full red-purple; the bouquet is quite fragrant, with moderately intense liquorice, cedar and spice aromas. On the palate cedar, chocolate and some red berry/cherry fruit is supported by soft tannins on quite a long finish. **rating:** 86

best drinking 2001–2007 **best vintages** '88, '90, '91, '92, '94, '97 **drink with** Mature cheddar, finer red meat dishes • $17

Chatsfield Cabernet Franc

The wine is challenging, because it is fermented in tank and taken straight to bottle, without any wood maturation at any stage.

ŸŸŸŸ **1998** Bright, full purple-red; the fresh berry aromas of the bouquet have a pleasing touch of bite; the palate offers lively, fresh cherry/berry fruit, with well-balanced acidity and tannin. Doesn't show the sweaty, tanky characters of some wines which have not been put in oak. **rating:** 87

best drinking 1999–2001 **best vintages** NA **drink with** Breast of duck • $15

chestnut grove ★★★☆

Chestnut Grove Road, Manjimup, WA 6258 **region** Pemberton

ph (08) 9771 4255 **fax** (08) 9772 4255 **open** By appointment

winemaker Kim Horton, John Griffiths (Consultant) **prod.** 10 000 **est.** 1988

prod. range ($16.95–23.95 CD) Chardonnay, Sauvignon Blanc, Verdelho, Pinot Noir, Cabernet Merlot, Merlot.

summary A joint venture between the Lange family of Alkoomi and Vic Kordic and his family, through to grandson Darren Cook. Initial vintages were slightly weak and dilute, but increasing vine age (and one suspects better viticulture) has resulted in a significant lift in wine quality, particularly with the Cabernet Merlot. Production, too, has been more than doubled, with distribution in the eastern States and exports to Germany.

Chestnut Grove Sauvignon Blanc

1998 was not a great vintage for Sauvignon Blanc anywhere in Australia, but this wine rises above the circumstances of the vintage. A surprise performer at the 1998 National Wine Show, but underlining the progress made by Chestnut Grove in building flavour in its wines over the past four or five years. It also suggests the Pemberton region may be difficult to put in any one pigeonhole.

ŸŸŸŸŸ **1998** Light straw-yellow; voluminous, fragrant passionfruit and gooseberry aromas which leap from the glass. The palate is crisp, clean and flavoursome, with excellent acidity and verve. **rating:** 94

best drinking 1999–2000 **best vintages** '98 **drink with** Smoked chicken • $17.95

Chestnut Grove Verdelho

Produced from 3 hectares of estate plantings established in 1988. Successive releases have shown increasing flavour and complexity, but right from the word go, clear varietal character has been evident. The 1998 wine topped its class at the 1999 *Winewise* annual wine competition.

ŸŸŸŸ **1998** Medium yellow-green; the bouquet has real character, with varietally-typical tropical/canned fruit salad aromas. The palate doesn't offer quite as much as the bouquet suggests, but it does have good length and acidity, and all in all, is a very good example of the variety. **rating:** 89

best drinking 1999–2000 **best vintages** '97 **drink with** Asian seafood • $16.95

chestnut hill vineyard NR

1280 Pakenham Road, Mount Burnett, Vic 3781 **region** Port Phillip Zone

ph (03) 5942 7314 **fax** (03) 5942 7314 **open** Weekends and public holidays 10.30–5.30, or by appointment

winemaker Charlie Javor **prod.** 1100 **est.** 1995

prod. range ($15–18 CD) Chardonnay, Sauvignon Blanc, Shiraz.

summary Charlie and Ivka Javor started Chestnut Hill with small plantings of chardonnay and shiraz in 1985, and have slowly increased the vineyards to their present total of a little over 3 hectares. The first wines were made in 1995, and all distribution is through the cellar door and direct to a few restaurants. Situated less than one hour's drive from Melbourne, the picturesque vineyard is situated among the rolling hills in the southeast of the Dandenongs near Mount Burnett. The wines reflect the cool climate.

chittering estate NR

Chittering Valley Road, Lower Chittering, WA 6084 **region** Other Wineries of WA
ph (08) 9273 6255 **fax** (08) 9273 6101 **open** Weekends and public holidays 11–4.30 (Apr–Dec)
winemaker Francois Jacquard **prod.** 12 000 **est.** 1982
prod. range ($14.90–21 R) Chardonnay, Hill Top Reserve Chardonnay, Semillon Sauvignon Blanc, Pinot Noir, Cabernet Merlot, Hill Top Reserve Cabernet Sauvignon.
summary Chittering Estate was sold in late 1997, and no information has been forthcoming about its new owner's intentions.

ciavarella NR

Evans Lane, Oxley, Vic 3678 **region** King Valley
ph (03) 5727 3384 **fax** (03) 5727 3384 **open** Mon–Sat 9–6, Sun 10–6
winemaker Cyril Ciavarella **prod.** 2000 **est.** 1978
prod. range ($10–14 CD) Chenin Blanc, Late Harvest Chenin Blanc, Chardonnay, Vendemmia (light red), Dolcino (medium-bodied red), Shiraz, Cabernet Sauvignon, Tawny Port.
summary The Ciavarellas have been grape growers in the King Valley for almost 20 years, selling their grapes to wineries such as Brown Brothers. Changes in the grape marketplace led to Ciavarella deciding to make limited quantities of wines, which were first offered for sale from the cellar door in early 1994. The '98 Chardonnay and '97 Cabernet Sauvignon are pleasant wines in lighter King Valley style.

clarendon hills winery ★★★★☆

Brookmans Road, Blewitt Springs, SA 5171 **region** McLaren Vale
ph (08) 8364 1484 **fax** (08) 8364 1484 **open** By appointment
winemaker Roman Bratasiuk **prod.** 10 000 **est.** 1989
prod. range ($60–175 R) Chardonnay (Kangarilla Vineyard and Norton Summit Vineyard), Pinot Noir, Merlot, Old Vines Grenache (Blewitt Springs Vineyard and Clarendon Vineyard), Astralis (Shiraz – and the flagship wine).
summary Clarendon Hills produces some of the most startlingly concentrated, rich and full-bodied red wines to be found in Australia, rivalled in this respect only by Wendouree. Roman Bratasiuk is a larger-than-life figure who makes larger-than-life wines. Technocrats may quibble about this or that aspect, but influential judges such as Robert Parker have neither reservations about nor problems with the immense, brooding red wines which Bratasiuk regularly produces from small patches of old, low-yielding vines which he ferrets out.

cleveland ★★★☆

Shannons Road, Lancefield, Vic 3435 **region** Macedon
ph (03) 5429 1449 **fax** (03) 5429 2017 **open** 7 days 9–6
winemaker Keith Brien **prod.** 4000 **est.** 1985
prod. range ($11–25 CD) Estate Chardonnay, Pinot Noir, Minus Five (Cabernets Merlot), Cabernet Sauvignon, Macedon Brut; Brien Family Selection Muscat Gordo Blanco, Chardonnay Gordo, Shiraz Cabernet.
summary The Cleveland homestead was built in 1889 in the style of a Gothic Revival manor house, but had been abandoned for 40 years when purchased by the Briens in 1983. It has since been painstakingly restored, and 3.5 hectares of surrounding vineyard established. Cleveland has done best with Pinot Noir and Chardonnay, but the occasional Cabernet Sauvignon attests to an unusually favourable vineyard site in a very cool region.

cliff house ★★★☆

57 Camms Road, Kayena, Tas 7270 **region** Northern Tasmania
ph (03) 6394 7454 **fax** (03) 6394 7454 **open** By appointment
winemaker Julian Alcorso (Contract) **prod.** 2500 **est.** 1983
prod. range ($18–20 R) Riesling, Chardonnay, Pinot Noir, Devil's Elbow (Pinot Cabernet blend), Cabernet Sauvignon.
summary Geoff and Cheryl Hewitt established 4 hectares of vineyard in the Tamar Valley area in 1983. After a slightly slow start, the Cliff House wines have really come into their own since 1994, having particular success at the 1999 Tasmanian Wines Show.

Cliff House Riesling

A gold medal winner at the 1999 Tasmanian Wines Show.
YYYYY **1998** Light yellow-green; the bouquet is shy and reserved, with more mineral than fruit characters showing as a young wine. However, the wine comes into its own on the palate with gentle lime flavours balancing the minerally characters and running through a long, pristine finish. **rating:** 94
best drinking 2002–2008 **best vintages** '98 **drink with** Fresh shellfish • $18

clonakilla ★★★★

Crisps Lane, Off Gundaroo Road, Murrumbateman, NSW 2582 **region** Canberra District
ph (02) 6227 5877 **fax** (02) 6227 5871 **open** 7 days 11–5
winemaker Tim Kirk **prod.** 1500 **est.** 1971
prod. range ($14–30 CD) Riesling, Chardonnay, Semillon Sauvignon Blanc, Viognier, Muscat, Shiraz, Shiraz Viognier, Cabernet.
summary The indefatigable Tim Kirk, who has many of the same personality characteristics as Frank Tate (of Evans & Tate), has taken over the management of Clonakilla from father and scientist Dr John Kirk. The quality of the wines is good, but none more so than the highly regarded Shiraz, which sells out quickly every year.

Clonakilla Riesling

Made in a consistent style throughout the 1990s, always showing considerable flavour, and quite often exhibiting a not unpleasant character which I can only describe as slightly cosmetic.
YYYY **1998** Light green-yellow; the bouquet is fresh with a mix of herbal/mineral/lime aromas. Those herb characters are the initial flavours, followed by lime and a hint of mineral towards the finish. **rating:** 86
best drinking 2000–2004 **best vintages** '91, '93, '94, '95, '97 **drink with** Spiced Asian dishes • $16

Clonakilla Shiraz Viognier

Now firmly established as Clonakilla's best wine, and indeed one of the best wines to come out of the Canberra District each year. Since 1994 the wine has been a blend of Shiraz and Viognier, with the Shiraz component typically accounting for between 80% and 85% of the blend. Interestingly, too, the grapes are not crushed, but are placed as whole bunches in open fermenters, foot-trodden, and the fermentation is completed in new and used French oak barriques. Originally simply labelled Shiraz with 'Viognier' added to the label since 1997.
YYYY **1997** Dense red-purple; the bouquet offers a mix of liquorice, blackberry and earthy fruit, with similar concentrated earthy and blackberry flavours. Overall gives the impression of dryness, and may well be going through a period of sulkiness. **rating:** 89
best drinking 2001–2007 **best vintages** '90, '92, '93, '94, '95, '97 **drink with** Jugged hare • $18

clos clare NR

Government Road, Watervale, SA 5452 **region** Clare Valley
ph (08) 8843 0161 **fax** (08) 8843 0161 **open** Weekends and public holidays 10–5
winemaker Various Contract **prod.** 1000 **est.** 1993
prod. range ($13.50–18 CD) Riesling, Shiraz.
summary Clos Clare is based on a small (1.5 hectares), unirrigated section of the original Florita Vineyard once owned by Leo Buring and which produces Riesling of extraordinary concentration and power.

clover hill ★★★★☆

Clover Hill Road, Lebrina, Tas 7254 **region** Northern Tasmania
ph (03) 6395 6114 **fax** (03) 6395 6257 **open** 7 days 10–5
winemaker Shane Clohesy, Dominique Portet, Chris Markell **prod.** 4000 **est.** 1986
prod. range ($29.50 R) Clover Hill (Sparkling).
summary Clover Hill was established by Taltarni in 1986 with the sole purpose of making a premium sparkling wine. Its 20 hectares of vineyards, comprising 12 hectares of chardonnay, 6.5 of pinot noir and 1.5 of pinot meunier, are still coming into bearing, and production is steadily increasing. Wine quality is excellent, combining finesse with power and length.

Clover Hill

A Chardonnay-predominant style given not less than 24 months on yeast lees, and invariably clean, fresh and finishing with pronounced acidity – a testament to the very cool climate in which the grapes are grown.

🍷🍷🍷🍷🍷 **1995** Medium straw-yellow, the bouquet is quite powerful with stylish biscuity/bready aromas; an elegant and fine wine on the palate with a crisp backbone of acidity which is better controlled than that of Jansz, another highly regarded Tasmanian. **rating:** 93

best drinking 1999–2002 **best vintages** '90, '91, '92, '95 **drink with** Caviar, shellfish • $29.50

clyde park NR

Midland Highway, Bannockburn, Vic 3331 **region** Geelong
ph (03) 5281 7274 **fax** (03) 5281 7274 **open** Not
winemaker Scott Ireland **prod.** 500 **est.** 1980
prod. range ($25 R) Chardonnay, Pinot Noir, Cabernet Sauvignon.
summary Sold by founder Gary Farr to leading Melbourne hotelier and restaurateur Donlevy Fitzpatrick in late 1994, but as at early 1998 once again on the market. In the meantime, the product range has been increased somewhat.

cobanov NR

Stock Road, Herne Hill, WA 6056 **region** Swan District
ph (08) 9296 4210 **open** Wed–Sun 9–5.30
winemaker Steve Cobanov **prod.** 10 000 **est.** 1960
prod. range ($6–10 CD) Chenin Blanc, Chardonnay, Sauvignon Blanc, Verdelho, Shiraz, Grenache, Cabernet Sauvignon.
summary A substantial family-owned operation producing a mix of bulk and bottled wine from 21 hectares of estate grapes. Part of the annual production is sold as grapes to other producers, including Houghton; part is sold in bulk; part sold in 2-litre flagons, and the remainder in modestly priced bottles.

cobaw ridge ★★★☆

Perc Boyer's Lane, East Pastoria via Kyneton, Vic 3444 **region** Macedon
ph (03) 5423 5227 **fax** (03) 5423 5227 **open** Weekends 10–5, or by appointment
winemaker Alan Cooper **prod.** 1300 **est.** 1985
prod. range ($18–25 CD) Chardonnay, Lagrein (375 ml), Shiraz, Shiraz Reserve.
summary Nelly and Alan Cooper have established Cobaw Ridge's 4-hectare vineyard at an altitude of 610 metres in the hills above Kyneton complete with self-constructed pole-framed mudbrick house and winery. Wine quality has been somewhat variable, but overall Alan Cooper has done extremely well. The plantings of cabernet sauvignon have been removed and partially replaced by lagrein, a variety which sent me scuttling to Jancis Robinson's seminal book on grape varieties, from which I learned it is a northeast Italian variety typically used to make delicate Rosé. Cooper believes it to be the only commercial planting in Australia, a claim disputed by at least one other winery, but no matter.

Cobaw Ridge Chardonnay

Produced from 2 hectares of estate plantings, and, like all the Cobaw Ridge wines, made and bottled on the estate.

🍷🍷🍷🍷 **1997** A surprisingly deep and developed yellow colour; the bouquet is a riotous mix of ripe, buttery diacetyl malolactic characters, oak and oatmeal aromas. The ripe, buttery butterscotch flavours, with a hint of nectarine, overpower the mouth with the aid of 14.5% alcohol. To my palate, over the top, but others will love it and feel my points are far too low. **rating:** 81

best drinking 1998–1999 **best vintages** NA **drink with** Honey chicken • $25

Cobaw Ridge Shiraz

This is by no means the first excellent Shiraz from Cobaw Ridge, but is the best for some years, reflecting the warm, dry and low-yielding vintage.

🍷🍷🍷🍷🍷 **1997** Youthful red-purple; the bouquet is clean and quite intense, with dark cherry/berry fruit and subtle oak. The palate is firm and well balanced, but needs time to soften and open up the complex fruit flavours which are certainly there. **rating:** 90

best drinking 2002–2007 **best vintages** '97 **drink with** Rare beef • $25

cobbitty wines NR

Cobbitty Road, Cobbitty, NSW 2570 **region** Other Wineries of NSW
ph (02) 4651 2281 **fax** (02) 4651 2671 **open** Mon–Sat 10–5, Sun 12–6
winemaker Giovanni Cogno **prod.** 5000 **est.** 1964
prod. range ($5–14 CD) A full range of generic table, fortified and sparkling wines under the Cobbitty Wines label; also cocktail wines.
summary Draws upon 10 hectares of estate plantings of muscat, barbera, grenache and trebbiano, relying very much on local and ethnic custom.

cockfighter's ghost vineyard ★★★

Lot 251 Milbrodale Road, Broke, NSW 2330 **region** Lower Hunter Valley
ph (02) 9667 1622 **fax** (02) 9667 1442 **open** Not
winemaker Various Contract **prod.** 12 000 **est.** 1994
prod. range ($16–22 ML) Semillon, Unwooded Chardonnay, Chardonnay, Pinot Noir, Shiraz.
summary Like Poole's Rock Vineyard, owned by eminent Sydney merchant banker David Clarke, but run and marketed as a separate venture, with lower wine prices. The wine has retail distribution throughout Australia, and is exported to the UK and New Zealand.

cofield ★★★☆

Distillery Road, Wahgunyah, Vic 3687 **region** Rutherglen
ph (02) 6033 3798 **fax** (02) 6033 3798 **open** Mon–Sat 9–5, Sun 10–5
winemaker Max Cofield, Damien Cofield **prod.** 5000 **est.** 1990
prod. range ($11.50–16.50 CD) Riesling, Chenin Blanc, Semillon, Semillon Chardonnay, Chardonnay, Late Harvest Tokay, Max's Blend Dry White, Max's Blend Dry Red, Shiraz, Merlot, Cabernet Sauvignon, Cabernet Merlot, Sparkling, Fortified.
summary District veteran Max Cofield, together with wife Karen and sons Damien, Ben and Andrew, is developing a strong cellar-door sales base by staging in-winery functions with guest chefs, and also providing a large barbecue and picnic area. The quality of the red wines, in particular, is good, and improving all the time. Limited retail distribution in Melbourne and Tasmania.

coldstream hills NR

31 Maddens Lane, Coldstream, Vic 3770 **region** Yarra Valley
ph (03) 5964 9410 **fax** (03) 5964 9389 **open** 7 days 10–5
winemaker James Halliday, Paul Lapsley **prod.** 50 000 **est.** 1985
prod. range ($24–42 CD) Pinot Gris, Sauvignon Blanc, Chardonnay, Reserve Chardonnay, Pinot Noir, Reserve Pinot Noir, Merlot, Reserve Merlot, Briarston (Cabernet Merlot), Reserve Cabernet Sauvignon; Limited Release Shiraz, Blanc de Noirs and Botrytis Chardonnay.
summary Founded by the author, who continues in charge of winemaking, but acquired by Southcorp in mid-1996. Expansion plans already then underway have been accelerated, with well in excess of 100 hectares of owned or managed estate vineyards as the base. Chardonnay and Pinot Noir continue to be the principal focus; varietal Merlot came on-stream from the 1997 vintage

Coldstream Hills Sauvignon Blanc

1997 was the first varietal Sauvignon Blanc produced by Coldstream Hills. In prior years the Sauvignon Blanc was blended with a greater volume of Semillon to produce a wine labelled Fumé Blanc. Twenty-five per cent of the wine was barrel-fermented in new French oak, the remainder fermented in stainless steel, both at low temperatures. The barrel-fermented portion was removed from oak immediately after the end of fermentation, and the components thereafter blended and bottled. It is a fruit-driven style, with the low crops and warm growing conditions of the 1997 vintage manifest in the wine, which has above-average length and concentration.
1998 Medium yellow-green; the bouquet is rich and full, with masses of sweet passionfruit and gooseberry aromas. On the palate the mouthfilling, rich tropical gooseberry fruit is balanced by acidity on the finish, the sweetness coming from glycerol, not residual sugar. The oak is perceptible more in terms of structure than either aroma or flavour.
best drinking 1999–2001 **best vintages** '97, '98 **drink with** Salad • $24

Coldstream Hills Chardonnay

Made from Yarra Valley grapes, part estate-grown and part purchased from other Yarra Valley growers. Largely barrel-fermented in a mix of new and used French oak under strictly controlled temperatures. Prolonged lees contact but no malolactic fermentation.
1998 Light to medium yellow-green; aromas of sweet fig and white peach on the bouquet, supported as always, by subtle oak. Similar, particularly sweet, fig and nectarine fruit flavours on the palate, no doubt deriving from the warm vintage. The wine has the hallmark length of flavour, a lingering aftertaste and good acidity.
best drinking 1999–2006 **best vintages** '86, '88, '91, '92, '93, '94, '96, '97 **drink with** Oven-roasted blue eye cod • $25.50

Coldstream Hills Reserve Chardonnay

Made primarily from estate-grown grapes which are 100% barrel-fermented in a mix of new (over 50%) and used French oak barriques, principally Vosges but with Troncais and Allier also used. Six months lees contact; 20% malolactic fermentation.

1997 Light yellow-green; melon, fig and stone fruit flavours are interwoven with a touch of spicy barrel-ferment character. By far the richest Reserve Chardonnay since 1994, and probably the richest ever. The integration of fruit and oak is near perfect, the mouth-coating flavours balanced by good acidity on the finish.

best drinking 1999–2006 **best vintages** '88, '91, '92, '93, '94, '96, '97 **drink with** Veal, chicken • $41.50

Coldstream Hills Pinot Noir

Part estate-grown and part sourced from other Yarra Valley growers, with a range of site climates. It is made using the full gamut of Burgundian techniques, including substantial use of whole bunches, foot-stamped and macerated. The primary fermentation is completed in a mix of new and used Troncais (French) oak.

1998 Medium red, with a touch of purple; has plum and forest characters on the bouquet, which will continue to grow and evolve over the next few years. There are pleasant plum and cherry flavours on the medium-weight palate. The wine has good structure and fine tannins running through a lingering finish.

best drinking 1999–2003 **best vintages** '87, '88, '91, '92, '94, '96, '97 **drink with** Seared or slow-cooked salmon, Asian cuisine • $26.80

Coldstream Hills Reserve Pinot Noir

Produced entirely from estate-grown grapes, in turn coming mainly from the Amphitheatre Block established in 1985. The same making techniques are used with the Reserve wine as with the standard, the difference being fruit selection and a much higher percentage of new Dargaud & Jaegle Troncais oak barriques.

1997 Medium purple-red; fine, with typical plum and black cherry fruit, balanced by gently chewy/spicy oak. Far more elegant than the very low yields would suggest, with complexity showing through the foresty/gamey nuances; excellent length, with soft tannins amplifying the finish.

best drinking 2001–2004 **best vintages** '87, '88, '91, '92, '94, '96, '97 **drink with** Quail, Asian cuisine • $42

Coldstream Hills Briarston

A blend of Cabernet Sauvignon, Cabernet Franc and Merlot. The percentages vary a little from year to year, but the Cabernet Sauvignon component is usually 80% or more. As with the other wines, part estate-grown, part purchased from other Yarra Valley growers. It is matured in a mix of new and used French oak (predominantly Nevers and Allier, with lesser amounts of Troncais) for 18 to 20 months before bottling.

1997 Medium red-purple; the bouquet is quite complex, of medium intensity, with red berry fruits running through hints of leaf, earth and forest. The palate is quite firm, with red berry and mint flavours to the mid-palate which will build as the wine matures; good grip to the finish.

best drinking 2000–2004 **best vintages** '88, '90, '91, '92, '94, '97 **drink with** Lamb with redcurrant sauce • $25.50

Coldstream Hills Reserve Cabernet Sauvignon

First introduced in 1992, and in fact will only be made in those years in which the quality of the cabernet sauvignon is outstanding. Entirely estate-produced from a single-vineyard block, the wine is matured in a high percentage of new Allier and Nevers oak barriques for 20 months. None made in 1996.

1997 Good, strong red-purple colour; cassis berry fruit and French oak are perfectly balanced and integrated on both bouquet and palate, with hints of sweet, earthy secondary Cabernet aromas adding interest. The palate has the hallmark elegance of the Coldstream Bordeaux-based reds, with fine, silky tannins woven through the length of the palate. A particular feature of the wine is the absence of any stalky, minty or herbaceous characters, being very much a reflection of a low-yielding, warm (for the Yarra Valley) vintage. Undoubtedly, the best Reserve Cabernet since 1992.
best drinking 2002–2012 **best vintages** '92, '93, '94, '97 **drink with** Rump steak • $42

connor park winery NR

59 Connors Road, Leichardt, Vic 3516 **region** Bendigo
ph (03) 5437 5234 **fax** (03) 5437 5204 **open** 7 days 10–6
winemaker Ross Lougoon **prod.** 2000 **est.** 1994
prod. range ($16–20 CD) Riesling, Semillon, Pinot Noir, Merlot, Shiraz, Cabernet Sauvignon, Port, Muscat.
summary The original planting of 2 hectares of vineyard dates back to the mid-1960s, and to the uncle of the present owners who had plans for designing an automatic grape harvester. The plans came to nothing, and when the present owners purchased the property in 1985, the vineyard had run wild. They resuscitated the vineyard (which formed part of a much larger mixed farming operation) and until 1994 were content to sell the grapes to other winemakers. Since then the vineyard has been expanded to 10 hectares, and while part of the grapes is sold to others, significant quantities are made under the Connor Park label and sold through cellar door and by mail order.

constable & hershon NR

1 Gillards Road, Pokolbin, NSW 2320 **region** Lower Hunter Valley
ph (02) 4998 7887 **fax** (02) 4998 7887 **open** 7 days 10–5
winemaker Neil McGuigan (Contract) **prod.** 3000 **est.** 1981
prod. range ($18.50–24.95 CD) Chardonnay, Unwooded Chardonnay, Pinot Noir, Cabernet Merlot, Reserve Cabernet Merlot.
summary Features four spectacular formal gardens, the Rose, Knot and Herb, Secret and Sculpture; a free garden tour is conducted every Monday to Friday at 10.30 am lasting 30 minutes. The 7-hectare vineyard is itself spectacularly situated under the backdrop of the Brokenback Range. Typically offers a range of several vintages of each variety ex-cellar door or by mailing list.

constables ★★★

Graphite Road, West Manjimup, WA 6258 **region** Pemberton
ph (08) 9772 1375 **open** 7 days 9–5
winemaker Houghton (Contract) **prod.** NFP **est.** 1988
prod. range ($10–15 CD) Riesling, Sauvignon Blanc, Chardonnay, Cabernet Sauvignon.
summary Father John and son Michael, together with other members of the Constable family, have established an 11-hectare vineyard at Manjimup. Most of the grapes are sold to Houghton under a long-term contract, and limited quantities are made for the Constable label by Houghton under contract.

cooinda vale NR

Bartonvale Road, Campania, Tas 7026 **region** Southern Tasmania
ph (03) 6260 4227 **open** By appointment
winemaker Andrew Hood (Contract) **prod.** 300 **est.** 1985
prod. range ($16–18 R) Riesling, Pinot Noir.
summary The tiny production means that the wines are not widely known, even in southern Tasmania, and quality has been somewhat variable. Retail distribution through Sutherland Cellars, Melbourne.

coolangatta estate ★★★☆

1335 Bolong Road, Shoalhaven Heads, NSW 2535 **region** Shoalhaven

ph (02) 4448 7131 **fax** (02) 4448 7997 **open** 7 days 10–5

winemaker Tyrrell's (Contract) **prod.** 5000 **est.** 1988

prod. range ($14–17 CD) Sauvignon Blanc, Semillon Sauvignon Blanc, Sauvignon Blanc Chardonnay, Unwooded Chardonnay, Alexander Berry Chardonnay, Verdelho, Chambourcin, Cabernet Shiraz, Merlot, Cabernet Sauvignon, Vintage Port.

summary Coolangatta Estate is part of a 150-hectare resort with accommodation, restaurants, golf course, etc, with some of the oldest buildings convict-built in 1822. It might be thought that the wines are tailored purely for the tourist market, but in fact the standard of viticulture is exceptionally high (immaculate Scott Henry trellising) and the winemaking is wholly professional (contract by Tyrrell's).

Coolangatta Estate Semillon

Previously used in blends, but in 1998 was made and released as a single varietal wine. It was picked early (at 10.2° baumé) in accordance with Tyrrell's instructions, and the outcome is a classic Semillon of high quality. Half of it will be held back for future release.

YYYYY **1998** Light to medium green-yellow; the bouquet shows excellent varietal character, with intense lemon and herb aromas. The palate is high toned and high flavoured, with more of those lemon/citrus notes. Will richly repay cellaring. **rating:** 90

best drinking 2004–2009 **best vintages** NA **drink with** Grilled fish • $17

Coolangatta Estate Sauvignon Blanc Chardonnay

The Scott Henry trellis system used at Shoalhaven is ideal for high vigour vineyards, particularly where (as in the case of Shoalhaven) the risk of summer rainfall is high. The canopy is split vertically, allowing sunlight and wind penetration, optimising air movement and reducing disease. For all that, the tempering effect of the nearby Shoalhaven River delays maturity, for vintage typically does not get underway until mid-March. Over the years several different blends involving Sauvignon Blanc have been made, initially incorporating Semillon, but in 1998 using Chardonnay.

YYYY **1998** Light green-yellow; the bouquet is not particularly distinctive in terms of varietal composition, tending firm, crisp and mineral. It is on the palate that the blend shows through, with sweeter fruit notes, starting with stone fruit and then moving through to ripe, sweet pampelmousse. The fruit and acidity are well-balanced. **rating:** 89

best drinking 1999–2000 **best vintages** '91, '94, '98 **drink with** Whitebait • $14

Coolangatta Estate Verdelho

An early-ripening variety, typically picked at the end of February (chardonnay follows in March) which is a distinct advantage in a climate in which summer rainfall always poses a threat. It is no doubt for the same reason that chambourcin has been planted, and likewise does well.

YYYY **1998** Medium yellow-green; the bouquet is clean and smooth with attractive fruit salad aromas. The palate follows down the same track, being very well balanced, zesty and lively, with a mix of citrus and stone fruit flavours. A gold medal winner at the 1998 Cairns Wine Show. **rating:** 87

best drinking 1999–2001 **best vintages** NA **drink with** Pasta • $15

Coolangatta Estate Alexander Berry Chardonnay

Named after Alexander Berry, who with Edward Wollstonecraft obtained a grant of 10 000 acres and 100 convicts in 1822, building their settlement on the foothills of a mountain named Coolangatta, an Aboriginal word meaning 'fine view'. Unlike the Verdelho, Chambourcin and Cabernet Shiraz, no claim to be estate-grown is made on the label.

🍷🍷🍷🍷 **1998** Medium yellow-green; the light melon fig fruit has a nice hint of oak to add a little complexity to the bouquet; on the palate, an elegant, unforced style which creeps up on you, with attractive melon fruit and unsuspected length to the flavour. Subtle oak. **rating:** 84

best drinking 1999–2001 **best vintages** '91, '94, '96, '97 **drink with** Avocado and seafood • $17

Coolangatta Estate Cabernet Shiraz

A surprise performer (on my score sheet at least) with the '96 vintage at the 1998 *Winewise* Small Makers Competition. However, came back to earth with a much lighter wine in 1997.

🍷🍷🍷 **1997** Light red; the bouquet is fairly plain, with faint nuances of leaf and spice. The palate delivers a little more, but is still short on fruit richness. **rating:** 78

best drinking 1998–1999 **best vintages** NA **drink with** Beef stroganoff • $17

coombend estate ★★★★

Coombend via Swansea, Tas 7190 **region** Southern Tasmania
ph (03) 6257 8256 **fax** (03) 6257 8484 **open** 7 days 9–6
winemaker Andrew Hood (Contract) **prod.** 1000 **est.** 1985
prod. range ($13–26 CD) Riesling, Sauvignon Blanc, Cabernet Sauvignon.

summary John Fenn Smith originally established 1.75 hectares of cabernet sauvignon, 2.2 hectares of sauvignon blanc and 0.5 hectare of riesling (together with a little cabernet franc) on his 2600-hectare sheep station, choosing that part of his property which is immediately adjacent to Freycinet. This slightly quixotic choice of variety has been justified by the success of the wine in limited show entries. In December 1998 Coombend opened a brand new, purpose-built cellar-door sales area, and has also significantly expanded its plantings to include riesling and sauvignon blanc.

Coombend Estate Riesling

Being graziers by background, the Fenn Smiths calculate their vineyard holdings not by area but by vine numbers, of which there are 4000 riesling.

🍷🍷🍷🍷 **1998** Light green-yellow; the bouquet is crisp, with herbal/mineral/talcum powder aromas, and no tropical fruit. The palate is much softer than the bouquet suggests, with moderate length and (unusually for Tasmania) low acidity. Perhaps, as the Fenn Smiths suggest, picked a tad too ripe. **rating:** 86

best drinking 1999–2002 **best vintages** NA **drink with** Avocado salad • $13

Coombend Estate Sauvignon Blanc

Produced from 3000 estate vines, and spends six weeks in French oak.

🍷🍷🍷🍷 **1998** Light green-yellow; light gooseberry fruit aromas are complexed by just a hint of oak on the bouquet; the palate is quite rich and ripe, with tropical fruit characters and soft – or ripe – acidity (if there can be such a thing) on the finish. **rating:** 87

best drinking 1999–2000 **best vintages** NA **drink with** Sugar-cured tuna • $20

Coombend Cabernet Sauvignon

Situated adjacent to Freycinet, and shares the same remarkable site climate, proving yet again how difficult it is to generalise about the Tasmanian climate. Both the 1990 and 1991 Cabernets have won trophies in Tasmanian shows; it is arguably the best Cabernet Sauvignon in Tasmania. The '94 vintage won a silver medal at the 1997 Tasmanian Wine Show. The '95 continues the tradition, with a richly deserved silver medal at the 1999 show.

🍷🍷🍷🍷🍷 **1995** Strong red-purple; the initial impression is of cedar, briar and cigar box, but berry fruit then emerges as the wine breathes. On the palate there is much the same impression, initially with soft, cedary flavours and ripe tannins, and then (almost perversely) sweet, dark berry fruit comes through on the finish. **rating:** 90

best drinking 1999–2006 **best vintages** '90, '91, '92, '94, '95 **drink with** Ragout of lamb • $26

coorinja ★★☆

Toodyay Road, Toodyay, WA 6566 **region** Swan District
ph (08) 9626 2280 **open** Mon–Sat 8–5
winemaker Michael Wood **prod.** 3200 **est.** 1870
prod. range ($8–10.50 CD) Dry White, Claret, Hermitage, Burgundy, Fortifieds; the latter account for 50% of Coorinja's production.
summary An evocative and historic winery nestling in a small gully which seems to be in a time-warp, begging to be used as a set for a film. A recent revamp of the packaging accompanied a more than respectable Hermitage, with lots of dark chocolate and sweet berry flavour, finishing with soft tannins.

cope-williams ★★★☆

Glenfern Road, Romsey, Vic 3434 **region** Macedon
ph (03) 5429 5428 **fax** (03) 5429 5655 **open** 7 days 11–5
winemaker Michael Cope-Williams **prod.** 7000 **est.** 1977
prod. range ($14–25 R) Chardonnay, Cabernet Merlot; d'Vine is second label, Riesling, Chardonnay and Cabernet Sauvignon; winery specialty sparkling wine Macedon R.O.M.S.E.Y.
summary One of the high country Macedon pioneers, specialising in sparkling wines which are full flavoured, but also producing excellent Chardonnay and Pinot Noir table wines in the warmer vintages. A traditional 'English Green'-type cricket ground is available for hire and booked out most days of the week from spring through till autumn.

coriole ★★★★☆

Chaffeys Road, McLaren Vale, SA 5171 **region** McLaren Vale
ph (08) 8323 8305 **fax** (08) 8323 9136 **open** Mon–Fri 10–5, weekends 11–5
winemaker Grant Harrison **prod.** 30 000 **est.** 1967
prod. range ($14–52 R) Lalla Rookh Semillon, Semillon Sauvignon Blanc, Chenin Blanc, Semillon, Chardonnay, Sangiovese, Shiraz, Redstone (Shiraz Cabernet Grenache), Diva (Sangiovese blend), Cabernet Sauvignon; Mary Kathleen (Cabernet blend), Lloyd Reserve Shiraz, Lalla Rookh Grenache Shiraz.
summary Justifiably best known for its Shiraz, which, both in the rare Lloyd Reserve, and also the standard form, is extremely impressive. It has spread its wings in recent years, being one of the first wineries to catch onto the Italian fashion with its Sangiovese, but its white varietal wines lose nothing by comparison. It is also a producer of high-quality olive oil distributed commercially through all Australian States. The wines are exported to the UK, US, Canada, Switzerland, Germany, Netherlands, Taiwan, Japan and New Zealand.

Coriole Lalla Rookh Semillon

A very interesting wine which over the past few years has shown strong passionfruit/gooseberry aromas when young, presumably due at least in part to the use of aromatic yeast during fermentation, but is possibly also partly vineyard character. Whatever the answer, it produces wines that are striking and attractive in their youth. The wine is partially fermented in French oak, and spends three months on yeast lees. A change of name in 1997 saw the addition of 'Lalla Rookh' to the name, but the wine remains unchanged. The '98 is, quite simply, outstanding.

🍷🍷🍷🍷🍷 **1998** Light to medium yellow-green; the bouquet is sophisticated and complex, with just the faintest whiff of smoky oak and exotic tropical fruit. The palate is flooded with citrus and herb flavours running through to a long finish. **rating:** 94

best drinking 1999–2004 **best vintages** '93, '94 '95, '97, '98 **drink with** Salad, seafood • $20.50

Coriole Semillon Sauvignon Blanc

A blend of 70% Semillon and 30% Sauvignon Blanc, fermented in stainless steel and early bottled without the intrusion of oak.

🍷🍷🍷🍷 **1998** Medium yellow-green; the bouquet is moderately complex and rich, with ripe, tangy fruit, the palate slightly more straightforward, but flavoursome and well balanced. **rating:** 86

best drinking 1999–2002 **best vintages** NA **drink with** Battered fish • $14

Coriole Shiraz

Produced from estate plantings on red loam over ironstone and limestone subsoils. While the vines are new in comparison to those used to make Lloyd Reserve, they are in fact old by any normal standards, dating back to the late 1960s. This has usually been a distinguished wine.

🍷🍷🍷 **1997** Medium to full purple-red; both the bouquet and palate show the influence of strong, sawdusty/pencilly oak which has not integrated with the wine, and I suspect never will. A pity, because the ripe cherry fruit and the prior history of this wine mean this is an aberrational outcome. **rating:** 84

best drinking 2001–2003 **best vintages** '90, '91, '96 **drink with** Steak and kidney pie • $23

Coriole Lloyd Reserve Shiraz

One of McLaren Vale's most distinguished Shiraz wines, with a track record going back to the establishment of Coriole in 1967, based upon 1.3 hectares of vines then 60 years old. Those vines provided the core of fruit over the intervening decades until additional plantings came first into bearing and then into maturity, allowing the introduction of a Reserve Shiraz from 1989 using only those original plantings. The vines typically produce 4 to 4.5 tonnes of grapes, enough to make about 250 cases per year.

🍷🍷🍷🍷🍷 **1996** Dark, dense red-purple; the bouquet offers a quite incredible essence of McLaren Vale chocolate, utterly delicious. The palate is likewise packed with dark Swiss chocolate, dark cherry, a hint of vanilla oak and long, lingering tannins. **rating:** 95

best drinking 2006–2016 **best vintages** '70, '74, '84, '88, '89, '90, '91, '92, '94, '95, '96 **drink with** Ragout of lamb • $52

Coriole Redstone

A blend of Shiraz, Cabernet Sauvignon, Grenache and Merlot which sounds very much like a marriage of convenience, but which has produced an attractive child.

🍷🍷🍷🍷 **1997** Medium red-purple; the moderately intense bouquet offers leaf, chocolate, berry fruit and subtle oak; there is an interesting range of flavours in the mouth which track over the bouquet before finishing with a typical touch of McLaren Vale chocolate and pleasant tannins. **rating:** 86

best drinking 2000–2005 **best vintages** NA **drink with** Lamb's fry • $16.50

cosham NR

101 Union Road, Carmel via Kalamunda, WA 6076 **region** Perth Hills

ph (08) 9293 5424 **fax** (08) 9293 5062 **open** Weekends and public holidays 10–5

winemaker Lyndon Crocker **prod.** 460 **est.** 1989

prod. range ($15–20 CD) Chardonnay, Pinot Noir, Cabernet Merlot, Methode Champenoise Brut.

summary The newest of the Perth Hills ventures, with a microscopic amount of wine available. Both the Chardonnay and Pinot Noir spend two years in French oak barriques before bottling – a long time by any standards. Improbable though it may seem, exports to the US.

cowra estate ★★★

Boorowa Road, Cowra, NSW 2794 **region** Cowra

ph (02) 6342 1136 **fax** (02) 6342 4286 **open** 7 days 9–6

winemaker Simon Gilbert (Contract) **prod.** 20 000 **est.** 1973

prod. range ($12–15 CD) Chardonnay, Unwooded Chardonnay, Cool Classic Chardonnay, Cool Classic Brut de Brut, Directors Reserve Merlot, Cabernet Rosé, Cabernet Shiraz, Cabernets. The Classic Bat series of Chardonnay, Pinot Noir and Cabernet Merlot is now at the head of the range.

summary Cowra Estate was purchased from the family of founder Tony Gray by South African-born food and beverage entrepreneur John Geber in 1995. A vigorous promotional campaign has gained a higher domestic profile for the once export-oriented brand. John Geber is very actively involved in the promotional effort, and rightly proud of the excellent value for money which the wines represent. The Quarry Wine Cellars and Restaurant offer visitors a full range of all of the Cowra Estate's wines, but also wines from the other producers in the region, including Richmond Grove, Hungerford Hill, Arrowfield, Mulyan and Chiverton.

crabtree of watervale ★★★

North Terrace, Watervale SA 5452 **region** Clare Valley

ph (08) 8843 0069 **fax** (08) 8843 0144 **open** 7 days 11–5

winemaker Robert Crabtree, Stuart Bourne **prod.** 4000 **est.** 1979

prod. range ($13–17 CD) Riesling, Late Harvest Riesling, Semillon, Watervale Dry Red, Grenache, Shiraz Cabernet, Cabernet Sauvignon, Muscat.

summary The gently eccentric Robert Crabtree has sold his winery to an Adelaide-based syndicate of wine lovers; the wines, and their styles, remain unchanged.

Crabtree of Watervale Riesling

Produced from hand-picked estate-grown grapes; the wine is partially whole-bunch pressed and conventionally fermented in stainless steel at low temperatures using two yeasts.

🍷🍷🍷🍷🍸 **1998** Medium yellow-green; a quite potent bouquet with powerful herb, lime and mineral fruit is followed by a palate of abundant flavour in the lime/citrus spectrum. Good acidity and length. **rating:** 90

best drinking 2000–2008 **best vintages** NA **drink with** Summer salads • $13

craig avon vineyard ★★★☆

Craig Avon Lane, Merricks North, Vic 3926 **region** Mornington Peninsula

ph (03) 5989 7465 **fax** (03) 5989 7615 **open** Weekends and public holidays 12–5

winemaker Ken Lang **prod.** 1000 **est.** 1986

prod. range ($24–32 CD) Chardonnay, Pinot Noir, Cabernet, Cabernet Merlot.

summary All of the wines are sold cellar door and by mailing list. The wines are competently made, clean and with pleasant fruit flavour.

craigie knowe ★★★

Glen Gala Road, Cranbrook, Tas 7190 **region** Southern Tasmania

ph (03) 6223 5620 **fax** (03) 6223 5009 **open** Weekends or by appointment

winemaker Dr John Austwick **prod.** 500 **est.** 1979

prod. range ($20–23 ML) Cabernet Sauvignon, Pinot Noir.

summary John Austwick makes a small quantity of full-flavoured, robust Cabernet Sauvignon in a tiny winery as a weekend relief from a busy metropolitan dental practice. The Pinot Noir is made in a style which will appeal to confirmed Cabernet Sauvignon drinkers, and John Austwick has a couple of barrels of 1998 Cabernet Sauvignon which, if bottled separately, would appeal to everyone who has ever lifted a wine glass.

craiglee ★★★★★

Sunbury Road, Sunbury, Vic 3429 **region** Sunbury
ph (03) 9744 4489 **fax** (03) 9744 4489 **open** Sun, public holidays 10–5, or by appointment
winemaker Patrick Carmody **prod.** 2000 **est.** 1976
prod. range ($17–30 CD) Chardonnay, Pinot Noir, Shiraz, Cabernet Sauvignon.
summary An historic winery with a proud nineteenth-century record which recommenced winemaking in 1976 after a prolonged hiatus. Produces one of the finest cool-climate Shirazes in Australia, redolent of cherry, liquorice and spice in the better (i.e. warmer) vintages, lighter-bodied in the cooler ones. Maturing vines and improved viticulture has made the wines more consistent (and even better) over the past ten years or so.

Craiglee Shiraz

Produced from 4 hectares of estate plantings, almost invariably producing wines of the highest imaginable quality, with wonderful cherry, pepper and spice aromas and flavours. The wines are fruit- rather than oak-driven; they are immaculately structured, having the fruit weight and vinous sweetness to balance the peppery/spicy tang.

🍷🍷🍷🍷🍷 **1996** The colour is not nearly as deep as one might expect, and the bouquet, too, is relatively light, albeit with attractive cherry spice fruit and a hint of charry oak. All of this leaves one unprepared for the marvellous flavour of the palate, with intense spice and cherry fruit, looking for all the world like a high-quality, lighter-year Northern Rhône Valley red. **rating:** 92

best drinking 2001–2006 **best vintages** '84, '86, '88, '91, '92, '93, '94, '96 **drink with** Italian cuisine • $30

craigmoor ★★★

Craigmoor Road, Mudgee, NSW 2850 **region** Mudgee
ph (02) 6372 2208 **fax** (02) 6372 4464 **open** Mon–Fri 9–4, weekends 10–4
winemaker Brett McKinnon **prod.** NFP **est.** 1858
prod. range ($12.95–13.95 R) Semillon, Chardonnay, Shiraz, Cabernet Sauvignon.
summary One of the oldest wineries in Australia to remain in continuous production, now subsumed into the Orlando/Wyndham group, with an inevitable loss of identity and individuality of wine style, although the technical quality of the wines cannot be faulted. Exports to UK, US, Asia.

craigow NR

Richmond Road, Cambridge, Tas 7170 **region** Southern Tasmania
ph (03) 6248 5379 **fax** (03) 6248 5482 **open** Not
winemaker Andrew Hood (Contract) **prod.** 200 **est.** 1989
prod. range ($15–19 ML) Pinot Noir.
summary Craigow has substantial vineyards, with 5 hectares of pinot noir and another 5 hectares divided between riesling, chardonnay and gewurztraminer. However, almost all of the grapes are sold, and only a little Pinot Noir is made each year.

craneford NR

Main Street, Springton, SA 5235 **region** Eden Valley
ph (08) 8568 2220 **fax** (08) 8568 2538 **open** Wed–Mon 11–5
winemaker Contract **prod.** 2400 **est.** 1978
prod. range ($11–15 CD) Riesling, Chardonnay, Shiraz, Cabernet Sauvignon.
summary A change of ownership may herald a revival in the fortunes of Craneford which had suffered from a range of winemaking problems over the past five years or so.

crane winery NR

Haydens Road, Kingaroy, Qld 4610 **region** Other Wineries of Qld
ph (07) 4162 7647 **fax** (07) 4162 7647 **open** Fri–Tues 9–4 or by appointment
winemaker John Crane **prod.** 3500 **est.** 1996
prod. range ($10–16 ML) Marsanne Verdelho Semillon, Chardonnay, Hillside White and Red, Cabernet Franc, Shiraz Cabernet Sauvignon, Sparkling Burgundy, Vintage Liqueur Shiraz, Liqueur Muscat.
summary Established by John and Sue Crane, Crane Winery is one of several in the burgeoning Kingaroy (or South Burnett) region in Queensland, drawing upon 3 hectares of estate plantings, but also purchasing grapes from other growers in the region. Interestingly, Sue Crane's great-grandfather established a vineyard planted to shiraz 100 years ago (in 1898) and which remained in production until 1970.

cranswick estate ★★★

Walla Avenue, Griffith, NSW 2680 **region** Riverina
ph (02) 6962 4133 **fax** (02) 6962 2888 **open** Mon–Fri 10–4.30, Weekends 10–4
winemaker Andrew Schulz, Tim Pearce **prod.** 750 000 **est.** 1976
prod. range ($7–22 R) There are three ranges in two price sectors; at the top come the Premium and Regional Selection ranges (with Autumn Gold Botrytis Semillon off to one side and higher priced again) comprising Barrel Fermented Semillon, Young Vine Chardonnay, Conlon Block Marsanne, McLaren Vale Sauvignon Blanc, Gnarled Vine Barossa Grenache, Cocoparra Vineyard Shiraz, Dry Country Cabernet Sauvignon and NV Sparkling Shiraz; then there is the volume-selling Vignette Range of Fruition (White Frontignac), Unoaked Chardonnay, Semillon Sauvignon Blanc, Semillon Cabernet Merlot, Shiraz.
summary Taking full advantage of the buoyant share market and the continuing export success of Australian wines, Cranswick Estate made a highly successful entry to the lists of the Australian Associated Stock Exchanges in 1997. The substantial capital raised will see the further expansion of an already thriving business, firmly aimed at the export market in the UK and Europe, US and Japan and southeast Asia.

Cranswick Estate Autumn Gold Botrytis Semillon (375 ml)

One hundred per cent Botrytis Semillon sourced from a single grower (Pat Zirilli) in Griffith, and given 12 months maturation in oak. Prior vintages have been prolific show medal winners; a style which the Riverina seems to achieve with almost effortless ease.

🍷🍷🍷🍷 **1996** Deep golden yellow; complex cumquat and lime fruit aromas swell into a cumquat, apricot and lime-flavoured palate, rescued from fatness by balanced acidity and subtle oak.
rating: 87

best drinking 1999–2003 **best vintages** '94, '95, '96 **drink with** Orange cake and King Island cream • $20

Cranswick Estate Cocoparra Vineyard Shiraz

As the name suggests, from the Cranswick Estate Cocoparra Vineyard and, in the quirky but far from uncommon nature of such things, was the first crop. The wine spent 12 months in American oak, oak which is barely noticeable. Winner of two trophies at the 1998 Griffith Wine Show, and a number of bronze medals at other shows including capital city wine shows.

🍷🍷🍷🍷🍸 **1997** Medium to full purple-red; a rich and concentrated bouquet with dark cherry varietal fruit is followed by a palate with plenty of concentration and power, with the fruit once again doing the talking. The only faint criticism is that it dips slightly towards the finish.
rating: 90

best drinking 1999–2003 **best vintages** '97 **drink with** Beef Wellington • $18

Cranswick Estate Vignette Range Cabernet Merlot

While most of the Cranswick Estate wines in the Vignette Range are no better than their price ($10-12) would suggest, the Cabernet Merlot does rise above its station. A blend of 52% Cabernet Sauvignon, 42% Merlot and 6% Ruby Cabernet, it is sourced from various vineyards in the Griffith region.

🍷🍷🍷🍷 **1996** Light to medium red-purple; the bouquet is sweet, with quite good fruit concentration, the palate with gentle dark fruit flavours and the supple, soft texture no doubt contributed by the Merlot component. The American oak is not intrusive. **rating:** 82

best drinking 1998–2001 **best vintages** NA **drink with** Pizza • $11

---

crawford river wines ★★★★★

Hotspur Upper Road, Condah, Vic 3303 **region** Far South West Victoria
ph (03) 5578 2267 **fax** (03) 5578 2240 **open** 7 days 10–4
winemaker John Thomson **prod.** 3500 **est.** 1975
prod. range ($12–25 CD) Riesling, Semillon Sauvignon Blanc, Chardonnay, Classic Dry White, Cabernet Merlot, Cabernet Sauvignon, Nektar.

summary Exemplary wines right across the range are made by full-time grazier, part-time winemaker John Thomson who clearly has the winemaker's equivalent of the gardener's green thumb. The Riesling is consistently outstanding, the Cabernet-based wines excellent in warmer vintages. Exports to the UK, Germany and Austria.

Crawford River Riesling

A wine of exemplary quality over the years, tight and reserved, and fully reflecting the very cool climate in which it is grown. It is 100% estate-grown (as are all of the Crawford River wines).

🍷🍷🍷🍷🍷 **1998** Light to medium yellow-green; the gently seductive aromas of lime and passionfruit are followed by a spotlessly clean palate, with similar flavours and a soft elegance and finesse all of its own. **rating:** 92

best drinking 2001–2007 **best vintages** '86, '88, '89, '91, '94, '96, '97, '98 **drink with** Antipasto • $19

Crawford River Cabernet Merlot

A blend of 50% Cabernet Sauvignon and 50% Merlot, and like the Cabernet Sauvignon, speaks as much of Bordeaux as it does of Australia. A wine that will inevitably reflect vintage conditions, for this is a distinctly cool region for these varieties.

🍷🍷🍷🍷 **1997** Medium red-purple; fragrant cedary, leafy, briary aromas are followed by an attractive palate with sweet berry fruit finishing with long, fine tannins. **rating:** 86

best drinking 2001–2006 **best vintages** NA **drink with** Rib of veal • $20

Crawford River Cabernet Sauvignon

Grown at the extreme edge for Cabernet Sauvignon, and always presents a rather austere, minerally, European cast, but in most years redeems itself with the length of its finish, and the way the flavour builds-up with the second glass.

🍷🍷🍷🍷 **1996** Light to medium red; the bouquet is clean, with rather light mint and cassis fruit, lacking concentration. The wine has better structure than the bouquet promises, helped by fine, almost milky tannins and cedary oak. **rating:** 84

best drinking 2000–2004 **best vintages** '86, '88, '90, '91, '93 **drink with** Roast veal • $25

---

crosswinds vineyard NR

10 Vineyard Drive, Tea Tree, Tas 7017 **region** Southern Tasmania
ph (03) 6268 1091 **fax** (03) 6268 1091 **open** 7 days 10–5 Oct–May, weekends Sept–June
winemaker Andrew Vasiljuk **prod.** NA **est.** 1990
prod. range ($15.85–20.60 R) Riesling, Chardonnay, Pinot Noir.
summary Crosswinds has two vineyards, with the 1-hectare Tea Tree Vineyard, and the 2-hectare Margate Vineyard. As well as cellar-door sales, has retail distribution through Sutherland Cellars, Melbourne, and small exports to UK and southeast Asia.

Crosswinds Non-Wooded Pinot Noir

A thoroughly unconventional approach to making Pinot Noir but which succeeds in this instance simply because of the exceptional quality of the grapes and of the vintage as a whole.

YYYY **1998** Medium to full red-purple; the bouquet exudes sweet plummy fruit, clean, but inevitably not complex. The palate has plenty of sweet, plummy fruit, all primary, and not complex, and one can only wonder how good the wine might have been had its fermentation been finished in high-quality French oak. **rating:** 88

best drinking 1999–2002 **best vintages** '98 **drink with** Braised duck • $18

cruickshank callatoota estate ★★☆

2656 Wybong Road, Wybong, NSW 2333 **region** Upper Hunter Valley
ph (02) 6547 8149 **fax** (02) 6547 8144 **open** 7 days 9–5
winemaker Hartley Smithers **prod.** 7000 **est.** 1973
prod. range ($8.50–13.50 CD) Cabernet Rosé, Cask 12, Two Cabernets, Show Reserve.
summary Owned by Sydney management consultant John Cruickshank and family. Wine quality has definitely improved in the 1990s, although the wines still show strong regional, rather earthy, characters, the label itself likewise doggedly remaining old-fashioned.

cullen wines ★★★★★

Caves Road, Cowaramup, WA 6284 **region** Margaret River
ph (08) 9755 5277 **fax** (08) 9755 5550 **open** 7 days 10–4
winemaker Vanya Cullen **prod.** 12 000 **est.** 1971
prod. range ($25–55 CD) Flagship wines: Chardonnay, Sauvignon Blanc, Reserve Cabernet Merlot; premium wines: Classic Dry White, Blanc de Noir, Velvet Red, Late Harvest Cabernet Sauvignon.
summary One of the pioneers of Margaret River which has always produced long-lived wines of highly individual style from the substantial and mature estate vineyards. Winemaking is now in the hands of Vanya Cullen, daughter of the founders; she is possessed of an extraordinarily good palate. The Cabernet Merlot goes from strength to strength; indeed, I would rate it Australia's best. The wines are distributed throughout Australia, and also make their way to significant export markets in the UK, the US, Europe and Asia.

Cullen Sauvignon Blanc Semillon

An extremely complex wine, a blend of 70% Sauvignon Blanc, 24% Semillon and 5% Chardonnay, the Chardonnay having been taken through malolactic fermentation in oak. A little over 40% is barrel-fermented, 15% utilising wild (or natural) yeast initiation of fermentation. This approach deliberately softens primary varietal characters.

YYYYY **1998** Light to medium yellow-green; the bouquet is moderately intense, with a nice mineral cut offset by just a touch of oak. The palate has exceptional length and intensity, with herb and lemon flavours, but its texture is its strongest point. **rating:** 94

best drinking 1999–2004 **best vintages** '97, '98 **drink with** Margaret River marron • $24

Cullen Cabernet Merlot

An estate-grown wine of the highest quality, that since 1990 has been arguably the best Margaret River Cabernet Merlot blend. Up to 1995 both a varietal and a Reserve version were made, but as from that vintage the decision was taken to only release one wine. Consistently outstanding. A blend of 75% Cabernet Sauvignon, 20% Merlot and 5% Cabernet Franc.

YYYYY **1997** Medium to full red-purple; powerful, but perfectly balanced, dark berry fruit and oak on the bouquet is logically followed by a palate with great depth to the dark berry fruit, lingering but balanced tannins, and exemplary oak. **rating:** 95

best drinking 2002–2012 **best vintages** '77, '84, '86, '90, '91, '92, '93, '95, '96, '97 **drink with** Lamb, strong cheddar • $55

currency creek wines ★★★

Winery Road, Currency Creek, SA 5214 **region** Other Wineries of SA

ph (08) 8555 4069 **fax** (08) 8555 4100 **open** 7 days 10–5

winemaker Phillip Tonkin **prod.** 8000 **est.** 1969

prod. range ($8.95–19.50 CD) Dry White, Semillon, Sauvignon Blanc, Chardonnay, Princess Alexandrina Noble Riesling, Gamay, Pinot Noir, Harmony (Shiraz Merlot), Ostrich Hill Shiraz, Cabernet Sauvignon, Sparkling, Fortifieds.

summary Constant name changes (Santa Rosa, Tonkins have also been tried) did not help the quest for identity or recognition in the marketplace, but the winery has nonetheless produced some outstanding wood-matured whites and pleasant, soft reds selling at attractive prices.

curtis NR

Foggo Road, McLaren Vale, SA 5171 **region** McLaren Vale

ph (08) 8323 8389 **open** Weekends 11–4.30

winemaker P Curtis **prod.** 1500 **est.** 1988

prod. range ($4–5.50 CD) Riesling, Moselle, Claret, Shiraz Grenache, Ruby Port, Tawny Port.

summary A small and relatively new producer in McLaren Vale, whose wines I have not tasted.

dalfarras ★★★★

PO Box 123, Nagambie, Vic 3608 **region** Goulburn Valley

ph (03) 5794 2637 **fax** (03) 5794 2360 **open** Not

winemaker Alister Purbrick **prod.** 35 000 **est.** 1991

prod. range ($13.95–49.95 R) Riesling, Unwooded Chardonnay, Barrel Fermented Chardonnay, Sauvignon Blanc, Marsanne, Shiraz, Cabernet Sauvignon, Reserve Cabernet Sauvignon.

summary The personal project of Alister Purbrick and artist-wife Rosa (née) Dalfarra, whose paintings adorn the labels of the wines. Alister, of course, is best known as winemaker at Chateau Tahbilk, the family winery and home, but this range of wines is intended to (in Alister's words) 'allow me to expand my winemaking horizons and mould wines in styles different to Chateau Tahbilk'. It now draws upon 37 hectares of its own plantings in the Goulburn Valley, and the business continues to grow year by year. In the 1998 show season the Dalfarras wines had outstanding success, with the 1991 Reserve Cabernet Sauvignon winning two trophies, a gold, two silver and seven bronze medals, backed up by the 1998 Sauvignon Blanc and 1997 Marsanne, both of which won gold medals, and a number of silver and bronze awards.

Dalfarras Shiraz

While some of the Dalfarras wines do not seem particularly different in style from those of Chateau Tahbilk, this wine certainly does. A blend of McLaren Vale and Goulburn Valley shiraz, which is matured in a mix of new American and French barrels for 18 months. Perhaps surprisingly, the wine does not show excessive oak pick-up.

🍷🍷🍷🍷 **1996** Medium to full red-purple; the bouquet is clean, smooth, of medium intensity with attractive black cherry varietal fruit and subtle oak. The same good balance of fruit and oak continues through the palate with a mix of black cherry, chocolate and vanilla flavours. Not over-extracted or heavy. **rating:** 89

best drinking 2000–2006 **best vintages** NA **drink with** Steak and kidney pie • $19.95

Dalfarras Cabernet Sauvignon

While the style of the Dalfarras wines is intended by Alister Purbrick to be different from those of Chateau Tahbilk, he obviously likes his red wines to have plenty of muscle and character. Made from grapes grown in Coonawarra, and spends 18 months in new American oak and French barrels. A pleasant wine, but one wonders where on earth all the oak went to.

🍷🍷🍷🍷 **1996** Medium red-purple, with some signs of early development. The bouquet is fragrant with a mix of peppermint and eucalypt aromas, the palate likewise light with a mix of sweet and leafy/minty fruit. **rating:** 84

best drinking 1998–1999 **best vintages** NA **drink with** Marinated beef • $22.95

Dalfarras Reserve Cabernet Sauvignon

An excellent wine which was rewarded with outstanding success at the 1998 Melbourne Wine Show, winning two of the top trophies including the Best Red Wine of Show. A blend of two-thirds Cabernet Sauvignon from Coonawarra and one-third Cabernet Sauvignon from the Goulburn Valley, it spent over two years in 100% new American oak in which it finished its primary fermentation. 785 cases made.

🍷🍷🍷🍷🍷 **1991** Strong, deep colour heralds a wine with strong cabernet varietal aromas and a powerful, structured palate offering lots of cherry fruit and tannins, without compromising the balance of the wine. Should age exceptionally well. **rating:** 93

best drinking 2001–2011 **best vintages** NA **drink with** Braised ox cheek • $49.95

dalrymple ★★★★

1337 Pipers Brook Road, Pipers Brook, Tas 7254 **region** Northern Tasmania
ph (03) 6382 7222 **fax** (03) 6382 7222 **open** 7 days 10–5
winemaker Bert Sundstrup **prod.** 3300 **est.** 1987
prod. range ($16–25 CD) Chardonnay, Unwooded Chardonnay, Sauvignon Blanc, Pinot Noir.
summary A partnership between Jill Mitchell and her sister and brother-in-law, Anne and Bertel Sundstrup, inspired by father Bill Mitchell's establishment of the Tamarway Vineyard in the late 1960s. In 1991 Tamarway reverted to the Sundstrup and Mitchell families, and it too, will be producing wine in the future, probably under its own label but sold ex the Dalrymple cellar door. As production has grown (significantly), so has that of wine quality across the board, often led by its Sauvignon Blanc.

Dalrymple Fumé Blanc

While the 1998 Dalrymple Sauvignon Blanc covered itself with glory at the Royal Hobart Wine Show, this wooded version was in better form at the 1999 Tasmanian Wines Show, indeed rating gold medal points on my score sheet.

🍷🍷🍷🍷🍷 **1998** Medium yellow-green; potent grassy fruit on the bouquet is only subtly influenced by oak, which is as it should be with this variety. The palate shows the same sensitive use of oak to produce a totally delicious wine in terms of flavour and balance. **rating:** 93

best drinking 1999–2001 **best vintages** '98 **drink with** Smoked trout • $20

Dalrymple Unwooded Chardonnay

Tasted at two wine shows in early 1999; on one occasion the wine appeared to show some sulphur dioxide (which will disappear with age in any event) and on the other occasion not. This tasting note is taken from the 1999 *Winewise* Small Makers Competition.

🍷🍷🍷🍷 **1998** Light yellow-green; a fragrant bouquet with tangy citrus fruit is followed by a fairly light-bodied, but pleasant palate with pure melon and citrus varietal fruit. Stood up well amongst bigger Chardonnays from warmer areas. **rating:** 85

best drinking 1999–2003 **best vintages** '98 **drink with** Scampi • $16

Dalrymple Chardonnay

Produced from 4.8 hectares of estate plantings, and a wine that always stands out amongst its Tasmanian fellows. Dalrymple winemaker Bert Sundstrup seems to be making better wines year by year. On my score sheet, at least, the '97 wine was unlucky not to do even better than the silver medal awarded to it at the 1999 Tasmanian Wines Show.

🍷🍷🍷🍷🍷 **1997** Bright yellow-green; strong citrus fruit is matched by strong but stylish barrel-ferment oak and some mlf characters. On the palate the power and character of the fruit carries the oak; long and mouthfilling. **rating:** 94

best drinking 1999–2003 **best vintages** '91, '92, '94, '97 **drink with** Braised pork neck • $18

Dalrymple Pinot Noir

Yet another of a range of very good wines coming from Dalrymple out of the 1998 vintage.

🍷🍷🍷🍷 **1998** Light to medium red-purple; while the bouquet is only of light to medium intensity, the fruit is sweet and complexed by a touch of forest. The same play occurs on the palate which ranges through plum, leaf and forest flavours in a delicate but appealing fashion. **rating:** 89

best drinking 1999–2001 **best vintages** '98 **drink with** Chinese barbecued pork • $25

dalwhinnie ★★★★★

Taltarni Road, Moonambel, Vic 3478 **region** Pyrenees

ph (03) 5467 2388 **fax** (03) 5467 2237 **open** 7 days 10–5

winemaker David Jones, Rick Kinzbrunner (Contract) **prod.** 4500 **est.** 1976

prod. range ($25–38 CD) Dalwhinnie Chardonnay and Pinot Noir, Moonambel Shiraz and Cabernet.

summary David and Jenny Jones have now acquired full ownership of Dalwhinnie from Ewan Jones, and have three children of their own to ensure the future succession. In the meantime, Dalwhinnie goes from strength to strength, making outstanding wines right across the board. The wines all show tremendous depth of fruit flavour, reflecting the relatively low-yielding but very well-maintained vineyards. It is hard to say whether the Chardonnay or the Shiraz is the more distinguished, the Pinot Noir (made with assistance from Rick Kinzbrunner) a startling arrival from out of nowhere. A further 8 hectares of shiraz (with a litle viognier) were planted in the spring of 1999 on a newly acquired block on Taltarni Road, permitting the further development of export markets already established in the UK, Asia, Germany and US.

Dalwhinnie Chardonnay

Produced from a little under 4 hectares of low-yielding, unirrigated estate-grown vines, with an ancestry going back to the clone introduced in the nineteenth century and discovered at Mudgee in the late 1960s. Barrel fermentation adds to the richness to produce an invariably extremely complex wine.

🍷🍷🍷🍷 **1997** Medium to full yellow-green; the bouquet is solid, of moderate to full intensity, subtle oak, and the fruit not aromatic. On the palate the wine lacks what might be termed fruit penetration, although there is no fault of any description. Perhaps time will help. **rating:** 85

best drinking 2000–2004 **best vintages** '87, '88, '90, '92, '93, '94, '96 **drink with** Turkey • $25

Dalwhinnie Pinot Noir

Produced from 2.4 hectares of estate plantings, and a wine that I tasted without any foreknowledge. It came as a total surprise, and it was only when I subsequently read that it was made by Rick Kinzbrunner (of Giaconda) that the pieces fell into place.

🍷🍷🍷🍷 **1997** Light to medium red; the bouquet is complex, with spicy nutmeg overtones followed by a long, sappy, spicy and strawberry-flavoured palate; subtle oak. Kinzbrunner's magic at work here yet again. **rating:** 88

best drinking 1999–2003 **best vintages** NA **drink with** Game • $30

Dalwhinnie Moonambel Shiraz

An exceptionally concentrated and powerful wine, fully reflecting the low yields and the influence of the quartz, clay and gravel soils. These are not wines for the faint-hearted, positively demanding long cellaring, but having the balance to repay patience.

🍷🍷🍷🍷🍸 **1997** Medium to full red-purple; the intense bouquet ranges through spice, berry, earth and vanilla; the palate is marked by very firm dark cherry fruit leavened by some spicy characters. The tannins, while not big, are quite firm, guaranteeing longevity. **rating:** 91

best drinking 2002–2012 **best vintages** '86, '88, '90, '91, '92, '94, '95, '97 **drink with** Potent cheeses, strong red meats • $38

Dalwhinnie Moonambel Cabernet

Predominantly Cabernet Sauvignon, but with a small percentage of estate-grown Cabernet Franc and Merlot. Every bit as powerful and concentrated as the Shiraz, and likewise needing many years in bottle.

🍷🍷🍷🍷 **1997** Medium to full purple-red; smooth, quite intense blackberry/cassis fruit is supported by subtle oak on the bouquet; powerful tannins run right through the palate, and one wonders whether they will ever soften. Certainly they unbalance the wine now, which is a pity. **rating:** 86

best drinking 2004–2010 **best vintages** '86, '88, '90, '91, '92, '93, '94 **drink with** Rare char-grilled rump steak • $34

dalyup river estate NR

Murrays Road, Esperance, WA 6450 **region** Other Wineries of WA
ph (08) 9076 5027 **fax** (08) 9076 5027 **open** Weekends 10–4
winemaker Tom Murray **prod.** 700 **est.** 1987
prod. range ($12–15 CD) Hellfire White, Esperance Sauvignon Blanc, Esperance Shiraz, Esperance Cabernet Sauvignon, Port.

summary Arguably the most remote winery in Australia other than Chateau Hornsby in Alice Springs. The quantities are as small as the cellar-door prices are modest; this apart, the light but fragrant wines show the cool climate of this ocean-side vineyard.

dal zotto wines NR

Edi Road, Cheshunt, Vic 3678 **region** King Valley
ph (03) 5729 8321 **fax** (03) 5729 8490 **open** By appointment
winemaker Otto Dal Zotto **prod.** 1500 **est.** 1987
prod. range ($15–16 CD) Chardonnay, Merlot, Cabernet Merlot, Cabernet Sauvignon.

summary Dal Zotto Wines remains primarily a contract grape grower, with almost 26 hectares of vineyards (predominantly chardonnay, cabernet sauvignon and merlot, but with trial plantings of sangiovese, barbera and marzemimo) but does make a small amount of wine for local sale (and by mail order).

d'arenberg ★★★★★

Osborn Road, McLaren Vale, SA 5171 **region** McLaren Vale
ph (08) 8323 8206 **fax** (08) 8323 8423 **open** 7 days 10–5
winemaker Chester Osborn, Phillip Dean **prod.** 120 000 **est.** 1912
prod. range ($10–50 R) Dry Dam Riesling, Dryland Sauvignon Blanc, Olive Grove Chardonnay, Other Side Chardonnay, Noble Riesling, Peppermint Paddock Sparkling Chambourcin, Twenty Eight Road Mourvedre, d'Arry's Original Shiraz Grenache, Custodian Grenache, Dead Arm Shiraz, Footbolt Old Vine Shiraz, Red Ochre, Ironstone Pressings, High Trellis Cabernet Sauvignon, The Coppermine Road Cabernet Sauvignon; Fortified.
summary d'Arenberg has adopted a much higher profile in the second half of the 1990s with a cascade of volubly worded labels and the opening of a spectacularly situated and high-quality restaurant, d'Arry's Verandah. Happily, wine quality has more than kept pace with the label uplifts. An incredible number of export markets spread across Europe, North America and Asia, with all of the major countries represented.

d'Arenberg The Noble Gewurztraminer

Packaged in a slender olive oil-type bottle and in 375 ml format, the back label still provides several minutes reading with the aid of a magnifying glass. From it I deduce that the first vintage of this wine was made in 1992, and that it has gained an international reputation. That, I have to admit, passed me by, but I also have to admit that the '97 is an exceptional wine, both in terms of interest and its sheer quality.

🍷🍷🍷🍷🍷 **1997** Full yellow-orange; the bouquet is wonderfully rich and complex, with lychee, peach and cumquat aromas which translate directly through to the luscious but extremely well-balanced palate. A magical combination of botrytis and varietal character. **rating:** 92

best drinking 1999–2004 **best vintages** '97 **drink with** Spiced apple tart and King Island cream • $30

---

d'Arenberg Dead Arm Shiraz

Part of a veritable cascade of new labels which tumbled out from d'Arenberg in the mid-1990s, and a decidedly strange name for the super-premium flagship of the winery. The name comes from a fungal disease (phomopsis viticola) that attacks vines in many parts of the world, and which can cause one half (or one side) of the vine to die, leaving the other side unaffected (at least for the time being). The older the vines, the more likely the incidence of attack. So much for the name, the wine itself can be absolutely outstanding, bringing together concentrated fruit with 12-18 months maturation in new American oak barriques.

🍷🍷🍷🍷🍷 **1997** Dense red-purple; an exceptional rich, full and concentrated bouquet with lushly ripe fruit married with oak. The palate effectively replicates the bouquet, finishing with soft tannins. Top gold 1998 National Wine Show. **rating:** 96

best drinking 2002–2012 **best vintages** '94, '95, '96, '97 **drink with** Marinated beef • $50

---

d'Arenberg The Footbolt Old Vine Shiraz

Produced from vines dating back as far as 1890 but harvested over a long period to achieve differing flavour and sugar levels to add complexity. Made traditionally in open fermenters, basket-pressed, and aged in a mix of French and American barriques and hogsheads for one year, all of which have been previously used, and which are not intended to impart excessive oak flavour. The adddition of the word 'Footbolt' on the front label has given rise to a world record back label, best read with a microscope and a spellcheck, the latter to tell you that the 'absinent' Joe Osborn was really abstinent.

🍷🍷🍷🍷🍷 **1997** Medium red-purple; the bouquet offers fresh, clean cherry fruit with touches of earth; the palate is nowhere near as concentrated as that of the Dead Arm, but does have fresh cherry, berry and mint fruit, and is supported by good tannins on the finish. **rating:** 90

best drinking 2002–2007 **best vintages** '82, '88, '90, '91, '94, '95, '96, '97 **drink with** Smoked lamb with redcurrant sauce • $18.50

---

d'Arenberg Ironstone Pressings

An intermittent release from d'Arenberg over the years, but likely to become a permanent part of the scene in the future, being made in both 1994 and 1995. Typically a blend of 85% Grenache and 15% Shiraz, it is not – as the name might suggest – made up of the pressings component of the various Grenache and Shiraz blends and wines made by d'Arenberg. In other words, it contains the usual mix of free-run and pressings material; to the extent it has extra weight, it comes from the late harvesting, with the ripest components nearing 16° baumé – terrorising stuff for the uninitiated.

1997 Medium red-purple; the moderately intense bouquet has cedar, plum and spicy Christmas cake aromas and avoids being jammy. The palate has a quite complex structure, with tannins running through the Christmas pudding spice flavours. **rating:** 90

best drinking 2002–2007 **best vintages** '91, '94, '95, '97 **drink with** Leave it in the cellar • $50

d'Arenberg d'Arry's Original Shiraz Grenache

A classic wine style with an illustrious show record and, as the classic wine entry indicates, dating back to 1961. Made from roughly equal proportions of low-yielding, old vine Shiraz and Grenache, held in a mix of large and small oak (principally old) for ten months prior to bottling. Quite deliberately made in a slightly old-fashioned style, fruit- rather than oak-driven, but none the worse for that.

1997 Excellent medium to full red-purple; the aromas are sweet, with blueberry and cherry, together with a hint of mint. The palate is very supple and elegant, with a silky, seamless mix of cherry, berry, mint and plum. Fine tannins on the finish. Top gold 1998 National Wine Show. **rating:** 94

best drinking 2000–2010 **best vintages** '63, '76, '86, '87, '88, '91, '95, '96, '97 **drink with** Jugged hare • $18.50

d'Arenberg The Custodian Grenache

As with all of the d'Arenberg wines, if you get to read the back label, these background notes will be as superfluous as they are brief. Suffice it to say that grenache has always been a major part of the d'Arenberg red wine production, and the best parcels of old vine grenache are used to produce the wine. It is kept in tank and in large wood for part of the time, before spending six months in new and old American oak barriques, thus putting the major emphasis on the luscious fruit – which is as it should be.

1997 Medium to full red-purple; cedar, tobacco and spice aromas of medium intensity are followed by a much more concentrated and weighty palate, with rich red berry fruit. **rating:** 87

best drinking 2000–2007 **best vintages** '91, '92, '94, '95 **drink with** Ragout of venison • $20

d'Arenberg The Coppermine Road Cabernet Sauvignon

The name comes from a road adjoining the vineyard (which in fact has four different names along parts of its not very great length, which is somehow apppropriate). It spends 18 months in new and used French and American oak barriques.

1997 Medium to full red-purple; rich, concentrated and ripe cassis berry aromas are followed by a rich, concentrated and voluptuous palate; some will see it as slightly extractive, others will enjoy the power and depth of the palate. **rating:** 91

best drinking 2005–2011 **best vintages** '97 **drink with** Leave it in the cellar • $50

dargo valley winery NR

Lower Dargo Road, Dargo, Vic 3682 **region** Gippsland

ph (03) 5140 1228 **fax** (03) 5140 1388 **open** Mon–Thur 12–8, weekends, holidays 10–8 (closed Fridays)

winemaker Hermann Bila **prod.** 200 **est.** 1985

prod. range ($12–14 CD) Traminer, Rhine Riesling, Sauvignon Blanc, Chardonnay, Pinot Noir, Cabernet Sauvignon, Port.

summary Two and a half hectares are situated in mountain country north of Maffra and looking towards the Bogong National Park. Hermann Bila comes from a family of European winemakers; there is also an on-site restaurant, and Devonshire teas and ploughman's lunches are provided – very useful given the remote locality. The white wines tend to be rustic, the sappy/earthy/cherry Pinot Noir the pick of the red wines.

darling estate ★★★

Whitfield Road, Cheshunt, Vic 3678 **region** King Valley

ph (03) 5729 8396 **fax** (03) 5729 8396 **open** By appointment

winemaker Guy Darling, Rick Kinzbrunner (Consultant) **prod.** 400 **est.** 1990

prod. range ($11–20 ML) Koombahla Riesling, Chenin Blanc, Nambucca Chenin Blanc, Koombahla Chardonnay, Pinot Noir, Koombahla Pinot Noir, Nambucca Gamay, Koombahla Shiraz, Koombahla Cabernets, Koombahla Cabernet Sauvignon.

summary Guy Darling was one of the pioneers of the King Valley when he planted his first vines in 1970. For many years the entire production was purchased by Brown Brothers, providing their well-known Koombahla Estate label. Much of the production from the 21 hectares is still sold to Brown Brothers (and others) but since 1991 Guy Darling has had a fully functional winery established on the vineyard, making a small portion of the production into wine – which was, in fact, his original motivation for planting the first vines. All the wines on sale have considerable bottle age.

darling park ★★★★

Red Hill Road, Red Hill, Vic 3937 **region** Mornington Peninsula

ph (03) 5989 2324 **fax** (03) 5989 2254 **open** 7 days Jan, Feb–Dec weekends and holidays 11–5

winemaker John Sargeant, Kevin McCarthy (Contract) **prod.** 1000 **est.** 1986

prod. range ($16–22 CD) Chardonnay, Pinot Gris, Querida (Rosé), Pinot Noir, Merlot, Cabernet Merlot, Reserve Cabernet Merlot.

summary John and Delys Sargeant have now opened their cellar door-cum-restaurant at Darling Park, which is open every day through January of each year. The labels are the most gloriously baroque of any to be found in Australia, and would give the American Bureau of Alcohol, Tobacco and Firearms (which governs such matters in the United States) total cardiac arrest. At the ripe young age of 70 John Sargeant took over the winemaking of the red wines as from the 1996 vintage.

darlington estate ★★★☆

Lot 39 Nelson Road, Darlington, WA 6070 **region** Perth Hills

ph (08) 9299 6268 **fax** (08) 9299 7107 **open** Thur–Sun and holidays 12–5

winemaker Caspar van der Meer **prod.** 2500 **est.** 1983

prod. range ($12–24 CD) Chardonnay, Semillon Sauvignon Blanc, Shiraz, Cabernet Sauvignon, Ruby Port; also cheaper Symphony (Chenin Blanc), Sonata (Sauvignon Blanc), Vin Primeur and Minuet (Rosé).

summary By far the largest producer in the Perth Hills region, and the best. Winemaking responsibilities have now passed to Caspar van der Meer, Balt's son, who graduated from Roseworthy in 1995 and, after a vintage at Chateau de Landiras in Bordeaux, joined the family business in 1996, returning to Languedoc in 1997 to make a large quantity of wine for the American market. Further improvements in wine quality seem highly probable.

david traeger ★★★

139 High Street, Nagambie, Vic 3608 **region** Goulburn Valley
ph (03) 5794 2514 **fax** (03) 5794 1776 **open** 7 days 10–5
winemaker David Traeger **prod.** 8000 **est.** 1986
prod. range ($16.50–25.50 R) Classic Verdelho, Shiraz, Cabernet, Reserve Cabernet (premium aged release); Helvetia range (available from cellar door only) Riesling, Late Harvest Riesling, Cabernet Dolce.
summary David Traeger learned much during his years as assistant winemaker at Mitchelton, and knows central Victoria well. The red wines are solidly crafted, the Verdelho interesting but more variable in quality. Reasonably active retail distribution through Melbourne, Sydney, Brisbane and Canberra, with exports to Germany, Hong Kong, UK and Singapore.

David Traeger Classic Verdelho

Something of a signature wine for David Traeger, although quality (and style) have varied somewhat over the years, suggesting the vineyard source is sensitive to climatic variation. The grapes, incidentally, are grown in the King Valley.

🍷🍷🍷🍷 **1997** Light to medium yellow-green; a very interesting wine with honey, mead and malt aromas, together with a rich, full-flavoured honeyed/meady palate. Seems far riper than the 12.5 degrees on the label; one suspects it came from very low-yielding vines. **rating:** 87

best drinking 1999–2000 **best vintages** '91, '94, '95, '97 **drink with** Sweet and sour pork • $16

David Traeger Shiraz

Produced from three vineyards in the Nagambie area, with a total of 10 hectares under vine. The American oak handling has been sensibly restrained.

🍷🍷🍷🍷 **1997** Medium to full red-purple; mint, briar and red berry fruit aromas are followed by a palate with plenty of concentration of ripe, cherry fruit. The oak handling is subtle but good throughout; the best from David Traeger for some years. **rating:** 88

best drinking 2002–2007 **best vintages** '88, '92, '97 **drink with** Wild duck • $25.50

deakin estate ★★★★

Kulkyne Way, via Red Cliffs, Vic 3496 **region** Murray Darling (Vic)
ph (03) 5029 1666 **fax** (03) 5024 3316 **open** Not
winemaker Mark Zeppel **prod.** 200 000 **est.** 1980
prod. range ($9.90–12.90 R) Colombard, Sauvignon Blanc, Chardonnay, Shiraz, Cabernet Sauvignon, Brut; Alfred Chardonnay and Alfred Shiraz are flag-bearers.
summary Effectively replaces the Sunnycliff label in the Yunghanns-owned Katnook Estate, Riddoch and (now) Deakin Estate triumvirate, which constitutes the Wingara Wine Group. Sunnycliff is still used for export purposes, but does not appear on the domestic market any more. Deakin Estate draws on 346 hectares of its own vineyards, making it largely self-sufficient, producing competitively priced wines of consistent quality and impressive value, getting better year by year. Exports to the UK, US, Canada, Switzerland, Hong Kong and the Philippines.

Deakin Estate Shiraz

Producing an early-bottled red wine from the Riverland and investing it with both flavour and structure is no easy task. In 1997 and 1998 Deakin Estate succeeded admirably, particularly given the modest price of the wine.

🍷🍷🍷🍷 **1998** Dense red-purple; youthful berry fruit aromas on the bouquet are followed by a wine with considerable substance and weight in the mouth; again, berry and earth flavours are supported by lingering tannins. Outstanding value at the price. **rating:** 85

best drinking 1999–2002 **best vintages** NA **drink with** Meat pie • $9.90

Deakin Estate Alfred Shiraz

1997 marked the inaugural release of Deakin Estate's Alfred Shiraz, paired with Alfred Chardonnay as the flag-bearers for Deakin Estate. It is a blend of 90% Shiraz and 10% Cabernet Sauvignon, matured in a mix of American (55%) and French (45%) oak hogsheads. One-third of the barrels are new, the remainder one and two years old, and the wine spends 14 months in barrel. It is an extremely impressive example of a wine made solely from fruit grown on the banks of the Murray River in northwestern Victoria – in other words, the Riverland.

🍷🍷🍷🍷 **1997** Bright purple-red; the bouquet is clean, with good earthy shiraz varietal character on the bouquet; an altogether impressive wine on the palate, with fresh cherry fruit, good acidity and length to the finish. The oak is well balanced and well integrated. **rating:** 88

best drinking 1999–2003 **best vintages** NA **drink with** Braised lamb shanks • $12.90

Deakin Estate Cabernet Sauvignon

Draws upon 43 hectares of estate plantings; while the region is far better suited to white grapes (and hence white wines) than reds, Deakin Estate winemaker Mark Zeppel has done a pretty good job with this wine. A big brother to the Shiraz, and similarly economically priced. In 1998 it produced an even better wine.

🍷🍷🍷🍷 **1998** Medium to full red-purple; the bouquet has above-average concentration with abundant dark berry fruit, the palate likewise with sweet, dark cassis berry flavours and subtle oak. Not callow or underworked, and outstanding at its price point. **rating:** 87

best drinking 1999–2003 **best vintages** NA **drink with** Steak and kidney pie • $9.99

de bortoli ★★★★

De Bortoli Road, Bilbul, NSW 2680 **region** Riverina

ph (02) 6964 9444 **fax** (02) 6964 9400 **open** Mon–Sat 9–5.30, Sun 9–4

winemaker Darren De Bortoli **prod.** 3 million **est.** 1928

prod. range ($4.90–42.50 CD) Noble One Botrytis Semillon is the flagship wine; Premium varietals under Deen De Bortoli label, and a low-priced range of varietal and generic wines under the Sacred Hill label. Substantial exports in bulk.

summary Famous among the cognoscenti for its superb Botrytis Semillon, that in fact accounts for only a minute part of its total production, this winery turns around low-priced varietal and generic wines which are invariably competently made and equally invariably of value for money. Financial and marketing acumen has made De Bortoli one of the fastest-growing large wineries in Australia in the 1990s. An export list that exceeds even that of D'Arenberg, with markets in 29 countries including such unlikely destinations as Latvia, Dubai and the whole of the Middle East.

De Bortoli Deen De Bortoli Vat 7 Chardonnay

A wine which has on occasions (such as 1993) been astonishingly good given its price, and the volume in which it is made. It is produced almost entirely from Griffith region grapes, but reflects careful selection of the best available material.

🍷🍷🍷½ **1997** Medium yellow-green; the bouquet is fairly straightforward, not showing a great deal of fruit or excitement, but the tangy palate comes as a surprise. A wine with length and some style, particularly at the price. **rating:** 83

best drinking 1998–2000 **best vintages** '90, '92, '93 **drink with** Pasta marinara • $10

De Bortoli Vat 5 Late Semillon

Effectively the third label behind Noble One and Rare Dry Botrytis Semillon, but a wine which shows the vast experience (and preferred access) the De Bortolis have to botrytised semillon. This really is a very useful wine.

ΥΥΥΥΥ **1996** Glowing golden yellow; the bouquet is quite rich, with peach and honey fruit, the palate showing good botrytis characters, with luscious fruit and nicely defined acidity. An outstanding bargain at the price. **rating:** 90

best drinking 1999–2003 **best vintages** NA **drink with** Peach tart • $10

De Bortoli Noble One Botrytis Semillon (375 ml)

A classic wine, without question the foremost example of barrel-fermented, wood-matured Botrytis Semillon in Australia. Every vintage made has won at least one trophy and innumerable gold medals.

ΥΥΥΥΥ **1997** Bright, light yellow-gold; the bouquet has slightly unusual honey and candle wax aromas over citrus fruit. The palate is very long, with excellent acidity, and still evolving and developing; tasted very early in its life, and the points may prove to be too low. **rating:** 92

best drinking 2001–2010 **best vintages** '82, '84, '87, '90, '91, '92, '93, '94, '95, '96 **drink with** Crème brûlée • $22

De Bortoli Black Noble

A quite fascinating wine, first released by De Bortoli in mid-1998, but coming from a lineage dating back to the 1930s, thence 1972 and finally 1982, the last being the year in which the first (conventional) Noble One Botrytis Semillon was released. This wine is made in the fashion of a fortified wine from northeast Victoria, and is very like a Rutherglen Tokay, except for its slightly lighter structure. Parcels of botrytised semillon have been set aside each year since 1982, and a blended solera system established incorporating a minute part of the 1972 Pedro Ximinez experiment. The wine is released at an average age of seven to eight years.

ΥΥΥΥΥ **NV** Mahogany brown with no red tints; you can literally see the raisins and the plum pudding swirling through the intense bouquet and through the classic plum pudding and spice-flavoured palate. Great length and balance, and not too heavy. **rating:** 94

best drinking 1998–1999 **best vintages** NA **drink with** On its own as an aperitif or with coffee • $29

de bortoli (victoria) ★★★★★

Pinnacle Lane, Dixons Creek, Vic 3775 **region** Yarra Valley

ph (03) 5965 2271 **fax** (03) 5965 2442 **open** 7 days 10–5

winemaker Stephen Webber, David Slingsby-Smith, David Bicknell **prod.** 150 000 **est.** 1987

prod. range ($12–55 R) At the top comes the premium Melba (Cabernet blend), followed by Yarra Valley Semillon, Chardonnay, Pinot Noir, Shiraz, Cabernet Sauvignon, Cabernet Merlot; then comes the intermediate Gulf Station range of Riesling, Chardonnay and Cabernet Sauvignon; the Windy Peak range of Riesling, Chardonnay, Pinot Noir, Cabernets, Prestige Cuvée; Montage White and Red.

summary The quality arm of the bustling De Bortoli group, run by Leanne De Bortoli and husband Stephen Webber, ex-Lindeman winemaker. The top label (De Bortoli), the second (Gulf Station) and the third label (Windy Peak) offer wines of consistently good quality and excellent value – the complex Chardonnay of outstanding quality. The omission of the (utterly justified) top rating for the winery in the previous edition of this book was entirely inadvertent. Shares the same amazing split of export markets with its parent.

De Bortoli Windy Peak Riesling

Made primarily from riesling grown in the Strathbogie Ranges and the Yarra Valley, often with a small percentage of Yarra Valley Gewurztraminer included. First made in 1989, the wine has garnered a constant stream of gold medals in national wine shows from that time.

ŸŸŸŸ **1998** Light green-yellow; the bouquet offers pleasant, soft lime and toast fruit with hints of more tropical characters. The palate is high flavoured, with a flick of residual sugar; very clever commercial winemaking. **rating:** 85

best drinking 1999–2003 **best vintages** '90, '92, '94, '95, '96, '97 **drink with** Fresh asparagus • $11.95

De Bortoli Gulf Station Semillon Sauvignon Blanc

A blend of predominantly Semillon with a substantial Sauvignon Blanc component, predominantly cold-fermented in stainless steel to preserve fruit flavour and freshness, but a small portion of ripe semillon was fermented in new American oak barriques for four weeks to add complexity.

ŸŸŸŸ **1998** Light green-yellow; the bouquet is light and clean, but not particularly aromatic. The balance of the wine is its strong point, as is its length. The fruit components (and the subliminal touch of vanilla oak) have come together so well no one character stands out, just delicate and fresh fruit. **rating:** 85

best drinking 1999–2001 **best vintages** NA **drink with** Gently spiced Thai • $14.95

De Bortoli Gulf Station Chardonnay

Achieved immediate fame for De Bortoli when it swept all before it at the 1996 Adelaide Wine Show. This new range, named after the historic Gulf Station on the outskirts of Yarra Glen, shows the winemaking skills of Stephen Webber to full advantage. It also shows just what can be achieved with American oak, a technique previously employed to full advantage by Yarra Ridge. Every vintage to date has been a gold medal winner, the wine a model of consistency.

ŸŸŸŸY **1998** Medium yellow-green; the bouquet is solid, with well-balanced and integrated tangy melon fruit and oak. The palate has good weight and mouthfeel, not alcoholic, and the oak is restrained. **rating:** 90

best drinking 1999–2003 **best vintages** '95, '96, '97, '98 **drink with** Yabbies • $14.95

De Bortoli Windy Peak Chardonnay

Sourced from vineyards in Victoria, but always containing a significant proportion of southern Victorian material, including the Yarra Valley. Over the years has consistently out-performed far more expensive wines in wine shows, and has always represented exceptionally good value for money. Some barrel-ferment and lees contact, together with the very clever use of American oak, are all contributors. Smart new packaging introduced for the 1997 vintage does the wine justice, making it appear far above its class.

ŸŸŸY **1998** Medium yellow-green; while the bouquet is quite solid, it is not particularly aromatic, and the palate follows in the same direction, solidly structured, yet with no particular highlights. May develop with a little time. **rating:** 84

best drinking 1999–2000 **best vintages** '90, '92, '93, '94, '97 **drink with** Sashimi • $11.95

De Bortoli Yarra Valley Chardonnay

Made from both estate-grown grapes and grapes purchased from other Yarra Valley vineyards. Usually in a fuller-bodied style, albeit with restrained oak from partial barrel fermentation. Richly repays cellaring; one of the Yarra Valley's most consistent and best Chardonnays.

ŸŸŸŸY **1998** Medium yellow-green; tangy melon and fig fruit is supported by smooth barrel-ferment oak aromas. A fruit-driven palate with melon and citrus fruit to the fore, hints of mlf and barrel-ferment characters on a long finish. **rating** 92

best drinking 1999–2004 **best vintages** '90, '92, '93, '94, '96, '97, '98 **drink with** Yabbies • $24

De Bortoli Yarra Valley Pinot Noir

In the manner of Socrates dissatisfied, Stephen Webber has until recently pronounced himself unhappy with the De Bortoli Pinot Noirs. However, the '95 won the prestigious Wine Press Club trophy for Best Mature Pinot Noir at the National Wine Show, and both the '96 and '97 wines are better than the '95. Partial cold maceration prior to fermentation, and partial barrel fermentation in 100% French oak are key techniques which have led to the improvement in style. The '97 is a multiple trophy winner, including at the 1999 Sydney Royal Wine Show.

🍷🍷🍷🍷🍷 **1997** Light to medium red-purple; fragrant cherry and plum fruit aromas, with a nice touch of sappiness and well-judged oak announce a wine with excellent structure and mouthfeel: lovely cherry and plum fruit on the mid-palate, followed by soft, very fine tannins on the finish. **rating:** 94

best drinking 1999–2003 **best vintages** '95, '96, '97 **drink with** Duck casserole • $30

De Bortoli Yarra Valley Shiraz

There are surprisingly few Yarra Valley varietal Shirazes on the market, and the fruit flavours vary significantly from one maker to the next, ranging from strong spice and pepper through to more minty aromas and flavours. The De Bortoli wine is usually in the latter spectrum, but hit the jackpot with the stunning '93 and an equally outstanding '94. The '97 outperformed all of the previous wines by winning the Jimmy Watson Trophy in 1998.

🍷🍷🍷🍷🍷 **1997** Medium to full red-purple; the bouquet is as complex as one might expect, with concentrated dark cherry fruit and sweet, cedary oak. The palate is very well balanced, with dark cherry fruit woven through sweet vanilla oak, finishing with soft, lingering tannins. **rating:** 94

best drinking 2002–2012 **best vintages** '88, '90, '91, '92, '94, '97 **drink with** Grilled calf's liver • $30

De Bortoli Melba Barrel Select

First made in the 1992 vintage, when winemaker Stephen Webber selected the very best parcels of grapes available in that vintage to make tiny quantities of the wine, named after Dame Nellie Melba, whose family home was in the heart of the Yarra Valley. It is a blend of Cabernet Sauvignon, Shiraz, Cabernet Franc and Merlot which spends two years in 100% new French oak and two years in bottle prior to release. Approximately 550 cases were made in 1994.

🍷🍷🍷🍷🍷 **1994** Excellent red-purple; the bouquet is smooth and sweet, with a mix of earthy, berry and cedary aromas. The elegant palate ripples through black fruits, cedar and vanilla, finishing with very fine but lingering tannins. Has blossomed over the past 12 months. **rating:** 95

best drinking 2000–2010 **best vintages** '93, '94 **drink with** Yarra Valley venison • $55

De Bortoli Yarra Valley Cabernet Sauvignon

Like the Shiraz, made from both estate-grown and purchased grapes. A 100% Cabernet Sauvignon (a Cabernet Merlot is also marketed) made using what might be described as traditional French techniques, and matured in a mix of new and used French barriques for 18 months before bottling. Can be outstanding; the 1995 vintage won five trophies and six gold medals in various Victorian wine shows.

🍷🍷🍷🍷 **1996** Medium red-purple; the bouquet is sweet, with varietal cassis berry aromas and no obvious green leaf characters. An elegant wine on the palate, strongly reminiscent of Bordeaux in style, with attractive cedary oak running through the palate. A very good outcome for a vintage which turned sour in its second half. **rating:** 85

best drinking 2000–2010 **best vintages** '88, '90, '91, '92, '94, '95 **drink with** Beef casserole • $30

deep dene vineyard NR

36 Glenisla Road, Bickley, WA 6076 **region** Perth Hills
ph (08) 9293 0077 **fax** (08) 9293 0077 **open** By appointment
winemaker Celine Rousseau (Contract) **prod.** 4000 **est.** 1994
prod. range ($25–28 R) Pinot Noir, Shiraz.
summary Deep Dene is the largest of the Perth Hills wineries, with French-born and trained Celine Rousseau as winemaker. It draws upon 4 hectares of pinot noir and half a hectare of shiraz continuing the near obsession of the Perth Hills vignerons with pinot noir in a climate which, to put it mildly, is difficult for the variety.

deep woods estate NR

Lot 10 Commonage Road, Yallingup, WA 6282 **region** Margaret River
ph (08) 9756 6066 **fax** (08) 9756 6066 **open** 7 days 11.30–5.30
winemaker Mark Lane **prod.** 4000 **est.** 1987
prod. range ($16–28 CD) Semillon, Sauvignon Blanc, Verdelho, Late Harvest (Verdelho), Cabernet Sauvignon, Reserve Cabernet Sauvignon.
summary The substantial-sized Deep Woods Estate has a convoluted history, with a number of changes of ownership since the first vines were planted in 1987. It is now owned by Malcolm and Margaret Gould who, with assistance from their children, are running both Deep Woods Estate and the Margaret River Marron Farm. In all, there are 24 hectares of vineyard planted to 14 different varieties. A winery was erected in 1998, and as from 1999 Mark Lane, former computer software designer-turned winemaker (with a vintage at Coldstream Hills) will be in charge of winemaking.

Deep Woods Estate Margaret River Semillon

Made in typically very ripe and full Margaret River style with a touch of what appears to be American oak.

🍷🍷🍷🍷 **1998** Medium yellow-green; the bouquet is clean, ripe and full, showing no herbal grass or citrus varietal character; some American oak. On the palate, a big, tropical, ripe pineapple and banana-flavoured wine which seems much riper than even the 12.8 degrees alcohol appearing on the label. **rating:** 84

best drinking 1999–2000 **best vintages** NA **drink with** Takeaway chicken • $19.50

Deep Woods Estate Margaret River Cabernet Reserve

A quantum leap up in quality over the varietal Cabernet Sauvignon of the same vintage from Deep Woods Estate.

🍷🍷🍷🍷 **1997** Medium to full red-purple; the bouquet is quite solid and clean, with sweet, dark blackberry fruit. A solidly rich and ripe wine on the palate with dark berry and chocolate flavours finishing with soft tannins and nice oak. **rating:** 89

best drinking 2001–2007 **best vintages** '97 **drink with** Roast beef • $28

de iuliis NR

Lot 1 Lovedale Road, Keinbah, NSW 2321 **region** Lower Hunter Valley
ph (02) 4930 7403 **fax** (02) 4968 8192 **open** By appointment
winemaker David Hook, Michael De Iuliis **prod.** 6500 **est.** 1990
prod. range ($12–18 CD) Semillon, Reserve Semillon Verdelho, Shiraz, Merlot, Cabernet Merlot.
summary Three generations of the De Iuliis family have been involved in the establishment of their 40-hectare vineyard at Keinbah in the Lower Hunter Valley. The family acquired the property in 1986, and planted the first vines in 1990, selling the grapes from the first few vintages to Tyrrell's, but retaining small amounts of grapes for release under the De Iuliis label. Michael

De Iuliis, the third-generation family member involved, is completing postgraduate studies in oenology at the Roseworthy Campus of Adelaide University, and assists David Hook (a long-term Hunter winemaker) in making the wine. As at February 1999, the wine was only available by mail order, but with this level of production, both retail and cellar-door sales will follow in the future.

delacolline estate ★★★

Whillas Road, Port Lincoln, SA 5606 **region** Other Wineries of SA
ph (08) 8682 5277 **fax** (08) 8682 4455 **open** Weekends 9–5
winemaker Andrew Mitchell (Contract) **prod.** 650 **est.** 1984
prod. range ($10–15 R) Riesling, Fumé Blanc, Cabernet Sauvignon.
summary Joins Boston Bay as the second Port Lincoln producer; the white wines are made under contract in the Clare Valley. The 3-hectare vineyard, run under the direction of Tony Bassett, reflects the cool maritime influence, with ocean currents that sweep up from the Antarctic.

delamere ★★★

Bridport Road, Pipers Brook, Tas 7254 **region** Northern Tasmania
ph (03) 6382 7190 **fax** (03) 6382 7250 **open** 7 days 10–5
winemaker Richard Richardson **prod.** 2500 **est.** 1983
prod. range ($12–24 CD) Chardonnay, White Pinot Noir, Pinot Noir (Standard, Dry Red and Reserve), Sparkling.
summary Richie Richardson produces elegant, rather light-bodied wines that have a strong following. The Chardonnay has been most successful, with a textured, complex, malolactic-influenced wine with great, creamy feel in the mouth. The Pinots typically show pleasant varietal fruit, but seem to suffer from handling problems with oak and a touch of oxidation. Retail distribution through fine wine outlets in Tasmania, Melbourne, Sydney and Brisbane.

delaney's creek winery NR

70 Hennessey Road, Delaneys Creek, Qld 4514 **region** D'Aguilar Range, Qld
ph (07) 5496 4925 **fax** (07) 5496 4926 **open** 7 days 10–5
winemaker Tom Weidmann **prod.** 3000 **est.** 1997
prod. range ($9.50–18 CD) Selection Chardonnay, Muscat Rosé, Late Shiraz
summary Tom Weidmann established Delaney's Creek Winery in 1997, and by doing so has expanded the vineyard map of Queensland yet further. Delaney's Creek is situated near the town of Woodford, itself not far northwest of Caboolture. In 1998 Weidmann planted an exotic mix of 1 hectare each of shiraz, chardonnay, sangiovese, touriga nacional and verdelho. In the meantime he is obtaining his grapes from 4 hectares of contract-grown fruit including cabernet sauvignon, cabernet franc, merlot, shiraz, chardonnay, marsanne and verdelho.

delatite ★★★★

Stoneys Road, Mansfield, Vic 3722 **region** Central Victorian High Country
ph (03) 5775 2922 **fax** (03) 5775 2911 **open** 7 days 10–4
winemaker Rosalind Ritchie **prod.** 14 000 **est.** 1982
prod. range ($11–27.50 CD) Unoaked Chardonnay, Chardonnay, Sauvignon Blanc, Riesling, Late Picked Riesling, Dead Man's Hill Gewürztraminer, Delmelza Pinot Chardonnay, Pinot Noir, Shiraz, Merlot, Malbec, Dungeon Gully, Devil's River (Cabernet Sauvignon Malbec Shiraz), Fortifieds.
summary With its sweeping views across to the snow-clad alps, this is uncompromising cool-climate viticulture, and the wines naturally reflect the climate. Light but intense Riesling and spicy Traminer flower with a year or two in bottle, and in the warmer vintages the red wines achieve flavour and mouthfeel, albeit with a distinctive mintiness.

demondrille vineyards NR

RMB 97 Prunevale Road, Kingsvale, NSW 2587 **region** Hilltops
ph (02) 6384 4272 **fax** (02) 6384 4292 **open** Sat–Sun 10.30–5
winemaker Pamela Gillespie, Tamburlaine (Contract), G Sissingh (Consultant) **prod.** NA **est.** 1979
prod. range ($15–20 CD) The Dove (Sauvignon Blanc Semillon), Purgatory (Pinot Noir), The Raven (Shiraz) and Black Rose (Cabernet Sauvignon Merlot Franc), together with lesser quantities of cellar-door-only wines (from the 'Tin Shed' range).
summary Pamela Gillespie and Robert Provan purchased the former Hercynia Vineyard and winery in 1995. Pam Gillespie has an Associate Diploma in Winemaking and Marketing from Adelaide University – Roseworthy, and has been in the hospitality industry since 1989. Her partner had a remarkable career, and is partway through a Bachelor of Science degree at Sydney University (as a mature-age student) majoring in agriculture. Most of the wines from Demondrille are made at Charles Sturt University, with smaller quantities made on site with assistance from Gerry Sissingh. Purgatory, though far from a great wine, is not so bad as to deserve its (bizarre) name. Specially created food platters featuring local produce are available at the winery each weekend.

dennis ★★★

Kangarilla Road, McLaren Vale, SA 5171 **region** McLaren Vale
ph (08) 8323 8665 **fax** (08) 8323 9121 **open** Mon–Fri 10–5, weekends, holidays 11–5
winemaker Peter Dennis **prod.** 10 000 **est.** 1970
prod. range ($13–30 CD) Sauvignon Blanc, Chardonnay, Shiraz, Cabernet Sauvignon, Merlot Cabernet, Mead, Egerton Vintage Port, Old Tawny Port.
summary A low-profile winery which has, from time to time, made some excellent wines, most notably typically full-blown, buttery/peachy Chardonnay.

d'entrecasteaux NR

Boorara Road, Northcliffe, WA 6262 **region** Pemberton
ph (08) 9776 7232 **open** By appointment
winemaker Alkoomi (Contract) **prod.** 600 **est.** 1988
prod. range ($NA) Chardonnay, Sauvignon Blanc, Pinot Noir, Cabernet Sauvignon.
summary Not to be confused with the now moribund Tasmanian winery of the same name, but likewise taking its name from the French explorer Admiral Bruni D'Entrecasteaux who visited both Tasmania and the southwest coast of Western Australia. Four hectares of estate vineyards, planted on rich Karri loam, produce grapes for the wines which are contract-made at Alkoomi.

derwent estate ★★★★

329 Lyell Highway, Granton, Tas 7070 **region** Southern Tasmania
ph (03) 6248 5073 **fax** (03) 6248 5073 **open** Not
winemaker Stefano Lubiana (Contract) **prod.** 300 **est.** 1993
prod. range ($17.50 ML) Riesling, Pinot Noir.
summary The Hanigan family has established Derwent Estate as part of a diversification programme for their 400-hectare mixed farming property. Five hectares of vineyard have been progressively planted since 1993, initially to riesling, followed by pinot noir.

Derwent Estate Riesling

Produced entirely from 1.5 hectares of estate plantings; typically picked at the end of the first week of May with yields of around 8 tonnes per hectare. Cold-fermented to dryness and early bottled.

🍷🍷🍷🍷🍷 **1998** Light to medium yellow-green; a mix of herb and mineral aromas (together with a touch of sulphur dioxide which rapidly dissipated) is followed by a wine with excellent lime/citrus flavour, and typical Tasmanian brisk acidity on the finish. **rating:** 94

best drinking 2001–2006 **best vintages** '98 **drink with** Fresh asparagus • $17.50

devil's lair NR

Rocky Road, Forest Grove via Margaret River, WA 6286 **region** Margaret River

ph (08) 9757 7573 **fax** (08) 9757 7533 **open** By appointment

winemaker Janice McDonald **prod.** 14 000 **est.** 1985

prod. range ($22–38 R) Chardonnay, Pinot Noir, Cabernet Merlot; Fifth Leg Dry White and Dry Red.

summary Having rapidly carved out a high reputation for itself through a combination of clever packaging and marketing allied with impressive wine quality, Devil's Lair was acquired by Southcorp (Penfolds, etc) wine group in December 1996, and production is projected to increase to 70 000 cases after the turn of the century. Readers should be aware that from mid-1997 I have assumed Group Winemaking responsibility (inter alia) for Devil's Lair, although Janice McDonald remains firmly in command of winemaking. Because of my involvement with the winery I neither rate the winery nor the wines.

Devil's Lair Chardonnay

Has always been made in an extremely rich, concentrated full-bodied Margaret River-style as befits its estate-grown origins. The generous use of oak goes well with the intense fruit. Low-yielding vineyards (never more than 2.5 tonnes per acre) and barrel fermentation and maturation in a mix of new and one-year-old Vosges and Alliers barriques from Dargaud & Jaegle and Francois Freres. Partial malolactic fermentation adds yet another level of complexity. Alcohol levels running up to 14.5 degrees complete a powerful picture.

1997 Light to medium green-yellow; complex, tangy melon and citrus fruit with Burgundian overtones and subtle oak on the bouquet introduce a rich, full, tangy palate with lots of flavour and balanced oak. Somewhere in the middle of the Margaret River Chardonnay style.

best drinking 1999–2003 **best vintages** '92, '94, '96, '97 **drink with** Rich white meat • $38

Devil's Lair Margaret River

An estate-grown Bordeaux-blend, with the percentage of Cabernet Sauvignon, Merlot and Cabernet Franc varying from year to year, but with the Cabernet component never less than 60%, and usually closer to 85%. Spends 18–20 months in French oak; a powerful, intense wine.

1996 Medium to full red-purple; clean, quite ripe dark berry fruits and subtle oak aromas are followed on the palate by dark berry and cassis fruit; well-structured, and pleasant oak.

best drinking 2000–2005 **best vintages** NA **drink with** Roast veal • $38

diamond valley vineyards ★★★★☆

2130 Kinglake Road, St Andrews, Vic 3761 **region** Yarra Valley

ph (03) 9710 1484 **fax** (03) 9710 1369 **open** Not

winemaker David Lance, James Lance **prod.** 6000 **est.** 1976

prod. range ($15–49.35 R) Estate Riesling, Chardonnay, Pinot Noir, Cabernets, Sparkling Diamond; Blue Label Semillon Sauvignon Blanc, Chardonnay, Pinot Noir, Cabernet Merlot; Close Planted Pinot Noir.

summary One of the Yarra Valley's finest producers of Pinot Noir, and an early pacesetter for the variety, making wines of tremendous style and crystal-clear varietal character. They are not Cabernet Sauvignon look-alikes, but true Pinot Noir, fragrant and intense. Much of the wine is sold through an informative and well-presented mailing list.

Diamond Valley Estate Chardonnay

Produced in very limited quantities from a little over 1 hectare of estate vines, and first released in 1990. Typically intense, the wine is fruit- rather than oak-driven. The wine has won numerous gold medals since 1990. The '96 was a richly deserved trophy and gold medal winner at the 1997 Royal Adelaide Wine Show, and has a long future.

🍷🍷🍷🍷 **1997** Medium yellow-green; the bouquet has unusual tobacco and spice edges to the aroma, possibly from oak. The palate straightens the ship, with nice melon and peach fruit of medium intensity in the usual Diamond Valley-style. Would have received higher points were it not for the strange bouquet. **rating:** 88

best drinking 2000–2005 **best vintages** '90, '92, '94, '96 **drink with** Cold smoked trout • $29.75

Diamond Valley Close Planted Pinot Noir

Produced from estate-grown grapes planted at a density of 1 metre between the rows, and 1.2 metres between the vines, a density which effectively precludes the use of anything other than hand labour in the vineyard. Only 125 dozen made, and the wine is not filtered.

🍷🍷🍷🍷🍷 **1997** Medium red, with just a touch of purple; a striking bouquet with intense and complex spicy, stalky, cedary, tobacco aromas. The palate is very long, powerful, with sappy/stalky/foresty characters and lingering acidity. **rating:** 91

best drinking 1999–2003 **best vintages** '97 **drink with** Braised duck • $49.35

Diamond Valley Estate Pinot Noir

Deserves its recognition as one of the greatest of the Yarra Valley Pinot Noirs, invariably generously flavoured, and invariably showing strong varietal character, often in a ripe mould A prolific trophy winner over the years, and seldom misses the mark. The secret lies in large part in the estate vineyard, with meticulous viticulture and low yields. I have to admit to a mild sense of disappointment with the '97, given the great vintage.

🍷🍷🍷🍷 **1997** Bright red-purple; the bouquet is surprisingly stalky/gamey/foresty, out of the usual style, and far more complex than is often the case. The palate is powerful, with more of the characters of the bouquet investing the cherry plum fruit with slightly sharp edges. All in all, very interesting. **rating:** 87

best drinking 1999–2003 **best vintages** '86, '90, '91, '92, '93, '94, '96 **drink with** Wild duck • $39.50

---

diggers rest NR

205 Old Vineyard Road, Sunbury, Vic 3429 **region** Sunbury

ph (03) 9740 1660 **fax** (03) 9740 1660 **open** By appointment

winemaker Peter Dredge **prod.** 1000 **est.** 1987

prod. range ($15–20 CD) Chardonnay, Pinot Noir, Shiraz, Cabernet Sauvignon.

summary Diggers Rest was purchased from the founders Frank and Judith Hogan in July 1998; the new owners Elias and Joseph Obeid intend to expand the vineyard resources and, by that means, significantly increase production.

djinta djinta winery NR

10 Stevens Road, Kardella South, Vic 3950 **region** Gippsland

ph (03) 5658 1163 **fax** (03) 5658 1863 **open** Weekends and public holidays 10–5 or by appointment

winemaker Peter Harley **prod.** 170 **est.** 1991

prod. range ($15–21 CD) Semillon, Sauvignon Blanc, Marsanne, Cabernet Sauvignon.

summary Is one of a group of wineries situated between Leongatha and Korumburra, the most famous being Bass Phillip. Vines were first planted in 1986 but were largely neglected until Peter and Helen Harley acquired the property in 1991, set about reviving the 2 hectares of sauvignon blanc and a little cabernet sauvignon, planting an additional 3 hectares (in total) of merlot, cabernet franc, cabernet sauvignon, semillon, marsanne, roussane and viognier. The first vintage was 1995, during the time that Peter Harley was completing a Bachelor of Applied Science (Wine Science) at Charles Sturt University. They are deliberately adopting a low-technology approach to both vineyard and winery practices, using organic methods wherever possible. They hope to finish the winery building and a 40-seat restaurant at some stage during 1999 or 2000. Naturally high levels of acidity may need watching, particularly on the evidence of the 1997 Semillon. The 1998 Sauvignon Blanc (84 points) is a nice wine with plenty of passionfruit flavour, the 1995 even better (86 points).

Djinta Djinta Sauvignon Blanc

The 1995 Sauvignon Blanc was the first wine produced under the Djinta Djinta label, and is a surprising wine in many ways. First, it has aged in bottle far better than most Sauvignon Blancs do; second, it is of quite excellent quality for a first attempt.

🍷🍷🍷🍷 **1995** Light yellow-green; a very complex, floral bouquet of herb, gooseberry and lantana is followed by similarly complex, high-flavoured tropical/gooseberry palate, part of these characters are doubtless from bottle development, but most have come from the style and quality of the grapes. **rating:** 86

best drinking 1999–2000 **best vintages** NA **drink with** Rich seafood • $21

domaine chandon ★★★★★

Maroondah Highway, Coldstream, Vic 3770 **region** Yarra Valley
ph (03) 9739 1110 **fax** (03) 9739 1095 **open** 7 days 10.30–4.30
winemaker Wayne Donaldson **prod.** 90 000 **est.** 1986
prod. range ($12.50–52 CD) Sparkling (Méthode Champenoise) specialist with five sparkling wines: Brut, Blanc de Blancs, Blanc de Noirs, Rosé, Yarra Valley Cuvée Riche; Green Point is export label, also used for still Pinot Noir (with a Reserve version now joining ranks) and Chardonnay; Colonnades is third label for table wines.
summary Wholly owned by Moet et Chandon, and the most important wine facility in the Yarra Valley, superbly located with luxurious tasting facilities (a small tasting charge is levied). The wines are exemplary, thought by many to be the best produced by Moet et Chandon in any of its overseas subsidiary operations, a complex blend of French and Australian style.

Domaine Chandon Green Point Chardonnay

The Green Point varietal table wines have slowly but steadily assumed greater significance in the Domaine Chandon winemaking portfolio. The Chardonnay and Pinot Noir are both sourced from the Yarra Valley, and predominantly (if not exclusively) from estate vineyards surrounding the winery. Right from the outset, the style has been generous and full flavoured.

🍷🍷🍷🍷 **1997** Light yellow-green; both the oak influence and the malolactic influence are very evident on the bouquet, and follow through into the complex nutty/cashew/melon-flavoured palate, with sweet spicy oak coming through on the finish. **rating:** 86

best drinking 1999–2003 **best vintages** '92, '93, '96 **drink with** Veal in white sauce • $22.25

Domaine Chandon Blanc de Blancs

Made entirely from Chardonnay sourced from all over Australia, with a significant contribution from Coonawarra and the Yarra Valley. Usually released between four and five years after vintage, and invariably shows just why French Champagne makers regard Chardonnay as particularly long-lived.

🍷🍷🍷🍷🍷 **1995** Brilliant green-yellow; fine, lemony, citrussy aromas with just a hint of bready autolysis is followed by a long, clean, crisp and still youthful palate. **rating:** 93

best drinking 1999–2004 **best vintages** '90, '92, '93, '95 **drink with** Aperitif • $26

Domaine Chandon Blanc de Noirs

The converse of the Blanc de Blancs, made entirely from Pinot Noir; like the Blanc de Blancs, held on lees for over three years before disgorgement. Sourced from numerous southern Australian vineyards, notably the Yarra Valley and Tasmania.

ŸŸŸŸŸ **1994** Pale straw bronze; the bouquet is complex, with nutty, strawberry aromas; there is excellent flavour, richness and balance, with an echo of strawberry on the palate. Perfect judgment of dosage. **rating:** 94

best drinking 1999–2004 **best vintages** '90, '92, '94 **drink with** Hors d'oeuvres • $26

Domaine Chandon Late Disgorged Brut

A relatively new direction for Domaine Chandon, but then not so surprising given the length of time such wines have to be left on yeast lees. A blend of 51% Pinot Noir, 47% Chardonnay and 2% Pinot Meunier. Spends five years on yeast lees.

ŸŸŸŸŸ **1993** Pale straw-yellow, although the bead (or mousse) seems bigger than expected; perhaps it was a quirk of the glass being used. The bouquet is finer than expected, with delicate yeasty/bready aromas. The palate is every bit as elegant as the bouquet promises, with a lingering but light finish. A very fine style. **rating:** 94

best drinking 1999–2002 **best vintages** '92, '93 **drink with** Caviar, of course • $36

Domaine Chandon Cuvée 2000

A special release made from 100% Pinot Noir (this, I must say, a surprise) laid down in 1994 to celebrate the millennium, and given three and a half years on yeast lees. It has thus also been given extended cork age, something that the former managing director of Domaine Chandon, Dr Tony Jordan, firmly believes to be necessary.

ŸŸŸŸŸ **1994** Pale straw-pink; aromas of strawberry and faintly earthy notes intermingle with bready autolysis characters on the bouquet. The palate is incredibly intense and long, with a lingering finish. A great success. **rating:** 96

best drinking 2000–2000 **best vintages** NA **drink with** Fireworks • $29.90

Domaine Chandon Brut Rosé

Made from a blend of 60% Pinot Noir and 40% Chardonnay, again sourced from vineyards across southern Australia. Without question, it is by far the most serious Rosé, albeit made in tiny quantities, and hard to find.

ŸŸŸŸY **1995** The colour is quite deep, strongly pink, the bouquet smooth with the benefit of over two years on yeast lees showing though in the autolytic bready notes. The mid-palate fruit flesh and relatively soft finish all contribute to the attractive Rosé style. **rating:** 92

best drinking 1999–2002 **best vintages** '90, '92, '93, '94 **drink with** Poached salmon, Asian cuisine • $33

donnelly river wines ★★★

Lot 159 Vasse Highway, Pemberton, WA 6260 **region** Pemberton

ph (08) 9776 2052 **fax** (08) 9776 2053 **open** 7 days 9.30–4.30

winemaker Blair Mieklejohn **prod.** 4000 **est.** 1986

prod. range ($15–19 CD) Chardonnay, Mist (white blend), Sauvignon Blanc, Pinot Noir, Cabernet Sauvignon, Port.

summary Donnelly River Wines draws upon 16 hectares of estate vineyards, planted in 1986 and which produced the first wines in 1990. It has performed consistently well with its Chardonnay.

donovan ★★☆

Main Street, Great Western, Vic 3377 **region** Grampians
ph (03) 5356 2288 **open** Mon–Sat 10–5.30, Sun 12–5
winemaker Chris Peters **prod.** 2000 **est.** 1977
prod. range ($12–17.50 CD) Riesling, Chardonnay, Shiraz, Cabernet Sauvignon, Chardonnay Brut.
summary Donovan quietly makes some attractively fragrant Riesling and concentrated, powerful Shiraz, most of which is sold cellar door and by mail order with considerable bottle age.

doonkuna estate ★★★☆

Barton Highway, Murrumbateman, NSW 2582 **region** Canberra District
ph (02) 6227 5811 **fax** (02) 6227 5085 **open** Sun–Fri 12–4
winemaker Malcolm Burdett **prod.** 2000 **est.** 1973
prod. range ($10–20 CD) Rhine Riesling, Sauvignon Blanc, Semillon Sauvignon Blanc, Chardonnay, Pinot Noir, Shiraz, Cabernet Merlot.
summary Following the acquisition of Doonkuna by Barry and Maureen Moran in late 1996, the plantings have been increased from a little under 4 hectares to 20 hectares (in 1998). The cellar-door prices remain modest, and increased production will follow in the wake of the new plantings.

Doonkuna Estate Riesling

As Doonkuna, Clonakilla, Lark Hill and Helm's most consistently show, Riesling is quite well suited to the Canberra district climate. This wine, produced from low-yielding, hand-pruned and hand-picked vines, has been consistently well made.

🍷🍷🍷🍷🍷 **1997** Light green-yellow; the bouquet is crisp and clean with focused citrus and herb aromas; the palate is quite delicious, still fresh as a daisy, with lovely passionfruit and lime flavours. Well-handled residual sugar is balanced by acidity. **rating:** 94

best drinking 1999–2003 **best vintages** '88, '90, '91, '92, '95, '97 **drink with** Antipasto • $10

Doonkuna Estate Chardonnay

Usually the better of the two white Doonkuna wines, and often the best Chardonnay made in the Canberra district, though sometimes shaded by Lark Hill. Invariably well made, with skilled use of barrel fermentation.

🍷🍷🍷🍷 **1997** Light to medium yellow-green; the bouquet is quite sophisticated, with deft oak handling; the palate is delicate and unforced, with cashew and fig fruit flavours, and an echo of that oak. **rating:** 85

best drinking 1999–2002 **best vintages** '88, '90, '91, '92, '95 **drink with** Snowy Mountains trout • $16

Doonkuna Estate Shiraz

At the lighter end of the spectrum, sometimes a little too leafy and minty, but not always.

🍷🍷🍷🍷 **1997** Medium to full red-purple; the bouquet is crammed full with American oak, but on the palate there is succulently rich and ripe plummy fruit which manages to express itself through the oak. Over-oaked, not under-fruited. **rating:** 86

best drinking 2000–2005 **best vintages** '92, '96 **drink with** Rack of lamb • $18

Doonkuna Estate Cabernet Merlot

The red wines of Doonkuna, and in particular the Cabernet Sauvignon, are generously flavoured wines which have improved markedly over recent vintages. The inclusion of 40% Merlot has seen a label change since 1996.

🍷🍷🍷🍷 **1997** Medium red-purple; there are distinctly minty overtones to the fresh but ripe berry fruit; balanced oak. Cassis berry fruits dominate the palate, with just a touch of mint showing; a well-balanced and structured wine. **rating:** 87

best drinking 2000–2005 **best vintages** '90, '91, '92, '96 **drink with** Lamb shanks • $20

dorrien estate NR

Cnr Barossa Valley Way/Siegersdorf Road, Tanunda, SA 5352 **region** Barossa Valley
ph (08) 8561 2200 **fax** (08) 8561 2299 **open** Not
winemaker R Bosward, W Dutschke, J Schwartzkopff **prod.** 750 000 **est.** 1982
prod. range ($9.95–24.95 CD) Produces a substantial number of wines under proprietary labels for the Cellarmaster Group; notable are Storton Hill Riesling, Di Fabio Shiraz and Mums Block Shiraz.

summary Dorrien Estate is the physical base of the vast Cellarmaster network which, wearing its retailer's hat, is by far the largest direct-sale outlet in Australia. It buys substantial quantities of wine from other makers either in bulk, or as cleanskin, i.e. unlabelled bottles, or with recognisable but subtly different labels of the producers concerned. It is also making increasing quantities of wine on its own account at Dorrien Estate. The Cellarmaster Group was acquired by Mildara Blass in 1997, and just how the winemaking activities of Dorrien Estate will be conducted in the future remains to be seen.

dowie doole ★★★☆

182 Main Road, McLaren Vale, SA 5171 **region** McLaren Vale
ph (08) 8323 7314 **fax** (08) 8323 7305 **open** Not
winemaker Brian Light (Contract) **prod.** 3500 **est.** 1996
prod. range ($15.50–21.50 R) Chenin Blanc, Semillon Sauvignon Blanc, Chardonnay, Merlot.

summary The imaginatively packaged and interestingly named Dowie Doole was a joint venture between two McLaren Vale grape growers: architect Drew Dowie and one-time international banker Norm Doole. Between them they have over 40 hectares of vineyards, and only a small proportion of their grapes are used to produce the Dowie Doole wines. In 1999 the partnership was expanded to include industry marketing veteran Leigh Gilligan, who returns to his native McLaren Vale after five years in Coonawarra. (Gilligan is also involved with Boar's Rock.) The wines have retail distribution in South Australia and the eastern States, and are exported to Ontario and British Columbia in Canada.

drayton's family wines ★★★☆

Oakey Creek Road, Cessnock, NSW 2321 **region** Lower Hunter Valley
ph (02) 4998 7513 **fax** (02) 4998 7743 **open** Mon–Fri 8–5, weekends and public holidays 10–5
winemaker Trevor Drayton **prod.** 100 000 **est.** 1853
prod. range ($7.50–22 CD) Several label ranges including budget-priced Oakey Creek, New Generation and Premium; Vineyard Reserve Chardonnay, Semillon, Shiraz, Merlot; Sparkling and Fortifieds; top-of-the-range Limited Release Chardonnay, Shiraz, Susanne Semillon, William Shiraz, Joseph Shiraz and Bin 5555 Shiraz.

summary A family-owned and run stalwart of the Valley, producing honest, full-flavoured wines which sometimes excel themselves, and are invariably modestly priced. The size of the production will come as a surprise to many, but it is a clear indication of the good standing of the brand. It is not to be confused with Reg Drayton Wines; national retail distribution with exports to New Zealand, US and southeast Asia.

Drayton's Vineyard Reserve Semillon

Produced entirely from grapes grown on the original Bellevue Vineyard, which was first planted in the 1860s. (These vines are much younger, of course.) Fermented in stainless steel and matured for three months on lees prior to bottling. 487 cases made.

🍷🍷🍷🍷 **1998** Medium to full yellow-green; a broad, early-developing style with honey and herb aromas; the palate is very rich with very unusual mid-palate fruit sweetness. All in all, a challenging and unconventional wine. **rating:** 84

best drinking 1999–2001 **best vintages** NA **drink with** Pasta • $19

Drayton's Vineyard Reserve Chardonnay

Produced from grapes grown on the family-owned Mangerton Estate vineyard, planted to chardonnay in 1976. Cold-fermented in stainless steel for 27 days, then taken to French oak barriques for nine months maturation. 493 cases made.

🍷🍷🍷🍷 **1998** Light to medium yellow-green; lemony oak is very evident on the bouquet, but fruit comes to the fore on the palate in a slightly grassy mould. The relatively high acidity (7.8 grams) jangles somewhat on the finish. **rating:** 85

best drinking 1999–2002 **best vintages** NA **drink with** Stir-fried chicken • $19

Drayton's Vineyard Reserve Pokolbin Shiraz

Produced from shiraz grown on the old Bellevue vineyards adjacent to the winery. Hand-picked, fermented for a little over one week, and spends nine months in one-year-old American oak barriques. 512 cases made.

🍷🍷🍷🍷 **1997** Medium red, with some purple hues. The bouquet is clean, of moderate intensity but not over-much varietal character or weight. The palate comes as a complete surprise after the bouquet, with far more depth and structure, and fruit-driven. Has undoubted long-term development potential. **rating:** 89

best drinking 2002–2012 **best vintages** NA **drink with** Stir-fried beef • $22

Drayton's Vineyard Reserve Pokolbin Merlot

Another instance of the high-quality wine which can be made from first-crop vines, coming as it does from a newly planted vineyard in the foothills of Mount Bright at Pokolbin. The wine spends eight months in two-year-old American oak barriques. 400 cases made.

🍷🍷🍷🍷 **1997** Bright red-purple; the bouquet is clean with attractive, faintly earthy fruit; the palate has considerable substance and good red wine structure, with particularly good ripe but fine tannins. **rating:** 88

best drinking 2002–2007 **best vintages** NA **drink with** Yearling steak • $22

drews creek wines NR

558 Wollombi Road, Broke, NSW 2330 **region** Lower Hunter Valley

ph (02) 6579 1062 **fax** (02) 6579 1062 **open** By appointment

winemaker David Lowe (Contract) **prod.** 300 **est.** 1993

prod. range ($10–14 R) Chardonnay, Unoaked Chardonnay, Merlot.

summary Graeme Gibson and his partners are developing Drews Creek step by step. The initial planting of 2 hectares of chardonnay and 3 hectares of merlot was made in 1991, and the first grapes produced in 1993. A further 2.5 hectares of sangiovese due to be planted in September 1999. Most of the grapes have been sold to contract-winemaker David Lowe, but a small quantity of wine has been made for sale to friends and through the mailing list. A cellar door is due to be opened some time before the end of 2000, and ultimately several holiday cabins overlooking the vineyard and Wollombi Brook will be opened.

driftwood estate ★★★★

Lot 13 Caves Road, Yallingup, WA 6282 **region** Margaret River
ph (08) 9755 6323 **fax** (08) 9755 6343 **open** 7 days 11–4.30
winemaker Lyndon Crockett, Nick Betts **prod.** 14 000 **est.** 1989
prod. range ($14–25 CD) Classic White, Semillon, Sauvignon Blanc Semillon, Chardonnay, Cane Cut Semillon, Late Harvest, Shiraz, Merlot, Cabernet Sauvignon, Sparkling, Tawny Port.
summary Driftwood Estate is yet another remarkable new entrant onto the vibrant Margaret River scene. Quite apart from offering a brasserie restaurant capable of seating 200 people (open 7 days for lunch and dinner) and a mock Greek open-air theatre, its wines feature striking and stylish packaging (even if strongly reminiscent of that of Devil's Lair) and opulently flavoured wines. The winery architecture is, it must be said, opulent but not stylish. The wines are exported to Japan, UK and Malaysia.

Driftwood Estate Sauvignon Blanc Semillon

Another Margaret River wine to fare very well in the 1999 *Winewise* Small Makers Competition, simply because it offered the fruit generosity which the majority of the wines lacked.

🍷🍷🍷🍷🍸 **1998** Light green-yellow; the bouquet is quite intense, with citrus, herb and a touch of stone aroma. The palate has considerable length and intensity in precisely the same flavour spectrum as the bouquet, all of which adds up to an appealing wine. **rating:** 91

best drinking 1999–2002 **best vintages** '98 **drink with** Antipasto • $14.10

Driftwood Estate Chardonnay

Produced from 3.2 hectares of estate plantings; right from the outset, has been made in what might be termed a 'Battlestar Galactica' style, even by the at-times exceedingly opulent and concentrated standards of the Margaret River.

🍷🍷🍷🍷🍸 **1998** Medium yellow-green; a complex, tangy bouquet with abundant grapefruit and melon is followed by a powerful, long and lingering palate; subtle oak use throughout. There is also an undertow of the character one finds in French White Burgundy, which adds interest. **rating:** 90

best drinking 2000–2004 **best vintages** '94, '95, '97 **drink with** Turkey, veal • $19.35

Driftwood Estate Cane Cut Semillon

The only such wine made in the Margaret River region; the severing of canes on the vine, allowing the fruit to raisin – thereby increasing both sugar and acid levels – is not uncommon in the Barossa and Clare Valleys, and is used in regions where botrytis does not occur. A most successful innovation for the Margaret River.

🍷🍷🍷🍷 **1998** Medium yellow-green; the honey and mead aromas are rich and luscious, logically leading an unctuous palate with the same honey/mead flavours, finishing with good acidity. **rating:** 86

best drinking 1999–2003 **best vintages** NA **drink with** Baked apple • $20

Driftwood Estate Shiraz

Estate-grown from vines which are now settling down. Well made, and further evidence of the suitability of Margaret River (or, at least, certain sites) to shiraz.

🍷🍷🍷🍷 **1997** Full purple-red; vanilla, marzipan and spice aromas intermingle on the bouquet the palate introduces cherry flavours as well as the characters from the bouquet. Subtle oak and fine tannins underpin a quite long finish. **rating:** 87

best drinking 2002–2007 **best vintages** NA **drink with** Lamb kebabs • $22.50

---

dromana estate ★★★★

Cnr Harrison's Road and Bittern-Dromana Road, Dromana, Vic 3936 **region** Mornington Peninsula

ph (03) 5987 3800 **fax** (03) 5981 0714 **open** 7 days 11–4

winemaker Garry Crittenden **prod.** 20 000 **est.** 1982

prod. range ($15–45 CD) Dromana Estate Sauvignon Blanc, Chardonnay, Reserve Chardonnay, Pinot Noir, Reserve Pinot Noir, Shiraz, Reserve Merlot, Cabernet Merlot; Second label Schinus, Riesling, Chenin Blanc, Sauvignon Blanc, Chardonnay, Merlot, Longest Lunch Brut, Rosé, Pinot Noir; and a newly packaged range of Italian generics Barbera, Dolcetto, Sangiovese, Granaccia, Riserva, Nebbiolo and Rosato under the Garry Crittenden i label.

summary Since it was first established, Dromana Estate has never been far from the headlines. The energetic marketing genius of Garry Crittenden has driven it hither and thither, launching a brief but very successful foray into the UK market, but since concentrating much of its efforts on a no less successful restaurant and cellar door, with a kaleidoscopic array of wines, first under the Dromana Estate label, then under the Schinus label and in late 1995, a strikingly revamped range of Italian-accented wines, the i wines. At the moment exports are restricted to the US and Asia.

Dromana Estate Chardonnay

Estate-grown from immaculately tended vineyards, barrel-fermented and given eight months in barriques on its lees. If malolactic fermentation has been used, it is not obvious.

▼▼▼▼▽ **1998** Medium yellow-green; the bouquet is clean and smooth, with a mix of citrus, melon and fig fruit; the nicely balanced and flavoured palate, with fruit dominant and oak virtually imperceptible looks sure to develop well. **rating:** 90

best drinking 1999–2004 **best vintages** '91, '92, '97, '98 **drink with** Crab • $25

Garry Crittenden Schinus Chardonnay

A blend of Chardonnay from the Goulburn Valley and McLaren Vale, which is lightly oaked (I imagine via the use of staves or chips), but well done.

▼▼▼▽ **1997** Light to medium yellow-green; a pleasantly made and balanced wine, with sweet peach and melon fruit, and the barest suggestion of oak. Well crafted. **rating:** 83

best drinking 1998–1999 **best vintages** NA **drink with** Pasta • $15

Dromana Estate Reserve Chardonnay

This wine has everything the smartest wine drinker could ever wish for: brilliant and innovative packaging, and the full kit and caboodle of winemaking (or rather, non-winemaking) tricks. Whole-bunch pressed straight to barrel, no yeasts were added, the wine fermenting naturally, and then spending 12 months on lees. It is not filtered, and winemaker Garry Crittenden says it may be slightly cloudy when poured but 'this is a positive attribute'. Why that should be so escapes me, but this is a high-quality wine nonetheless.

▼▼▼▼▼ **1997** Light to medium yellow-green; the bouquet is fine and aromatic, with melon, citrus and stone fruit supported by subtle oak, and no obvious malolactic-fermentation characters. The palate is equally fine, complexed by excellent oak handling producing subtly spicy overtones to the fruit. Very impressive. **rating:** 94

best drinking 1999–2003 **best vintages** '91, '94, '96, '97 **drink with** Kassler • $45

Dromana Estate Pinot Noir

Like the Chardonnay, estate-grown. Also like the Chardonnay, showing to the full the impact of a magnificent vintage in the region.

🍷🍷🍷🍷 **1997** Medium red, already showing some development. The bouquet is in a sappy/briary/foresty spectrum, quite stylish, and which comes together very well with the richer, plummy fruit of the mid-palate. Subtle oak handling and some foresty characters add complexity to a well-structured palate. **rating:** 89

best drinking 1999–2002 **best vintages** '97 **drink with** Grilled quail • $25

Dromana Estate Reserve Pinot Noir

I cannot do better than quote the marvellous piece of self-aggrandising propaganda produced by Garry Crittenden, however much I disagree with it. 'Pinot Noir continues to provide the greatest quest for Australia's premium quality red wine makers. In my personal view both California and New Zealand are leading Australia in the Pinot stakes. With this wine I have accepted the challenge and believe it to be the equal of anything produced in these countries. It has a clear varietal definition and manages to capture the complexity and silky texture found in great French Burgundies.'

🍷🍷🍷🍷🍷 **1997** Medium to full red-purple; the bouquet is complex, with ripe foresty aromas, followed by an exceptionally concentrated palate with very good structure. The flavours are principally forest floor together with some plum, and the wine has a long, persistent finish. What it lacks is that touch of fruit vinosity to provide a little more sweetness. **rating:** 90

best drinking 1999–2003 **best vintages** NA **drink with** Braised duck • $45

Dromana Estate Shiraz

1997 was the second vintage for Dromana Estate Shiraz, sourced from the Mornington Peninsula but not, so far as I know, estate-grown. There is no question that in years such as 1997 the wine can exhibit quite lovely cool-climate varietal character.

🍷🍷🍷🍷🍷 **1997** Medium red-purple; the question is not so much what aromas are present in the bouquet as what are missing. To be found are spicy, woody, leafy, savoury steak and animal/gamey/Cote Rotie characters. The palate is neither extractive nor tannic, simply showing good cool-climate Shiraz flavours of medium intensity. **rating:** 90

best drinking 1999–2003 **best vintages** '97 **drink with** Beef stroganoff • $30

Garry Crittenden i Rosato

The wine is a blend of Pinot Noir, Shiraz, Sangiovese and Nebbiolo, made by the Saignée method, involving the run-off of juice from the crushed must of the varieties concerned before fermentation commences. A wine which catches the ground swell of interest in Rosé-style wines in Australia.

🍷🍷🍷🍷 **1998** Bright pink; the bouquet is distinctly fruity, with pronounced strawberry aromas, bordering on confection. The palate has good mouthfeel, again with sweet fruit flavours, and is well balanced. A cut above many such wines. **rating:** 87

best drinking 1999–2000 **best vintages** NA **drink with** Brasserie food • $20

Dromana Estate Cabernet Merlot

As with all of the wines under the Dromana Estate label, entirely estate-grown (and, of course, made and bottled). A blend of 90% Cabernet Sauvignon and 10% Merlot.

🍷🍷🍷🍷 **1997** Medium red-purple; the bouquet is unashamedly varietal with leaf, tobacco and spice aromas. The palate opens with attractive black cherry fruit, finishing with a slightly green varietal kick. An interesting wine from a very good vintage on the Mornington Peninsula. **rating:** 88

best drinking 2000–2005 **best vintages** '91, '93, '97 **drink with** Beef in red wine sauce • $25

dulcinea ★★★★

Jubilee Road, Sulky, Ballarat, Vic 3352 **region** Ballarat
ph (03) 5334 6440 **fax** (03) 5334 6828 **open** 7 days 9–5
winemaker Rod Stott **prod.** 800 **est.** 1983
prod. range ($14–18 CD) Chardonnay, Sauvignon Blanc, Pinot Noir, Shiraz, Cabernet Sauvignon.
summary Rod Stott is a part-time but passionate grape grower and winemaker who chose the name Dulcinea from 'The Man of La Mancha' where only a fool fights windmills. With winemaking help from various sources, he has produced a series of very interesting and often complex wines, the current releases being exceptionally impressive.

Dulcinea Pinot Noir

A great achievement, and in many ways a landmark wine for Ballarat Pinot Noir.

YYYY **1998** Medium to full purple-red; the bouquet is particularly attractive, opening with plum and briar but with lots of spice then coming through. The palate has excellent plummy pinot fruit, again with Christmas cake spice which is not at all at odds with the variety. To be hypercritical, the acid is slightly hard, which kept it out of even higher points. **rating:** 89

best drinking 1999–2003 **best vintages** '96, '97, '98 **drink with** Roast duck • $18

Dulcinea Shiraz

Produced not from Ballarat-grown grapes, but from Bendigo.

YYYY **1997** Deep purple-red; the bouquet is quite scented, with attractive spice and black cherry fruit leading into a palate in which mint, spice, cherry and a hint of leaf all add up to a high-flavoured yet unforced wine. **rating:** 87

best drinking 2001–2007 **best vintages** NA **drink with** Daube of beef • $16

duncan estate ★★★

Spring Gully Road, Clare, SA 5453 **region** Clare Valley
ph (08) 8843 4335 **fax** (08) 8843 4335 **open** 7 days 10–4
winemaker John Duncan **prod.** 2500 **est.** 1968
prod. range ($11–15 CD) Riesling, Chardonnay, Semillon Sauvignon Blanc, Spätlese, Shiraz, Cabernet Merlot Shiraz.
summary The Duncan family has been growing grapes in the Clare Valley since 1968, and first produced wines from its 7.4 hectares of vineyards in 1984. Over the years some attractive wines have been produced, with the Cabernet Merlot and Shiraz usually good.

dyson wines NR

Sherriff Road, Maslin Beach, SA 5170 **region** McLaren Vale
ph (08) 8386 1092 **fax** (08) 8327 0066 **open** 7 days 10–5
winemaker Allan Dyson **prod.** 2000 **est.** 1976
prod. range ($14–18 CD) Chardonnay, Ella Rose Sauvignon Blanc Semillon, Cabernet Sauvignon, White Port.
summary Owned by district veteran Allan Dyson, drawing on 6.5 hectares of estate vineyard. Sauvignon Blanc Semillon to one side, full-bodied Chardonnay, typical of the region, is the most reliable wine.

east arm vineyard NR

111 Archers Road, Hillwood, Tas 7250 **region** Northern Tasmania
ph (03) 6334 0266 **fax** (03) 6334 1405 **open** By appointment
winemaker Andrew Hood, Bert Sundstrup (Contract) **prod.** NA **est.** 1993
prod. range ($18–21 R) Riesling, Chardonnay, Pinot Noir.

summary East Arm Vineyard was established by Launceston gastroenterologist Dr John Wettenhall and partner Anita James who also happens to have completed the Charles Sturt University Diploma in Applied Science (wine growing). The 2 hectares of vineyard which came into full production in 1998 are more or less equally divided between riesling, chardonnay and pinot noir. It is established on an historic block, part of a grant made to retired British soldiers of the Georgetown garrison in 1821, and slopes down to the Tamar River. The property is 25 hectares, and there are plans for further planting, and somewhere down the track, a winery. The 1998 Riesling was the top gold medal winner in its class at the 1999 Tasmanian Wines Show.

East Arm Riesling

Produced from a small estate planting of about two-thirds of a hectare of riesling established in 1993. Tiny amounts of wine were made in 1996 and 1997, with the first commercial vintage coming in 1998. Voted the Best Tasmanian Riesling of 1998 at the Tasmanian Wine Centre in November 1998, and on its second showing, was the top gold medal in the very strong 1998 Riesling class at the 1999 Tasmanian Wines Show.

YYYYY **1998** Pale straw-green; the wine has a piercing bouquet with aromas of lemon, lime, stone fruit and some quite spicy lift. The palate is uncompromising, with the flavours of the bouquet supported and reinforced by lingering, potent acidity. A wine which begs to be cellared.

rating: 94

best drinking 2000–2008 **best vintages** '98 **drink with** Scallops • $18

eastbrook estate NR

Lot 3 Vasse Highway, Eastbrook, WA 6260 **region** Pemberton

ph (08) 9776 1251 **fax** (08) 9776 1251 **open** Fri–Sun, public holidays 11–4

winemaker Kim Skipworth **prod.** 2000 **est.** 1990

prod. range ($10–17 CD) Chardonnay, Pinot Noir, Cabernet Sauvignon, Port.

summary Established on part of the same former grazing property which also accommodates Salitage, Phoenicia, and Dr Bill Pannell's vineyard. A jarrah pole, limestone and cedar weatherboard winery and restaurant have been built on the site by former Perth real estate agent Kim Skipworth, who is also a shareholder in one of the major Margaret River cheese factories. The wines come from 7 hectares of estate plantings of pinot noir, chardonnay, sauvignon blanc and shiraz.

eastern peake NR

Clunes Road, Coghills Creek, Vic 3364 **region** Ballarat

ph (03) 5343 4245 **fax** (03) 5343 4365 **open** 7 days 10–5

winemaker Norman Latta **prod.** 1500 **est.** 1983

prod. range ($15–20 CD) Chardonnay, Alba, Persuasion (Pinot Rosé), Pinot Noir.

summary Norm Latta and Di Pym commenced the establishment of Eastern Peake, situated 25 kilometres northeast of Ballarat on a high plateau overlooking the Creswick Valley almost 15 years ago. In the early years the grapes were sold to Trevor Mast of Mount Chalambar and Mount Langi Ghiran, but the 4.5 hectares of vines are now dedicated to the production of Eastern Peake wines. The Pinot Noir is on the wirey/minerally/stemmy side; earlier bottling might preserve more of the sweet fruit.

Eastern Peake Chardonnay

Just how cool Eastern Peake vineyards are can be gauged from the fact that this white was picked in early May, and yet achieved only 12.8 degrees alcohol. A wine which speaks ever so clearly of the very cool growing conditions, but which has been sensitively handled in the winery to maximise its character and quality, being taken through partial malolactic fermentation to soften the acidity. An extremely interesting and well-made debut.

🍷🍷🍷🍷 **1998** Medium yellow-green; the bouquet is light and clean, with subtle oak in support of the slightly green citrus fruit aromas. Precisely the same characters appear on the palate, with lemony citrus fruit and nicely controlled oak. **rating:** 86

best drinking 1999–2002 **best vintages** NA **drink with** Sushi • $20

eden valley wines NR

Main Street, Eden Valley, SA 5235 **region** Eden Valley
ph (08) 8564 1111 **fax** (08) 8564 1110 **open** 7 days 10–5
winemaker Vickie Bartier **prod.** 4000 **est.** 1994
prod. range ($10–20 CD) Riesling, Chardonnay, Shiraz, Matara, Cabernet Sauvignon, White Port, Tawny Port.
summary Eden Valley Wines has waxed and waned over the years, but seems now very much in the ascendant. The venture now has 30 hectares each of recently planted riesling, cabernet sauvignon, shiraz, with 5 hectares of much older mourvedre. A major part of the production is sold as grapes to others; the wines currently on sale have varied (non-estate) backgrounds.

elan vineyard NR

17 Turners Road, Bittern, Vic 3918 **region** Mornington Peninsula
ph (03) 5983 1858 **fax** (03) 5983 2821 **open** First weekend of month, public holidays 11–5 or by appointment
winemaker Selma Lowther **prod.** 600 **est.** 1980
prod. range ($13–16 CD) Olive's Paddock Riesling, Chardonnay, Shiraz, Gamay, Cabernet Merlot.
summary Selma Lowther, fresh from Charles Sturt University (as a mature-age student) made an impressive debut with her spicy, fresh, crisp Chardonnay. Most of the grapes from the 2.9 hectares of estate vineyards are sold; production remains minuscule.

Elan Vineyard Olive's Paddock Riesling

Riesling is one of nature's conundrums, with no particular climatic pattern evident in Australia to indicate where it should or shouldn't be planted. On balance the evidence suggests the Mornington Peninsula is an appropriate place.

🍷🍷🍷🍷 **1998** Light to medium yellow-green; a fairly shy, minerally bouquet showing a touch of gun flint when first poured but which cleaned up progressively with aeration. The palate, likewise, is relatively austere and minerally, but does have excellent length. Patience will be richly rewarded. **rating:** 85

best drinking 2000–2008 **best vintages** '98 **drink with** Grilled fish • NA

elderton ★★★★

3 Tanunda Road, Nuriootpa, SA 5355 **region** Barossa Valley
ph (08) 8562 1058 **fax** (08) 8562 2844 **open** Mon–Fri 8.30–5, weekends, holidays 11–4
winemaker James Irvine **prod.** 32 000 **est.** 1984
prod. range ($13–45 R) Riesling, Golden Riesling, Semillon, Chardonnay, Sparkling, Old Hayshed Red, Shiraz, Command Shiraz, Cabernet Sauvignon Merlot, Cabernet Sauvignon, Merlot; the Elderton Domain range is in fact the second label.
summary The wines are based around some old, high-quality Barossa floor estate vineyards, and all are driven to a lesser or greater degree by lashings of American oak; the Command Shiraz is at the baroque end of the spectrum, and has to be given considerable respect within the parameters of its style. National retail distribution, with exports to the UK, US, Europe and Asia.

eldredge ★★★☆

Spring Gully Road, Clare, SA 5453 **region** Clare Valley
ph (08) 8842 3086 **fax** (08) 8842 3086 **open** 7 days 11–5
winemaker Leigh Eldredge **prod.** 4000 **est.** 1993
prod. range ($12–20 CD) Watervale Riesling, Semillon Sauvignon Blanc, Late Harvest Riesling, New Age Grenache, Blue Chip Shiraz, Cabernet Sauvignon, Sparkling, Port.
summary Leigh and Karen Eldredge have established their winery and cellar-door sales area in the Sevenhill Ranges, at an altitude of 500 metres, above the town of Watervale. Contract-winemaking by Tim Adams has ensured a solid start to the business. The popular Leigh Eldredge collected the trophy for Best Cabernet Sauvignon at the 1997 Royal Adelaide Wine Show. The wines are exported to the UK, US and Canada, and southeast Asia.

Eldredge Watervale Riesling

The long experience of consultant winemaker Tim Adams shows through in this wine, grown in the heartland of Australia's riesling.

🍷🍷🍷🍷🍷 **1998** Light yellow-green; the bouquet is moderately intense, with crisp, herbal, tangy fruit giving an attractive cut to the aromas. Attractively tangy fruit comes through on a very well-balanced palate, with good length and intensity. **rating:** 91

best drinking 1999–2008 **best vintages** NA **drink with** Bruschetta • $13

eldridge estate NR

Red Hill Road, Red Hill, Vic 3937 **region** Mornington Peninsula
ph (03) 5989 2644 **fax** (03) 5989 2644 **open** Weekends, public holidays and January 1–26 11–5
winemaker David Lloyd **prod.** 700 **est.** 1985
prod. range ($20–28 CD) Semillon Sauvignon Blanc, Chardonnay, Pinot Noir, Cabernet Merlot, Sparkling.
summary The Eldridge Estate vineyard, with seven varieties included in its 3.5 hectares, was purchased by Wendy and David Lloyd in 1995. Major retrellising work has been undertaken, changing to Scott Henry, and all of the wines will now be estate-grown and made. The wines are available at the Victorian Wine Centre and Richmond Hill Cellars in Melbourne, and a few leading restaurants in Melbourne and Sydney.

Eldridge Estate Pinot Noir

David Lloyd has had many years experience in working with cool-climate grapes, principally from the Yarra Valley, which shows through in this wine.

🍷🍷🍷🍷 **1997** Light red; the bouquet is light, with lots of forest/briar/forest floor aromas and needing a little more sweet fruit to give balance. However, there is a nice touch of strawberry and violets at the heart of the palate which lifts the wine considerably. Looks very much as if it is at its best now. **rating:** 85

best drinking 1999–2000 **best vintages** NA **drink with** Antipasto • $28

elgee park NR

Wallaces Road, Merricks North, Vic 3926 **region** Mornington Peninsula
ph (03) 5989 7338 **fax** (03) 5989 7553 **open** One day a year – Sunday of Queen's Birthday weekend
winemaker Various (Contract) **prod.** 1500 **est.** 1972
prod. range ($12–25 R) Chardonnay, Pinot Noir, Cabernet, Cuvée Brut.
summary The pioneer of the Mornington Peninsula in its twentieth-century rebirth, owned by Baillieu Myer and family. The wines are now made at Stonier's, Elgee Park's own winery having been closed, and the overall level of activity decreased. Melbourne retail distribution through Flinders Wholesale.

Elgee Park Riesling

The Elgee Park vineyards are the oldest on the Mornington Peninsula, with vines over 20 years of age. These invest the wine with an extra degree of concentration and flavour.

🍷🍷🍷🍷 **1998** Light green-yellow; mineral, herb and lime aromas and flavours run through a wine with well-above-average concentration and length. The lime flavours on the palate persist on a clean aftertaste. **rating:** 89

best drinking 1999–2005 **best vintages** NA **drink with** Crab and asparagus • $12.50

Elgee Park Chardonnay

Produced in very limited quantities from the small estate plantings, part of the chardonnay now going to make the Cuvée Brut. In recent times released under the Family Reserve label, and as competently made by Tod Dexter as one would imagine.

🍷🍷🍷🍷 **1997** Light green-yellow; the bouquet is firm and direct, with slightly herbal overtones. The palate, likewise, is in a no-nonsense, no-frills style, but is very lively, with good length and acidity. Will age nicely. **rating:** 88

best drinking 2000–2005 **best vintages** '91, '92, '94, '95 **drink with** Tasmanian salmon • $22

eling forest winery NR

Hume Highway, Sutton Forest, NSW 2577 **region** Southern New South Wales Zone
ph (02) 4878 9499 **fax** (02) 4878 9499 **open** 4 days 10–5
winemaker Leslie Fritz **prod.** 3500 **est.** 1987
prod. range ($13.50–20 CD) Eling Forest Riesling, Traminer Riesling, Chardonnay, Chardonnay Blend, Catherine Hill, Botrytis Riesling, Furmint, Cabernet Sauvignon, Peach Brandy, Cherry Brandy, Peach Ambrosia, Cherry Ambrosia.

summary Eling Forest's mentally agile and innovative founder Leslie Fritz celebrated his 80th birthday not long after he planted the first vines at his Sutton Forest vineyard in 1987. He proceeded to celebrate his 88th birthday by expanding the vineyards from 3 hectares to 4, primarily with additional plantings of the Hungarian varieties. He has also developed a Cherry Port, and is using the spinning cone technology to produce various peach-based liqueurs utilising second class peach waste. The '97 Riesling and '97 Chardonnay (both 83 points) are both well made table wines with plenty of flavour and varietal character.

elmslie ★★☆

Upper McEwans Road, Legana, Tas 7277 **region** Northern Tasmania
ph (03) 6330 1225 **fax** (03) 6330 2161 **open** By appointment
winemaker Ralph Power **prod.** 600 **est.** 1972
prod. range ($18 ML) Pinot Noir, Cabernet Sauvignon.

summary A small, specialist red winemaker, from time to time blending Pinot Noir with Cabernet. The fruit from the now fully mature vineyard (half a hectare of pinot noir and 1.5 hectares of cabernet sauvignon) has depth and character, but operational constraints mean that the style of the wine is often somewhat rustic.

elsewhere vineyard ★★★★☆

40 Dillons Hill Road, Glaziers Bay, Tas 7109 **region** Southern Tasmania
ph (03) 6295 1509 **fax** (03) 6295 1509 **open** Not
winemaker Andrew Hood (Contract), Steve Lubiana (Contract) **prod.** 3000 **est.** 1984
prod. range ($18–25 ML) Riesling, Dry White, Chardonnay, Pinot Noir, Méthode Champenoise.

summary Eric and Jette Phillips' evocatively named Elsewhere Vineyard jostles for space with a commercial flower farm also run by the Phillips. It is a mark of the success of the wines that in 1993 some of the long-established flower areas made way for additional chardonnay and riesling, although it is Elsewhere's long-lived Pinot Noirs that are so stunning. The estate-produced range comes from 4 hectares of pinot noir, 3 hectares of chardonnay and 2 hectares of riesling constituting the immaculately-tended vineyard. Retail distribution through Sutherland Cellars, Melbourne.

Elsewhere Vineyard Riesling

A typical Tasmanian Riesling, fragrant and with wild herb overtones.

ΨΨΨΨ **1998** Light yellow-green; a fragrant, floral bouquet with a mix of lime, herb and thyme aromas; the palate is crisp and spicy, with more of those floral lime characters, followed by typical Tasmanian acidity on the finish. **rating:** 86

best drinking 2002–2008 **best vintages** NA **drink with** Prawn salad • $18

eltham vineyards ★★★

225 Shaws Road, Arthurs Creek, Vic 3099 **region** Yarra Valley
ph (03) 9439 4688 **fax** (03) 9439 5121 **open** By appointment
winemaker George Apted, John Graves **prod.** 850 **est.** 1990
prod. range ($14.95–18.95 ML) Chardonnay, Pinot Noir, Cabernet Sauvignon.
summary Drawing upon vineyards at Arthurs Creek and Eltham, John Graves (brother of David Graves of the illustrious Californian Pinot producer Saintsbury) produces tiny quantities of quite stylish Chardonnay and Pinot Noir, the former showing nice barrel-ferment characters.

elysium vineyard NR

393 Milbrodale Road, Broke, NSW 2330 **region** Lower Hunter Valley
ph (02) 9664 2368 **fax** (02) 9664 2368 **open** By appointment
winemaker Tyrrell's (Contract) **prod.** 500 **est.** 1990
prod. range ($19.95 CD) Verdelho.
summary Elysium was once part of a much larger vineyard established by John Tulloch. John Tulloch (not part of the Tulloch operation owned by Southcorp) continues to look after the viticulture, with the 1 hectare of verdelho being vinified at Tyrrell's. The Elysium Vineyard cottage, large enough to accommodate six people, has won a number of tourism awards, and proprietor Victoria Foster conducts wine education weekends on request, with meals prepared by a chef brought in for the occasion. As at the time of writing, the cost per person for a gourmet weekend was $300.

emerald estate NR

Main North Road, Stanley Flat, SA 5453 **region** Clare Valley
ph (08) 8842 3296 **fax** (08) 8842 2220 **open** Mon, Tues, Thur, Fri 11–5, weekends 10–5
winemaker Tim Adams (Consultant), Frank Sheppard **prod.** 1670 **est.** 1990
prod. range ($9–15 CD) Riesling, Unwooded Chardonnay, Classic Dry White, Shiraz, Cabernet Sauvignon.
summary Don and Gwen Carroll purchased a 33-hectare property at Stanley Flat in 1990. A small existing vineyard was pulled out, and since 1990 20 hectares of vines have been established. Most of the production is sold to leading wineries in the region, a portion being retained for the Emerald Estate wine range, with limited retail distribution in Melbourne.

eppalock ridge NR

633 North Redesdale Road, Redesdale, Vic 3444 **region** Bendigo
ph (03) 5425 3135 **fax** (03) 5425 3135 **open** 7 days 10–6 by appointment
winemaker Rod Hourigan **prod.** 1000 **est.** 1979
prod. range ($23.50 CD) Cabernet Merlot, Shiraz.
summary A low-key operation now focusing almost entirely on Shiraz produced from the 4 hectares of this variety; the other wine in the portfolio comes from the 2.7 hectares of cabernet sauvignon, merlot and cabernet franc.

ermes estate NR

2 Godings Road, Moorooduc, Vic 3933 **region** Mornington Peninsula
ph (03) 5978 8376 **open** Weekends and public holidays 11–5
winemaker Ermes Zucchet **prod.** 500 **est.** 1989
prod. range ($10–15 CD) Riesling Malvasia, Chardonnay Pinot Grigio, Cabernet Merlot.
summary Ermes and Denise Zucchet commenced planting of the 2-hectare estate in 1989 with chardonnay, riesling, cabernet sauvignon and merlot, adding pinot gris in 1991. In 1994 an existing piggery on the property was converted to a winery and cellar-door area (in the Zucchets' words, the pigs having been evicted) and the modestly priced wines are on sale during the weekends. No recent tastings.

eurunderee flats winery NR

Henry Lawson Drive, Mudgee, NSW 2850 **region** Mudgee
ph (02) 6373 3954 **fax** (02) 6373 3750 **open** Sun–Fri 10–4, Sat 9–5
winemaker Peter Knights **prod.** 1500 **est.** 1985
prod. range ($10–15 CD) Sauvignon Blanc, Chardonnay, Shiraz, Merlin Rouge, Liqueur Muscat, Tawny Port.
summary Sometimes called Knights Vines, although the wines are marketed under the Eurunderee Flats label. There are 5 hectares of vineyards producing white wines of variable quality, and rather better dry red table wines.

evans family ★★★★☆

Palmers Lane, Pokolbin, NSW 2321 **region** Lower Hunter Valley
ph (02) 4998 7333 **fax** (02) 4998 7798 **open** By appointment
winemaker Contract **prod.** 3000 **est.** 1979
prod. range ($16–22.50 ML) Pinchem Chardonnay, Howards Chardonnay, Statue Vineyard Sparkling Pinot, Chapel Gamay, Hillside Pinot Noir.
summary In the wake of the acquisition of Rothbury by Mildara Blass, Len Evans' wine interests now focus on Evans Family (estate-grown and produced from vineyards around the family home) and on the Evans Wine Company (a quite different, part-maker, part-negociant business). Len Evans continues to persist with the notion that the Hunter Valley can produce Gamay and Pinot Noir of quality, and irritatingly occasionally produces evidence to suggest he may be half right. There is, of course, no such reservation with the Chardonnay.

evans & tate ★★★★☆

Metricup Road, Willyabrup, WA 6280 **region** Margaret River
ph (08) 9296 4666 **fax** (08) 9296 1148 **open** 7 days 10.30–4.30
winemaker Brian Fletcher **prod.** 100 000 **est.** 1970
prod. range ($14–40 R) There are two basic echelons: an expanding range of commercial wines including the newly released Tate White (an extraordinary blend of Chardonnay and Grenache) and Tate Red; Western Australia Classic Sauvignon Blanc, Two Vineyards Chardonnay, Gnangara Shiraz Cabernet and Barrique 61 Cabernet Merlot. The top echelon is based on the Margaret River vineyards which produce Chardonnay, Semillon, Sauvignon Blanc, Shiraz, Merlot and Cabernet Sauvignon.
summary Single-handedly changed perceptions of the Swan Valley red wines in the '70s before opening its highly successful Margaret River operation which goes from strength to strength. The most recent expansion has been the establishment of a large vineyard in the new Jindong subregion of the Margaret River, precipitating a flood of other arrivals in that area. The continuing rapid growth of the business has taken edge off the wines which, while immaculately crafted and ever-reliable, lack the concentration and complexity of the very best wines of the region.

Evans & Tate Margaret River Sauvignon Blanc Semillon

Now produced predominantly from grapes grown at Evans & Tate's Jindong Vineyard.

🍷🍷🍷🍷 **1998** Light yellow-green; the bouquet is clean, fairly light, with herbal/grassy/minerally notes. The wine is helped by some sweetness on the mid-palate but does not make much of a statement about its varietal origins. **rating:** 86

best drinking 1998–1999 **best vintages** NA **drink with** Shellfish • $19.55

Evans & Tate Margaret River Classic

Previously sourced from various vineyards throughout Western Australia, the 1998 Classic comes back to the Margaret River (hence the change in name) and, indeed, to largely estate-grown grapes from Evans & Tate's new venture at Jindong, where it has established Lionels Vineyard. The area is fertile, and there is abundant water, resulting in vigorous growth and generous yields. While Jindong may not produce the same concentration as traditional Margaret River subregions, it is doubtless well suited to lighter, cheaper wines such as Margaret River Classic. For the record, it is predominantly Semillon and Sauvignon Blanc, with lesser amounts of Chardonnay and Verdelho.

🍷🍷🍷🍷 **1998** Light yellow-straw; the bouquet is fragrant and intensely fruity, with strong passionfruit aromas, perhaps partially yeast-derived. The palate is crisp and lively, with passionfruit and tropical fruit flavours, finishing with crisp acid. **rating:** 87

best drinking 1998–1999 **best vintages** NA **drink with** Asian seafood • $15.95

Evans & Tate Margaret River Shiraz

A 100% estate-grown wine from the Redbrook Vineyard (using the Hermitage name until 1992, but of course made from shiraz). The Margaret River region has not done a great deal with Shiraz overall, preferring to concentrate on other varieties, but this wine shows the potential that the region has for this variety. Matured in a French oak for a quite miraculous two years, given that the wine was bottled and packaged by January 1998.

🍷🍷🍷🍷🍷 **1996** Full red-purple; the bouquet is dense, complex with distinctly gamey Rhône Valley boot polish and liquorice aromas. The palate is no less rich and concentrated, crammed full with ripe black cherry, liquorice and spice flavours. Remarkably uninhibited. **rating:** 92

best drinking 2001–2011 **best vintages** '86, '88, '90, '91, '92, '93, '95, '96 **drink with** Strong red meat dishes • $40

Evans & Tate Margaret River Merlot

Evans & Tate were one of the first producers in Australia to market a varietal Merlot. The grapes are grown at Evans & Tate's Redbrook Vineyard from now fully mature vines. Partial (10%) barrel fermentation and maturation in a mix of one and two-year-old French oak barriques for 18 months produces an elegant wine. It is perfectly obvious that winemakers and wine writers alike in Australia are still trying to reach agreement on the desirable characters of and descriptors for Merlot. Brian Fletcher thinks the '96 has the flavours he is after, but I see it in the contrary light. Only time will tell who is right.

🍷🍷🍷🍷 **1996** Medium red-purple; the bouquet is clean, smooth and sweet, but does not seem to have particularly distinctive varietal character. The palate undoubtedly has lots of flavour, with sweet plum and raspberry fruit, and touches of dark chocolate. **rating:** 84

best drinking 2000–2005 **best vintages** NA **drink with** Osso buco • $40

Evans & Tate Margaret River Cabernet Sauvignon

Sometimes produced from 100% Cabernet Sauvignon, but in some vintages with a little Merlot blended in, produced from Evans & Tate's Redbrook Vineyard in the Margaret River. Whether coincidence or not, the arrival of Brian Fletcher lifted this wine into another class and dimension. The wine is matured in French oak barriques for 22 months.

🍷🍷🍷🍷 **1996** Dense red-purple; the bouquet shows extremely ripe and concentrated blackberry/blackcurrant fruit with an almost resinous overtone. The palate is concentrated and essency with luxuriant blackberry fruit – and which borders on going over the top. **rating:** 89

best drinking 2005–2015 **best vintages** '86, '88, '90, '91, '92, '95, '96 **drink with** Braised ox cheek in red wine sauce • $40

evelyn county estate NR

35 New Road, Kangaroo Ground, Vic 3097 **region** Yarra Valley
ph (03) 9437 1668 **fax** (03) 9437 1232 **open** By appointment
winemaker David Lance (Contract) **prod.** NA **est.** 1994
prod. range ($16.50–23.50 CD) Black Paddock range of Sauvignon Blanc, Chardonnay, Pinot Noir, Merlot, Cabernet Sauvignon.
summary The 7-hectare Evelyn County Estate has been established by former Coopers & Lybrand managing partner Roger Male and his wife Robyn, who is currently halfway through a degree in Applied Science (Wine Science) at Charles Sturt University. David Lance (of Diamond Valley) is currently making the wines, and an architect-designed cellar-door sales and gallery is due to open in the year 2000. As one would expect, the quality of the wines is good.

Evelyn County Estate Black Paddock Chardonnay

Picked very ripe (as almost all 1998 Yarra Valley Chardonnays were) at 13.5° baumé and barrel-fermented. Looks as if it may develop reasonably quickly.

🍷🍷🍷🍷 **1998** Glowing yellow-green; the oak imparts a slightly unusual aroma, vaguely reminiscent of tobacco ash, but the palate throws off that uncertainty, with attractive melon, citrus and clingstone peach fruit; subtle barrel-ferment characters run through the finish. **rating:** 85

best drinking 1999–2001 **best vintages** NA **drink with** Yarra Valley trout • $18

excelsior peak NR

22 Wrights Road, Drummoyne, NSW 2047 (postal address) **region** Tumbarumba
ph (02) 9719 1916 **fax** (02) 9719 2752 **open** Not
winemaker Charles Sturt University (Contract) **prod.** 1150 **est.** 1980
prod. range ($18–22 ML) Chardonnay, Pinot Noir, Méthode Champenoise.
summary Excelsior Peak proprietor Juliet Cullen established the first vineyard in Tumbarumba in 1980. That vineyard was thereafter sold to Southcorp, and Juliet Cullen subsequently established another vineyard, now releasing wines under the Excelsior Peak label. Plantings total 9 hectares, with 6.5 hectares in production.

eyton-on-yarra ★★★☆

Cnr Maroondah Highway and Hill Road, Coldstream, Vic 3370 **region** Yarra Valley
ph (03) 5962 2119 **fax** (03) 5962 5319 **open** 7 days 10–5
winemaker Matthew Aldridge **prod.** 15 000 **est.** 1991
prod. range ($17–30 R) There are now three labels in the range: at the top NDC Reserve Merlot and Shiraz, a tribute to the late Newell Cowan, who effectively founded Eyton-on-Yarra; the main varietal range under the Eyton label; and the second label range of Dalry Road, the name of the second vineyard owned by Eyton.
summary Now owned and run by the energetic and innovative Deidre Cowan, overseeing an excellent and capacious restaurant, a sound shell for concerts of every shape and hue, and – of course – the winemaking side of a substantial business which is able to draw on 40 hectares of estate vineyards. Retail distribution through Victoria, New South Wales and ACT.

Eyton-on-Yarra Chardonnay

Produced from estate-grown fruit; Eyton-on-Yarra draws on two substantial vineyards, one around the winery and the other some distance away on flatter country closer to the Yarra River. Much of the grape production has been sold to others, with the best reserved for the Eyton-on-Yarra label. The wine spends 18 months in French oak and is given partial malolactic fermentation.

🍷🍷🍷🍷 **1997** Medium to full yellow-green; the bouquet is solid, smooth, with well-balanced and integrated oak. A seamless wine on the palate, stacked with creamy, nutty characters surrounding the peach and melon fruit. Very much the product of a warm, dry year. **rating:** 87

best drinking 1998–2003 **best vintages** NA **drink with** Smoked salmon pasta • $22

Eyton-on-Yarra Dalry Road Cabernets

A blend of 52% Cabernet Franc harvested from the Eyton Vineyard, and 48% Cabernet Sauvignon from the Dalry Road Vineyard, with a full month between the picking of the cabernet franc (end March) and cabernet sauvignon (end April). The two portions are separately made, the Cabernet Franc receiving no oak, and the Cabernet Sauvignon three months maturation in older French oak barriques. This approach results in a very fresh, elegant early-drinking style.

🍷🍷🍷 **1996** Light to medium red-purple; the bouquet is clean, of light to medium intensity with a mix of leaf, earth and red berry fruits. The palate is quite attractive, with sweet cassis berry flavours on the mid-palate, almost no oak and likewise very low tannins. A successful drink-now style. **rating:** 84

best drinking 1999–2000 **best vintages** NA **drink with** Light meat or pasta dishes • $17

fairfield vineyard NR

Murray Valley Highway, Browns Plains via Rutherglen, Vic 3685 **region** Rutherglen

ph (02) 6032 9381 **open** Mon–Sat 10–5, some Sun 12–5

winemaker Andrew Sutherland-Smith **prod.** 4200 **est.** 1959

prod. range ($8.50–15 CD) White Hermitage, Riesling, Moselle, Rosé, Light Red, Shiraz, Durif, Cabernet Sauvignon, Fortified.

summary Specialist in red and fortified wines made with nineteenth-century wine equipment housed in the grounds of the historic Fairfield Mansion built by G F Morris. A tourist must. Offers a wide range of back vintages.

faisan estate NR

Amaroo Road, Borenore, NSW 2800 **region** Orange

ph (02) 6365 2380 **open** Not

winemaker Col Walker **prod.** 500 **est.** 1992

prod. range ($8–13 ML) Chardonnay, Canobolas Classic White, Pinot Noir, Britton's Block Cabernet Sauvignon, Old Block Cabernet Sauvignon.

summary Faisan Estate, within sight of Mount Canobolas, and 20 kilometres west of the city of Orange, has been established by Trish and Col Walker. They now have almost 7 hectares of vineyards coming into bearing, and have purchased grapes from other growers in the region in the interim. The 1998 Chardonnay (84 points) offers lots of flavour (both fruit and oak) at the bargain price of $12 per bottle.

Faisan Estate Chardonnay

A well-made wine which shows very competent oak handling, and represents exceptional value for money.

🍷🍷🍷 **1998** Light to medium yellow-green; the bouquet is quite complex, with tangy fruit and what appears to be barrel-ferment oak. That spicy, toasty oak comes to the fore throughout the palate, with slight crème brûlée overtones. **rating:** 84

best drinking 1999–2002 **best vintages** NA **drink with** Chinese honey chicken • $12

farrell's limestone creek ★★★☆

Mount View Road, Mount View, NSW 2325 **region** Lower Hunter Valley
ph (02) 4991 2808 **fax** (02) 4991 3414 **open** Weekends, public holidays 10–5
winemaker Neil McGuigan (Consultant) **prod.** 1000 **est.** 1982
prod. range ($11–18 CD) Semillon, Chardonnay, Late Harvest Verdelho, Shiraz, Merlot, Cabernet Sauvignon Merlot.
summary The Farrell family purchased 50 acres on Mount View in 1980, and gradually established 18 acres of vineyards planted to semillon, verdelho, chardonnay, shiraz, cabernet sauvignon and merlot. Most of the grapes are sold to McWilliam's, which contract-makes a small amount for cellar-door sales. The quality of the wines is as good as one would expect, with a range of back vintages available at cellar door.

felsberg winery NR

Townsends Road, Glen Aplin, Qld 4381 **region** Granite Belt
ph (07) 4683 4332 **fax** (07) 4683 4377 **open** 7 days 9–5
winemaker Otto Haag **prod.** 1600 **est.** 1983
prod. range ($12–25 CD) Rhine Riesling, Traminer, Sylvaner, Chardonnay, Traminer Rosé, Merlot, Shiraz, Cabernet Sauvignon, Mead.
summary Felsberg has been offering wine for sale via cellar door, made by former master brewer Otto Haag. I had problems with the first few vintages, but after a gap in tasting of several years, renewed my acquaintance with Felsberg through the attractive '97 vintage reds, notably the Shiraz (85 points) and Merlot (86 points).

Felsberg Winery Shiraz

Shiraz (along with Semillon) has proved itself to be well suited to the Granite Belt climate, and has produced most of the best reds to come from the region.
🍷🍷🍷🍷 **1997** Medium to full red-purple; the bouquet has some slightly lifted characters, but there is plenty of fruit there. Similarly, the palate has strong cherry fruit, subtle oak and just a hint of that lifted character of the bouquet. **rating:** 85
best drinking 2000–2004 **best vintages** NA **drink with** Braised beef • $15

Felsberg Winery Merlot

Merlot is a relatively new arrival on the scene in the Granite Belt, but holds much promise if the wines of Adam Champman's Violet Cane and this Felsberg version are anything to go by.
🍷🍷🍷🍷 **1997** Medium red-purple; the bouquet is clean, with aromas of cedar, cigar and a touch of leaf; the palate has nice flavour, balance and structure, with more of those cedary/leafy flavours, finishing with dusty tannins. **rating:** 86
best drinking 2000–2005 **best vintages** NA **drink with** Veal ragout • $25

ferguson falls estate NR

Pile Road, Dardanup, WA 6236 **region** Geographe
ph (08) 9728 1083 **fax** (08) 9728 1083 **open** By appointment
winemaker James Pennington (Contract) **prod.** 320 **est.** 1983
prod. range ($15–17 CD) Chardonnay, Cabernet Sauvignon.
summary Peter Giumelli and family are dairy farmers in the lush Ferguson Valley, 180 kilometres south of Perth. In 1983 they planted 3 hectares of cabernet sauvignon, chardonnay and merlot, making their first wines for commercial release from the 1995 and 1996 vintages. Both confirm the suitability of the region for the production of premium wine.

fergusson ★★★★

Wills Road, Yarra Glen, Vic 3775 **region** Yarra Valley
ph (03) 5965 2237 **fax** (03) 5965 2405 **open** 7 days 11–5
winemaker Christopher Keyes, Peter Fergusson **prod.** 10 000 **est.** 1968
prod. range ($14.50–28.50 CD) There are two basic ranges: the lower-priced Tartan Range sourced from grapes grown outside the Yarra Valley, with Sauvignon Blanc, Chardonnay, Semillon Sauvignon Blanc, Shiraz, Sparkling; and three wines in the Estate Range, Victoria Chardonnay, Jeremy Shiraz and Benjamin Cabernet Sauvignon.
summary Best known as a favoured tourist destination, particularly for tourist coaches, and offering hearty fare in comfortable surroundings accompanied by wines of non-Yarra Valley origin. For this reason the limited quantities of its estate wines are often ignored, but should not be.

Fergusson Victoria Chardonnay

Confusingly, Victoria Chardonnay does not indicate an appellation – in other words, that the wine comes from various regions within Victoria – but is named after the Fergussons' daughter, and is made from estate-grown grapes.

YYYY **1997** Medium to full yellow-green; the bouquet is quite developed, with rich, honeyed, toasty overtones to buttery/peachy fruit. The palate is round and ripe, with plenty of mouthfeel, showing the rich buttery/peachy and slightly charry oak characters promised by the bouquet. Retasted March 1999, the oak has settled down nicely. **rating:** 87

best drinking 1999–2002 **best vintages** '90, '92, '93, '94, '97 **drink with** Seafood salad • $28.50

Fergusson Jeremy Shiraz

Named after Peter and Louise Fergusson's son Jeremy, and made from estate-grown grapes. The '93 vintage was a knockout: when the subject of Yarra Valley Shiraz comes up, Fergusson is more often than not forgotten. It should not be.

YYYY **1997** Medium to full red-purple; the bouquet has hints of liquorice, boot polish and earth in a Rhône mode. The palate is very powerful, with slightly astringent tannins which might well have been fined out before bottling, but which may soften with time. **rating:** 85

best drinking 2004–2009 **best vintages** '88, '91, '92, '93 **drink with** Osso buco • $28.50

fermoy estate NR

Metricup Road, Willyabrup, WA 6280 **region** Margaret River
ph (08) 9755 6285 **fax** (08) 9755 6251 **open** 7 days 11–4.30
winemaker Michael Kelly **prod.** 15 000 **est.** 1985
prod. range ($10.50–26 CD) Sauvignon Blanc, Semillon, Chenin Blanc, Chardonnay, Reserve Chardonnay, Cabernet Sauvignon, Reserve Cabernet.
summary Consistently produces wines with a particular character and style, with the focus away from primary fruit and into secondary flavours, with strong structure; the Americans would call them 'food styles'. Quite deliberately out of the mainstream.

fern hill estate ★★☆

Ingoldby Road, McLaren Flat, SA 5171 **region** McLaren Vale
ph (08) 8383 0167 **fax** (08) 8383 0107 **open** Mon–Fri 10–5, weekends 10–5
winemaker Grant Burge (Contract) **prod.** 5000 **est.** 1975
prod. range ($14.95–17.95 CD) Semillon, Chardonnay, Shiraz, Cabernet Sauvignon.
summary One suspects there have been significant changes since Wayne Thomas sold Fern Hill to the Hill International Group, not all for the better. The wines are now exported to the UK, US, Canada, Japan, New Zealand, Singapore and Switzerland.

fishburn & o'keefe ★★★☆

16 Pioneer Avenue, New Norfolk, Tas 7140 **region** Southern Tasmania
ph (03) 6286 1238 **fax** (03) 6261 4029 **open** 7 days at Meadowbank Vineyard
winemaker Greg O'Keefe **prod.** 3000 **est.** 1991
prod. range ($15–25 CD) Riesling, Chardonnay, Trout White (White Pinot), Pinot Noir, Cabernet Sauvignon, Trout Brut.
summary Wine consultant and contract-winemaker Greg O'Keefe, one time winemaker at Normans, has joined forces with Hutchins schoolteacher Mike Fishburn to produce wines made from grapes purchased from various growers across Tasmania, but with an estate vineyard in the course of establishment. Greg O'Keefe also has an active consultancy and contract-winemaking business in his own right. Has also managed to produce a hard to find Sparkling Shiraz from grapes grown in the Tamar Valley. The details provided here are current and correct at the time of going to print, but changes are likely before the end of 1999.

five oaks vineyard NR

Aitken Road, Seville, Vic 3139 **region** Yarra Valley
ph (03) 5964 3704 **fax** (03) 5964 3064 **open** Weekends and public holidays 10–5
winemaker Wally Zuk, Michael Zitzlaff (Consultant) **prod.** 1500 **est.** 1997
prod. range ($18–25 CD) Riesling, Merlot, Cabernet Merlot, Cabernet Sauvignon.
summary Wally and Judy Zuk purchased the Five Oaks Vineyard in Aitken Road, Seville from Oakridge Estate, which has moved to its new premises on the other side of the Yarra Valley. Wally Zuk, with a background in physics, has completed his Wine Science degree at Charles Sturt University, spending part of his time working in Sydney and Canberra as a physicist and part as winemaker (with help from Michael Zitzlaff of Oakridge) at Five Oaks. The '97 Cabernet Sauvignon is well made and has powerful varietal Cabernet flavour.

fox creek wines ★★★★☆

Malpas Road, Willunga, SA 5172 **region** McLaren Vale
ph (08) 8556 2403 **fax** (08) 8556 2104 **open** 7 days 11–5
winemaker Njal (Sparky) Marquis, Sarah Marquis **prod.** 15 000 **est.** 1995
prod. range ($15–40 CD) Verdelho, Chardonnay, Botrytis Chenin Blanc, Vixen Sparkling Burgundy, Reserve Shiraz, JSM Shiraz Cabernets, Merlot, Reserve Cabernet Sauvignon.
summary Fox Creek has made a major impact since coming on-stream late in 1995. It is the venture of a group of distinguished Adelaide doctors (three of them professors) with particular input from the Watts family, which established the vineyard back in 1985 (selling the grapes) and whose daughter Sarah is now married to winemaker Sparky Marquis. The Reserve red wines, and especially the Reserve Shiraz, are outstanding and have enjoyed considerable show success. As well as comprehensive distribution throughout Australia, the wines are exported to the UK, US, Germany, Switzerland and Singapore.

Fox Creek Reserve Shiraz

Estate-grown; a gold medal winner at the 1997 Liquorland National Wine Show.
🍷🍷🍷🍷🍷 **1996** Strong red-purple; there is masses of powerful, earthy shiraz on the bouquet and a mix of mint, black cherry and earthy varietal fruit on the palate, supported by strong tannins. The oak influence is subtle. **rating:** 95

best drinking 2002–2010 **best vintages** '96 **drink with** Grilled beef • $40

Fox Creek JSM Shiraz Cabernets

JSM are the initials of James Stanley Malpas (born 1873) and are cut into the stone lintel above the present-day tasting room, but which was his house after he had graduated from Roseworthy Agricultural College. The wine is a blend of Shiraz, Cabernet Sauvignon and Cabernet Franc,

70% of which is matured for 12 months in second-use American oak and 30% for 12 months in second-use French oak.

🍷🍷🍷🍷 **1997** Youthful purple-red; a fresh, powerful, juicy earthy berry bouquet is followed by a similarly vibrantly youthful palate offering plenty of fruit backed by tannins and subtle oak. Tasted early in its life, and should merit significantly higher points in the future. **rating:** 86

best drinking 2002–2007 **best vintages** NA **drink with** Rib of beef • $19

Fox Creek Reserve Cabernet Sauvignon

An imperious wine, which spent an awesome 52 days on its skins during the fermentation and post-fermentation maceration period, followed by maturation in new French oak for 12 months. The '96 was a gold medal winner at the 1997 Adelaide Wine Show, the '97 of similar style and quality.

🍷🍷🍷🍷🍷 **1997** Dense purple-red; the bouquet has massive weight and extract, with powerful, if not forbidding, earthy aromas; the palate provides more of the same, but there is enough cassis/blackcurrant fruit to hold the whole thing together. **rating:** 93

best drinking 2007–2017 **best vintages** '96, '97 **drink with** Game pie • $30

frankland estate ★★★★

Frankland Road, Frankland, WA 6396 **region** Great Southern
ph (08) 9855 1544 **fax** (08) 9855 1549 **open** By appointment
winemaker Barrie Smith, Judi Cullam **prod.** 12 000 **est.** 1988
prod. range ($18–29 R) Riesling, Sauvignon Blanc, Chardonnay, Isolation Ridge (Shiraz), Olmo's Reward (Bordeaux-blend of Cabernet Franc, Merlot, Malbec, Cabernet Sauvignon, with Petit Verdot in future vintages), Cabernet.

summary A rapidly growing Frankland River operation, situated on a large sheep property owned by Barrie Smith and Judi Cullam. The 26-hectare vineyard has been established progressively since 1988, and a winery built on the site for the 1993 vintage. The Riesling, Isolation Ridge and Olmo's Reward are consistently good. Exports to US, UK, Taiwan, Japan, Belgium, Switzerland and Singapore.

Frankland Estate Isolation Ridge

First made in 1991, Isolation Ridge has moved from being predominantly made from Shiraz to 100% Shiraz, entirely estate-grown. It is normally the richest and strongest of the Frankland Estate reds.

🍷🍷🍷🍷 **1997** Medium red-purple; the moderately intense bouquet has attractive cedar, chocolate and berry aromas, characters which flow into the fore and middle palate. Then quite pronounced tannins, coupled with cedary/briary notes, follow on, giving the wine a certain austerity, but which is far from unpleasant. **rating:** 88

best drinking 2002–2007 **best vintages** '92, '93, '97 **drink with** Braised lamb shanks • $21

freycinet ★★★★★

Tasman Highway via Bicheno, Tas 7215 **region** Southern Tasmania
ph (03) 6257 8384 **fax** (03) 6257 8454 **open** Mon–Fri 9–5, weekends 10–4
winemaker Claudio Radenti **prod.** 5000 **est.** 1980
prod. range ($14–35 CD) Riesling Muller Thurgau, Chardonnay, Pinot Noir, Cabernet Sauvignon, Cabernet Franc.

summary The 4-hectare Freycinet vineyards are beautifully situated on the sloping hillsides of a small valley. The soils are podsol and decaying granite with a friable clay subsoil, and the combination of aspect, slope, soil and heat summation produce red grapes of unusual depth of colour and ripe flavours. One of Australia's foremost producers of Pinot Noir, with a wholly enviable track record of consistency – rare with such a temperamental variety.

Freycinet Chardonnay

Freycinet has produced a remarkable string of vintages between 1993 and 1996 inclusive, with wines of exemplary quality and a richness of texture not often encountered in Tasmania. The wines reinforce the message of the Freycinet Pinot Noir: that the amphitheatre/bowl in which the vineyard is situated provides a unique site climate, and with it, an extra dimension of weight and flavour.

🍷🍷🍷🍷🍷 **1998** Medium to full yellow-green; the bouquet is extremely rich, with nutty malolactic-fermentation characters supported by masses of melon, grapefruit and fig flavours. The palate is exceptionally rich, with mealy/nutty/cashew/creamy flavour and texture, all attesting to the fact that it has had a total malolactic fermentation and spent a long time on lees. **rating:** 94

best drinking 2000–2005 **best vintages** '93, '94, '95, '96, '98 **drink with** Abalone • $24

Freycinet Pinot Noir

The remarkable site climate of the Freycinet vineyard is primarily responsible for the outstanding quality of the Pinot Noir, although the experience and skills of Geoff Bull, daughter Lindy Bull and Claudio Radenti ensure that the potential quality is maximised. Interestingly, the only red fermenter the winery possesses is a rotary fermenter, which in turn helps in the extraction of both colour and flavour. Year in, year out one of Australia's (let alone Tasmania's) best Pinot Noirs, the '97 being no exception.

🍷🍷🍷🍷🍷 **1997** Light to medium red-purple; in the manner of many great Pinot Noirs, tends to creep up on you, initially a little light on the bouquet, but then first the stylish character of the wine starts to manifest itself with delicious plum fruit supported by oak swelling on the bouquet. The palate builds the pace and weight further, highly flavoured, with the clever use of oak with marvellously sweet plummy fruit. **rating:** 94

best drinking 1999–2004 **best vintages** '91, '92, '94, '95, '96, '97 **drink with** Duck, hare, venison • $35

Freycinet Cabernet Sauvignon

Estate-grown, and a testament to the exceptional qualities of the amphitheatre-shaped Freycinet vineyard which is a sun and heat trap, well protected from the least desirable winds.

🍷🍷🍷🍷🍷 **1994** Full red-purple; a very complex bouquet with hints of mint, which quickly move into rich dark chocolate and blackberry flavours on the palate. The wine has excellent texture, and ripe tannins. **rating:** 94

best drinking 1999–2010 **best vintages** NA **drink with** Double-thick lamb loin chops • $28

fyffe field NR

Murray Valley Highway, Yarrawonga, Vic 3730 **region** Goulburn Valley

ph (03) 5748 4282 **fax** (03) 5748 4284 **open** 7 days 10–5

winemaker David Traeger (Contract) **prod.** 1300 **est.** 1993

prod. range ($7.95–13.50 CD) Traminer, Riesling, Diamond White, Chardonnay, Brut, Big Rivers Rosé, 2nd Vintage Shiraz, Cabernet (from Langhorne Creek), Shiraz, Tokay, Tawny Port, Tokay and 'Tawny Snort' (a fearsome blend of Tawny Port, Reserve Port and Muscat).

summary Fyffe Field has been established by Graeme and Liz Diamond near the Murray River between Cobram and Yarrawonga in a mudbrick and leadlight tasting room opposite an historic homestead. A highlight is the ornamental pig collection on display, a display set up long before Babe was born. Wine quality is adequate.

galafrey ★★★☆

145 Lower Sterling Terrace, Albany, WA 6330 **region** Great Southern

ph (08) 9841 6533 **fax** (08) 9851 2324 **open** Mon–Sat 10–5

winemaker Ian Tyrer **prod.** 4000 **est.** 1977

prod. range ($10–27 CD) Riesling, Chardonnay, Müller Thurgau, Pinot Noir, Shiraz, Cabernet Sauvignon.

summary Relocated to a new purpose-built but utilitarian winery after previously inhabiting the exotic surrounds of the old Albany wool store, Galafrey makes wines with plenty of character, drawing grapes in the main from 12 hectares of estate plantings at Mount Barker. The wines always have character and flavour, but are not necessarily technically precise. No recent tastings.

galah wine ★★★☆

Tregarthen Road, Ashton, SA 5137 **region** Adelaide Hills

ph (08) 8390 1243 **fax** (08) 8390 1243 **open** Available at Ashton Hills

winemaker Stephen George **prod.** 750 **est.** 1986

prod. range ($7.50–35 ML) Barossa Valley Riesling, Barossa Valley Fume Blanc, Marlborough Chardonnay, McLaren Vale Shiraz, SE Aust Shiraz Cabernet, Barossa Valley Grenache, Clare Valley Cabernet Malbec, Clare Valley Shiraz, Sparkling Shiraz, Galah Brut, Vintage Port.

summary Over the years, Stephen George has built up a network of contacts across South Australia from which he gains some very high-quality small parcels of grapes or wine for the Galah label. These are all sold direct at extremely low prices for the quality.

Galah Clare Valley Shiraz

Although the label makes no mention of it, Stephen George is consultant winemaker at Wendouree, and it is well known that he has first option on any small parcels not required by Wendouree for its own label. All the hallmarks of having come from that marvellous source.

🍷🍷🍷🍷🍸 **1997** Medium to full red-purple; the bouquet promises a wine of immense power and earthy extract, and that is precisely what the palate delivers, with the fruit flavours running through chocolate, earth and blackberry, presently in the iron clasp of austere tannins. Ten years in the cellar, and a little prayer (or a little luck) should result in a great full-bodied dry red.

rating: 90

best drinking 2007–2017 **best vintages** '88, '89, '90, '91, '92, '95, '97 **drink with** Venison, kangaroo • $16

garbin estate NR

209 Toodyay Road, Middle Swan, WA 6056 **region** Swan District

ph (08) 9274 1747 **fax** (08) 9274 1747 **open** 7 days 10–5.30

winemaker Peter Garbin **prod.** 1000 **est.** 1956

prod. range ($10–15 CD) Chenin Blanc, Chardonnay, Shiraz, Cabernet Merlot, Dessert Wine, Ruby Port.

summary Peter Garbin, winemaker by weekend and design draftsman by week, decided in 1990 that he would significantly upgrade the bulk fortified winemaking business commenced by his father in 1956. The vineyards have been replanted, the winery re-equipped, and the first of the new generation wines produced in 1994.

garden gully vineyards ★★★☆

Garden Gully, Great Western, Vic 3377 **region** Grampians

ph (03) 5356 2400 **fax** (03) 5356 2400 **open** Mon–Fri 10.30–5.30, weekends 10–5.30

winemaker Brian Fletcher, Warren Randall **prod.** 2000 **est.** 1987

prod. range ($12.50–27 CD) Riesling, Shiraz, Sparkling Burgundy, Sparkling Chardonnay, Sparkling Pinot Noir.

summary Given the skills and local knowledge of the syndicate which owns Garden Gully, it is not surprising that the wines are typically good: an attractive stone cellar-door sales area is an additional reason to stop and pay a visit. Shiraz produced from the 100-year-old vines adjoining the cellar door is especially good. The 3 hectares of shiraz is complemented by 3 hectares of riesling. Another good wine.

Garden Gully Riesling

The label does not indicate the origin of the grapes, but it really doesn't matter: this is a truly excellent wine at the price.

🍷🍷🍷🍷🍷 **1998** Light green-yellow; the bouquet is light but intense with lime and herb aromas, and the palate has the same paradoxical fine but flavoursome, delicate yet intense lime juice flavour. Very pure, very appealing. **rating:** 92

best drinking 1999–2005 **best vintages** '98 **drink with** Light seafood • $13

geebin wines NR

3729 Channel Highway, Birchs Bay, Tas 7162 **region** Southern Tasmania
ph (03) 6267 4750 **fax** (03) 6267 5090 **open** 7 days 10–5
winemaker Andrew Hood (Contract) **prod.** 60 **est.** 1983
prod. range ($15–16.50 CD) Riesling, Cabernet Sauvignon.
summary Although production is minuscule, quality has been consistently high. The Riesling is well made, but the interesting wine from this far southern vineyard is Cabernet Sauvignon: clearly, the vineyard enjoys favourable ripening conditions. With 0.7 hectare of vineyards (including 0.3 hectare of chardonnay yet to come into bearing) Geebin claims to be the smallest commercial producer in Australia, but isn't: Scarp Valley and (temporarily) Jollymont are smaller.

gehrig estate ★★★

Cnr Murray Valley Highway. and Howlong Road, Barnawartha, Vic 3688 **region** Rutherglen
ph (02) 6026 7296 **fax** (02) 6026 7424 **open** Mon–Sat 9–5, Sun 10–5
winemaker Brian Gehrig **prod.** 5000 **est.** 1858
prod. range ($12–18 CD) Chenin Blanc, Riesling, Chardonnay, Trebbiano, Autumn Riesling, Late Harvest, Pinot Noir, Shiraz, Cabernet Sauvignon, Fortifieds.
summary An historic winery (and adjacent house) are superb legacies of the nineteenth century. Progressive modernisation of the winemaking facilities and operations has seen the quality of the white wines improve significantly, while the red wines now receive a percentage of new oak. Another recent innovation has been the introduction of the Gourmet Courtyard serving lunch on weekends, public holidays and Victorian school holidays. A wide-ranging tasting of dry white, red and fortified wines in March 1999 showed Brian Gehrig to be equally at home with all three styles, the wines very reasonably priced for their quality.

Gehrig Estate Old Tawny Port

The rich Tawny Ports of northeast Victoria are quite unique, infinitely richer and sweeter than their Portuguese counterparts.

🍷🍷🍷🍷 **NV** The colour is dark, with a few red hues still there indicating intermediate age. The bouquet is solid, with sweet fruit and some rancio; a big, rich Rutherglen style on the palate which, like the bouquet, does not show any signs of staleness. **rating:** 86

best drinking 1999–2000 **best vintages** NA **drink with** Coffee • $18

gembrook hill ★★★★

Launching Place Road, Gembrook, Vic 3783 **region** Yarra Valley
ph (03) 5968 1622 **fax** (03) 5968 1699 **open** By appointment
winemaker Ian Marks, Martin Williams **prod.** 2000 **est.** 1983
prod. range ($28–35 R) Sauvignon Blanc, Chardonnay, Pinot Noir.
summary The 6-hectare Gembrook Hill Vineyard is situated on rich, red volcanic soils 2 kilometres north of Gembrook in the coolest part of the Yarra Valley. The vines are not irrigated, with consequent natural vigour control. The Sauvignon Blanc is invariably good, sometimes outstanding. Ian Marks is presently engaged in building an on-site winery, to be completed in time for vintage 2000.

Gembrook Hill Sauvignon Blanc

The wine for which Gembrook Hill first came into prominence, and suited (in terms of wine style) both to the site and climate. It has proved to be a very difficult variety to grow, or at least to crop well, with a tiny production from the 2 hectares of vines. Devotees of the style are pleased that Dr Ian Marks has persevered.

🍷🍷🍷🍷 **1998** Light green-yellow; a crisp, light and clean bouquet with an utterly correct mix of light gooseberry and more herbaceous aromas is followed by a commensurately delicate and fresh palate, with flavours tracking those of the bouquet. **rating:** 87

best drinking 1999–2000 **best vintages** '90, '92, '93, '94, '95, '98 **drink with** Lobster bisque • $28

Gembrook Hill Chardonnay

Immaculate viticulture, with great attention to detail and to canopy management and manipulation, neatly offsets the ultra-cool site climate at Gembrook. The Chardonnay always shows its cool-climate origins, but is never thin or herbal.

🍷🍷🍷🍷 **1997** Light green-yellow; the bouquet is very light, with citrussy aromas veering more towards Sauvignon Blanc; subtle oak. The palate is similar, with a nice twist of lemon drop acidity on the finish. Will very probably evolve well in bottle. **rating:** 84

best drinking 1999–2002 **best vintages** '90, '91, '93, '94, '97 **drink with** Crab, prawns • $28

Gembrook Hill Pinot Noir

Tends to reflect the very cool climate, and the rich, red volcanic soils rather more than the two white wines, but nonetheless has style and genuine appeal in a lighter mould.

🍷🍷🍷🍷 **1997** Medium red-purple; the bouquet is quite complex with a range of gamey/foresty/tobacco aromas. The palate is relatively long and intense, with the slightly feral/forest characters of the bouquet again manifesting themselves. Does have over 13% alcohol and there is a core of ripe fruit there. **rating:** 86

best drinking 1999–2002 **best vintages** '97 **drink with** Asian seafood dishes • $35

geoff merrill ★★★★

291 Pimpala Road, Woodcroft, SA 5162 **region** McLaren Vale

ph (08) 8381 6877 **fax** (08) 8322 2244 **open** Mon–Fri 10–5, Sun 12–5

winemaker Geoff Merrill, Goe DiFabio, Scott Heidrich **prod.** 35 000 **est.** 1980

prod. range ($14–40 R) A change in brand structure has resulted in the Geoff Merrill range becoming Geoff Merrill Reserve representing the ultra-premium wines; the former Premium range is now released under the Geoff Merill label.

summary In 1998 the product range was rearranged into two tiers: premium (in fact simply varietal) and reserve, the latter being the older (and best) wines, reflecting the desire of this otherwise exuberant winemaker for elegance and subtlety. As well as national retail distribution, significant exports to UK, Europe, US and Asia.

Geoff Merrill Reserve Chardonnay

As at 1998, no less than five vintages, from 1991 to 1995, were available under the Reserve label, all priced at $26. There is considerable consistency across all vintages, with oak (and bottle age) the major factors in shaping that style.

🍷🍷🍷🍷🍷 **1994** Light to medium green-yellow; there are fresh and lively citrus and stone fruit aromas followed by a wine with considerable elegance, and which is ageing particularly well. Citrus and melon fruit is beautifully balanced with sensitive oak. A wine which should have been on the market in 1998, rather than the '92 vintage. **rating:** 94

best drinking 1999–2002 **best vintages** NA **drink with** Pan-fried chicken breast • $26

Geoff Merrill Premium Shiraz

The fruit source for this wine sometimes contains a portion of Goulburn Valley fruit, but is always predominantly based in McLaren Vale. In 1996 it was 100% McLaren Vale, and spent 21 months in new and used American oak.

🍷🍷🍷🍷 **1996** Medium red-purple; a clean, fresh red cherry/berry bouquet leads into a palate with more weight and complexity than the usual Merrill style, quite ripe and rich, with flavours of chocolate and cherry. **rating:** 87

best drinking 1999–2003 **best vintages** '96 **drink with** Ragout of veal • $14.95

Geoff Merrill Reserve Cabernet Sauvignon

Effectively takes over from the wine previously simply labelled Cabernet Sauvignon, sharing with it the multi-regional source and the occasional dash of Cabernet Franc or Merlot. Those sources span Coonawarra, McLaren Vale and the Goulburn Valley.

🍷🍷🍷🍷 **1995** Medium purple-red; the bouquet is quite pungent, with slightly strange aromas in addition to the earth and leaf found in many of these Cabernets. The palate has much more tannin and grip than the older vintages, although it is far from certain how well it will age, because there seems to be a slightly green streak running through it. **rating:** 86

best drinking 2000–2005 **best vintages** NA **drink with** Rack of lamb • $27

geoff weaver ★★★★★

2 Gilpin Lane, Mitcham, SA 5062 **region** Adelaide Hills

ph (08) 8272 2105 **fax** (08) 8271 0177 **open** Not

winemaker Geoff Weaver **prod.** 5000 **est.** 1982

prod. range ($21–32 ML) Riesling, Chardonnay, Sauvignon Blanc, Cabernet Merlot.

summary This is now the full-time business of former Hardy Group chief winemaker Geoff Weaver. He draws upon a little over 11 hectares of vineyard established between 1982 and 1988; for the time being, at least, the physical winemaking is carried out by Geoff Weaver at Petaluma. He produces an invariably immaculate Sauvignon Blanc, and one of the longest-lived Chardonnays to be found in Australia, with intense grapefruit and melon flavour. The beauty of the labels ranks supreme with that of Pipers Brook. The wines are exported to the US, UK and to five European destinations.

Geoff Weaver Lenswood Riesling

Produced from 1 hectare of riesling planted in 1982 at an altitude of 540 metres. The wines are unequivocally long-lived.

🍷🍷🍷🍷🍷 **1998** Light green-yellow; the bouquet is fragrant, with distinct herb/herbal notes together with more minerally/kerosene characters. The palate is very tight and intense, with mineral, lime and herb flavours on a long finish. In its absolute infancy. **rating:** 90

best drinking 2002–2010 **best vintages** '90, '93, '94, '96, '98 **drink with** Fresh asparagus • $21

Geoff Weaver Lenswood Chardonnay

The 3.5 hectares of estate chardonnay are now over 15 years old, producing grapes with great intensity of flavour, but with a particular grapefruit citrus character which has been present since the very first vintage. The wine matures slowly and gracefully, and is released with several years bottle age.

🍷🍷🍷🍷🍷 **1997** Glowing yellow-green; a complex bouquet in which all of the parts are relatively subtle, with gently tangy fruit and controlled barrel-ferment aromas. The palate has intense melon and grapefruit flavours running through a long finish; oak plays a junior role. **rating:** 91

best drinking 2000–2005 **best vintages** '95, '97 **drink with** Sweetbreads • $32

giaconda ★★★★★

McClay Road, Beechworth, Vic 3747 **region** Ovens Valley
ph (03) 5727 0246 **fax** (03) 5727 0246 **open** By appointment
winemaker Rick Kinzbrunner **prod.** 1000 **est.** 1985
prod. range ($36–60 R) Chardonnay, Pinot Noir, Cabernet Sauvignon.
summary Wines which have a super-cult status and which, given the tiny production, are extremely difficult to find, sold chiefly through restaurants and mail order. All have a cosmopolitan edge befitting Rick Kinzbrunner's international winemaking experience. The Chardonnay and Pinot Noir are made in contrasting styles: the Chardonnay tight and reserved, the Pinot Noir usually opulent and ripe.

Giaconda Chardonnay

Four hundred and fifty cases of hand-crafted wines are produced from a little under 1 hectare of estate vineyard every year. The style of the wine is entirely different from mainstream Australian Chardonnay, relying far more on texture and structure, and far less on primary fruit. An exceptionally distinguished and consistent wine which is the very deliberate product of Rick Kinzbrunner's winemaking philosophy. The label bears the ultimate politically correct statement 'Unfiltered. Wild yeast.'

🍷🍷🍷🍷🍷 **1997** Glowing yellow-green; a typically complex bouquet, very Burgundian, with an array of secondary fruit aromas supported by subtle oak. The palate is no less interesting than the bouquet, with characters ranging all the way from creamy to minerally, and a touch of spicy oak emerging on the finish. Juicy fruity it isn't. **rating:** 95

best drinking 2000–2007 **best vintages** '86, '88, '90, '92, '93, '94, '95, '96, '97 **drink with** Slow-roasted Tasmanian salmon • $60

Giaconda Pinot Noir

As fastidiously produced and as full of character as the Chardonnay. It comes from a little over half a hectare of estate plantings, and is made in tiny quantities. The style has been quite different from the Pinot Noirs of southern Victoria, being much fuller and more robust, with the obvious potential to age well. In recent vintages, the style and flavour has veered more towards Burgundy, and become better and better.

🍷🍷🍷🍷🍸 **1997** Light to medium red, with just a hint of purple; there are amazing anise, French cashew and spice aromas which are decidedly unconventional, but the silky, long texture restores some of the balance and conventionality. The flavours and feel on the finish are excellent. **rating:** 93

best drinking 1999–2003 **best vintages** '85, '86, '88, '90, '91, '92, '95, '96, '97 **drink with** Tea-smoked duck • $49

---

gilbert ★★★★☆

RMB 438 Albany Highway, Kendenup via Mount Barker, WA 6323 **region** Great Southern
ph (08) 9851 4028 **fax** (08) 9851 4021 **open** Wed–Mon 10–5
winemaker Plantagenet (Contract) **prod.** 2000 **est.** 1980
prod. range ($14–23 CD) Riesling, Alira (medium sweet), Chardonnay, Shiraz.
summary A part-time occupation for sheep and beef farmers Jim and Beverly Gilbert, but a very successful one. The now mature vineyard, coupled with contract-winemaking at Plantagenet, has produced small quantities of high-quality Riesling and Chardonnay; the small production sells out quickly each year with retail distribution through New South Wales, Victoria and Western Australia, and exports to Japan.

Gilbert Chardonnay

Chardonnay, with a little under 2.5 hectares, accounts for the largest proportion of the Gilbert vineyard plantings, but invariably sells out first. It is made in a very disciplined style, without much new oak input, and with an unusual capacity to age gracefully.

ΨΨΨΨ **1997** Medium yellow-green; a clean bouquet with subtle and well-integrated oak running through the melon fruit. The innate power of the wine becomes obvious on the palate; old vines and 14 degrees alcohol produce ripe melon flavours and a hint of cashew on a powerful finish. **rating:** 90

best drinking 1999–2004 **best vintages** '90, '92, '94, '95 **drink with** Grilled fish • $18

Gilbert Shiraz

All the wines of Gilbert are made in tiny quantities, none more so than the Shiraz – which is a pity.

ΨΨΨΨΨ **1997** Medium to full red-purple; quite intense cherry/plum fruit is supported by a touch of toasty oak on the bouquet. The wine has lovely flavour, with masses of sweet cherry fruit, sophisticated oak treatment and attractive, ripe tannins. **rating:** 94

best drinking 2001–2011 **best vintages** '97 **drink with** Seared beef • $23

gilgai winery NR

Tingha Road, Gilgai, NSW 2360 **region** Other Wineries of NSW

ph (02) 6723 1204 **open** 7 days 10–6

winemaker Keith Whish **prod.** 400 **est.** 1968

prod. range ($NA) Semillon, Shiraz Cabernet, Chandelier (Fortified White), Port.

summary Inverell medical practitioner Dr Keith Whish has been quietly producing wines from his 6-hectare vineyard for almost 30 years. All of the production is sold through cellar door.

glaetzer wines NR

34 Barossa Valley Way, Tanunda, SA 5352 **region** Barossa Valley

ph (08) 8563 0288 **fax** (08) 8563 0218 **open** Mon–Sat 10.30–4.30, Sunday and public holidays 11.30–4.30

winemaker Colin Glaetzer, Ben Glaetzer **prod.** 3000 **est.** 1995

prod. range ($16-45 CD) Bush Vine Semillon, Semillon Ratafia, Grenache Mourvedre; Sparkling Pinot Noir, Bishop Shiraz, Malbec Cabernet Sauvignon, Shiraz, Sparkling Shiraz.

summary Colin and Ben Glaetzer are almost as well known in South Australian wine circles as Wolf Blass winemaker John Glaetzer, and, needless to say, they are all related. Glaetzer Wines purchases its grapes from Barossa Valley growers, and makes an array of traditional Barossa styles. Wholesalers have been appointed in each State, and the wines are already exported to four destinations in Europe and to the US.

glenara wines ★★★☆

126 Range Road North, Upper Hermitage, SA 5131 **region** Adelaide Hills

ph (08) 8380 5056 **fax** (08) 8380 5056 **open** Mon–Fri 11–5 (closed public holidays)

winemaker Trevor Jones **prod.** 6000 **est.** 1971

prod. range ($16–24 CD) Riesling, Chardonnay, Sauvignon Blanc Semillon, Unwooded Chardonnay, Pinot Noir, Shiraz, Cabernet Rosé, Cabernet Merlot, Cabernet Sauvignon, Sparkling, Old Tawny Port.

summary Glenara has been owned by the Verrall family since 1924; the first vines were planted in 1971, the first wine made in 1975, and the winery built in 1988. Has proceeded to produce many good wines, particularly the full-flavoured Rieslings, but also with creditable full-bodied reds. The wines have limited retail distribution in all States, and are exported to Japan and Canada.

glenayr ★★★☆

Back Tea Tree Road, Richmond, Tas 7025 **region** Southern Tasmania
ph (03) 6260 2388 **fax** (03) 6260 2691 **open** Mon–Fri 8–5
winemaker Chris Harrington (at Stoney Vineyard) **prod.** 350 **est.** 1975
prod. range ($18–20 CD) Riesling, Chardonnay, Pinot Noir, Cabernet Shiraz Merlot; Tolpuddle Vineyards Chardonnay and Pinot Noir.
summary The principal occupation of Chris Harrington is as viticultural manager of the substantial Tolpuddle Vineyard, the grapes of which are sold to Domaine Chandon. Tiny quantities of wine are made from an adjacent 1-hectare vineyard for mailing list sales under the GlenAyr label; chardonnay and pinot noir grapes are also purchased from Tolpuddle Vineyards.

GlenAyr Pinot Noir

The Coal River region rivals the Tamar Valley in its capacity to produce full-flavoured red wines of a style one does not traditionally associate with Tasmania. The '97 was a silver medal winner at the 1999 Tasmanian Wines Show, and also scored well at the 1999 *Winewise* Small Makers Competition.

ΥΥΥΥ **1997** Light to medium red, quite developed. The bouquet has a mix of earthy/leafy/red cherry/mint aromas; the palate is relatively light and soft with fresh cherry/strawberry fruit, offset by more cedary, foresty notes. **rating:** 89

best drinking 1999–2001 **best vintages** '91, '94, '96 **drink with** Braised duck • $20

glen erin vineyard retreat NR

Woodend Road, Lancefield, Vic 3435 **region** Macedon
ph (03) 5429 1041 **fax** (03) 5429 2053 **open** Weekends, public holidays 10–6
winemaker Brian Scales **prod.** 1000 **est.** 1993
prod. range ($14–26 CD) Gewurztraminer, Chardonnay, Pinot Noir, Mystic Park Macedon Sparkling, Cabernet Merlot; Deep Creek Riesling, Colombard Chardonnay and Shiraz Grenache.
summary Brian Scales acquired the former Lancefield Winery and has renamed it Glen Erin Grange. The accompanying restaurant is open on Friday and Saturday evenings à la carte and for Saturday and Sunday lunch.

glenfinlass NR

Elysian Farm, Parkes Road, Wellington, NSW 2820 **region** Other Wineries of NSW
ph (02) 6845 2011 **fax** (02) 6845 3329 **open** Sat 9–5 or by appointment
winemaker Brian G Holmes **prod.** 500 **est.** 1971
prod. range ($10 CD) Sauvignon Blanc, Shiraz, Hill Vineyard Shiraz Cabernet, Cabernet Sauvignon.
summary The weekend and holiday hobby of Wellington solicitor Brian Holmes, who has wisely decided to leave it at that. I have not tasted the wines for many years, but the last wines I did taste were competently made. Brian Holmes says that the wines are currently in unusually short supply owing to drought and frost damage, but there are three vintages of Shiraz ('92, '96 and '97) available at cellar door.

glenguin NR

River Oaks Vineyard, Lot 8 Milbrodale Road, Broke, NSW 2330 **region** Lower Hunter Valley
ph (02) 6579 1009 **fax** (02) 6579 1009 **open** At Boutique Wine Centre, Broke Road, Pokolbin
winemaker Robin Tedder **prod.** 6000 **est.** 1993
prod. range ($19–24.50 R) Individual Vineyard Semillon, Chardonnay, Shiraz, Merlot.

summary Glenguin's vineyard has been established along the banks of the Wollombi Brook by Robin, Rita and Andrew Tedder, Robin and Andrew being the grandsons of Air Chief Marshal Tedder, who was made Baron of Glenguin by King George VI in recognition of his wartime deeds. Glenguin was in fact a Scottish distillery which continues to produce a single malt, but there is no other connection between the two Glenguins. Glenguin has 16 hectares of vineyard at Broke, and another 5 hectares at Orange (cabernet and merlot). The wines are imported into the UK by Bibendum.

Glenguin Individual Vineyard Merlot

One of a series of Individual Vineyard wines from Glenguin, the white wines coming from the Hunter Valley and the reds variously from Mudgee and, in this case, from Orange. It comes from a vineyard situated close to the town of Orange at an altitude of 800 metres on a gravelly clay soil.

🍷🍷🍷🍷 **1997** Medium red-purple; the bouquet is clean, with firm red berry fruit and minimal oak influence. The palate has sweet, small red berry fruit flavours supported by lingering, silky tannins. A fruit-driven wine with good concentration of flavour. **rating:** 86

best drinking 2000–2005 **best vintages** NA **drink with** Grilled chops • $24.50

gloucester ridge vineyard ★★★

Burma Road, Pemberton, WA 6260 **region** Pemberton

ph (08) 9776 1035 **fax** (08) 9776 1390 **open** 7 days 10–5, later on Saturdays

winemaker John Wade **prod.** 5000 **est.** 1985

prod. range ($14–22.50 CD) Pemberton White, Late Harvest Riesling, Sauvignon Blanc, Aurora, Chardonnay, Pemberton Red, Pinot Noir, Cabernets.

summary Gloucester Ridge is the only vineyard located within the Pemberton town boundary, within easy walking distance. It is owned and operated by Don and Sue Hancock; quality has varied, but as the Sauvignon Blanc shows, can be good.

gnadenfrei estate NR

Seppeltsfield Road, Marananga via Nuriootpa, SA 5355 **region** Barossa Valley

ph (08) 8562 2522 **fax** (08) 8562 3470 **open** Tues–Sun 10–5.30

winemaker Malcolm Seppelt **prod.** 1500 **est.** 1979

prod. range ($12–20 CD) Riesling, Semillon, Traminer Riesling, Shiraz Grenache, Tawny Port, Sparkling.

summary A strictly cellar-door operation, which relies on a variety of sources for its wines, but has a core of 2 hectares of estate shiraz and 1 hectare of grenache. A restaurant presided over by Joylene Seppelt is open for morning teas, lunches and afternoon teas. Small quantities of the wines make their way to Pennsylvania, US.

golden grape estate NR

Oakey Creek Road, Pokolbin, NSW 2321 **region** Lower Hunter Valley

ph (02) 4998 7588 **fax** (02) 4998 7730 **open** 7 days 10–5

winemaker Neil McGuigan (Consultant) **prod.** NFP **est.** 1985

prod. range ($14.95–29.90 CD) Premier Semillon, Gewurztraminer, Sauvignon Blanc, Semillon Verdelho, Happy Valley Chardonnay, Five Star (light fruity), Frizzante Rosé, Mount Leonard (Cabernet Sauvignon), Domaine Springton (Shiraz), Classic Red, Fortifieds.

summary German-owned and unashamedly directed at the tourist, with a restaurant, barbecue and picnic areas, wine museum and separate tasting room for bus tours. The substantial range of wines are of diverse origins and style. The operation now has over 42 hectares of Hunter Valley plantings.

golden grove estate ★★★☆

Sundown Road, Ballandean, Qld 4382 **region** Granite Belt
ph (07) 4684 1291 **fax** (07) 4684 1247 **open** 7 days 9–5
winemaker Sam Costanzo **prod.** NA **est.** 1993
prod. range ($10–15 CD) Accommodation Creek Classic White and Classic Dry Red, Muscadean, Rosé, Shiraz, Cabernet Merlot, Liqueur Muscat.
summary Golden Grove Estate was established by Mario and Sebastiana Costanzo in 1945, producing stone fruits and table grapes for the fresh fruit market. The first wine grapes (shiraz) were planted in 1972, but it was not until 1985, when ownership passed to son Sam Costanzo and wife Grace that the use of the property started to change. In 1993 chardonnay and merlot joined the shiraz, followed by cabernet sauvignon, sauvignon blanc and semillon. Wine quality has steadily improved, with 14 medals in regional shows awarded up to July 1998, leading to national (though limited) retail distribution.

Golden Grove Estate Classic White

Yet another Golden Grove wine to come near the top of its class in the 1999 *Winewise* Small Makers Competition, and deservedly so.
▼▼▼▼▽ **1998** Medium yellow-green; the bouquet is very full, with sweet tangy fruit and a nice hint of oak. The palate is complex and powerful, with lots of presence and character. **rating:** 92
best drinking 1999–2001 **best vintages** '98 **drink with** Coquilles St Jacques • $12

Golden Grove Estate Chardonnay

A wine which simply underlines the overall quality of this exciting newcomer to the Queensland scene.
▼▼▼▼ **1998** Medium yellow-green; the bouquet is quite stylish, with complex barrel-ferment characters which also come through strongly on the palate. A well-made wine, even if in a full-on oak style. **rating:** 88
best drinking 1999–2000 **best vintages** NA **drink with** Smoked salmon risotto • $14

golders vineyard ★★★☆

Bridport Road, Pipers Brook, Tas 7254 **region** Northern Tasmania
ph (03) 6395 4142 **open** By appointment
winemaker Richard Crabtree **prod.** 400 **est.** 1991
prod. range ($20 R) Pinot Noir.
summary The initial plantings of 1.5 hectares of pinot noir have been supplemented by 1 hectare of chardonnay. The quality of the Pinot Noir has been good from the initial vintage in 1995.

Golders Pinot Noir

Winemaking runs in the family, it seems, for Richard Crabtree is Robert Crabtree's (of Watervale in the Clare Valley) brother. His 1995 Pinot Noir, sourced from Craig Hogarth's small vineyard at Pipers Brook, was his first wine, and a creditable effort, followed up by an even better '96, and an equally good effort in 1997.
▼▼▼▼▽ **1997** Medium red-purple; the bouquet offers an array of briary, sappy, stemmy, foresty aromas; the palate has excellent length and intensity, progressively building plum, briar and forest flavours running through to a finish with fine tannins. **rating:** 90
best drinking 1999–2002 **best vintages** '95, '96, '97 **drink with** Quail • $20

goona warra vineyard ★★★☆

Sunbury Road, Sunbury, Vic 3429 **region** Sunbury
ph (03) 9740 7766 **fax** (03) 9744 7648 **open** 7 days 10–5
winemaker John Barnier **prod.** 2500 **est.** 1863

prod. range ($18–23 R) Semillon, Chardonnay, Black Cygnet Chardonnay, Pinot Noir, Cabernet Franc, Black Cygnet Cabernets, Black Widow Brut, Tawny Port.

summary An historic stone winery, established under this name by a nineteenth-century Victorian premier. Excellent tasting facilities; an outstanding venue for weddings and receptions; Sunday lunch also served. Situated 30 minutes drive from Melbourne (10 minutes north of Tullamarine airport). Berry Bros & Rudd import the wines into the UK.

Goona Warra Chardonnay

Produced from estate-grown grapes, and given full-blown barrel-ferment and oak maturation. The cool-climate fruit character holds the wine in check notwithstanding the expansive winemaking techniques.

🍷🍷🍷🍷 **1997** Glowing yellow-green; a big, rich wine with lots of character on the bouquet, and richness on the strong barrel-ferment-influenced palate. Toasty characters are evident through the wine, which does however finish a fraction short. **rating:** 88

best drinking 1999–2004 **best vintages** NA **drink with** Pasta carbonara • $20

goundrey ★★★★

Muir Highway, Mount Barker, WA 6324 **region** Great Southern

ph (08) 9851 1777 **fax** (08) 9851 1997 **open** Mon–Sat 10–4.30, Sun 11–4.30

winemaker Keith Brown **prod.** 187 000 **est.** 1978

prod. range ($16–26 R) Chenin Blanc, Classic White, Unwooded Chardonnay, Cabernet Merlot, Cabernet Sauvignon; Reserve range of Riesling, Sauvignon Blanc, Chardonnay, Pinot Noir, Shiraz, Cabernet Sauvignon; second label Fox River Classic White, Chardonnay, Classic Red.

summary Under the ownership of Perth businessman Jack Bendat, not to mention the injection of many millions of dollars into vineyard and winery expansion, Goundrey grows apace. There seems to be a widening gap between the quality of the Reserve wines (usually, but not invariably, outstanding) and the varietal range (workmanlike). This may be no bad thing from a commercial viewpoint, particularly if the differential is reflected in the price, but does make an overall rating difficult.

Goundrey Reserve Chardonnay

A wine which has had considerable success over the years, hitting a high point with the '94 vintage. For a while it seemed to me to be unnecessarily oaky, and the 1997 vintage marks a return to better balance. It was matured in 80% French and 20% American Calistoga oak for 12 months, with six months on lees.

🍷🍷🍷½ **1997** Medium to full yellow-green; quite intense citrus and stone fruit aromas are supported by well-handled, gentle oak on the bouquet. The palate opens with delicate but quite complex peach and stone fruit flavours, although oak spice phenolics come on the attack at the finish. A wine which very nearly hit the mark; I suspect the damage was done by the American oak component. **rating:** 84

best drinking 1998–2001 **best vintages** '91, '94, '95 **drink with** Coquilles St Jacques • $29.60

gralyn cellars ★★★★

Caves Road, Willyabrup, WA 6280 **region** Margaret River

ph (08) 9755 6245 **fax** (08) 9755 6245 **open** 7 days 10.30–4.30

winemaker Graham Hutton, Merilyn Hutton **prod.** 1700 **est.** 1975

prod. range ($12–36 CD) Riesling, Classic Dry White, Late Harvest Riesling, Shiraz, Old Vine Shiraz, Late Harvest Cabernet, Cabernet Nouveau, Cabernet Shiraz, Cabernet Sauvignon, and an extensive range of fortifieds including White Port and Tawny Port.

summary The move from primarily fortified wine to table wine production continues, and does so with considerable success. The red wines are made in a distinctively different style from most of those from the Margaret River region, with a softness and sweetness (in part from American oak) which is reminiscent of some of the better-made wines from the eastern States.

Gralyn Old Vine Shiraz

Originally named after son Michael Hutton, produced from 24-year-old shiraz vines, and matured in a mix of new French and American oak. The 1994 vintage won a silver medal at the Perth Show in open competition against wines from all over Australia; the '95 did even better, winning silver at both the Perth Show and – even more significantly a gold medal – at the Sheraton Wine Awards. The '96 repeated almost exactly the performance of the '95. The '97 has been renamed Old Vine Shiraz, and will retain this title in the future.

🍷🍷🍷🍷 **1997** Old Vine. Medium to full red-purple; nice earthy berry Shiraz varietal fruit together with a touch of vanilla oak on the bouquet introduce a pleasantly flavoured wine with red berry fruit flavours, sweet American oak and soft tannins all coming together convincingly. **rating:** 87

best drinking 2000–2005 **best vintages** '94, '95, '96 **drink with** Strong red meat dishes • $36

Gralyn Cabernet Shiraz

Obviously enough, a blend of Cabernet Sauvignon and Shiraz matured in predominantly new French and American oak barrels.

🍷🍷🍷🍷 **1997** Medium to full red-purple with just a touch of purple remaining; a complex bouquet offers sweet berry fruit and a generous helping of vanilla oak; the same mix of generously sweet berry fruit and vanilla oak, together with soft tannins make for an excellent show style. **rating:** 88

best drinking 2001–2007 **best vintages** '95 **drink with** Smoked lamb • $36

granite ridge wines NR

Sundown Road, Ballandean, Qld 4382 **region** Granite Belt
ph (07) 4684 1263 **fax** (07) 4684 1250 **open** By appointment
winemaker Dennis Fergusson **prod.** 850 **est.** 1995
prod. range ($10–20 CD) Semillon Chardonnay, Unwooded Chardonnay, Chardonnay, Shiraz, Shiraz Cabernet, Cabernet Sauvignon, Liqueur Muscat.
summary Formerly known as Denlana Fergusson Estate Wines, Granite Ridge has had considerable success, with both the 1995 and 1996 Cabernet Sauvignon being judged Queensland's Best Cabernet (though quite by whom I am not sure), but continues to be run by Dennis Fergusson.

grant burge ★★★★☆

Jacobs Creek, Tanunda, SA 5352 **region** Barossa Valley
ph (08) 8563 3700 **fax** (08) 8563 2807 **open** 7 days 10–5
winemaker Grant Burge **prod.** 108 000 **est.** 1988
prod. range ($12–70 R) Has recently moved to a series of vineyard-designated varietal wines including Thorn Vineyard Riesling, Kraft Vineyard Sauvignon Blanc, Zerk Vineyard Semillon, Barossa Ranges Chardonnay, Lily Farm Frontignac, Filsell Shiraz, Hillcott Merlot, and Cameron Vale Cabernet Sauvignon. Top-of-the-range red are Meshach Shiraz, The Holy Trinity (Grenache Shiraz Mourvedre) and Shadrach Cabernet Sauvignon; also Rubycid and Virtuoso.
summary As one might expect, this very experienced industry veteran makes consistently good, full-flavoured and smooth wines chosen from the pick of the crop of his extensive vineyard holdings, which total an impressive 200 hectares; the immaculately restored/rebuilt stone cellar-door sales buildings are another attraction. The provocatively named The Holy Trinity (a Grenache Shiraz Mourvedre blend) joins Shadrach and Meshach at the top of the range. Grant Burge enjoyed great success in the 1998 and 1999 wine shows, with gold medals going to wines right across the range. The wines are exported to the UK, Europe, US, Canada and Asia.

Grant Burge Thorn Vineyard Riesling

The Thorn Vineyard is situated in the Eden Valley, one of South Australia's two classic riesling areas. Traditionally, Eden Valley Rieslings took many years to develop their characteristic lime aroma and flavour, but modern winemaking seems to be bringing out those characters earlier in the life of the wines.

🍷🍷🍷🍷🍷 **1998** Light yellow-green; powerful aromatics leap from the glass the moment the wine is swirled, and the palate precisely replicates the bouquet in attractively full-frontal style. Top gold medal Class 1 1998 National Wine Show. **rating:** 94

best drinking 1998–2001 **best vintages** '88, '90, '92, '93, '94, '96, '98 **drink with** Smoked trout mousse • $14.85

Grant Burge Zerk Semillon

The 1998 vintage won a gold medal in Class 19 at the 1999 Sydney Royal Wine Show, part of panoply of award-winning wines from Grant Burge.

🍷🍷🍷🍷🍸 **1998** Medium yellow-green; the bouquet offers above-average complexity and intensity, as does the palate, which has considerable length. It does, however, look as if it will thicken up pretty quickly, and is best drunk sooner rather than later. **rating:** 90

best drinking 1999–2000 **best vintages** NA **drink with** Grilled chicken • $13.45

Grant Burge Kraft Vineyard Sauvignon Blanc

Produced from grapes grown at Val and Dennis Kraft's vineyard at Tanunda. Eighty-five per cent of the wine is cold-fermented in stainless steel, 15% barrel-fermented and kept on yeast lees for three months prior to blending and release.

🍷🍷🍷🍷 **1998** Medium yellow-green; the bouquet is full of rich, ripe passionfruit/tropical character, quite striking given the vintage. The palate carries on in much the same vein, with an abundance of ripe tropical/sweet gooseberry fruit. With all this, it is hardly surprising the finish is slightly soft. **rating:** 88

best drinking 1999–2000 **best vintages** '94 **drink with** Rich seafood • $14.85

Grant Burge Virtuoso

A blend of Sauvignon Blanc (around 55%) and Semillon (around 45%), with avant garde packaging deliberately aimed at the brasserie market. The wine is cold-fermented in stainless steel, and relatively early-bottled. Arguably an altogether better wine than the jolly packaging suggests it will be. The 1998 won a gold medal in Class 2 at the 1999 Sydney Royal Wine Show.

🍷🍷🍷🍷🍸 **1998** Light yellow-green; the bouquet is very aromatic and tangy, having some almost Riesling-like floral characters. The wine has a tangy, grassy palate with the Sauvignon Blanc coming through clearly, and considerable length to the flavour. **rating:** 91

best drinking 1999–2000 **best vintages** '98 **drink with** Brasserie food • $15.50

Grant Burge Rubycind

A blend of 80% Pinot Noir and 20% Ruby Cabernet, fermented off skins at relatively low temperatures. It is a far better wine than the novelty packaging suggests it will be.

🍷🍷🍷🍷 **1998** Bright, light rosé; sweet, lifted cherry aromas follow through onto a flavoursome, cherry-filled palate with well-balanced acidity and no reliance on residual sugar. **rating:** 86

best drinking 1999–2000 **best vintages** NA **drink with** Antipasto • $15.50

Grant Burge Filsell Shiraz

Grant Burge's Filsell Vineyard is situated on the deep alluvial soils of the Lyndoch Valley; many of the vines are 80 years old. The wine finishes its fermentation in new American oak barrels, where it spends the next two years before final blend adjustments are made.

🍷🍷🍷🍷 **1996** Medium to full red-purple; a traditional style, with plenty of stuffing and concentration, and – as ever – lots of American oak wrapped around the fruit. The palate is no less rich and concentrated with black fruits and vanilla in what I call the magimix style. **rating:** 86

best drinking 2001–2006 **best vintages** NA **drink with** Pizza • $22.25

Grant Burge The Holy Trinity

A new super-premium wine from Grant Burge, presented in a custom-made bottle with the Grant Burge seal embossed in the glass. Wine and the church have had a long history together, which is just as well given the name of this wine, and its launch date on 7 June 1998, which was Holy Trinity Sunday. The Trinity in the wine is Grenache, Shiraz and Mourvedre.

🍷🍷🍷🍷 **1996** Youthful red-purple; there are aromas of dark cherry/berry, a hint of chocolate and subtle oak, the palate repeating those flavours but then adding lots of mint. Something of a chameleon. **rating:** 86

best drinking 1999–2004 **best vintages** NA **drink with** Jugged hare • $29.95

Grant Burge Shadrach Cabernet Sauvignon

Introduced in 1993 as a companion to Meshach. A gold medal winner at both the Melbourne and Barossa Valley Wine Shows, and made in the baroque oaky style, but with a fair degree of fruit to carry that oak.

🍷🍷🍷🍷🍷 **1996** Medium red-purple; there is abundant, ripe, dark berry, earthy varietal fruit and subtle oak on the bouquet. The palate has attractive juicy cassis berry fruit, good tannins and balanced oak. **rating:** 91

best drinking 2001–2007 **best vintages** '93, '96 **drink with** Beef Bordelaise • $42.50

great lakes wines NR

Herivals Road, Wootton, NSW 2423 **region** Hastings River

ph (02) 4997 7255 **fax** (02) 4997 7255 **open** 7 days 8.30–6

winemaker John Webber, Ian Lindeman (Consultant) **prod.** 1200 **est.** 1990

prod. range ($8–10 CD) Chardonnay, Semillon Classic Dry White, Mellow White, Chambourcin, Mellow Red, Cabernets.

summary John Webber and family began planting their vineyard on the mid north coast of New South Wales in 1990. There is now approximately half a hectare of each of chardonnay, semillon, verdelho, shiraz, cabernet sauvignon and chambourcin and the Webbers are proud of the fact that (to use their words) 'we are a fair dinkum winery where we "grow the grapes" and "we make the wine"'. All of the wines on current release have won at least one medal in regional wine shows, the 1998 Chambourcin winning silver medals at both the 1998 Cowra Show and 1998 New South Wales Small Winemakers Show.

greenock creek cellars NR

Radford Road, Seppeltsfield, SA 5360 **region** Barossa Valley

ph (08) 8562 8103 **fax** (08) 8562 8259 **open** Wed–Mon 11–5

winemaker Michael Waugh **prod.** 1500 **est.** 1978

prod. range ($13–16.50 CD) Chardonnay, Shiraz, Cabernet Sauvignon.

summary Michael and Annabelle Waugh are disciples of Rocky O'Callaghan of Rockford Wines, and have deliberately accumulated a series of old dryland, low-yielding Barossa vineyards, aiming to produce wines of unusual depth of flavour and character. They have handsomely succeeded in this aim. They also offer superior accommodation in the ancient but beautifully restored two-bedroom cottage 'Miriam's'; Michael Waugh is a highly skilled stonemason.

green valley vineyard NR

3137 Sebbes Road, Forest Grove, WA 6286 **region** Margaret River
ph (08) 9384 3131 **open** Weekends and public holidays 10–6 or by appointment
winemaker Clive Otto **prod.** 3000 **est.** 1980
prod. range ($16.50–27.50 CD) Chardonnay, Riesling, Müller Thurgau, Dolce (Chenin Blanc), Cabernet Sauvignon.
summary Owners Ed and Eleanore Green commenced the development of Green Valley Vineyard in 1980. It is still a part-time operation, with the wines made by contract, but production has grown steadily from the 7.7 hectares of vines, and the Cabernet Sauvignon has been a consistent medal winner. Exports to Singapore.

grevillea estate NR

Buckajo Road, Bega, NSW 2550 **region** Other Wineries of NSW
ph (02) 6492 3006 **fax** (02) 6492 5330 **open** 7 days 9–5
winemaker Nicola Collins **prod.** 4000 **est.** 1980
prod. range ($14–16 CD) Daisy Hill Rhine Riesling, Lunatic Hill Sauvignon Blanc, Unoaked Chardonnay, Gewurztraminer, Rougon, Grosse's Creek Merlot, Edmund Kirby Cabernet Sauvignon.
summary A tourist-oriented winery which successfully sells all of its surprisingly large production through cellar door and to local restaurants. The best of the current release wines is the 1996 Edmund Kirby Tribute Cabernet Sauvignon; this apart, all of the wines have very attractive labels.

grosset ★★★★★

King Street, Auburn, SA 5451 **region** Clare Valley
ph (08) 8849 2175 **fax** (08) 8849 2292 **open** Wed–Sun 10–5 from 1st week of September for approx 6 weeks
winemaker Jeffrey Grosset **prod.** 8000 **est.** 1981
prod. range ($23.95–48 R) Watervale Riesling, Polish Hill Riesling, Semillon Sauvignon Blanc, Piccadilly Chardonnay, Gaia (a Cabernet blend), Noble Riesling, Reserve Pinot Noir.
summary Jeffrey Grosset served part of his apprenticeship at the vast Lindeman Karadoc winery, moving from the largest to one of the smallest when he established Grosset Wines in its old stone winery. He now crafts the wines with the utmost care from grapes grown to the most exacting standards; all need a certain amount of time in bottle to achieve their ultimate potential, not the least the Rieslings and Gaia, among Australia's best examples of their kind. At a Riesling Summit held in Hamburg in the latter part of 1998, Grosset was voted Riesling Winemaker of the Year. Exports to UK, US, Japan, Belgium, Germany mean a continuous shortage of the wines in all markets.

Grosset Polish Hill Riesling

A finer, crisper and more elegant wine than the Watervale, with more lime and citrus fruit, albeit less generous. Since 1985 the Molloy Vineyard has been the major source, but as from 1994 estate plantings also contribute. Like the Watervale, made with neutral yeasts and without the use of enzymes. Always brilliant.

🍷🍷🍷🍷🍷 **1998** Light green-yellow; intense mineral, lime, spice and powdery aromas are followed by a similarly intensely-flavoured palate veering more towards lime and spice. A classically austere wine with a bone-dry finish. As good as ever. **rating:** 96

best drinking 2003–2013 **best vintages** '82, '86, '87, '90, '93, '94, '96, '97, '98 **drink with** Grilled South Australian whiting • $28

Grosset Watervale Riesling

Made from hand-picked grapes grown on a single vineyard established on red clay over limestone at an altitude of 450 metres. It is a richer, fuller style than the Polish Hill River wine, and tends to be slightly earlier maturing. All of the recent vintages have been made bone-dry, with deliberately neutral yeast influence.

🍷🍷🍷🍷🍷 **1998** Glowing yellow-green; as ever, a pure and intense celebration of Clare Riesling with lime, citrus and a hint of mineral on the bouquet. The palate is perfectly balanced, again showing pure varietal fruit character, with fine lime, citrus and herb flavours; perfectly balanced, with a lingering finish. **rating:** 97

best drinking 2001–2008 **best vintages** '81, '86, '90, '93, '94, '95, '96, '97, '98 **drink with** Thai soup • $23.95

Grosset Piccadilly Chardonnay

Since 1994 Jeffrey Grosset has sourced his Chardonnay from Piccadilly in the Adelaide Hills, and labelling it as such. In a far finer style than the preceding Clare Valley wines, and – one would imagine – a longer future. Forty per cent of the wine is taken through malolactic fermentation.

🍷🍷🍷🍷🍷 **1997** Medium yellow-green; the bouquet is already starting to develop some secondary characters with a range of peach, melon and fig fruit leading the way. The palate takes a while to open up, initially folded in, but ultimately yielding an attractive array of creamy/nutty characters. **rating:** 93

best drinking 2001–2007 **best vintages** '96, '97 **drink with** Gravlax • $36.95

Grosset Gaia

A blend of 85% Cabernet Sauvignon, 10% Cabernet Franc and 5% Merlot, typically made in amounts of less than 1000 cases. Shot to stardom with the initial vintage of 1990, and has not faltered since.

🍷🍷🍷🍷🍷 **1996** Strong, deep red-purple; the bouquet, as ever, is profound and complex with dark briary berry fruit. The palate is formidable, with plum, dark berry, briar and relatively assertive tannins. Does not quite seem to have the brilliant edge of purity of previous years, but it may also be simply a question of time. **rating:** 93

best drinking 2006–2016 **best vintages** '90, '91, '92, '94, '95 **drink with** Game pie • $48

grove hill NR

120 Old Norton Summit Road, Norton Summit, SA 5136 **region** Adelaide Hills
ph (08) 8390 1437 **fax** (08) 8390 1437 **open** Sunday 11–5
winemaker Roman Bratasiuk (Contract) **prod.** 500 **est.** 1978
prod. range ($18–30 ML) Riesling, Chardonnay, Marguerite Pinot Chardonnay.
summary Grove Hill is situated on the site of a heritage property established in 1846 with the original homestead and outbuildings and held by the same family since that time. The wines from the 3 hectares of vineyards are made in the full-frontal (and unpredictable) style one expects from Roman Bratasiuk. Exports to the US (fuelled by the *Wine Spectator* awarding the 1996 Chardonnay 91 points).

haan wines NR

Siegersdorf Road, Tanunda, SA 5352 **region** Barossa Valley
ph (08) 8562 2122 **fax** (08) 8562 3034 **open** Not
winemaker James Irvine (Contract) **prod.** 4000 **est.** 1993
prod. range ($18–40 ML) Merlot Prestige, Cabernet Rosé.
summary Hans and Fransien Haan established their business in 1993, when they acquired a 16-hectare vineyard near Tanunda. The primary focus is on Merlot, and in particular on the luxury Merlot Prestige, and they understandably chose James Irvine as their contract-winemaker.

There are no cellar-door sales; the wines are sold through distributors in the eastern States, and Australian Prestige Wines also acting as export distributor.

Haan Merlot Prestige

Contract/consultant-winemaker James Irvine was one of the first in Australia to realise the potential for varietal Merlot, and likewise to focus on the export, rather than the domestic, market. This is very much in the James Irvine Grand Merlot style, and was selected in the Top 100 in the 1999 Sydney International Wine Competition. It is estate-grown, and matured in a mix of new, once-used and twice-used French oak for 18 months. It is then held in bottle for a further six to 12 months before release.

▼▼▼▼▽ **1996** Medium to full red; the bouquet is ripe and clean, with sweet chocolate and blackcurrant aromas, the palate rich and supple, with sweet mid-palate chocolate, mulberry and blackcurrant flavours running through to cedary oak tannins on the finish. **rating:** 90

best drinking 2000–2008 **best vintages** '96 **drink with** Game pie • $20

habitat NR

Old Canobolas Road, Nashdale, NSW 2800 **region** Orange
ph (02) 6365 3294 **fax** (02) 6362 3257 **open** At Ibis Wines
winemaker Phil Stevenson (Contract) **prod.** 120 **est.** 1989
prod. range ($18–20 CD) Pinot Noir, Merlot.
summary The 2.5-hectare Habitat vineyard is situated on the northern slope of Mount Canobolas on deep-red basalt soil at an altitude of 1100 metres, making it one of the highest – if not the highest – vineyards in Australia. In prior vintages the grapes were sold to Charles Sturt University to make sparkling (and table) wines.

haig NR

Square Mile Road, Mount Gambier, SA 5290 **region** Mount Gambier
ph (08) 8725 5414 **fax** (08) 8725 0252 **open** 7 days 11–5
winemaker Katnook (Contract) **prod.** 500 **est.** 1982
prod. range Chardonnay, Late Harvest Chardonnay, Pinot Noir, Shiraz Pinot, Fortifieds.
summary The 4 hectares of estate vineyards are planted on the rich volcanic soils near the slopes of the famous Blue Lake of Mount Gambier. I have neither seen nor tasted the wines.

hainault ★★★

255 Walnut Road, Bickley, WA 6076 **region** Perth Hills
ph (08) 9328 6728 **fax** (08) 9328 6895 **open** Weekends 10–5
winemaker Celine Rousseau, Gary Dixon **prod.** 2300 **est.** 1980
prod. range ($14–21 CD) The Terroir Range of Gewurztraminer, Pinot Noir, Merlot, Shiraz; Barking Owl Sauvignon Blanc Semillon, Chardonnay, Fruity Muscat, Old Vine Red, Cabernets, Vineyard Port; and Talus Sparkling.
summary Under the energetic ownership of public affairs consultant and businessman Bill Mackey and wife Vicki, the changes have come thick and fast at Hainault. Plantings have increased to 11 hectares; Celine Rousseau, a highly qualified French-born and trained oenologist, has been installed as winemaker; and the Barking Owl range, attractively packaged, and sourced from Pemberton, the Bickley and Swan Valleys introduced to sit under the Hainault Terroir range. Limited distribution in Victoria and the ACT.

Barking Owl Old Vine Red

Produced from sections of the Bickley Valley plantings which are now 20 years old. It is an unspecified varietal mix.

🍷🍷🍷🍷 **1997** Medium red; the bouquet is light, with pleasant, soft, earthy fruit. The palate has similar sweet earthy fruit balanced by lingering tannins and subtle oak. **rating:** 85

best drinking 1999–2002 **best vintages** NA **drink with** Mushroom risotto • $15

Hainault Terroir Merlot

Sourced from Bickley Valley fruit, and a silver medal winner at the Rutherglen Wine Show.

🍷🍷🍷🍷 **1997** Medium red-purple; sweet vanilla oak dominates the bouquet and makes a major contribution to the palate. Overall it is a quite rich wine, with soft, persistent tannins; well made although not particularly varietal. **rating:** 85

best drinking 1999–2004 **best vintages** NA **drink with** Baby lamb • $21

halcyon daze ★★★

19 Uplands Road, Lilydale, Vic 3140 **region** Yarra Valley

ph (03) 9726 7111 **fax** (03) 9726 7111 **open** By appointment

winemaker Richard Rackley **prod.** 500 **est.** 1982

prod. range ($20–25 ML) Riesling, Chardonnay, Pinot Noir, Cabernets; also Sparkling.

summary One of the lower-profile wineries with a small, estate-grown production which in fact sells the major part of its output of grapes from its 6.5 hectares of vines to others. Immaculate viticulture ensures that the grapes have a strong market.

half mile creek ★★★☆

George Campbell Drive, Mudgee, NSW 2850 **region** Mudgee

ph (02) 6372 3880 **fax** (02) 6372 2977 **open** 7 days 10–4

winemaker Rob Guadagnini **prod.** 10 000 **est.** 1918

prod. range ($10 R) Hunter Valley Verdelho, Cowra Chardonnay, Mudgee Shiraz, Mudgee Cabernet Merlot.

summary Half Mile Creek was once the Augustine Vineyard, established by the Roth family in the nineteenth century, and was purchased by Dr Thomas Fiaschi (one of the great unsung heroes of the Australian wine industry) in 1917. It has had a chequered career ever since, but has a great history, and is still a substantial producer (and part of the Mildara Blass empire).

hamelin bay ★★★★

Five Ashes Vineyard, RMB 116 McDonald Road, Karridale, WA 6288 **region** Margaret River

ph (08) 9389 6020 **fax** (08) 9389 6020 **open** By appointment

winemaker Eddie Price, Greg Tilbrook (Contract) **prod.** 15 000 **est.** 1992

prod. range ($15.40–24.95 CD) Sauvignon Blanc, Semillon Sauvignon Blanc, Chardonnay, Merlot, Shiraz, Cabernet Sauvignon.

summary The 25-hectare Hamelin Bay vineyard, established by the Drake-Brockman family, has enjoyed outstanding success with its first wine releases from the 1996 and 1997 vintages. For the time being its wines are contract-made, but a winery with cellar-door sales facility is due to be opened in the year 2000. In the meantime, production has soared from 5000 to 15 000 cases.

Hamelin Bay Semillon Sauvignon Blanc

A blend of 50% Semillon and 50% Sauvignon Blanc, with 5% barrel-fermented in American oak, the remainder in stainless steel. 1997 was the first vintage, winning a gold, two silver and two bronze medals in wine shows in 1997, the 1998 good but not quite in the same class.

🍷🍷🍷🍷 **1998** Light to medium yellow-green; the bouquet is crisp, moderately intense with mineral and citrus aromas. An elegant style with mineral and tangy citrus flavours, finishing with a nice bite. **rating:** 86

best drinking 1999–2000 **best vintages** NA **drink with** Tempura • $19.95

Hamelin Bay Chardonnay

Between 25% and 50% of the wine is barrel-fermented in French oak, the remainder in stainless steel, lees contact and partial malolactic fermentation then follow. A light-bodied but well-constructed style.

🍷🍷🍷🍷 **1998** Light to medium green-yellow; the bouquet has attractive, light melon citrus fruit, fruit which drives the palate from start to finish, although subtle oak is there in support. Should develop well. **rating:** 87

best drinking 1999–2003 **best vintages** NA **drink with** Lemon chicken • $22.95

hamilton ★★★★

Main Road, Willunga, SA 5172 **region** McLaren Vale

ph (08) 8556 2288 **fax** (08) 8556 2868 **open** 7 days 10–5

winemaker Phillipa Treadwell **prod.** 45 000 **est.** 1837

prod. range ($12–45 R) Synergy Dry White, Natural Chardonnay, Chenin Semillon, Cabernet Merlot; Richard Hamilton Chenin Blanc, Muscat Blanc, McLaren Vale Shiraz, Hut Block Cabernet Sauvignon, Merlot; Hamilton Ewell Reserve Marion Vineyard Grenache Shiraz, Reserve Merlot, Old Vines Shiraz; Burtons Vineyard Old Bush Vine Grenache Shiraz is premium release; also Gumpers Block Shiraz, Ayliffe's Orchard Sauvignon Blanc, Almond Grove Chardonnay, The Hills Chardonnay and Pinot Noir.

summary The quality and character of the Richard Hamilton wines have grown in leaps and bounds over the past five years or so, no doubt due to the skills of winemaker Phillipa Treadwell, former winemaker Ralph Fowler and support from owner Dr Richard Hamilton. The wines are boldly styled, full of flavour and character.

Hamilton Ayliffe's Orchard Sauvignon Blanc

Imitation being the sincerest form of flattery, Richard Hamilton (now shortened to Hamilton Estate as a winery brand name) has copied the d'Arenberg penchant for garrulous label names and stories. Violet Ayliffe, for those who do not know, was Richard Hamilton's grandmother who married Frank Hamilton in 1895.

🍷🍷🍷🍷 **1998** Light to medium yellow-green; the bouquet is moderately intense with tropical peach and passionfruit aromas, the palate likewise positively flavoured with ripe, tropical pineapple and passionfruit. Particularly commendable at the price. **rating:** 85

best drinking 1999–2000 **best vintages** NA **drink with** Blue swimmer crab • $11

Hamilton Almond Grove Chardonnay

Almonds were an important crop in McLaren Vale from the early 1840s onwards, and in the early 1930s, shortly before the arrival of Richard Hamilton's father Burton in McLaren Vale, the Almond Grove Company was formed and plantings increased further. The connection between this interesting historical fact and Chardonnay is not immediately apparent. Incidentally, the wine was barrel-fermented in French oak casks (not new) but is fruit-driven.

🍷🍷🍷🍷 **1998** Light to medium yellow-green; the bouquet is of moderate intensity, quite smooth, with subtle oak, but not a great deal of aromaticity. A pleasant McLaren Vale-style on the palate, again with soft chardonnay fruit providing the major taste activity, and only minimal oak. **rating:** 84

best drinking 1999–2001 **best vintages** NA **drink with** Pasta • $13

Richard Hamilton Signature Chardonnay

As the name attests, sits at or near the top of the Hamilton white wines.

🍷🍷🍷🍷 **1997** Light green-yellow; the bouquet is fresh, smooth, relatively undeveloped with light melon fruit and subtle oak. The palate is a logical corollary, elegant and understated; the question is whether it will build character with time in bottle, or whether, in the final analysis, it lacks fruit intensity. **rating:** 88

best drinking 2000–2004 **best vintages** NA **drink with** Abalone • $18.50

Hamilton The Hills Chardonnay

One of two new wines from the Hamilton Group, sourced from the Stirling district in the Adelaide Hills. Barrel-fermented in new French oak barriques, and given extended time on yeast lees.

🍷🍷🍷🍷 **1998** Light yellow-green; tangy fruit leads the bouquet with light toasty, charry barrel-ferment oak in support. The palate once again has tangy melon fruit to the fore, exhibiting good length and style. Elegant and unforced. **rating:** 87

best drinking 1999–2003 **best vintages** NA **drink with** White-fleshed fish • $22

Hamilton The Hills Pinot Noir

Sourced from the same vineyard as The Hills Chardonnay, the vines being grown by the Sampson family. Hamilton says it is a 'Burgundian style made using traditional techniques'.

🍷🍷🍷🍷 **1998** Quite strong red-purple; there are quite intense gamey characters, partly deriving from the vineyard, partly from whole-bunch maceration. The palate has powerful plum cherry fruit but not so much texture. May develop interesting secondary characters in bottle. **rating:** 86

best drinking 1999–2004 **best vintages** NA **drink with** Kangaroo fillet • $22

Hamilton Gumper's Block Shiraz

James Gumprs (spelt without an 'e') was the first licensee (in 1851) of the building which these days houses McLaren Vale's best and best-known restaurant, The Salopian Inn. One assumes the 'e' has been put into the name and it has been singularised for fear of jibes about misspelling.

🍷🍷🍷🍷 **1997** Strong red-purple; the bouquet is slightly closed, but potent earthy shiraz varietal character is there. The palate has powerful, potent and intense fruit with spikes of slightly green mint, and lingering tannins. In an utterly idiosyncratic style which is evident in a number of the Hamilton old vine Shiraz offerings from McLaren Vale. **rating:** 85

best drinking 2002–2007 **best vintages** NA **drink with** Kebabs • $18.95

Hamilton McLaren Vale Shiraz

Introduced in 1996 to sit underneath the Hamilton Ewell Reserve Old Vines Shiraz.

🍷🍷🍷🍷🍸 **1997** Full red-purple; the bouquet is clean, with pristine black cherry fruit and subtle oak. The palate is particularly well balanced, with sweet cherry fruit, oak and tannins all playing their part. **rating:** 90

best drinking 2002–2007 **best vintages** NA **drink with** Devilled kidneys • $17.95

Hamilton Hut Block Cabernet Sauvignon

The original vines of the Hut Block were planted by Richard Hamilton's father, Burton Hamilton, in 1947.

🍷🍷🍷🍷 **1997** Medium to full red-purple; an interesting mix of ripe cassis, liquorice and a touch of leaf on the bouquet; the palate is fresh, with cassis fruit and persistent tannins; overall, seems slightly callow and underworked. **rating:** 85

best drinking 2001–2006 **best vintages** '86, '90, '91, '93 **drink with** Beef Provençale • $19

---

hamiltons bluff NR

Ryegates Lane, Canowindra, NSW 2804 **region** Cowra

ph (02) 6344 2079 **fax** (02) 6344 2165 **open** Weekends and holidays 10–4, Mon–Fri by appointment

winemaker Andrew Margan **prod.** 2000 **est.** 1995

prod. range ($13–18 CD) Cowra Chardonnay, Canowindra grossi Chardonnay, Chairman's Reserve Chardonnay.

summary Hamiltons Bluff is owned and operated by the Andrews family, which planted 45 hectares of vines in 1995. 1998 produced the first crop, and three different Chardonnays were contract-made by Andrew Margan. The Cowra Chardonnay and Canowindra grossi Chardonnay received medals at the 1998 Cowra Wine Show. Cellar-door sales opened in early 1999, heralding a new stage of development for the Cowra region.

hanging rock winery ★★★☆

The Jim Jim, Jim Road, Newham, Vic 3442 **region** Macedon
ph (03) 5427 0542 **fax** (03) 5427 0310 **open** 7 days 10–5
winemaker John Ellis **prod.** 15 000 **est.** 1982
prod. range ($10–44 CD) Macedon Cuvée V, Colemans Gully Riesling, The Jim Jim Sauvignon Blanc, Victoria Semillon Sauvignon Blanc, Victoria Chardonnay, Reserve Swan Hill Chardonnay, Late Harvest Riesling, Late Harvest Gewurztraminer, Central Highlands Pinot Noir, Victoria Cabernet Merlot, Gralaine Merlot, Victoria Shiraz, Picnic Red and White.
summary The Macedon area has proved very marginal in spots, and the Hanging Rock vineyards, with their lovely vista towards the Rock, are no exception. John Ellis has thus elected to source additional grapes from various parts of Victoria, to produce an interesting and diverse style of wines. The low-priced Picnic White and Picnic Red, with the striking label, have been particularly successful. Exports to the UK, US, Asia, Italy and Japan.

Hanging Rock Macedon Cuvée

A blend of Pinot Noir and Chardonnay from vintages falling between 1987 and 1995, the older vintages being used as Reserve wines. It spent 22 months on lees, and was disgorged July 1998.
🍷🍷🍷🍷 **NV** VI. Medium to full straw-yellow; a rich, concentrated and complex bouquet in a deliberate would-be-Bollinger style. The aldehyde characters are controlled, and despite the richness of the palate, the wine has a nice, dry finish. **rating:** 88
best drinking 1999–2002 **best vintages** NA **drink with** Oysters Kilpatrick • $44

Hanging Rock Winery Heathcote Shiraz

Without question the most distinguished wine in the Hanging Rock portfolio, made from dry-grown, low-yielding vines in what many regard as one of Australia's best Shiraz regions.
🍷🍷🍷🍷🍷 **1997** Youthful purple-red; there is vibrant, sweet dark cherry fruit and relatively subtle oak on the bouquet; a wine with very good structure thanks to fine tannins which run through sweet, dark cherry fruit on the palate, again with sensitive oak use. **rating:** 92
best drinking 2002–2012 **best vintages** '97 **drink with** Roast venison • $38

hankin estate NR

Johnsons Lane, Northwood via Seymour, Vic 3660 **region** Goulburn Valley
ph (03) 5792 2396 **fax** (03) 9353 2927 **open** Weekends 10–5
winemaker Dr Max Hankin **prod.** 600 **est.** 1975
prod. range ($7–22 CD) Semillon, Sauvignon Blanc, Premium Dry White, Rosé, Shiraz, Shiraz Cabernet Malbec, Cabernet Sauvignon.
summary Hankin Wines is now the principal retirement occupation of Dr Max Hankin, who has retired from full-time medical practice. He also has to contend with phylloxera, which decimated the original plantings, with the replanting process still underway.

hanns creek estate NR

Kentucky Road, Merricks North, Vic 3926 **region** Mornington Peninsula
ph (03) 5989 7266 **fax** (03) 5989 7500 **open** 7 days 11–5
winemaker Tony Aubrey-Slocock **prod.** 1500 **est.** 1987
prod. range ($18–25 CD) Chardonnay, Rosé, Pinot Noir, Cabernet Shiraz, Cabernet Sauvignon.
summary Denise and Tony Aubrey-Slocock have established a 3-hectare vineyard on the slopes of Merricks North. After an uncertain start, with contract-winemaking moving around, Kevin McCarthy took control, and wine-style steadied.

hanson ★★★

340 Old Healesville Road, Yarra Glen, Vic 3775 **region** Yarra Valley
ph (03) 9439 7425 **fax** (03) 9435 9183 **open** Not
winemaker Dr Ian Hanson **prod.** 800 **est.** 1983
prod. range ($23.55–25.55 R) Pinot Noir, Cabernets, Cabernet Franc, Cabernet Sauvignon.
summary Dental surgeon Ian Hanson planted his first vines in the late 1960s close to the junction of the Yarra and Plenty Rivers; in 1983 those plantings were extended (with 3000 vines), and in 1988 the Tarrahill property at Yarra Glen was established with 10 further acres. The wines all now bear the Hanson label.

happs ★★★☆

Commonage Road, Dunsborough, WA 6281 **region** Margaret River
ph (08) 9755 3300 **fax** (08) 9755 3846 **open** 7 days 10–5
winemaker Erl Happ, Frank Kittler **prod.** 14 000 **est.** 1978
prod. range ($13–28 CD) Dry table wines are Classic (Semillon Chardonnay), Marrime (Semillon Chenin Blanc), Chardonnay, Margaret River Red, Shiraz, Merlot and Cabernet Merlot; sweet table wines are Fuschia, Topaz and Late Picked Verdelho; fortifieds are Fortis (Vintage Port), 10 Year Fortis (Tawny), Garnet (from Muscat à Petit Grains), Pale Gold (White Port) and Old Bronze (Muscat). In 1994 a Preservative Free Red was also made.
summary Former schoolteacher turned potter and winemaker Erl Happ is an iconoclast and compulsive experimenter. Many of the styles he makes are very unconventional, the future likely to be even more so: the Karridale vineyard planted in 1994 has no less than 28 different varieties established. Merlot has been a winery specialty for a decade.

harcourt valley vineyards NR

Calder Highway, Harcourt, Vic 3453 **region** Bendigo
ph (03) 5474 2223 **fax** (03) 5474 2293 **open** 7 days 10–6
winemaker John Livingstone **prod.** 2000 **est.** 1976
prod. range ($12.50–25 CD) Chardonnay, Riesling, Barbara's Shiraz, Cabernet Sauvignon.
summary Traditional producer of rich, full-bodied red wines typical of the district, but sporadic (and largely outdated) tastings since ownership changed preclude evaluation. No recent tastings; however 1996 Barbara's Shiraz has won two gold and two silver medals during 1997, strongly suggesting the quality of this lovely wine has been maintained.

hardys ★★★★☆

Reynella Road, Reynella, SA 5161 **region** McLaren Vale
ph (08) 8392 2222 **fax** (08) 8392 2202 **open** 7 days 10–4.30
winemaker Peter Dawson, Stephen Pannell, Tom Newton, Ed Carr **prod.** NFP **est.** 1853
prod. range ($8–55 R) Starts with Old Castle Rhine Riesling, St Vincent Chablis; then McLaren Vale Hermitage, Classic Dry White; Nottage Hill Riesling, Chardonnay, Cabernet Sauvignon; then the generic Bird series, then Insignia Chardonnay Sauvignon Blanc, and Cabernet Sauvignon Shiraz, followed by Siegersdorf Chardonnay and Rhine Riesling. No Preservative Added Chardonnay and Cabernet Sauvignon are available for allergy sufferers. The Sir James range and the Bankside wines fill in the middle; at the very top Eileen Hardy Chardonnay, Shiraz and Thomas Hardy Cabernet Sauvignon. There is a full range of sparkling wines, superior quality brandies and ports including Australia's finest Vintage Port.
summary Since the 1992 merger of Thomas Hardy and the Berri Renmano group, the business has flourished, and the shareholders have profited greatly. The merged group has confounded expectations by aggressively, and very successfully, pushing the premium end of the business, making a number of acquisitions and investments across the length and breadth of Australia, all aimed at the upper end of the market. A high level of winemaking expertise and commitment has been an essential part of this success. It is basically for these wines that the winery rating is given.

Hardys Eileen Hardy Chardonnay

Prior to the release of the '94 vintage, one might have argued whether or not the wine has a sufficient track record or distinction to justify its rating as a classic, but is a more than usually elegant wine, and some of the early vintages have aged with far greater distinction than most Australian Chardonnays. However, the shift from Padthaway to a Yarra Valley base, coupled with refinements to the winemaking used, has resulted in a great wine.

▼▼▼▼▽ **1997** Medium yellow-green; the bouquet is rich and complex, although the oak was still to fully integrate when the wine was 18 months old. The palate, likewise, offers fine, long fruit flavour, again with the oak being a touch intrusive, and needing time. A blend of Adelaide Hills and Yarra Valley material. **rating:** 92

best drinking 2002–2007 **best vintages** '85, '87, '90, '91, '93, '94, '96 **drink with** Fresh Atlantic salmon • $32.95

Hardys Sir James Vintage

In response to Hardys ever-growing wine empire, this top-of-the-range sparkling wine is now sourced almost entirely from the Yarra Valley and Tasmania, and knowing sparkling winemaker Ed Carr's preference for these areas, is likely to remain so in the future. A blend of Pinot Noir, Chardonnay and a little Pinot Meunier, and which spends almost three years on yeast lees. Some strong winemaker inputs are evident, not the least being the apparent adoption of higher than normal fermentation temperatures with what may be unclarified juice. Whatever be the truth, a challenging style.

▼▼▼▼▽ **1994** Glowing yellow-green, the bouquet is smooth, with a nice touch of bready/creamy autolysis, followed by a flavoursome but elegant citrus and melon-flavoured palate. A fine, gently crisp finish balances the mid-palate flavour. **rating:** 93

best drinking 1999–2000 **best vintages** '94 **drink with** Hors d'oeuvres • $24

Hardys Tintara Shiraz

A new superbly packaged and presented premium range from Hardys, simultaneously replacing and upgrading the now discontinued Bankside range. Made from McLaren Vale shiraz, open-fermented and basket-pressed, it spends two years in a combination of French and American oak.

▼▼▼▼▼ **1996** Very dark yet bright purple-red; exceptionally concentrated and complex dark berry and chocolate fruit soaks up the considerable oak doubtless invested in the wine. The palate follows the bouquet, powerful, complex and concentrated, with tannins to spare. **rating:** 94

best drinking 2006–2016 **best vintages** NA **drink with** Beef in black bean sauce • $25

Hardys Tintara Grenache

A marvellously packaged wine, with the best super-deluxe bottle on the Australian market at the present time, but the wine is much more than a pretty face. It is made from unirrigated old vine McLaren Vale grenache, with a percentage of shiraz incorporated at the time of crushing (around 10%) and spends 12 months in French oak, or American, depending on whether you believe the back label on the wine or the separate information provided by the brand manager. All of this reflects some refinement of the first release in 1995, and in my view is the best Grenache in Australia, challenged only by the top d'Arenberg releases.

▼▼▼▼▼ **1997** Strong purple-red; there is a cascade of rich, ripe fruit aromas ranging through raspberry, plum, black cherry and spice, with the French oak playing a subtle support role. The wine is satin-smooth on entry to the mouth, with the fruit flavours representing a continuum of the bouquet. Fine tannins and subtle oak. **rating:** 94

best drinking 2000–2007 **best vintages** '95, '97 **drink with** Very rich meat dishes • $25

Hardys Vintage Port

Made from very ripe, low-yielding McLaren Vale shiraz, fortified using a very particular Brandy spirit made by Hardys. Just as with Portuguese Vintage Port, there is much debate about the relative input and importance of the base wine on the one hand and the fortifying spirit (and fortifying techniques) on the other. What is certain is that this is Australia's finest Vintage Port, the best vintages of which age superbly.

🍷🍷🍷🍷🍷 **1984** Dense purple-red; potent dark berry chocolate with ever so distinctive brandy spirit aromas followed by a palate of chocolate, blackberry and liquorice shot through with lingering spirit and finishing with appropriate tannins. A trophy and gold medal winner. **rating:** 93

best drinking 2000–2015 **best vintages** '45, '51, '54, '56, '71, '73, '75, '81, '82, '84 **drink with** Coffee • $22.90

hardys (padthaway) NR

Stonehaven Winery, Riddoch Highway, Padthaway, SA 5271 **region** Padthaway
ph (08) 8765 6140 **fax** (08) 8765 6137 **open** Not
winemaker Tom Newton, Duncan McGillivray, Robert Mann **prod.** NFP **est.** 1998
prod. range ($15.95 R) Unwooded Chardonnay, Cabernet Sauvignon.

summary The $18 million Stonehaven winery, the largest single new winery to be built in Australia in 20 years, was opened in March 1998. It has a capacity of 10 000 tonnes and will process all of the Hardy intake from the Limestone Coast Zone. A cellar-door and tasting facility will open in 1999.

Hardys Padthaway Unwooded Chardonnay

A new product introduced for the '95 vintage, with the inevitable new, avant garde label. More importantly, it is one of the better unwooded Chardonnays, and a powerful testament to the quality of Chardonnay from Padthaway.

🍷🍷🍷🍷 **1998** Medium yellow-green; very intense, fragrant passionfruit and grapefruit aromas are effectively reproduced on the palate, although it tails off slightly on the finish. Gold medal Cowra Wine Show 1998. **rating:** 86

best drinking 1999–2000 **best vintages** NA **drink with** Mussels • $15.95

harewood estate ★★★☆

Scotsdale Road, Denmark, WA 6333 **region** Great Southern
ph (08) 9840 9078 **fax** (08) 9840 9053 **open** By appointment
winemaker John Wade (Contract) **prod.** 400 **est.** 1988
prod. range ($30 R) Chardonnay, Pinot Noir.

summary Keith and Margie Graham have established a showpiece vineyard at Binalong. The majority of the grapes are sold to Howard Park and Domaine Chandon, but gradually increasing amounts of wine are being made under the Harewood Estate label. The wines have retail distribution in Perth, and are exported to the UK, but are otherwise only available by mail order.

Harewood Estate Chardonnay

Immaculate viticulture, with a mixture of high-tech trellis systems (Scott Henry, Smart Dyson and Geneva Double Curtain) and varied row orientation pay dividends in this lush vineyard setting. The wine is barrel-fermented and matured in French oak for ten months.

🍷🍷🍷🍷🍷 **1997** Medium to full yellow-green; the moderately intense bouquet features nectarine/clingstone peach fruit which is quite complex. The palate has been very well put together, with the many components perfectly balanced: fruit, oak and a nice hint of mlf cashew. **rating:** 90

best drinking 2000–2004 **best vintages** '97 **drink with** Seafood risotto • $30

hartzview wine centre ★★☆

RSD 1034 Off Cross Road, Gardners Bay, Tas 7112 **region** Southern Tasmania
ph (03) 6295 1623 **open** 7 days 9–5
winemaker Andrew Hood (Contract), Robert Patterson **prod.** NFP **est.** 1988
prod. range ($18 CD) Chardonnay, Pinot Noir; also a range of Pig and Whistle Hill fruit wines.
summary A combined wine centre offering wines from a number of local Huon Valley wineries and also newly erected and very comfortable accommodation for six people in a separate, self-contained house. Hartzview table wines (produced from 3 hectares of estate plantings) are much to be preferred to the self-produced Pig and Whistle Hill fruit wines.

haselgrove ★★★★

Foggo Road, McLaren Vale, SA 5171 **region** McLaren Vale
ph (08) 8323 8706 **fax** (08) 8323 8049 **open** Mon–Fri 9–5, weekends 10–5
winemaker Nick Haselgrove **prod.** 40 000 **est.** 1981
prod. range ($9.90–40 R) McLaren Vale Pictures Series Sauvignon Blanc, Chardonnay, Chenin Blanc, Frontignac, Grenache Shiraz, Cabernet Merlot Shiraz; Futures Shiraz; premium releases under 'H' Reserve label; Sparkling, Port; lesser priced varietals under Sovereign Series.
summary Haselgrove Wines became a wholly owned subsidiary of the publicly listed Australian Premium Wines Limited in mid-1997. Under Nick Haselgrove's direction, the premium red wines, and in particular the 'H' Reserve range, have gone from strength to strength, the 1996 'H' Reserve Shiraz winning multiple trophies, including Best Wine of Show, at the 1998 Sydney International Wine Competition (otherwise known as the 'Top 100').

Haselgrove Grenache

The wine is made from old vine grenache, some of the vines being over 70 years old, together with a small percentage of shiraz from Wrattonbully, the new region near Coonawarra and which has had such a difficult time agreeing on its name (Koppamurra was the preferred alternative). The wine is matured in older oak to keep the focus firmly on the fruit.

🍷🍷🍷🍷🍷 **1998** Medium to full red-purple; the bouquet offers ripe, dense mulberry/blackberry fruit which is not the least bit jammy. The palate has virtually the same flavours in the mulberry/blackberry spectrum, and there is an abundance of tannin on the finish. Said to be an early-drinking style, but I would prefer to regard it as a food style. **rating:** 90

best drinking 1999–2002 **best vintages** NA **drink with** Steak and kidney pie • $16

hastwell & lightfoot NR

Foggo Road, McLaren Vale, SA 5171 **region** McLaren Vale
ph (08) 8323 8692 **fax** (08) 8323 8098 **open** Not
winemaker Nick Haselgrove (Contract) **prod.** 700 **est.** 1990
prod. range ($10–12 ML) Chardonnay, Cabernet.
summary Hastwell & Lightfoot is an offshoot of a rather larger grape growing business with the majority of the grapes from the 15 hectares of vineyard being sold to others; the vineyard was planted in 1988 and the first grapes produced in 1990. More recently Nick Haselgrove has made small quantities of wine for Hastwell & Lightfoot, the 1996 Cabernet winning a gold medal at the 1998 Australian Small Winemakers Show, no mean feat.

hawley vineyard NR

Hawley Beach, Hawley, Tas 7307 **region** Northern Tasmania
ph (03) 6428 6221 **fax** (03) 6428 6844 **open** 7 days
winemaker Andrew Pirie (Contract) **prod.** 1000 **est.** 1988
prod. range ($18–25 R) Rubicon Chardonnay, Unwooded Chardonnay, Rubicon Pinot Noir.

summary Hawley Vineyard overlooks Hawley Beach, and thence northeast to Bass Strait. It is established on an historic 200-hectare farming property, with Hawley House offering dining and accommodation in a grand style. There are no other vineyards in what is a unique winegrowing region, and few hoteliers-cum-viticulturists as flamboyant as owner Simon Hawley. Limited distribution in Sydney.

Hawley Vineyard Unwooded Chardonnay

Together with most other wine-writers and wine judges in Australia, I shrink at the very mention of unwooded Chardonnay, but every now and then one comes along which suggests the style has legitimacy. This is one such wine. Re-entered in the 1999 Tasmanian Wines Show and has blossomed in the intervening 12 months since first tasted.

🍷🍷🍷🍷🍸 **1997** Medium yellow-green; very attractive sweet melon fruit on the bouquet swells into even more attractive melon and fig flavours on a long palate. **rating:** 92

best drinking 1999–2002 **best vintages** NA **drink with** White-fleshed fish • $18

hay shed hill ★★★☆

RSM 398 Harmans Mill Road, Willyabrup, WA 6280 **region** Margaret River

ph (08) 9755 6234 **fax** (08) 9755 6305 **open** 7 days 10.30–5

winemaker Peter Stanlake **prod.** 12 000 **est.** 1987

prod. range ($13–40 CD) Sauvignon Blanc, Semillon, Chardonnay, Pitchfork Pink (Rose), Group 20 Cabernet Sauvignon (light, unwooded), Cabernet Franc, Cabernet Sauvignon, Pinot Noir.

summary A landmark on the Margaret River scene, with a striking new 200-tonne winery, a carefully devised business plan by the Morrison family, energetic marketing, and innovative label design. Wine quality has been a touch inconsistent, but the 'sold-out' sign so often displayed speaks for itself. At their best, tangy and incisive. The wines are distributed through retail outlets in Perth, Melbourne and Sydney, and are exported to the UK and the US.

Hay Shed Hill Cabernet Franc

One of barely a dozen varietal Cabernet Francs made in Australia, where, overall its performance has not encouraged makers to produce a varietal wine. It must be said that much the same view is held in California and Bordeaux.

🍷🍷🍷🍷 **1996** Medium red-purple, holding brightness and hue. The bouquet has pronounced cedar and leaf aromas, moving into berry and raspberry flavours on the palate, which has some length and persistence on the finish. A more than creditable effort. **rating:** 85

best drinking 1999–2004 **best vintages** NA **drink with** Veal chops • $40

Hay Shed Hill Cabernet Sauvignon

Made from estate-grown grapes produced from vines which are now almost 20 years old, and indicative of the quality to be expected from the winery under this label. The wine spends 18 months in French oak, and is typically produced in a lighter, elegant style.

🍷🍷🍷🍷🍸 **1996** Medium to full red-purple; there is quite sweet and powerful dark berry fruit on the bouquet, followed logically by a powerful palate, with dark cassis varietal fruit, plenty of concentration, and fine, persistent tannins on the finish. **rating:** 90

best drinking 2001–2011 **best vintages** '92, '94, '96 **drink with** Beef casserole • $30

hayward's whitehead creek NR

Lot 18A Hall Lane, Seymour, Vic 3660 **region** Goulburn Valley

ph (03) 5792 3050 **open** Mon–Sat 9–6, Sun 10–6

winemaker Sid Hayward, David Hayward **prod.** 600 **est.** 1975

prod. range ($7.50–12.50 CD) Riesling, Shiraz, Cabernet Sauvignon.

summary The last tastings, some years ago, were of somewhat rustic but full-flavoured wines. The ultra low-yielding vineyards (4.5 hectares) are used solely to produce the Whitehead Creek wines.

heathcote winery ★★★☆

183-185 High Street, Heathcote, Vic 3523 **region** Bendigo
ph (03) 5433 2595 **fax** (03) 5433 3081 **open** 7 days 10–6 summer, rest of year 11–5
winemaker Mark Kelly **prod.** 5000 **est.** 1978
prod. range ($19–33.50 CD) Heathcote Winery range of Viognier, Chardonnay, Cane Cut, Mail Coach Shiraz; 7th Horse range of Chardonnay and Shiraz (early-drinking styles predominantly sourced from Padthaway).
summary The Heathcote Winery is now back in business with a vengeance. The wines are being produced predominantly from the 12 hectares of estate vineyard, and some from local growers under long-term contracts.

Heathcote Winery Viognier

A wine as interesting for its future as for the future, for this is one of the earlier plantings of viognier in the Bendigo region, the climate of which should be well suited to this relatively rare grape. The wine is made by fermentation and short-term maturation on lees in French oak barriques, but the oak influence is not overdone.

🍷🍷🍷🍷 **1997** Medium yellow-green; the bouquet is clean, with faint tropical fruit characters, but not a great deal of distinctive varietal character. The palate has good mouthfeel and weight, although the fruit flavours are amorphous, and it is inherently unlikely any stronger definition will emerge with time. **rating:** 86

best drinking 1999–2003 **best vintages** NA **drink with** Stir-fried chicken • $33

Heathcote Winery 7th Horse Shiraz

Part of the rebirth of Heathcote Winery, and which had a dream debut winning a trophy at the 1998 Royal Adelaide Wine Show. It is a blend of Bendigo and Padthaway Shiraz.

🍷🍷🍷🍷½ **1997** Brightly coloured; the bouquet is intense, with a mix of spice, earth and berry characters; strong berry, mint and earth fruit is supported by well-handled oak on the palate. **rating:** 92

best drinking 2000–2007 **best vintages** NA **drink with** Kebabs • $14.80

Heathcote Winery Mail Coach Shiraz

Typically has a small percentage of Viognier added; spends 21 months in barriques, predominantly French-coopered American oak, 25% of which is new.

🍷🍷🍷🍷 **1997** Full red-purple; a massive wine with dark berry and smoky oak aromas. The palate shows similar massive concentration (a drought year) with strong tannins and some spice. **rating:** 87

best drinking 2002–2012 **best vintages** NA **drink with** Barbecued meat • $22.50

heathfield ridge wines NR

Cnr Caves Road and Penola Highway, Naracoorte, SA 5271 **region** Wrattonbully
ph (08) 8762 4133 **fax** (08) 8762 0141 **open** Opening in October 1999
winemaker Pat Tocaciu, Neil Doddridge **prod.** 20 000 **est.** 1998
prod. range Riesling, Sauvignon Blanc, Chardonnay, Shiraz, Merlot and Cabernet Sauvignon.
summary Heathfield Ridge Wines is the major winery in the Naracoorte region. Opened in time for the 1998 vintage, its major function was a contract crush facility, but it offers full winemaking facilities for others and will also release wines under the Heathfield Ridge label. The cellar-door sales area will be opened after October 1999. Exports to the UK.

heemskerk ★★★★

40 Baxters Road, Pipers River, Tas 7252 **region** Northern Tasmania
ph (03) 6382 7122 **fax** (03) 6382 7231 **open** Nov–Apr 10–5
winemaker Fiona West **prod.** 16 000 **est.** 1974
prod. range ($18–31 R) Riesling, Semillon Sauvignon Blanc, Chardonnay, Pinot Gris, Botrytis Riesling, Pinot Noir, Cabernet Merlot.
summary In February 1998 the ownership of Heemskerk changed once again, and it is now part of the Pipers Brook Group. It will continue as an important cellar-door site, with the probability of vineyard-linked Chardonnay and Pinot Noir being made. With the sale of the prestige Jansz Méthode Champenoise brand to Yalumba in April 1998, the sparkling wine focus of the group will be firmly placed on the Pipers Brook Pirie Méthode Champenoise released mid-1998.

heggies vineyard ★★★★

Heggies Range Road, Eden Valley, SA 5235 **region** Eden Valley
ph (08) 8565 3203 **fax** (08) 8565 3380 **open** Not
winemaker Simon Adams **prod.** 12 500 **est.** 1971
prod. range ($15–25 R) Riesling, Viognier, Chardonnay, Botrytis Riesling, Cabernets.
summary Heggies was the second of the high-altitude (570 metres) vineyards established by S Smith & Sons (Yalumba), with plantings on the 120-hectare former grazing property commencing in 1973. The once simple view of Heggies as a better white than red wine producer has become more complicated, with the pendulum swinging backwards and forwards according to vintage. Exports to all major markets.

Heggies Riesling

The wine upon which the Heggies fame, with its ever so distinctive label, was founded. As with Pewsey Vale, hasn't always lived up to its early reputation, but wines such as the '95, '96 and '98 restore the faith.

1998 Light straw-green; quite fragrant mineral and lime aromas are replicated on the powerful palate, which has length, balance and intensity. Shy now, but will blossom with age. **rating:** 90

best drinking 2000–2008 **best vintages** '95, '96, '98 **drink with** Seafood salad • $15.95

Heggies Viognier

While viognier has been grown on the Mornington Peninsula, at Elgee Park, for many years, Yalumba (with a separate importation of clonal material) has been the first to commercialise viognier in Australia. The early vintages produced fairly thin, ordinary wines, but no one could accuse the wines made since 1993, '95 or '97 of lacking character.

1997 Glowing yellow-green; interesting honeycomb and mead aromas which are rich and striking. The palate exhibits more of the same, with honeycomb and pastille fruit flavours, all of which are in the mainstream of viognier varietal character. **rating:** 90

best drinking 1999–2000 **best vintages** '93, '95, '97 **drink with** Pork fillet • $21.95

Heggies Chardonnay

Made, needless to say, from 100% estate-grown grapes, barrel-fermented and always given time in bottle before it is brought onto the market. The result is a complex, rich wine, often sweet and toasty.

1996 Medium to full yellow-green; strong toasty/charry barrel-ferment oak aromas dominate the bouquet, and the same strong oak influence carries on through the wine. Needlessly assertive, for there is some pleasant fruit there. **rating:** 80

best drinking 1998–1999 **best vintages** '86, '91, '93 **drink with** Veal, turkey • $21.95

Heggies Botrytis Riesling (375 ml)

Some exceptionally good wines have appeared under this label over the years; the Yalumba Group as a whole were one of the pioneers of botrytis riesling. Not made every vintage, but when it does appear, is of high quality.

🍷🍷🍷🍷🍷 **1997** Medium to full yellow-green; abundant apricot and mandarin fruit aromas lead into a flavourful and high-quality wine, with similar fruit tastes balanced by good acidity. **rating:** 90

best drinking 1998–2001 **best vintages** '92, '94, '97 **drink with** Poached fruit or fruit tart • $14.95

Heggies Merlot

First made its mark with the '93 vintage, repeating the performance in even more splendid fashion in 1994, and following up with a solid wine in 1995.

🍷🍷🍷🍷 **1995** Medium red-purple; the bouquet is soft, with an attractive mix of earthy berry fruit and gentle oak. The palate is essentially savoury, with good flavour and texture, finishing with lingering, fine tannins. **rating:** 86

best drinking 2000–2005 **best vintages** '93, '95 **drink with** Rack of veal • $22.95

helm ★★★☆

Butt's Road, Murrumbateman, NSW 2582 **region** Canberra District

ph (02) 6227 5536 **fax** (02) 6227 5953 **open** Thur–Mon 10–5

winemaker Ken Helm **prod.** 2500 **est.** 1973

prod. range ($12–25 CD) Rhine Riesling Classic Dry, Cowra Riesling Classic Dry, Traminer Riesling, Chardonnay (Non Oaked), Cabernet Merlot, Cabernet Sauvignon, Helm (Cabernet blend).

summary Ken Helm is well known as one of the more stormy petrels of the wine industry, and is an energetic promoter of his wines and of the Canberra district generally. His wines have been consistent bronze medal winners, with silvers and the occasional gold dotted here and there, such as the gold medal to the 1997 Cabernet Merlot at the 1999 Sydney Royal Wine Show. The wines have limited retail distribution in New South Wales and Victoria.

henke ★★☆

175 Henke Lane, Yarck, Vic 3719 **region** Central Victorian High Country

ph (03) 5797 6277 **fax** (03) 5797 6277 **open** By appointment

winemaker Tim Miller, Caroline Miller **prod.** 250 **est.** 1974

prod. range ($18 CD) Shiraz, Shiraz Cabernet.

summary Produces tiny quantities of deep-coloured full-flavoured, minty red wines known only to a chosen few. Typically, a range of back vintages up to five years of age are available at cellar door; as at 1999, both the '90 and '91 vintages of each of the two wines were available.

henley park wines NR

149 Swan Street, West Swan, WA 6055 **region** Swan District

ph (08) 9296 4328 **fax** (08) 9296 1313 **open** Tues–Sun 10–5

winemaker Claus Petersen, Lisbet Petersen **prod.** 3500 **est.** 1935

prod. range ($9.95–15.95 CD) Semillon, Chenin Blanc, Classic White, Muscat Gordo Blanco (late picked), Autumn Harvest (Sauternes style), Mousse Rosé Brut (Méthode Champenoise), Shiraz, Cabernet Sauvignon, Shiraz Cabernet Merlot, Tawny Port.

summary Henley Park, like so many Swan Valley wineries, was founded by a Yugoslav family, but is now jointly owned by Danish and Malaysian interests, a multicultural mix if ever there was one. Majority owner and winemaker Claus Petersen arrived in 1986, and had his moment of glory in 1990 when Henley Park was the Most Successful Exhibitor at the Mount Barker Wine Show. Much of the production is sold through cellar door (and exported to Denmark and Malaysia).

henschke ★★★★★

Henschke Road, Keyneton, SA 5353 **region** Eden Valley
ph (08) 8564 8223 **fax** (08) 8564 8294 **open** Mon–Fri 9–4.30, Sat 9–12, public holidays 10–3
winemaker Stephen Henschke **prod.** 40 000 **est.** 1868
prod. range ($14–170 R) From the Henschke Eden Valley sources, Sauvignon Blanc, Chardonnay, Chenin Blanc, Dry White Frontignac, Joseph Hill Gewurztraminer, Louis Semillon, Tilly's Vineyard, Julius Riesling, Keyneton Estate, Mount Edelstone, Cyril Henschke Cabernet Sauvignon, Hill of Grace. From the Lenswood Vineyard in the Adelaide Hills, Green's Hill Riesling, Croft Chardonnay, Giles Pinot Noir, Abbott's Prayer Cabernet Merlot. Also Barossa Ranges Eden Valley Chardonnay.
summary Unchallenged as the best medium-sized red wine producer in Australia, and has gone from strength to strength over the past 14 years or so under the guidance of Stephen and Prue Henschke. The red wines fully capitalise on the very old, low-yielding, high-quality vines, and are superbly made with sensitive but positive use of new small oak; the same skills are evident in the white winemaking. Hill of Grace is second only to Penfolds Grange as Australia's red wine icon. Exports to UK, US, Canada, New Zealand, Switzerland, France, Germany, Denmark, Holland, Austria and Japan.

Henschke Green's Hill Riesling

In the year 1982 the Henschke family acquired a 14-hectare apple orchard at Lenswood, high in the southern end of the Adelaide Hills at an altitude of 550 metres. The first significant vintage was 1989, but production overall is limited. The Green's Hill Riesling, which overlooks the apple orchards operated by the Green family since 1893, is a marvellous example of cool-climate Riesling, intense yet generous.

🍷🍷🍷🍷 **1998** Light green-yellow; the bouquet is quite tight, with apple blossom and citrus aromas; the lively palate offers more apple blossom and citrussy fruit, finishing with good acidity. **rating:** 87

best drinking 1999–2004 **best vintages** '94, '95 **drink with** Smoked trout pâté • $19.50

Henschke Croft Chardonnay

Another distinguished wine from the cool-climate Lenswood vineyards of Henschke, sitting alongside the Giles Pinot Noir.

🍷🍷🍷🍷 **1998** Light green-yellow; a fine, elegant melon-accented bouquet attesting to the cool climate is followed by a long but relatively restrained and understated palate. The oak has been sensitively used, and the wine will unquestionably develop well. **rating:** 86

best drinking 2000–2004 **best vintages** NA **drink with** Gravlax • $33

Henschke Hill Of Grace

Made entirely from 100-year-old shiraz vines on the Hill of Grace Vineyard, planted in the late 1860s by a Henschke ancestor, Nicholas Stanitzki. Is second only to Penfolds Grange, which it rivals in terms of quality, scarcity and (almost) price. The wine has never been entered in wine shows, nor will it ever be: it has its own standards. Its soaring retail price bears testament to the public esteem in which the wine is rightly held.

🍷🍷🍷🍷🍷 **1994** Medium purple-red; vanillin oak is immediately evident on the bouquet, but without subduing the smooth supple fruit, showing none of the 'sauvage' characters of the '93. A lovely wine on the palate, although the oak is very evident, perhaps supporting, perhaps distracting, from the fruit. A mild, dry summer with a slow start to vintage was followed by two weeks of hot weather which ripened everything at once. Stephen Henschke described it as the year of the 'black ferments', and regards it as having the potential of being one of the best ever vintages. **rating:** 94

best drinking 2004–2019 **best vintages** '59, '61, '62, '66, '78, '82, '85, '86, '88, '90, '91, '93 '96
drink with Rich casserole dishes • $170

Henschke Mount Edelstone

Made entirely from shiraz grown on the Mount Edelstone Vineyard, planted in the 1920s and acquired by Henschke in 1974, although the wine was first made (and labelled as such) in 1952. A wine of tremendous character and quality. The price is rapidly being pulled upwards by the Hill of Grace, but, here too, the wine quality justifies the price.

🍷🍷🍷🍷🍷 **1996** Medium to full red-purple; complex aromas run through the bouquet with nuances of pepper, liquorice and mint to the core of red cherry fruit. The palate is smooth, with cherry, mint and a touch of pepper all showing; the oak is sweet but not forceful, the tannins soft and supple. Top gold 1998 National Wine Show. **rating:** 95

best drinking 2001–2011 **best vintages** '52, '56, '61, '66, '67, '78, '82, '86, '88, '90, '92, '93, '94, '95, '96 **drink with** Beef bourguignon • $50

Henschke Keyneton Estate Shiraz Cabernet Malbec

Not, as the label might half suggest, a single vineyard or estate wine in the classic sense of that term, but rather a blend of 70% Shiraz, 25% Cabernet Sauvignon and 5% Malbec grown in the Eden and Barossa Valleys. It is made in the traditional Henschke fashion in open fermenters, and matured in new and used American and French oak for 12 months. It may not be the greatest of the Henschke red wines, but frequently offers the best value for money.

🍷🍷🍷🍷 **1996** Medium purple-red; the complex bouquet opens up with vanilla and cedar aromas, then moves into berry and chocolate, and it is the fruit which carries the palate; here mint, red berry and a touch of leaf all marry with subtle oak and fine tannins. **rating:** 93

best drinking 2001–2011 **best vintages** '82, '84, '86, '88, '90, '92, '93, '94, '96 **drink with** Veal chops • $26.60

henty brook estate NR

Box 49, Dardanup, WA 6236 **region** Geographe

ph (08) 9728 1459 **fax** (08) 9728 1459 **open** Not

winemaker James Pennington (Contract) **prod.** NA **est.** 1994

prod. range Semillon, Sauvignon Blanc, Shiraz.

summary One hectare each of shiraz and sauvignon blanc, and half a hectare of semillon were planted in the spring of 1994, and are still coming into bearing. James Pennington is the contract-winemaker; the first releases will follow over the next few years.

heritage farm wines NR

RMB 1005 Murray Valley Highway, Cobram, Vic 3655 **region** Goulburn Valley

ph (03) 5872 2376 **open** 7 days 9–5

winemaker Kevin Tyrrell **prod.** 5000 **est.** 1987

prod. range ($5–11 CD) Riesling, Traminer Riesling, Chardonnay are varietal releases; there are a considerable number of generic releases and fortified wines on sale at cellar door.

summary Heritage Farm claims to be the only vineyard and orchard in Australia still using horsepower, with Clydesdales used for most of the general farm work. The winery and cellar-door area also boasts a large range of restored horse-drawn farm machinery and a bottle collection. All of the wines are sold by mailing list and cellar door.

heritage wines ★★★☆

106a Seppeltsfield Road, Marananga, SA 5355 **region** Barossa Valley

ph (08) 8562 2880 **fax** (08) 8562 2692 **open** 7 days 11–5

winemaker Stephen Hoff **prod.** 6000 **est.** 1984

prod. range ($11.50–19 CD) Riesling, Semillon, Chardonnay, Shiraz, Cabernet Franc, Cabernet Malbec, Cabernet Sauvignon, Rosscos Shiraz.

summary A little-known winery which deserves a far wider audience, for Stephen Hoff is apt to produce some startlingly good wines. At various times the Chardonnay, Riesling (from old Clare Valley vines) and Rosscos Shiraz (now the flag-bearer) have all excelled, at other times not

heritage wines of stanthorpe ★★☆

New England Highway, Cottonvale, Qld 4375 **region** Granite Belt
ph (07) 4685 2197 **fax** (07) 4685 2112 **open** 7 days 9–5
winemaker Jim Barnes **prod.** 6000 **est.** 1992
prod. range ($10.50–18.50 CD) Semillon Chardonnay, Sauvignon Blanc, Classic White (Chardonnay Semillon Sauvignon Blanc), Chardonnay, Simply Red, Club Red, Shiraz, Fortified and flavoured wines.
summary A tourist-oriented venture, as are many of the Stanthorpe region wineries, established in a painstakingly restored cool-storage shed. The estate plantings comprise chardonnay (2 hectares), merlot (2 hectares), shiraz (1 hectare) and cabernet sauvignon (1 hectare).

hermes morrison wines NR

253 Swan Ponds Road, Woodstock, NSW 2793 **region** Cowra
ph (02) 6345 0153 **fax** (02) 6345 0153 **open** 7 days 10–5 summer, weekends and public holidays, winter by appointment
winemaker Jill Lindsay (Contract) **prod.** 750 **est.** 1990
prod. range ($9–15 CD) Riesling, Semillon, Sauvignon Blanc, Chardonnay, Pinot Noir, Shiraz Cabernet.
summary The Morrison family established their Hermes Pol Dorset Stud in 1972, which continues, but has now been joined by Hermes Morrison wines. The cellar door has been established by the side of a large lake fed by cold, clear water welling up from subterranean caves, and a 10-minute walk takes you to the summit of Mount Palatine, one of the highest peaks in the shire and with a spectacular view of the Canobolas Mountains 80 kilometres away.

herons rise vineyard NR

Saddle Road, Kettering, Tas 7155 **region** Southern Tasmania
ph (03) 6267 4339 **fax** (03) 6267 4245 **open** By appointment
winemaker Andrew Hood **prod.** 300 **est.** 1984
prod. range ($12.50–20 CD) Dry White, Pinot Noir.
summary Sue and Gerry White run a small stone country guesthouse in the D'Entrecasteaux Channel area, and basically sell the wines produced from the surrounding 1 hectare of vineyard to those staying at the guesthouse. The postal address for bookings is PO Box 271, Kettering, Tas 7155.

hesperos wines NR

Corner West Calgardup Road and Bussell Highway, Witchcliffe, WA 6285 **region** Margaret River
ph (08) 9757 3302 **fax** (08) 9757 3302 **open** By appointment
winemaker Jürg Muggli **prod.** 1000 **est.** 1993
prod. range ($18.50–30 CD) Sauvignon Blanc, Shiraz.
summary Hesperos is the venture of Jürg Muggli and Sandra Hancock. It supplies Jürg Muggli's winemaking skills to Xanadu, where Muggli has been resident winemaker for many years. It also has a 30-hectare property near Witchcliffe between Cape Mentelle and Devil's Lair, with the potential of 15 hectares of vineyard, with planting due to commence in the winter of 1999. In the meantime the Hesperos wines are made from purchased grapes; the first wine produced under the Hesperos label was a 1993 Shiraz, followed by a 1995 Sauvignon Blanc. Shiraz and Sauvignon Blanc have been produced in each successive vintage.

hewitson ★★★★

4 Mann St, Hyde Park, SA 5061 **region** Warehouse
ph (08) 8271 5755 **fax** (08) 8271 5570 **open** Not
winemaker Dean Hewitson **prod.** 4000 **est.** 1996
prod. range ($17–37 R) La Source Eden Valley Riesling, L'Oizeau Shiraz, Barossa Valley Mourvedre, Barossa Valley Shiraz, Barossa Valley Grenache.
summary Dean Hewitson was a Petaluma winemaker for ten years, and during that time, managed to do three vintages in France, one in Oregon as well as undertaking his Masters at UC Davis, California. It is hardly surprising that the Hewitson wines are immaculately made from a technical viewpoint. However, he has also managed to source 30-year-old riesling from the Eden Valley and 70-year-old shiraz from McLaren Vale for his first two releases, following on with a 1998 Barossa Valley Mourvedre produced from 145-year-old vines at Rowland Flat, and a Barossa Valley Shiraz and Grenache, coming from 60-year-old vines at Tanunda. The vineyards are now under long-term contracts to Ian Hewitson.

Hewitson La Source Eden Valley Riesling

Produced from grapes grown on 30-year-old vines from the Eden Valley, and does full justice to the quality of the source.
🍷🍷🍷🍷🍷 **1998** Medium yellow-green, with above-average depth for such a young wine. The bouquet, too, is quite forward and full, with lime tropical fruit cut by more chalky/mineral characters. The palate is well-balanced, with smooth, gently tropical lime fruit. Appealing mouthfeel. **rating:** 90

best drinking 1999–2003 **best vintages** NA **drink with** Summer salads • $18

Hewitson L'Oizeau Shiraz

Produced from dry-grown, 70-year-old vines in McLaren Vale. It takes its name from the brigantine Josephine L'Oizeau which was shipwrecked on the Fleurieu Peninsula in 1856, taking its cargo of wine, spirits and tobacco down with it. Why anyone would wish to be reminded of that, I don't know, but I am not a brand manager.
🍷🍷🍷🍷 **1996** Medium to full red-purple; ripe, concentrated plum, prune and earth fruit on the bouquet is repeated on the palate, with touches of mint. Overall, very ripe, and seemingly higher than the 13.5 degrees alcohol. To be hypercritical, a fraction extractive. **rating:** 89

best drinking 2001–2006 **best vintages** NA **drink with** Hearty red meat dishes • $37

hickinbotham NR

Nepean Highway (near Wallaces Road), Dromana, Vic 3936 **region** Mornington Peninsula
ph (03) 5981 0355 **fax** (03) 5981 0355 **open** 7 days
winemaker Andrew Hickinbotham **prod.** 2000 **est.** 1981
prod. range ($14–26 CD) King Valley Riesling, Chardonnay, Classic White, Taminga, Sparkling (Strawberry Kiss and Futures), Pinot Noir, Merlot, Shiraz, Cabernets.
summary After a peripatetic period, and a hiatus in winemaking, Hickinbotham established a permanent vineyard and winery base at Dromana. It now makes only Mornington Peninsula wines, drawing in part on 10 hectares of estate vineyards, and in part on contract-grown fruit. The wines are principally sold through cellar door and mail order. 1998 The Taminga (84 points) is an interesting wine, unique to Hickinbotham and an Australian-bred variety created by the CSIRO. Grapey, spicy, floral aromas and flavours are the key.

hidden creek NR

Eukey Road, Ballandean, Qld 4382 **region** Granite Belt
ph (07) 4684 1383 **fax** (07) 4684 1383 **open** Saturday 10–4, Sunday 8.30–3 or by appointment
winemaker Andrew Vasiljuk (Contract) **prod.** 750 **est.** 1998
prod. range ($13–15 CD) Classic Dry White, Shiraz.

summary A beautifully-located vineyard and winery on a ridge overlooking the Ballandean township and the Severn River valley, separated from Girraween National Park by Doctors Creek. The granite boulder-strewn hills mean that the 70-hectare property will only provide a little over 6 hectares of vineyard, in turn divided into six different blocks. The two wines tasted ('97 Dry White and '95 Shiraz) were of modest but acceptable quality.

highbank NR

Main Penola–Naracoorte Road, Coonawarra, SA 5263 **region** Coonawarra
ph (08) 8736 3311 **open** By appointment
winemaker Dennis Vice, Trevor Mast **prod.** 4000 **est.** 1986
prod. range ($20–25 CD) Chardonnay, Basket Pressed Cabernet Blend, Basket Pressed Cabernet Sauvignon.
summary Mount Gambier lecturer in viticulture Dennis Vice makes a tiny quantity of smooth, melon-accented Chardonnay and stylish Coonawarra Cabernet blend of good quality which are sold through local restaurants and cellar door, with limited Melbourne distribution. The wines have retail distribution through the eastern States, and are exported to the US, Taiwan, Japan, Malaysia and Hong Kong.

Highbank Basket Pressed Cabernet Sauvignon

Produced from a small vineyard in Coonawarra which is now fully mature. As the name suggests, produced in small quantities and pressed in a basket press at the end of fermentation.
ŸŸŸŸ **1997** Medium to full red-purple; the bouquet is quite complex and generous with dark berry fruit and earth aromas backed by plenty of oak. A firmly structured wine showing pure varietal cabernet flavours in an overall dry style. **rating:** 87
best drinking 2002–2007 **best vintages** NA **drink with** Braised ox cheek • $35

highland heritage estate ★★★☆

Mitchell Highway, Orange, NSW 2800 **region** Orange
ph (02) 6361 3612 **fax** (02) 6361 3613 **open** Mon–Fri 9–3, weekends 9–5
winemaker John Hordern, Rex D'Aquino **prod.** 3500 **est.** 1984
prod. range ($10–30 CD) Under the Mount Canobolas label: Chardonnay, Sauvignon Blanc, Pinot Noir; Gosling Creek Chardonnay; and the newly released Wellwood Estate label.
summary The estate plantings have increased from 4 hectares to over 15 hectares, with new plantings in 1995 and 1997 to come into full production by 2001. The tasting facility is unusual: a converted railway carriage overlooking the vineyard.

highway wines NR

612 Great Northern Highway, Herne Hill, WA 6056 **region** Swan District
ph (08) 9296 4354 **open** Mon–Sat 8.30–6
winemaker Tony Bakranich **prod.** 4000 **est.** 1954
prod. range ($5–14.50 CD) Exclusively Fortified wines, of which 20 are available, including six different styles of Sherry, six Muscats, three Ports, and so forth.
summary A survivor of another era, when literally dozens of such wineries plied their business in the Swan Valley. It still enjoys a strong local trade, selling much of its wine in fill-your-own-containers, and 2-litre flagons, with lesser quantities sold by the bottle.

hill of hope NR

Cobcroft Road, Broke, NSW 2330 **region** Lower Hunter Valley
ph (02) 6579 1161 **fax** (02) 6579 1373 **open** 7 days 10–5
winemaker Peter Howland **prod.** 30 000 **est.** 1996

prod. range ($15–20 CD) Verdelho, Semillon, Unwooded Chardonnay, Chardonnay, Botrytised Semillon, Blanc de Noir, Cabernet Merlot.
summary The Hill of Hope is the reborn Saxonvale Winery purchased by Michael Hope, a Broke/Fordwich grape grower, with 90 hectares of vineyards. Ever-increasing amounts of wine are made and sold under the Hill of Hope label.

hill-smith estate ★★★★

c/o Yalumba Winery, Angaston, SA 5353 **region** Eden Valley
ph (08) 8561 3200 **fax** (08) 8561 3393 **open** Not
winemaker Robert Hill-Smith **prod.** 6000 **est.** 1973
prod. range ($16 R) Sauvignon Blanc, Chardonnay.
summary Part of the Yalumba stable, drawing upon its own estate plantings comprising 23 hectares of chardonnay and sauvignon blanc. Over the years, has produced some excellent wines, but quality does seem to vary significantly with vintage, and the winery rating is a compromise between the best and the least. Exports to all major markets.

hillstowe ★★★★

104 Main Road, Hahndorf, SA 5245 **region** Adelaide Hills
ph (08) 8388 1400 **fax** (08) 8388 1411 **open** 7 days 10–5
winemaker Chris Laurie **prod.** 12 000 **est.** 1980
prod. range ($14.95–31.95 R) A range of vineyard and varietal-designated wines of ascending price and quality, being Buxton McLaren Vale Sauvignon Blanc, Buxton Sauvignon Blanc, Chardonnay, Buxton Shiraz, Buxton Cabernet Merlot; and at the top end Adelaide Hills Udy's Mill Chardonnay, Adelaide Hills Carey Gully Pinot Noir, Yarra Valley Hoddles Pinot Noir.
summary Founded by renowned viticulturist David Paxton and Chris Laurie, but now owned by the latter and his family. Its principal vineyard, Udy's Mill, at Lenswood has 17 hectares planted, supplementing McLaren Vale grapes coming from the Buxton Vineyard. The wines are exported to the UK, Canada, US, Europe and Asia.

Hillstowe Udy's Mill Chardonnay

Produced from 3.2 hectares of chardonnay grown in the Carey Gully subdistrict of the Adelaide Hills, adjacent to McLaren Vale. The climate is distinctly cooler, the grapes harvested later, and clearly reflect the climate. The '97 was selected in the 1999 Top 100 at the Sydney International Wine Competition.

ΥΥΥΥ **1997** Glowing yellow-green; a big wine with peachy fruit and some charry oak on the bouquet is followed by a solid, fruit-driven palate in which the oak is less evident, and which is a little less complex than the bouquet suggests. **rating:** 87

best drinking 2000–2004 **best vintages** '90, '92, '93, '95, '96 **drink with** Trout or salmon mousse • $29

Hillstowe Mary's Hundred Shiraz

Mary's Hundred has a history to put the most voluble of the d'Arenberg stories to shame. Mary Laurie has been fairly described as South Australia's first woman winemaker, taking over winemaking responsibility from her husband Buxton Laurie when he died in 1876, continuing that role until 1892 when she suffered a stroke. This wine, produced from very old vines (some more than 100 years old) at the Laurie family's McLaren Vale vineyard, is named in her honour.

ΥΥΥΥΥ **1997** Strong red-purple; a rich wine with berry and chocolate fruit matched with sweet vanilla oak on the bouquet and the palate alike, the latter being rounded off with nice tannins on the finish. **rating:** 90

best drinking 2002–2012 **best vintages** '96, '97 **drink with** Rack of lamb • $31.95

---

hjt vineyards NR

Keenan Road, Glenrowan, Vic 3675 **region** Glenrowan

ph (03) 5766 2252 **fax** (03) 5765 3260 **open** Fri, Sat, 10–5 and Sunday during school holidays

winemaker Wendy Tinson **prod.** 1200 **est.** 1979

prod. range ($11.50–17.50 CD) A varietal range, with occasional use of bin numbers denoting winemaking approaches, Bin 4 being more delicate, Bin 19 fuller-bodied. Wines include Riesling Bins 4 and 19, Chardonnay, Chenin Blanc Bin 19, Late Picked Riesling, Pinot Noir, Cabernet Pinot, Shiraz, Cabernet Sauvignon, Merlot, Tawny Port.

summary Founded by the late Harry Tinson after he left Baileys after a long and illustrious stewardship, and now run by his daughter Wendy Tinson, with tiny production all sold from the cellar door.

hoffmann's NR

Ingoldby Road, McLaren Flat, SA 5171 **region** McLaren Vale

ph (08) 8383 0232 **fax** (08) 8383 0232 **open** 7 days 10–5

winemaker Nick Holmes (Consultant) **prod.** 500 **est.** 1996

prod. range ($13–16 CD) Chardonnay, Shiraz, Cabernet Sauvignon.

summary Peter and Anthea Hoffmann have been growing grapes at their property in Ingoldby Road since 1978, and Peter Hoffmann has worked at various wineries in McLaren Vale since 1979. Both he and Anthea have undertaken courses at the Regency TAFE Institute in Adelaide, and (in Peter Hoffmann's words) 'in 1996 we decided that we knew a little about winemaking and opened a small cellar door'. Only small quantities of wine are made for the Hoffmann's label; the balance of the production is sold to Mildara Blass (for the Ingoldby label).

hollick ★★★★

Riddoch Highway, Coonawarra, SA 5263 **region** Coonawarra

ph (08) 8737 2318 **fax** (08) 8737 2952 **open** 7 days 9–5

winemaker Ian Hollick, Matt Pellew **prod.** 25 000 **est.** 1983

prod. range ($12–50 R) A very disciplined array of products with Sauvignon Blanc Semillon, Unoaked Chardonnay and Shiraz Cabernet Malbec at the lower end of the price range; Reserve Chardonnay and Cabernet Sauvignon Merlot in the middle, along with Sparkling Merlot; Ravenswood, the deluxe Cabernet Sauvignon at the top end. Also small range of limited cellar-door releases.

summary Hollick has, if it were possible, added to the reputation of Coonawarra since it released its first wines in the mid-1980s. Winner of many trophies (including the most famous of all, the Jimmy Watson), its wines are well crafted and competitively priced, although sometimes a little on the light side. National distribution in all States; exports to the UK, US, New Zealand, Canada, Hong Kong, Singapore and Switzerland.

hollyclare NR

Lot 6 Milbrodale Road, Broke, NSW 2330 **region** Lower Hunter Valley

ph (02) 6579 1193 **fax** (02) 6579 1269 **open** Weekends 10–5

winemaker Tamburlaine (Contract) **prod.** 1500 **est.** 1987

prod. range ($14.50–16.50 CD) Chardonnay, Unoaked Chardonnay, Chardonnay Semillon, Semillon, Shiraz.

summary John Holdsworth established the Hollyclare Vineyard (now totalling 3 hectares of chardonnay, semillon and shiraz) ten years ago, but the Hollyclare label is a relatively new one on the market. While the wines are made under contract at Tamburlaine, Hollyclare has its own dedicated wine tanks and all of the wines are estate-grown.

holm oak ★★★☆

RSD 256 Rowella, West Tamar, Tas 7270 **region** Northern Tasmania
ph (03) 6394 7577 **fax** (03) 6394 7350 **open** 7 days 12–5
winemaker Nick Butler **prod.** 1800 **est.** 1983
prod. range ($18.50 R) Pinot Noir Chardonnay (still table wine), Pinot Noir, Cabernet Sauvignon.
summary The Butler family produces tremendously rich and strongly flavoured red wines from the vineyard situated on the banks of the Tamar River, and which takes its name from the grove of oak trees planted around the turn of the century and originally intended for the making of tennis racquets. Together with Marion's Vineyard, it suggests that this section of the Tamar Valley may even be too warm for pinot noir in some vintages (not '97); certainly it is best suited to cabernet sauvignon and chardonnay.

Holm Oak Pinot Noir

Produced from estate-grown grapes grown on the banks of the Tamar River. The '97 shows excellent varietal character, and by far the best from Holm Oak to date.

🍷🍷🍷🍷🍷 **1997** Medium to full red-purple; dark briary/plummy fruit aromas with subtle oak. There is abundant dark plum fruit on the palate with a nice touch of forest floor; good weight and balance. **rating:** 92

best drinking 1998–2001 **best vintages** NA **drink with** Squab • $25

home hill NR

73 Nairn Street, Ranelagh, Tas 7109 **region** Southern Tasmania
ph (03) 6228 0128 **fax** (03) 6264 1028 **open** Not
winemaker Michael Vishacki **prod.** 1600 **est.** 1994
prod. range ($16–20 ML) Chardonnay, Dry White, Pinot Noir.
summary Terry and Rosemary Bennett planted their first half hectare of vines in 1994 on gentle slopes in the beautiful Huon Valley. The plantings were quickly extended to 3 hectares, with another hectare being planted in 1999. The 3 hectares are in production, providing significant quantities of wine in 1998, the first commercial release. Both the Chardonnay (85 points) and the Pinot Noir (86 points) are full of promise.

Home Hill Chardonnay

The first commercial vintage of Home Hill Chardonnay was 1998, with a yield of 4 tonnes per hectare from 1993 plantings. Contract-made at Panorama.

🍷🍷🍷🍷 **1998** Medium to full yellow-green; a very rich, exotic bouquet with ripe fruit and extremely strong barrel-ferment oak characters. The palate will polarise opinion, some reacting adversely to the very strong oak, others accepting its Burgundian overtones. My points are something of a compromise between the views. **rating:** 85

best drinking 1999–2002 **best vintages** NA **drink with** Veal goulash • $18

Home Hill Pinot Noir

Like the Chardonnay, the '98 was the first commercial vintage for Home Hill, and also like the Chardonnay, from very small crops, with only 400 cases made. Once again like the Chardonnay, the oak is very strong.

🍷🍷🍷🍷 **1998** Light to medium red; there is a mix of sappy cherry pinot noir varietal fruit underlying very strong oak on the bouquet. The palate provides more of the same, very savoury, and with considerable (oaky) length. **rating:** 86

best drinking 1999–2002 **best vintages** NA **drink with** Smoked quail • $18

honeytree estate NR

16 Gillards Road, Pokolbin, NSW 2321 **region** Lower Hunter Valley
ph (02) 4998 7693 **fax** (02) 4998 7693 **open** Fri–Mon 10–5
winemaker Garry Reid **prod.** 3600 **est.** 1970
prod. range ($10–22 CD) Semillon, Traminer, Semillon Chardonnay, Clairette, Shiraz, Cabernet Sauvignon.
summary The Honeytree Estate vineyard was first planted in 1970, and for a period of time wines were produced under the Honeytree Estate label. It then disappeared, but has since been revived. Its 10 hectares of vines are of shiraz, cabernet sauvignon, semillon and a little clairette, known in the Hunter Valley as blanquette, and a variety which has been in existence there for well over a century. Jancis Robinson comments that the wine 'tends to be very high in alcohol, a little low in acid and to oxidise dangerously fast' but in a sign of the times, the first Honeytree Clairette sold out so quickly (in four weeks) that 2.2 hectares of vineyard has been grafted over to additional clairette.

Honeytree Estate Semillon

A classic Hunter Semillon, in large measure reflecting the maturity of the vines, now 30 years of age.

🍷🍷🍷🍷 **1998** Light yellow-green; the bouquet is light and clean, with herbal/grassy aromas. The palate has grapefruit and herb flavours running through to a long, lingering and dry finish.

rating: 85

best drinking 2000–2005 **best vintages** NA **drink with** Thai chicken • $19

hoppers hill vineyards NR

Googodery Road, Cumnock, NSW 2867 **region** Other Wineries of NSW
ph (02) 6367 7270 **open** Weekends 11–5
winemaker Robert Gilmore **prod.** NFP **est.** 1990
prod. range ($10–12 CD) Chardonnay, Sauvignon Blanc, Dry White, Cabernet Franc Merlot, Cabernet Sauvignon.
summary The Gilmores planted their vineyard in 1980, using organic growing methods and using no preservatives or filtration in the winery which was established in 1990. Not surprisingly, the wines cannot be judged or assessed against normal standards, but may have appeal in a niche market.

horseshoe vineyard NR

Horseshoe Road, Horseshoe Valley via Denman, NSW 2328 **region** Upper Hunter Valley
ph (02) 6547 3528 **open** Weekends 9–5
winemaker John Hordern **prod.** NFP **est.** 1986
prod. range ($13–18 CD) Classic Hunter Semillon, Chardonnay Semillon, Chardonnay, Pinot Noir.
summary Seems to have fallen by the wayside after its wonderful start in 1986, with rich, full-flavoured, barrel-fermented Semillons and Chardonnays. The '87 Semillon was exhibited in the Museum Class at the 1996 Hunter Valley Wine Show and was still drinking beautifully, winning a strong silver medal. Younger vintages do not have the same magic.

hotham valley estate ★★★★

South Wandering Road, Wandering, WA 6308 **region** Other Wineries of WA
ph (08) 9884 1525 **fax** (08) 9884 1079 **open** By appointment
winemaker James Pennington, Garry Baldwin (Consultant) **prod.** 12 000 **est.** 1987
prod. range ($12–18 CD) Semillon, Semillon Sauvignon Blanc, Chenin Blanc, Classic Dry White, Chardonnay, Cabernet Merlot, Classic Red.

summary An impressive newcomer to the scene, situated in a region of its own making, 120 kilometres southeast of Perth. It has a continental climate with cold winters and hot summer days, but cool nights, tempered by the altitude of 350 metres. Some exceptionally good wines have been made by former science teacher and now Charles Sturt University graduate James Pennington, on whose family property Hotham Valley Estate is established, albeit by way of a subdivision with outside investment. A state-of-the-art winery was built in 1993. James Pennington has also created considerable interest with a patented development of oak treatment using sandalwood. Exports to the UK.

houghton ★★★★★

Dale Road, Middle Swan, WA 6056 **region** Swan District
ph (08) 9274 5100 **fax** (08) 9274 5372 **open** 7 days 10–5
winemaker Larry Cherubino **prod.** 300 000 **est.** 1836
prod. range ($9.90–65 R) At the bottom end come the Wildflower Ridge range; then White Burgundy, Chablis, Frankland River Riesling, Semillon Sauvignon Blanc, Cabernet Sauvignon are the basic wines; next the Crofters range introduced in 1996; at the top end Gold Reserve Verdelho, Chardonnay and Cabernet Sauvignon with occasional special releases of aged Show Reserve wines including White Burgundy, Riesling and Verdelho, always of high quality. Finally, the super-premium Jack Mann (a Cabernet blend) was introduced in July 1997.
summary The five-star rating may seem extreme, but is very deliberate and is in no small measure justified by Houghton White Burgundy, one of Australia's largest selling white wines, almost entirely consumed within days of purchase, but which is superlative with seven or so years bottle age. To borrow a phrase of the late Jack Mann, 'There are no bad wines here'. Former winemaker Paul Lapsley's abundant winemaking skills brought the Jack Mann red and the Houghton Reserve Shiraz to the very forefront of Australian wine quality.

Houghton Show Reserve Riesling

One of the quartet of Show Reserve white wines first released in 1997, and the best of an outstanding bunch. Produced from grapes grown on Houghton's Frankland River vineyard and prior vintages (notably the '91) have had outstanding success in wine shows.

🍷🍷🍷🍷🍷 **1995** Medium yellow-green; there is obvious bottle development showing on the bouquet with lime and touches of kerosene and toast. The palate has lots of flavour and style with toasty lime characters; it has developed more quickly than the '91 did. **rating:** 90

best drinking 1999–2003 **best vintages** '91, '95 **drink with** Avocado salad • $24.95

Houghton Crofters Verdelho Chardonnay

Blends of this kind, while strange by eastern States' standards, are common in Western Australia, even with a premium wine such as this. Moreover, one cannot deny that it works well.

🍷🍷🍷🍷🍷 **1996** Medium to full yellow-green; very complex barrel-ferment characters surround the tangy fruit of the bouquet; on the palate the impact is reversed, with citrus, fig and nectarine fruit foremost, and then spicy barrel-ferment characters coming through on the finish. **rating:** 90

best drinking 1999–2003 **best vintages** '96 **drink with** Weiner Schnitzel • $16

Houghton White Burgundy

A wine with an extraordinary pedigree over its 60-year history, made from a blend of Chenin Blanc, Muscadelle, Semillon, Verdelho and Chardonnay primarily grown in the Swan Valley and at Gingin. Released and almost entirely consumed within 12 months of vintage, it invariably matures wonderfully well in bottle over a six- to eight-year period, leading to tiny releases of the Show Reserve Wines which accumulate innumerable gold medals and trophies.

1998 Light to medium yellow-green; a lively, attractive and aromatic bouquet with hints of passionfruit is replicated on the fresh, relatively light palate which will build weight and complexity over the next five years. **rating:** 87

best drinking 1999–2004 **best vintages** '83, '87, '89, '91, '93, '95 **drink with** Fish, chicken, veal • $9.90

Houghton Show Reserve White Burgundy

A re-release of 500 cases of the best bottling of the standard Houghton White Burgundy of 1991, the wine is a blend of Chenin Blanc, Muscadelle, Semillon, Verdelho and Chardonnay.

1991 Medium yellow-green; the bouquet is remarkably elegant and fine, still tight and youthful given the age of the wine. The palate is supple, with smooth, faintly tropical fruit, and has excellent length. Top gold 1998 National Wine Show. **rating:** 94

best drinking 1998–2001 **best vintages** '91 **drink with** West Australian marron • $25

Houghton Show Reserve Shiraz

Produced from 30-year-old, low-yielding shiraz grown in some of Houghton's best vineyards in the Frankland River region. The wine spends 18 months in new American oak barriques before further maturation on cork for an additional 18 months. The 1995 vintage won the prestigious Tucker Seabrook Trophy for the Champion Wine Exhibited at all National Wine Shows in the 12 months to February 1998, as well as winning trophies at the Sydney Wine Show including Best Wine of Show. Prior vintages have also been prolific trophy and gold medal winners.

1995 The colour is so dense that it stains the glass as the wine is swirled, signalling a wine which is massive and powerful in every respect. The aromas and flavours are somewhere between a Shiraz and a Cabernet Sauvignon in character, having many of the dark berry/blackcurrant and mulberry flavours of Cabernet but the soft structure of Shiraz. As is the case with so many show reserve wines, the oak contribution is substantial, and the wine really needs to be tamed with ten years bottle age. **rating:** 95

best drinking 2005–2020 **best vintages** '95 **drink with** Aged beef • $65

Houghton Jack Mann

Released in July 1997 in honour of the late, great Jack Mann. It is a blend of Cabernet Sauvignon, Malbec and Shiraz primarily sourced from Houghton's Frankland River vineyard, with a lesser component from the Mount Barker Omrah vineyard. It spent two years in a mix of Nevers and Allier French oak; I suspect the cooper was Dargaud & Jaegle. An absolutely superb wine in every respect.

1995 Dark, dense red; an immensely powerful and deep bouquet with earth, briary, chocolate and vanilla oak is followed by a very rich and wonderfully sweet palate, luscious yet not jammy, with the fruit and oak totally welded together. **rating:** 95

best drinking 2002–2017 **best vintages** '94, '95 **drink with** Fillet of lamb • $62

Houghton Crofters Cabernet Merlot

A blend of Cabernet Sauvignon and Merlot drawn from the Margaret and Frankland River regions, but with no percentages specified in either instance. The wine is matured for 16 months in a mix of new and one-year-old French oak barriques, and the inaugural release (from the 1994 vintage) won three gold medals, the '95 following in its footsteps with a gold medal at the 1997 Sydney Wine Show; the '97 is another excellent wine.

1997 Medium to full red-purple; the bouquet is very rich and concentrated with abundant ripe, dark berry fruit, the palate no less powerful with dark plum, briar and powerful tannins. Happily, the oak has been sensitively handled. **rating:** 91

best drinking 2002–2007 **best vintages** '94, '95, '97 **drink with** Kangaroo • $22

Houghton Red Stripe Cabernet Sauvignon

A cheaper version of the very good Gold Reserve, and sourced from what is described as the cool southern regions of Western Australia, no doubt meaning Margaret River, Great Southern and Pemberton. Very often outstanding value.

ΨΨΨΨ **1996** Medium purple-red; the moderately intense bouquet offers nicely ripened fruit and soft oak. The palate has good feel and balance, with cassis/berry fruit supported by good acidity and gentle tannin. **rating:** 87

best drinking 1999–2004 **best vintages** '90, '92, '94 **drink with** Rack of lamb • $11

howard park ★★★★★

Lot 377 Scotsdale Road, Denmark, WA 6333 **region** Great Southern
ph (08) 9848 2345 **fax** (08) 9848 2064 **open** 7 days 10–4
winemaker John Wade, Michael Kerrigan **prod.** 70 000 **est.** 1986
prod. range ($17–66 R) Madfish Premium Dry White and Red provide low-priced volume; limited quantities of Howard Park Riesling, Chardonnay and Cabernet Merlot.
summary John Wade, one of the most talented winemakers in Western Australia, is poised on the edge of a new venture; he has joined forces with Jeff and Amy Burch to build a winery near Denmark (Mount Shadforth Drive) which will make the Howard Park and Madfish Bay wines, but also act as contract-winemaker to a dozen Great Southern vignerons. It will be a singularly important centre of winemaking for the region. The crystal pure Riesling is one of the three best in Australia, the Cabernet Merlot likewise in vintages such as 1994. Production has soared since the establishment of the new winery, with exports to the UK, US and Singapore.

Howard Park Riesling

First made in 1986, and in my view the greatest Riesling made in the Great Southern region, itself home of many of Australia's finest examples of the style. It ages superbly, the '86 still with years in front of it.

ΨΨΨΨΨ **1998** Light to medium green-yellow; a beautifully intense wine with lime/citrus/tropical aromas, and a delicate, crisp palate in which passionfruit, lime and mineral flavours intermingle. Quite flawless. **rating:** 95

best drinking 2002–2008 **best vintages** '86, '87, '88, '91, '93, '94, '95, '96, '97, '98 **drink with** Fresh asparagus, Asian seafood • $22

Howard Park Madfish Premium Dry White

The strikingly labelled and named Madfish Bay was designed by Maxine Fumagalli, a Noongar artist who lives in Denmark, Western Australia. The name itself comes from a favourite fishing spot 15 kilometres west of Denmark. The white is an unwooded blend of 85% Chardonnay, 15% Semillon and 5% Sauvignon Blanc from the Great Southern region.

ΨΨΨΨ **1998** Quite strong green-yellow; the bouquet is smooth, moderately intense with nectarine, melon and citrus fruit; the nicely balanced palate offers an abundance of the flavours promised by the bouquet. A far better than average unwooded style. **rating:** 88

best drinking 1999–2001 **best vintages** NA **drink with** Chicken • $17

Howard Park Chardonnay

John Wade long resisted the temptation to make a Chardonnay, arguing that he was not happy to do so until he was assured of grapes of the highest quality. In 1993 he realised that a component of Madfish Bay met his requirements, and the wine was made from a blend of 50% Chardonnay grown in the Denmark area, and 50% at Pemberton. In subsequent vintages the sources have varied according to the quality of the available material.

🍷🍷🍷🍷 **1998** Medium yellow-green; the bouquet is clean, but surprisingly light and not (yet) complex. The palate is delicate, indeed almost painfully shy and reserved, but does have a core of citrus/nectarine fruit, subtle oak and a long finish. **rating:** 89

best drinking 2001–2006 **best vintages** '93, '94, '95, '96, '97 **drink with** Pan-fried veal • $25

Howard Park Cabernet Sauvignon Merlot

Like the Riesling, first made in 1986. The regional and varietal mix has changed a little over the years, with the Cabernet component ranging from between 70% and 85% and the wine being labelled Cabernet Merlot. Always a great wine, it hit new heights in 1994, placing it in the top half-dozen red wines in the country, slipping in '95.

🍷🍷🍷🍷🍷 **1996** Medium red, with some softening in the purple hue. The bouquet is fragrant and complex, with that typical spicy/cedary oak of Howard Park; the palate has admirable structure with cedary, foresty flavours, soft, fine tannins, and sweetness deriving partly from the fruit and partly from the Troncais oak. A welcome return to form. **rating:** 93

best drinking 2001–2011 **best vintages** '86, '88, '89, '90, '92, '93, '94, '96 **drink with** Lamb fillets, mature cheddar • $66

howards way vineyard NR

Cobcroft Road, Broke, NSW 2330 **region** Lower Hunter Valley

ph (02) 4998 1336 **fax** (02) 4938 3775 **open** Not

winemaker Andrew Margan (Contract) **prod.** 3500 **est.** NA

prod. range ($15–20 ML) Semillon, Pinot Noir, Shiraz.

summary Yet another of the dozens of new vineyards and labels that have appeared in the Hunter Valley in the latter part of the 1990s. Eight hectares of shiraz, 3 of pinot noir and 2 of semillon provide a substantial base, and it is expected that retail distribution will commence in 1999.

howarth's pycnantha hill NR

Benbournie Road, Clare, SA 5453 **region** Clare Valley

ph (08) 8842 2137 **fax** (08) 8842 2137 **open** Not

winemaker Jim Howarth **prod.** 600 **est.** 1997

prod. range ($9–10.75 ML) Riesling, Chardonnay, Shiraz, Cabernet Sauvignon.

summary The Howarth family has progressively established 2 hectares of vineyard since 1987, making its first commercial vintage ten years later in 1997. Acacia pycnantha is the botanic name for the golden wattle which grows wild over the hills of the Howarth farm, and they say it was 'a natural choice to name our vineyards Pycnantha Hill'. I am not too sure that marketing gurus would agree, but there we go. The 1997 Chardonnay won a silver medal at the Australian Small Winemakers Show at Stanthorpe in 1998, which certainly suggests that there are no problems with the quality of the wines, for this is a tough show at which to succeed. The tiny production (which will peak at around 1300 cases) is sold by mail order or wholesale on a first-come, first-served basis.

hugh hamilton ★★★

McMurtrie Road, McLaren Vale, SA 5171 **region** McLaren Vale

ph (08) 8323 8689 **fax** (08) 8323 9488 **open** Mon–Fri 10–5, weekends and public holidays 11–5

winemaker Hugh Hamilton **prod.** 8500 **est.** 1992

prod. range ($14.50–19.50 R) Chenin Blanc, Unwooded Chardonnay, Shiraz, Merlot, Cabernet Sauvignon, Sparkling Shiraz.

summary Hugh Hamilton is a member of the famous Hamilton winemaking family, there being an intensely (and well-known) competitive spirit existing between those various members – notably between Richard and Hugh – which can only be good for the consumer. The '98 Unwooded Chardonnay is pleasant, the '96 Cabernet Sauvignon powerful and austere. A new cellar-door facility was planned for opening in late 1999.

Hugh Hamilton Chardonnay

A wine which catches the attention because, unlike the majority of unwooded Chardonnays, powerful, high-quality fruit has been used to make it.

🍷🍷🍷🍷 **1998** Medium yellow-green; the bouquet is lifted and tangy; the curious thing is that both the bouquet and palate seem to have underlying hints of oak. Perhaps a small percentage was indeed barrel-fermented. Whatever, ripe fruit comes through on the palate, investing it with generous flavour and length. **rating:** 87

best drinking 1999–2000 **best vintages** NA **drink with** Pasta • $14.50

hugo ★★★

Elliott Road, McLaren Flat, SA 5171 **region** McLaren Vale
ph (08) 8383 0098 **fax** (08) 8383 0446 **open** Sun–Fri 10.30–5, Sat 12–5
winemaker John Hugo **prod.** 9000 **est.** 1982
prod. range ($14.95–19.50 CD) Sauvignon Blanc, Chardonnay, Unwooded Chardonnay, Shiraz, Cabernet Sauvignon, Port.

summary A winery which came from relative obscurity to prominence in the late 1980s with some lovely ripe, sweet reds which, while strongly American oak influenced, were quite outstanding. Subsequent red releases have continued in the same style, albeit slightly less exciting. There are 32 hectares of estate plantings, with part of the grape production sold to others. The wines are exported to the US and Hong Kong/China.

Hugo Shiraz

The 12 hectares of estate vineyards drawn upon for the label are situated in the foothills of the Mount Lofty Ranges, and were established in 1950 by Colin Hugo. The vineyard is not irrigated, and the fruit flavours are intense.

🍷🍷🍷🍷 **1997** Medium red-purple; the clean bouquet, with an attractive array of varietal fruit aromas ranging through berry, earth and a hint of liquorice is followed by a palate which has good fruit weight and concentration in the dark cherry flavour spectrum, supported by fine tannins on the finish. **rating:** 88

best drinking 2001–2007 **best vintages** '86, '88, '90, '92, '94 **drink with** Lamb shashlik • $19.50

hungerford hill NR

McDonalds Road, Pokolbin, NSW 2321 **region** Lower Hunter Valley
ph (02) 4998 7666 **fax** (02) 4998 7682 **open** Mon–Fri 9–4.30, weekends 10–4.30
winemaker Ian Walsh **prod.** 18 000 **est.** 1967
prod. range ($10–22.50 R) Tumbarumba Chardonnay, Tumbarumba Sauvignon Blanc, Tumbarumba Pinot Gris, Cowra Chardonnay, Cowra Verdelho, Young Chardonnay, Young Semillon, Late Picked Semillon, Late Picked Riesling, Cabernet Merlot, Hunter Shiraz, Hilltops Cabernet Sauvignon, Adelaide Hills–McLaren Vale Cabernet, Tumbarumba Pinot Noir.

summary Now purely a brand owned by Southcorp, with the wines being made at Tulloch. However, eye-catching new labels and a new range of regionally sourced wines (all from New South Wales) have substantially elevated the status of the brand on the ever parochial Sydney market and raised wine quality to a significant degree. It should be noted that since 1997 I have had some responsibilities for this brand, and hence rate neither the winery nor the wines.

Hungerford Hill Tumbarumba Sauvignon Blanc

Produced from sauvignon blanc grown at Tumbarumba in the Snowy Mountains, the first vintage being 1995. The wine is stainless steel-fermented and early bottled, without any oak input. Note: I had no part in the making or blending of this wine.

1998 Light green-yellow; the bouquet is discreet and relatively light but with sufficient grassy herbal varietal character to please the purists. On the palate the wine has good balance and mouthfeel, with a long finish; it is not souped-up as are many of the '98 vintage white wines.

best drinking 1998–1999 **best vintages** NA **drink with** Seafood risotto • $19.50

Hungerford Hill Tumbarumba Pinot Noir

Back in the 1980s, Hungerford Hill released several Pinot Noirs from Coonawarra, long forgotten and unlamented. In 1997 it made its first Pinot Noir from an entirely new and infinitely more appropriate source for the grape, Tumbarumba in the foothills of the New South Wales Snowy Mountains. It is an exciting source, and it will be interesting to watch the future releases of the wines as vintage conditions permit. Disclosure: I played no part in the making of this wine. Available cellar door (in the Hunter Valley) and through restaurants, mainly in Sydney and Melbourne.

1997 Medium to full red-purple, strong for a variety such as this. There is distinctive plummy pinot noir fruit on the bouquet matched by subtle oak. A full-bodied Pinot on the palate, a little too much so for my taste, perhaps, but authentic nonetheless, and an excellent entry point for those comfortable with Australian dry reds and not so comfortable with Pinot Noir.

best drinking 1999–2001 **best vintages** NA **drink with** Braised duck • $22.50

Hungerford Hill Hilltops Cabernet Sauvignon

The first Hungerford Hill Cabernet Sauvignon from the Hilltops region was released in 1994, then using the now-discarded town name of Young for the region. The reasons for changing are doubtless self-evident. The wine is matured in a mix of new, one- and two-year-old French oak barrels for 14 months and given additional bottle age prior to release. Disclosure of interest as Group Winemaker Southcorp Regional Wineries (including Hungerford Hill).

1996 Dense red, with just a touch of purple. Attractive and sweet chocolate, berry and vanilla aromas are followed by a luscious and sweet palate with cassis berry, chocolate and vanilla all present in a harmonious and neatly packaged wine.

best drinking 2001–2006 **best vintages** NA **drink with** Designer meat pie • $18.50

hunt's foxhaven estate NR

Canal Rocks Road, Yallingup, WA 6282 **region** Margaret River

ph (08) 9755 2232 **fax** (08) 9255 2249 **open** Weekends, holidays 11–5

winemaker David Hunt **prod.** 1000 **est.** 1978

prod. range ($12–15 CD) Riesling (dry and sweet), Semillon, Semillon Sauvignon Blanc, Sauvignon Blanc Riesling, Cabernet Sauvignon.

summary Draws upon 4 hectares of vines progressively established, the oldest being 20-year-old riesling. All of the wine is sold through cellar door and by mail order.

hunter ridge NR

Hermitage Road, Pokolbin, NSW 2320 **region** Lower Hunter Valley

ph (02) 4998 7500 **fax** (02) 4998 7211 **open** 7 days 10–5

winemaker Stephen Pannell **prod.** NFP **est.** 1996

prod. range ($14.50–22 R) Semillon, Verdelho, Chardonnay, Shiraz, Cabernet Sauvignon.

summary Hunter Ridge is effectively a joint venture between BRL Hardy and McGuigan Wines Limited. The grapes come from the vineyards surrounding the Hunter Ridge cellar door and which are owned by McGuigan Wines. The wines are fermented and partially matured in the Hunter Valley, but are finally blended, finished and bottled by BRL Hardy in South Australia.

huntington estate ★★★★☆

Cassilis Road, Mudgee, NSW 2850 **region** Mudgee

ph (02) 6373 3825 **fax** (02) 6373 3730 **open** Mon–Fri 9–5, weekends 10–4

winemaker Susie Roberts **prod.** 25 000 **est.** 1969

prod. range ($10–20 CD) Semillon, Semillon Chardonnay, Chardonnay Non Wooded, Rosé, sundry sweet whites; red wines are released under bin numbers (FB = full-bodied, MB = medium-bodied) comprising Shiraz, Cabernet Merlot and Cabernet Sauvignon.

summary The remarkable Roberts family members have a passion for wine which is equalled only by their passion for music, with the Huntington Music Festival a major annual event. The red wines of Huntington Estate are outstanding, and sell for absurdly low prices. The wines are not exported; almost all are sold via cellar door and mailing list.

Huntington Estate Shiraz

In very much the same style and quality class as the Cabernet Sauvignon, made from estate-grown grapes and producing wines of great longevity. As the wines age, cherry/berry fruits gradually soften and take on that typically, gently earthy Shiraz character, and the tannins soften at the same rate as the fruit rounds off and develops. Thus the balance of the wine is never threatened, and patience is rewarded.

🍷🍷🍷🍷🍸 **1995** Bin FB27. Medium to full red-purple; quite delicious rich and ripe chocolate-accented fruit provides the bouquet, followed by mouthfilling chocolate and berry fruit on the palate with lingering, soft tannins. Exemplary oak handling. **rating:** 92

best drinking 2001–2010 **best vintages** '74, '75, '78, '79, '84, '90, '91, '93, '94, '95 **drink with** Kangaroo fillet • $14

Huntington Estate Special Reserve Shiraz

Small parcels of wine which Susan Roberts considers to be specially ageworthy are held back for release several years after the varietal releases, themselves given more bottle age than most.

🍷🍷🍷🍷🍷 **1993** Bin FB29. Medium to full red-purple; the bouquet is complex and concentrated with rich, dark berry/dark chocolate fruit, just starting to show a touch of softening. The palate is redolent of plum, chocolate and black cherry fruit, with the first sign of the earthy regional characters which will build as the wine slowly ages. Retasted March 1999 with identical notes and points. **rating:** 95

best drinking 1999–2013 **best vintages** '93 **drink with** Venison • $14

huntleigh vineyards ★★☆

Tunnecliffes Lane, Heathcote, Vic 3523 **region** Bendigo

ph (03) 5433 2795 **open** 7 days 10–5.30

winemaker Leigh Hunt **prod.** 425 **est.** 1975

prod. range ($14.50–16 CD) Riesling, Traminer, Shiraz, Cabernet Sauvignon; Leckie Shiraz.

summary A retirement hobby, with robust, rather astringent red wines which need time in bottle to lose some of the rough edges. Part of the production from the 5.2 hectares of vines is sold to others.

ibis wines NR

237 Kearneys Drive, Orange, NSW 2800 **region** Orange

ph (02) 6362 3257 **fax** (02) 6362 5779 **open** Weekends and public holidays 11–5 or by appointment

winemaker Phil Stevenson **prod.** 700 **est.** 1988

prod. range ($16–22 CD) Riesling, Chardonnay, Pinot Noir, Kanjara Shiraz, Cabernet Sauvignon, Cabernet Franc.

summary Phil Stevenson constructed a small winery in 1995, but his winemaking in fact extends back to 1993. He also acts as winemaker for Habitat. The Ibis production comes from 1 hectare of estate plantings, half of which are Cabernet Sauvignon, which is the best of the wines, powerful and concentrated when young, and gaining a vaguely Italianate feel with bottle age. Four vintages of Cabernet Sauvignon ('93 – '96 inclusive) were available at cellar door in early 1999, along with an eclectic range of other wines.

idyll vineyard ★★★

265 Ballan Road, Moorabool, Vic 3221 **region** Geelong
ph (03) 5276 1280 **fax** (03) 5276 1537 **open** Tues–Sun, holidays 10–5
winemaker Dr Daryl Sefton **prod.** 4500 **est.** 1966
prod. range ($13–18.50 R) Idyll Blush, Gewurztraminer, Chardonnay, Bone Idyll (Lighter style Shiraz), Shiraz, Cabernet Shiraz; with Sefton Estate as budget-priced second label Dry White and Cabernet Shiraz.
summary A stalwart of the region, producing wines in an individual style (pungent, assertive Traminer, long-vatted reds) which are almost as well known and appreciated overseas as they are in Australia.

Idyll Vineyard Geelong Dry Gewurztraminer

Idyll Vineyard has always produced Gewurztraminer with an unusual depth of flavour and varietal character. This has shone through even when the wine has been oak-matured, although I much prefer the straightforward approach adopted in more recent years.

🍷🍷🍷🍷 **1997** Medium to full yellow-green; the bouquet is flooded with powerful floral lychee aromas; the palate is similarly powerful and lychee-dominated, the finish as dry as the name suggests. Lots of character and flavour; perhaps fractionally hard on the finish. **rating:** 86

best drinking 1999–2004 **best vintages** NA **drink with** Asian cuisine • $15

Idyll Vineyard Geelong Shiraz

Estate-grown and produced. In bygone years the red wines were left in barrel for an unusually long time (two to three years), but that time has now been shortened, resulting in a marked change in style.

🍷🍷🍷🍷 **1997** Dense red-purple; the bouquet is concentrated, but shows slightly callow, unworked earthy/briary fruit. The palate is very rich and super-concentrated, with earthy black fruit flavours and soft tannins. All fingers point to a very low yield. **rating** 88

best drinking 2002–2012 **best vintages** '97 **drink with** Lamb Provencale • $18.50

Idyll Vineyard Geelong Cabernet Sauvignon

As with all the Idyll wines, estate-grown. Here, too, there have been winemaking changes which have resulted in the elimination of the volatile acidity problems which once bedevilled Idyll.

🍷🍷🍷🍷 **1997** Medium red; the bouquet is soft, with some chocolate and vanilla over red fruits. Sweet cassis berry and chocolate is accompanied by pleasant vanilla oak on the palate; a pleasant and unchallenging wine. **rating:** 87

best drinking 2001–2006 **best vintages** '97 **drink with** Fillet steak • $18.50

indigo ridge NR

Icely Road, Orange, NSW 2800 **region** Orange
ph (02) 6362 1851 **fax** (02) 6362 1851 **open** By appointment
winemaker Contract **prod.** NA **est.** 1995
prod. range ($15–25 ML) Sauvignon Blanc, Cabernet Sauvignon.
summary The newly established Indigo Ridge has 2 hectares each of sauvignon blanc and cabernet sauvignon; production is still very small, and all of the wines are sold by cellar door and mail order. No tastings to date.

inglewood vineyards ★★★

Yarrawa Road, Denman, NSW 2328 **region** Upper Hunter Valley
ph (02) 6547 2556 **fax** (02) 6547 2546 **open** Not
winemaker Simon Gilbert (Contract) **prod.** 18 000 **est.** 1988
prod. range ($9–19 R) Broken into three levels: at the bottom the Rivers label Classic Red and White; then under the Two Rivers label Chardonnay, Unwooded Chardonnay, Semillon Sauvignon Blanc, Verdelho, Cabernet Sauvignon; at the top end is Inglewood Show Reserve Chardonnay, Semillon, Verdelho.
summary A very significant addition to the viticultural scene in the Upper Hunter Valley, with almost 170 hectares of vineyards established, involving a total investment of around $7 million. Much of the fruit is sold to Southcorp under long-term contracts, but part is made under contract for the expanding winemaking and marketing operations of Inglewood. The emphasis is on Chardonnay and Semillon, and the wines have been medal winners in the wine show circuit.

ingoldby ★★★★

Kangarilla Road, McLaren Flat, SA 5171 **region** McLaren Vale
ph (08) 8323 8853 **fax** (08) 8323 8550 **open** 7 days 10–5
winemaker Phil Reschke, Charles Hargreaves **prod.** 15 000 **est.** 1972
prod. range ($14 CD) Colombard, Hugo's Hill Riesling, Sauvignon Blanc, Shiraz, Grenache, Cabernet Sauvignon, Meteora Tawny Port.
summary Acquired by Mildara Blass in 1995, apparently for the grapes and grape contracts controlled by Ingoldby. Phil Reschke is making some excellent wines, just as he is for the Andrew Garrett label. Exports and distribution via Mildara Blass.

Ingoldby Shiraz

Made under the direction of Phil Reschke, who consistently brings out the best in McLaren Vale fruit.

ΥΥΥΥ **1997** Bright red-purple; the aromas are of medium intensity in the red berry range, supported by subtle oak. The palate has attractive fruit, quite sweet and rich, together with hints of chocolate, all contributing to the good mouthfeel. **rating:** 87

best drinking 1999–2004 **best vintages** NA **drink with** Steak and kidney pie • $14

innisfail vineyards NR

Cross Street, Batesford, Vic 3221 **region** Geelong
ph (03) 5276 1258 **fax** (03) 5221 8442 **open** Not
winemaker Ron Griffiths **prod.** 1800 **est.** 1980
prod. range ($13–24 ML) Riesling, Chardonnay, Pinot Noir, Cabernet Sauvignon.
summary This 6-hectare vineyard released its first wines in 1988, made in a small but modern winery on site with a chewy, complex Chardonnay from both 1989 and 1990 attesting to the quality of the vineyard. No recent tastings, however.

ironbark ridge vineyard NR

Middle Road Mail Service 825, Purga, Qld 4306 **region** Other Wineries of Qld
ph (07) 5464 6787 **fax** (07) 5464 6858 **open** By appointment
winemaker Mark Ravenscroft (Contract) **prod.** 250 **est.** 1984
prod. range ($18–25 ML) Chardonnay, Vintage Port.
summary Ipswich is situated on the coastal side of the Great Dividing Range, and the high summer humidity and rainfall will inevitably provide challenges for viticulture here. On the evidence of the '98 Chardonnay, Ironbark Ridge is capable of producing Chardonnay equal to the best from Queensland.

Ironbark Ridge Vineyard Chardonnay

Estate-grown, and a gold medal winner at the Australian Small Makers Show 1998, also taking the title of Best Queensland White at that show.

YYYY 1998 Medium yellow-green; the bouquet proclaims a well-made wine, clean, moderately intense, with smooth melon and fig fruit. The palate is even more attractive, with citrus and melon fruit running through a palate of considerable length; imperceptible oak. **rating:** 89

best drinking 1999–2002 **best vintages** NA **drink with** Morton Bay bugs • $20

iron pot bay wines ★★★☆

West Bay Road, Rowella, Tas 7270 **region** Northern Tasmania
ph (03) 6394 7320 **fax** (03) 6394 7346 **open** By appointment
winemaker Andrew Hood, Jim Chatto (Contract) **prod.** 2500 **est.** 1988
prod. range ($20 R) Unwooded Chardonnay, Sauvignon Blanc, Pinot Grigio.
summary Iron Pot Bay is now part of the syndicate which has established Rosevears Estate, with its large, state-of-the-art winery erected on the banks of the Tamar. The vineyard takes its name from a bay on the Tamar River, and is strongly maritime-influenced, producing delicate but intensely flavoured unwooded white wines.

ironwood estate NR

RMB 1316 Bolganup Road, Porongurup, WA 6324 **region** Great Southern
ph (08) 9853 1126 **fax** (08) 9853 1172 **open** By appointment
winemaker Steve Pester **prod.** NA **est.** 1996
prod. range ($NA) Riesling, Chardonnay, Shiraz, Cabernet Sauvignon.
summary Ironwood Estate was established in 1996 when the first wines were made from purchased grapes. In the same year chardonnay, shiraz and cabernet sauvignon were planted on a northern slope of the Porongurup Range. The twin peaks of the Porongurups rise above the vineyard and provide the basis for the label design. The first estate-grown grapes were vinified at the new Porongurup Winery erected for the 1999 vintage, and jointly owned by Jingalla and Chatsfield Wines. The fragrant passionfruit-accented lime juice-flavoured '96 Riesling (87 points) and the cassis berry '96 Cabernet Sauvignon (86 points) should get Ironwood away to a flying start.

irvine ★★★☆

Roeslers Road, Eden Valley, SA 5235 **region** Eden Valley
ph (08) 8564 1046 **fax** (08) 8564 1046 **open** Not
winemaker James Irvine **prod.** 4000 **est.** 1980
prod. range ($14–90 R) Under the cheaper Eden Crest label: Unwooded Chardonnay, Pinot Merlot, Merlot Cabernet, Meslier Brut, Pinot Chardonnay Brut; under the premium James Irvine label: Brut Royale, Merlot Brut and (at the top of the tree) Grand Merlot.
summary Industry veteran Jim Irvine, who has successfully guided the destiny of so many South Australian wineries, quietly introduced his own label in 1991, although the vineyard from which the wines are sourced was commenced in 1980, and now comprises 1 hectare of petit meslier, 1 hectares of merlot and 7 hectares of chardonnay. Between one-third and two-thirds of the production is exported each year, principally to Europe and Taiwan.

island brook estate NR

Lot 817 Bussell Highway, Metricup, WA 6280 **region** Margaret River
ph (08) 9755 7501 **fax** (08) 9755 7501 **open** 7 days 10–5
winemaker Stuart Pym **prod.** 700 **est.** 1985
prod. range ($15 CD) Semillon, Summer Garden (white).

summary Ken and Judy Brook operate a vineyard café offering brunch, lunch, cappuccinos and Devonshire teas throughout the day, with a most unusual maze (built with rammed earth walls) to occupy children (and perhaps adults). Much of the production from the 8-hectare estate is sold to other makers, with limited quantities of wine made for sale from the restaurant and cellar door.

ivanhoe wines NR

Cnr Oakey Creek and Marrowbone Roads, Pokolbin, NSW 2320 **region** Lower Hunter Valley
ph (02) 4998 7325 **fax** (02) 4998 7848 **open** 7 days 10–5
winemaker Stephen Drayton, Tracy Drayton **prod.** NA **est.** 1995
prod. range Various varietal wines under the Ivanhoe, Stephen Drayton Signature Series and Lancelot brands, including Chardonnay and Shiraz.
summary Stephen Drayton is the son of the late Reg Drayton, and with wife Tracy is the third branch of the family to be actively involved in winemaking in the Hunter Valley. The property on which the vineyard is situated has been called Ivanhoe for over 140 years, and 25 hectares of 30-year-old vines provide high-quality fruit for the label. The plans are to build a replica of the old homestead (burnt down, along with much of the winery in the 1968 bushfires) to operate as a sales area.

jackson's hill ★★★

Mount View Road, Mount View, NSW 2321 **region** Lower Hunter Valley
ph (02) 4990 1273 **fax** (02) 4991 3233 **open** Thur–Mon and public holidays 10–5
winemaker Mike Winborne **prod.** 1000 **est.** 1984
prod. range ($16–18 CD) Semillon, Oak Fermented Semillon, Late Harvest Semillon, Cabernet Franc.
summary One of the newer arrivals on the spectacularly scenic Mount View Road, making tiny quantities of wine sold exclusively through the cellar door, and specialising in Cabernet Franc. The '97 Semillon and Cabernet Franc were not exhilarating, but Jackson's Hill does produce the most marvellous home-made chocolates I have tasted in a long time.

jadran NR

445 Reservoir Road, Orange Grove, WA 6109 **region** Perth Hills
ph (08) 9459 1110 **open** Mon–Sat 10–8, Sun 11–5
winemaker Steve Radojkovich **prod.** NFP **est.** 1967
prod. range ($6–12 CD) Riesling, Hermitage, generic red and white table wines, Sparkling, Fortifieds.
summary A quite substantial operation which basically services local clientele, occasionally producing wines of quite surprising quality from a variety of fruit sources.

james estate NR

Mudgee Road, Baerami via Denman, NSW 2333 **region** Upper Hunter Valley
ph (02) 6547 5168 **fax** (02) 6547 5164 **open** 7 days 10–4.30
winemaker Peter Orr **prod.** 20 000 **est.** 1971
prod. range ($12–16 CD) Sylvaner, Late Harvest Sylvaner, Semillon, Chardonnay, Reserve Chardonnay, Merlot, Cabernet Shiraz.
summary A very substantial viticultural enterprise with 60 hectares of vineyards equally divided between sylvaner, semillon, chardonnay, shiraz and cabernet sauvignon. Overall quality has been there or thereabouts. Sold in 1997, and after a transition period, Peter Orr has been appointed as winemaker and the wines are now sold under the James Estate label.

jane brook estate ★★★

229 Toodyay Road, Middle Swan, WA 6056 **region** Swan District
ph (08) 9274 1432 **fax** (08) 9274 1211 **open** 7 days 12–5
winemaker Julie White, David Atkinson **prod.** 20 000 **est.** 1972
prod. range ($16.50–18.50 CD) Wood Aged Chenin Blanc, Chardonnay, Sauvignon Blanc, Late Harvest Cabernet Merlot, Shiraz, Elizabeth Jane Méthode Champenoise, Fortifieds.
summary An attractive winery which relies in part on substantial cellar-door trade and in part on varying export markets, with much work having been invested in the Japanese market in recent years. The white wines are usually best, although the quality of the oak isn't always up to the mark. Exports to Japan, Malaysia, Singapore and Taiwan, with distribution in each of the Australian States.

Jane Brook Estate Sauvignon Blanc

A blend of 60% Pemberton grapes and 40% Swan Valley, the latter picked almost two months before the Pemberton component. The parcels were separately fermented, and blended prior to bottling.
🍷🍷🍷🍷 **1998** Light green-yellow; the bouquet is clean, with moderately intense herb and mineral aromas and a faintly earthy cut. The palate has distinct presence, length and grip, although for those who are not enamoured of Sauvignon Blanc it will not be user-friendly. **rating:** 86
best drinking 1999–2001 **best vintages** NA **drink with** Marinated scallops • $16.50

---

jardee NR

Old School House, Jardee, WA 6258 **region** Pemberton
ph (08) 9777 1552 **fax** (08) 9777 1552 **open** Not
winemaker Barrie Smith **prod.** 510 **est.** 1994
prod. range ($18–22 R) Chardonnay, Pinot Noir.
summary Jardee is a pioneering mill town, the wines are in fact made in tiny quantities from purchased fruit, the operation being a part-time interest for proprietor Steve Miolin.

jasper hill ★★★★★

Drummonds Lane, Heathcote, Vic 3523 **region** Bendigo
ph (03) 5433 2528 **fax** (03) 5433 3143 **open** By appointment
winemaker Ron Laughton **prod.** 3000 **est.** 1975
prod. range ($17–66 R) Georgia's Paddock Riesling, Georgia's Paddock Shiraz, Emily's Paddock Shiraz Cabernet Franc.
summary The red wines of Jasper Hill are highly regarded and much sought after, invariably selling out at cellar door and through the mailing list within a short time after release. These are wonderful wines, reflecting the very low yields and the care and attention given to them by Ron Laughton. The oak is not overdone, the fruit flavours showing Central Victoria at its best.

Jasper Hill Georgia's Paddock Shiraz

Georgia's Paddock is by far the larger of the two vineyard blocks, with 9.5 hectares of shiraz and 3 hectares of riesling (the latter of course going to make the varietal wine of that name). The vines are unirrigated, and are normally low-yielding, but yields were reduced even further in 1996 as a result of drought followed by the first frost in 22 years, which destroyed the bottom third of the paddock in December 1995. These calamities to one side, a perfect vintage which produced wines with 14.5 degrees alcohol.
🍷🍷🍷🍷🍷 **1997** Full, deep purple-red; the bouquet is quite overwhelming, with scented, spicy, exotic aromas tumbling out of the glass. The palate is no less rich and ripe with plum and cherry fruit still locked in arm-to-arm combat with the tannins. The rating is, to a degree, an article of faith. **rating:** 94
best drinking 2003–2013 **best vintages** '90, '91, '93, '95, '96, '97 **drink with** Wild duck • $41

Jasper Hill Emily's Paddock Shiraz Cabernet Franc
Produced from 3 hectares of shiraz and 0.2 hectare of cabernet franc, the latter having an impact on style disproportionate to the area of grapes. Ron Laughton also suggests since 1992 he has endeavoured to move the wine towards a more elegant style, and I think he has succeeded admirably. In 1996, due mainly to low bunch numbers and the impact of the drought, only 100 cases of Emily's Paddock were produced.

🍷🍷🍷🍷🍷 **1997** Dense, youthful purple-red. The extremely rich, ripe and concentrated bouquet offers liquorice, cardamom and mint; the palate, likewise, has an amazing concentration of exotic spices (cardamom, etc) and sweet liquorice fruit, fruit which has gobbled up the oak. **rating:** 95

best drinking 2002–2012 **best vintages** '90, '91, '93, '95, '96, '97 **drink with** Rare eye fillet of beef • $30

jasper valley NR

RMB 880 Croziers Road, Berry, NSW 2535 **region** Shoalhaven
ph (02) 4464 1596 **fax** (02) 4464 1596 **open** 7 days 9.30–5.30
winemaker Contract **prod.** 1100 **est.** 1976
prod. range ($4.20–12 CD) White Burgundy, Riesling, Traminer Riesling, Moselle, Summer Red, Cabernet Sauvignon, Port; also non-alcoholic fruit wines.
summary A strongly tourist-oriented winery with most of its wine purchased as cleanskins from other makers. Features about 1 hectare of lawns, barbecue facilities, and sweeping views.

jeanneret wines NR

Jeanneret Road, Sevenhill, SA 5453 **region** Clare Valley
ph (08) 8843 4308 **fax** (08) 8843 4251 **open** Mon–Fri 11–5, weekends and public holidays 10–5
winemaker Ben Jeanneret, Denis Jeanneret **prod.** 5000 **est.** 1992
prod. range ($14–20 CD) Riesling, Semillon, Sparkling Shiraz, Shiraz, Cabernet Sauvignon.
summary Jeanneret's fully self-contained winery has a most attractive outdoor tasting area and equally attractive picnic facilities situated on the edge of a small lake surrounded by bushland. While it did not open the business until October 1994, its first wine was in fact made in 1992 (Shiraz) and it has already established a loyal following. Export markets have been established in Canada and the US.

jeir creek ★★★

Gooda Creek Road, Murrumbateman, NSW 2582 **region** Canberra District
ph (02) 6227 5999 **fax** (02) 6227 5900 **open** Fri–Sun, holidays 10–5
winemaker Rob Howell **prod.** 3500 **est.** 1984
prod. range ($16–18 CD) Riesling, Late Harvest Riesling, Botrytis Semillon Sauvignon Blanc, Semillon Sauvignon Blanc, Sauvignon Blanc, Chardonnay, Pinot Noir, Shiraz, Cabernet Merlot.
summary Rob Howell came to part-time winemaking through a love of drinking fine wine, and is intent on improving both the quality and consistency of his wines. It is now a substantial (and still growing) business, with the vineyard plantings increased to 11 hectares by the establishment of more cabernet sauvignon, shiraz and merlot.

jenke vineyards ★★★★

Barossa Valley Way, Rowland Flat, SA 5352 **region** Barossa Valley
ph (08) 8524 4154 **fax** (08) 8524 5044 **open** 7 days 10–4.30
winemaker Kym Jenke **prod.** 7000 **est.** 1989
prod. range ($10–22 CD) Semillon, Chardonnay, Late Harvest Riesling, Mourvedre, Merlot, Cabernet Franc, Cabernet Sauvignon, Shiraz.

summary The Jenkes have been vignerons in the Barossa since 1854, and have over 25 hectares of vineyards; a small part of the production is now made and marketed through a charming restored stone cottage cellar door. The red wines have been particularly impressive, notably the recently introduced Old Vine Mourvedre.

jim barry wines ★★★★☆

Main North Road, Clare, SA 5453 **region** Clare Valley
ph (08) 8842 2261 **fax** (08) 8842 3752 **open** Mon–Fri 9–5, weekends, holidays 9–4
winemaker Mark Barry **prod.** 50 000 **est.** 1959
prod. range ($6.50–120 R) Watervale Riesling, Lodge Hill Riesling, Personal Selection Semillon Sauvignon Blanc and Chardonnay, Semillon, Unwooded Chardonnay, Lavendar Hill, Cabernet Shiraz, Personal Selection Cabernet Sauvignon, McCrae Wood Shiraz and Cabernet Malbec, The Armagh (Shiraz).
summary The Armagh and the McCrae Wood range continue to stand out as the very best wines from Jim Barry, exceptionally concentrated and full flavoured. The remainder are seldom less than adequate, but do vary somewhat from one vintage to the next. Has an exceptional viticultural resource base of 160 hectares of mature Clare Valley vineyards. Exports to the UK, much of Europe, North America, Japan and southeast Asia.

Jim Barry The Armagh

First made in 1985 from very old, unirrigated, low-yielding vines. Not produced in 1986, and not exhibited in wine shows until the 1987 vintage was made. Since that time every vintage since the '87 has received at least one gold medal, with numerous trophies bestowed on the '89 and '90 vintages, respectively. Unashamedly a Grange pretender, and succeeding well in its aim.

🍷🍷🍷🍷🍷 **1996** Full red-purple; lovely dark cherry and blackberry fruit dominates a complex bouquet, with only subtle oak in evidence. A big, succulently dense black cherry and chocolate-flavoured palate follows, finishing with soft tannins. Notwithstanding all that flavour, not over-extracted. **rating:** 94

best drinking 2006–2011 **best vintages** '89, '90, '91, '92, '93, '95, '96 **drink with** The richest game dish possible • $120

Jim Barry McCrae Wood Cabernet Malbec

The blend of Cabernet and Malbec is a Clare Valley specialty, pioneered by Leasingham in the vinous dawn of time. Malbec flourishes in the Clare Valley as in few parts of Australia, adding a juicy sweetness to the formidable power of Clare Valley Cabernet.

🍷🍷🍷🍷 **1996** Medium to full red-purple; the bouquet has solid, sweet dark fruit plus that hallmark touch of mint which is always evident in the McCrae Wood wines. The palate has a mix of berry, earth and mint flavours, finishing with slightly hard-edged acidity. All in all, a style I often find difficult to come to grips with. **rating:** 85

best drinking 2001–2006 **best vintages** NA **drink with** Oxtail • $39.95

---

jindalee wines NR

PO Box 5146, North Geelong, Vic 3215 **region** Geelong
ph (03) 5277 2836 **fax** (03) 5277 2840 **open** Not
winemaker Contract **prod.** 8000 **est.** 1997
prod. range ($8–15 CD) Chardonnay, Colombard Chardonnay, Shiraz, Merlot, Cabernet Sauvignon.
summary Jindalee Wines made its debut with the 1997 vintage. It is part of the Littore Group, which currently has 400 hectares of premium wine grapes in wine production and under development in the Riverland. The first wines were contract-made, but it is planned to have a winery constructed by the year 2000, and to expand the product range to include Cabernet

Sauvignon, Merlot and Colombard Chardonnay. The first release wines did not rise above their price station.

jingalla ★★★★

RMB 1316 Bolganup Dam Road, Porongurup, WA 6324 **region** Great Southern
ph (08) 9853 1023 **fax** (08) 9853 1023 **open** 7 days 10.30–5
winemaker Steven Pester **prod.** 3000 **est.** 1979
prod. range ($11–20 CD) Great Southern White and Red, Riesling, Semillon, Verdelho, Reserve Shiraz, Cabernet Rouge, Cabernet Sauvignon, Late Harvest Semillon, Tawny Port, Liqueur Muscat.
summary Jingalla is a family-run business, owned and run by Geoff and Nita Clarke and Barry and Shelley Coad, the latter the ever-energetic wine marketer of the business. The 8 hectares of hillside vineyards are low-yielding, with the white wines succeeding best. Consistently competent winemaking at Goundrey has resulted in a range of very reliable, positively flavoured wines. While best known for its wooded and unwooded whites, it also produces some lovely red wines.

Jingalla Riesling

Produced from 2 hectares of riesling, with two clones, including the so-called Geisenheim clone. Consistent performer, always exhibiting abundant fruit flavour. The 1998 contains 10% Traminer.
🍷🍷🍷🍷🍷 **1998** Full yellow-green, surprisingly developed for a young Riesling. The bouquet is rich and generous with sweet, tropical lime fruit, the palate tighter than the bouquet, but still with abundant fruit in typical Jingalla style. Eminently drinkable right now. **rating:** 90
best drinking 1999–2003 **best vintages** '84, '86, '90, '93, '96, '98 **drink with** Chinese or Thai cuisine • $13

Jingalla Semillon

There are 1.5 hectares of semillon on the hillside vineyard, situated on the northern slopes of the Porongurup Ranges, but with very limited production: less than 300 cases were made in 1997. In the manner of Hunter Semillons, stainless steel fermented and bottled within a few months of vintage.
🍷🍷🍷🍷 **1997** Light to medium yellow-green; the bouquet is clean, of only moderate intensity and with faintly grassy overtones. The palate is delicate and crisp, but well balanced, and will very probably develop considerable character as it ages. **rating:** 86
best drinking 2000–2005 **best vintages** NA **drink with** Smoked eel • $13

jinks creek winery NR

Tonimbuk Road, Tonimbuk, Vic 3815 **region** Gippsland
ph (03) 5629 8502 **fax** (03) 5629 8551 **open** By appointment
winemaker Andrew Clarke **prod.** NA **est.** 1981
prod. range ($14 CD) Sauvignon Blanc, Chardonnay, Pinot Noir.
summary Jinks Creek Winery is situated between Gembrook and Bunyip, bordering the evocatively named Bunyip State Park. While the winery was not built until 1992, planting of the 2.5-hectare vineyard started back in 1981, and all of the wines are estate-grown. The 'sold out' sign goes up each year.

joadja vineyards NR

Joadja Road, Berrima, NSW 2577 **region** Other Wineries of NSW
ph (02) 4878 5236 **fax** (02) 4878 5236 **open** 7 days 10–5
winemaker Kim Moginie **prod.** 2000 **est.** 1983
prod. range ($13–21 CD) Classic Dry White, Sauvignon Blanc, Chardonnay, Botrytis Autumn Riesling, Sauternes, Classic Dry Red, Cabernet Malbec, Christopher Tawny Port.

summary The strikingly labelled Joadja Vineyards wines, first made in 1990, are principally drawn from 7 hectares of estate vineyards situated in the cool hills adjacent to Berrima. Both the red and whites have a consistent eucalypt/peppermint character which is clearly a product of the climate and (possibly) soil. Joadja is well worth a visit.

john gehrig wines ★★☆

Oxley–Milawa Road, Oxley, Vic 3678 **region** King Valley
ph (03) 5727 3395 **fax** (03) 5727 3699 **open** 7 days 9–5
winemaker John Gehrig **prod.** 5600 **est.** 1976
prod. range ($7–14.50 CD) Oxley Dry White, Riesling, Chenin Blanc, Verdelho, Chardonnay, Sparkling, Pinot Noir, King River Red, Merlot, Cabernet Merlot, Fortifieds.
summary Honest, if seldom exciting, wines; the occasional Chardonnay, Pinot Noir, Merlot and Cabernet Merlot have, however, risen above their station.

jollymont NR

145 Pullens Road, Woodbridge, Tas 7162 **region** Southern Tasmania
ph (03) 6267 4594 **fax** (03) 6267 4594 **open** Not
winemaker Andrew Hood (Contract) **prod.** 10 **est.** 1990
prod. range Chardonnay, Pinot Noir.
summary However briefly, Jollymont has displaced Scarp Valley as the smallest producer in Australia, its 1998 vintage (the first) producing ten cases. The vines are not irrigated, nor will they be, and Peter and Heather Kreet do not intend to sell any wine younger than three to four years old. Their aim is to produce wines of maximum intensity and complexity.

jones winery NR

Jones Road, Rutherglen, Vic 3685 **region** Rutherglen
ph (02) 6032 8496 **fax** (02) 6032 8495 **open** Fri–Sat 10–5, Sun and holidays 10–4
winemaker Mandy Jones **prod.** 7000 **est.** 1864
prod. range Chablis, Riesling, White Burgundy, Light Red, Dry Red, Fortifieds.
summary Late in 1998 the winery was purchased by Leanne Schoen and Mandy and Arthur Jones (nieces and nephew of founder Les Jones). They are planning to redevelop the property the next few years, but to concentrate on Shiraz and the styles of wine that the winery is known for. All wine is sold through the cellar door.

kaesler ★★★

Barossa Valley Way, Nuriootpa, SA 5355 **region** Barossa Valley
ph (08) 8562 2711 **fax** (08) 8562 2788 **open** 7 days 10–5
winemaker Contract **prod.** 3000 **est.** 1990
prod. range ($9–30 CD) Prestige Semillon, Cottage Block White, Late Harvest Semillon, Bush Vine Grenache, Old Vine Shiraz, Beerenauslese, Prestige Cuvee, Méthode Champenoise, Old Vine Shiraz, Fortifieds.
summary Toby and Treena Hueppauff purchased Kaesler Farm, with its 12 hectares of vines in 1985, and since 1990 have had the wines made under contract by others. The winery has an à la carte restaurant offering both indoor and outdoor dining; there is also accommodation.

kaiser stuhl ★★☆

Tanunda Road, Nuriootpa, SA 5355 **region** Barossa Valley
ph (08) 8560 9389 **fax** (08) 8562 1669 **open** Mon–Sat 10–5, Sun 1–5
winemaker Nigel Logos **prod.** 1.3 million **est.** 1931
prod. range ($4–12 R) Black Forest, generic whites under bin numbers, Claret Bin 33, Bin 44 Riesling, Bin 55 Moselle, Bin 66 Burgundy, Bin 77 Chablis, Sparkling; also extensive cask and flagon range.

summary Part of the Southcorp Wines empire, but a shadow of its former self, with its once-famous Green Ribbon Riesling and Red Ribbon Shiraz no more. Essentially provides flagon-quality wines in bottles at competitive prices.

kalari vineyards NR

120 Carro Park Road, Cowra, NSW 2794 **region** Cowra
ph (02) 6342 1465 **fax** (02) 6342 1465 **open** 7 days 10–4
winemaker Jill Lindsay (Contract) **prod.** NA **est.** 1995
prod. range ($14–17 ML) Verdelho, Chardonnay, Shiraz.
summary Kalari Vineyards is yet another of the new brands to appear in the Cowra region. Fourteen and a half hectares of vines have been established, with a Verdelho, Chardonnay and Shiraz being included in the initial release.

kangarilla road vineyard & winery ★★★☆

Kangarilla Road, McLaren Vale, SA 5171 **region** McLaren Vale
ph (08) 8383 0533 **fax** (08) 8383 0044 **open** Mon–Fri 9–5, weekends 11–5
winemaker Kevin O'Brien **prod.** 10 000 **est.** 1975
prod. range ($10–30 CD) Chardonnay, Zinfandel, Shiraz, Cabernet, Tawny Port, Vintage Port.
summary Kangarilla Road Vineyard & Winery was formerly known as Stevens Cambrai. Long-time industry identity Kevin O'Brien and wife Helen purchased the property in July 1997, and are continuing to sell the existing stocks of Cambrai wines, but with the strikingly labelled Kangarilla Road brand to progressively replace it.

Kangarilla Road Shiraz

Produced from grapes grown on three McLaren Vale vineyards, the first in the hills behind Coriole, the second on sandy soil at Blewitt Springs, and the third (and latest-ripening) from Kangarilla Road's own vineyard. All the parcels are open-fermented, and were separately matured in various combination of American and French and American oak. The wine spends months in barrel before being bottled.

🍷🍷🍷🍷 **1997** Medium red-purple; there is an interesting array of aromas, opening with hints of pine needle, then spice, leather and black cherry. The palate is archetypal McLaren Vale, with rich red berry and chocolate flavours flooding the mouth on entry, finishing with soft, almost milky, tannins; lurking somewhere in the wine is an almost subliminal hint of green leaf, a variant of the pine needle on the bouquet. **rating:** 89

best drinking 2002–2007 **best vintages** NA **drink with** Barbecued beef • $18

kangaroo island vines NR

c/o 413 Payneham Road, Felixstow, SA 5070 **region** Other Wineries of SA
ph (08) 8365 3411 **fax** (08) 8336 2462 **open** Not
winemaker Caj Amadio **prod.** 600 **est.** 1990
prod. range ($19–23 ML) Island Sting, Cabernet Merlot, Special Reserve Cabernet Merlot.
summary Kangaroo Island is another venture of Caj and Genny Amadio, with the wines being sold through the Chain of Ponds cellar door. The Amadios have been the focal point of the development of vineyards on Kangaroo Island, producing the wines not only from their own tiny planting of 450 vines on quarter of an acre, but buying grapes from other vignerons on the island. The tiny quantities of wine so far produced strongly support the notion that Kangaroo Island has an excellent climate for Bordeaux-style reds, particularly the excellent Special Reserve Cabernet Merlot.

Kangaroo Island Vines Cabernet Merlot

Produced from grapes grown on the Florance Vineyard at Cygnet River. The first vintage was 1991, and the vines are now achieving mature flavours. The '97 caused a major surprise by winning a gold medal at the 1998 Royal Adelaide Wine Show, both because of the tiny production of the wine and because of its quality.

🍷🍷🍷🍷🍷 **1997** Medium red-purple; the bouquet is smooth, moderately intense, with excellent balance and integration of olive and berry fruit with cedary oak. The palate, likewise, is elegant, with a nice touch of cassis followed by fine, lingering tannins on the finish. **rating:** 92

best drinking 2001–2007 **best vintages** '97 **drink with** Kangaroo fillet • $22

kangderaar vineyard NR

Melvilles Caves Road, Rheola, Vic 3517 **region** Bendigo
ph (03) 5438 8292 **fax** (03) 5438 8292 **open** Mon–Sat 9–5, Sun 10–5
winemaker James Nealy **prod.** 500 **est.** 1980
prod. range ($12–15 CD) Chardonnay, Vintage Reserve Chardonnay, Cabernet Sauvignon.
summary The 4.5-hectare vineyard is situated at Rheola, near the Melville Caves, said to have been the hideout of the bushranger Captain Melville in the 1850s, and surrounded by the Kooyoora State Park. It is owned by James and Christine Nealy.

kara kara vineyard ★★★

Sunraysia Highway, St Arnaud, Vic 3478 (10 km south of St Arnaud) **region** Pyrenees
ph (03) 5496 3294 **fax** (03) 5496 3294 **open** Mon–Fri 10.30–6, weekends 9–6
winemaker John Ellis (Contract) **prod.** 2500 **est.** 1977
prod. range ($15–19.50 CD) Fumé Blanc, Sauvignon Blanc, Chardonnay Semillon, Late Harvest Semillon, Shiraz Cabernet.
summary Hungarian-born Steve Zsigmond comes from a long line of vignerons, and sees Kara Kara as the eventual retirement occupation for himself and wife Marlene. The first step has been the decision to have their production contract-made (first by Mitchelton, then John Ellis – previously the grapes were sold) with predictably consistent results over the first few years. Draws upon 9 hectares of estate plantings.

karina vineyard ★★★★

RMB 4055 Harrisons Road, Dromana, Vic 3936 **region** Mornington Peninsula
ph (03) 5981 0137 **fax** (03) 5981 0137 **open** Weekends 11–5, 7 days in January
winemaker Gerard Terpstra **prod.** 1500 **est.** 1984
prod. range ($11–19 CD) Riesling, Sauvignon Blanc, Chardonnay, Cabernet Merlot, Bald Hill Creek (Cabernet Sauvignon).
summary A typical Mornington Peninsula vineyard, situated in the Dromana/Redhill area on rising, north-facing slopes, just 3 kilometres from the shores of Port Phillip Bay, immaculately tended and with picturesque garden surrounds. Fragrant Riesling and cashew-accented Chardonnay are its best wines.

Karina Riesling

Made from half a hectare of estate plantings employing the neatly trimmed vertical canopy so necessary to obtain fruit ripeness in this cool region. Always elegant and crisp, and has improved as the vines have aged.

🍷🍷🍷🍷🍷 **1998** Light to medium yellow-green; aromas of citrus, passionfruit and stone fruit run through a moderately intense bouquet, leading into a spotlessly clean and bright palate with lovely fruit. The wine is of only moderate weight but doesn't need any more. **rating:** 91

best drinking 1999–2004 **best vintages** '94, '97, '98 **drink with** Crab, mussels • $15

Karina Vineyard Sauvignon Blanc

Cold-fermented in stainless steel and early bottled in traditional fashion. Sauvignon Blanc did well on the Mornington Peninsula in 1998. A cleverly made wine precisely aimed at the palate of the majority of cellar-door customers, and none the worse for that.

🍷🍷🍷🍷🍷 **1998** Light green-yellow; the bouquet is clean, with quite intense and stylish citrus and herb fruit. The palate swirls into a rounder, fleshy style supported by what seems to be a well-judged touch of residual sugar. **rating:** 90

best drinking 1999–2000 **best vintages** '98 **drink with** Saffron mussels • $17

Karina Chardonnay

Has been a particularly successful wine for Karina, and in particular avoids going over the top in the fashion of some Mornington Chardonnays. The wine is barrel-fermented, with partial malolactic-fermentation, and spends 12 months in Vosges oak barriques.

🍷🍷🍷🍷 **1997** Bright, light to medium green-yellow; the bouquet is lifted, with a faintly herbal edge to straw-tinged malolactic aromas. The palate, however, provides bell-clear cool-climate fruit, with tangy, citrus and melon flavours. Of almost piercing intensity. **rating:** 87

best drinking 1998–2001 **best vintages** '91, '92, '94, '96, '97 **drink with** Grilled spatchcock • $19

karl seppelt ★★★☆

Ross Dewells Road, Springton, SA 5235 **region** Eden Valley

ph (08) 8568 2378 **fax** (08) 8568 2799 **open** 7 days 10–5

winemaker Karl Seppelt, Petaluma (Contract) **prod.** 2500 **est.** 1981

prod. range ($18–22 R) Riesling, Springton Chardonnay, Springton Cabernet Sauvignon, Langhorne Creek Shiraz, Chardonnay Brut, Sparkling Shiraz, Brut Sauvage, Fino Sherry, Vintage Port, Tawny Port.

summary After experimenting with various label designs and names, Karl Seppelt (former marketing director of Seppelt) has decided to discontinue the brand name Grand Cru (although retaining it as a business name) and henceforth market the wines from his estate vineyards under his own name. The quality is very consistent across the range, and the wines are exported to Canada, Germany and Japan.

Karl Seppelt Springton Cabernet Sauvignon

Drawn from 2.75 hectares of estate plantings. The vines are now fully mature, and, in better vintages, produce grapes of tremendous concentration, and wine of commensurate power and extract.

🍷🍷🍷🍷 **1996** Medium to full red-purple; the bouquet is quite intense, with blackberry/earthy cabernet sauvignon varietal fruit supported by subtle oak. A powerful palate ranges through dark berry and bitter chocolate flavours finishing with firm tannins and subliminal oak. A purist style perhaps, but a good one. **rating:** 88

best drinking 2001–2011 **best vintages** '90, '91, '96 **drink with** Aged parmesan • $21.50

karrivale ★★★★☆

Woodlands Road, Porongurup, WA 6324 **region** Great Southern

ph (08) 9853 1009 **fax** (08) 9853 1129 **open** Wed–Sun 10–5

winemaker Gavin Berry (Contract) **prod.** 1170 **est.** 1979

prod. range ($12–18 CD) Riesling, Chardonnay.

summary A tiny Riesling specialist in the wilds of the Porongurups forced to change its name from Narang because Lindemans felt it could be confused with its Nyrang Shiraz brand; truly a strange world. The viticultural skills of owner Campbell McGready and ultra-competent contract-winemaking fulfil the promise of this beautifully sited vineyard, and its long-lived Riesling.

karriview ★★★★

RMB 913 Roberts Road, Denmark, WA 6333 **region** Great Southern

ph (08) 9840 9381 **fax** (08) 9840 9381 **open** Summer school holidays 7 days 11–4, Feb–Dec Fri–Tues 11–4

winemaker John Wade (Contract) **prod.** 800 **est.** 1986

prod. range ($23–21 CD) Chardonnay, Late Harvest Riesling, Pinot Noir.

summary One hectare each of immaculately tended pinot noir and chardonnay on ultra-close spacing produce tiny quantities of two wines of remarkable intensity, quality and style. Available only from the winery, but worth the effort. There is some vintage variation; the winery rating is based upon the successes, not the disappointments. Typically, back vintages are available, with age, the Pinot Noir acquires strong foresty characters which are quite Burgundian.

katnook estate ★★★★☆

Riddoch Highway, Coonawarra, SA 5263 **region** Coonawarra

ph (08) 8737 2394 **fax** (08) 8737 2397 **open** Mon–Fri 9–4.30, weekends 10–4.30

winemaker Wayne Stehbens **prod.** 70 000 **est.** 1979

prod. range ($8.50–75 R) Under the premium Katnook label: Riesling, Sauvignon Blanc, Chardonnay, Botrytis Riesling, Cabernet Sauvignon, Merlot, Odyssey (super-premium Cabernet) and Chardonnay Brut; under the Riddoch label: Chardonnay, Sauvignon Blanc, Shiraz and Cabernet Shiraz Merlot; also Woolshed Chardonnay and Cabernet Shiraz Merlot.

summary Still the largest contract-grape grower and supplier in Coonawarra, selling 60% of its grape production to others. The historic stone woolshed in which the second vintage in Coonawarra (1896) was made and which has served Katnook since 1980 is being restored. Together with the 1997 launch of the flagship Odyssey, points the way for a higher profile for the winemaking side of the venture. Exports to UK, Northern Ireland, US, Canada, Switzerland, Hong Kong and the Philippines.

Katnook Estate Sauvignon Blanc

One of the signature wines of Katnook Estate, now made from 8 remaining hectares of plantings of this variety. A chance tasting of the '82 vintage in late 1994 emphasised how well the wines can age, even if the change in character is quite radical. On balance, best drunk young. The '98 vintage, incidentally, marked the nineteenth release, putting it in the forefront of Australian Sauvignon Blanc in terms of history.

▼▼▼▼▽ **1998** Light straw-green; initially the aromas are of herb and mineral, but more tropical gooseberry characters then open up. A powerful, strongly structured wine on the palate with abundant weight. The finish grabs ever so slightly. Gold medal 1998 National Wine Show. **rating:** 93

best drinking 1999–2002 **best vintages** '84, '86, '90, '92, '94, '95, '96, '98 **drink with** Grilled whiting • $25

Riddoch Sauvignon Blanc

Part of the newly packaged and identified Riddoch range, which delicately distances itself (or perhaps vice versa) from the Katnook Estate wines. Straightforward cold fermentation in stainless steel emphasises varietal character. Comparatively speaking, the 1998 was a success.

▼▼▼▼ **1998** Medium green-yellow; the bouquet is surprisingly rich, ripe and full with tropical gooseberry fruit, the palate almost heavily ripe with gooseberry, baked apple and lychee flavours. **rating:** 85

best drinking 1999–2000 **best vintages** NA **drink with** Goat's cheese • $15

Riddoch Chardonnay

Forty-five per cent of the wine is fermented in oak, 20% new American and 25% two-year-old. It is then left in barrel on yeast lees for 12 months prior to bottling.

🍷🍷🍷🍷 **1997** Medium to full yellow; spicy nutmeg oak gives lift and focus to the bouquet, but raises the question whether there is sufficient fruit on the palate. The answer here is equivocal; it is a substantial wine, with plenty of total flavour, but I fancy the fruit will drop away before the oak. **rating:** 85

best drinking 1999–2000 **best vintages** NA **drink with** Pasta carbonara • $15

Riddoch Shiraz

The wine is matured in small American oak barrels (20% new, 30% one-year-old, the balance older) for 15 months.

🍷🍷🍷🍷 **1997** Medium to full red-purple; a solid wine on the bouquet, with plenty of earthy/plummy fruit and the sensible use of oak. The rich flavours carry through onto the palate, with cherry/plum fruit flavour, and it is only on the finish that the oak becomes fractionally assertive. **rating:** 88

best drinking 1999–2004 **best vintages** NA **drink with** Barbecued steak • $15

Katnook Estate Odyssey Cabernet Sauvignon

The first vintage (1991) was released in February 1997, having been launched at Katnook's centenary celebrations in December of the preceding year. One hundred per cent Cabernet Sauvignon matured in new French oak barriques for 30 months, and then given an additional three years bottle age before release, it is produced in tiny quantities and represents the best Katnook is able to produce. The 1994 is the third release of Odyssey.

🍷🍷🍷🍷🍷 **1994** Medium to full red-purple; the bouquet is ripe and complex with masses of cedar and vanilla oak wrapped around intense mulberry and blackberry fruit. The palate is opulently sweet, ripe and rich, particularly for the 1994 vintage, with soft, lingering tannins. Fruit and oak are here in abundance, providing a head-on clash with John Riddoch Cabernet Sauvignon, and succeeding well. **rating:** 94

best drinking 2000–2010 **best vintages** '91, '92, '94 **drink with** Yearling beef • $75

kay bros amery ★★★

Kay Road, McLaren Vale, SA 5171 **region** McLaren Vale

ph (08) 8323 8211 **fax** (08) 8323 9199 **open** Mon–Fri 9–5, weekends and public holidays 12–5

winemaker Colin Kay **prod.** 6500 **est.** 1890

prod. range ($15–36 R) Sauvignon Blanc, Late Harvest Frontignac, Chardonnay, Shiraz, Block 6 Shiraz, Grenache, Cabernet Sauvignon; Port Liqueur, Muscat.

summary A traditional winery with a rich history and 16 hectares of priceless old vines; while the white wines have been variable, the red wines and fortified wines can be very good. Of particular interest is Block 6 Shiraz, made from 100-year-old vines; both vines and wine are going from strength to strength.

kellermeister ★★★

Barossa Valley Highway, Lyndoch, SA 5351 **region** Barossa Valley

ph (08) 8524 4303 **fax** (08) 8524 4880 **open** 7 days 9–6

winemaker Trevor Jones **prod.** 8000 **est.** 1970

prod. range ($8.50–26.50 CD) High Country Riesling, Show Reserve Riesling, Abendlese, Frontignan Spatlese and Auslese, Late Harvest Sylvaner, Cabernet Rosé, Black Sash Shiraz, Cabernet Sauvignon, Cabernet Shiraz, Sparkling and Fortifieds; also the Trevor Jones range (under his own label) of Virgin Chardonnay, Riesling, Cabernet Merlot and Dry Grown Barossa Shiraz.

summary Specialises in older vintage wines made in traditional fashion, an extraordinary array of which are on offer at enticing prices. There is always a range of vintages available; the wines are soft and generous, if very traditional, in style.

kellybrook ★★★☆

Fulford Road, Wonga Park, Vic 3115 **region** Yarra Valley
ph (03) 9722 1304 **fax** (03) 9722 2092 **open** Mon–Sat 9–6, Sun 11–6
winemaker Darren Kelly, Philip Kelly **prod.** 3000 **est.** 1960
prod. range ($16–40 CD) Chardonnay, Riesling, Gewürztraminer, Pinot Noir, Shiraz, Cabernet Merlot, Cabernet Shiraz, Méthode Champenoise, Champagne Cider, Apple Brandy, Liqueur Muscat, Old Vintage Tawny Port.
summary The 8-hectare vineyard is situated at Wonga Park at the entrance to the principal winegrowing areas of the Yarra Valley, replete with picnic area and a full-scale restaurant. As well as table wine, a very competent producer of both cider and apple brandy (in Calvados style). Retail distribution through Victoria, New South Wales, Queensland and South Australia.

kennedys keilor valley NR

Lot 3 Overnewton Road, Keilor, Vic 3036 **region** Sunbury
ph (03) 9311 6246 **fax** (03) 9331 6246 **open** By appointment
winemaker Peter Dredge **prod.** 300 **est.** 1994
prod. range ($16 CD) Chardonnay.
summary A newly established estate-based Chardonnay specialist, producing its only wine from 1.8 hectares of relatively newly established vineyards.

kevin sobels wines NR

Cnr Broke and Halls Roads, Pokolbin, NSW 2321 **region** Lower Hunter Valley
ph (02) 4998 7766 **fax** (02) 4998 7475 **open** 7 days 10–5
winemaker Kevin Sobels **prod.** 5000 **est.** 1992
prod. range ($12–18 CD) Chardonnay, Semillon, Traminer, Pinot Noir.
summary Veteran winemaker Kevin Sobels has found yet another home, drawing upon 8 hectares of vineyards (originally planted by the Ross Jones family) to produce wines sold almost entirely through cellar door and mail order, with limited retail representation. The cellar door offers light meals and picnic and barbecue facilities.

kilikanoon NR

Penna Lane, Skillogalee Valley, Penwortham, SA 5453 **region** Clare Valley
ph (08) 8843 4377 **fax** (08) 8843 4377 **open** Thur–Sun and public holidays 11–5
winemaker Kevin Mitchell **prod.** 2500 **est.** 1997
prod. range ($14.95–22.95 R) Morts Block Riesling, Blocks Road Riesling, Prodigal Grenache, Shiraz, Cabernet Sauvignon.
summary Kilikanoon has 6 hectares of estate vineyards at Leasingham and Penwortham. It also has a restaurant in a restored 1880s cottage which is open from Thursday to Sunday and on public holidays.

Kilikanoon Morts Block Riesling

A classic, no-frills Watervale Riesling which will blossom with age.
🍷🍷🍷🍷 **1998** Light to medium yellow-green; finer and more delicate than the Blocks Road Riesling, with lime and passionfruit aromas. The palate, likewise, is more delicate yet also intense, with lingering citrus and passionfruit on the long finish. **rating:** 87
best drinking 1999–2006 **best vintages** NA **drink with** Japanese tempura • $15.95

Kilikanoon Cabernet Sauvignon

Produced from 2.5 hectares of estate plantings, and has all the character one expects of Clare Valley Cabernet.

🍷🍷🍷🍷 **1997** Medium purple-red; the bouquet is solid, with sweet berry fruits, notes of sweet leather and earth, and a touch of vanilla oak. Ripe cassis berry fruit, vanilla oak and persistent, milky tannins run through the very attractive palate. **rating:** 87

best drinking 2002–2007 **best vintages** NA **drink with** Roast lamb • $22.95

killawarra ★★★★

Tanunda Road, Nuriootpa, SA 5355 **region** Barossa Valley
ph (08) 8560 9389 **fax** (08) 8562 1669 **open** See Penfolds
winemaker Steve Goodwin **prod.** 205 000 **est.** 1975
prod. range ($8.95–13.95 R) Only Sparkling wines: Non Vintage Brut, Vintage Brut, Brut Cremant, Premier Brut and Reserve Brut; also Non Vintage Sparkling Burgundy.
summary Purely a Southcorp brand, without any particular presence in terms of either vineyards or winery, but increasingly styled in a mode different from the Seaview or Seppelt wines. As one would expect, the wines are competitively priced, and what is more, regularly sweep all before them in national wine shows.

Killawarra Premier Brut

A blend of Pinot Noir and Chardonnay grown in the Adelaide Hills, Coonawarra and the Eden Valley, which has been a prolific show trophy and medal winner over the years.

🍷🍷🍷🍷 **1995** Medium to full straw-yellow; the bouquet has pleasant bready/yeasty overtones to delicate citrus/stone fruit aromas, and the palate is lively, with almost peachy fruit flavours, and an appropriately lingering finish. **rating:** 86

best drinking 1999–2000 **best vintages** '90, '91, '93, '94 **drink with** Hors d'oeuvres, Asian food • $13

killerby ★★★☆

Lakes Road, Capel, WA 6230 **region** Geographe
ph 1800 655 722 **fax** 1800 679 578 **open** 7 days 10–4.30
winemaker Paul Boulden **prod.** 10 000 **est.** 1973
prod. range ($20–28 CD) Semillon, Selection Sauvignon Blanc, Chardonnay, Shiraz, Cabernet Sauvignon and budget-priced April Classic White (Traminer Semillon Chardonnay blend) and April Classic Red (Shiraz Pinot Cabernet blend).
summary The members of the Killerby family are long-term residents of the southwest; Ben Killerby is the fourth generation. The 21 hectares of vines were established by Ben's father, the late Dr Barry Killerby, in 1973, and are now fully mature. The Chardonnay, in particular, is very highly rated by some critics, but I would like to see a little more succulence and concentration in the wines.

Killerby Selection Sauvignon Blanc

A wine which surprised with its exceptionally strong showing at the 1999 *Winewise* Small Makers Competition, being highly scored by all judges including myself.

🍷🍷🍷🍷🍸 **1998** Light to medium yellow-green; the bouquet is ripe and complex, with a mix of stone fruit and gooseberry aromas. The palate, too, offers ripe, tropical fruit which stood out against the numerous '98 Sauvignon Blancs from southeastern Australia and New Zealand. **rating:** 90

best drinking 1999–2000 **best vintages** '98 **drink with** Sugar-cured tuna • $20

kings creek winery ★★★★

237 Myers Road, Bittern, Vic 3918 **region** Mornington Peninsula
ph (03) 5983 2102 **fax** (03) 5983 5153 **open** 7 days 11–5
winemaker Brien Cole **prod.** 3000 **est.** 1981
prod. range ($18–24 R) Black Label Pinot Noir, Chardonnay, Cabernet Sauvignon; White Label Pinot Gris, Unwooded Chardonnay, Sauvignon Blanc, Pinot Noir.
summary Kings Creek is owned and operated by the Bell, Glover and Perraton families. Planting commenced in 1981, and the vines are now fully mature. Since 1990 the quality of the wines, particularly of the Pinot Noir and Chardonnay, has been beyond reproach.

Kings Creek Pinot Noir

Estate-grown, and vinified using a range of Burgundian techniques including 20% whole bunches in the fermentation. Matured for ten months in French oak and, like the Chardonnay from Kings Creek, a wine of real complexity and style, with many medals to its credit.

🍷🍷🍷🍷 **1998** Strong purple-red; the bouquet is bright and clean with fresh cherry, a touch of strawberry and subtle oak. The palate shows attractive ripe cherry and plum flavour with the sensible use of oak; a clean and moderately long finish. **rating:** 89

best drinking 1999–2004 **best vintages** '88, '89, '91, '92, '94, '95, '97, '98 **drink with** Chinese duck with mushrooms • $18

kingsley ★★★

6 Kingsley Court, Portland, Vic 3305 **region** Far South West Victoria
ph (03) 5523 1864 **fax** (03) 5523 1644 **open** 7 days 1–4
winemaker Contract **prod.** 1900 **est.** 1983
prod. range ($12–14 CD) Riesling, Botrytis Riesling, Late Harvest Riesling, Chardonnay, Cabernet Sauvignon.
summary Only a small part of the 10 hectares is made into wine under contract, the remainder being sold as grapes. Older vintages are usually available at cellar door. In early 1999, wines spanning the 1995 to 1998 vintages were on sale at low prices.

kingston estate ★★★☆

PO Box 67, Kingston-on-Murray, SA 5331 **region** Riverland
ph (08) 8583 0599 **fax** (08) 8583 0304 **open** Not
winemaker Bill Moularadellis, Rod Chapman **prod.** 100 000 **est.** 1979
prod. range ($12–35 R) Soft Press Chardonnay, Chardonnay, Semillon Sauvignon Blanc, Shiraz, Cabernet Sauvignon, Merlot; Reserve range of Chardonnay, Shiraz, Merlot and Petit Verdot.
summary Kingston Estate is a substantial and successful Riverland winery, crushing 10 000 tonnes a year, and exporting 80% of its production. It is only in recent years that it has turned its attention to the domestic market with national distribution. The wines are modestly priced and offer exceptionally good value for money, with some particularly good wines in the Reserve range. In more recent years Kingston Estate has set up long-term purchase contracts with growers in the Clare Valley, Adelaide Hills, Langhorne Creek and Mount Benson.

Kingston Estate Reserve Chardonnay

Only limited quantities of the Reserve wines are made, typically between 500–1000 cases, representing but a tiny fraction of the Kingston Estate output. The '91 Reserve Chardonnay won the prestigious Hyatt Advertiser Trophy for Best South Australian Chardonnay in 1994, and at its peak was a quite lovely wine. The wine is given some skin contact and is fermented in one-year-old French oak puncheons, and given extended lees contact. Whether the style is to be preferred to the more restrained varietal wine, and whether the price differential is always justified, will depend partly on the year and partly on the eye of the beholder.

🍷🍷🍷🍷 **1997** Medium to full yellow-green; quite rich, with positive fruit and subtle oak. The moderately full and smooth palate has plenty of buttery/nutty flavours in a drink-me-up style. **rating:** 87

best drinking 1999–2000 **best vintages** '91, '97 **drink with** Roast stuffed chicken • $22.95

Kingston Estate Reserve Shiraz

The winemakers of the Riverina have at last realised what is necessary to produce good red wines: far lower crop levels than they have traditionally obtained. Pruning, reduced irrigation and careful canopy management are all part of the equation, the reduction in irrigation being the most significant.

🍷🍷🍷🍷½ **1997** Medium to full red-purple; the bouquet offers solid fruit with attractive oak. There is abundant, sweet, dark cherry fruit on the palate, again showing good handling of French and American oak, and finishing with good tannins. **rating:** 90

best drinking 1999–2003 **best vintages** '97 **drink with** Spaghetti bolognese • $26.95

Kingston Estate Reserve Merlot

I doubt we shall ever see a great Merlot out of the Riverina, no matter what viticultural skills are used, simply because the climate is too warm to allow the grapes to retain full varietal character. Nonetheless, this is a creditable red wine.

🍷🍷🍷🍷 **1997** Medium to full red-purple; the bouquet is clean, smooth and sweet with subtle oak, the palate showing pleasant red berry fruit and, once again, balanced oak – the key to the success of all of the Kingston Estate Reserve wines. **rating:** 86

best drinking 1999–2003 **best vintages** NA **drink with** Veal parmigiana • $26.95

kingtree wines NR

Kingtree Road, Wellington Mills via Dardanup, WA 6326 **region** Geographe

ph (08) 9728 3050 **fax** (08) 9728 3113 **open** 7 days 12–5.30

winemaker Contract **prod.** 1000 **est.** 1991

prod. range ($16–20 CD) Riesling, Sauvignon Blanc, Gerrasse White, Cabernet Merlot.

summary Kingtree Wines, with 2.5 hectares of estate plantings, is part of the Kingtree Lodge development, a four and a half-star luxury retreat in dense Jarrah forest.

kinvarra estate NR

RMB 5141, New Norfolk, Tas 7140 **region** Southern Tasmania

ph (03) 6286 1333 **fax** (03) 6286 2026 **open** Not

winemaker Andrew Hood (Table Wine), Greg O'Keefe (Sparkling) **prod.** 90 **est.** 1990

prod. range ($13.50–15 ML) Riesling, Pinot Noir.

summary Kinvarra is the part-time occupation of David and Sue Bevan, with their wonderful 1827 homestead depicted on the label. There is only 1 hectare of vines, half riesling and half pinot noir, and most of the crop is sold to Wellington Wines and Fishburn & O'Keefe.

kirkham estate NR

3 Argyle Street, Camden, NSW 2570 **region** Other Wineries of NSW

ph (02) 4655 7722 **fax** (02) 4655 7722 **open** 7 days 11–5

winemaker Stan Aliprandi **prod.** 5000 **est.** 1993

prod. range ($9.50–15.50 CD) Traminer Riesling, Semillon, Semillon Chardonnay, Botrytis Semillon, Pinot Noir, Merlot, Cabernet Sauvignon, Tawny Port.

summary Kirkham Estate is one of six or so wine producers near Camden, a far cry from the 18 producers of the mid-nineteenth century, but still indicative of the growth of vineyards and winemakers everywhere. It is the venture of Stan Aliprandi, a former Riverina winemaker with an interesting career going back over 30 years. It draws upon 10 hectares of vineyards, planted to chardonnay, semillon, shiraz, merlot, pinot noir and cabernet sauvignon, supplemented, it would seem, by grapes (and wines) purchased elsewhere.

knappstein wines ★★★★☆

2 Pioneer Avenue, Clare, SA 5453 **region** Clare Valley
ph (08) 8842 2600 **fax** (08) 8842 3831 **open** Mon–Fri 9–5, Sat 11–5, Sun and public holidays 11–4
winemaker Andrew Hardy **prod.** 40 000 **est.** 1976
prod. range ($16.25–27.90 R) Riesling, Gewurztraminer, Fumé Blanc, Chardonnay, Sauvignon Blanc Semillon, Botrytis Riesling, Shiraz, Enterprise Reserve Shiraz, The Franc (Cabernet Franc), Cabernet Merlot, Enterprise Cabernet Sauvignon.
summary The bell has tolled, and Tim Knappstein (together with wife Annie) are now involved full-time in their Lenswood Vineyard. Petaluma stalwart Andrew Hardy has been placed in charge at Knappstein Wines, and will no doubt place his stamp on the brand. Clever label redesign and a subtle label name change (dropping the word 'Tim') are physical signs of the new order. Exports to the UK and much of Europe, Japan, Hong Kong and much of Asia.

Knappstein Semillon Sauvignon Blanc

As from 1998 sourced entirely from the Clare Valley, and with Semillon the dominant partner. That component is fermented in new Dargaud & Jaegle barriques, and retained in those barrels for three months with weekly lees stirring. Fifty per cent of the Sauvignon Blanc is cold-fermented in stainless steel, the remainder in one-year-old French oak, and matured in that oak for two months, once again with weekly stirring. The parcels are finally blended, clarified and bottled in December. All in all, very sophisticated and time-consuming making.

🍷🍷🍷🍷🍷 **1998** Bright yellow-green; the bouquet is very aromatic and tangy, with complex high-toned fruit, at once interesting and striking. The palate is rich, offering a combination of riper and more tangy/citrussy/faintly grassy characters. **rating:** 92

best drinking 1999–2003 **best vintages** '98 **drink with** Rich fish dishes • $19.90

Knappstein Shiraz

Sourced from five small vineyards in four separate subdistricts of the Clare Valley, reflected in a five-week span in picking dates. Partly fermented in open fermenters, and part in closed (Potter) fermenters, with the wine in the latter being kept on skins for one month after the conclusion of fermentation. Fifteen per cent of the wine is matured in new Nevers oak, the balance in one and two-year-old French oak; in all the wine spends between 12 and 15 months in wood before being bottled in or about October of the year following vintage.

🍷🍷🍷🍷 **1997** Medium red-purple; the bouquet is clean, moderately intense, with touches of vanilla bean to the faintly earthy fruit. The smooth, harmonious palate offers more sweet fruit than the bouquet in a delicate sweet cherry spectrum; finishes with fine tannins. **rating:** 89

best drinking 2000–2007 **best vintages** NA **drink with** Moroccan lamb • $20.90

Knappstein Enterprise Shiraz

Enterprise is the flagship label for Knappstein, first introduced with the 1994 vintage. It comes from a single 32-year-old block of vines on the eastern slopes of the Clare Valley; the terra rossa soil produces small crops of intense fruit using minimal irrigation. The wine spends 15 months in a mix of one and two-year-old French oak before being eggwhite-fined and bottled without filtration. The outstanding quality of the wine is a direct reflection of the vineyard and the minimalist winemaker intervention.

🍷🍷🍷🍷🍷 **1996** Dense red-purple; the powerful and concentrated bouquet and palate are packed with dark berry/blackberry fruit with an ever so faintly spicy finish. The tannins are beautifully modulated, and the oak appropriately subtle. **rating:** 94

best drinking 2001–2016 **best vintages** '94, '96 **drink with** Rich red meat dishes • $27.90

Knappstein Cabernet Merlot

A blend of Cabernet Sauvignon (dominant) and Merlot, 82% from the Clare Valley and 18% from the Adelaide Hills. The wine is fermented in a mix of open and closed fermenters, and spends between 11 and 14 months in a mix of French oak and a small percentage of American oak.

🍷🍷🍷🍷 **1997** Medium red-purple; the bouquet is clean, moderately intense, and with no outstanding or obvious fruit or oak characters, other than a faintly olive-like aroma. The palate is quite long, with a touch of herbaceous fruit which is Bordeaux-like, but which is less striking than most of the Knappstein wines. **rating:** 85

best drinking 1999–2003 **best vintages** NA **drink with** Roast veal • $20.90

Knappstein Enterprise Cabernet Sauvignon

This is the second Enterprise bottling of Cabernet Sauvignon, coming from the best parcels of cabernet grown on the Knappstein Vineyard which was planted in 1971. It also contains a small percentage of Malbec, a traditional Clare blend. It is fermented in a mix of open and closed fermenters, and spends between 18 and 24 months in oak, before being blended and bottled more than two years after vintage.

🍷🍷🍷🍷🍷 **1996** Dense red; the bouquet is concentrated and complex with a mix of earth, bitter chocolate and blackberry fruit; the palate offers a similar rich melange of blackcurrant, cassis and chocolate fruit supported by lingering tannins. The oak has been deliberately underplayed throughout. **rating:** 92

best drinking 2001–2011 **best vintages** '96 **drink with** Thick-cut lamb loin chops • $27.90

knight granite hills ★★★☆

Burke and Wills Track, Baynton RSD 391, Kyneton, Vic 3444 **region** Macedon

ph (03) 5423 7264 **fax** (03) 5423 7288 **open** Mon–Sat 10–6, Sun 12–6

winemaker Llew Knight **prod.** 6000 **est.** 1970

prod. range ($14–28.50 R) Riesling, Chardonnay, Pinot Noir, Shiraz, Cabernet Sauvignon; also MICA Unwooded White and Cabernet Sauvignon.

summary Knight Granite Hills was one of the early pacesetters, indeed the first pacesetter, for cool-climate, spicy Shiraz and intense Riesling. Revived marketing in a buoyant market, and the introduction of the lesser-priced MICA range, has resulted in greater activity; plantings remain the same at 9 hectares of mature, low-yielding vineyards.

Knight Granite Hills Riesling

The 2 hectares of estate plantings on the bare, windswept hills are low-yielding and inevitably late-ripening, producing a wine with considerable intensity of flavour in most vintages. The '92 was re-released as a mature wine at the end of 1997, marking the start of a mature-release programme.

🍷🍷🍷🍷🍷 **1998** Light to medium yellow-green; the aromas are crisp and clean, with delicate lime and toast; the palate has excellent length and intensity with lingering lime/citrus flavours. Will richly repay cellaring. **rating:** 91

best drinking 2003–2008 **best vintages** '86, '90, '93, '94, '98 **drink with** Sugar-cured tuna • $15

knowland estate NR

Mount Vincent Road, Running Stream, NSW 2850 **region** Mudgee
ph (02) 6358 8420 **fax** (02) 6358 8423 **open** By appointment
winemaker Peter Knowland **prod.** 250 **est.** 1990
prod. range ($12.50–18 CD) Mt Vincent Sauvignon Blanc, Orange Pinot Noir, Mt Vincent Pinot Noir, Wellington Cabernet.
summary The former Mount Vincent Winery which sells much of its grape production from the 3.5 hectares of vineyards to other makers, but which proposes to increase production under its own label.

kominos ★★★

New England Highway, Severnlea, Qld 4352 **region** Granite Belt
ph (07) 4683 4311 **fax** (07) 4683 4291 **open** 7 days 9–5
winemaker Tony Comino **prod.** 4000 **est.** 1976
prod. range ($10–13 CD) Riesling, Sauvignon Blanc, Chardonnay, Vin Doux, White Shiraz, Nouveau, Shiraz, Cabernet Merlot, Cabernet Sauvignon.
summary Tony Comino is a dedicated viticulturist and winemaker; and together with his father, battled hard to prevent ACI obtaining a monopoly on glass production in Australia, foreseeing many of the things which have in fact occurred. However, Kominos keeps a very low profile, selling all of its wine through cellar door and mailing list. No recent tastings.

kongwak hills winery NR

1030 Korumburra–Wonthaggi Road, Kongwak, Vic 3951 **region** South Gippsland
ph (03) 5657 3267 **open** Weekends and public holidays 10–5
winemaker Peter Kimmer **prod.** 350 **est.** 1989
prod. range ($10–25 CD) Riesling, Pinot Noir, Shiraz, Cabernet Malbec.
summary Peter and Jenny Kimmer started the development of their vineyard in 1989, and now have half a hectare each of cabernet sauvignon, shiraz and pinot noir, together with lesser quantities of malbec, merlot and riesling. Most of the wines are sold at cellar door, with limited distribution in Melbourne through Woods Wines Pty Ltd of Fitzroy.

Kongwak Hills Pinot Noir

The best of the Kongwak Hills current releases, even if slightly rustic.
🍷🍷🍷🍷 **1997** Full red-purple; the bouquet offers a mix of foresty, briary and plummy fruit, with distinct varietal character. The palate is surprisingly ripe and soft for the 12 degrees alcohol, again with positively flavoured plummy, foresty fruit. **rating:** 84
best drinking 1999–2001 **best vintages** NA **drink with** Barbecued quail • $21

koppamurra ★★★☆

Joanna via Naracoorte, SA 5271 **region** Wrattonbully
ph (08) 8271 4127 **fax** (08) 8271 0726 **open** By appointment
winemaker John Greenshields **prod.** 2500 **est.** 1973
prod. range ($8.50–16 ML) Riesling, Autumn Pick Riesling, Botrytis Riesling, Chardonnay, Pinot Meunier, Cabernet Merlot, Merlot, Cabernet Sauvignon, Two Cabernets, McLaren Vale Muscat.
summary Which Hollywood actress was it who said 'I don't care what they say about me, as long as they spell my name right'? This might be the motto for Koppamurra, which became embroiled in a bitter argument over the use of the name Koppamurra for the region in which its vineyards are situated, and which through what seems to be sheer bloody mindedness by various of the parties involved, is now known as Wrattonbully. The wines have limited retail distribution in the eastern States, and are exported to the US, UK and EU.

kraanwood NR

8 Woodies Place, Richmond, Tas 7025 **region** Southern Tasmania
ph (03) 6260 2540 **open** Not
winemaker Frank van der Kraan, Alan Bird (Consultant) **prod.** 150 **est.** 1994
prod. range ($16.20 ML) Schonburger, Montage, Pinot Noir.
summary Frank van der Kraan and wife Barbara established their half-hectare vineyard Kraanwood between 1994 and 1995, with approximately equal plantings of pinot noir, chardonnay and cabernet sauvignon. Frank van der Kraan also manages the 1-hectare Pembroke Vineyard, and procures from it small quantities of schonberger, chardonnay, riesling and sauvignon blanc. The Kraanwood wines are made with help from Alan Bird of Palmara.

krondorf ★★★☆

Krondorf Road, Tanunda, SA 5352 **region** Barossa Valley
ph (08) 8563 2145 **fax** (08) 8562 3055 **open** 7 days 10–5
winemaker Nick Walker **prod.** 30 000 **est.** 1978
prod. range ($9–18 R) Barossa Valley Riesling, Barossa Valley Chablis, Family Reserve Chardonnay, Semillon, Frontignac Spätlese, Coonawarra Shiraz, Shiraz Cabernet, Family Reserve Cabernet Sauvignon; Show Reserve Chardonnay and Cabernet Sauvignon are top-end wines.
summary Part of the Mildara Blass Group, with a tightly focused and controlled range of wines. The Show Reserve Chardonnay is of particular merit, but all of the wines exhibit the technical gloss and represent the value-for-money expected of one of Australia's foremost wine-producing groups. Distribution via Mildara Blass.

Krondorf Family Reserve Cabernet Sauvignon

A new range, and I have to admit to a strong allergic reaction to the concept of Family Reserve being produced by a winery owned by one of the largest brewing companies in the world, Fosters Brewing Group. Brand manager and label quibbles to one side, this takes Cabernet Sauvignon from McLaren Vale, Coonawarra and Barossa according to the vintage, part barrel-fermented in French oak and then matured in a mix of French and American. All in all, deserves better than the name it has been given. The inaugural 1995 vintage was very good, the follow-on vintages of '96 and '97 quite superb.

🍷🍷🍷🍷🍷 **1997** Medium to full red-purple; there is abundant ripe chocolate and berry fruit with well-handled oak on the bouquet. The palate is quite outstanding, very rich, with lovely sweet fruit and sensuous oak handling. Barossa-sourced. **rating:** 94

🍷🍷🍷🍷🍷 **1996** Medium to full red-purple. Both on bouquet and palate this is a wonderfully elegant, fruit-driven wine showing very pure and correct cabernet varietal character. Even though I find the name derisory, the wine itself is excellent. Gold medal 1998 National Wine Show. **rating:** 94

best drinking 2002–2010 **best vintages** '95, '96, '97 **drink with** Tongue • $17

---

kulkunbulla ★★★★

Brokenback Estate, Cnr Broke and Hermitage Roads, Pokolbin, NSW 2320 **region** Lower Hunter Valley
ph (02) 9954 1873 **fax** (02) 9954 1930 **open** Not
winemaker David Lowe (Contract) **prod.** 2400 **est.** 1996
prod. range ($21–29.50 ML) Hunter Valley Semillon, The Glandore Semillon, Hunter Valley Chardonnay, Nullabor Chardonnay, The Brokenback Chardonnay.
summary Kulkunbulla is owned by a relatively small Sydney-based company, headed by Gavin Lennard, and which has purchased part of the Brokenback Estate in the Hunter Valley formerly owned by Rothbury. For the time being all Kulkunbulla's wines are sold by mail order, with a sophisticated brochure entitled Vinsight. The company got away to a flying start with its first

vintage in 1997; notwithstanding the rain, the Kulkunbulla Brokenback won the trophy for Best 1997 Chardonnay at the Hunter Valley Wine Show of that year, a dream debut.

Kulkunbulla Hunter Valley Semillon

Produced from three rows of hand-picked 30-year-old semillon; only the free-run juice is used, and cold-fermented in stainless steel.

ΥΥΥΥ **1998** Medium yellow-green; the bouquet is quite generous and full with touches of thyme, a hint of honey, and also touches of mineral. The palate has developed considerable flavour early in its life, with honey offset by a twist of citrus. Good length, but will develop relatively quickly. **rating:** 85

best drinking 1999–2004 **best vintages** NA **drink with** Seafood salad • $25

Kulkunbulla The Glandore Semillon

Glandore was the original name of the Brokenback vineyard in the 1940s before it went out of production. (It was replanted in the late 1960s and given its present name.) This is the top-end Semillon of Kulkunbulla.

ΥΥΥΥ **1998** Medium yellow-green; the bouquet offers a varietally correct mix of grass, mineral and citrus aromas. The palate is quite tight and powerful, with flavour persistence; will develop well over the medium term. **rating:** 88

best drinking 2001–2008 **best vintages** '98 **drink with** Balmain bugs • $25.50

Kulkunbulla Hunter Valley Chardonnay

Given the usual barrel-ferment and lees contact treatment. I in fact prefer the balance of this wine to the more expensive The Brokenback Chardonnay.

ΥΥΥΥΥ **1998** Medium to full yellow-green; the bouquet is quite complex, with well-handled spicy oak which does not overwhelm the fruit. There is considerable flavour, style and length on the palate, once again showing an appropriate balance between fruit and oak, finishing with good acidity. **rating:** 91

best drinking 1999–2003 **best vintages** '98 **drink with** Rich seafood • $25

Kulkunbulla The Brokenback Chardonnay

Barrel-fermented in a mix of new and one-year-old Vosges oak barriques, with lees stirring and contact, and the usual bag of tricks for top-end Chardonnay.

ΥΥΥΥΥ **1998** Medium to full yellow-green; the bouquet is concentrated, more aggressive and more complex than the Hunter Chardonnay, as is the palate, with obvious winemaker inputs all over it. In the end, I slightly prefer the gentler touch with the Hunter Valley Chardonnay. **rating:** 90

best drinking 1999–2002 **best vintages** '98 **drink with** Pan-fried chicken • $29

kyeema estate ★★★★

PO Box 282, Belconnen, ACT 2616 **region** Canberra District

ph (02) 6254 7536 (ah) **fax** (02) 6254 7536 **open** Not

winemaker Andrew McEwin **prod.** 700 **est.** 1986

prod. range ($12–16 ML) Semillon, Chardonnay, Shiraz, Cabernet Merlot.

summary Part-time winemaker, part-time wine critic (with *Winewise* magazine) Andrew McEwin produces wines full of flavour and character; every wine released under the Kyeema Estate label has won a show award of some description.

laanecoorie ★★★

Bendigo Road, Betley, Vic 3472 **region** Pyrenees

ph (03) 5468 7260 **fax** (03) 5468 7388 **open** Not

winemaker John Ellis (Contract) **prod.** 1500 **est.** 1982

prod. range ($20 R) A single Bordeaux-blend dry red of Cabernet Franc, Cabernet Sauvignon and Merlot in roughly equal proportions.

summary John McQuilten's 7.5-hectare vineyard produces grapes of consistently high quality, and competent contract-winemaking by John Ellis at Hanging Rock has done the rest.

ladbroke grove NR

Coonawarra Road, Penola, SA 5277 **region** Coonawarra

ph (08) 8737 2082 **fax** (08) 8762 3236 **open** 7 days 10–4

winemaker Ken Ward **prod.** 800 **est.** 1982

prod. range ($8–14 CD) Riesling, Late Picked Riesling, Shiraz, Premium Shiraz.

summary Relaunched with both standard and Premium Shiraz after a hiatus; wine quality has been variable, but it does have 2 hectares of hand-pruned shiraz planted by John Redman in the 1960s upon which to draw.

lake barrington estate ★★★☆

1133-1136 West Kentish Road, West Kentish, Tas 7306 **region** Northern Tasmania

ph (03) 6491 1249 **fax** (03) 6334 2892 **open** Wed–Sun 10–5 (Nov–Apr)

winemaker Steve Lubiana (Sparkling), Andrew Hood (Table), both Contract **prod.** 600 **est.** 1988

prod. range ($14–24 CD) Previously, Riesling, Chardonnay, Pinot Noir, Cabernet Sauvignon; henceforth only sparkling, including Alexandra Méthode Champenoise.

summary Lake Barrington Estate is owned by the vivacious and energetic Maree Taylor, and takes its name from the adjacent Lake Barrington, 30 kilometres south of Devonport, on the northern coast of Tasmania. There are picnic facilities at the vineyard, and, needless to say, the scenery is very beautiful.

lake breeze wines ★★★★☆

Step Road, Langhorne Creek, SA 5255 **region** Langhorne Creek

ph (08) 8537 3017 **fax** (08) 8537 3267 **open** 7 days 10–5

winemaker Greg Follett **prod.** 10 000 **est.** 1987

prod. range ($10–32 CD) Chardonnay, White Frontignac, Grenache, Cabernet Sauvignon, Shiraz, Bernoota (Cabernet Shiraz), Tawny Port. The premium Winemakers Selection range was introduced in 1996.

summary The Folletts have been farmers at Langhorne Creek since 1880, grape growers since the 1930s. Since 1987 a small proportion of their grapes has been made into wine, and a cellar-door sales facility was opened in early 1991. The quality of the releases has been exemplary, the new Winemakers Selection red wines particularly striking. Retail distribution in Melbourne and Sydney is now augmented by exports to the US and UK.

Lake Breeze Winemakers Selection Shiraz

Sold only from cellar door and by mail order. First tasted early in 1998, when it was still settling down after bottling, and the components still coming together. Nine months later it was virtually unrecognisable, having improved enormously – albeit consistently with the performance of many of the '96 vintage reds.

🍷🍷🍷🍷🍷 **1996** Full red-purple; very concentrated and rich black cherry fruit and lots of spicy oak. The palate shows the same abundant mix of sweet spicy fruit and oak interwoven. The texture is soft, rich and chewy, and notwithstanding the masses of sweet vanilla oak, the total flavour is irresistible. **rating:** 94

best drinking 2001–2010 **best vintages** '96 **drink with** Leave it in the cellar • $32

lake george winery ★★★

Federal Highway, Collector, NSW 2581 **region** Canberra District
ph (02) 4848 0039 **fax** (02) 4848 0039 **open** Not
winemaker Dr Edgar F Riek **prod.** 500 **est.** 1971
prod. range ($25–27 R) Chardonnay, Semillon, Sauternes, Pinot Noir, Cabernet Sauvignon, Merlot, Fortifieds.
summary Dr Edgar Riek is an inquisitive, iconoclastic winemaker who is not content with his role as Godfather and founder of the Canberra district, forever experimenting and innovating. His fortified wines, vintaged in northeastern Victoria but matured at Lake George, are very good. By 1998, however, Edgar Riek was looking for a successor, and threatening to hang up his boots. We shall see.

lake's folly ★★★★★

Broke Road, Pokolbin, NSW 2321 **region** Lower Hunter Valley
ph (02) 4998 7507 **fax** (02) 4998 7322 **open** Sat 10–4 when wine is available
winemaker Stephen Lake **prod.** 4000 **est.** 1963
prod. range ($32 CD) Simplicity itself: Chardonnay and Cabernets (with occasional small releases of Reserve Cabernets).
summary The first of the weekend wineries to produce wines for commercial sale, long revered for its Cabernet Sauvignon and thereafter its Chardonnay. Very properly, terroir and climate produce a distinct regional influence, and thereby a distinctive wine style. Some find this attractive, others are less tolerant. The winery continues to enjoy an incredibly loyal clientele, with much of each year's wine selling out quickly by mail order. A little of the wine finds its way to the UK.

Lake's Folly Chardonnay

Only 1200 cases a year (with some seasonal variation) are made from estate-grown grapes. The wine is invariably correct in style, and – unlike the Cabernets – should cause no discussion or argument, except that it seems to be getting better year by year.

🍷🍷🍷🍷🍷 **1997** Medium yellow-green; the bouquet is very smooth, with seamless integration of fruit and oak and gentle peach/melon aromas. The wine has an unexpected delicacy and elegance on entering the mouth, and gently builds on this to provide a really attractive mouthfeel. Bang on the quality of the '96. **rating:** 94

best drinking 2000–2007 **best vintages** '81, '82, '83, '84, '86, '89, '92, '94, '96, '97 **drink with** Sweetbreads • $32

Lake's Folly Cabernets

Like Max Lake himself, never far from controversy; again like Max Lake, full of earthy personality. It is not a wine which can or should be judged by conventional standards; if it were to be so treated, the judgment would not do the wine justice. A slightly varying blend of 65% Cabernet Sauvignon, 15% Petit Verdot, 10% Shiraz and 10% Merlot. Cork taint affecting many older bottles is an unpredictable hazard.

1997 Strong and bright red-purple; a powerful wine, but the bottle tasted was indubitably corked and the replacement did not arrive in time for publication.

best drinking 2002–2012 **best vintages** '69, '75, '81, '87, '89, '91, '93, '96 **drink with** Rabbit, hare • $32

---

lalla gully wines ★★★☆

PO Box 377, Launceston, Tas 7250 **region** Northern Tasmania
ph (03) 6331 2325 **fax** (03) 6331 2325 **open** At Ripples The River Cafe in Launceston
winemaker Andrew Hood (Contract), Kim Seagram **prod.** 3000 **est.** 1988
prod. range ($18–19 ML) Chardonnay, Sauvignon Blanc, Pinot Noir.

summary Former owners Rod and Kim Ascui sold the vineyard recently, but continue to sell Lalla Gully wines made from the 1997 and 1998 vintages at their beautifully situated Water's Edge restaurant.

lamont wines ★★★☆

85 Bisdee Road, Millendon, WA 6056 **region** Swan District
ph (08) 9296 4485 **fax** (08) 9296 1663 **open** Wed–Sun 10–5
winemaker Mark Warren **prod.** 6000 **est.** 1978
prod. range ($9–25 CD) Riesling, Verdelho, Barrel Fermented Semillon, Chardonnay, Barrel Fermented Chardonnay, WB (White Burgundy), Sweet White, Light Red Cabernet, Cabernet, Shiraz, Merlot; Fortifieds, including Flor Fino, Amontillado and Reserve Sherry (Oloroso style).
summary Corin Lamont is the daughter of the late Jack Mann, and makes her wines in the image of those her father used to make, resplendent in their generosity. Lamont also boasts a superb restaurant, with a gallery for the sale and promotion of local arts.

langanook wines NR

Faraday Road RSD 1, Castlemaine, Vic 3450 **region** Bendigo
ph (03) 5474 8250 **fax** (03) 5474 8250 **open** Not
winemaker Matt Hunter **prod.** 600 **est.** 1985
prod. range ($15–20 ML) Chardonnay, Cabernet Sauvignon, Candlebark Cabernet Franc.
summary The Langanook vineyard was established back in 1985 (the first wines coming much later), at an altitude of 450 metres on the slopes of Mount Alexander. The wines are available through a mailing list and limited Victorian distribution through Rathdowne Cellars, Armadale Cellars, the Victorian Wine Centre and Castlemaine Cellars.

langmeil winery ★★★☆

Cnr Para and Langmeil Roads, Tanunda, SA 5352 **region** Barossa Valley
ph (08) 8563 2595 **fax** (08) 8563 3622 **open** 7 days 10–5
winemaker Paul Lindner **prod.** 6000 **est.** 1996
prod. range ($10.50–25 CD) White Frontignac, Barossa Riesling, Chardonnay, Cabernet Rosé, Shiraz, Barossa Grenache, Selwin's Lot (Cabernet blend), Fortifieds.
summary Vines were first planted at Langmeil in the 1840s, and the first winery on the site, known as Paradale Wines, opened in 1932. In 1996 cousins Carl and Richard Lindner along with brother-in-law Chris Bitter formed a partnership to acquire and refurbish the winery and its 5-hectare vineyard, planted to shiraz, including 2 hectares planted in 1846. This vineyard has now been supplemented by another vineyard acquired in 1998, taking total plantings to 14.5 hectares and including cabernet sauvignon and grenache.

Langmeil Shiraz

The team at Langmeil describe this wine as 'a classic example of the honest traditional Barossan Shiraz'. No hyperbole here; that is exactly what the wine is.

YYYY 1997 Medium to full red-purple; the bouquet shows deep, dark cherry fruit and subtle oak. In the style of the winery, the palate is powerful and concentrated, with masses of fruit supported by tannins which run through the length of the palate. Oak in happy restraint. **rating:** 86

best drinking 1999–2006 **best vintages** '96, '97 **drink with** Kangaroo fillet • $19.50

Langmeil Cabernet Sauvignon

Another wine made in traditional Barossa fashion, with the primary emphasis on ripe fruit.

YYYY 1997 Medium to full red-purple; there is a lot happening on the bouquet, with ripe blackcurrant and earth aromas. The palate is likewise ripe, with masses of blackberry and blackcurrant fruit on the entry, and big earthy tannins then taking over. **rating:** 85

best drinking 2000–2007 **best vintages** NA **drink with** Barbecued sausages • $16.50

---

lark hill ★★★★☆

RMB 281 Gundaroo Road, Bungendore, NSW 2621 **region** Canberra District
ph (02) 6238 1393 **fax** (02) 6238 1393 **open** 7 days 10–5
winemaker Dr David Carpenter, Sue Carpenter **prod.** 6000 **est.** 1978
prod. range ($15–27 R) Rhine Riesling, Sauvignon Blanc Semillon, Chardonnay, Late Harvest (dessert wine), Pinot Noir, Cabernet Merlot, The Canberra Fizz Methode Champenoise.
summary The Lark Hill vineyard is situated at an altitude of 860 metres, level with the observation deck on Black Mountain Tower, and offering splendid views of the Lake George Escarpment. Right from the outset, David and Sue Carpenter have made wines of real quality, style and elegance, but achieved extraordinary success at the 1997 Sydney Wine Show, topping two classes (with gold medals in each). At the 1998 Canberra Regional Wine Show, Lark Hill won two trophies, five silver and four bronze medals. All of the wines are sold ex-winery.

Lark Hill Chardonnay

Chardonnay occupies a major portion of the 4-hectare estate vineyard (with a 2-hectare satellite on a neighbouring property). The wines are barrel-fermented, and are taken through malolactic fermentation to produce a textured, complex style away from primary fruit flavours.
🍷🍷🍷🍷▽ **1997** Medium yellow-green; a rich and complex bouquet exhibits ripe fig and melon fruit married with positive, well-handled oak. That oak is a fraction assertive on the palate, but overall, a wine of real character, grip and length. **rating:** 91
best drinking 1999–2001 **best vintages** '88, '91, '92, '93, '95 **drink with** Crispy chicken • $24

Lark Hill Pinot Noir

A wine which forces me to drink my words, for I have hitherto held the view that the climate of the Canberra district is not suited to Pinot Noir. In 1996, at least, David and Sue Carpenter achieved a minor miracle with this gold medal winner from the 1997 Sydney Wine Show. The '97 and '98 vintages, while not in the class of the '96, still show good varietal character.
🍷🍷🍷🍷 **1998** The colour shows ominous development for such a young wine, already losing its purple hues. The bouquet is ripe, with plummy fruit and a touch of spice, the palate likewise showing sweet plummy fruit in distinct varietal mould. However, it shows all the signs of a wine which will age quickly; enjoy it now. **rating:** 85
best drinking 1999–2000 **best vintages** '96, '97 **drink with** Venison • $24

Lark Hill Cabernet Merlot

A blend of Cabernet Sauvignon, Cabernet Franc and Merlot which does best in warmer vintages. Overall, the move at Lark Hill is towards the earlier-ripening varieties, with the wine showing leafy/minty characters in less warm years. The '95 vintage produced the top gold medal at the 1997 Sydney Wine Show (class 56 Small Producers Dry Red Table Wine), rounding off a marvellous show for Lark Hill; the '97 is another good wine.
🍷🍷🍷🍷 **1997** Dense purple-red; a massive wine on both bouquet and palate, with black fruits, a hint of liquorice (from where I do not know) and lingering tannins. **rating:** 89
best drinking 2002–2012 **best vintages** '88, '91, '92, '93, '94 '95, '97 **drink with** Rabbit, hare • $24

latara NR

Cnr McDonalds and Deaseys Roads, Pokolbin, NSW 2320 **region** Lower Hunter Valley
ph (02) 4998 7320 **open** Sat 9–5, Sun 9–4
winemaker Iain Riggs (Contract) **prod.** 250 **est.** 1979
prod. range ($9.50–11 CD) Semillon, Cabernet Sauvignon, Shiraz.

summary The bulk of the grapes produced on the 5-hectare Latara vineyard, which was planted in 1979, are sold to Brokenwood. A small quantity is vinified for Latara and sold under its label. As one would expect, the wines are very competently made, and are of show medal standard.

latitude wines (or 2 bud spur) NR

Postal address 252 Strickland Avenue, South Hobart, Tas 7004 **region** Southern Tasmania
ph (03) 6224 1639 **fax** (03) 6233 3477 **open** Not
winemaker Andrew Hood, Michael Vishacki **prod.** NA **est.** 1996
prod. range ($NA) Sauvignon Blanc, Chardonnay, Pinot Noir.
summary Phil Barker and Anne Lasala commenced establishing 2.5 hectares of vineyard in 1996. Phil Barker has the most extraordinary qualifications, having worked as a chef for over ten years after acquiring a PhD in Botany and is now a botanist with the Tasmanian Parks and Wildlife Department. There is still much agonising about the name; Latitude may well become 2 Bud Spur, a term viticulturists are very familiar with, but which will completely confuse the average wine drinker.

laurel bank ★★★☆

130 Black Snake Lane, Granton, Tas 7030 **region** Southern Tasmania
ph (03) 6263 5977 **fax** (03) 6263 3117 **open** By appointment
winemaker Andrew Hood (Contract) **prod.** 500 **est.** 1987
prod. range ($17–23 R) Sauvignon Blanc, Pinot Noir, Cabernet Merlot.
summary Laurel (hence Laurel Bank) and Kerry Carland planted their 2-hectare vineyard in 1986. They delayed the first release of their wines for some years, and (by virtue of the number of entries they were able to make) won the trophy for Most Successful Exhibitor at the 1995 Royal Hobart Wine Show. Things have settled down since, wine quality is solid and reliable. Retail distribution through Sutherland Cellars, Melbourne.

Laurel Bank Pinot Noir

Produced from half a hectare of estate plantings. The 1997 is the best Pinot to so far come from Laurel Bank, and offers much for the future.

🍷🍷🍷🍷🍷 **1997** Medium to full red-purple; a massively rich bouquet with plum, spice and mint aromas, almost into the liquorice of Shiraz. The palate, properly but inevitably, follows the bouquet, with a depth and structure reminiscent of Paringa Estate. **rating:** 92

best drinking 2000–2004 **best vintages** '94, '95 **drink with** Tasmanian venison • $23

lauren brook ★★★☆

Eedle Terrace, Bridgetown, WA 6255 **region** Other Wineries of WA
ph (08) 9761 2676 **fax** (08) 9761 1879 **open** 7 days 11–4
winemaker Stephen Bullied **prod.** 1300 **est.** 1993
prod. range ($14.40–20.90 CD) Riesling, Bridgetown Blend, Fume, Late Harvest, Shiraz, Cabernet Sauvignon, Fortissimo.
summary Lauren Brook is established on the banks of the beautiful Blackwood River, and is the only commercial winery in the Bridgetown subregion of Mount Barker. An 80-year-old barn on the property has been renovated to contain a micro-winery and a small gallery. There is 1 hectare of estate chardonnay, supplemented by grapes purchased locally.

lavender bay NR

39 Paringa Road, Red Hill South, Vic 3937 **region** Mornington Peninsula
ph (03) 9869 4405 **fax** (03) 9869 4423 **open** Not
winemaker Garry Crittenden (Contract) **prod.** NA **est.** 1988
prod. range Chardonnay, Pinot Noir.

summary Marketing consultant Kevin Luscombe established Lavender Bay in 1988 on a spectacular 4-hectare property in Red Hill South, with its view of the Bay to Phillip Island. Tiny quantities of the first three vintages were progressively released onto the market in mid-1997, distributed through Flinders Wholesale Wines.

lawson's hill ★★★

Henry Lawson Drive, Eurunderee, Mudgee, NSW 2850 **region** Mudgee
ph (02) 6373 3953 **fax** (02) 6373 3948 **open** Mon, Thur, Fri, Sat 10–4.30, Sun 10–4
winemaker Various Contract and José Grace **prod.** 3300 **est.** 1985
prod. range ($11–39 CD) Chardonnay, Verdelho, Sauvignon Blanc, Riesling, Traminer Riesling, Louisa Rose, Cabernet Merlot, Pinot Noir Gamay, Reserve Dryland Cabernet Sauvignon, Port.
summary Former music director and arranger (for musical acts in Sydney clubs) José Grace and wife June run a strongly tourist-oriented operation situated nextdoor to the Henry Lawson Memorial, offering a kaleidoscopic array of wines, produced from 8 hectares of vineyard, and made under contract. The red wines are richly representative of the deeply coloured, flavoursome Mudgee-style.

leasingham ★★★★☆

7 Dominic Street, Clare, SA 5453 **region** Clare Valley
ph (08) 8842 2555 **fax** (08) 8842 3293 **open** Mon–Fri 8.30–5.30, weekends 10–4
winemaker Kerri Thompson **prod.** 95 000 **est.** 1893
prod. range ($12–36 R) Classic Clare Riesling, Shiraz, and Cabernet Sauvignon at the top end; mid-range Bin 7 Riesling, Bin 37 Chardonnay, Bin 42 Semillon Sauvignon Blanc, Bin 56 Cabernet Malbec, Bin 61 Shiraz; finally low-priced Hutt Creek Riesling, Sauvignon Blanc, Shiraz Cabernet.
summary Successive big-company ownerships and various peregrinations in labelling and branding have not resulted in any permanent loss of identity or quality. With a core of high-quality, aged vineyards to draw on, Leasingham is in fact going from strength to strength under BRL Hardy's direction. The stentorian red wines take no prisoners, compacting densely rich fruit and layer upon layer of oak into every long-lived bottle.

Leasingham Classic Clare Riesling

Leasingham is following in the footsteps of noted Riesling producers such as Leo Buring in holding back part of its Riesling from better vintages and blocks for re-release when partially mature.

🍷🍷🍷🍷🍷 **1996** Medium yellow-green; fragrant toast and lime aromas precede a very intense, powerful palate with the lime juice and toast flavours promised by the bouquet running through a long finish. **rating:** 92

best drinking 2000–2010 **best vintages** '95, '96 **drink with** Braised neck of pork • $20

Leasingham Classic Clare Bin 7 Riesling

The wine has a rich history; the vineyards were planted in the 1940s against contemporary thinking in the 1960s when a newly graduated Tim Knappstein arrived at Leasingham, and proceeded to make Bin 7 one of Australia's most highly rated white wines. Thirty years later, the 50-year-old vines have been reworked with a new trellis and vertical spur positioning to produce even better grapes.

🍷🍷🍷🍷 **1997** Medium yellow-green; the bouquet is relatively soft with a mix of lime, toast and honey aromas, the palate similarly soft and round yet finishing with nicely balanced acidity. Indeed a classic Clare Riesling and, for that matter, a classic Australian Riesling. **rating:** 88

best drinking 1999–2007 **best vintages** NA **drink with** Vegetable terrine • $12

Leasingham Bin 61 Shiraz

A junior brother to the Classic Clare, with less weight and extract, and seemingly relying upon the use of some oak chips as well as barrels. The quality of Clare Valley Shiraz is still very evident, and the wine is exceptionally well-priced.

🍷🍷🍷🍷 **1997** Medium to full red-purple; a powerful, rich bouquet with masses of dark berry and oak evident; a head and shoulders wine in which the initial impact is substantial, but which seems to hollow out on the back palate. May correct this with time, who knows. **rating:** 85

best drinking 2002–2007 **best vintages** '88, '90, '91, '93, '94, '96 **drink with** Spiced lamb kebabs • $19

Leasingham Classic Clare Shiraz

The best Clare Valley Shiraz available to Leasingham is matured in (real) new American oak barrels, consistently producing a wine of tremendous depth and richness. Whether less would be better depends on one's personal perspective; certainly the wine has had great success in wine shows, none more so than the '94 which won the Jimmy Watson Trophy in 1995, but all vintages have won gold medals in the past ten years.

🍷🍷🍷🍷🍷 **1996** Deep purple; an exceptionally complex and exceptionally rich wine with liquorice and black fruit aromas interwoven with sweet oak on the bouquet. The palate is textured and ripe, with all of the flavours promised by the bouquet. Top gold 1998 National Wine Show. **rating:** 95

best drinking 2001–2010 **best vintages** '88, '90, '91, '92, '94, '95, '96 **drink with** Kangaroo, strong red meat, strong cheese • $36

Leasingham Bin 56 Cabernet Malbec

The blend of 85% Cabernet Sauvignon and 15% Malbec has been a Clare Valley specialty for decades; anyone lucky enough to have the '71 Bin 56 or virtually any of Wendouree's Cabernet Malbecs will need no persuasion of the merits of the blend. Its show record is second to none.

🍷🍷🍷🍷🍷 **1996** Medium to full purple-red; very sweet berry fruit aromas with that unique slightly jammy note of Malbec contributing. The palate, likewise, is rich and sweet, with layered mint and berry flavours. Cleverly handled oak is also a feature. Multiple trophy winner 1998 National Wine Show. **rating:** 96

best drinking 2001–2011 **best vintages** '88, '90, '91, '94, '95, '96 **drink with** Jugged hare • $19

Leasingham Classic Clare Cabernet Sauvignon

Like peas in a pod with the Classic Clare Shiraz, a massively – at times dauntingly – powerful wine with layer upon layer of fruit and layer upon layer of American oak, bound together with lashings of tannin and extract. In many ways, a throwback to older times.

🍷🍷🍷🍷🍷(half) **1996** Medium to full red-purple; the solid blackberry/dark berry fruit is supported by the inevitable scoop of oak always ladled into the wine. The rich, ripe, luscious blackberry/blackcurrant/mulberry palate is surrounded by appropriately balanced tannin and oak. Patience is required, but not too much else. **rating:** 92

best drinking 2006–2016 **best vintages** '88, '90, '91, '92, '93, '96 **drink with** Rich red meat dishes • $36

leconfield ★★★★☆

Penola Road, Coonawarra, SA 5263 **region** Coonawarra

ph (08) 8737 2326 **fax** (08) 8737 2285 **open** 7 days 10–5

winemaker Phillipa Treadwell **prod.** 17 000 **est.** 1974

prod. range ($14.40–28.95 CD) Riesling, Twelve Rows Commemorative Riesling, Noble Riesling, Chardonnay (Wooded and Unwooded), Merlot, Shiraz, Cabernet.

summary A distinguished estate with a proud, even if relatively short, history. Long renowned for its Cabernet Sauvignon, its repertoire has steadily grown with the emphasis on single varietal wines. The style overall is fruit- rather than oak-driven.

Leconfield Chardonnay

Made in two styles, the unwooded version released earlier than the barrel-fermented (and typically elegant) wooded release from Leconfield.

🍷🍷🍷🍷 **1998** Medium yellow-green; the bouquet shows clever use of gently spicy oak on melon and fig fruit. The palate is fruit-driven, with quite sweet fruit on the mid-palate, yet not at all heavy. Subtle oak; nicely balanced. **rating:** 88

best drinking 1999–2003 **best vintages** '98 **drink with** Robe lobster • $17.50

Leconfield Shiraz

A wine which is not produced every vintage, but only in those years in which the fruit is judged to have gained sufficient ripeness and depth. In 1995 the wine won vinous Tattslotto, walking away with a truckload of trophies from the Adelaide Wine Show, including the title of 'Winemaker of the Year' for Ralph Fowler (and a business-class trip around the world for two).

🍷🍷🍷🍷 **1997** Light to medium red-purple; the bouquet is spicy, with pronounced gamey characters right on the edge of acceptability and some mint. The palate reassures the taster that gamey notes of the bouquet are in fact varietally derived, with light spicy berry Rhône characters, and again a touch of mint. **rating:** 86

best drinking 1999–2003 **best vintages** '88, '90, '91, '94, '95, '96 **drink with** Beef casserole • $24

Leconfield Coonawarra Merlot

A relatively early entrant in the Merlot stakes; Coonawarra has the appropriate climate for the variety, but winemakers everywhere seem to have different ideas about style.

🍷🍷🍷🍷🍷 **1997** Medium red-purple; the bouquet offers very ripe plummy fruit and subtle oak, with similar ripe, plummy fruit flavours the major force on the palate, fleshed out with some cedary oak and soft tannins. Yet another of the infinitely variable faces of Merlot. **rating:** 90

best drinking 2002–2009 **best vintages** NA **drink with** Duck casserole • $28.95

leeuwin estate ★★★★★

Stevens Road, Margaret River, WA 6285 **region** Margaret River

ph (08) 9757 6253 **fax** (08) 9430 5687 **open** 7 days 10.30–4.30

winemaker Bob Cartwright **prod.** 40 000 **est.** 1974

prod. range ($15–67 CD) Art Series Chardonnay, Riesling, Sauvignon Blanc, Pinot Noir, Cabernet Sauvignon; Prelude Classic Dry White, Chardonnay, Pinot Noir, Cabernet Sauvignon are lower-priced alternatives, with a non-vintage Prelude blended white the cheapest wine on the list.

summary Leeuwin Estate's Chardonnay is, in my opinion, Australia's finest example based on the wines of the last 15 years. The Cabernet Sauvignon, too, is an excellent wine with great style and character. Almost inevitably, the other wines in the portfolio are not in the same Olympian class, although the Prelude Chardonnay and Sauvignon Blanc are impressive at their lower price level. The '96 Chardonnay and '95 Cabernet Sauvignon were released as this book was going to print; I wish to suspend judgment until I have had the opportunity of tasting (as opposed to drinking) the wines.

lefroy brook NR

Glauder Road, Pemberton, WA 6260 **region** Pemberton

ph (08) 9386 8385 **open** Not

winemaker Peter Fimmel (Contract) **prod.** 350 **est.** 1982

prod. range ($21.95 R) Chardonnay, Pinot Noir.

summary Owned by Perth residents Pat and Barbara Holt, the former a graduate in biochemistry and microbiology working in medical research, but with a passion for Burgundy. The 1.5 hectares of vines are now both netted and fenced with steel mesh, producing wines which, on tastings to date, are outside the mainstream.

leland estate ★★★★

PO Lenswood, SA 5240 **region** Adelaide Hills
ph (08) 8389 6928 **open** Not
winemaker Robb Cootes **prod.** 1000 **est.** 1986
prod. range ($13–20 CD) Sauvignon Blanc (piercingly pure and fragrant), Pinot Noir, Adele (Sparkling).
summary Former Yalumba senior winemaker Robb Cootes, with a Master of Science degree, deliberately opted out of mainstream life when he established Leland Estate, living in a split-level, one-roomed house built from timber salvaged from trees killed in the Ash Wednesday bushfires. The Sauvignon Blanc is usually superb. A tiny part of the production finds its way to Victoria where it is distributed by Dilettarre.

lengs & cooter NR

24 Lindsay Terrace, Belair, SA 5042 **region** Other Wineries of SA
ph (08) 8278 3998 **fax** (08) 8278 3998 **open** Not
winemaker Contract **prod.** 2500 **est.** 1993
prod. range ($14–24 ML) Watervale Riesling, Clare Valley Semillon, Old Bush Vines Grenache, Clare Valley Old Vines Shiraz, Victor (Grenache Shiraz), Swinton (Cabernet blend)
summary Carel Lengs and Colin Cooter began making wine as a hobby in the early 1980s. Each had (and has) a full-time occupation outside the wine industry, and it was all strictly for fun. One thing has led to another, and although they still possess neither vineyards nor what might truly be described as a winery, the wines have graduated to big boy status, winning gold medals at national wine shows and receiving critical acclaim from writers across Australia. However, I am not sure the current releases are so exciting.

Lengs & Cooter Watervale Riesling

Watervale riesling is the base, competent winemaking does the rest.

🍷🍷🍷🍷 **1998** Medium yellow-green; the bouquet is clean, of medium intensity, with a mix of grassy herbal, mineral and faintly spicy fruit; the light, fresh and delicate palate picks up the aromas of the bouquet, finishing crisply dry. **rating:** 87

best drinking 1999–2004 **best vintages** '98 **drink with** Summer salads • $14

Lengs & Cooter Clare Valley Semillon

Produced from relatively old vine semillon grown in the Clare Valley; wisely, the makers have been content to let the fruit speak for itself without over-extraction or distracting oak.

🍷🍷🍷🍷 **1998** Light to medium green-yellow; the bouquet is quite aromatic, with a faintly spicy scent and more traditional herb and mineral notes. There is plenty of flavour on the palate running through to and building on the finish. **rating:** 86

best drinking 2001–2007 **best vintages** NA **drink with** Yabbies • $14

Lengs & Cooter Swinton

Swinton takes its name from a property established by the forefathers of the proprietors in 1849 in the Sturt River Gorge of the Coromandel Valley. The property was subdivided in the late 1960s, but a small planting of cabernet sauvignon was established in 1971, and it is this planting, blended with Shiraz and Merlot from Blewitt Springs, which produces Swinton.

🍷🍷🍷🍷 **1997** Medium red-purple; very pronounced high toast oak provides the initial impact on the bouquet, with minty berry fruit following along in its wake. The palate shows that there is, in fact, good depth to the fruit, although the tannins are a little rough and that oak needs to settle down. All of these things should happen with time. **rating:** 87

best drinking 2007–2012 **best vintages** '96 **drink with** Leave it in the cellar • $20

lenswood vineyards ★★★★★

3 Cyril John Court, Athelstone, SA 5076 **region** Adelaide Hills

ph (08) 8389 8111 **fax** (08) 8389 8555 **open** Not

winemaker Tim Knappstein **prod.** 10 000 **est.** 1981

prod. range ($24.75–41.25 R) Semillon, Sauvignon Blanc, Chardonnay, Pinot Noir, The Palatine, Cabernets.

summary Lenswood Vineyards is now the sole (and full-time) occupation of Tim and Annie Knappstein, Tim Knappstein having retired from the winery which bears his name, and having sold most of the Clare vineyards to Petaluma (along with the wine business). With 25.5 hectares of close-planted, vertically trained vineyards maintained to the exacting standards of Tim Knappstein, the business will undoubtedly add to the reputation of the Adelaide Hills as an ultra-premium area. Complex Chardonnay, intense Sauvignon Blanc and broodingly powerful yet stylish Pinot Noir are trailblazers. The wines are exported to the UK, US, Germany and Singapore.

Lenswood Vineyards Semillon

An extension to the Lenswood range of wines, and an impressive addition at that. Three-quarters of the wine is barrel-fermented in a mix of new and one-year-old French oak barriques, and kept on its lees for five months with fortnightly stirring. The balance is stainless steel-fermented and kept in steel until blending and bottling.

🍷🍷🍷🍷🍸 **1998** Medium to full yellow-green; a sophisticated style, with spicy barrel-ferment oak lifting the bouquet; the palate is well balanced, with the oak less assertive than it is on the bouquet, showing grass/citrus varietal fruit and good length. **rating:** 90

best drinking 2000–2005 **best vintages** '98 **drink with** Pork rillettes • $24.75

Lenswood Vineyards Sauvignon Blanc

Produced from 7.6 hectares, which is in fact the largest varietal planting at Lenswood Vineyards (22 hectares in all) and which makes a quite superlative wine.

🍷🍷🍷🍷🍸 **1998** Medium yellow-green; the bouquet is not particularly intense, but has clearly defined gooseberry and passionfruit aromas, the palate at once delicate yet flavoursome, with continuing varietal definition. Good balance and length; a success for the vintage. **rating:** 90

best drinking 1999–2000 **best vintages** '94, '95, '97, '98 **drink with** Shellfish • $23

Lenswood Vineyards Chardonnay

Produced from 2.2 hectares of estate vineyards planted in 1981 and 1984. Only the best wine is chosen for release under the Lenswood Vineyards label, the remainder being disposed of elsewhere. The releases to date have been of very high quality.

🍷🍷🍷🍷🍸 **1997** Medium to full yellow-green; there is lots of power and concentration on the bouquet, with strong but stylish barrel-ferment influences. The palate has tangy melon and citrus fruit complexed both by barrel-ferment and mlf influences. **rating:** 93

best drinking 1999–2005 **best vintages** '93, '94, '95, '96, '97 **drink with** Terrine of smoked salmon • $32.20

Lenswood Vineyards Pinot Noir

Since a stellar debut in 1990, has been quite outstanding, significantly outperforming other producers in the region, and leaving no doubt that the Lenswood area will in time become a most important producer of Pinot Noir. As Tim Knappstein himself observes, the style is fuller, riper and more robust than that of the Yarra Valley. It has found much favour, and the Lenswood

Pinots have established themselves as leading examples of the fuller style of Australian Pinot Noir. The '96 won a number of major awards and gold medals in 1997, and the '97 is an even better wine.

▼▼▼▼▼ **1997** Light to medium red-purple; the bouquet opens up with moderately intense plum, forest floor and spice aromas; it is the intense and lingering palate which lifts the wine into the highest class. **rating:** 97

best drinking 1999–2004 **best vintages** '91, '93, '94, '95, '96, '97 **drink with** Quail, hare • $38

Lenswood Vineyards The Palatine

A new wine for Lenswood Vineyards, produced from its estate plantings of cabernet sauvignon, merlot and malbec. I suspect it will do best in vintages such as 1997 and 1998, but time alone will show. This is certainly a very promising start.

▼▼▼▼▽ **1997** Dense purple-red; solid, dark berry plum fruit has soaked up the fine oak on the bouquet. A powerful wine, with fully ripe fruit, and long but persistent tannins. A most interesting outcome for the Adelaide Hills. **rating:** 90

best drinking 2002–2012 **best vintages** '97 **drink with** Roast venison • $41.25

lenton brae estate NR

Willyabrup Valley, Margaret River, WA 6285 **region** Margaret River
ph (08) 9755 6255 **fax** (08) 9755 6268 **open** 7 days 10–6
winemaker Edward Tomlinson **prod.** NFP **est.** 1983
prod. range ($16–21 CD) Chardonnay, Semillon Sauvignon Blanc, Sauvignon Blanc, Late Harvest, Cabernet Sauvignon, Cabernet Merlot.
summary Former architect, town-planner and political wine activist Bruce Tomlinson built a strikingly beautiful winery, but would not stand for criticism of his wines. Son Edward is more relaxed and is in fact making wines which require no criticism.

Lenton Brae Semillon Sauvignon Blanc

In a similar mould to the Sauvignon Blanc, although more complex. The '95 thoroughly deserved the gold medal it won at the Mount Barker Wine Show of that year, but the '98 is a return to less exalted heights.

▼▼▼▽ **1998** Medium yellow-green; the bouquet is clean and smooth with some tropical fruit characters, the palate having good length but a slightly grippy finish. Solid but unspectacular. **rating:** 82

best drinking 2000–2003 **best vintages** '94, '95 **drink with** Asian dishes • $16

Lenton Brae Chardonnay

One of a number of very impressive white wines coming from Lenton Brae over the past few years, benefiting both from mature vines and astute winemaking influences.

▼▼▼▼▽ **1997** Light to medium yellow-green; a stylish and elegant wine from start to finish, with a subtle yet complex bouquet of melon, fig and a touch of cream. The fine-flavoured palate, with fig and melon flavours comes together deliciously. **rating:** 93

best drinking 1999–2003 **best vintages** '97 **drink with** Smoked chicken • $20

Lenton Brae Cabernet Sauvignon

Another wine to mark the emphatic change for the better in the wines of Lenton Brae. Estate-grown and skilfully made.

▼▼▼▼ **1996** Medium red, still with a touch of purple. The bouquet is in mainstream Margaret River-style, clean with some leafy/briary hints to the fruit. The palate is firm, well structured and flavoured in traditional regional style. Has good length, and will age well. **rating:** 88

best drinking 2001–2008 **best vintages** NA **drink with** Grilled beef • $21

leo buring ★★★★

Tanunda Road, Nuriootpa, SA 5355 **region** Barossa Valley
ph (08) 8563 2184 **fax** (08) 8563 2804 **open** Mon–Sat 10–5, Sun 1–5
winemaker Geoff Henriks **prod.** 23 000 **est.** 1931
prod. range ($9.90–23.50 R) A very much simplified range of Clare Valley Riesling, Late Picked Clare Valley Riesling, Clare Valley Chardonnay, Clare Valley Semillon and Barossa Valley/Coonawarra Cabernet Sauvignon, all under the split label introduced in 1996; the Aged Show Releases are now under the Leonay Eden Valley label.
summary Earns its high rating by virtue of being Australia's foremost producer of Rieslings over a 30-year period, with a rich legacy left by former winemaker John Vickery. But it also has the disconcerting habit of bobbing up here and there with very good wines made from other varieties, even if not so consistently.

Leo Buring Clare Valley Riesling

Sourced from a range of vineyards in the Clare Valley, and extremely well priced.

🍷🍷🍷🍷 **1998** Light green-yellow; the bouquet is of medium intensity, with tight herb and citrus aromas, leading into a well-balanced and constructed palate. The flavours run through a similar lime and herb spectrum; good length and excellent development potential. **rating:** 89

best drinking 2003–2008 **best vintages** NA **drink with** South Australian whiting • $9.90

Leo Buring Leonay Riesling

In a move seemingly directed to making the position of the Leo Buring brand as difficult as humanly possible to follow, there are now two levels of release. The first is the commercial Riesling simply called 'Leo Buring', typically released at around 9–12 months age. The other is the Leonay Riesling, either with Watervale or Eden Valley appended. This is typically released when five years old, and is in effect the Show Reserve of years gone by with the bin number omitted. In some years there is a dual release from both Watervale and the Eden Valley, in other years there is no release at all.

🍷🍷🍷🍷🍷 **1994** Eden Valley. Medium yellow-green; the bouquet is starting to unfold with sweet lime fruit at the core of still-building complexity. The palate is amazingly fresh, lively and youthful with lingering delicate lime flavours on the finish. Will go on for years; it is not surprising this is only a partial release, with a portion held back for further release in the future. **rating:** 94

🍷🍷🍷🍷🍷 **1991** Watervale. Strong yellow-green; there are classic toast, honey and lime aromas in abundance on the bouquet, the palate showing lots of bottle development but still with years and years in front of it. It is well balanced, with lovely toast and lime flavours running through to the finish. Gold medal 1998 National Wine Show. **rating:** 94

best drinking 2000–2010 **best vintages** '70, '72, '75, '77, '79, '90, '91, '92, '94 **drink with** Asparagus and salmon salad • $23.50

Leo Buring Clare Valley Shiraz

A newcomer to the Leo Buring range, and a welcome one, showing Clare Valley Shiraz to full advantage.

🍷🍷🍷🍷 **1997** Medium to full purple-red; the bouquet suggests a traditional full-bodied big extract style with a deal of oak, but the palate is a surprise packet, with opulent dark cherry fruit the leading force, supported by well-balanced oak. **rating:** 87

best drinking 2002–2007 **best vintages** '97 **drink with** Rare roast beef • NA

---

liebich wein NR

Steingarten Road, Rowland Flat, SA 5352 **region** Barossa Valley
ph (08) 8524 4543 **fax** (08) 8524 4543 **open** Weekends 11–5, Mon–Fri by appointment

winemaker Ron Liebich **prod.** 600 **est.** 1992
prod. range ($9–18 CD) Riesling of the Valleys (a blend of Barossa and Clare Valley Riesling), Riesling Traminer, Chardonnay, Cabernet Sauvignon, Bush Vine Grenache, Classic Old Barossa Tawny Port, Benno Port; bulk port constitutes major sales.
summary Liebich Wein is Barossa Deutsch for 'Love I wine'. The Liebich family has been grape growers and winemakers at Rowland Flat since 1919, with Ron 'Darky' Liebich one of the great local characters. He himself commenced making wine in 1969, but it was not until 1992 that together with wife Janet he began selling wine under the Liebich Wein label.

lillydale vineyards ★★★☆

Lot 10, Davross Court, Seville, Vic 3139 **region** Yarra Valley
ph (03) 5964 2016 **fax** (03) 5964 3009 **open** 7 days 11–5
winemaker Jim Brayne **prod.** NFP **est.** 1976
prod. range ($14–17 R) Sauvignon Blanc, Chardonnay, Pinot Noir, Cabernet Merlot.
summary Acquired by McWilliam's Wines in 1994; Alex White has departed, and Max McWilliam is in charge of the business. With a number of other major developments, notably Coonawarra and Barwang, on its plate, McWilliam's has so far adopted a softly, softly approach to Lillydale Vineyards, although a winery restaurant was opened in February 1997.

Lillydale Sauvignon Blanc

Lillydale Vineyards was one of the first producers of Sauvignon Blanc in the Yarra Valley, finding the variety very difficult to handle from a viticultural viewpoint, with inconsistent and often very low yields. Production has increased somewhat in recent years, the wine normally being made without the use of oak. It is situated in a relatively cool subregion, and the wine has always been delicate.

🍷🍷🍷🍷🍷 **1998** Light green-yellow; excellent passionfruit and gooseberry aromas, while of only light to medium intensity, are followed by a fresh, light and lively wine, with quite lovely flavour. A major success in a difficult vintage for aromatic varieties. **rating:** 93

best drinking 1998–2000 **best vintages** NA **drink with** Delicate fish dishes • $15.95

Lillydale Chardonnay

One of the pioneers of Chardonnay, with Alex White then making a fruit-driven rather than oak-driven style. The early vintages were very good by the standards of their time, but were outpaced by many others in the latter part of the 1980s. The decision has been taken to continue the lightly oaked style established by Alex White.

🍷🍷🍷🍷 **1997** Light to medium green-yellow; the bouquet is light, fresh and elegant, with gentle citrus fruit and minimal oak. The palate is light, with tangy citrus fruit almost verging on herbal. In typical light, restrained Lillydale style; will slowly evolve and grow in bottle. **rating:** 85

best drinking 2000–2004 **best vintages** '86, '88, '90, '91, '94 **drink with** Avocado • $15.95

lillypilly estate ★★★☆

Lillypilly Road, Leeton, NSW 2705 **region** Riverina
ph (02) 6953 4069 **fax** (02) 6953 4980 **open** Mon–Sat 10–5.30, Sun by appointment
winemaker Robert Fiumara **prod.** 10 000 **est.** 1982
prod. range ($9.70–21.50 CD) Riesling, Chardonnay, Sauvignon Blanc, Pound Hill Classic Dry White, Tramillon® (Traminer Semillon), Noble Riesling, Noble Traminer, Noble Semillon, Noble Muscat of Alexandria, Red Velvet® (medium sweet red), Cabernet Sauvignon, Shiraz, Vintage Port.
summary Apart from occasional Vintage Ports the best wines by far are the botrytised white wines, with the Noble Muscat of Alexandria unique to the winery; these wines have both style and intensity of flavour, and can age well. The Noble Semillon and Noble Traminer add strings to the bow.

Lillypilly Estate Semillon

Lillypilly is one of the participants in the Semillon of the Riverina promotion. A particular strength of this wine is the extremely subtle use of oak.

🍷🍷🍷 **1998** Light yellow-green; fresh, gentle slightly tropical fruit aromas lead logically into the delicate palate, where once again tropical/passionfruit flavours appear. Nicely balanced acidity; the oak influence is subliminal. **rating:** 84

best drinking 1999–2000 **best vintages** NA **drink with** Pan-fried scallops • $13

lindemans (coonawarra) ★★★★

Main Penola–Naracoorte Road, Coonawarra, SA 5263 **region** Coonawarra

ph (08) 8736 2613 **fax** (08) 8736 2959 **open** 7 days 10–5

winemaker Phillip John, Greg Clayfield **prod.** 15 000 **est.** 1908

prod. range ($9.50–37.70 R) Under the new Coonawarra Vineyard label, Riesling, Sauvignon Blanc; then come the premium red trio of Pyrus (Cabernet blend), Limestone Ridge (Shiraz Cabernet), and St George (Cabernet Sauvignon).

summary Lindemans is clearly the strongest brand other than Penfolds in the Southcorp Group, with some great vineyards and a great history. The Coonawarra vineyards are of ever-increasing importance because of the move towards regional identity in the all-important export markets, which has led to the emergence of a new range of regional/varietal labels. Whether the fullest potential of the vineyards (from a viticultural viewpoint) is being realised is a matter of debate. Worldwide distribution.

Lindemans Limestone Ridge

Arguably the most distinguished of the Coonawarra trio. A variable blend of Shiraz and Cabernet Sauvignon, varying from as much as 80% Shiraz to as little as 55%. The wine is matured in new American oak barrels for 20 months, and is given additional bottle age prior to release. The 1997 vintage was a multiple trophy winner at the 1999 Sydney Royal Wine Show (not yet released).

1997 Medium to full red-purple; lusciously ripe plummy/cherry fruit is swathed in high-quality oak on the bouquet. The same ripe cherry and plum fruit runs through a seductive palate with a long, soft finish. Oak certainly makes its contribution, but does so in balance and harmony with the fruit. **rating:** NR

best drinking 2002–2012 **best vintages** '86, '88, '90, '91, '93, '94 **drink with** Beef casserole • $36

Lindemans St George

Made entirely from cabernet sauvignon grown on the 12-hectare St George Vineyard, first produced in 1973. A very distinguished label with an at-times brilliant show record, but which has seen style vary and quality fluctuate over the years. The oak can dominate in lighter vintages, not surprising given that it spends up to 25 months in new French hogsheads and barriques.

🍷🍷🍷🍷🍷 **1995** Medium to full red-purple; the bouquet is sweet and smooth, with well-balanced fruit and oak; the palate is even more attractive, with sweet chocolate and mint-flavoured fruit, well-above-average tannin structure, and nicely judged oak. The best of the trio from 1995. **rating:** 91

best drinking 2000–2010 **best vintages** '86, '88, '90, '91 **drink with** Shoulder of lamb • $37.70

lindemans (hunter valley) NR

McDonalds Road, Pokolbin, NSW 2320 **region** Lower Hunter Valley

ph (02) 4998 7684 **fax** (02) 4998 7682 **open** Mon–Fri 9–4.30, weekends and public holidays 10–4.30

winemaker Patrick Auld **prod.** 12 000 **est.** 1843

prod. range ($7.50–95 R) Standard wines under annually changing Bin numbers of Semillon, Chablis, White Burgundy, Semillon Chardonnay, Chardonnay, Shiraz Burgundy, Hermitage, deluxe releases under Reserve Bin label, individual vineyard label (e.g. Steven) and revitalised older Classic Release label.

summary I have to declare an interest in Lindemans Hunter Valley: not only did I cut my vinous teeth on it, but since 1997 I have had group winemaker responsibility for it within Southcorp. I have long been on record in saying that its crown had slipped somewhat. A major winery upgrade in 1997–98 will help to restore things, as will the renovations to the historic Ben Ean facility. Worldwide distribution. As with Tulloch and Hungerford Hill, I rate neither the winery nor the wines with which I have been involved.

Lindemans Hunter River Semillon

As from 1998, the Lindemans Hunter Valley white wines are being made in what is to all intents and purposes a brand new winery (at Tulloch, across the road from Ben Ean). New airbag presses, temperature-controlled stainless steel fermentation and a new approach to some aspects of the handling of the wine should be evident in both this and future wines. Disclosure of interest: I have a direct influence in the making of the wine in my position as Group Winemaker Regional Wineries. However, the hands-on making is that of the Hunter Valley team.

1998 Very pale straw-yellow; the bouquet is clean, light and fresh, with attractive lemony fruit. A stylish wine, crisp and clean on the palate with nice structure. The only possible criticism is a slightly short finish. For all that, has many years improvement in front of it. **rating:** NR

best drinking 2003–2008 **best vintages** '63, '65, '68, '70, '79, '86, '87, '91, '93, '94, '98 **drink with** Yabbies • $17

Lindemans Hunter River Reserve Bin 8650 Semillon

Part of the saga of ever-changing labels and bin numbers, the 'Reserve' tag being relatively new, but which will likely be retained for the best wines from the better vintages. Incidentally, just for the record, I had nothing whatsoever to do with the making or selection of this wine, which predated my involvement.

1995 Glowing yellow-green; the bouquet is clean, with nice honeyed characters to the varietal fruit. An attractive wine on the palate with very good mouthfeel and balance, the ripeness and richness being balanced by good acidity on the finish. The problem I have with the wine is that it looks far older and more developed than a normal four-year-old Semillon, and I cannot see how it has any cellaring potential. **rating:** NR

best drinking 1999–2000 **best vintages** NA **drink with** Sautéed veal Swiss style • $22

Lindemans (Hunter Valley) Hunter River Reserve Bin 8680 Chardonnay

The wine is barrel-fermented and given extended lees contact, but bottled quite early. Shortly before its release in January 1999 it won trophies at the 1998 Perth Wine Show for Best Chardonnay and for Best Table Wine of Show, a pleasing result for Lindemans no doubt but lightning only struck once. Incidentally, I had no involvement in the making of this wine.

1996 Glowing yellow-green, very developed, looking rather like a ten-year-old New Zealand Chardonnay. The toasty/buttery bouquet leads into a soft, lush palate without any truly distinctive Chardonnay characteristics; it is simply a regional (Hunter) white wine which in the good old days would have been called White Burgundy. **rating:** NR

best drinking 1999–2000 **best vintages** NA **drink with** Vegetarian lasagne • $22

Lindemans Hunter River Shiraz

This is the 'standard' release, sourced from vineyards in the Pokolbin region, and matured in a mix of one and two-year-old French and American oak barrels for 12 months.

1995 Bin 9003. Medium red; the bouquet is of medium intensity with earthy, distinct varietal character surrounded by soft vanilla oak. The palate is likewise of medium weight, with earthy/chocolatey flavours and soft tannins running through to the finish. Pleasant but not great.
rating: NR

best drinking 2000–2005 **best vintages** '86, '87, '91, '94 **drink with** Marinated beef • $19

Lindemans Steven Vineyard Shiraz

From the long-established Lindemans-owned Steven Vineyard, which is a consistent producer of very typical but generously flavoured Shiraz (or Hermitage as it is still called on the label). The wine is matured in American oak hogsheads for 12 months, but typically shows little impact from the oak. Occasionally, as in 1994, the wine is released under a Reserve Bin number. For the record, this wine was matured in new American oak, and shows it.

1996 Medium red-purple; the soft, leathery/earthy fruit aromas are regional, but not aggressively or unpleasantly so. An attractive wine on the palate with good fruit weight and a supple, silky texture, subtle oak and a long finish. **rating:** NR

best drinking 2001–2011 **best vintages** '79, '83, '86, '87, '90, '91, '96 **drink with** Mild mature cheese • $17.95

lindemans (karadoc) ★★★★

Edey Road, Karadoc via Mildura, Vic 3496 **region** Murray Darling (Vic)
ph (03) 5051 3333 **fax** (03) 5051 3390 **open** Mon–Sun 10–4.30
winemaker Phillip John (Chief) **prod.** 10 million **est.** 1974

prod. range ($7–40 R) Bin 23 Riesling, Bin 65 Chardonnay (one of the largest selling Chardonnay brands in the world), Bin 95 Sauvignon Blanc, Bin 99 Pinot Noir, Bin 50 Shiraz Cabernet, Bin 60 Merlot are the most important in terms of volume; Cawarra range of Colombard Chardonnay, Classic Dry White, Traminer Riesling and Shiraz Cabernet; also Nyrang Semillon and Shiraz. Karadoc also produces the great fortified wines, including the premium Fino, Amontillado and Oloroso Sherries, Old Liqueur Muscat, Tokay and Madeira and fine Tawny Ports.

summary Now the production centre for all of the Lindemans and Leo Buring wines, with the exception of special lines made in the Coonawarra and Hunter wineries. The biggest and most modern single facility in Australia allowing all-important economies of scale, and the major processing centre for the beverage wine sector (casks, flagons and low-priced bottles) of the Southcorp empire. Its achievement in making several million cases of Bin 65 Chardonnay a year is extraordinary given the quality and consistency of the wines. Worldwide distribution.

Lindemans Bin 65 Chardonnay

A winemaking tour de force, and one of the world's leading brands of Chardonnay. It is sourced from no less than 14 different wine-growing regions across southeastern Australia, and shows no sign of buckling under the ever-increasing production volumes. It has been praised by wine critics around the world, and is the only wine to have ever been rated a 'Best Buy' by the *Wine Spectator* for nine consecutive vintages. Notwithstanding the ever-increasing volume (1.5 million cases in 1998) the '97 and '98 vintages have been of even better quality than the preceding years.

🍷🍷🍷🍷 **1998** Like Old Man River, just keeps rollin' along, winning show medals and plaudits from around the world. Light to medium yellow-green; there is fruit complexity on the bouquet, and cleverly infused oak. The technical brilliance shows through in the palate, with above-average length and complexity, fruit-driven but with oak lurking in the background. Has won a medal at every show in which is has been entered. **rating:** 87

best drinking 1999–2000 **best vintages** NA **drink with** Virtually anything you choose • $8.90

Lindemans Bin 50 Shiraz

Bin 50 Shiraz may not be in precisely the same league as Bin 65 Chardonnay, but is up there nonetheless in world terms. It is the second-largest selling varietal Shiraz in the US and comes in fourth in Australia. Improbable though it may seem, it is sourced largely from South Australia's Barossa Valley, and does spend time in French and American oak.

🍷🍷🍷🍷 **1997** Medium red-purple; the bouquet is clean and smooth with dark berry fruit aromas and a hint of oak. There is attractive plummy fruit on the pleasantly sweet mid-palate, and although it has a slightly short finish, it is very well priced. **rating:** 81

best drinking 1998–2001 **best vintages** NA **drink with** Pasta with meat sauce • $8.90

lindemans (padthaway) ★★★★☆

Naracoorte Road, Padthaway, SA 5271 **region** Padthaway
ph (08) 8765 5155 **fax** (08) 8765 5073 **open** Not
winemaker Phillip John, Greg Clayfield **prod.** 68 000 **est.** 1908
prod. range ($11–17 R) Sauvignon Blanc, Verdelho, Chardonnay, Pinot Noir, Cabernet Merlot, also Winemakers Reserve Chardonnay, Limestone Coast Chardonnay, Limestone Coast Shiraz and Padthaway Merlot.

summary Lindemans Padthaway Chardonnay could be said to be the best premium Chardonnay on the market in Australia, with an exceptional capacity to age. Back vintages win gold medals seemingly at will, the performance of the '94 vintage winning the Aged Chardonnay Trophy at the Liquorland National Wine Show in both 1996 and 1997 being quite remarkable. Worldwide distribution.

Lindemans Padthaway Chardonnay

A wine with a long and at times very illustrious history. The style has changed somewhat over the years, starting off as a fruit-driven wine which aged well (winning a major national wine show trophy for mature Chardonnay along the way) then becoming very oaky, but with the '94 vintage returning to its very best, and providing one of the great bargains of 1995, before going on to win the most prestigious Chardonnay trophy in Australia at the Liquorland National Wine Show for two years in succession (1996 and 1997), recalling the feats of Wolf Blass and Cape Mentelle with the Jimmy Watson Trophy in Melbourne. For the record, it is entirely barrel-fermented in a mix of new and one-year-old French (Allier and Troncais) oak, and given extended lees contact.

🍷🍷🍷🍷🍷 **1997** Medium yellow-green; the palate is in typically, opulently oaked style with nice melon fruit. The palate is rich and complex with spicy/toasty oak set against rounded melon and citrus fruit flavours. What you see is what you get. **rating:** 90

best drinking 1999–2002 **best vintages** '84, '85, '90, '94, '96, '97 **drink with** Chinese prawns with cashew nuts • $13.50

Lindemans Padthaway Cabernet Merlot

A blend of about 70% Cabernet Sauvignon and 30% Merlot which shows that, if yields are controlled, Padthaway can produce red wines of equivalent quality to its white wines. The wine was matured in a mix of new and used French and American oak, and has an abundance of fruit substance and flavour.

🍷🍷🍷🍷 **1996** Medium purple-red; the bouquet is clean, of moderate to full intensity, with very pleasant sweet mulberry/berry aromas and no green edges. A substantial wine on the palate, with red and dark berry fruit flavours and tannins which run through its length. **rating:** 88

best drinking 2000–2006 **best vintages** '91, '94, '95 **drink with** Fillet steak • $16.95

lirralirra estate ★★★

Paynes Road, Lilydale, Vic 3140 **region** Yarra Valley
ph (03) 9735 0224 **fax** (03) 9735 0224 **open** Weekends and holidays 10–6, Jan 7 days
winemaker Alan Smith **prod.** 400 **est.** 1981
prod. range ($14–20 CD) Semillon, Wooded Semillon, Semillon Sauvignon Blanc, Sauvignon Blanc, Pinot Noir, Yarra Valley Cabernets.
summary Off the beaten track, and one of the lesser-known Yarra Valley wineries; owner Alan Smith originally intended to make a Sauternes-style wine from Semillon, Sauvignon Blanc and Muscadelle, but has found the conditions do not favour the development of botrytis, and is hence producing dry red and white wines.

little river wines NR

Cnr West Swan and Forest Roads, Henley Brook, WA 6055 **region** Swan District
ph (08) 9296 4462 **fax** (08) 9296 1022 **open** Fri–Wed 10–5
winemaker Bruno de Tastes **prod.** 4000 **est.** 1934
prod. range ($15–28 CD) Chenin Blanc, Viognier, Chardonnay, Cabernet Sauvignon, Old Vines Shiraz, Cabernet Sauvignon, Grenache Shiraz, Brut de Brut, Vin Doux Late Harvest, Noble Classic.
summary Following several quick changes of ownership (and of consultant winemakers) the former Glenalwyn has gone through a period of change. It now has as its winemaker the eponymously named Count Bruno de Tastes. No recent tastings. Exports to Malaysia, Taiwan and Japan.

little's winery ★★☆

Lot 3 Palmers Lane, Pokolbin, NSW 2321 **region** Lower Hunter Valley
ph (02) 4998 7626 **fax** (02) 4998 7867 **open** 7 days 10–4.30
winemaker Ian Little **prod.** 6000 **est.** 1984
prod. range ($12–18 CD) Chardonnay, Semillon, Gewurztraminer, Late Harvest Semillon, Pinot Noir Blanc de Noir, Shiraz, Cabernet Sauvignon, Vintage Port.
summary A successful cellar-door operation with friendly service and friendly wines: aromatic, fresh and sometimes slightly sweet white wines and light, inoffensive red wines.

Little's Chardonnay

A full-bodied wine made in the mainstream Hunter fashion, opulent and ready to drink when young.
🍷🍷🍷🍷🍷 **1998** The bright green-yellow is a welcoming start; the bouquet is full, and the oak tends to sit on the top of the fruit like a floppy hat, and there is lots and lots of oak on the palate. Notwithstanding this, there is also delicious, tangy stone fruit and citrus flavour which ends up winning the war with the oak. **rating:** 90
best drinking 1999–2000 **best vintages** '98 **drink with** Pasta carbonara • $16.95

lochvie wines NR

28 Lavender Park Road, Eltham, Vic 3095 **region** Yarra Valley
ph (03) 9439 9444 **open** Weekends 9.30–5.30, weekdays by appointment
winemaker John Lewis **prod.** NFP **est.** 1985
prod. range ($10 CD) Cabernet Merlot.
summary A tiny home winery producing a single 65%/35% Cabernet Merlot blend. Since 1993 the grapes have been sold to others, and no further wines made, but the '90 to '92 wines were available at cellar door on last advice.

long gully estate ★★★

Long Gully Road, Healesville, Vic 3777 **region** Yarra Valley
ph (03) 9807 4246 **fax** (03) 9807 2213 **open** Weekends, holidays 11–5
winemaker Peter Florance **prod.** 30 000 **est.** 1982
prod. range ($8.30–29 CD) Riesling, Chardonnay, Semillon, Sauvignon Blanc, Merlot, Irmas Cabernets, Pinot Noir, Shiraz; Limited Edition Reserve Chardonnay and Cabernet.
summary One of the larger (but by no means largest) of the Yarra Valley producers which has successfully established a number of export markets over recent years. Wine quality is consistent rather than exhilarating; it is able to offer a range of wines with two to three years bottle age. Recent vineyard extensions underline the commercial success of Long Gully.

Long Gully Chardonnay

Produced from estate-grown grapes grown on 7 hectares of gentle north-facing slopes. A mini-vertical tasting in February 1997 covering the '92 to '96 vintages showed wines of extraordinarily consistent style which are ageing slowly and gracefully. The '97, by contrast, reflects the small crop and warm vintage.

🍷🍷🍷🍷🍷 **1997** Medium to full yellow-green; the bouquet is complex and powerful with barrel-ferment characters over strong fruit. The palate is even more powerful, highly flavoured, and with a faintly hot finish, but has lots of character. **rating:** 90

best drinking 1999–2003 **best vintages** '89, '92, '93, '94 **drink with** Abalone • $18

longleat ★★★

Old Weir Road, Murchison, Vic 3610 **region** Goulburn Valley
ph (03) 5826 2294 **fax** (03) 5826 2510 **open** 7 days 10–5
winemaker Alister Purbrick (Consultant) **prod.** 1500 **est.** 1975
prod. range ($10.50–17 CD) Riesling, Sauvignon Blanc, Chardonnay, Shiraz, Cabernet Sauvignon, Liqueur Muscat, Sparkling.
summary Longleat has long had a working relationship with Chateau Tahbilk, which makes the Longleat wines under contract, and buys significant quantities of grapes surplus to Longleat's requirements. The wines are always honest and full-flavoured.

longview creek vineyard NR

150 Palmer Road, Sunbury, Vic 3429 **region** Sunbury
ph (03) 9744 1050 **fax** (03) 9744 1050 **open** Sunday 11–5
winemaker David Hodgson **prod.** 150 **est.** 1988
prod. range ($16–20 CD) Chardonnay, Chenin Blanc, Pinot Noir, Cabernet Sauvignon.
summary A relatively new arrival in the Sunbury subdistrict of the Macedon region, owned by Ron and Joan Parker. A total of 3.3 hectares of chardonnay, pinot noir and chenin blanc are in production with an additional 2 hectares of shiraz coming into bearing.

lost valley winery NR

35 Yamby Road, Strath Creek, Vic 3141 **region** Central Victorian High Country
ph (03) 5797 0212 **fax** (03) 9351 2005 **open** Not
winemaker Alex White (Contract) **prod.** 1600 **est.** 1995
prod. range ($25–35 ML) Verdleho, Shiraz, Merlot, Cortese.
summary Dr Robert Ippaso planted the Lost Valley vineyard at an elevation of 450 metres on the slopes of Mount Tallarook, with 1.25 hectares of shiraz and 0.85 hectare each of merlot, verdelho and cortese, the last the only such planting in Australia. It pays homage to Dr Ippaso's birthplace in Savoie in the Franco-Italian Alps, where cortese flourishes.

lovey's estate NR

1548 Melba Highway, Yarra Glen, Vic 3775 **region** Yarra Valley
ph (03) 5965 2444 **fax** (03) 5965 2460 **open** Wed–Sun 12–5
winemaker Brian Love **prod.** 1000 **est.** 1989
prod. range ($15–24.50 CD) Sauvignon Blanc, Chardonnay, Muscadelle, Pinot Noir, Shiraz, Cabernet Sauvignon.
summary Lovey's Estate is part of a restaurant and accommodation complex situated prominently on the Melba Highway, just on the far side of Yarra Glen. The majority of the production from the 11-hectare vineyard is sold; part is made under contract at Tarrawarra, and is available through cellar door and the restaurant.

lowe family wines ★★★

9 Paterson Road, Bolwarra, NSW 2320 **region** Mudgee
ph (02) 4930 0233 **fax** (02) 4930 0233 **open** Not
winemaker David Lowe, Jane Wilson **prod.** 6000 **est.** 1987
prod. range ($18–25 ML) Semillon, Peacock Hill Chardonnay, Lawless Chardonnay, Peacock Hill Shiraz.
summary Former Rothbury winemaker David Lowe and Jane Wilson make the Lowe Family Wines at the Oakvale Winery, drawing upon 15 hectares of family-owned vineyards in Mudgee, supplemented by purchases from Orange and the Hunter Valley. Interestingly, the plantings include sangiovese, barbera and zinfandel. David and Jane also run an increasingly important contract-winemaking business.

lyre bird hill ★★★

Inverloch Road, Koonwarra, Vic 3954 **region** Gippsland
ph (03) 5664 3204 **fax** (03) 5664 3206 **open** Weekends, holidays 10–5 or by appointment
winemaker Owen Schmidt **prod.** 1000 **est.** 1986
prod. range ($11–25 CD) Riesling, Traminer, Chardonnay (wooded and unwooded), Pinot Noir, Pinot Noir Cellar Reserve, Cabernet Sauvignon, Shiraz, Rhapsody (Sparkling), Phantasy (Sparkling).
summary Former Melbourne professionals Owen and Robyn Schmidt make small quantities of estate-grown wine (the vineyard is 2.4 hectares in size), offering accommodation for three couples (RACV four-star rating) in their newly built spacious house. Shiraz has been the most successful wine to date, although the Pinot Noir has pronounced varietal character in a sappy/tomato wine style which will appeal to some. The Cabernet Sauvignon is also a pleasant, well-made wine.

Lyre Bird Hill Traminer

Over the years, the Traminer has been one of the better wines from Lyre Bird Hill, usually showing strong varietal character. This is no exception.

🍷🍷🍷🍷 **1998** Medium yellow-green; there is intense lime, lychee and rose petal varietal fruit on the bouquet, which cascades through onto the powerful, all-encompassing palate. Almost claustrophobic, and certainly very varietal. **rating:** 85

best drinking 1999–2002 **best vintages** NA **drink with** Strongly flavoured Chinese dishes • $15

Lyre Bird Hill Pinot Noir Cellar Reserve

The Lyre Bird Hill Pinot Noir comes in two forms: varietal and Cellar Reserve. The Cellar Reserve comes from selected barrels and is similar to (although slightly better than) the varietal wine.

🍷🍷🍷🍷 **1997** Medium red; the bouquet is quite solid, with a mix of briary, foresty, gamey canopy characters. The palate has some length, with touches of cherry and plum coming through the more sappy flavours. Has been handled well once the grapes came into the winery. **rating:** 86

best drinking 1999–2002 **best vintages** NA **drink with** Wild duck • $25

macquariedale estate NR

40 Rusty Lane, Branxton, NSW 2335 **region** Lower Hunter Valley
ph (02) 4938 1408 **fax** (02) 4938 1408 **open** Not
winemaker Contract **prod.** 2000 **est.** 1993
prod. range ($15–20 ML)
summary Macquariedale Estate dates back to 1993, when Ross McDonald commenced the planting of a small vineyard at Branxton. One thing led to another, and he and his family have moved from Sydney to the Hunter Valley, and significantly extended the scope of the operations by purchasing a 25-year-old vineyard to supplement production from the home vineyard. The wines are contract-made, and are sold through the mailing list and the Boutique Wine Centre in Broke Road, Pokolbin.

madew wines NR

Westering, Federal Highway, Lake George, NSW 2581 **region** Canberra District
ph (02) 4848 0026 **open** Weekends, public holidays 11–5
winemaker David Madew **prod.** 2500 **est.** 1984
prod. range ($13–20 CD) Riesling, Reserve Riesling, Semillon, Chardonnay, Phoenix (Botrytis Chardonnay), Dry Red, Merlot, Cabernets.
summary Madew Wines bowed to the urban pressure of Queanbeyan, and purchased the Westering Vineyard from Captain G P Hood some years ago. Plantings there have now increased to 9.5 hectares, with 1 hectare each of shiraz and pinot gris coming into bearing.

maglieri ★★★★☆

Douglas Gully Road, McLaren Flat, SA 5171 **region** McLaren Vale
ph (08) 8383 0177 **fax** (08) 8383 0136 **open** Mon–Sat 9–4, Sun 12–4
winemaker John Loxton **prod.** 110 000 **est.** 1972
prod. range ($18 R) While still billing itself as the 'House of Lambrusco', and still producing a range of Italian-derived styles for specialty markets within Australia, is increasingly known for the quality of its varietal table wines, spearheaded by Semillon, Cabernet Sauvignon and Shiraz, the last released in two guises: as a simple varietal, and the top-end Steve Maglieri. Typically several vintages available at any one time.
summary One of the better-kept secrets among the wine cognoscenti, but not among the many customers who drink thousands of cases of white and red Lambrusco every year, an example of niche marketing at its profitable best. Its dry red wines are invariably generously proportioned and full of character, the Shiraz particularly so – and of the highest quality. It is for these wines that the winery rating is given. Exports and distribution via Mildara Blass.

Maglieri Shiraz

Sourced primarily from estate-grown grapes in McLaren Vale and Blewitt Springs, but from time to time incorporating a 10% Coonawarra component. Matured in American oak, and invariably rich and full flavoured in style. A consistent show medal winner over the years, particularly in the better vintages. The '95 won the Shiraz Trophy at the very important and enormously influential 1997 International Wine Challenge in London. It was also a gold medal winner in the 1997 Great Australian Shiraz Challenge, and the '96 and '97 are in the same class.

🍷🍷🍷🍷🍷 **1997** Dense purple-red, promising much; rich, dark berry, dark chocolate fruit aromas on the bouquet start to fulfil that promise, the wine swelling further with wonderful luscious berry and cherry fruit on the palate. The wine is not at all heavy, the oak is subtle and the tannins fine and long. A quintessential example of McLaren Vale Shiraz. **rating:** 96

best drinking 2002–2012 **best vintages** '90, '91, '93, '95, '96, '97 **drink with** Ravioli • $18

Maglieri Merlot

I cannot remember previously tasting a Maglieri Merlot; if this is indeed the first, it's a more than creditable start.

TTTT **1997** Medium to full red-purple; solid, clean ripe berry fruit with some earth and green leaf overtones attesting to the variety. The palate offers sweet berry fruit in a legitimate varietal manifestation; good texture, fine tannins and a silky finish. **rating:** 87

best drinking 2002–2007 **best vintages** NA **drink with** Roast veal • $18

Maglieri Cabernet Sauvignon

Like the Shiraz, made primarily from grapes grown on estate vineyards in McLaren Vale and Blewitt Springs, and with an intermittent component from Coonawarra. Another excellent wine from Maglieri, matured for 18 months in a mix of French and American oak, but as with all the Maglieri wines, fruit- rather than oak-driven.

TTTTY **1997** Medium to full red-purple; clean, earthy cabernet varietal fruit aromas together with subtle oak introduce a wine with abundant cassis berry fruit, a touch of vanilla oak and well-balanced tannins on the finish. **rating:** 90

best drinking 2002–2010 **best vintages** '90, '91, '95, '97 **drink with** Parmesan cheese • $18

main ridge estate ★★★★

Lot 48 William Road, Red Hill, Vic 3937 **region** Mornington Peninsula

ph (03) 5989 2686 **fax** (03) 5931 0000 **open** Mon–Fri 12–4, weekends 12–5

winemaker Nat White **prod.** 1000 **est.** 1975

prod. range ($32–38 CD) Chardonnay, Pinot Noir, Half Acre Pinot Noir, Cabernet Merlot.

summary Nat White gives meticulous attention to every aspect of his viticulture and winemaking, doing annual battle with one of the coolest sites on the Peninsula. The same attention to detail extends to the winery and the winemaking. Not surprisingly, Nat White took full advantage of the excellent, warm 1997 vintage. Incidentally, with such minuscule production, the wines are available only through cellar door and mail order.

Main Ridge Chardonnay

Since 1991 Nat White has moved away from conventional Australian techniques to a far more French-influenced regimen, using barrel fermentation, malolactic fermentation and lees contact. As is inevitably the case in the Mornington Peninsula, style and quality reflect vintage variations, with the warmer, drier years tending to produce the best wines. 1997 was such a vintage.

TTTT **1997** Medium to full yellow-green; the bouquet is rich and complex, with smooth, ripe peach and melon fruit. A very concentrated, full-bodied wine on the palate with some faintly malty/meady flavours. Subtle oak. **rating:** 88

best drinking 1998–2000 **best vintages** '91, '92, '94, '96, '97 **drink with** Sweetbreads • $32

Main Ridge Half Acre Pinot Noir

The cool-climate Main Ridge vineyard does peculiar things to Pinot Noir, sometimes producing wines with spicy pepper characteristics which makes them look for all the world like light-bodied Shiraz. In other years the flavours are more towards the stalky/tobacco end; never does the wine become jammy or heavy. Then a vintage like 1997 comes along, and blows away all preconceptions; this is an absolutely gorgeous wine.

TTTTT **1997** Excellent full red-purple; the bouquet is complex, with wonderfully sweet plum, cherry and spice aromas. The wine has much more concentration and power on the palate than usual, with neither spice nor green sappy characters the least bit evident; instead there is that lovely sweet fruit promised by the bouquet which carries through on a long finish. **rating:** 94

best drinking 1999–2007 **best vintages** '97 **drink with** Grilled salmon • $38

mair's coalville ★★★

Moe South Road, Moe South, Vic 3825 **region** Gippsland
ph (03) 5127 4229 **fax** (03) 5127 2148 **open** By appointment
winemaker Dr Stewart Mair **prod.** 250 **est.** 1985
prod. range ($15 ML) A single wine, predominantly Cabernet Sauvignon with a little Cabernet Franc, Malbec and Merlot, labelled Coalville Red.
summary Dr Stewart Mair has fashioned a remarkably consistent wine from his small vineyard, on the lean side perhaps, but with the elegance which comes from very cool-grown fruit.

majella ★★★★

Lynn Road, Coonawarra, SA 5263 **region** Coonawarra
ph (08) 8736 3055 **fax** (08) 8736 3057 **open** 7 days 10–4.30
winemaker Bruce Gregory **prod.** 6000 **est.** 1969
prod. range ($20–40 CD) Shiraz, Cabernet Sauvignon, Malleea (Shiraz Cabernet), Sparkling Shiraz.
summary Majella is one of the more important contract grape growers in Coonawarra, with 55 hectares of vineyard, principally shiraz and cabernet sauvignon, and with a little riesling and merlot, in production and now fully mature. Common gossip has it that part finds its way into the Wynns John Riddoch Cabernet Sauvignon and Michael Shiraz, or their equivalent within the Southcorp Group. Production under the Majella label is increasing as long-term supply contracts expire. Production under the Majella label has increased substantially over the past few years, rising from 2000 to 6000 cases, with exports to Asia and the US.

Majella Shiraz

First made in 1991, although the estate vineyards are almost 30 years old. The concentration has varied somewhat over the years, the '91 and '93 vintages being quite outstanding, the '97 in much the same class.

🍷🍷🍷🍷🍸 **1997** Dense purple-red; full, dark berry, plum and mulberry fruit is woven through abundant oak on the bouquet. A concentrated palate with dark berry fruit, soft, persistent tannins and oak in balance. **rating:** 90

best drinking 2000–2010 **best vintages** '97 **drink with** Lamb chops • $20

Majella The Malleea Shiraz Cabernet

A new prestige cuvée for Majella. If the gossip that some of the best grapes had been sold to Wynns for inclusion in John Riddoch Cabernet and Michael Shiraz is correct, this wine may well represent Majella's gain and Wynns' loss. In any event, it represents a singularly impressive debut for a new super-premium Coonawarra red.

🍷🍷🍷🍷🍷 **1996** Dense red-purple; ripe, rich dark berry fruit aromas, with hints of chocolate and of excellent concentration lead into a palate with classic Coonawarra flavours; dark cherry, blackberry, mulberry and cassis. The oak has not been overdone, and the wine has quite marvellous length. **rating:** 95

best drinking 2001–2010 **best vintages** '96 **drink with** Marinated beef • $40

Majella Cabernet Sauvignon

The '93 vintage was the first Cabernet to be released from Majella, produced from the House Block. Aged in French oak hogsheads for two years, and deservedly accumulated five gold medals. The '96 marked a return to form after a pedestrian '95 vintage; the '97 falling somewhere in the middle.

🍷🍷🍷🍷 **1997** Medium to full red-purple; there are ripe, slightly jammy characters with whispers of hay/straw on the bouquet; the palate is very ripe but throws off the jammy characters of the bouquet, and both the tannins and oak are appropriately balanced. **rating:** 86

best drinking 2002–2007 **best vintages** '93, '94, '96 **drink with** Rack of lamb • $24

---

malcolm creek ★★★☆

Bonython Road, Kersbrook, SA 5231 **region** Adelaide Hills
ph (08) 8389 3235 **fax** (08) 8389 3235 **open** Weekends, public holidays 11–5
winemaker Reg Tolley **prod.** 650 **est.** 1982
prod. range ($16 R) Chardonnay, Cabernet Sauvignon.
summary Malcolm Creek is the retirement venture of Reg Tolley, and keeps a low profile. However, the wines are invariably well made and age gracefully; they are worth seeking out.

Malcolm Creek Chardonnay

Produced from 0.6 hectare of chardonnay planted in 1982, with a further 0.2 hectare planted in 1996 and to come into bearing in future years. A wine which ages with considerable grace; the '94 was the currently available wine in 1997. The '96 and '97 vintages were available in 1999.

🍷🍷🍷🍷🍷 **1996** Medium yellow-green; the bouquet shows good bottle-developed aromas with complex tangy fruit and a touch of honey. The palate has lovely flavours, with a mix of nectarine, citrus and again a flick of that honey; considerable length. **rating:** 90

best drinking 1999–2003 **best vintages** '88, '90, '91, '94, '95, '96 **drink with** South Australian whiting • $16

Malcolm Creek Cabernet Sauvignon

A 100% Cabernet Sauvignon produced from 0.6 hectare of vines planted in 1982, although the area was doubled in 1996.

🍷🍷🍷🍷 **1996** Medium red-purple; the bouquet is clean, with quite sweet blackberry fruit and a touch of chocolate; the palate is relatively austere compared to the bouquet, firm and with fine tannins, but probably has its best years in front of it. **rating:** 86

best drinking 2000–2006 **best vintages** '86, '88, '93, '96 **drink with** Yearling beef • $16

mann NR

105 Memorial Avenue, Baskerville, WA 6056 **region** Swan District
ph (08) 9296 4348 **fax** (08) 9296 4348 **open** Weekends 10–5 and by appointment
winemaker Dorham Mann **prod.** 500 **est.** 1988
prod. range ($15 CD) Méthode Champenoise.
summary Industry veteran Dorham Mann has established a one-wine label for what must be Australia's most unusual wine: a dry, only faintly pink, sparkling wine made exclusively from cabernet sauvignon grown on the 2.4-hectare estate surrounding the cellar door. Dorham Mann explains, 'our family has made and enjoyed the style for more than 30 years, although just in a private capacity until recently'.

manning park NR

Cnr Olivers and Chalk Hill Roads, McLaren Vale, SA 5171 **region** McLaren Vale
ph (08) 8323 8209 **fax** (08) 8323 9474 **open** 7 days 10–5
winemaker Warren Randall **prod.** 5000 **est.** 1979
prod. range ($11–20 CD) Great White (Sauvignon Blanc/Semillon), Colombard, Chenin Blanc, Savage Grenache, Native Cabernet, Wild Shiraz, Stormy Shiraz (Sparkling Burgundy), Tawny Port.
summary Once a low-key operation focusing primarily on fortified wines, has been revived by former Seppelt and Andrew Garrett (wunderkind) Warren Randall, now offering a range of exclusively McLaren Vale-sourced table wines. A barbecue and paved garden area are available for the use of cellar-door visitors.

mansfield wines NR

204 Eurunderee Road, Mudgee, NSW 2856 **region** Mudgee
ph (02) 6373 3871 **fax** (02) 6373 3708 **open** Mon–Sat 9–5, Sun 10–5
winemaker Bob Heslop **prod.** 3680 **est.** 1975
prod. range ($4.50–25 CD) Chardonnay, Golden Gully (semi-sweet), Colonial Gold (sweet white), Sparkling Muscat, Cabernet Sauvignon Merlot, and a selection of fortified wines.
summary Mansfield Wines is one of the old-style wineries, offering a mix of varietal and generic table wines at low prices, and an even larger range of miscellaneous fortified wines, not all of which are locally produced. Distribution is through cellar door and various regional outlets.

margan family winegrowers NR

1238 Milbrodale Road, Broke, NSW 2330 **region** Lower Hunter Valley
ph (02) 6579 1317 **fax** (02) 6579 1317 **open** 9–5 Mon–Fri
winemaker Andrew Margan **prod.** 15 000 **est.** 1989
prod. range ($15–20 R) Semillon, Verdelho, Chardonnay, Merlot, Shiraz, Cabernet Sauvignon.
summary Andrew Margan followed in his father's footsteps by entering the wine industry 20 years ago, and has covered a great deal of territory since, working as a Flying Winemaker in Europe, then for Tyrrell's, first as a winemaker then as marketing manager. His wife Lisa, too, has had many years of experience in restaurants and marketing. They now have 10 hectares of fully yielding vines at their 50-hectare Ceres Hill homestead property at Broke, and lease the nearby Vere Vineyard of 13 hectares. The first stage of a 700-tonne on-site winery was completed in 1998, the first wines having been made elsewhere in 1997. Wine quality (and the packaging) is good.

Margan Family Chardonnay

An impressive wine which proved its class at the 1999 Sydney Royal Wine Show.
🍷🍷🍷🍷🍷 **1998** Medium to full yellow-green; the bouquet is quite complex, with attractive barrel-ferment characters imposed on smooth melon fruit. The palate follows on convincingly, with length and style. **rating:** 91
best drinking 1999–2001 **best vintages** '98 **drink with** Breast of chicken • $15

marienberg ★★★

2 Chalk Hill Road, McLaren Vale, SA 5171 **region** McLaren Vale
ph (08) 8323 9666 **fax** (08) 8323 9600 **open** 7 days 10–5
winemaker Grant Burge (Contract) **prod.** 30 000 **est.** 1966
prod. range ($9.95–24.95 R) Cottage Classic range of Riesling, Sauvignon Blanc Semillon, Unwooded Chardonay, Cabernet Grenache Mourvedre; Reserve Chardonnay, Shiraz and Cabernet Sauvignon; Limeburner's Chardonnay, Botrytis Semillon, Cabernet Sauvignon; also Late Picked Frontignac, Nicolle Méthode Champenoise and Tawny Port.
summary The Marienberg brand was purchased by the Hill International group of companies in late 1991 following the retirement of Ursula Pridham. Releases under the new regime have been honest, if unashamedly commercial, wines. The newly-introduced Reserve 1998 Chardonnay and 1996 Reserve Shiraz do offer a significant lift in quality above the basic range. The wines are exported to the UK, US, Canada, Hong Kong, Germany, Holland, New Zealand, Philippines, Switzerland and Thailand.

mariners rest NR

Jamakarri Farm, Roberts Road, Denmark, WA 6333 **region** Great Southern
ph (08) 9840 9324 **fax** (08) 9840 9324 **open** 7 days 10–5
winemaker Ron Cocking, Peter Cocking **prod.** 450 **est.** 1996
prod. range ($13.50–19 R) Chardonnay, Autumn Gold, Autumn Red, Southern Red, Pinot Noir.

summary Mariners Rest is the reincarnation of the now defunct Golden Rise winery. A new 2.5-hectare vineyard was planted in the spring of 1997, and in the meantime some excellent wines from the Golden Rise days are being marketed under the Mariners Rest label.

marion's vineyard ★★★☆

Foreshore Drive, Deviot, Tas 7275 **region** Northern Tasmania
ph (03) 6394 7434 **fax** (03) 6394 7434 **open** 7 days 10–5
winemaker Mark Semmens, Marion Semmens **prod.** 2000 **est.** 1980
prod. range ($15–30 ML) Chardonnay, Müller Thurgau, Pinot Noir, Cabernet Sauvignon.
summary The irrepressible Mark Semmens and indefatigable wife Marion have one of the most beautifully situated vineyards and wineries in Australia on the banks of the Tamar River. As well as an outdoor restaurant and accommodation, there is a jetty and a stage – indeed, life is a stage for Mark Semmens.

maritime estate NR

Tucks Road, Red Hill, Vic 3937 **region** Mornington Peninsula
ph (03) 9848 2926 **fax** (03) 9882 8325 **open** Weekends and public holidays 11–5, Dec 27–Jan 26 7 days
winemaker T'Gallant (Contract) **prod.** 1000 **est.** 1988
prod. range ($18–25 CD) Unwooded Chardonnay, Chardonnay, Pinot Noir, Cabernet Sauvignon.
summary John and Linda Ruljancich have enjoyed great success since their first vintage in 1994, no doubt due in part to skilled contract-winemaking, but also to the situation of their vineyard looking across the hills and valleys of the Red Hill subregion.

markwood estate NR

Morris Lane, Markwood, Vic 3678 **region** King Valley
ph (03) 5727 0361 **fax** (03) 5727 0361 **open** 7 days 9–5
winemaker Rick Morris **prod.** 900 **est.** 1971
prod. range ($15–21 CD) Rhine Riesling, Chardonnay, Cabernet Sauvignon, Shiraz, Muscat, Tokay, Port.
summary A member of the famous Morris family, Rick Morris shuns publicity, and relies virtually exclusively on cellar-door sales for what is a small output. Of a range of table and fortified wines tasted in February 1997, the Old Tawny Port (a cross between Port and Muscat, showing more of the character of the latter than the former) and a White Port (seemingly made from Muscadelle) were the best. Agglomerate corks may present a threat for prolonged cellaring, however.

marribrook ★★★

Rocky Gully Road, Frankland, WA 6396 **region** Great Southern
ph (08) 9457 7885 **fax** (08) 9457 7885 **open** Not
winemaker Gavin Berry (Contract) **prod.** 2000 **est.** 1990
prod. range ($13–16 CD) Unwooded Chardonnay, Botanica Chardonnay, Marsanne, Cabernet Sauvignon, Cabernet Malbec.
summary The Brooks family purchased the former Marron View 5.6-hectare vineyard from Kim Hart in 1994, and renamed the venture Marribrook Wines. Those wines are now made by Gavin Berry at Plantagenet, having been made at Alkoomi up to 1994. An interesting wine on the roster is Western Australia's only Marsanne, although it has to be said that the wine in the bottle is perhaps less interesting than the label. Retail distribution in Victoria, Queensland and South Australia; exports to the UK.

Marribrook Botanica Chardonnay

Unwooded Chardonnay is just that; Marribrook does the right thing in releasing the wine when it is young and retains maximum life.

🍷🍷🍷🍷 **1998** Light to medium green-yellow; the bouquet is clean, but neutral in the manner of many unwooded Chardonnays. The palate has more intensity, weight and length than the bouquet suggests, particularly on the tangy melon back palate. **rating:** 86

best drinking 1999–2004 **best vintages** NA **drink with** Marron, of course • $13

Marribrook Reserve Chardonnay

There is a praiseworthy gap between the Botanica Chardonnay and the Reserve Chardonnay; not only has the better fruit gone into the Reserve, but the skilful use of barrel fermentation and lees contact has added complexity.

🍷🍷🍷🍷🍷 **1998** Medium to full yellow-green; the tangy, stylish bouquet shows complex barrel-ferment characters leading into a palate in which the oak is evident but not intrusive, supporting attractive citrus/melon fruit on a long finish. **rating:** 91

best drinking 1999–2005 **best vintages** NA **drink with** Fricassee of veal • $15

marsh estate ★★★

Deasey Road, Pokolbin, NSW 2321 **region** Lower Hunter Valley

ph (02) 4998 7587 **fax** (02) 4998 7884 **open** Mon–Fri 10–4.30, weekends 10–5

winemaker Peter Marsh **prod.** 4000 **est.** 1971

prod. range ($15–21 CD) High Hill Semillon, Semillon, Private Bin Semillon, Chardonnay (oaked and unoaked), Traminer, Semillon Sauternes, Shiraz (Private Bin, Vat S and Vat R), Cabernet Merlot, Champagne Brut, Andrew IV Vintage Port.

summary Through sheer consistency, value-for-money and unrelenting hard work, the Marsh family (who purchased the former Quentin Estate in 1978) has built up a sufficiently loyal cellar-door and mailing list clientele to allow all of the considerable production to be sold direct. Wine style is always direct, with oak playing a minimal role, and prolonged cellaring paying handsome dividends.

martins hill wines NR

Sydney Road, Mudgee, NSW 2850 **region** Mudgee

ph (02) 6373 1248 **fax** (02) 6373 1248 **open** Not

winemaker Pieter Van Gent (Contract) **prod.** 800 **est.** 1985

prod. range ($13–16 R) Sauvignon Blanc, Pinot Noir.

summary Janette Kenworthy and Michael Sweeny are committed organic grape growers, and are members of the Organic Vignerons Association. It is a tiny operation at the moment, with only half a hectare each of sauvignon blanc and pinot noir in production, but with an additional hectare of cabernet sauvignon to produce its first grapes in the year 2000, and 1.5 hectares of shiraz in 2001. While there is no cellar door (only a mailing list), organic vineyard tours and talks can be arranged by appointment.

marybrook vineyards NR

Vasse–Yallingup Road, Marybrook, WA 6280 **region** Margaret River

ph (08) 9755 1143 **fax** (08) 9755 1112 **open** Fri–Mon 10–5, 7 days 10–5 school holidays

winemaker Mike Lemmes (Contract) **prod.** 2000 **est.** 1986

prod. range ($12.50–22.95 CD) Chardonnay, Verdelho, Classic White, Nectosia (sweet), Cabernets, Cabernet Franc, Temptation (sweet red), Jetty Port.

summary It is easy to confuse Marybrook Vineyards with Marybrook Estate; they are in fact separate operations. Marybrook Vineyards is owned by Aub and Jan House. Seven hectares of vineyards are in production. As at March 1999, five vintages of Cabernets or Cabernet Franc ('92–'96) were available cellar door.

massoni main creek ★★★★

Mornington–Flinders Road, Red Hill, Vic 3937 **region** Mornington Peninsula
ph (03) 5989 2352 **fax** (03) 5989 2014 **open** By appointment
winemaker Ian Home **prod.** 2500 **est.** 1984
prod. range ($32–39 R) Chardonnay, Pinot Noir, Ian Home Lectus Cuvée, Shiraz are the top wines; also Homes Chardonnay, Pinot Noir and Cabernet Merlot.
summary The changes have continued to flow after Ian Home (best known as the founder of Yellowglen) acquired the remaining 50% of Massoni from former restaurateur Leon Massoni. There are now two ranges of wines, and (quite sensibly) the Shiraz and Cabernet Merlot releases are in whole or in part sourced from outside the Mornington Peninsula (from Langhorne Creek). The flagships continue to be the Chardonnay, Pinot Noir and Lectus Cuvée, supported by Shiraz. Exports to US and Germany.

Massoni Main Creek Chardonnay

Produced from 1.5 hectares of estate plantings, and made using the full gamut of barrel fermentation, malolactic fermentation and lees contact. In many ways the most striking example of Mornington Peninsula Chardonnay, with tremendous character, body and richness.

🍷🍷🍷🍷🍷 **1997** Medium yellow-green; an exceptionally ripe, full and complex bouquet with tangy melon/vegetal fruit strongly reminiscent of Burgundy. The usual blockbuster flavour on the palate, yet miraculously smooth and supple, with a long finish. Subtle oak. **rating:** 94

best drinking 1998–2002 **best vintages** '89, '90, '92, '93, '94, '96, '97 **drink with** Veal, pork • $33

Massoni Main Creek Pinot Noir

The mirror image of the Chardonnay, always rich, full-bodied and opulent, where so many of the Mornington wines – Pinot and Chardonnay alike – tend to be prettier and more elegant.

🍷🍷🍷🍷🍷 **1997** Strong red-purple; the bouquet is quite powerful, with abundant dark plum fruit which comes through as the centrepiece of the palate. Overall the wine has excellent flavour, concentration and length, and will undoubtedly build further complexity with age. This is no shrinking violet. 400 cases made. **rating:** 94

best drinking 1999–2003 **best vintages** '91, '92, '93, '94, '96, '97 **drink with** Breast of duck • $39

---

matilda's meadow ★★★

Eladon Brook Estate, RMB 654 Hamilton Road, Denmark, WA 6333 **region** Great Southern
ph (08) 9848 1951 **fax** (08) 9848 1957 **open** Wed–Mon 10–4
winemaker Brenden Smith **prod.** 1500 **est.** 1990
prod. range ($12.50–23.50 CD) Riesling, Late Picked Riesling, Unwooded Chardonnay, Semillon Chardonnay, Autumn Amethyst (light red), Pinot Noir, Cabernet Sauvignon Cabernet Franc.
summary Former hotelier Don Turnbull and oil-industry executive Pamela Meldrum have quickly established a thriving business at Matilda's Meadow, based on 6 hectares of estate plantings, and with a restaurant offering morning and afternoon teas and lunches every day. The '98 Semillon and '97 Cabernet Sauvignon are technically sound, but do show rather green fruit characters.

maxwell wines ★★★★

Olivers Road, McLaren Vale, SA 5171 **region** McLaren Vale
ph (08) 8323 8200 **fax** (08) 8323 8900 **open** 7 days 10–5
winemaker Mark Maxwell **prod.** 8000 **est.** 1979
prod. range ($10–30 R) Under the Maxwell Wines brand, Semillon, Sauvignon Blanc, Chardonnay, Cabernet Merlot, Reserve Shiraz; Ellen Street Shiraz and Lime Cave Cabernet Sauvignon; and excellent Honey Mead, Spiced Mead and Liqueur Mead.

summary Maxwell Wines has come a long way since opening for business in 1979 using an amazing array of Heath Robinson equipment in cramped surroundings. A state-of-the-art and infinitely larger winery was built on a new site in time for the 1997 vintage, appropriate for a brand which has produced some excellent white and red wines in recent years. Exports to US, UK, Switzerland, Austria, Hong Kong and New Zealand.

mcalister vineyards NR

Golden Beach Road, Longford, Vic 3851 **region** Gippsland
ph (03) 5149 7229 **fax** (03) 5149 7229 **open** By appointment
winemaker Peter Edwards **prod.** 550 **est.** 1975
prod. range ($NA) A single wine, The McAlister, a blend of Cabernet Sauvignon, Cabernet Franc and Merlot.
summary The McAlister Vineyards actively shun publicity or exposure which, on the basis of prior tastings, is a pity.

mcguigan brothers ★★★

PO Box 31, Cessnock, NSW 2335 **region** Lower Hunter Valley
ph (02) 4998 7400 **fax** (02) 4998 7401 **open** 7 days 10–5
winemaker Brian McGuigan **prod.** 400 000 **est.** 1992
prod. range ($9.50–25 R) The wines are sold in three price brackets: at the bottom, Harvest Range Semillon Chardonnay, Night Harvest Graves, Autumn Harvest Traminer Riesling, Black Shiraz; then the Bin range 2000 Hermitage, 3000 Merlot, 4000 Cabernet Sauvignon, 6000 Verdelho, 7000 Chardonnay; finally Shareholder Reserve Chardonnay, Cabernet Merlot, Sauternes; also Personal Reserve recently added to the range.
summary A public-listed company which is the ultimate logical expression of Brian McGuigan's marketing drive and vision, on a par with that of Wolf Blass in his heyday. Highly successful in its chosen niche market notwithstanding exceedingly garish labels. Has been particularly active in export markets, notably the US, and more recently in China. Wine quality seems less important than marketing magic.

McGuigan Brothers Shareholders Chardonnay

These days you can take your choice between McGuigan Brothers Chardonnay, McGuigan Brothers Shareholders Chardonnay, and McGuigan Brothers Personal Reserve Chardonnay. It all sounds a bit like Gallo, but this is a pleasant wine at a modest price.

🍷🍷🍷🍸 **1997** Full yellow; developed buttery/peachy aromas are followed by a big, rich fast-developing style of Chardonnay with a jab of oak to help it move along. **rating:** 84

best drinking 1999–2000 **best vintages** NA **drink with** Chicken • $16

McGuigan Brothers Shareholders Shiraz

No region or origin is claimed for this wine, which is almost certainly a southeast Australian blend.

🍷🍷🍷🍸 **1997** Medium red-purple; the bouquet is earthy, with slightly astringent characters, but the palate redeems the wine, with solid, flavoursome fruit supported by vanillin American oak. Traditional Australian Shiraz. **rating:** 84

best drinking 2000–2004 **best vintages** NA **drink with** Big Mac • $16

mcivor creek NR

Costerfield Road, Heathcote, Vic 3523 **region** Bendigo
ph (03) 5433 3000 **fax** (03) 5433 3456 **open** 7 days 10–5.30
winemaker Peter Turley **prod.** 5000 **est.** 1973
prod. range ($9.95–14.95 CD) Riesling, Auslese Riesling, Shiraz, Cabernet Shiraz, Fine Old Tawny Port.

summary The beautifully situated McIvor Creek winery is well worth a visit, and does offer wines in diverse styles of which the red wines are the most regional. Peter Turley has 5 hectares of cabernet sauvignon together with 2.5 hectares of cabernet franc and merlot and supplements his intake with grapes from other growers. No recent tastings.

mcmanus NR

Rogers Road, Yenda, NSW 2681 **region** Riverina
ph (02) 6968 1064 **open** 7 days 9–5
winemaker Dr David McManus **prod.** 500 **est.** 1972
prod. range ($4–8 CD) Chardonnay, Chardonnay Semillon, Malbec, Merlot, Shiraz, Pinot Malbec Shiraz; many named after family members.
summary An extremely idiosyncratic winery run by Griffith GP Dr David McManus, his sister and other family members. Natural winemaking methods lead to considerable variation in quality, but the prices are from another era, some of the vintages likewise.

mcwilliam's ★★★★

Winery Road, Hanwood, NSW 2680 **region** Riverina
ph (02) 6963 0001 **fax** (02) 6963 0002 **open** Mon–Sat 9–5
winemaker Jim Brayne **prod.** NFP **est.** 1877
prod. range ($6–40 R) A disciplined and easy-to-follow product range (all varietally identified) commencing with Hillside casks; Inheritance Range; Hanwood; Charles King; JJ McWilliams (first released 1996), finally Limited Release Hunter Valley Chardonnay, Eden Valley Riesling and JJ McWilliam Riverina Botrytis Semillon. Also superb fortified wines including MCW11 Liqueur Muscat and 10-Year-Old Hanwood Tawny Port heading a much larger range of Sherries which still form an important part of the business.
summary Some of the best wines to emanate from the Hanwood winery are from other regions, notably the Barwang Vineyard at Hilltops in New South Wales, Coonawarra and Eden Valley; on the other side of the coin as it were, the critical mass of the business continues to come from the Murrumbidgee Irrigation Area, which provides the bulk of the rapidly growing export business of the company. The rating is a compromise between the best and the least of the wide range. Exports to many countries, the most important being the UK, US, Germany and New Zealand. (Exports currently account for 30% of Mcwilliam's sales.)

McWilliam's Hanwood Cabernet Sauvignon

There is no doubt that the winemakers of the Riverina, big and small, are achieving what was once considered impossible with their best red wines, investing them with far greater depth of fruit flavour. This, in turn, has been achieved by improved viticulture and limiting yields.

🍷🍷🍷🍷 **1998** Medium red, with a touch of purple; the bouquet is clean and smooth, with an attractive amalgam of berry, earth and vanilla. Smooth, cherry/berry fruit and subtly balanced oak mark a well-made wine at a bargain basement price. **rating:** 86

best drinking 1999–2002 **best vintages** NA **drink with** Pizza • $9.95

McWilliam's MCW11 Liqueur Muscat

This wine, blended to a constant quality and style, has amassed four trophies and 42 gold medals in its show career. Very long ageing in barrel is the secret; the wines do not have the intense Muscat fruit character of those of northeastern Victoria when young, but do attain tremendous complexity with age.

🍷🍷🍷🍷🍷 **NV** Very deep dark mahogany brown; extreme wood age shows with concentrated plum pudding aromas leading on to a multiflavoured palate ranging from malt to toffee to plum pudding, finishing with cleansing acidity to balance the intense sweetness of the mid-palate. **rating:** 94

best drinking 1999–2000 **best vintages** NA **drink with** A winter's night • $39.95

mcwilliam's mount pleasant ★★★★★

Marrowbone Road, Pokolbin, NSW 2320 **region** Lower Hunter Valley
ph (02) 4998 7505 **fax** (02) 4998 7761 **open** 7 days 10–4.30
winemaker Phillip Ryan **prod.** NFP **est.** 1880

prod. range ($10–40 R) Much simplified and rationalised over the past year. The base range now comprises Mount Pleasant Elizabeth, Philip, Late Harvest Dessert Wine, Chardonnay and Verdelho; then individual vineyard wines, Rosehill Shiraz, Old Paddock & Old Hill Shiraz, Lovedale Semillon (previously known as Anne), then Maurice O'Shea Chardonnay and Shiraz; finally Museum releases of Elizabeth.

summary McWilliam's Elizabeth and the glorious Lovedale Semillon are now the only mature Hunter Semillons generally commercially available, and are undervalued and underpriced treasures, with a consistently superb show record. The three individual vineyard wines, together with the Maurice O'Shea memorial wines, add to the lustre of this proud name. Exports to many countries, the most important being the UK, US, Germany and New Zealand.

McWilliam's Mount Pleasant Elizabeth

A wine with an exceptional pedigree and deserved reputation for consistency, yet chronically underpriced and hence underrated (or the reverse, I am not sure which). Changes to the packaging, notably the bottle shape, and the hand of McWilliam's chief executive Kevin McLintock may well see the wine gradually being repositioned in the market to assume its rightful place. Even without this, an undoubted classic, having won 13 trophies and 129 gold medals since 1981.

🍷🍷🍷🍷🍷 **1994** Glowing yellow; there is a mix of honey, toast and a lesser amount of herb on the bouquet; the palate has marvellous zest and life, with a twist of lemon peel, and brisk acid on the finish. Has already accumulated four gold medals, with many more in store for it. **rating:** 94

best drinking 1999–2009 **best vintages** '75, '81, '82, '83, '86, '89, '90, '91, '93, '94 **drink with** Rich seafood • $14.95

McWilliam's Mount Pleasant O'Shea Chardonnay

This is the flagship Chardonnay from the Hunter Valley, sitting alongside the O'Shea Shiraz. Only limited quantities are made, and indeed is not necessarily produced every vintage.

🍷🍷🍷🍷 **1996** Medium yellow-green; the oak is rather strong and not integrated on the bouquet, but high-toned, powerful and slightly angular chardonnay fruit comes through on the palate. A wine which may be on an upward curve. **rating:** 86

best drinking 1999–2003 **best vintages** NA **drink with** Roast pork • $26.95

McWilliam's Mount Pleasant O'Shea (OP & OH) Shiraz

OP & OH is the sort of obscure designation which so delighted Maurice O'Shea. The letters in fact stand for Old Paddock and Old Hill, planted respectively, in 1880 and 1920, and which provide the bulk of the grapes which go into this distinguished wine. To confuse matters a little, since 1987 it has been sold as O'Shea Hermitage, in honour of Maurice O'Shea, in outstanding vintage years. So sometimes you will see it sold as O'Shea, sometimes as OP & OH.

🍷🍷🍷🍷🍷 **1996** O'Shea. Medium red-purple; a wonderfully scented and aromatic bouquet with cedar/smoky characters introducing a fine, long cherry-flavoured palate, finishing with supple tannins. Gold medal 1998 National Wine Show. **rating:** 94

best drinking 2001–2011 **best vintages** '65, '66, '67, '79, '85, '87, '90, '91, '94, '95, '96 **drink with** Roast veal • $23.95

McWilliam's Mount Pleasant Rosehill Shiraz

The Rosehill Vineyard shares a hill of terra rossa soil bisected in the middle by the Broke Road, with Lake's Folly on the opposite side to Rosehill. Over the years it has produced many great wines (the '59 is particularly memorable) and, it must be said, a few disappointments too. Recent vintages have been more consistent.

▼▼▼▼▽ **1996** Light to medium red-purple; a fragrant but strongly regional bouquet with cherry, earth and leather aromas. The palate follows down precisely the same track, being an excellent example of Hunter Valley Shiraz in full-on regional style, yet which is not the least extractive or tannic. Its very life and liveliness is deceptive, for it will cellar well. **rating:** 90

best drinking 2006–2016 **best vintages** '59, '65, '66, '67, '75, '95, '96 **drink with** Marinated spatchcock • $23.95

meadowbank wines ★★★☆

'Meadowbank', Glenora, Derwent Valley, Tas 7410 **region** Southern Tasmania
ph (03) 6286 1269 **fax** (03) 6286 1133 **open** 7 days 11–5
winemaker Andrew Hood (Contract) **prod.** 3000 **est.** 1974
prod. range ($16–22 CD) Riesling, Chardonnay, Grace Elizabeth Chardonnay, Pinot Noir, Cabernet Sauvignon.
summary Now an important part of the Ellis family business on what was once (but no more) a large grazing property on the banks of the Derwent. Increased plantings are being established under contract to BRL Hardy.

Meadowbank Riesling

Produced from estate plantings on the banks of the Derwent River which are now more than 20 years old.

▼▼▼▼ **1998** Light yellow-green; the bouquet is quite fragrant, with floral, lime and passionfruit aromas. The palate follows the same track; elegant, well-balanced and with pleasing citrus/passionfruit flavours. **rating:** 87

best drinking 1999–2005 **best vintages** NA **drink with** Asparagus • $17

Meadowbank Pinot Noir

One of the wines to fulfil the promise for so long held out by Tasmania. Made from estate-grown grapes, with 4 hectares of pinot in production.

▼▼▼▼▽ **1998** Medium red-purple; an aromatic bouquet with spice, cherry and plum fruit is followed by a well-balanced and flavoured palate; the fruit flavours are those of the bouquet, and oak handling has been sensitive throughout. **rating:** 90

best drinking 1999–2001 **best vintages** '94, '98 **drink with** Breast of duck • $20

meerea park ★★★☆

2 Denton Close, Windella via Maitland, NSW 2321 **region** Upper Hunter Valley
ph 0417 693 310 **fax** (02) 4930 7100 **open** At The Boutique Wine Centre, Broke Road, Pokolbin
winemaker Rhys Eather **prod.** 15 000 **est.** 1991
prod. range ($17–35 ML) Sauvignon Blanc Semillon, Semillon, Lindsay Hill Verdelho, Unoaked Chardonnay, Chardonnay, Alexander Munro Chardonnay, Alexander Munro Shiraz, Cabernet Merlot, Cabernet Sauvignon.
summary An interesting operation, selling its substantial production primarily through The Boutique Wine Centre, Broke Road, Pokolbin and by mailing list. All of the wines are produced from grapes purchased from growers, primarily in the Broke/Fordwich region, but also from as far afield as McLaren Vale, the Barossa Valley, Mudgee and Orange. It is the brainchild of Rhys Eather, great-grandson of Alexander Munro, a leading vigneron in the middle of the nineteenth

century, and who makes the wine in Simon Gilbert's contract winery at Muswellbrook. Retail distribution through the principal States, and the wines are exported to the UK and Asia.

Meerea Park Alexander Munro Chardonnay

Fermented using so-called wild yeast in 100% new French barriques, and left in barrel for nine months.

▼▼▼▼ **1997** Medium yellow-green; the bouquet is still quite fresh with stone fruit aromas and very subtle oak. The palate is tight, still crisp, with brisk acidity; does not show the oak one would expect, and is an altogether unusual style for the Hunter Valley. **rating:** 85

best drinking 2000–2003 **best vintages** NA **drink with** Marinated octopus • $25

Meerea Park Alexander Munro Shiraz

Produced from grapes from the 40-year-old shiraz vines on the Tallawanta Vineyard once owned by Elliots. Part finishes its fermentation in French and American oak.

▼▼▼▼▽ **1997** Youthful purple-red; an excellent palate with well-weighted dark cherry/berry fruit is followed by a firm, youthful dark cherry-flavoured palate and lingering tannins. Would appear to have outstanding potential. **rating:** 90

best drinking 2002–2007 **best vintages** NA **drink with** Braised beef • $35

Meerea Park Cabernet Merlot

Released under the Art label, and a blend of 70% Cabernet Sauvignon and 30% Merlot grown in the Hunter Valley, Mudgee and Orange regions. The wine is given two weeks post-fermentation maceration, and matured in a mix of new and used American oak barrels. The first release from 1996 won a gold, and five medals in capital city wine shows in 1997. The follow-on vintage is also pleasant.

▼▼▼▼ **1997** Medium red, with just a touch of purple; the bouquet is only of light to medium intensity, with some leaf and mint aromas balanced by subtle oak. The pleasant, fairly light-bodied palate has an attractive mix of berry, mint and leaf, all pointing to early drinking. **rating:** 85

best drinking 1999–2000 **best vintages** NA **drink with** Washed rind cheese • $18

merrebee estate ★★★☆

Lot 3339 St Werburghs Road, Mount Barker, WA 6234 **region** Great Southern

ph (08) 9851 2424 **fax** (08) 9851 2425 **open** Weekends and public holidays 10–4 and by appointment

winemaker Brendan Smith (Contract) **prod.** 1000 **est.** 1986

prod. range ($14–22 CD) Riesling, Chardonnay, Unwooded Chardonnay, Shiraz.

summary The 3.5-hectare Merrebee Estate vineyards were established in 1985, and the first wines released from the 1995 and 1996 vintages. The wines are available from selected retailers in Western Australia and from Rathdowne Cellars, Melbourne. These and subsequent releases have, at the very least, been good, some excellent.

Merrebee Estate Riesling

More proof, if proof were needed, of the symbiotic relationship between riesling and the Great Southern, particularly when you take into account the youth of the vines.

▼▼▼▼▽ **1995** Glowing yellow-green; the bouquet is quite powerful, with a mix of toast, lime and mineral aromas. The palate surpasses the promise of the bouquet with concentrated, smooth, round lime juice and honey flavours running through to the very finish. **rating:** 90

best drinking 1999–2003 **best vintages** NA **drink with** Roast pork • $18

Merrebee Estate Shiraz

A nicely-handled wine which has not been forced to be something which it is not, either by over-extraction or the over-compensating use of oak.

🍷🍷🍷🍷 **1996** Medium red, with a touch of purple remaining; the moderately intense bouquet offers earthy, cedary aromas, the palate hints of spice, liquorice and sweet vanilla oak. Both the structure and texture are good. **rating:** 86

best drinking 1999–2004 **best vintages** NA **drink with** Lasagne • $20

merricks estate ★★★☆

Thompsons Lane, Merricks, Vic 3916 **region** Mornington Peninsula
ph (03) 5989 8416 **fax** (03) 9627 4035 **open** First weekend of each month 12–5
winemaker Michael Zitzlaff **prod.** 2500 **est.** 1977
prod. range ($20–27.95 CD) Chardonnay, Shiraz, Cabernet Sauvignon, Pinot Noir.
summary Melbourne solicitor George Kefford, together with wife Jacquie, runs Merricks Estate as a weekend and holiday enterprise as a relief from professional practice. Right from the outset it has produced very distinctive, spicy, cool-climate Shiraz which has accumulated an impressive array of show trophies and gold medals.

Merricks Estate Pinot Noir

Like so many other Mornington producers, had a warm and dry 1997 vintage to remember for a long time.

🍷🍷🍷🍷 **1997** Medium to full red-purple; there is the classic array of complex Pinot aromas: briary, foresty and with some plum. Dark plum and cherry fruit comes through quite strongly on the palate, as does acidity on the finish. Very restrained oak use throughout. **rating:** 87

best drinking 1999–2003 **best vintages** '97 **drink with** Squab • $27.95

Merricks Estate Shiraz

Always an intense and striking wine which, right from the word go, has shown strong cool-climate characteristics with lots of spice and lots of leafy characters.

🍷🍷🍷🍷🍷 **1997** Quite deep red-purple; there are no green or herbaceous notes whatsoever on the sweet, plummy, cherry-accented bouquet. The palate opens with most attractive plum and cherry fruit, followed by a welcome hint of spice on the very finish. **rating:** 90

best drinking 1999–2005 **best vintages** '84, '88, '90, '93, '97 **drink with** Risotto • $27.95

merrivale wines ★★★

Olivers Road, McLaren Vale, SA 5171 **region** McLaren Vale
ph (08) 8323 9196 **fax** (08) 8323 9746 **open** 7 days 11–5
winemaker Contract **prod.** 10 000 **est.** 1971
prod. range ($10–18 CD) Under the Tapestry label Riesling, Chardonnay, Spaetlese, Shiraz, Cabernet, Muscat of Alexandria, Old Tawny Port.
summary After a relatively brief period of ownership by Brian Light and family, was then acquired by the Gerard family, owners of Chapel Hill.

Merrivale Tapestry Spaetlese

Spaetlese is a term Australia has agreed to cease using within a few years. It denotes a wine which is in fact neither dry on the one hand nor lusciously sweet on the other.

🍷🍷🍷🍷 **1996** Light green-yellow; the bouquet is intensely perfumed and grapey, the flavours promised by the bouquet coming on the palate together with an intriguing touch of lavender. The crisp acid on the finish reduces the amount of apparent sweetness almost to the point of dryness. **rating:** 86

best drinking 2001–2007 **best vintages** NA **drink with** Lightly poached fruit • $8

michelini wines NR

Great Alpine Road, Myrtleford, Vic 3737 **region** King Valley
ph (03) 5751 1990 **fax** (03) 5751 1990 **open** Thur–Mon 10–5
winemaker Contract **prod.** 3000 **est.** 1982
prod. range ($13–18.50 CD) Riesling, Unwooded Chardonnay, Chardonnay, Marzemo, Merlot, Shiraz, Cabernet Sauvignon.
summary The Michelini family are among the best known grape growers in the Buckland Valley of northeast Victoria. Having migrated from Italy in 1949, the Michelinis originally grew tobacco, diversifying into vineyards in 1982. A little over 42 hectares of vineyard have been established on terra rossa soil at an altitude of 300 metres, mostly with frontage to the Buckland River. The major part of the production is sold (to Orlando and others), but since 1996 an on-site winery has permitted the Michelinis to vinify part of their production. The winery in fact has capacity to handle 1000 tonnes of fruit, thereby eliminating the problem of moving grapes out of a declared phylloxera area. The quality of the initial releases was modest, but the area does have the potential to produce pleasant wine, and doubtless better things are in store.

middlebrook NR

Sand Road, McLaren Vale, SA 5171 **region** McLaren Vale
ph (08) 8383 0600 **fax** (08) 8383 0557 **open** Mon–Fri 9–5, weekends 10–5
winemaker Walter (Bill) Clappis **prod.** 5000 **est.** 1947
prod. range ($14–20 CD) In the course of redevelopment, with a top-end range under the Walter Clappis label, and a mid-range under the Middlebrook label.
summary Middlebrook has been acquired and is being redeveloped by industry veteran Bill Clappis after his former winery, Ingoldby, was acquired by Mildara Blass. The Middlebrook winery and restaurant are currently under renovation, and reopened in the second-half of 1998 with an appropriate fanfare of trumpets.

middleton estate NR

Flagstaff Hill Road, Middleton, SA 5213 **region** McLaren Vale
ph (08) 8555 4136 **fax** (08) 8555 4108 **open** Fri–Sun 11–5
winemaker Nigel Catt **prod.** 3000 **est.** 1979
prod. range ($9–16 CD) Riesling, Sauvignon Blanc, Semillon Sauvignon Blanc, Cabernet Hermitage.
summary Nigel Catt has demonstrated his winemaking skills at Andrew Garrett and elsewhere, so wine quality should be good; despite its decade of production, I have never seen or tasted its wines. A winery restaurant helps the business turnover.

mildara (coonawarra) ★★★★

Penola–Naracoorte Road, Coonawarra, SA 5263 **region** Coonawarra
ph (08) 8736 3380 **fax** (08) 8736 3307 **open** Mon–Fri 9–4.30, weekends 10–4
winemaker David O'Leary, Toni Stockhausen **prod.** 90 000 **est.** 1955
prod. range ($10–40 R) The volume is driven by Jamiesons Run Red and Jamiesons Run Chardonnay, then comes Robertson's Well Chardonnay (in fact a blend of Yarra Valley and King Valley fruit) and Cabernet Sauvignon and Flanagan's Ridge Cabernet Sauvignon. Jamiesons Run has been 'brand-extended' with the introduction of a smartly packaged Reserve, but Alexanders remains the top-end wine for the time being.
summary The quality jewel in the crown of the Fosters Brewing Group's Mildara Blass wine empire, but a jewel which has always been put to industrial use, with the emphasis on volume brands such as Jamiesons Run, Robertson's Well and so forth. For all that, it has to be said the quality of Jamiesons Run has been zealously protected, notwithstanding the growth in volume of its production (and the 1998 introduction of Jamiesons Run Reserve). Worldwide distribution.

Mildara Jamiesons Run Reserve Red

Given Mildara's penchant for creating new labels, the brand extension of Jamiesons Run into a Reserve Red comes as something of a surprise, but then Ray King has pulled many rabbits from the hat in his time as chief executive. The wine comes in its own branded, embossed bottle, and there is no doubt the first vintage ('95) was significantly better than the standard release of that year, as is the '96.

1996 Medium red-purple; the moderately intense bouquet is clean, with attractive earthy/berry fruit and subtle oak. The palate is firm but not harsh or aggressive, with a similar mix of earth, mint and red berry flavours; subtle oak, soft tannins. **rating:** 90

best drinking 2000–2005 **best vintages** '95, '96 **drink with** Porterhouse steak • $39

Mildara Jamiesons Run Coonawarra Merlot

Everyone with Merlot is hopping onto the bandwagon, it seems. I am not too sure what to make of this brand extension for Jamiesons Run; one might have expected a better end use for Coonawarra Merlot.

1997 Light to medium red-purple; the bouquet is fairly light, with clean berryish fruit and subtle oak. The palate offers some of the expected structure of Merlot, being soft and round, the flavours offer touches of mint and leaf, but do not send the pulse racing. **rating:** 84

best drinking 1999–2002 **best vintages** NA **drink with** Pan-fried veal • $15

Mildara Robertson's Well

Introduced in 1992, and named after a local landmark which (slightly embarrassingly) turned out to be outside the tentative boundaries for Coonawarra. The problem was resolved by bestowing the name on one of the Coonawarra vineyards within the boundary. Had an auspicious start winning the Stodart Trophy and three gold medals in national wine shows. A similar blend to Jamiesons Run but with more weight.

1997 Medium red-purple; the bouquet is clean, of medium intensity, with relatively straightforward berry fruit and controlled oak. The palate is very lively, with crisp cherry fruit and well-balanced and integrated oak. Gold medal winner Melbourne Show 1998. **rating:** 87

best drinking 2000–2005 **best vintages** '92, '93, '96 **drink with** Roast beef • $19

mildara (murray darling) ★★★

Wentworth Road, Merbein, Vic 3505 **region** Murray Darling and Swan Hill
ph (03) 5025 2303 **fax** (03) 5025 3300 **open** Mon–Fri 9–5, weekends 10–4
winemaker Steve Guy **prod.** 80 000 **est.** 1888

prod. range ($8–10 R) Church Hill Chardonnay, Fumé Blanc, Cabernet Merlot; Jimmy Watson Chardonnay and Cabernet Sauvignon; also makes fine Sherries (Chestnut Teal, George and Supreme) and superb Pot Still Brandy.

summary A somewhat antiquated Merbein facility remains the overall group production centre following its acquisition of Wolf Blass, although all of its premium wines are sourced from and made at Coonawarra.

milford vineyard ★★★★

Richmond Road, Cambridge, Tas 7170 **region** Southern Tasmania
ph (03) 6248 5029 **fax** (03) 6224 2331 **open** Not
winemaker Andrew Hood (Contract) **prod.** 250 **est.** 1984

prod. range ($20 R) Pinot Noir.

summary Given the tiny production, Milford is understandably not open to the public, the excellent Pinot Noir being quickly sold by word of mouth. The 150-hectare grazing property (the largest Southdown sheep stud in Australia) has been in Charlie Lewis's family since 1830. Only 15 minutes from Hobart, and with an absolute water frontage to the tidal estuary to the

Coal River, it is a striking site. The vineyard is established on a patch of 5-foot-deep sand over a clay base, with lots of lime impregnation.

Milford Vineyard Pinot Noir

Produced from a little over 1 hectare of estate-grown pinot noir, a size chosen by the partnership which runs the vineyard (with property owner Charlie Lewis) because it represents one day's picking – and only one day. The 1996 Pinot Noir won a gold medal at the Royal Hobart Wine Show.

🍷🍷🍷🍷🍷 **1996** Medium red-purple; a voluminous bouquet initially displaying cherry, plum and spice, and ultimately touches of mint. The palate likewise shows a mix of plum, spice and hints of mint, but no foresty/herby characters. Amazing for a tough vintage. **rating:** 90

best drinking 1999–2003 **best vintages** '96 **drink with** Char-grilled salmon • $20

millers samphire NR

Watts Gully Road, Cnr Robertson Road, Kersbrook, SA 5231 **region** Adelaide Hills
ph (08) 8389 3183 **open** 7 days 9–6 by appointment
winemaker Tom Miller **prod.** 130 **est.** 1982
prod. range ($9 CD) Riesling.
summary Next after Scarp Valley, one of the smallest wineries in Australia offering wine for sale; pottery also helps. Tom Miller has one of the more interesting and diverse CVs, with an early interest in matters alcoholic leading to the premature but happy death of a laboratory rat at Adelaide University and his enforced switch from biochemistry to mechanical engineering. The Riesling is a high-flavoured wine with crushed herb and lime aromas and flavours.

millinup estate NR

RMB 1280 Porongurup Road, Porongurup, WA 6324 **region** Great Southern
ph (08) 9853 1105 **fax** (08) 9853 1105 **open** Weekends 10–5
winemaker Gavin Berry (Contract) **prod.** 220 **est.** 1989
prod. range ($12–16 CD) Twin Peaks Riesling, Late Harvest Riesling, Cabernet Sauvignon Franc Merlot.
summary The Millinup Estate vineyard was planted in 1978, when it was called Point Creek. Owners Peter and Lesley Thorn purchased it in 1989, renaming it and having the limited production (from half a hectare each of riesling and merlot, supplemented by purchased cabernet sauvignon) vinified at Plantagenet.

minimbah NR

Minimbah House, Whittingham, NSW 2330 **region** Upper Hunter Valley
ph (02) 6572 4028 **fax** (02) 6572 1513 **open** Not
winemaker Simon Gilbert (Contract) **prod.** NA **est.** 1996
prod. range Chardonnay, Shiraz.
summary Four hectares of chardonnay and 1 hectare of shiraz were planted in 1996, and the first vintage, to be made by Simon Gilbert, was expected in 1999.

minot vineyard NR

PO Box 683, Margaret River, WA 6285 **region** Margaret River
ph (08) 9757 3579 **fax** (08) 9757 2361 **open** By appointment
winemaker Various Contract **prod.** 1000 **est.** 1600
prod. range ($11–15 ML) Semillon Sauvignon Blanc, Cabernet Sauvignon.
summary Minot, which takes its name from a small Chateau in the Loire Valley in France, is the husband and wife venture of the Miles family, producing just two wines from the 4-hectare plantings of semillon, sauvignon blanc and cabernet sauvignon. Both wines won silver medals at the 1997 Mount Barker Wine Show.

mintaro wines NR

Leasingham Road, Mintaro, SA 5415 **region** Clare Valley
ph (08) 8843 9046 **fax** (08) 8843 9050 **open** 7 days 9–5
winemaker Peter Houldsworth **prod.** 4000 **est.** 1984
prod. range ($9–20 CD) Dry Riesling, Late Picked Riesling, Shiraz, Shiraz Cabernet Franc Cabernet Sauvignon, Cabernet Sauvignon Cabernet Franc.
summary Has produced some very good Riesling over the years, developing well in bottle. The red wines are formidable, massive in body and extract, built for the long haul.

miramar ★★★★

Henry Lawson Drive, Mudgee, NSW 2850 **region** Mudgee
ph (02) 6373 3874 **fax** (02) 6373 3854 **open** 7 days 9–5
winemaker Ian MacRae **prod.** 8000 **est.** 1977
prod. range ($11–30 CD) Semillon, Chardonnay, Fumé Blanc, Riesling, Eurunderee Rosé, Cabernet Sauvignon, Shiraz, Encore (Sparkling).
summary Industry veteran Ian MacRae has demonstrated his skill with every type of wine over the decades, ranging through Rosé to Chardonnay to full-bodied reds. All have shone under the Miramar label at one time or another, although the Ides of March are pointing more to the red than the white wines these days. A substantial part of the production from the 38 hectares of estate vineyard is sold to others, the best being retained for Miramar's own use.

miranda ★★★☆

57 Jondaryan Avenue, Griffith, NSW 2680 **region** Riverina
ph (02) 6962 4033 **fax** (02) 6962 6944 **open** 7 days 9–5
winemaker Doug Wilson **prod.** 1.5 million **est.** 1939
prod. range ($4.95–23.95 R) Top of the range is Show Reserve Chardonnay, Old Vine Shiraz, Shiraz Cabernet; Golden Botrytis; followed by the High Country series (from the King Valley) of Riesling, Chardonnay, Merlot, Shiraz and Cabernet; Mirool Creek Dry White, Chardonnay and Cabernet Shiraz; Somerton Riesling Traminer, Semillon Chardonnay and Shiraz Cabernet; also lower priced Christy's Land and assorted varietals, generics, sparkling and ports.
summary Miranda Wines continues its aggressive and successful growth strategy, having opened a new winery in the King Valley in 1998, and previously expanded winemaking operations into the Barossa Valley. A veritable cascade of wines now appear under the various brand names, the majority representing good value for money.

Miranda High Country Merlot

Miranda has taken a significant position in Victoria's King Valley, from whence this wine comes. The region does not produce concentrated or weighty wine, but is clearly well suited in climatic terms to merlot – particularly at higher elevations.

YYYY **1997** Bright red-purple; while the bouquet is not complex, it is clean, and offers quite sweet red berry fruit; there is more fresh red fruit on the palate, with a flick of tannin and a whisk of American oak. Well put together. **rating:** 85

best drinking 1999–2000 **best vintages** NA **drink with** Pasta • $13

miranda rovalley estate ★★★☆

Barossa Highway, Rowland Flat, SA 5352 **region** Barossa Valley
ph (08) 8524 4537 **fax** (08) 8524 4066 **open** 7 days 9–4.30
winemaker David Norman **prod.** NA **est.** 1919
prod. range ($9–25 CD) Premium Late Harvest Riesling; the Grey Series of Riesling, Semillon, Sauvignon Blanc, Chardonnay, Shiraz, Bush Vine Grenache and Cabernet Sauvignon; followed by Show Reserve range of Chardonnay, Old Vine Shiraz and Shiraz Cabernet.

summary Increasingly absorbed into the Miranda Wine Group since its acquisition, drawing on grapes produced both in the Barossa Valley and throughout other parts of southeast Australia. The accent is on value for money, with consistent show success underlining the quality.

mistletoe wines NR

Lot 1 Hermitage Road, Pokolbin, NSW 2335 **region** Lower Hunter Valley
ph (02) 4998 7770 **fax** (02) 4998 7792 **open** Mon–Fri 10–6 or by appointment
winemaker Jon Reynolds (Contract) **prod.** 2000 **est.** 1967
prod. range ($15.50–17 R) Semillon, Barrel Fermented Chardonnay, Shiraz.
summary Mistletoe Wines, owned by Ken and Gwen Sloan, can trace its history back to 1909, when a substantial vineyard was planted on what was then called Mistletoe Farm. The Mistletoe Farm brand made a brief appearance in the late 1970s, but disappeared, and has now been revived under the Mistletoe Wines label by the Sloans, with contract-winemaking by Jon Reynolds. The Shiraz is the best of the wines, with pleasant, straightforward varietal character, the white wines adequate. The wines won a number of medals in wine shows in 1997 and are available both at the McGuigan Bros cellar door and at the old Hungerford Hill Village site on the corner of McDonalds and Broke Roads.

mitchell ★★★★☆

Hughes Park Road, Sevenhill via Clare, SA 5453 **region** Clare Valley
ph (08) 8843 4258 **fax** (08) 8843 4340 **open** 7 days 10–4
winemaker Andrew Mitchell **prod.** 30 000 **est.** 1975
prod. range ($15–30 CD) Watervale Riesling, Peppertree Vineyard Shiraz, The Growers Semillon, The Growers Grenache, Sevenhill Cabernet, Sparkling Peppertree.
summary For long one of the stalwarts of the Clare Valley, producing long-lived Rieslings and Cabernet Sauvignons in classic regional style, but having extended the range with very creditable Semillon and Shiraz. A lovely old stone apple shed provides the cellar door and upper section of the compact winery. Production has increased by almost 50% over the past few years and as well as national retail distribution, the wines are exported to the UK, US, Canada, New Zealand, Switzerland and Hong Kong.

Mitchell Watervale Riesling

First made in 1977 and produced from the estate vineyards in the Watervale region. This is a classic Clare Riesling which can age magnificently for up to 15 years in good vintages.

YYYY 1998 Light green-yellow; the bouquet is powerful, full and slightly broad, the palate equally powerful and rich with some fruit sweetness (not residual sugar); an honest, generous style. **rating:** 85

best drinking 1999–2005 **best vintages** '78, '84, '90, '92, '93, '94, '95 **drink with** Grilled fish • $15

Mitchell The Growers Semillon

Introduced in 1996 and replaced the former Barrel Fermented Semillon. As with its predecessor, an each-way drinking style, good when young, but even better with some age. The grapes are grown by the Pulford, Haig and Howard families.

YYYY 1998 Light green-yellow; the bouquet is firm, quite intense, with attractive, slightly grassy fruit. The palate is well balanced and flavoured, complexed by a touch of Sauvignon Blanc and a whisper of oak. **rating:** 87

best drinking 1999–2003 **best vintages** NA **drink with** Sugar-cured tuna • $15

Mitchell Peppertree Vineyard Shiraz

Like the Semillon, first made in 1984, and which takes its name from the old peppertree which grows in the shiraz vineyard at Watervale. The wine is aged for 18 months in small French and American oak; in some years it shows minty characters, in other years more spice and cherry.

ŸŸŸŸ **1997** Medium red-purple; while clean and fairly ripe, the bouquet is tending light. The palate offers more, with gently ripe berry fruit and soft, fine tannins. **rating:** 85

best drinking 2000–2005 **best vintages** '84, '86, '87, '94, '95, '96 **drink with** Devilled kidneys • $22

Mitchell Sevenhill Cabernet Sauvignon

First made in 1976; from then to 1983 fashioned entirely from Cabernet Sauvignon, but between then and 1985 first Cabernet Franc and then Merlot were added, now contributing 5–15% of the finished wine, which is aged in a mix of new and older French oak.

ŸŸŸŸ **1996** Medium to full red-purple; an attractively rich bouquet with blackberry, chocolate and earth follows through into a solid, ripe palate with sweet blackcurrant fruit, a nice touch of oak and balanced tannins on the finish. **rating:** 89

best drinking 2000–2008 **best vintages** '78, '80, '84, '86, '90, '92, '94, '96 **drink with** Roast lamb • $22

mitchelton ★★★★☆

Mitchellstown via Nagambie, Vic 3608 **region** Goulburn Valley
ph (03) 5794 2710 **fax** (03) 5794 2615 **open** 7 days 10–5
winemaker Don Lewis **prod.** 200 000 **est.** 1969

prod. range ($12.95–42.95 CD) Top-of-the-range is Print Label Red; then come Chardonnay, Cabernet Sauvignon, Marsanne; next Mitchelton III wines, White (Marsanne, Grenache, Viognier), Red (Shiraz, Grenache, Mourvedre); Chinaman's Bridge Merlot, Blackwood Park Riesling. Preece Chardonnay, Sauvignon Blanc, Merlot and Cabernet Sauvignon are volume sellers; Goulburn Valley Shiraz introduced in 1996. Finally, intermittent aged classic releases.

summary Acquired by Petaluma in 1994, having already put the runs on the board in no uncertain fashion with a gifted team of Stephen Shelmerdine and winemaker Don Lewis. Boasts an impressive array of wines across a broad spectrum of style and price, but each carefully aimed at a market niche. The wines are exported throughout the UK and Europe, Singapore, Taiwan and US.

Mitchelton Blackwood Park Riesling

Over the past few years Blackwood Park has laid claim to being the best commercial Riesling in Australia, even though its origins go back to 1978 under different labels. Since that time various vintages have won eight trophies, 51 gold, 66 silver and 102 bronze medals.

ŸŸŸŸ **1998** Yet another wine under this label which seems to hit the spot year in, year out. As always, has that extra dimension of tropical lime aroma and flavour on a pure but intense palate; good acidity on a lingering finish. **rating:** 87

best drinking 1999–2003 **best vintages** '85, '90, '91, '92, '94, '95 **drink with** Sashimi • $12.95

Mitchelton Viognier Roussanne

A new and very interesting release from Mitchelton, held back for bottle maturation prior to release.

ŸŸŸŸ **1996** Bright green-yellow; a highly scented and distinctive bouquet with a mix of dried flowers and honeysuckle is followed by a palate with attractive flavours initially, but which ends up being slightly loose. Perhaps a touch more acid would have helped, but nonetheless a very interesting wine. **rating:** 88

best drinking 1999–2000 **best vintages** NA **drink with** Turkey breast • $24.95

Mitchelton Airstrip Marsanne

The wine was made using techniques dating back to the eighth century BC, when the Greeks described the technique, basically devolving around drying grapes on straw mats in the sun or in specially-constructed drying houses. In modern-day France it is used to make Vin de Paille (straw wine) in the Jura and in the Rhône Valley and Vin Santo in Tuscany and elsewhere. Mitchelton winemaker Don Lewis used the oldest of methods, laying the grapes on mats in the sun to dehydrate them. The wine is not unlike young Vin de Paille from the Rhône Valley, and one hopes and assumes it will gain considerable additional complexity in bottle. The question is largely academic, for only 900 375-ml bottles were made, and the wine was not sold commercially.

▼▼▼▼ **1997** Brilliant yellow-green; the bouquet offers touches of honeycomb and honey, characters which come through in the slightly off-dry palate. You have to wonder whether the effort was worthwhile, but it is an interesting wine. For the record it has 15 degrees alcohol. **rating:** 85

best drinking 2000–2005 **best vintages** NA **drink with** Rich pasta • $11.95

Mitchelton Goulburn Valley Marsanne

Mitchelton adopts precisely the opposite approach to Chateau Tahbilk by seeking to invest its Marsanne with complexity from the word go. For many years the approach was to give the wine substantial oak, but the approach is now more subtle. The Marsanne is barrel-fermented, and also has 15% Viognier and Roussane included.

▼▼▼▽ **1997** Light to medium yellow-green; the bouquet is quite complex, but the oak is pretty obvious. The wine is cleverly made, but really needs more fruit or more acidity to lift the fruit profile of the palate above that of the oak. **rating:** 84

best drinking 1999–2001 **best vintages** NA **drink with** Creamy pasta • $13.95

Mitchelton Chardonnay

This wine has undergone a number of name changes (not necessarily coincident with the arrival of new brand managers) in the 1990s. First called Reserve Chardonnay, the name was then changed to Victoria Chardonnay, and now simply Chardonnay: it is the best and yet the only Chardonnay released under the Mitchelton label. The same changes have taken place with the Marsanne and Cabernet Sauvignon.

▼▼▼▼▽ **1997** Medium yellow-green; the bouquet is very smooth and clean with fig and melon fruit, and just a hint of barrel-ferment character. The gentle palate is harmonious and smooth, with similar fruit flavours to the bouquet, and subtle oak. Trophy winner 1998 National Wine Show: Best Chardonnay. **rating:** 92

best drinking 1999–2002 **best vintages** '81, '85, '90, '91, '92 **drink with** Crumbed brains • $21.95

Mitchelton Blackwood Park Late Harvest Riesling

In some years Mitchelton produces a Botrytis Riesling from Blackwood Park and in others (when botrytis does not intervene) a late harvest style. This sector of the market is not oversupplied with medium-sweet Rieslings of quality. Given five to ten years bottle age, such wines can attain wonderful bottle-developed characters, difficult to visualise when they are young. Available only from cellar door.

▼▼▼▼ **1997** Very light green-yellow colour; a relatively straightforward, crisp lime/lemon-accented bouquet is precisely replicated on the palate. Deserves to be given a chance in the cellar. **rating:** 87

best drinking 2003–2010 **best vintages** NA **drink with** Scallops in bechamel sauce • $12.95

Mitchelton Goulburn Valley Shiraz

Yet another new label from Mitchelton, using a blend of 85% Shiraz and 15% Grenache and Mourvedre. I suspect this has replaced the former Mitchelton III Red.

YYYY **1996** Light to medium red-purple; the bouquet is relatively light, and not particularly rich, with vanilla oak doing much of the work. However, attractive, bright, juicy cherry/berry fruit makes its presence felt on the refreshing palate. **rating:** 87

best drinking 1999–2002 **best vintages** NA **drink with** Braised duck • $14.95

Mitchelton Print Label Shiraz

In 1981 Mitchelton conceived the idea of staging a Print Exhibition and making an annual purchase of the best print in the exhibition for subsequent use as the label of the best red wine of the vintage. Both the '90 and '91 vintages were outstanding, the former winning the Jimmy Watson Trophy. Having started life as a Cabernet Sauvignon, Shiraz has ruled the roost for most recent vintages. However, the best wine of the 1994 vintage was a Cabernet Sauvignon, but with 1995 the wine reverted to Shiraz.

YYYYY **1995** Medium to full purple-red; a complex, rich bouquet with liquorice and dark plum/cherry fruit marries well with sweet oak. A wine which starts well and gets even better on the palate, with abundant liquorice and black cherry fruit, well-balanced and integrated oak, and finishing with good tannins. **rating:** 94

best drinking 2001–2011 **best vintages** '81, '82, '90, '91, '92, '95 **drink with** Marinated venison • $42.95

Mitchelton Preece Merlot

The wine comes from contract growers in the King Valley, and from Mitchelton's vineyards in the Goulburn Valley. It is aged in two- to three-year-old oak barrels for 17 months prior to release.

YYYY **1997** Medium red-purple; the bouquet is clean, of light to medium intensity with soft, slightly anonymous fruit. The wine has gentle mouthfeel, with pleasant, soft and sweet berry fruit and soft, fine tannins. **rating:** 85

best drinking 2000–2004 **best vintages** NA **drink with** Veal • $15.95

molly morgan vineyard ★★★☆

Talga Road, Lovedale, NSW 2321 **region** Lower Hunter Valley

ph (02) 4930 7695 **fax** (02) 9235 1876 **open** Weekends, public holidays 10–5

winemaker Contract **prod.** 1400 **est.** 1963

prod. range ($13.50–25 CD) Joe's Block Semillon, Old Vines Semillon, Semillon Chardonnay, Shiraz.

summary Molly Morgan been acquired by Andrew and Hady Simon, who established the Camperdown Cellars group in 1971, becoming the largest retailer in Australia before passing on to other pursuits, and John Baker, one of Australia's best known fine wine retailers who owned or managed Quaffers, Double Bay Cellars, the Newport Bottler and Grape Fellas (Epping) at various times. The property is planted to 14 acres of 25-year-old unirrigated semillon, which goes to make the Old Vines Semillon, 2 acres for Joe's Block Semillon, 6 acres of chardonnay and 3 acres of shiraz. The wines are contract-made (as has always been the case, in fact, but to a high standard).

Molly Morgan Joe's Block Semillon

A single vineyard wine, named after the third husband of the redoubtable Molly Morgan.

YYYY **1998** Considerable colour, suggesting the use of some skin contact. The bouquet is rich and full, with ripe Semillon varietal character, tending a little broad, and reinforcing the likelihood of some skin contact prior to fermentation. The palate is quite rich, with an interesting balance of ripe fruit and a distinct vein of mineral acid running through it. **rating:** 85

best drinking 1999–2003 **best vintages** NA **drink with** Seafood • $19

Molly Morgan Old Vines Semillon

Produced from 25-year-old, unirrigated vines which provide the best grapes from the estate.

🍷🍷🍷🍷🍷 **1998** Lighter colour than the Joe's Block Semillon. The bouquet has more intensity, yet more restraint than the Joe's Block wine. It has a full-bodied but smooth and stylish palate, with rounded lemony fruit, good structure and finish. **rating:** 92

best drinking 1999–2008 **best vintages** '98 **drink with** Calamari • $19

monbulk winery ★★☆

Macclesfield Road, Monbulk, Vic 3793 **region** Yarra Valley

ph (03) 9756 6965 **fax** (03) 9756 6965 **open** Weekends and public holidays 12–5, or by appointment

winemaker Paul Jabornik **prod.** 800 **est.** 1984

prod. range ($10–15 CD) Chardonnay, Riesling, Pinot Noir, Cabernet Sauvignon, Shiraz; also Kiwifruit wines.

summary Originally concentrated on Kiwifruit wines, but now extending to table wines; the very cool Monbulk subregion should be capable of producing wines of distinctive style, but the table wines are (unfortunately) not of the same standard as the Kiwifruit wines, which are quite delicious.

monichino wines ★★☆

1820 Berrys Road, Katunga, Vic 3640 **region** Goulburn Valley

ph (03) 5864 6452 **fax** (03) 5864 6538 **open** Mon–Sat 9–5, Sun 10–5

winemaker Carlo Monichino, Terry Monichino **prod.** 14 000 **est.** 1962

prod. range ($8–22 CD) Riesling, Semillon Sauvignon Blanc, Sauvignon Blanc, (Strathbogie) Botrytis Semillon, Orange Muscat, Lexia, Rose Petals Spätlese, Shiraz, Merlot, Malbec, Cabernet Franc, Cabernet Sauvignon; various Ports and fortifieds; bulk sales also available.

summary A winery which has quietly made some clean, fresh wines in which the fruit character is carefully preserved.

montagne view estate NR

555 Hermitage Road, Pokolbin, NSW 2335 **region** Lower Hunter Valley

ph (02) 4998 7822 **fax** (02) 6574 7276 **open** 7 days 10–5

winemaker Greg Silkman (Contract) **prod.** 500 **est.** 1993

prod. range ($20–25 CD) Edith Margaret Chardonnay, Vivian Laurie Merlot.

summary The major investment and principal business of Montagne View is the eight-studio-suite guesthouse sitting among the 5 hectares of vines. There is also a high-quality restaurant (Brents) offering the prospect of all-inclusive gourmet weekends for around $550 per couple. The estate wines are sold through the restaurant and cellar door, with other local wines available in the restaurant.

montara ★★★

Chalambar Road, Ararat, Vic 3377 **region** Grampians

ph (03) 5352 3868 **fax** (03) 5352 4968 **open** Mon–Sat 9.30–5, Sun 12–4

winemaker Mike McRae **prod.** NFP **est.** 1970

prod. range ($10–19.50 CD) New Wave Riesling, Pinot Noir, Shiraz, Cabernet Sauvignon, Shiraz Port.

summary Achieved considerable attention for its Pinot Noirs during the 1980s, but other regions (and other makers) have come along since. A recent tasting of the currently available wines ('97 and '96 vintages) was interesting, with a number of the wines having fragrance and elegance, but lacking richness and concentration. Smart new label designs do help. Limited national distribution; exports to the UK, Switzerland, Canada and Hong Kong.

montgomery's hill NR

Hassell Highway, Lower King, Albany, WA 6330 **region** Great Southern
ph (08) 9844 7177 **fax** (08) 9844 1104 **open** Not
winemaker Steve Pester (Porongurup Winery) **prod.** 2000 **est.** 1996
prod. range Chardonnay, Cabernet Sauvignon Franc.

summary Montgomery's Hill is situated 16 kilometres northeast of Albany on a north-facing slope on the banks of the Kalgan River. The vineyard is situated on an area which was previously an apple orchard, and is a diversification for the third generation of the Montgomery family which owns the property. Chardonnay, cabernet sauvignon and cabernet franc were planted in 1996, followed by sauvignon blanc, shiraz and merlot in 1997. The 1998 wines were contract-made by Brenden Smith at West Cape Howe Wines, but as from 1999 Montgomery's Hill will be made at the new Porongurup Winery.

montrose ★★★☆

Henry Lawson Drive, Mudgee, NSW 2850 **region** Mudgee
ph (02) 6373 3853 **fax** (02) 6373 3795 **open** Mon–Fri 9–4, weekends 10–4
winemaker Brett McKinnon **prod.** 50 000 **est.** 1974
prod. range ($9.95–20.95 R) Poet's Corner Semillon Sauvignon Blanc Chardonnay, Unwooded Chardonnay and Shiraz Cabernet Sauvignon Cabernet Franc are at the inexpensive end; premium varietals are Chardonnay, Barbera, Sangiovese, Black Shiraz and Cabernet Sauvignon.

summary A small piece of the Orlando/Wyndham empire, acting partly as a grape and bulk wine source for that empire, and partly as a quality producer in its own right, making typically full-flavoured whites and deep-coloured reds. Poet's Corner always provides excellent value for money. Exports to UK, US and Asia.

Montrose Mudgee Chardonnay

Montrose followed closely in the footsteps of neighbour Craigmoor (both now being part of the Wyndham empire, but not then) in producing some of the first Chardonnays in Australia. From the outset, the quality has been high, with then winemaker Carlo Corino anticipating the future in first making unwooded Chardonnay (in the 1970s). Over the years, various vintages of this wine have been trophy winners.

🍷🍷🍷🍷 **1997** Medium to full yellow-green; the bouquet is of moderate intensity, with well-balanced fruit and oak, the palate with plenty of substance and flavour, aided by clever oak infusion. Gold medal winner at the Mudgee Wine Show in 1997. **rating:** 86

best drinking 1999–2001 **best vintages** '95 **drink with** Pasta • $14.95

Montrose Black Shiraz

Black Shiraz is simply a name given to the wine; it is not a particular form of Shiraz, nor even a special clone. However, it is not a bad descriptor of the strength of Mudgee Shiraz.

🍷🍷🍷🍷 **1996** Excellent, strong red-purple; rich and concentrated dark plum and berry fruit on the bouquet floods into sweet plum, cherry and rippling, soft tannins on the palate. Oak is there but you barely notice it. **rating:** 89

best drinking 2000–2010 **best vintages** NA **drink with** Beef with olives and red wine • $18.95

Montrose Barbera

Montrose was established by Italian-born engineers Carlo Salteri and Franco Belgiorno-Nettis (of Transfield Corporation), who in turn hired Italian winemaker Carlo Corino as winemaker. Inevitably, Italian grape varieties were planted and were briefly made, but disappeared from the scene for over a decade (although the plantings were not removed). The Italian craze has led to their reappearance, and not before time. In both 1996 and 1997 I preferred the Barbera to the Sangiovese.

TTTT **1997** Youthful medium to full purple-red; the bouquet is sweet, ripe and lush with shiraz-like dark cherry fruit. The palate is generously flavoured, with good tannins, a nice touch of vanilla oak, all making for a very pleasant wine. I am not so sure what is says about varietal character. **rating:** 87

best drinking 2002–2007 **best vintages** NA **drink with** Bistecca fiorentina • $20.95

moondah brook ★★★★

c/o Houghton, Dale Road, Middle Swan, WA 6056 **region** Swan District

ph (08) 9274 5372 **fax** (08) 9274 5372 **open** Not

winemaker Larry Cherubino **prod.** 90 000 **est.** 1968

prod. range ($13–17 R) Chardonnay, Chenin Blanc, Verdelho, Sauvignon Blanc, Shiraz, Cabernet Sauvignon, Maritime (Sparkling); also occasional Show Reserve releases of Chenin Blanc and Verdelho.

summary Part of the BRL Hardy wine group which has its own special character as it draws part of its fruit from the large Gingin vineyard, 70 kilometres north of the Swan Valley, and part from the Margaret River and Great Southern. In recent times it has excelled even its own reputation for reliability with some quite lovely wines, in particular honeyed, aged Chenin Blanc and finely structured Cabernet Sauvignon.

Moondah Brook Chenin Blanc

Stakes a strong claim as Australia's best Chenin Blanc. Produced entirely from Moondah Brook's Gingin Vineyard, the '92 won the White Wine of the Year Award in the 1993 *Wine* magazine International Wine Challenge in London. The wines are cold-fermented and then matured in new American oak for six months.

TTTY **1996** Medium yellow-green; the bouquet is soft, with tropical fruit salad aromas, the palate no less soft with similar flavours together with a touch of passionfruit. The barrel fermentation component is all but imperceptible, but does add some length to the palate. **rating:** 83

best drinking 1998–2003 **best vintages** '80, '87, '89, '91, '93, '94 **drink with** Prosciutto and figs • $13

Moondah Brook Verdelho

Sourced predominantly from the Gingin Vineyard, but with a smaller contribution from the Margaret River. It is cold-fermented in stainless steel, and not given any oak maturation. The wine can develop wonderful secondary characters given time in bottle, with attractive tropical and citrus fruit aromas and flavours.

TTTY **1997** Light green-yellow; the bouquet is clean, with a touch of fruit salad, but tending neutral. There is rather more to the palate, with nice tropical fruit salad flavours; slow-developing. **rating:** 84

best drinking 2000–2004 **best vintages** '80, '87, '89, '90, '91, '93 **drink with** Sugar-cured tuna • $13

Moondah Brook Chardonnay

A wine which is sometimes plain, sometimes quite brilliant. In the latter category was the '95, the top gold medal in the 70-strong Commercial Dry White Table Wine (full-bodied) C class at the 1997 Sydney Wine Show, but in vintages such as '93 and '97 simply a pleasant commercial white wine.

TTTT **1997** Medium yellow-green; the bouquet is highly aromatic with peachy fruit and just a hint of passionfruit. The palate has similar high fruit flavours which track the bouquet; no oak evident. Has developed well over the past 12 months, throwing off the green characters it showed when young. **rating:** 85

best drinking 1999–2000 **best vintages** '95 **drink with** Marron • $13

Moondah Brook Shiraz

Sourced from various parts of Western Australia, but predominantly from Houghton's Frankland River vineyard and given generous – though not excessive – oak treatment.

🍷🍷🍷🍷 **1997** Very good purple-red; fresh, strongly earthy, shiraz aromas are supported by a nice touch of sweet oak. The palate is firm, youthful and earthy, needing much time, but should handsomely repay patience. **rating:** 88

best drinking 2002–2007 **best vintages** NA **drink with** Beef stew • $17

Moondah Brook Cabernet Sauvignon

The Moondah Brook label is something of a misnomer, as fruit sources have gradually extended across the State of Western Australia. So it is with the Cabernet Sauvignon, which burst into prominence with the '91 vintage, winning gold medals at national shows, made from Margaret River and Great Southern fruit. The '96 and '97 were excellent sequels, other vintages less so.

🍷🍷🍷🍷 **1997** Medium red-purple; quite strong oak inputs surround blackberry fruit on the bouquet. The palate is more fruit-driven, with sweet cassis berry fruit, soft tannins and not over much oak. A good outcome for a less than brilliant vintage. **rating:** 87

best drinking 2000–2005 **best vintages** '82, '88, '91, '93, '94, '96, '97 **drink with** Veal chops Italian-style • $17

moonshine valley winery NR

374 Mons Road, Forest Glen, Buderim, Qld 4556 **region** Other Wineries of Qld
ph (07) 5445 1198 **fax** (07) 5445 1799 **open** 7 days 10–5
winemaker Frederick Houweling **prod.** 1700 **est.** 1985

prod. range ($10.40–32.95 CD) A kaleidoscopic array of basically fruit-based wines including Le Dry (Mulberries and Blueberries, oak-matured), Moonlight White (West Indian Limes), Sunshine Nouveau (locally grown Jaboticabas), Old Buderim Ginger (fortified base with Honey and Ginger added), Strawberry Wine, Exporto (Mulberry and Blueberry-based Port Wine).

summary Frederick Houweling brings a European background to his making of these fruit-based wines. The winery is situated on a large property among natural lakes and forest, and also offers a restaurant, cafeteria, and souvenir shop.

moorebank vineyard NR

Palmers Lane, Pokolbin, NSW 2320 **region** Lower Hunter Valley
ph (02) 4998 7610 **fax** (02) 4998 7367 **open** Fri–Mon 10–5 or by appointment
winemaker Iain Riggs (Contract) **prod.** 1750 **est.** 1977

prod. range ($19.50–24.50 CD) Chardonnay, Summar Semillon, Gewurztraminer, Merlot, now sold in the narrow 500-ml Italian glass bottle known as Bellissima.

summary Ian Burgess and Debra Moore own a mature 5.5-hectare vineyard with a small cellar-door operation offering immaculately packaged wines in avant-garde style. The peachy Chardonnay has been a medal winner at Hunter Valley Wine Shows.

moorilla estate ★★★★☆

655 Main Road, Berriedale, Tas 7011 **region** Southern Tasmania
ph (03) 6249 2949 **fax** (03) 6249 4093 **open** 7 days 10–5
winemaker Alain Rousseau **prod.** 9000 **est.** 1958

prod. range ($19–38 R) Riesling, Chardonnay, Gewurztraminer, Winter Collection Cabernet Merlot, Cabernet Sauvignon, Vintage Brut; Pinot Noir in three tiers: White Label, Black Label and Reserve.

summary Moorilla Estate is an icon in the Tasmanian wine industry and, under the ownership of David Walsh, is thriving. Alain Rousseau is an exceptionally gifted winemaker producing better wines than have ever previously come from Moorilla (particularly Pinot Noir and Merlot) while the opening of the Museum in the marvellous Alcorso house designed by Sir Roy Grounds adds even more attraction for visitors to the estate, a mere 15–20 minutes from Hobart.

Moorilla Estate Riesling

In my view, consistently the best of the Moorilla Estate wines, with a long track record of excellence. Produced from 1.5 hectares of estate vineyards. In addition, the occasional releases of Botrytised Riesling (by mailing list and through cellar door), are sensational. The 1994 won the trophy for Best Museum Wine in the 1998 Tasmanian Wines Show. The '98 sharply split the judges a the 1999 Tasmanian Wines Show; the points given here are equivalent to those which I gave the wine, my fellow judges being much less impressed with it. I can't help but think I might have been closer to the mark.

🍷🍷🍷🍷🍷 **1998** Light green-yellow; the bouquet is clean and powerful, with potent lime and herb fruit aromas. On the palate it has lots of lovely lime and herb fruit, with perfect length and balance. **rating:** 94

best drinking 1999–2005 **best vintages** '81, '82, '90, '91, '93, '94, '95, '97, '98 **drink with** Asparagus • $21

Moorilla Estate Gewurztraminer

A wine which has always been on the fringe of inclusion in the *Wine Companion*, with the delicacy of youth telling against it in prior vintages. Given a few years in bottle, it would undoubtedly have made the cut. In this instance, the '98 vintage (and very likely the skills of winemaker Alain Rousseau) has resulted in a wine which cannot be ignored because of its youth.

🍷🍷🍷🍷 **1998** Medium yellow-green; the bouquet has powerful lychee fruit on the bouquet supported by an unusually powerful, almost thick, palate which is quite different from the usual delicate Moorilla style, but which will appeal to many simply because of the amount of flavour. **rating:** 87

best drinking 1999–2003 **best vintages** NA **drink with** Chinese stir-fried prawns • $26

Moorilla Estate Chardonnay

A complex wine which is, as one would expect, very much in cool-climate style. It is produced from 2.5 hectares of immaculately trained estate vineyards, and given what might loosely be called 'the full treatment' in the winery. The '95 won the trophy for Best Wine of Show at the 1997 Tasmanian Wines Show, bettering the trophy (Best White Wine) won by the '94 the previous year. The '97 is right up there, winning a gold medal at the 1999 show.

🍷🍷🍷🍷🍸 **1997** Light green-yellow; a very fragrant and tangy bouquet with a mix of citrus, melon and apple fruit touched with subtle oak is followed by a commensurately lively, fresh, crisp and tangy palate. Unmistakably cool-grown, with those apple fruit flavours lingering, and an appropriately gentle touch with oak. **rating:** 90

best drinking 1999–2003 **best vintages** '81, '82, '90, '91, '92, '94, '95 **drink with** Tasmanian salmon • $24

Moorilla Estate Black Label Pinot Noir

The third tier of the Moorilla Pinot Noir structure, coming underneath the White Label Pinot Noir and ultimately the Reserve Pinot, a three-tiered classification which was introduced in 1997.

🍷🍷🍷🍷 **1998** Bright, youthful purple-red. A quite fragrant bouquet with a mix of cherry and more leafy/minty overtones. The palate shows a mix of strong cherry/cherry pip fruit and more stalky/stemmy characters. Good early release, early-drinking style. **rating:** 85

best drinking 1999–2001 **best vintages** NA **drink with** Milk-fed veal • $21

Moorilla Estate Reserve Pinot Noir

First introduced into the Moorilla Estate range in 1996, and on the evidence of the '96 and '97 vintages, significantly superior to the varietal releases.

🍷🍷🍷🍷🍷 **1997** Excellent purple-red; the bouquet is fragrant, with an exotic mix of red berry and pine needle aromas; the striking palate is marvellously concentrated, with powerful plummy fruit and a long finish. **rating:** 91

best drinking 2000–2004 **best vintages** '96, '97 **drink with** Venison • $38

Moorilla Estate White Label Pinot Noir

As if to underline the vast changes for the better since the arrival of Alain Rousseau, the '98 was the top (and only) gold medal winner in the 1998 Pinot Noir Class at the 1999 Tasmanian Wines Show.

🍷🍷🍷🍷🍷 **1998** Medium red-purple; the bouquet is complex, with cool-grown gamey varietal pinot noir aromas. A lovely wine on the palate, with sweet plummy fruit at the core surrounded by foresty nuances. **rating:** 94

best drinking 1999–2004 **best vintages** '84, '86, '90, '91, '92 **drink with** Quail • $19.50

Moorilla Estate Cabernet Sauvignon

One of two silver medal winners in its class in the 1999 Tasmanian Wines Show, the other silver (higher-pointed) going to the Reserve Cabernet Sauvignon from Moorilla Estate. A convincing demonstration of the all-round skills of Alain Rousseau.

🍷🍷🍷🍷 **1997** Medium to full red-purple, bright and strong. A wine desperately needing time for the powerful, youthful fruit and slightly unintegrated oak on both bouquet and palate to come together, but the balance is there. **rating:** 88

best drinking 2003–2010 **best vintages** NA **drink with** Marinated venison • $22.80

Moorilla Estate Reserve Cabernet Sauvignon

There is no doubt the Tamar Valley (St Matthias) Vineyard is playing its part here. Equally, there is no doubt that the Reserve Cabernet Sauvignon deserves its status.

🍷🍷🍷🍷🍷 **1997** Outstanding purple-red colour; like the '97 standard wine of the same year, powerful, solid and yet to build the secondary aromas which will come with time, but has masses of flavour and well-balanced extract. **rating:** 90

best drinking 2004–2010 **best vintages** NA **drink with** Rare beef • $35.95

mooroodcuc estate ★★★★☆

501 Derril Road, Moorooduc, Vic 3936 **region** Mornington Peninsula

ph (03) 9696 4130 **fax** (03) 9696 2841 **open** First weekend each month 12–5

winemaker Dr Richard McIntyre **prod.** 7000 **est.** 1983

prod. range ($19–45 R) Sauvignon Blanc, Chardonnay, Pinot Noir, Cabernet; also Devil Bend Creek Chardonnay and Pinot Noir.

summary Dr Richard McIntyre regularly produces one of the richest and most complex Chardonnays in the region, with grapefruit/peach fruit set against sumptuous spicy oak, and that hallmark soft nutty/creamy/regional texture. As well as retail distribution, the wines are exported to Japan.

Moorooduc Estate Chardonnay

Produced from 2.5 hectares of estate plantings, and made by Dr Richard McIntyre in a very consistent style, which is always strongly influenced by the secondary malolactic fermentation. The resultant nutty/buttery wine tends to stand out even in the background of the Mornington Peninsula, and certainly against other wine styles.

🍷🍷🍷🍷🍷 **1997** Light to medium yellow-green; the bouquet is clean, with an attractive balance of primary fruit and secondary malolactic fermentation inputs. The palate is similarly fresh and crisp, with nectarine, mineral and more nutty characters sparking off each other. Outside the mainstream Mornington style, and all the better for that. **rating:** 91

best drinking 2000–2008 **best vintages** '88, '90, '91, '92, '93, '94, '95, '97 **drink with** Grilled spatchcock • $30

Moorooduc Estate Pinot Noir

Produced from 1 hectare of estate plantings with two clones: D2V5 and D5V12. Typically very fragrant, sappy and spicy, excelling in the warmer, drier vintages.

🍷🍷🍷🍷 **1997** Medium red, with touches of purple; a fragrant and stylish bouquet, moderately intense with an attractive sappy edge to the fruit. The palate revolves around small red berry fruit flavours at the core, with lingering sappy/foresty characters providing style and bite. **rating:** 89

best drinking 1999–2002 **best vintages** '88, '91, '92, '93, '97 **drink with** Asian meat dishes • $45

morning cloud wines NR

15 Ocean View Avenue, Red Hill South, Vic 3937 **region** Mornington Peninsula
ph (03) 5989 2762 **fax** (03) 5989 2700 **open** By appointment
winemaker Ken Lang (Contract), Lindsay McCall (Contract) **prod.** 500 **est.** 1983
prod. range ($17–20 R) Chardonnay, Cabernet Sauvignon.
summary Morning Cloud Wines (previously Cloud Valley) is a joint venture between Kathy and Bill Allen and Peter and Judy Maxwell. Each family has its own vineyard at Red Hill South, and the grapes are pooled and the wine made under contract at Stonier's Winery. The Cabernet Sauvignon tends to be very leafy in Chinon-style; the Chardonnay, medium-bodied crisp and citrus-tinged.

morningside wines ★★★★

RMB 3002 Middle Tea Tree Road, Tea Tree, Tas 7017 **region** Southern Tasmania
ph (03) 6268 1748 **fax** (03) 6268 1748 **open** Not
winemaker Peter Bosworth **prod.** 400 **est.** 1980
prod. range ($20–25 ML) Riesling, Pinot Noir, Cabernet Sauvignon.
summary The name 'Morningside' was given to the old property on which the vineyard stands because it gets the morning sun first. The property on the other side of the valley was known as 'Eveningside', and, consistently with the observation of the early settlers, the Morningside grapes achieve full maturity with good colour and varietal flavour. Production is as yet tiny, but will increase as the 1.5-hectare vineyard matures. Retail distribution through Sutherland Cellars, Melbourne.

Morningside Chardonnay

1997 was the first vintage of Chardonnay made by Peter Bosworth, with two barrels produced. It spontaneously went through malolactic fermentation, and spent considerable time on lees. An outstanding wine for a first vintage.

🍷🍷🍷🍷🍷 **1997** Light yellow-green; the bouquet is spotlessly clean, with attractive melon and apple fruit supported by subtle oak. A wine with remarkable length and freshness, especially the acidity remaining after the malolactic fermentation. **rating:** 91

best drinking 1999–2006 **best vintages** '97 **drink with** Tasmanian lobster • $20

mornington vineyards estate ★★★

Moorooduc Road, Moorooduc South, Vic 3931 **region** Mornington Peninsula
ph (03) 5974 2097 **fax** (03) 5974 2097 **open** Weekends and public holidays 11–5
winemaker Kim Hart **prod.** 2500 **est.** 1989
prod. range ($18–28 CD) Chardonnay, Sauvignon Blanc, Pinot Noir, Shiraz.

summary As with so many Mornington Peninsula vineyards, a high degree of viticultural expertise, care and attention, coupled with skilled contract-winemaking has paid dividends. With a little over 20 hectares in production, it is one of the larger vineyards on the Peninsula, but sells a significant portion of its grapes. In cool, wet vintages (such as 1996), the vineyard does struggle to achieve full fruit ripeness. However, to be honest, did not do a great deal better in 1997.

morris ★★★★★

Mia Mia Road, Rutherglen, Vic 3685 **region** Rutherglen
ph (02) 6026 7303 **fax** (02) 6026 7445 **open** Mon–Sat 9–5, Sun 10–5
winemaker David Morris **prod.** NFP **est.** 1859
prod. range ($12.95–41.95 R) A limited range of table wines sparingly distributed, the most important of which is the red wine Durif, and also Shiraz; then fortified wines comprising Mick Morris Old Tawny Port, Liqueur Tokay, Old Premium Liqueur Tokay, Mick Morris Muscat, Old Premium Liqueur Muscat, Mick Morris Commemorative Liqueur Muscat; Old Premium Liqueur Muscat at the top end of the range; tiny quantities of Show Reserve are released from time to time, mainly ex-winery.
summary One of the greatest of the fortified winemakers, some would say the greatest. If you wish to test that view, try the Old Premium Muscat and Old Premium Tokay, which are absolute bargains given their age and quality and which give rise to the winery rating. The table wines are dependable, the white wines all being made by owner Orlando.

Morris Liqueur Tokay

Made from the muscadelle grape, used as a minor component in Sauternes, but nowhere else in the world used to make fortified dessert wines. In this version, which contains a greater percentage of younger material (two to four years old) than the premium labels, the accent is thrown firmly onto the very distinctive varietal character of muscadelle.
🍷🍷🍷🍷🍷 **NV** Light to medium golden-brown; a fragrant bouquet with fresh tea-leaf varietal aroma. There is masses of flavour on the palate, yet the wine is quite fresh with archetypal cold tea and butterscotch flavours, with the mid-palate sweetness followed by a cleansing, crisp finish. **rating:** 90
best drinking 1999–2009 **best vintages** NA **drink with** Either aperitif or at the end of the meal • $16.95

Morris Liqueur Muscat

As with the Tokay in the same range, the accent is thrown onto the varietal character of Muscat, otherwise known as Brown Frontignac. The style differs from that of Baileys, which tends to be sweet and in some ways more complex, but with less clarity of varietal character. Which of the two one prefers is very much a matter of personal taste.
🍷🍷🍷🍷🍷 **NV** Light to medium red-brown; clearly articulated, lively, raisiny muscat varietal aromas. In the mouth you can literally taste the grapes, as if one were chewing on an explosively rich raisin; great length, and perfect balance. **rating:** 91
best drinking 1999–2009 **best vintages** NA **drink with** Aperitif or digestif • $16.95

Morris Old Premium Liqueur Muscat

Here the component of old material is much greater than in the less-expensive standard Liqueur range. The blending of young, middle-aged and very old wines lies at the heart of the style, with a tiny percentage of 30- and 40-year-old wood-aged material having a disproportionate impact on the blend.
🍷🍷🍷🍷🍷 **NV** Medium to full tawny, with a hint of olive on the rim. A rich bouquet with complex caramel, toffee and coffee aromas intermingling with the raisins. The palate shows more of the raisiny varietal fruit, although the complexity of the bouquet does repeat itself. A great example of blending. **rating:** 95
best drinking 1999–2009 **best vintages** NA **drink with** Coffee, petits fours • $45.95

moss brothers ★★★

Caves Road, Willyabrup, WA 6280 **region** Margaret River
ph (08) 9755 6270 **fax** (08) 9755 6298 **open** 7 days 10–5
winemaker Jane Moss, David Moss **prod.** 14 000 **est.** 1984
prod. range ($16–25 R) Semillon, Sauvignon Blanc, Unwooded Chardonnay, Barrel Fermented Chardonnay, Moses Rock White, Sauvignon Blanc, Cabernet Merlot, Cellar Door Red, Moses Rock Red (the last two unusual blends, Moses Rock including Merlot, Pinot Noir, Grenache, and Cabernet Franc).
summary Established by long-term viticulturist Jeff Moss and his family, notably sons Peter and David and Roseworthy graduate daughter Jane. A 100-tonne rammed-earth winery was constructed in 1992, and draws upon both estate-grown and purchased grapes. Seemingly without warning, Moss Brothers produced two very good white wines in 1998 which suggest more can be expected of it in the future.

Moss Brothers Semillon

A quite outstanding wine, the quality of which was largely determined in the vineyard, for it was made in the traditional style, stainless steel-fermented at 13 degrees, and bottled in July of the year of vintage. It was justly rewarded with a gold medal at the 1998 Sheraton Wine Awards, a gold medal at the Sydney International Wine Competition, and was one of the top wines in the 1999 *Winewise* Small Makers Competiton.

ŸŸŸŸŸ **1998** Medium yellow-green; the bouquet shows intense varietal character, with fine lemony, herbal aromas reminiscent of the best Hunter Valley Semillons. The palate fulfils all of the promise of the bouquet, with great flavour, balance and mouthfeel. **rating:** 95

best drinking 1999–2001 **best vintages** '98 **drink with** Fried chicken • $22

Moss Brothers Sauvignon Blanc

Another very good wine from the 1998 vintage for Moss Brothers, emphasising the difference between the best wines and the least wines from this winery. Like the Semillon, made without artifice or elaboration.

ŸŸŸŸ **1998** Light green-yellow; the clean, light and crisp bouquet has minerally overtones; the quite delicate but well-balanced palate adds touches of passionfruit and gooseberry to the mix. All in all, reminiscent of Marlborough Sauvignon Blanc. **rating:** 88

best drinking 1999–2002 **best vintages** NA **drink with** Calamari • $17

moss wood ★★★★★

Metricup Road, Willyabrup, WA 6280 **region** Margaret River
ph (08) 9755 6266 **fax** (08) 9755 6303 **open** By appointment
winemaker Keith Mugford **prod.** 5000 **est.** 1969
prod. range ($25–75 R) Semillon, Wood Matured Semillon, Chardonnay, Cabernet Sauvignon, Pinot Noir.
summary Widely regarded as one of the best wineries in the region, capable of producing glorious Semillon (the best outside the Hunter Valley) in both oaked and unoaked forms, unctuous Chardonnay and elegant, gently herbaceous, superfine Cabernet Sauvignon which lives for many years. Exports to the UK, US, Switzerland, Germany, Denmark, Belgium, France, Japan, Hong Kong, Singapore and New Zealand.

Moss Wood Semillon

As with all the Moss Wood wines, entirely estate-grown and produced from 1.35 hectares of fully mature vines (the first vintage was 1977). Produced in both unoaked and oaked versions, both superb and age-worthy.

🍷🍷🍷🍷🍷 **1998** Light to medium yellow-green; the bouquet is powerful, with interesting, European overtones, perhaps suggesting the deliberate use of some solids in the juice. In the mouth, the wine has that extra dimension of weight (without sweetness) of Margaret River, easily carrying the massive 14.5 degrees alcohol. **rating:** 90

best drinking 2000–2005 **best vintages** '81, '82, '83, '84, '86, '87, '92, '94, '95, '97, '98 **drink with** Crab, lobster • $23

Moss Wood Chardonnay

To be honest, I have not the faintest idea why prior vintages of Moss Wood Chardonnay have not been included in the *Wine Companion*. It is always a richly seductive wine, with the typical Margaret River fruit concentration, offset by that special softness which the Moss Wood vineyard site imparts.

🍷🍷🍷🍷🍷 **1998** Medium yellow-green; a typically powerful bouquet with concentrated fruit and smoky oak is followed by a power-packed palate in which the fruit and oak have been seamlessly welded together, neither predominating. **rating:** 91

best drinking 1999–2004 **best vintages** '90, '91, '95, '98 **drink with** Smoked chicken • $45

Moss Wood Cabernet Sauvignon

First made in 1973, and has established itself as one of the classic wine styles not only of the Margaret River, but of Australia. The wines have a distinctive suppleness and softness which sets them apart from other Margaret River Cabernets, and which Dr John Gladstones firmly attributes to the particular terroir of the vineyard.

🍷🍷🍷🍷🍷 **1996** Medium to full red-purple; a concentrated bouquet with soft, berry and earth flavours supported by subtle oak is, like the palate, in typical Moss Wood style. The palate has complex yet fine secondary flavours, with earth, berry and chocolate merged with oak and finishing with lingering tannins. **rating:** 93

best drinking 2000–2010 **best vintages** '77, '80, '86, '87, '90, '91, '93, '94, '95, '96 **drink with** Beef with olives • $75

mountadam ★★★★☆

High Eden Road, High Eden Ridge, SA 5235 **region** Eden Valley
ph (08) 8564 1101 **fax** (08) 8361 3400 **open** 7 days 11–4
winemaker Adam Wynn, Andrew Ewart **prod.** 12 000 **est.** 1972
prod. range ($12–60 R) Under the premium Mountadam label, Riesling, Chardonnay, Pinot Noir, The Red (50% Merlot, 50% Cabernet), Merlot, Cabernet Sauvignon, Pinot Noir Chardonnay Brut; under the David Wynn label Riesling, Sauvignon Blanc, Chardonnay, Pinot Noir, Shiraz, Patriarch Shiraz, Cabernet Sauvignon; under organically grown Eden Ridge label, Sauvignon Blanc, Cabernet Sauvignon; also Ratafia Riesling and Chardonnay.
summary One of the leading small wineries, founded by David Wynn and run by winemaker son Adam Wynn, initially offering only the Mountadam range at relatively high prices. The subsequent development of the three ranges of wines has been very successful, judged both by the winemaking and wine-marketing viewpoint. Mountadam has built up an extensive export network over many years, with the US, Canada, Hong Kong, Japan and the UK being the major markets, but extending across the breadth of Europe and most Asian markets.

Mountadam Chardonnay

The Mountadam Chardonnay accounts for around 70% of the production under the Mountadam label. It is 100% estate-grown from low-yielding, high-altitude vineyards, the resulting fruit concentration fully expressing itself in the wine. Although Adam Wynn says that the changes in winemaking since 1984 have essentially been limited to finetuning, there is no doubt that there has been a sea change in the end product since 1989.

YYYYY **1997** Medium yellow-green; the bouquet is very complex with nutty minerally notes, logically introducing a wine in which the primary fruit has been softened by the deliberate introduction of secondary characters, an approach which works exceptionally well. **rating:** 92

best drinking 2000–2005 **best vintages** '86, '89, '90, '91, '92, '93, '94 **drink with** Salmon terrine • $32

Mountadam Patriarch Shiraz

In my view, has been the outstanding red wine from Mountadam, year in, year out. The '93 was magnificent, the '94 very good. The grapes are produced from low-yielding Eden Valley vineyards established on sandy soils, and the wine is matured in a mix of American and French oak barrels for 12 months. The '95 apparently fell victim to an ordinary vintage, but the '96 is a return to form.

YYYY **1996** Medium to full red-purple; a pungently complex bouquet ranging through red berry, green leaf and earth varietal fruit characters is followed by a powerful palate with dark fruit flavours, relatively fine tannins and subtle oak. **rating:** 89

best drinking 2001–2010 **best vintages** '90, '91, '93, '94 **drink with** Barbecued beef • $40

Mountadam Merlot

Merlot has been planted at Mountadam since 1984, but in all vintages prior to 1996, the Merlot was blended into The Red. Increased plantings have now resulted in the release of a 100% Merlot, aged for 12 months in French barriques and given additional maturation time prior to release.

YYYY **1995** Light to medium red, with the faintest touch of brick on the rim. The bouquet offers a somewhat unusual mix of jammy and sappy characters, yet is definitively Merlot. The palate has a distinctively sappy/vegetal edge to go with the plum, leather and mint flavours. All in all, something quite differenet. **rating:** 88

best drinking 1999–2004 **best vintages** NA **drink with** Milk-fed lamb • $60

Mountadam The Red

A strangely-named wine, although it is not hard to see what Mountadam is getting at. A blend of Cabernet and Merlot, typically 60% Cabernet Sauvignon and 40% Merlot, and aged in oak for 18 months. I had never much fancied the style, but the '96 builds on the much-improved '95.

YYYYY **1996** Medium red-purple; the bouquet opens with ripe, cassis fruit, with more cedary/briary notes thereafter. The palate has considerable concentration, with clean, rich fruit, lingering tannins and subtle oak. Certainly the best vintage so far under this label. **rating:** 90

best drinking 2000–2010 **best vintages** '95, '96 **drink with** Braised lamb • $50

Mountadam Cabernet Sauvignon

Estate-grown, and aged in French Troncais oak barriques for 18 months. The yields are always low, but were particularly small in 1995. Mountadam regards the vintage as one of its best ever.

YYYY **1995** Medium to full red-purple; the aromas are strongly reminiscent of Bordeaux, quite olivaceous and tangy, with subtle oak. Those Bordeaux overtones come again on the palate with distinctive tastes of olive and woody/foresty characters (fruit-derived, not oak-derived). All in all, a sophisticated wine. **rating:** 88

best drinking 2000–2005 **best vintages** '90, '91, '94, '95 **drink with** Lamb cutlets • $50

mountain creek wines NR

Mountain Creek Road, Moonambel, Vic 3478 **region** Pyrenees
ph (03) 5467 2230 **fax** (03) 5467 2230 **open** Weekends, holidays 10–7
winemaker Contract **prod.** 500 **est.** 1973

prod. range ($10–16 CD) Sauvignon Blanc, Sparkling Passion, Cabernet Shiraz, Frontignac, Muscat.

summary Brian Cherry acquired the Mountain Creek Vineyard in 1975 and has extended it to a total of 13 hectares. The first wine was made in 1987, all or part of the grapes before and in some years since then sold to other Pyrenees wineries. The wine is made under contract, and shows all of the substance and weight for which the district is renowned.

mount alexander vineyard ★★

Calder Highway, North Harcourt, Vic 3453 **region** Bendigo
ph (03) 5474 2262 **fax** (03) 5474 2553 **open** 7 days 10–5.30
winemaker Keith Walkden **prod.** 6000 **est.** 1984

prod. range ($10–14 CD) A wide range of various table wines, sparkling, fortifieds, meads and liqueurs.

summary A substantial operation with large vineyards with 12 hectares planted to all the right varieties. It is several years since I have tasted the wines, but I have no reason to suppose they have changed much.

mount anakie wines ★★☆

Staughton Vale Road, Anakie, Vic 3221 **region** Geelong
ph (03) 5284 1452 **fax** (03) 5284 1405 **open** Tues–Sun 11–6
winemaker Otto Zambelli **prod.** 6000 **est.** 1968

prod. range ($10–18 R) Biancone, Riesling, Semillon, Chardonnay, Dolcetto, Shiraz, Cabernet Franc, Cabernet Sauvignon.

summary Also known as Zambelli Estate, and once produced some excellent wines (under its various ownerships and winemakers), all distinguished by their depth and intensity of flavour. No recent tastings; prior to that, the wines tasted were but a shadow of their former quality. The level of activity seems relatively low; the current price list was issued in 1995, and spans the 1992 to 1995 vintages.

mount avoca vineyard ★★★☆

Moates Lane, Avoca, Vic 3467 **region** Pyrenees
ph (03) 5465 3282 **fax** (03) 5465 3544 **open** Mon–Fri 9–5, weekends 10–5
winemaker Matthew Barry **prod.** 17 000 **est.** 1970

prod. range ($10–18 R) Sauvignon Blanc, Chardonnay, Rhapsody, Trioss, Shiraz, Cabernets

summary A substantial winery which has for long been one of the stalwarts of the Pyrenees region, and steadily growing, with 23.7 hectares of vineyards. There has been a significant refinement in the style and flavour of the red wines over the past few years.

Mount Avoca Shiraz

Produced from a little over 6 hectares of relatively low-yielding estate plantings. The wine spends 8–12 months in a mix of older French and new American oak barrels.

🍷🍷🍷🍷 **1997** Medium red; the bouquet is clean, with earthy/chocolatey fruit and no hint of the bacterial/brettanomyces characters which dogged some of the earlier vintages. The palate is similarly clean, with red berry and earth flavours of light to medium intensity. **rating** 84

best drinking 2000–2004 **best vintages** NA **drink with** Marinated beef • $17

Mount Avoca Cabernet

Produced from 4 hectares of estate-grown grapes. Since 1995 the wine has moved from a straight Cabernet Sauvignon to 85% Cabernet Sauvignon with 15% Merlot and Cabernet Franc included. This change has also coincided with the eradication of some errant barrel-derived characters.

ŸŸŸŸ **1996** Medium to full red-purple; the bouquet is dense but clean, with chocolate and dark fruit aromas. An attractive palate, with full chocolate and dark berry fruit supported by fine, lingering tannins. **rating:** 87

best drinking 2000–2006 **best vintages** '88, '90, '91, '92, '93, '95, '96 **drink with** Char-grilled steak • $18

mount beckworth ★★★☆

RMB 915 Learmonth Road, Tourello via Ballarat, Vic 3363 **region** Ballarat
ph (03) 5343 4207 **fax** (03) 5343 4207 **open** Weekends 10–6 and by appointment
winemaker Paul Lesock **prod.** 630 **est.** 1984
prod. range ($14–16 CD) Chardonnay, Pinot Noir, Shiraz, Cabernets.
summary The 4-hectare Mount Beckworth vineyard was planted between 1984 and 1985, but it was not until 1995 that the full range of wines under the Mount Beckworth label appeared. Until that time much of the production was sold to Seppelt Great Western for sparkling wine use. It is owned and managed by Paul Lesock, who studied viticulture at Charles Sturt University, and his wife Jane. The wines reflect the very cool climate. Limited Victorian retail distribution.

Mount Beckworth Unwooded Chardonnay

A wine which was entered in the unwooded section of the 1999 *Winewise* Small Makers Competition, but which seemed to show quite evident spicy oak. Whatever, a good wine.

ŸŸŸŸ **1998** Light to medium yellow-green; the fruit is light and clean, supported by what appears to be spicy French oak. The palate shows similarly delicate fruit, once again with some oak influence. **rating:** 86

best drinking 1999–2003 **best vintages** NA **drink with** Grilled fish • $14

mount charlie winery NR

Mount Charlie Road, Riddells Creek, Vic 3431 **region** Macedon
ph (03) 5428 6946 **fax** (03) 5428 6946 **open** Not
winemaker Trefor Morgan **prod.** 700 **est.** 1991
prod. range ($15–22 ML) Sauvignon Blanc, Chardonnay, Red (Cabernet Shiraz blend).
summary Mount Charlie's wines are sold principally through mail order (no cellar door sales) and through selected restaurants. A futures programme encourages mailing list sales with a discount of over 25% on the ultimate release price. Trefor Morgan is particularly pleased with the quality of the 1998 wines, the product of a great year in Macedon.

mount duneed ★★☆

Feehan's Road, Mount Duneed, Vic 3216 **region** Geelong
ph (03) 5264 1281 **fax** (03) 5264 1281 **open** Public holidays and weekends 11–5 or by appointment
winemaker Ken Campbell, John Darling **prod.** 1000 **est.** 1970
prod. range ($10–18 CD) Semillon, Sauvignon Blanc, Riesling, Botrytis Semillon, Malbec, Cabernet Malbec, Cabernet Sauvignon.
summary Rather idiosyncratic wines are the order of the day, some of which can develop surprisingly well in bottle; the Botrytis Noble Rot Semillon has, from time to time, been of very high quality. A significant part of the production from the 7.5 hectares of vineyards is sold to others.

mountford ★★★

Bamess Road, West Pemberton, WA 6260 **region** Pemberton
ph (08) 9776 1439 **fax** (08) 9776 1439 **open** Fri–Sun 10–4
winemaker Andrew Mountford **prod.** 5000 **est.** 1987

prod. range ($15–50 CD) Sauvignon Blanc, Chardonnay, Blanc de Noir, Pinot Noir, Merlot Cabernet Sauvignon.

summary English-born and trained Andrew Mountford and wife Sue migrated to Australia in 1983, first endeavouring to set up a winery at Mudgee, and thereafter moving to Pemberton with far greater success. Their strikingly packaged wines (complete with beeswax and paper seals) have been well-received on eastern Australian markets, being produced from 6 hectares of permanently netted, dry-grown vineyards.

mount gisborne wines NR

5 Waterson Road, Gisborne, Vic 3437 **region** Macedon
ph (03) 5428 2834 **fax** (03) 5428 2834 **open** By appointment
winemaker Stuart Anderson **prod.** 1200 **est.** 1987

prod. range ($10–24 CD) Chardonnay, Dessert Chardonnay, Pinot Noir, Pinot Noir Limited Release.

summary Mount Gisborne Wines is very much a weekend and holiday occupation for proprietor David Ell, who makes the wines from the 6-hectare vineyard under the watchful and skilled eye of industry veteran Stuart Anderson, now living in semi-retirement high in the Macedon Hills.

mount horrocks ★★★★

The Old Railway Station, Curling Street, Auburn, SA 5451 **region** Clare Valley
ph (08) 8849 2243 **fax** (08) 8849 2243 **open** Weekends and public holidays 11–5
winemaker Stephanie Toole **prod.** 5000 **est.** 1982

prod. range ($18.75–34.50 R) Watervale Riesling, Chardonnay, Semillon, Semillon Sauvignon Blanc, Cordon Cut Riesling, Cabernet Merlot, Shiraz.

summary Mount Horrocks has well and truly established its own identity in recent years, aided by positive marketing and, equally importantly, wine quality which has resulted in both show success and critical acclaim.

Mount Horrocks Watervale Riesling

Mount Horrocks was originally the wine-production end of the Ackland Brothers' extensive vineyard holdings. The two have now effectively been split, and Mount Horrocks is dependent on contract-grown grapes through the Watervale region. Sourced from two growers in the Watervale subdistrict of the Clare Valley who continue to both hand-prune and hand-pick their vines.

🍷🍷🍷🍷 **1998** Light yellow-green; herb, spice and chalk aromas intermingle with some bottling sulphur dioxide which will hopefully disappear with age. The palate has a very attractive fruit core of citrus and lime flavours, although the effects of the sulphur dioxide were still noticeable when the wine was tasted. **rating:** 85

best drinking 1999–2005 **best vintages** '86, '87, '90, '93, '94, '97 **drink with** Thai or Chinese soup • $18.75

Mount Horrocks Cordon Cut Riesling (375 ml)

The term cordon cut describes a wine made from grapes which have reached high sugar levels as the result of the canes being severed from the trunk of the vine, and being allowed to hang on the trellis with the leaves and grapes attached. A process of dehydration follows, involving the loss of water but the increase in sugar and acid levels. The one thing missing is the complexity which the botrytis mould (the alternative way of achieving similar levels of dehydration) bestows. The wine, incidentally, is made from unspecified varieties. The '97 was a trophy winner for Best Sweet White Wine at the 1997 Clare Valley Wine Show.

🍷🍷🍷🍷 **1998** Medium yellow-green; the aromas are as intriguing as the flavours, with slightly cosmetic overtones to hints of dried apricot. There is extremely sweet fruit on the palate, richly textured, and almost as if not fermented (although it undoubtedly is). **rating:** 89

best drinking 1998–2000 **best vintages** NA **drink with** Rich, sweet cake • $22.95

mount hurtle ★★★☆

291 Pimpala Road, Woodcroft, SA 5162 **region** McLaren Vale
ph (08) 8381 6877 **fax** (08) 8322 2244 **open** Mon–Fri 10–5, Sun 12–5
winemaker Geoff Merrill, Goe DiFabio, Scott Heidrich **prod.** 30 000 **est.** 1897
prod. range ($9.99–12 R) Sauvignon Blanc Semillon, Grenache Shiraz, Grenache Rosé. The cheekily named Who Cares White (66% Chenin Blanc, 34% Sauvignon Blanc) and Who Cares Red (90% Grenache, 10% Shiraz) were introduced in 1996.
summary Now a brand sold exclusively through Liquorland. The wines are of consistent quality and well priced.

mount ida ★★★★

Northern Highway, (vineyard only) Heathcote, Vic 3253 **region** Bendigo
open Not
winemaker Toni Stockhausen **prod.** 3000 **est.** 1978
prod. range ($30 R) Shiraz.
summary Established by the famous artist Leonard French and Dr James Munro, but purchased by Tisdall after the 1987 bushfires and thereafter by Mildara Blass when it acquired Tisdall. Up to the time of the fires, wonderfully smooth, rich red wines with almost voluptuous sweet, minty fruit were the hallmark. After a brief period during which the name was used as a simple brand (with various wines released) has returned to a single estate-grown wine, sharing similar packaging with Mount Helen.

mountilford NR

Mount Vincent Road, Ilford, NSW 2850 **region** Mudgee
ph (02) 6358 8544 **fax** (02) 6358 8544 **open** 7 days 10–4
winemaker Don Cumming **prod.** NFP **est.** 1985
prod. range ($10–16 CD) Rhine Riesling, Gewürztraminer, Chardonnay, Highland White, Windamere, Sylvaner, Cabernet Shiraz, Pinot Shiraz, Sir Alexander Port.
summary Surprisingly large cellar-door operation which has grown significantly over the past few years. I have not, however, had the opportunity of tasting the wines.

mount langi ghiran vineyards ★★★★★

Warrak Road, Buangor, Vic 3375 **region** Grampians
ph (03) 5354 3207 **fax** (03) 5354 3277 **open** Mon–Fri 9–5, weekends 12–5
winemaker Trevor Mast, Andrew McLoughney **prod.** 25 000 **est.** 1969
prod. range ($16–45 R) Chardonnay, Riesling, Pinot Grigio; under Langi label Shiraz and Cabernet Merlot; Billi Bill Creek Shiraz Cabernet.
summary A maker of outstanding cool-climate peppery Shiraz, crammed with flavour and vinosity, and very good Cabernet Sauvignon. The Shiraz points the way for cool-climate examples of the variety, for weight, texture and fruit richness all accompany the vibrant pepper-spice aroma and flavour. Now partly owned by Trevor Mast, and partly by German wine entrepreneur Riquet Hest; the most tangible sign of the partnership has been the erection of a totally new, state-of-the-art winery, not to mention the expansion of the estate vineyards to over 70 hectares and the establishment of an export network throughout the US, UK, Asia, New Zealand and Canada.

Mount Langi Ghiran Pinot Grigio

This is the third release of Pinot Grigio from Mount Langi Ghiran; I cannot recollect having tasted the previous two releases. Although the label is not absolutely explicit on the point, I assume it does come from estate-grown grapes at Mount Langi Ghiran. If so, it certainly extends the area in which pinot grigio can flourish, for this is a good wine.

🍷🍷🍷🍷 **1998** Medium yellow-green; there is good varietal character on the bouquet with that faintly musky fruit salad character of pinot gris. In the mouth, the strength of the wine is in the middle palate flavour, softer and fuller than most white wine varieties, yet not the least bit flabby.

rating: 87

best drinking 1999–2001 **best vintages** NA **drink with** Turkey • $19

Langi Shiraz

One of the top half-dozen Shirazes in Australia. The site climate of the Mount Langi Ghiran vineyards produces wines which have tremendous depth and complexity: there are pepper and spice notes, but there are also all of the lush ripe fruit flavours running from cherry to liquorice which a top Rhône Valley maker would immediately recognise and appreciate.

🍷🍷🍷🍷🍷 **1997** Medium to full red-purple; the bouquet has abundant ripe black cherry and liquorice fruit, supported by very subtle oak. The palate has abundant ripe fruit flavours in the same spectrum as the bouquet, again with subtle oak. Just misses out on that 'sauvage' spark of the very best Langis.

rating: 92

best drinking 2001–2007 **best vintages** '86, '88, '90, '92, '93, '94, '96 **drink with** Kangaroo, venison • $45

Langi Cabernet Merlot

Another good, estate-grown wine from Mount Langi Ghiran, perhaps lacking the sheer brilliance of the Shiraz, but that is hardly surprising.

🍷🍷🍷🍷 **1996** Medium red-purple; the bouquet offers a mix of sweet chocolate and more leafy notes, characters which come through on the chocolate, leaf and red berry-flavoured palate; fine tannins and minimal oak impact.

rating: 88

best drinking 2000–2005 **best vintages** NA **drink with** Moroccan lamb • $35

mount macedon NR

Bawden Road, Mount Macedon, Vic 3441 **region** Macedon

ph (03) 5427 2735 **fax** (03) 5427 1071 **open** 7 days 10–6

winemaker Peter Dredge **prod.** 1500 **est.** 1989

prod. range ($15–20 CD) Chardonnay, Unwooded Chardonnay, Macedon Ranges Chardonnay, Winemaker's Reserve Chardonnay, Pink Reflections, Pinot Noir, Macedon Ranges Pinot Noir, Hay Hill Shiraz.

summary Don and Pam Ludbey have established a substantial operation at Mount Macedon drawing upon two separate vineyards, Mount Macedon and Hay Hill. In all, they have 12.5 hectares under vine, and also operate a restaurant during the weekend. The '96 Chardonnay and Pinot Noir lacked the necessary fruit ripeness.

mount mary ★★★★★

Coldstream West Road, Lilydale, Vic 3140 **region** Yarra Valley

ph (03) 9739 1761 **fax** (03) 9739 0137 **open** Not

winemaker Dr John Middleton, Mario Marson **prod.** 3000 **est.** 1971

prod. range ($32–60 ML) Chardonnay, Triolet (Sauvignon Blanc, Semillon, Muscadelle), Pinot Noir, Cabernets Quintet (Bordeaux-blend).

summary Superbly refined, elegant and intense Cabernets, and usually outstanding and long-lived Pinot Noirs, fully justify Mount Mary's exalted reputation. The Triolet blend is very good, more recent vintages of Chardonnay likewise. However, John Middleton does not believe in wine critics, and, least of all myself, so no (official) tastings. Limited quantities of the wines are sold through the wholesale/retail distribution system in Victoria, New South Wales, Queensland and South Australia.

mount panorama winery NR

117 Mountain Straight, Mount Panorama, Bathurst, NSW 2795 **region** Central Ranges Zone
ph (02) 6331 5368 **fax** (02) 6331 5368 **open** By appointment
winemaker Bill Stuart, Deborah Stuart **prod.** 600 **est.** 1991
prod. range ($11.50–20 CD) Forrest Elbow Riesling, Chase Chardonnay, Hell Corner Shiraz, The Esses Cabernet Sauvignon, The Cutting (Cabernet blend), Thunder Mountain Port.
summary For all the obvious reasons, Mount Panorama Winery makes full use of its setting on Mountain Straight, after the 'Hell Corner' on the inside of the famous motor racing circuit. Bill and Deborah Stuart are wholly responsible for the production of the wine, from picking, using the hand-operated basket press through to bottling, labelling, etc. The '97 wines have had some success at the local (Yass Valley) wine show. They are gradually extending both the size and scope of the cellar-door facilities to take advantage of the tourist opportunities of the site.

mount prior vineyard ★★★

Cnr River Road and Popes Lane, Rutherglen, Vic 3685 **region** Rutherglen
ph (02) 6026 5591 **fax** (02) 6026 7456 **open** 7 days 9–5
winemaker Richard Langford **prod.** 15 000 **est.** 1860
prod. range ($11–23 CD) Chardonnay, Chenin Blanc, Classic Ibis White, Semillon Chardonnay, Late Picked Riesling, Noble Gold, Classic Ibis Dry Red, Cabernet Merlot, Shiraz, Durif, Sparkling Shiraz/Durif, Brut Cuvée, Port, Muscat, Tokay.
summary A full-scale tourist facility, with yet more in the pipeline. Full accommodation packages at the historic Mount Prior House, a restaurant operating weekends under the direction of Trish Hennessy, with four consecutive *Age Good Food Guide* awards to its credit, picnic and barbecue facilities, and a California-style gift shop. The wines are basically sold through cellar door and an active mailing list. The already substantial 40 hectares of vineyards were expanded by a further 5 hectares of durif planted in 1998, a mark both of the success of Mount Prior and of the interest in Durif.

Mount Prior Chardonnay

Produced from 15 hectares of estate plantings; a truly excellent example of barrel-fermented Chardonnay and was the equal top-pointed wine in the 1998 Chardonnay Class at the 1999 *Winewise* Small Makers Competition.

🍷🍷🍷🍷🍷 **1998** Medium yellow-green; the bouquet is complex, with attractively tangy fruit offset by smoky bacon oak, giving a Burgundian feel to the wine. The palate has abundant, quite sweet melon and grapefruit flavours, with good oak integration and balance. **rating:** 93

best drinking 1999–2002 **best vintages** '98 **drink with** Weiner schnitzel • $14.50

mount tamborine winery NR

32 Hartley Road, Mount Tamborine, Qld 4272 **region** Other Wineries of Qld
ph (07) 5545 3981 **fax** (07) 5545 3311 **open** 7 days 10–4
winemaker Craig Robinson, Kevin Watson **prod.** 4000 **est.** 1993
prod. range ($7.50–25 ML) Sauvignon Blanc Chardonnay, Semillon, Sauvignon Blanc Semillon, Cedar Ridge Chardonnay, Merlot Emily Cuvee de Rouge, Cedar Ridge Blanc de Blanc, Flaxton Nouveau Shiraz, Tehembrin Merlot, Shiraz Cabernet Merlot, Mountain Muscat, Bush Turkey (Port).
summary Mount Tamborine Winery draws upon 3 hectares of estate plantings adjacent to the winery, 30 hectares in Stanthorpe, and also purchases wine from the King Valley, Cowra and the Riverland to produce a wide range of wine styles. The Chardonnay and Merlot have both had

success in Queensland wine shows and competitions, and the wines are sold both locally and exported to southeast Asia.

mount trio vineyard ★★★★

Cnr Castle Rock and Porongurup Roads, Porongurup WA 6324 **region** Great Southern
ph (08) 9853 1136 **open** By appointment
winemaker Gavin Berry (Contract) **prod.** 1000 **est.** 1989
prod. range ($15–19 R) Sauvignon Blanc, Chardonnay, Pinot Noir, Cabernet Shiraz, Cabernet Merlot.
summary Mount Trio was established by Gavin Berry and Gill Graham shortly after they moved to the Mount Barker district in late 1988. Gavin Berry was assistant winemaker to John Wade, and Gill managed the cellar-door sales. Gavin is now senior winemaker and managing director of Plantagenet and Gill the mother of two young children. In the meantime they have slowly built up the Mount Trio business, based in part upon estate plantings of 2 hectares of pinot noir and half a hectare of chardonnay and in part on purchased grapes. An additional 6 hectares was planted in the spring of 1999, and plans are to ultimately increase production to around 5000 cases.

Mount Trio Chardonnay

Barrel-fermented in French oak, but very much a fruit-driven style.
🍷🍷🍷🍷🍸 **1998** Medium yellow-green; highly perfumed passionfruit and grapefruit aromas are repeated on the intensely flavoured palate. This is a regional character which is either more or less intense according to the vintage. Some may like it, others not; I am in the former camp. **rating:** 91
best drinking 1999–2004 **best vintages** '98 **drink with** West Australian marron • $15.50

Mount Trio Cabernet Merlot

A blend of 90% Cabernet Sauvignon and 10% Merlot, which effectively replaces the prior Cabernet Shiraz Merlot blend.
🍷🍷🍷🍷 **1997** Youthful purple-red; the bouquet is fresh and clean in fruit terms, although the oak is slightly raw and unintegrated. The palate is dominated by attractive juicy berry cassis fruit, and throws off the uncertainty of the bouquet. Does, however, need time for the parts to come together. **rating:** 86
best drinking 2002–2007 **best vintages** NA **drink with** Beef in red wine sauce • $17.30

mount view estate ★★★★

Mount View Road, Mount View, NSW 2325 **region** Lower Hunter Valley
ph (02) 4990 3307 **fax** (02) 4991 1289 **open** Mon–Fri 10–4, weekends, holidays 10–5
winemaker Keith Tulloch **prod.** 3000 **est.** 1971
prod. range ($12–26 CD) Verdelho, Reserve Verdelho, Kester Chardonnay, Verdelho Chardonnay, Kester Shiraz, Reserve Shiraz, Cabernet Sauvignon, Cabernet Port, Liqueur Verdelho, Trophy Muscat.
summary Since the involvement of former Rothbury winemaker Keith Tulloch, the overall quality of the wines has improved considerably. Harry Tulloch continues to zealously guard the vineyards, and the combination of the skills of the two, together with the expenditure on some better oak, has taken the focus from the fortified wines to an array of table wines, with Verdelho a winery specialty. Not for the first time, the winery teased me by sending its samples minutes before the text of this book was delivered to the printer.

Mount View Estate Verdelho

Produced both under a straight varietal label, and in a reserve configuration. The difference between the two is minimal, and in 1998 I had a marginal preference for the varietal. 13.6 degrees alcohol and stainless steel-fermented.

ŸŸŸŸ **1998** Medium to full yellow-green; the bouquet is clean and soft, with a mix of honey and fruit salad aromas underpinned by a trace of minerally characters. A mouthfilling wine, with far more interest and focus than the average rather bland Verdelho coming from the majority of wine producers. Very ripe fruit salad and peach fruit flavour really help drive the wine along. **rating:** 87

best drinking 1999–2003 **best vintages** '98 **drink with** Creamy pasta • $15

Mount View Estate Kester Shiraz

The Kester range (Chardonnay and Shiraz) sits at the top of the quality tree, and overall it is not hard to see why.

ŸŸŸŸ **1997** Medium red-purple; the bouquet is moderately intense, with sweet, earthy Hunter Shiraz coming through on both bouquet and palate. The stylish and restrained use of French oak gives ample opportunity for Hunter Valley typicité to express itself. **rating:** 87

best drinking 2002–2007 **best vintages** NA **drink with** Baby rack of lamb • $26

Mount View Estate Reserve Shiraz

Made from estate-grown fruit coming from vines which are now more than 25 years old. Good grapes have been well handled in the winery. The Reserve Shiraz is at a lower price point than the Kester Shiraz, and (perhaps perversely) I preferred the bigger Reserve to the more elegant Kester. However, both are excellent wines.

ŸŸŸŸ **1997** Strong red-purple; full dark berry/plum fruit and a vanilla slice of oak on the bouquet leads into a palate with plenty of depth to the sweet, round fruit; good oak balance and integration, and pleasantly soft tannins on the finish. **rating:** 88

best drinking 2000–2010 **best vintages** '97 **drink with** Loin of lamb • $22

mountview wines ★★★☆

Mount Stirling Road, Glen Aplin, Qld 4381 **region** Granite Belt
ph (07) 4683 4316 **fax** (07) 4683 4111 **open** Wed–Sat 9–5, Sun 10–4
winemaker David Price **prod.** 1200 **est.** 1990
prod. range ($10–25 CD) Chardonnay Semillon Sauvignon Blanc, Chardonnay Royal (sparkling), Bianco (sweet white), Cerise (light red), Shiraz, Cabernet Merlot, Sparkling Shiraz.
summary David and Linda Price are refugees from the Sydney rat-race operating a small, neat, red cedar farm-style winery. Various vintages of Mountview Shiraz have deservedly won trophies and gold medals at Queensland wine shows. Half of the 2-hectare vineyard is planted to shiraz, and Mountview also makes a Sparkling Shiraz from estate-grown grapes.

mount vincent mead NR

Common Road, Mudgee, NSW 2850 **region** Mudgee
ph (02) 6372 3184 **fax** (02) 6372 3184 **open** Mon–Sat 10–5, Sun 10–4
winemaker Jane Nevell **prod.** 2000 **est.** 1972
prod. range ($3–32 CD) Does make a Shiraz and a Liqueur Muscat, but is essentially a meadery, with White Box Honey Wine, Napunyah Dry Mead, White Box Dry Mead, Napunyah Medium Sweet, Stringy Bark Sweet, Thistle Sweet Metheglin and Stringy Bark Liqueur Mead. Each of these is vintage dated.
summary Forget the table wines, and concentrate on the meads, which can be absolutely outstanding, dramatically reflecting the impact of the different plants from which the bees have collected their honey.

mount william winery NR

Mount William Road, Tantaraboo, Vic 3764 **region** Macedon
ph (03) 5429 1595 **fax** (03) 5429 1998 **open** 7 days 11–5
winemaker Murray Cousins, Michael Cope-Williams (Contract) **prod.** 1500 **est.** 1987

prod. range ($15–25 CD) Bedbur's Riesling, Chardonnay Semillon, Chardonnay, Pinot Noir Cabernet Franc, Louis Clare Sparkling Red.

summary Adrienne and Murray Cousins established 6 hectares of vineyards between 1987 and 1992, planted to pinot noir, cabernet franc, semillon and chardonnay. The wines are made under contract (Cope-Williams), and are sold through a stone tasting room cellar-door facility which was completed in 1992, and also through a number of fine wine retailers around Melbourne.

Mount William Bedbur's Riesling

Showed well at the 1999 *Winewise* Small Winemakers Competition, whence this note comes.

ΨΨΨΨ **1998** Light to medium yellow-green; the bouquet is light with faint mineral and chalk aromas, but the pleasant palate moves into lime and stone fruit, and is very nicely balanced.

rating: 87

best drinking 1999–2003 **best vintages** NA **drink with** Chicken salad • $15

mudgee wines NR

Henry Lawson Drive, Mudgee, NSW 2850 **region** Mudgee
ph (02) 6372 2258 **open** Thur–Mon 10–5, holidays 7 days
winemaker Jennifer Meek **prod.** 1000 **est.** 1963

prod. range ($9–15 CD) Chardonnay, Gewurztraminer, Crouchen, Riesling, Rosé, Shiraz, Pinot Noir, Cabernet Sauvignon.

summary All of the wines are naturally fermented with wild yeasts and made without the addition of any chemicals or substances including sulphur dioxide, a very demanding route, particularly with white wines. For some consumers, any shortcoming in quality will be quite acceptable.

munari wines ★★★★

1129 Northern Highway, Heathcote, Vic 3523 **region** Bendigo
ph (03) 5433 3366 **fax** (03) 5433 3095 **open** 7 days 10–5
winemaker Adrian Munari, Deborah Munari **prod.** 1000 **est.** 1993

prod. range ($15–25 CD) Chardonnay, Shiraz, Schoolhouse Red, Reserve Merlot.

summary Adrian and Deborah Munari have made a singularly impressive entry into the winemaking scene, each of their 1997 vintage wines winning medals at various Victorian wine shows, including the Royal Melbourne and Victorian Wines Show. With a little over 4 hectares of estate vines, production will be limited, but the wines are well worth seeking out. The only retail distribution is through Vintage Cellars South Melbourne; otherwise from cellar door or mailing list.

Munari Shiraz

There is a powerful case for the proposition that Heathcote (which is to receive its own status under the Geographic Indications legislation) is the foremost area in Australia for the production of full-bodied Shiraz. This first vintage does nothing to challenge that case.

ΨΨΨΨ **1997** Dense red-purple; the bouquet is dominated by strong, pungent high-toast oak, which also comes through on the palate, but is there justified by the fruit concentration and the powerful tannins. Like all four new Munari releases, does not hide its light under a bushel.

rating: 87

best drinking 2002–2012 **best vintages** NA **drink with** Barbecued lamb • $25

Munari Reserve Merlot

The most powerful of a powerhouse range of wines from Munari, obviously picked ripe and equally obviously, from very low-yielding vines. The only question mark is whether this wine really exhibits varietal character.

🍷🍷🍷🍷🍷 **1997** Medium to full red-purple; an exceptionally concentrated, powerful and dense bouquet full of sweet fruit and notes of briar is followed by a lush, succulent, blackcurrant-flavoured palate which has swallowed up the oak. **rating:** 92

best drinking 2002–2012 **best vintages** '97 **drink with** Lamb Provençale • $25

murray robson wines NR

'Bellona' Old North Road, Rothbury, NSW 2335 **region** Lower Hunter Valley

ph (02) 4938 3577 **fax** (02) 4938 3577 **open** 7 days 9–5

winemaker Murray Robson **prod.** 4000 **est.** 1993

prod. range ($18–24 CD) Traminer, Semillon, Chardonnay, Shiraz, Merlot Cabernet, Cabernet Sauvignon.

summary Like a phoenix from the ashes, Murray Robson Wines rises once again, having reopened in its new location in February 1997. Four hectares of estate plantings are supplemented by grapes purchased from other growers in the valley; the initial releases from 1996 were all produced in tiny quantities of 150 cases or less, produced and packaged with the irrepressible flair of Murray Robson and using the same label which appeared back in the early 1970s, each one hand-signed – as ever – by Murray Robson. Exports to US, UK and New Zealand.

murrindindi ★★★★

Cummins Lane, Murrindindi, Vic 3717 **region** Central Victorian High Country

ph (03) 5797 8217 **fax** (03) 5797 8422 **open** Not

winemaker Alan Cuthbertson, Hugh Cuthbertson **prod.** 2000 **est.** 1979

prod. range ($22 R) Chardonnay, Cabernets Merlot.

summary Situated in an unequivocally cool climate which means that special care has to be taken with the viticulture to produce ripe fruit flavours. In more recent vintages, Murrindindi has succeeded handsomely in so doing. Limited Sydney and Melbourne distribution through Wine Source.

Murrindindi Chardonnay

The original plantings of 2.5 hectares of chardonnay have been dramatically extended in the mid-1990s with an additional 12 hectares, attesting to the suitability of the climate for this variety. In July 1998 a mini vertical tasting of the '97, '96, '94 and '92 vintages emphasised the extraordinary fidelity and constancy of the style, and also its longevity.

🍷🍷🍷🍷 **1997** Medium yellow-green; the bouquet is complex though elegant, with toasty/nutty/mineral overtones to the melon fruit. The palate is similarly subtle, well balanced and harmonious. All the wine needs is time. **rating:** 87

best drinking 1999–2005 **best vintages** '84, '90, '91, '92, '93, '96 **drink with** Mussels • $22

murrumbateman winery NR

Barton Highway, Murrumbateman, NSW 2582 **region** Canberra District

ph (02) 6227 5584 **open** Thur–Mon 10–5

winemaker Duncan Leslie **prod.** 1000 **est.** 1972

prod. range ($12–18 CD) Sauvignon Blanc, Unoaked Chardonnay, Sally's Sweet White, Shiraz, Cabernet Sauvignon, Fortifieds and Sparkling.

summary Revived after a change of ownership, the Murrumbateman Winery draws upon 4.5 hectares of vineyards, and also incorporates an à la carte restaurant, function room, together with picnic and barbecue areas.

narkoojee ★★★★

1110 Francis Road, Glengarry, Vic 3854 **region** Gippsland

ph (03) 5192 4257 **fax** (03) 5192 4257 **open** By appointment

winemaker Harry Friend, Axel Friend **prod.** 700 **est.** 1981
prod. range ($15–28 CD) Chardonnay, Cabernets Merlot, The Rose, The Athelstan Merlot.
summary Narkoojee Vineyard is within easy reach of the old goldmining town of Walhalla, and looks out over the Strzelecki Ranges. The wines are produced from 4 hectares of estate vineyards. Harry Friend was an amateur winemaker of note before turning to commercial winemaking with Narkoojee, his skills showing through with all the wines.

Narkoojee Gippsland Chardonnay

An opulently flavoured wine, showing the full range of winemaking techniques. In recent years, the style has been tightened somewhat, with the promise of greater longevity.

🍷🍷🍷🍷 **1997** Medium yellow-green; archetypal, ripe melon fruit merged with a touch of honeyed toast runs through the bouquet; a quite long and powerful palate with the strength to age well.
rating: 89

best drinking 2000–2005 **best vintages** '87, '89, '92, '93, '94, '97 **drink with** Salmon pizza • $20

Narkoojee The Athelstan Merlot

A special bottling of limited quantities of varietal Merlot, named in the memory of Harry Friend's father, Athelstan Hardcastle Friend. Another delicious wine, and true to variety.

🍷🍷🍷🍷 **1996** Medium red-purple; the soft bouquet shows the clever use of sweet oak to introduce a hint of spice to the berry fruit. The same sweet berry/cherry fruit and sweet oak is softly interwoven with soft, fine tannins on the palate. **rating:** 86

best drinking 2000–2005 **best vintages** NA **drink with** Veal chops • $28

Narkoojee Cabernets

A blend of Cabernet Sauvignon, Cabernet Franc and Merlot, with the vines dating back to 1981.

🍷🍷🍷🍷 **1996** Light to medium red-purple; the bouquet has fresh, gently ripe berry fruit and subtle oak, the palate offering similarly attractive berry fruit backed up by persistent but not aggressive tannins. **rating:** 85

best drinking 2000–2006 **best vintages** NA **drink with** Gippsland lamb • $21

---

nashdale wines NR

Borenore Lane, Nashdale, NSW 2800 **region** Orange
ph (02) 6365 2463 **fax** (02) 6361 4495 **open** Weekends 2–6
winemaker Mark Davidson (Contract) **prod.** 1000 **est.** 1990
prod. range ($10–25 CD) Riesling, Sauvignon Blanc, Chardonnay, Pinot Noir, Cabernet Sauvignon.
summary Orange solicitor Edward Fardell commenced establishing the 10-hectare Nashdale Vineyard in 1990. At an elevation of 1000 metres, it offers panoramic views of Mount Canobolas and the Lidster Valley, with a restaurant-café open on weekends.

nepenthe vineyards ★★★★☆

Vickers Road, Lenswood, SA 5240 **region** Adelaide Hills
ph (08) 8389 8039 **fax** (08) 8389 8019 **open** By appointment
winemaker Peter Leske **prod.** 12 000 **est.** 1994
prod. range ($15–38 R) Riesling, Sauvignon Blanc, Unwooded Chardonnay, Chardonnay, Pinot Gris, Pinot Noir, Zinfandel, Fugue (Cabernet Merlot).
summary The Tweddell family established 21 hectares of close-planted vineyards at Lenswood between 1994 and 1997 with an exotic array of varieties reflected in the wines. In late 1996 it obtained the second licence to build a winery in the Adelaide Hills, Petaluma being the only other successful applicant back in 1978. A 500-tonne winery has been constructed, with Peter

Leske in charge of winemaking. Nepenthe has quickly established itself as one of the most exciting new wineries in Australia. Distribution through most States, and exports to the UK.

Nepenthe Vineyards Riesling

Produced from 1.2 hectares of estate vineyards, and almost inevitably well made in the highly disciplined Australian style.

🍷🍷🍷🍷🍷 **1998** Light to medium green-yellow; the floral bouquet has a mix of lime juice and herb, the palate adding a mineral note or two; long, intense and lingering. Gold medal winner 1998 Adelaide Hills Wine Show. **rating:** 94

best drinking 2001–2008 **best vintages** '98 **drink with** Sashimi • $20

Nepenthe Vineyards Sauvignon Blanc

Yet another high-class wine from Nepenthe, which has 4 hectares of estate plantings.

🍷🍷🍷🍷🍷 **1998** Light to medium yellow-green; the bouquet is quite intense, with gooseberry and more grassy/herbal characters; in a not dissimilar fashion, the long, clean and crisp palate has a substrate of mineral to sustain the intense fruit sweetness. **rating:** 93

best drinking 1999–2000 **best vintages** NA **drink with** Poached scallops in white sauce • $20

Nepenthe Vineyards Unwooded Chardonnay

Nepenthe, it seems, does not know how to make a bland, let alone a bad, wine. Unwooded Chardonnay does not come better than this.

🍷🍷🍷🍷🍷 **1998** Light green-yellow; very fragrant fruit, primarily citrus and with a dash of passionfruit is followed by an intensely flavoured palate running through melon, citrus and passionfruit flavours. **rating:** 90

best drinking 1999–2000 **best vintages** NA **drink with** Mussel soup • $15

newstead winery NR

Tivey Street, Newstead, Vic 3462 **region** Bendigo

ph (03) 5476 2733 **fax** (03) 5476 2536 **open** By appointment

winemaker Ron Snep, Cliff Stubbs **prod.** 900 **est.** 1994

prod. range ($15–16 CD) Welshmans Reef Semillon, Barrel Fermented Semillon, Unwooded Chardonnay, Cabernet Sauvignon; Burnt Acre Riesling, Shiraz.

summary Newstead Winery is established in the old Newstead Butter Factory, drawing upon two distinct vineyards at Welshmans Reef (near Maldon) and Burnt Acre Vineyard at Marong, west of Bendigo. Vineyard designations are used for each of the wines.

nicholson river ★★★★

Liddells Road, Nicholson, Vic 3882 **region** Gippsland

ph (03) 5156 8241 **fax** (03) 5156 8433 **open** 7 days 10–4

winemaker Ken Eckersley **prod.** 2000 **est.** 1978

prod. range ($15-39 CD) Gippsland White (Semillon Riesling Chardonnay), Semillon, Cuvée, Gippsland Red (Pinot Shiraz), Montview Cabernet Merlot, Tawny Port. The Chardonnays come in four levels: at the bottom Gippsland, then Montview, then Montview Special, and finally Nicholson River.

summary The fierce commitment to quality in the face of the temperamental Gippsland climate and frustratingly small production has been handsomely repaid by some stupendous Chardonnays, mostly sold through cellar door, and a little is exported to the UK and US. Ken Eckersley does not refer to his Chardonnays as white wines, but as gold wines, and lists them accordingly in his newsletter.

Nicholson River Chardonnay

Produced from half a hectare of 18-year-old vines, producing 150–180 cases of wine a year. Arguably the most powerful and complex Chardonnay made in Australia today. The wine has everything: concentrated fruit, abundant use of French oak, deliberately challenging winemaking techniques which place the emphasis on structure and weight, and away from primary fruit characters. It is a style which polarises opinion. The wine is matured in a mix of French-coopered barrels, 30% new. For seven years in a row, it was the most popular white wine at the Victorian Winemakers Exhibition.

▼▼▼▼▽ **1997** Deep yellow; the bouquet has a cascade of tangerine, malolactic-fermentation and oak influences, but nonetheless the massive sweet nectarine/tangerine fruit flavours which coat the mouth on the palate are unexpected. As ever, a very difficult wine to assess by conventional standards. **rating:** 90

best drinking 1999–2003 **best vintages** '86, '87, '92, '94, '97 **drink with** Pheasant with truffles • $39

---

noon winery ★★★★

Rifle Range Road, McLaren Vale, SA 5171 **region** McLaren Vale
ph (08) 8323 8290 **fax** (08) 8323 8290 **open** Thur–Mon 10–5
winemaker Drew Noon **prod.** 2000 **est.** 1976
prod. range ($12.50–23 CD) Solaire, One Night (Rosé), Twelve Bells, Eclipse (Grenache Shiraz), The Reles, Grenache Shiraz, Reserve Shiraz, Reserve Cabernet Sauvignon, Solaire Reserve Grenache, Chalice (Vintage Port).

summary Drew Noon has returned to McLaren Vale and purchased Noon's from his parents, having spent many years as a consultant oenologist and viticulturist in Victoria, thereafter as winemaker at Cassegrain. Some spectacular and unusual wines have followed, such as the 16.8 degrees alcohol Solaire Grenache, styled like an Italian Amarone. In 1998 Drew Noon gained the coveted Master of Wine (MW) award.

Noon Reserve Shiraz

Drew Noon seems intent on beating Roman Bratasiuk of Clarendon Hills at his own game. He is picking his grapes at exceptionally high sugar levels, and then neither fining nor filtering the wine before it goes to bottle. In 1997 this resulted in a wine with 15.8% alcohol.

▼▼▼▼ **1997** Dense, impenetrable purple; the bouquet seems underworked and slightly callow, though it is undeniably potent with earthy undertones. Inevitably, launches a massive assault on the mouth, and a wine which needs infinite patience. The fruit is actually hidden at this level and you can wonder whether it will ever escape the net. **rating:** 87

best drinking 2010–2020 **best vintages** NA **drink with** Don't • $19.50

Noon Eclipse Grenache Shiraz

Noon Winery has always made a specialty of very ripe, late-picked red wine styles, but it is hard to imagine anything eclipsing this Eclipse. A blend of 65% Grenache and 35% Shiraz, the '97 vintage had a mind-boggling alcohol of 16.3%.

▼▼▼▼▽ **1997** Strong purple-red; the bouquet has attractive blackberry/earthy fruit and minimal oak evident. The palate has abundant but pleasantly chewy chocolate and red fruit flavours; the tannins are not aggressive and the wine has excellent length. Amazingly, the alcohol does not result in a hot finish. **rating:** 92

best drinking 2000–2007 **best vintages** '97 **drink with** Venison • $15.50

---

normans ★★★☆

Grant's Gully Road, Clarendon, SA 5157 **region** McLaren Vale
ph (08) 8383 6138 **fax** (08) 8383 6089 **open** Mon–Fri 9–5, weekends and public holidays 11–5
winemaker Roger Harbord, Peter Fraser **prod.** 1.1 million **est.** 1853
prod. range ($3.95–35 R) A spread of wines starting with the Riverland-sourced Lone Gum range of Chardonnay and Shiraz Cabernet and Jesse's Blend White and Red; Chandlers Hill Chardonnay Semillon, Chenin Blanc, Fumé Blanc and Shiraz; Conquest and Pinot Noir Brut; then the White Label series of Bin number-identified Chardonnay, Verdelho, Merlot, Pinot Noir and Shiraz; and at the top of the scale Chais Clarendon Chardonnay, Shiraz and Cabernet Sauvignon.
summary In late 1994 Normans raised $6 million in new share capital, joining the lists of Australian Associated Stock Exchanges. The issue reflected the success Normans has enjoyed in recent years in establishing its brand both in domestic and export markets. The quality of the Chais Clarendon range is exemplary. Exports to the UK, much of Europe, US, Canada, Japan, southeast Asia and New Zealand.

Normans Chais Clarendon Shiraz

Although the Chais Clarendon label is strictly a brand, the majority of the grapes for the wines under the label does come from the Clarendon district at the southern end of the Adelaide Hills. Normans has a 6.6-hectare vineyard here which produces fruit of high quality.
🍷🍷🍷🍷🍷 **1996** Medium to full red-purple; both on bouquet and palate the wine has a powerful structure surrounding the mint, cherry and berry flavours. A highlight are the lovely milky tannins which add to the depth of the wine. Gold medal 1998 National Wine Show. **rating:** 94
best drinking 2001–2006 **best vintages** '82, '86, '90, '91, '92, '94, '95, '96 **drink with** Shoulder of lamb • $35

Normans Langhorne Creek Cabernet Sauvignon Cabernet Franc

One of three Signature Series wines first released by Normans in 1998. A blend of 75% Cabernet Sauvignon and 25% Cabernet Franc from a single vineyard with 25- to 30-year-old vines. The wine was aged in new and second-use French hogsheads for 18 months prior to bottling.
🍷🍷🍷🍷 **1996** Medium red-purple; there is good concentration of quite sweet fruit on the bouquet with well-handled vanillin oak. The palate, too, has attractive sweet, almost juicy, fruit flavours with soft tannin and subtle oak. A nice example of Langhorne Creek, offering more than the other wines in the initial Signature Series release. **rating:** 84
best drinking 2001–2006 **best vintages** NA **drink with** Yearling steak • $24

notley gorge ★★★★☆

West Tamar Highway, Rosevears, Tas 7277 **region** Northern Tasmania
ph (03) 6344 1114 **fax** (03) 6344 1114 **open** By appointment
winemaker James Chatto **prod.** 15 000 **est.** 1983
prod. range ($16–20 R) Sauvignon Blanc, Chardonnay, Tamar Dry White, Pinot Noir, Glengarry Cabernet Sauvignon, Cabernet Merlot (light red).
summary Notley Gorge has been transformed overnight from a vineyard with its wine contract-made for it into a large and modern winery which will not only make the wine for the Notley Gorge label, Ironpot Bay and Rosevears Estate (all associated entities) but will also contract-make wine for others. From poacher to gamekeeper indeed. The $2.2 million investment has been funded by a syndicate headed by Dr Michael Beamish, and is an important part of the rapidly changing face of the Tasmanian industry.

nuggetty ranges winery NR

Maldon-Shelbourne Road, Nuggetty, Vic 3463 **region** Bendigo
ph (03) 5475 1347 **fax** (03) 5475 1647 **open** Weekends 10–4 or by appointment
winemaker Greg Dedman **prod.** 500 **est.** 1993
prod. range ($20–28 ML) Semillon, Shiraz, Cabernet Sauvignon.
summary Draws upon 5 hectares of estate plantings, with mailing list and cellar door sales available while stocks last.

oakridge estate ★★★★

864 Maroondah Highway, Coldstream, Vic 3770 **region** Yarra Valley
ph (03) 9739 1920 **fax** (03) 9739 1923 **open** 7 days 10–5
winemaker Michael Zitzlaff, Paul Evans **prod.** 15 000 **est.** 1982
prod. range ($20–38 R) Riesling, Chardonnay, Pinot Noir, Cabernet Merlot; Reserve Chardonnay, Merlot and Cabernet Sauvignon.
summary The 1997 capital raising by Oakridge Vineyards Limited was successful, and a new winery was built and officially opened on 31 January 1998. Production of the Oakridge Estate wines is projected to increase in leaps and bounds during the remainder of this decade and into the first few years of the next millennium.

oakvale ★★★

Broke Road, Pokolbin, NSW 2321 **region** Lower Hunter Valley
ph (02) 4998 7520 **fax** (02) 4998 7747 **open** 7 days 10–5
winemaker Barry Shields, David Lowe **prod.** 5000 **est.** 1893
prod. range ($16–25 CD) Peach Tree Semillon, Selections Chardonnay (unwooded), Peach Tree Unwooded Chardonnay, Frontignac, Peppercorn Shiraz, Cabernet Merlot and a handful of fortified wines.
summary Former Sydney solicitor Barry Shields seems content with the change in his lifestyle as he presides over the historic Oakvale winery which he purchased from the Elliott family over a decade ago. The emphasis is on Semillon, Chardonnay and a blend of the two in both oaked and unoaked versions, most of which are offered with three to five years bottle age, with deep-yellow colour and rich, sometimes heavy, buttery/nutty flavours. The best wine, however, is the Peppercorn Shiraz.

Oakvale Peppercorn Shiraz

A good example of Hunter Shiraz, which while bottled and released prior to the following vintage, does not appear to have been rushed into bottle but, rather, well manipulated in the winery.

🍷🍷🍷🍷 **1998** Medium to full red-purple; the bouquet is clean, with sweet cherry fruit and subtle oak. The palate shows the same pleasant cherry fruit and subtle oak, finishing with soft, fine tannins. **rating:** 87

best drinking 2000–2005 **best vintages** NA **drink with** Chinese stir-fried beef • $20

old barn NR

Langmeil Road, Tanunda, SA 5352 **region** Barossa Valley
ph (08) 8563 0111 **open** Mon–Fri 9–5, weekends 10–5
winemaker Trevor Jones **prod.** 2200 **est.** 1990
prod. range ($6–13 CD) Riesling, Semillon, Traminer Riesling, Fumé Blanc, Old Fashioned Hock, Barossa Dry Red, Cabernet Shiraz, Light Sweet Red, and a range of other sweet and fortified wines.
summary Owned by a partnership of Barry and Elizabeth Chinner and Janet Hatch, with an honorary establishment date of 1861, being the date of construction of the old stone barn from

which the wines are exclusively sold. The wines are made by Trevor Jones, also winemaker at Kellermeister and Glenara, and are available only from cellar door, where back vintages are available.

old caves NR

New England Highway, Stanthorpe, Qld 4380 **region** Granite Belt
ph (07) 4681 1494 **fax** (07) 4681 2722 **open** Mon–Sat 9–5, Sun 10–5
winemaker David Zanatta **prod.** 2200 **est.** 1980
prod. range ($7.50–13.50 CD) Chardonnay, Classic Dry White, Light Red, Shiraz, Cabernet Sauvignon and a range of generic wines in both bottle and flagon, including fortifieds.
summary Has a strictly local, relatively uncritical and evidently loyal clientele.

old kent river ★★★☆

Turpin Road, Rocky Gully, WA 6397 **region** Great Southern
ph (08) 9855 1589 **fax** (08) 9855 1589 **open** By appointment
winemaker Alkoomi (Contract), Michael Staniford **prod.** 2000 **est.** 1985
prod. range ($16–54 CD) Chardonnay, Sauvignon Blanc, Pinot Noir, Reserve Pinot Noir, Shiraz, Diamontina (Sparkling).
summary Mark and Debbie Noack have done it tough all of their relatively young lives, but have earned respect from their neighbours and from the other producers to whom they sell more than half the production from their 10-hectare vineyard established on their sheep property. 'Grapes', they say 'saved us from bankruptcy'.

Old Kent River Pinot Noir

Replete with the striking label (a photograph of a dead jarrah tree taken by Mark's brother Ken) the wine invariably has strong colour and solid, ripe flavours, notwithstanding the generous yields.

🍷🍷🍷🍷🍸 **1997** Light to medium red-purple; the bouquet is clean and smooth, with pleasant cherry varietal fruit, if a fraction simple and straightforward. The palate picks up the pace, with plenty of fruit weight and well-structured plum and cherry fruit supported by subtle oak. **rating:** 90
best drinking 1999–2003 **best vintages** NA **drink with** Pasta, cheese • $25

olive farm ★★★

77 Great Eastern Highway, South Guildford, WA 6055 **region** Swan District
ph (08) 9277 2989 **fax** (08) 9279 4372 **open** Mon–Tues and Thur–Fri 10–5, weekends 11–3
winemaker Ian Yurisich **prod.** 4000 **est.** 1829
prod. range ($12.90–24.50 CD) Verdelho, Chenin Blanc, Classic White, Chardonnay, Gewurztraminer, Sauterne Style, Pinot Noir, Cabernet Sauvignon, Cabernet Shiraz Merlot, Sherry, Port, Sparkling.
summary The oldest winery in Australia in use today, and arguably the least communicative. The ultra-low profile tends to disguise the fact that wine quality is by and large good. The wines come from 12 hectares of estate plantings of 11 different varieties.

oliverhill NR

Seaview Road, McLaren Vale, SA 5171 **region** McLaren Vale
ph (08) 8323 8922 **open** 7 days 10–5
winemaker Stuart Miller **prod.** 1300 **est.** 1973
prod. range ($5–11 CD) Great Outdoors White and Red, Chardonnay, Shiraz Cabernet, Port, Muscat.
summary Oliverhill has changed hands, but otherwise continues an operation aimed almost entirely at the local tourist trade.

olssens of watervale NR

Government Road, Watervale, SA 5452 **region** Clare Valley
ph (08) 8843 0065 **fax** (08) 8843 0065 **open** 7 days 11–5 or by appointment
winemaker Contract **prod.** 1000 **est.** 1994
prod. range ($13–15 CD) Riesling, Semillon, Cabernet Sauvignon Cabernet Franc Merlot.
summary Kevin and Helen Olssen first visited the Clare Valley in December 1986. Within two weeks they and their family had decided to sell their Adelaide home and purchase a property in a small, isolated valley 3 kilometres north of the township of Watervale. Between 1987 and 1993 production from the 5-hectare vineyard was sold to other makers, but in 1993 the decision was taken to produce wine under the Olssen label. The 1998 Riesling (85 points) is the best of the current releases, intense, tight and long, and bound to repay cellaring.

orani vineyard NR

Arthur Highway, Sorrel, Tas 7172 **region** Southern Tasmania
ph (03) 6225 0330 **fax** (03) 6225 0330 **open** Weekends and public holidays 9.30–6.30
winemaker Various Contract **prod.** NA **est.** 1986
prod. range ($15.60–18.35 R) Riesling, Chardonnay, Pinot Noir.
summary The first commercial release from Orani was of a 1992 Pinot Noir, with Chardonnay and Riesling following in the years thereafter. The solidly constructed '96 Orani Pinot Noir, with abundant plum and mint fruit, is the best yet, well deserving of its bronze medal at the 1998 Tasmanian Wines Show. Retail distribution through Sutherland Cellars, Melbourne.

orlando ★★★★☆

Barossa Valley Way, Rowland Flat, SA 5352 **region** Barossa Valley
ph (08) 8521 3111 **fax** (08) 8521 3100 **open** Mon–Fri 10–5, weekends 10–4
winemaker Philip Laffer **prod.** NFP **est.** 1847
prod. range ($8.95–52 R) The table wines are sold in four ranges: first the national and international best-selling Jacobs Creek Semillon Sauvignon Blanc, Chardonnay, Riesling, Shiraz Cabernet and Grenache Shiraz and special Limited Releases; then the Gramps range of Chardonnay, Botrytis Semillon, Grenache, Cabernet Merlot; next the Saint range, St Helga Eden Valley Riesling, St Hilary Padthaway Chardonnay, St Hugo Coonawarra Cabernet Sauvignon; finally the premium range of Steingarten Riesling, Jacaranda Ridge Cabernet Sauvignon and Lawsons Padthaway Shiraz. Also Russet Ridge Coonawarra Chardonnay and Cabernet Shiraz Merlot off to one side; sparkling wines under the Trilogy and Carrington labels.
summary Jacobs Creek is one of the largest-selling brands in the world, and is almost exclusively responsible for driving the fortunes of this French (Pernod Ricard) owned company. The super-premium wines in the range continue to improve; Orlando had conspicuous trophy success at the 1997 National Wine Show. A colossus in the export game, chiefly to the UK and Europe, but also to the US and Asia.

Orlando Jacobs Creek Riesling

Jacobs Creek is one of the marketing miracles of the modern world, with more than one million cases a year sold in the UK market alone. It has been brand-extended from its original Shiraz Cabernet Malbec base to a range of four, also including Riesling, Chardonnay and Semillon Chardonnay. Notwithstanding the very large volume in which the Jacobs Creek Riesling is made, recent vintages have been consistent gold medal winners in national wine shows, competing against far more expensive (and more prestigious) wines.

YYYY 1998 Light yellow-green; a clean, crisp and very correct bouquet with hints of lime juice; the palate, too, is clean, again with a touch of lime balanced by crispness; moderate length and will probably build. **rating:** 85

best drinking 1999–2001 **best vintages** NA **drink with** Salads, seafood • $8.95

Orlando St Helga Eden Valley Riesling

Produced from grapes grown on Orlando's St Helga Vineyard in the Eden Valley and first made in 1980. That initial release was not put onto the market until it was four years old, and was initially priced as a premium product. It is now conventionally marketed in the year of production, and in real terms the price has come down significantly.

▼▼▼▼ **1998** Light yellow-green; by the end of 1998 already showing some richness in typical style, but with the near guarantee of improvement in front of it. The palate has plenty of flavour, quite rich, with soft lime fruit. **rating:** 87

best drinking 2000–2003 **best vintages** '90, '92, '94, '95, '96 **drink with** Asian cuisine • $15.95

Orlando Steingarten Riesling

In 1962 Colin Gramp, then Orlando's managing director, embarked on one of the great romantic follies: establishing a riesling vineyard on a vineyard entirely composed of rock schist (hence 'steingarten', or garden of stones) at a height of 490 metres on the East Barossa Ranges. Planted at a density of 6000 vines per hectare, the yields were nonetheless minuscule, and during the latter part of the '60s and '70s this was one of the most celebrated of all Rieslings. Commercial reality has since turned the wine into a brand, the back label freely admitting that Steingarten is used, where possible, as a base for the wine. In other words, in some years there may be a contribution from the Steingarten Vineyard, but in other years not necessarily so.

▼▼▼▼▽ **1998** Bright light green-yellow; the bouquet is fresh, clean and lively, with lime and touches of apple. A very pretty wine in the mouth, with lime, apple and passionfruit flavours which cannot help but please. **rating:** 90

best drinking 1999–2004 **best vintages** '98 **drink with** Sautéed prawns • $21

Orlando Gramp's Botrytis Semillon (375 ml)

Produced from semillon grown in the Griffith region, from whence all the great Australian botrytised Semillons are made, led conspicuously by De Bortoli's Noble One. Gramps has been producing the wine since 1988 (although not every vintage, for sometimes the botrytis is not present) and over the years the wines have won seven trophies and 19 gold medals. The '97 vintage is best yet, with a trophy and four gold medals to its credit, including an Intervin gold medal. It also happens to be ludicrously underpriced.

▼▼▼▼▼ **1997** Full yellow; intense, complex apricot cumquat and clingstone peach aromas flood the bouquet. Excellent winemaking has resulted in perfectly balanced acidity and subtle oak to balance the intensely luscious fruit on the mid-palate. **rating:** 94

best drinking 1999–2003 **best vintages** '97 **drink with** Baked apple • $14.95

Orlando Centenary Hill Shiraz

A new Barossa-based Shiraz from very old, dry-grown vines from a single vineyard which has had great success throughout 1998 and early 1999 in Australian wine shows.

▼▼▼▼▽ **1996** Medium to full red-purple; the bouquet is really quite elegant, with well-balanced and integrated fruit and oak. The palate retains those keys of elegance, berry fruit and vanilla oak; a quite firm tannin finish will stand the wine in good stead in the years ahead. **rating:** 92

best drinking 2001–2011 **best vintages** '96 **drink with** Braised quail • $50

Orlando Russet Ridge Coonawarra Cabernet Shiraz Merlot

First released in 1991, and continuing the move towards rather tighter and more concentrated fruit in the Orlando red wines. It is made entirely from Coonawarra grapes, a blend of approximately 85% Cabernet Sauvignon, 10% Shiraz and 5% Merlot. The wine is matured in a mixture of new and used French and American oak hogsheads for between 12 and 18 months.

YYYY 1996 Medium red-purple; the bouquet is fresh with typical Cabernet Merlot earthy/leafy/berry aromas, the palate sweetened by touches of chocolate and oak-derived vanilla. **rating:** 85

best drinking 1998–2003 **best vintages** '91, '92 **drink with** Hot, honey-glazed ham • $15.95

Orlando Gramp's Cabernet Merlot

Overall, Australian Cabernet Merlot blends have provided more disappointments than successes over the past ten years, but this is a success, particularly at the price.

YYYY 1996 Medium red-purple; the bouquet has plenty of weight and aroma, ranging through earth, blackberry and chocolate fruit, and the palate delivers precisely what the bouquet promises: attractive sweet, dark berry fruit contained in a firm structure. Lots of flavour. **rating:** 87

best drinking 1999–2006 **best vintages** NA **drink with** Braised oxtail • $14.95

Orlando Jacaranda Ridge Cabernet Sauvignon

A Coonawarra Cabernet Sauvignon which has been the red wine flagship for Orlando, sitting alongside Lawson's Shiraz. As with Lawson's, the American oak treatment has seemed rather heavy-handed, but good things are happening at Orlando, and recent vintages are much improved.

YYYYY 1996 Full red-purple; very pure and very correct cabernet varietal character is prominent on both bouquet and palate. Cassis and earth aromas and flavours intermingle, with extremely well-handled oak. Top gold 1998 National Wine Show. (Not yet released.) **rating:** 94

best drinking 2006–2011 **best vintages** '86, '88, '90, '94, '96 **drink with** Grilled beef • $52

Orlando St Hugo Cabernet Sauvignon

When first released about 1980, a benchmark, but which seemed to lose its way in a competitive field. Once again, resurgent, with particularly good outcomes in 1994 and 1996.

YYYY 1996 Medium to full red-purple; the bouquet is smooth, with sweet cassis berry fruit and nicely handled oak. The palate is of medium weight, with slightly more leafy/earthy characters (legitimately varietal) than the bouquet suggests. **rating:** 88

best drinking 2000–2005 **best vintages** '86, '88, '90, '91, '92, '94, '96 **drink with** Mixed grill • $31.95

orlando (padthaway) NR

c/o Barossa Valley Way, Rowland Flat, SA 5352 **region** Padthaway
ph (08) 8521 3111 **fax** (08) 8521 3100 **open** Not
winemaker Philip Laffer **prod.** NFP **est.** NA
prod. range ($14.95–55 R) St Hilary Padthaway Chardonnay, Jacobs Creek Limited Release Chardonnay, Lawsons Padthaway Shiraz.

summary While Padthaway serves the same purpose for Orlando as it does for the other major companies in the region by providing good-quality wine for cross-regional blends, it also produces three of Orlando's most distinguished wines. The first to appear was Lawson's Shiraz; St Hilary Chardonnay is a more recent arrival, the special Jacob's Creek Limited Release Chardonnay even more noteworthy. Other than its large vineyard, Orlando has no physical presence in Padthaway, but its Padthaway wines are sold at Padthaway Estate (and, of course, through retail shops).

Orlando Lawson's Padthaway Shiraz

Named after a nineteenth-century pioneer surveyor, Robert Lawson. It is matured for two years in new Nevers oak hogsheads, with further bottle maturation before release. This expansive oak input made its mark on the wine: the regular shower of gold and silver medals that descend on the wine shows that many judges did not regard the input as excessive. I have to admit that the vintages of the 1990s seem to me to be getting better and better, particularly so far as the oak regime is concerned, reaching a height of perfection with the '94 vintage.

🍷🍷🍷🍷🍷 **1994** Medium red-purple; there are lovely sweet fruit aromas on the bouquet running through mint and red berry; the palate is fine and smooth, still exceptionally youthful, with very well-balanced and integrated oak. Soft tannins, delicious wine. Gold medal 1998 National Wine Show. Retasted March 1999 with similar description and marginally higher points. **rating:** 95

🍷🍷🍷🍷🍷 **1993** Medium red-purple; a stylish and tangy bouquet offers lifted fruit with leafy nuances offset by a neat touch of vanillin oak. On the palate, sweet, dark cherry and berry fruit is more evident, although once again, there is plenty of well-handled vanillin oak influence. Three gold medals to its credit. **rating:** 91

best drinking 2000–2009 **best vintages** '88, '90, '91, '93, '94 **drink with** Beef stroganoff • $55

osborns ★★★★

RMB 5935 Ellerina Road, Merricks North, Vic 3926 **region** Mornington Peninsula
ph (03) 5989 7417 **fax** (03) 5989 7510 **open** By appointment
winemaker Richard McIntyre (Consultant) **prod.** 1600 **est.** 1988
prod. range ($20–23 CD) Chardonnay, Pinot Noir, Cabernet Merlot.
summary Frank and Pamela Osborn are now Mornington Peninsula veterans, having purchased the vineyard land in Ellerina Road in 1988, and (with help from son Guy) planted the vineyard over the following four years. The first release of wines in 1997 offered six vintages each of Chardonnay and Pinot Noir and five vintages of Cabernet Sauvignon, quite a debut. Part of the production from the 5.5 hectares of vineyards is sold to others, but increasing amounts are made and marketed under the Osborns label.

Osborns Chardonnay

Produced from 2.3 hectares of chardonnay situated on the north-facing slope of a small valley in the warmer part of the Mornington Peninsula. The wine style is absolutely in the mainstream of Mornington Peninsula-style, with the hallmark malolactic influence.

🍷🍷🍷🍷🍷 **1997** Medium yellow-green; the bouquet is at once subtle yet complex, gradually revealing nutty butterscotch mlf alongside melon fruit and a hint of mineral. The palate is still relatively youthful, with a smooth, creamy texture from the mlf contribution, yet retains fruit freshness and definition. **rating:** 91

best drinking 2000–2005 **best vintages** '97 **drink with** Sautéed scallops • $20

Osborns Pinot Noir

Produced from 1.6 hectares of estate plantings on well-drained duplex soils.

🍷🍷🍷🍷 **1997** Medium to full red-purple; the aromas are predominantly of soft, sweet plum, but there are some more sappy characters there to aid complexity. The palate has good flavour, and even better structure; plum and briar fruit is matched with soft, fine tannins. **rating:** 87

best drinking 2000–2004 **best vintages** '97 **drink with** Grilled quail • $23

padthaway estate ★★★☆

Keith–Naracoorte Road, Padthaway, SA 5271 **region** Padthaway
ph (08) 8765 5039 **fax** (08) 8765 5097 **open** 7 days 10–4.30
winemaker Nigel Catt **prod.** 6000 **est.** 1980
prod. range ($15–20 R) Eliza Pinot Chardonnay Cuvée, Eliza Pinot Noir Brut, Eliza Sparkling Burgundy, Eliza Chardonnay (wooded and unwooded).
summary The only functioning winery in Padthaway, set in the superb grounds of the Estate in a large and gracious old stone woolshed; the homestead is in the Relais et Chateaux mould, offering luxurious accommodation and fine food. Sparkling wines are the specialty of the Estate. Padthaway Estate also acts as a tasting centre for other Padthaway-region wines.

palmara ★★★☆

1314 Richmond Road, Richmond, Tas 7025 **region** Southern Tasmania
ph (03) 6260 2462 **open** Summer 12–6
winemaker Allan Bird **prod.** 250 **est.** 1985
prod. range ($14.50–17.50 CD) Chardonnay, Semillon Ehrenfeltzer, Montage Blend, Exotica (Siegerrebe), Pinot Noir, Cabernet Sauvignon.
summary Allan Bird makes the Palmara wines in tiny quantities. (The vineyard is slightly less than 1 hectare in total). The Pinot Noir has performed consistently well since 1990. The Exotica Siegerrebe blend is unchallenged as Australia's most exotic and unusual wine, with amazing jujube/lanolin aromas and flavours.

palmer wines ★★★★

Caves Road, Willyabrup, WA 6280 **region** Margaret River
ph (08) 9797 1881 **fax** (08) 9797 0534 **open** By appointment
winemaker Eddie Price, Amberley Estate (Contract) **prod.** 6000 **est.** 1977
prod. range ($15–19.50 R) Sauvignon Blanc, Semillon, Classic White, Chardonnay, Merlot, Cabernet Sauvignon, Cabernet Merlot.
summary Stephen and Helen Palmer planted their first hectare of vines way back in 1977, but a series of events (including a cyclone and grasshopper plagues) caused them to lose interest and instead turn to thoroughbred horses. But, with encouragement from Dr Michael Peterkin of Pierro, and after a gap of almost ten years, they again turned to viticulture, and now have 15 hectares planted to the classic varieties. The Cabernet Merlot is especially good.

Palmer Cabernet Merlot

A newly introduced wine into the Palmer range, and a great success. A blend of 77% Cabernet Sauvignon and 23% Merlot.

🍷🍷🍷🍷🍷 **1997** Medium red-purple; the bouquet opens with sweet berry fruit, but then some cedary/savoury nuances appear. A most attractive wine on the palate with lots of cedar and no less blackberry fruit; good oak handling. **rating:** 90

best drinking 2002–2007 **best vintages** '97 **drink with** Fillet mignon • $15

pankhurst NR

Old Woodgrove, Woodgrove Road, Hall, NSW 2618 **region** Canberra District
ph (02) 6230 2592 **fax** (02) 6230 2592 **open** Sundays, public holidays and by appointment
winemaker Sue Carpenter (Contract) **prod.** 4000 **est.** 1986
prod. range ($13–20 ML) Chardonnay, Semillon, Semillon Sauvignon Blanc, Pinot Noir, Cabernet Merlot.
summary Agricultural scientist and consultant Allan Pankhurst and wife Christine (with a degree in pharmaceutical science) have established a 3-hectare, split canopy vineyard. Tastings of the first wines produced showed considerable promise. In recent years Pankhurst has shared success with Lark Hill in the production of surprisingly good Pinot Noir – surprising given the climatic limitations. Says Christine Pankhurst 'the result of good viticulture here and great winemaking at Lark Hill', and she may well be right. The '98 Chardonnay (85 points) is a pleasant, well put together wine which seems to be unoaked.

panorama ★★★

RSD 297 Lower Wattle Grove, Cradoc, Tas 7109 **region** Southern Tasmania
ph (03) 6266 3409 **fax** (03) 6266 3409 **open** 6 days 10–5
winemaker Michael Vishacki **prod.** 250 **est.** 1974
prod. range ($10–30 CD) Chardonnay, Sauvignon Blanc, Pinot Noir, Cabernet Sauvignon

summary Michael and Sharon Vishacki purchased Panorama from Steve Ferencz two years ago, and have since spent considerable sums in building a brand new winery, an attractive cellar-door sales outlet, and in trebling the vineyard size.

Panorama Sauvignon Blanc

Picked very ripe and stainless steel-fermented.

ŸŸŸŸ 1998 Light green-yellow; ripe gooseberry aromas are the dominant force in the bouquet, although there are some grassy notes as well. There is a similar interplay between the lush, ripe gooseberry fruit which marks the forepalate, and the acidity which comes on the finish. A quite dramatic if rambunctious style. **rating:** 85

best drinking 1999–2000 **best vintages** NA **drink with** Asian dishes • $19

Panorama Pinot Noir

Produced entirely from pinot noir grown on the 3 hectares of estate vineyards, which are close planted and part organic, with no pesticides of any description used.

ŸŸŸŸ 1998 Light to medium red, showing development already; plum, spice and cherry fruit aromas, together with subtle oak, are followed by a relatively light-bodied but quite complex wine on the palate, with more plum, spice and cherry flavours. Very pleasant, but looks as if it will be short-lived. **rating:** 85

best drinking 1999–2000 **best vintages** '90, '91, '92, '93 **drink with** Antipasto • $24

paracombe wines ★★★★

Main Road, Paracombe, SA 5132 **region** Adelaide Hills

ph (08) 8380 5058 **fax** (03) 8380 5488 **open** Not

winemaker Paul Drogemuller (Overseeing Contract) **prod.** 1500 **est.** 1983

prod. range ($18–22 R) Chardonnay, Sauvignon Blanc, Pinot Noir, Cabernet Franc, Shiraz Cabernet, Pinot Chardonnay Méthode Champenoise, Sparkling Shiraz.

summary The Drogemuller family have established 6 hectares of vineyards at Paracombe, reviving a famous name in South Australian wine history. The wines are in fact contract-made at Petaluma, and are sold by mail order and through retailers in South Australia. It has had particular success with its Sauvignon Blanc (1997 Adelaide Wine Show trophy) but all of the wines in the range are worth chasing. A substantial portion of the production from the 13 hectares of estate plantings is sold to others.

paradise enough NR

Stewarts Road, Kongwak, Vic 3951 **region** Gippsland

ph (03) 5657 4241 **fax** (03) 5657 4229 **open** Sun, public holidays 12–5

winemaker John Bell, Sue Armstrong **prod.** 600 **est.** 1987

prod. range ($13–25 CD) Chardonnay, Reserve Chardonnay, Pinot Noir, Cabernet Merlot, Pinot Chardonnay.

summary Phillip Jones of Bass Phillip persuaded John Bell and Sue Armstrong to establish their small vineyard on a substantial dairy and beef cattle property.

paringa estate ★★★★★

44 Paringa Road, Red Hill South, Vic 3937 **region** Mornington Peninsula

ph (03) 5989 2669 **fax** (03) 5989 2669 **open** Mon, Wed–Fri 12–5, weekends, public holidays 11–5

winemaker Lindsay McCall **prod.** 2000 **est.** 1985

prod. range ($20–35 R) Chardonnay, Pinot Noir, Shiraz, Cabernet Sauvignon.

summary No longer a rising star, but a star shining more brightly in the Mornington Peninsula firmament than any other. As recent vintages have emphasised, the Mornington Peninsula region is very sensitive to growing season conditions, with problems in 1995 and 1996, but having wonderful years in 1997 and 1998. Paringa shines most brightly in the warmer years. Construction of a restaurant was completed in 1998.

Paringa Estate Chardonnay

Produced entirely from half a hectare of estate plantings, and 100% barrel-fermented in 100% new Vosges and Allier French oak barriques of the highest quality. Typically, 10% (and no more) is taken through malolactic fermentation, and the wine spends almost a year on yeast lees before bottling. The style is opulent and concentrated.

▼▼▼▼ **1997** Medium yellow-green; a solid bouquet with substantial oak and what appear to be slightly burnt aromas, but which may simply be derived from the oak, is followed by a full-bodied wine, showing its 14 degrees alcohol, with ripe melon, peach and fig fruit. Overall, fractionally heavy in the mouth. **rating:** 89

best drinking 1999–2003 **best vintages** '91, '92, '93, '94, '96, '97 **drink with** Pan-fried veal with abalone mushrooms • $30

Paringa Estate Pinot Noir

The most strikingly individual style in the Paringa range, and arguably the most controversial, although its formidable show record, strewn with trophies and gold medals, would not suggest any great element of controversy. It is very pungent, with herb and spice aromas and flavours attesting to the very cool climate, yet to my palate it has the requisite vinosity and sweet fruit.

▼▼▼▼▼ **1997** Amazingly deep red-purple; a hugely powerful and complex wine in its youth, looking like a cross between Burgundy and Cote Rotie in terms of its aromas. An extraordinarily powerful wine on the palate which literally blossomed in the nine months after it was bottled. Tasted three times between October 1998 and January 1999, on each occasion looking better, and with further improvement in front of it. Could be one of the greatest Australian Pinots. **rating:** 96

best drinking 1999–2006 **best vintages** '88, '90, '91, '92, '93, '95, '97 **drink with** Wild duck, game • $35

Paringa Estate Shiraz

The Mornington Peninsula is an unlikely place to produce first class Shiraz, particularly given that it is usually picked on the Queen's Birthday in June. And indeed, this extremely late-harvest date indicates that the growing conditions must be good (and in particular, warm) for the wine to really succeed – as it did brilliantly in a run of vintages from '90 to '94 inclusive. In this time the wine won numerous gold medals and trophies, the most prestigious being the trophy for Best Australian Wine in the 1995 Australia v South Africa Wine Challenge. In '95 and '96 the less favourable growing conditions have made their mark; the '97 is back to near top form, and will improve with time in bottle.

▼▼▼▼▽ **1997** Vivid purple-red; the aromatic qualities of the bouquet are somewhat closed, but dark berry fruits are certainly there in abundance. On the palate leaf, berry and spice fruit flavours are supported by a degree of sweet oak; the lingering tannins grip slightly on the finish, but the wine has obvious development potential. **rating:** 90

best drinking 2000–2005 **best vintages** '91, '92, '93, '94, '97 **drink with** Stir-fried beef • $30

---

parker coonawarra estate ★★★★

Penola Road, Coonawarra, SA 5263 **region** Coonawarra

ph (02) 9357 3376 **fax** (02) 9358 1517 **open** Not

winemaker Chris Cameron **prod.** 4000 **est.** 1985

prod. range ($25–65 R) Cabernet Sauvignon under two labels, Parker Coonawarra Estate First Growth and Parker Coonawarra Estate Terra Rossa.

summary Parker Coonawarra Estate is now a 50/50 joint venture between founder John Parker and family and James Fairfax. It is by this mechanism that Pepper Tree in the Hunter Valley (controlled by James Fairfax) has its Coonawarra stake. It has also led to the highly regarded wines being made by Pepper Tree winemaker Chris Cameron, albeit using the Balnaves winery in Coonawarra to do so. A maturation cellar and cellar-door sales facility in Coonawarra was planned to be functional in 1999.

passing clouds ★★★★

RMB 440 Kurting Road, Kingower, Vic 3517 **region** Bendigo
ph (03) 5438 8257 **fax** (03) 5438 8246 **open** 7 days 12–5 by appointment
winemaker Graeme Leith, Greg Bennett **prod.** 3000 **est.** 1974
prod. range ($23–30 CD) Red wine specialist; principal wines include Pinot Noir, Grenache, Shiraz, Shiraz Cabernet, Graeme's Blend (Shiraz Cabernet), Angel Blend (Cabernet), Glenalbyn Cabernet; also Onidne (Sparkling Shiraz Cabernet); Chardonnay and Sauvignon Blanc from the Goulburn Valley.

summary Graeme Leith is one of the great personalities of the industry, with a superb sense of humour, and makes lovely regional reds, with cassis, berry and mint fruit. His smiling, bearded face has adorned the front cover of many of the Victorian Tourist Bureau's excellent tourist publications over the past few years. The cellar in which he is seen dispensing wine is not his, incidentally; it is that of Chateau Tahbilk.

Passing Clouds Shiraz

Contrary to what one might expect, 1992 marked the first release of a 100% Shiraz from Passing Clouds; prior to that time the Shiraz was used in the various Shiraz Cabernet blends. It was an extremely auspicious start, albeit from a very good vintage. Subsequent releases have continued to impress.

🍷🍷🍷🍷 **1997** Medium to full red-purple; the bouquet is ripe, quite intense, with a mix of gamey/earthy fruit and a nice touch of vanilla oak. The palate offers sweet berry fruit, a dash of mint, soft, fine tannins and subtle oak; a nicely put together wine. **rating:** 86

best drinking 2001–2007 **best vintages** '92, '94, '97 **drink with** Oxtail • $23

Passing Clouds Graeme's Blend Shiraz Cabernet

The blend of 60% Shiraz and 40% Cabernet Sauvignon was first made in 1980, and has since become the linchpin in the Passing Clouds repertoire.

🍷🍷🍷🍷½ **1997** Medium to full red-purple; rich and sweet dark berry fruit swirls up from the bouquet, cascading into a rich dark cherry and chocolate-flavoured palate, smooth and long, closing with just a hint of vanilla oak. A lovely wine. **rating:** 92

best drinking 2001–2010 **best vintages** '81, '82, '86, '90, '91, '92, '94, '97 **drink with** Yearling steak or veal • $23

Passing Clouds Angel Blend

A blend of 90% Cabernet Sauvignon and 5% each of Cabernet Franc and Merlot aged in predominantly new American oak.

🍷🍷🍷🍷 **1997** Medium to full red-purple; the bouquet is concentrated, with some earthy/briary notes over dark fruit aromas; mouth-coating ripe, dense prune and mulberry fruit – with a touch of mint – finishes with soft tannins. **rating:** 87

best drinking 2002–2007 **best vintages** '90, '91, '92, '94, '96, '97 **drink with** Wild duck • $30

paternoster NR

17 Paternoster Road, Emerald, Vic 3782 **region** Yarra Valley
ph (03) 5968 3197 **open** Weekends 10.30–5.30
winemaker Philip Hession **prod.** 600 **est.** 1985

prod. range ($12–30 CD) Semillon, Chardonnay, Pinot Noir, Shiraz, Cabernets, Vintage Port.
summary The densely planted, non-irrigated vines (at a density of 5000 vines to the hectare) cascade down a steep hillside at Emerald in one of the coolest parts of the Yarra Valley. Pinot Noir is the specialty of the winery, producing intensely flavoured wines with a strong eucalypt mint overlay reminiscent of the wines of Delatite. No recent tastings; there also seems to be some dispute as to whether Paternoster falls within the Yarra Valley.

patrick creek vineyard NR

Springfield Park, North Down, Tas 7307 **region** Northern Tasmania
ph (03) 6424 6979 **fax** (03) 6424 6380 **open** By appointment
winemaker Andrew Hood (Contract) **prod.** 350 **est.** 1990
prod. range ($15 CD) Semillon, Chardonnay, Classic Dry White, Pinot Noir.
summary Patrick Creek Vineyard came into being in 1990 when Pat and Kay Walker established high-density plantings of chardonnay, pinot noir, semillon and sauvignon blanc in a 1-hectare vineyard. Patrick Creek produced a commendably sturdy Pinot Noir in 1996 which won a bronze medal at the 1997 Tasmanian Wines Show, repeating the dose with its '97 Pinot Noir at the 1998 Tasmanian Wines Show.

patritti wines ★★☆

13-23 Clacton Road, Dover Gardens, SA 5048 **region** Other Wineries of SA
ph (08) 8296 8261 **fax** (08) 8296 5088 **open** Mon–Sat 9–6
winemaker G Patritti, J Patritti **prod.** 65 000 **est.** 1926
prod. range ($4.50–8 CD) A kaleidoscopic array of table, sparkling, fortified and flavoured wines (and spirits) offered in bottle and flagon. The table wines are sold under the Blewitt Springs Estate, Patritti and Billabong Wines brands.
summary A traditional, family-owned business offering wines at modest prices, but with impressive vineyard holdings of 40 hectares in Blewitt Springs and another 40 hectares at Aldinga.

pattersons ★★★☆

St Werburghs Road, Mount Barker, WA 6234 **region** Great Southern
ph (08) 9851 2063 **fax** (08) 9851 2063 **open** Sun–Wed 10–5
winemaker Plantagenet (Contract) **prod.** 2000 **est.** 1982
prod. range ($17–25 CD) Chardonnay, Unwooded Chardonnay, Pinot Noir, Shiraz, Sparkling Shiraz.
summary Schoolteachers Sue and Arthur Patterson have grown chardonnay, shiraz and pinot noir and grazed cattle as a weekend relaxation for a decade. The cellar door is in a recently completed and very beautiful rammed-earth house, and a number of vintages are on sale at any one time. Good Chardonnay and Shiraz have been complemented by the occasional spectacular Pinot Noir.

Pattersons Unwooded Chardonnay

It is difficult to say much about unwooded Chardonnay in general; this wine, however, does have that little bit extra.
TTTT **1998** Light to medium yellow-green; the bouquet is quite aromatic with melon, nectarine and citrus fruit aromas, the palate providing the same flavours, with above-average length.
rating: 85

best drinking 1999–2001 **best vintages** NA **drink with** Roast pork • $17

paul conti ★★★☆

529 Wanneroo Road, Woodvale, WA 6026 **region** Swan District
ph (08) 9409 9160 **fax** (08) 9309 1634 **open** Mon–Sat 9.30–5.30, Sun by appointment
winemaker Paul Conti, Jason Conti **prod.** 8000 **est.** 1948

prod. range ($9.50–22 CD) Chenin Blanc, Chardonnay, Carabooda Chardonnay, Late Harvest Fronti, Grenache, Shiraz, Mariginiup Shiraz, Cabernet Sauvingon, White Port, Reserve Port.

summary In 1968 Paul Conti succeeded his father in the business the latter had established in 1948, and since assuming responsibility has quietly gone about making some outstanding wines, doing much to help pioneer the Mount Barker region, even though most of the wines are made from grapes grown on the southwest coastal plains around Perth. However, it has to be said this one-time leader of the field has been caught by the pack. The wines are exported to Japan and the UK and are distributed through the Australian east coast by Domaine Wine Shippers.

paul osicka ★★★★

Majors Creek Vineyard at Graytown, Vic 3608 **region** Bendigo
ph (03) 5794 9235 **fax** (03) 5794 9288 **open** Mon–Sat 10–5, Sun 12–5
winemaker Paul Osicka **prod.** NFP **est.** 1955
prod. range ($14–21.95 CD) Chardonnay, Riesling, Cabernet Sauvignon, Shiraz.

summary A low-profile producer but reliable, particularly when it comes to its smooth but rich Shiraz. The wines are distributed in Melbourne and Sydney by Australian Prestige Wines, with exports to the UK, Hong Kong and Japan.

Paul Osicka Cabernet Sauvignon

Paul Osicka has been a remarkably consistent producer of elegant Shiraz over the past 20 years or so. In 1997 it was the turn of the Cabernet Sauvignon, albeit in the same elegant style.

ỲỲỲỲ 1997 Medium red; the bouquet is only of moderate intensity, but has sweet berry fruit with a nicely judged touch of American oak. Berry and raspberry flavours on the palate are supported by soft, lingering tannins. Once again, the oak is delicate. **rating:** 86

best drinking 2001–2006 **best vintages** NA **drink with** Yearling steak • $21.95

paulett ★★★★

Polish Hill Road, Polish Hill River, SA 5453 **region** Clare Valley
ph (08) 8843 4328 **fax** (08) 8843 4202 **open** 7 days 10–5
winemaker Neil Paulett **prod.** 14 000 **est.** 1983
prod. range ($13–17 CD) Riesling, Sauvignon Blanc, Chardonnay, Late Harvest Riesling, Shiraz, Andreas Shiraz, Cabernet Merlot, Trillians (Sparkling Riesling).

summary The completion of the winery and cellar-door sales facility in 1992 marked the end of a development project which began back in 1982 when Neil and Alison Paulett purchased a 47-hectare property with a small patch of old vines and a house in a grove of trees (which were almost immediately burnt by the 1983 bushfires). The beautifully situated winery is one of the features of the scenic Polish Hill River region, as is its Riesling and its Cabernet Merlot. Exports to New Zealand, UK and Thailand.

Paulett Polish Hill River Riesling

Produced from 4 hectares of estate plantings on the Polish Hill River slopes. While in classic Clare-style, often develops a little more quickly than many of the other Rieslings of the district, which is no bad thing.

ỲỲỲỲ 1998 Medium yellow-green; the bouquet is rich and full with tropical lime fruit, the palate likewise full bodied and relatively advanced, but has good length and grip from a touch of minerally acidity on the finish. **rating:** 88

best drinking 1999–2003 **best vintages** '84, '90, '92, '93, '95, '96, '98 **drink with** Quiche Lorraine • $13

peacock hill vineyard NR

Cnr Branxton Road and Palmers Lane, Pokolbin, NSW 2320 **region** Lower Hunter Valley
ph (02) 4998 7661 **fax** (02) 4998 7661 **open** Fri–Mon, public and school holidays, 10–5
winemaker David Lowe, Jane Wilson-Lowe **prod.** 2000 **est.** 1969
prod. range ($14.50–24 CD) Absent Friends Chardonnay, Reserve Chardonnay, Jason Shiraz, Cabernet Sauvignon.

summary The Peacock Hill Vineyard was first planted in 1969 as part of the Rothbury Estate, originally being owned by a separate syndicate, but then moving under the direct control and ownership of Rothbury. After several further changes of ownership as Rothbury sold many of its vineyards, George Tsiros and Silvi Laumets acquired the 8-hectare property in October 1995. Since that time they have rejuvenated the vineyard and built a small but attractive accommodation lodge for two people, and have a tennis court and petanque rink for their exclusive enjoyment. Over the years, Peacock Hill has been a consistent medal winner in local wine shows.

pearson vineyards NR

Main North Road, Penwortham, SA 5453 **region** Clare Valley
ph (08) 8843 4234 **fax** (08) 8843 4141 **open** Mon–Fri 11–5, weekends 10–5
winemaker Jim Pearson **prod.** 800 **est.** 1993
prod. range ($13–18 CD) Riesling, Late Harvest Riesling, Cabernet Franc, Cabernet Sauvignon.

summary Jim Pearson makes the Pearson Vineyards wines at Mintaro Cellars. The 1.5-hectare estate vineyards surround the beautiful little stone house which acts as a cellar door and which appears on the cover of my book *The Wines, The History, The Vignerons of the Clare Valley*. Both the '98 Riesling (85 points) and the '96 Cabernet Sauvignon (84 points) are well made wines.

Pearson Late Harvest Riesling

Made in traditional Clare Valley late harvest style, with no botrytis influence, and the sweetness evident but no more than that. These wines can be superb at ten years of age or more.

🍷🍷🍷🍷 **1996** Light green-yellow; rich lime and toast fruit aromas are followed by a palate with considerable richness to identical lime/toasty fruit, the style being shaped as much by the fruit as by residual sugar. **rating:** 86

best drinking 2000–2004 **best vintages** NA **drink with** Conversation • $13

peel estate ★★★★

Fletcher Road, Baldivis, WA 6171 **region** South West Coast
ph (08) 9524 1221 **fax** (08) 9524 1625 **open** 7 days 10–5
winemaker Will Nairn **prod.** 8000 **est.** 1974
prod. range ($13.50–34 R) Chardonnay, Wood Matured Chenin Blanc, Medium Dry Chenin Blanc, Unwooded Chardonnay, Classic White, Verdelho, Shiraz, Zinfandel, Cabernet Sauvignon.

summary The winery rating is given for its Shiraz, a wine of considerable finesse and with a remarkably consistent track record. Every year Will Nairn holds a Great Shiraz tasting for six-year-old Australian Shirazes, and pits Peel Estate (in a blind tasting attended by 60 or so people) against Australia's best. It is never disgraced. The white wines are workmanlike, the wood-matured Chenin Blanc another winery specialty, although not achieving the excellence of the Shiraz. At five years of age it will typically show well, with black cherry and chocolate flavours, a strong dash of American oak, and surprising youth. There is limited retail distribution through each Australian State, and exports to Ireland.

Peel Estate Shiraz

Produced from 20-year-old estate vines which naturally produce high levels of ripeness and hence alcohol. The wine is open-fermented and spends two years in oak, 75% French, 25% American. The wine is then held a further ten months in bottle prior to release. The '94 is a simply marvellous wine, a reflection of the best vintage ever at Peel Estate.

🍷🍷🍷🍷🍷 **1994** Medium red-purple; a lovely, rich, sweet, dark cherry/black cherry-scented bouquet is followed by a richly fleshy palate where limitless amounts of dark cherry and chocolate are woven through a filigree of sweet tannins and subtle oak. Good now or in 20 years. **rating:** 95

🍷🍷🍷🍷 **1993** Light to medium red-purple; the bouquet is sweet and soft, with light chocolate, earth and berry aromas, and the palate is more fragrant and savoury than the '94. On the other side of the coin it lacks the richness and complexity of the younger wine. **rating:** 90

best drinking 2000–2020 **best vintages** '93, '94 **drink with** Leg of lamb • $34

peerick vineyard NR

Wild Dog Track, Moonambel, Vic 3478 **region** Pyrenees
ph (03) 9817 1611 **fax** (03) 9817 1611 **open** Weekends and public holidays 10–4
winemaker Contract **prod.** 3000 **est.** 1990
prod. range ($12.75–16.50 CD) Sauvignon Blanc, Semillon Sauvignon Blanc, Cabernet Sauvignon.

summary Peerick is the venture of Melbourne lawyer Chris Jessup and wife Meryl. They have mildly trimmed their Joseph's coat vineyard by increasing the plantings to 5.95 hectares and eliminating the malbec and semillon, but still manage to grow cabernet sauvignon, shiraz, cabernet franc, merlot, sauvignon blanc, and viognier. Don't ask me where the semillon for the Semillon Sauvignon Blanc blend comes from; I don't know. The '98 Sauvignon Blanc (85 points) is a clean, elegant light-bodied wine with nice varietal character.

pendarves estate ★★★☆

110 Old North Road, Belford, NSW 2335 **region** Lower Hunter Valley
ph (02) 6574 7222 **fax** (02) 9970 6152 **open** Weekends 11–5, Mon–Fri by appointment
winemaker Greg Silkman (Contract) **prod.** 9000 **est.** 1986
prod. range ($16–20 CD) An unusual portfolio of Verdelho, Sauvignon Blanc, Semillon, Chardonnay, Pinot Noir, Chambourcin, Shiraz, Merlot Malbec Cabernet.

summary The perpetual-motion general practitioner and founder of the Australian Medical Friends of Wine, Dr Philip Norrie, is a born communicator and marketer, as well as a wine historian of note. He also happens to be a passionate advocate of the virtues of Verdelho, inspired in part by the high regard held for that variety by vignerons around the turn of the century. His ambassadorship for the cause of wine and health in both Australia and overseas has no doubt indirectly contributed to the significant rise in production, and to the establishment of export markets in Singapore, the UK and the US (as well as national distribution).

penfolds ★★★★★

Tanunda Road, Nuriootpa, SA 5355 **region** Barossa Valley
ph (08) 8560 9389 **fax** (08) 8562 2494 **open** 7 days 10.30–4.30
winemaker John Duval **prod.** 1.1 million **est.** 1844
prod. range ($9–250 R) Kalimna Bin 28 Shiraz, 128 Coonawarra Shiraz, 389 Cabernet Shiraz, 407 Cabernet Sauvignon, 707 Cabernet Sauvignon and Special Show Bin reds. Brands include Minchinbury Sparkling; Rawson's Retreat Semillon Chardonnay and Cabernet Shiraz; Penfolds The Valleys Chardonnay, Old Vine Barossa Valley Semillon and Old Vine Barossa Valley Shiraz Grenache Mourvedre; Koonunga Hill Shiraz Cabernet, Semillon Sauvignon Blanc and Chardonnay; Magill Estate; Clare Estate and Clare Estate Chardonnay; St Henri Cabernet Shiraz;

Yattarna, Grange. Also various export-only labels. Finally, Grandfather Port and Great Grandfather Port.

summary Senior among the 17 wine companies or stand-alone brands in Southcorp Wines, and undoubtedly one of the top wine companies in the world in terms of quality, product range and exports. The consistency of the quality of the red wines and their value for money can only be described as breathtaking. In 1998 it released its long-awaited super-premium white wine, Yattarna Chardonnay, intended to sit alongside Grange, Australia's greatest red wine. It is also increasingly giving a regional identity to its fighting varietal brands. Worldwide distribution.

Penfolds Rawson's Retreat Chardonnay

Launched with the 1998 vintage, and a signpost for the direction in which the 'surplus' of Australian Chardonnay is headed. It also shows Penfolds' growing skill in white winemaking, being far less heavy-handed than earlier wines at this price point.

🍷🍷🍷🍷 **1998** Light to medium green-yellow; the bouquet is clean, and more elegant than one would expect, reflecting the continuous refinement of these cheaper wines in the Penfolds portfolio. The palate is well balanced and structured, with sweet melon fruit and a light touch of oak flavour. **rating:** 84

best drinking 1998–1999 **best vintages** NA **drink with** KFC • $9.95

Penfolds Reserve Bin Chardonnay

Keeping up with the twists and turns of Penfolds development of a super-premium white wine to sit alongside Grange is not easy. Between 1992 and 1997 more than 12 experimental wines have been made by Penfolds, some commercially released, some not. In November 1997 there was a one-off release of Bin 94A, an extraordinarily complex wine produced from 90% Chardonnay and 10% Sauvignon Blanc grown in the Adelaide Hills (52%), Eden Valley (24%), Tumbarumba (21%) and Macedon (3%). It was totally barrel-fermented in French oak (72% new) and matured on lees in those barrels for ten months, with 90% of the wine undergoing malolactic fermentation. The '96 is very nearly as complex in its make-up.

🍷🍷🍷🍷🍷 **1996** Light to medium yellow-green; the complex bouquet brings together elegant melon fruit and sophisticated, toasty barrel-ferment oak characters. On the palate fine, tight melon and citrus fruit runs through to a long finish, with the oak far less evident than it is on the bouquet. **rating:** 94

best drinking 2001–2006 **best vintages** '94, '96 **drink with** Smoked chicken • $42

Penfolds The Valleys Chardonnay

Made from chardonnay grown in the Clare and Barossa Valleys, and now 100% barrel-fermented in a mixture of new and one-year-old French oak, in which it spends eight months. A fast-maturing style which derives a considerable amount of its impact from the oak. One cannot quibble about the value of the wine, but I do wonder whether in this instance more is indeed better. 'The Valleys' was added to the label name in late 1995, part of a broad-ranging move towards greater regional identification for the wines in this price bracket.

🍷🍷🍷🍷 **1997** Light to medium yellow-green; there is a faintly earthy substrate running through the bouquet although the fruit is smooth and quite solid. A wine very much made in the traditional style, solid and dependable. **rating** 84

best drinking 1998–1999 **best vintages** '93 **drink with** Pasta, chicken, turkey • $14.50

Penfolds Bin 128 Coonawarra Shiraz

Curiously, the only wine made by Penfolds every vintage from Coonawarra grapes and sold in the Australian market (there are some export labels). No less curious has been the lacklustre performance (by Penfolds standards and given the reputation of Coonawarra) of the wine over the years. However, since 1986 the wine has taken a distinct turn for the better, which is as it should be. It is one of the few wines in the Penfolds range which is matured entirely in French oak, 20% new and 80% one and two years old.

🍷🍷🍷🍷🍸 **1996** Medium to full red-purple; generous, ripe dark berry/plum, a hint of spice and nice oak on the bouquet announce a concentrated and powerful wine, with dark berry fruit in abundance, lingering tannins and subtle oak. **rating:** 91

best drinking 2001–2011 **best vintages** '63, '66, '80, '86, '89, '90, '91, '93, '94, '96 **drink with** Veal; mild cheddar • $20

Penfolds Grange

Australia's greatest red wine, with a turbulent early history chronicled in Huon Hooke's book entitled *Max Schubert: Winemaker* (1994). What can be said in a few words? The bare bones are: first made 1955; sourced from low-yielding, old shiraz vines (and up to 15% Cabernet Sauvignon) chiefly from the Barossa but also from the Clare Valley and McLaren Vale. Fermentation finished in and matured in new American oak. The 1990 was rated 'Number One Wine in the World' by the all-powerful *Wine Spectator* in 1995. The '94 comes from a relatively unheralded vintage; it will be interesting to see whether others think it is as great as I do.

🍷🍷🍷🍷🍷 **1994** Dense red-purple; the typically rich and powerful bouquet has layers of dark berry fruit and integrated and balanced oak. What makes it unusual is the distinct hint of chocolate, which appears again on the silkily powerful palate. Here there is an array of all of the fruit flavours one could possibly expect, but once again that touch of dark Swiss chocolate adding an intriguing note. The oak balance and integration is better than any five-year-old Grange I can remember, the tannins perfectly balanced into the bargain. **rating:** 97

best drinking 2004–2024 **best vintages** '52, '53, '55, '62, '66, '67, '71, '76, '78, '80, '85, '86, '90, '91, '92, '93, '94 **drink with** Rich game dishes • $250

Penfolds Kalimna Bin 28 Shiraz

A multi-region blend sourced from the Barossa and Clare Valleys, McLaren Vale and Langhorne Creek regions. There is a good argument to be made that it should be Bin 28, rather than Bin 389, which is 'Poor Man's Grange'. The emphasis is on the same lush fruit; the oak used is second- and third-use American barrels handed down through the Grange and Bin 389 programmes.

🍷🍷🍷🍷🍷 **1996** Dense red-purple; the very rich and concentrated bouquet is full of dark berry fruit and dark Swiss chocolate; the palate is rich and full, with luscious dark plum and rippling tannins. The best Bin 28 for years. **rating:** 94

best drinking 2001–2016 **best vintages** '64, '66, '71, '80, '81, '83, '86, '90, '91, '94, '95, '96 **drink with** Lamb or beef casserole • $20

Penfolds Magill Estate

Made solely from the 5.2-hectare vineyard established on the site where Dr Christopher Rawson Penfold built his house in 1844. It is an estate wine in the fullest sense of the term, made at the historic Magill winery, which has been restored and refurbished. Situated within ten minutes of the centre of Adelaide, Magill Estate will become one of Australia's foremost viticultural landmarks.

🍷🍷🍷🍷🍷 **1996** Medium red-purple; the bouquet is full of sweet, dark cherry fruit, the palate similarly full of delicious dark cherry flavours, fine, lingering tannins and perfectly integrated and balanced oak. **rating:** 95

best drinking 2001–2011 **best vintages** '83, '86, '89, '90, '91, '93, '95, '96 **drink with** Rack of lamb • $39.95

Penfolds Old Vine Barossa Valley Shiraz Mourvedre Grenache

The first vintage of this wine (1992) was released in 1995 and market demand has since shortened the time between vintage and release. It is typically a blend of around 55% Shiraz, 25% Mourvedre and 20% Grenache, the last two components sourced from old, low-yielding vineyards in the Barossa Valley. Although it spends 14 months in barrel, no new oak is used, and the oak influence is minimal.

🍷🍷🍷🍷 **1995** Medium red-purple; the bouquet is quite intense and solid, with nuances of black cherry, earth and liquorice. The palate is sustained by a gently sweet fruit core of black cherry and small berry fruit, finishing with soft tannins. Significantly less Teutonic than preceding vintages, and all the better for that. **rating:** 85

best drinking 1999–2004 **best vintages** NA **drink with** Rich stews • $20

Penfolds Koonunga Hill Shiraz Cabernet

First made in 1976, it is sourced from premium areas across South Australia (not including the Riverland) from a varying blend, with the Cabernet component usually ranging between 30% and 40%, but with a high of 48% and a low of 5% according to vintage. It is matured in real oak (no chips) and is not batch-bottled. Outstanding value for money, its integrity protected by the introduction at a lower price point of Rawson's Retreat.

🍷🍷🍷🍷 **1997** Medium to full red-purple; quite rich and ripe blackberry and chocolate fruit aromas with a hint of vanilla on the bouquet leads into a palate as honest as the day is long, with solid ripe fruit, good tannins and the barest touch of American oak. This is not a wine about sophistication or elegance, simply straightforward flavour. **rating:** 86

best drinking 1999–2007 **best vintages** '82, '87, '90, '91, '92, '94, '96 **drink with** Hearty fare • $12

Penfolds Clare Estate

Made from a blend of Merlot, Cabernet Sauvignon, Cabernet Franc and Malbec, grown entirely on Penfolds 180-hectare vineyard in the Polish Hill River Valley, east of the main Clare Valley region. The wine is matured in new French oak for one year. I have often had difficulty with this wine, typically finding it over-oaked or under-fruited, depending on which side of the coin you look. The '96 seems to represent a breakthrough, although it did come from a particularly good vintage. It will be interesting to see whether the same results are obtained in the future.

🍷🍷🍷🍷½ **1996** Medium purple-red; the bouquet opens with pleasant cedar and vanilla oak, but then sweet dark berry fruit swells up. An elegant wine on the palate, medium-bodied, with cedar and dark fruit flavours, finishing with soft, lingering tannins. Surely the best yet. **rating:** 90

best drinking 2000–2006 **best vintages** '88, '90, '91, '92, '94, '96 **drink with** Yearling steak; mild cheese • $21

Penfolds Bin 389 Cabernet Shiraz

First made in 1960, and promptly dubbed 'Poor Man's Grange'. A blend of 50% to 60% Cabernet Sauvignon and 40% to 50% Shiraz from the Barossa Valley, Coonawarra, Padthaway, McLaren Vale, Langhorne Creek and the Clare Valley. Matured in a mixture of new (20%) and older American oak barrels.

🍷🍷🍷🍷🍷 **1996** Medium to full red-purple; a dense wine on the bouquet with layers of earth, chocolate, berry and oak. Wonderfully sweet blackberry fruit is offset with sweet chocolate and rounded off with sweet tannins and oak on the finish mark an exceptionally good Bin 389. **rating:** 95

best drinking 2001–2016 **best vintages** '66, '70, '71, '86, '90, '93, '94, '96 **drink with** Double lamb loin chops • $26

Penfolds Rawson's Retreat Cabernet Shiraz

Introduced in the mid-1990s to fill the volume gap underneath Koonunga Hill, and made in very substantial quantities. One of Australia's top-selling red wines under $10 and, for this reason if no other, cannot be ignored. Typically a blend of 50% Cabernet with the balance more or less equally split between Shiraz and Ruby Cabernet, the wine is given some maturation in used American and French oak barrels.

🍷🍷🍷🍷 **1997** Medium red-purple; there is sweet berry fruit on the pleasantly balanced bouquet, and likewise on the palate. Overall, appealingly soft, with controlled oak influence. **rating:** 84

best drinking 1999–2000 **best vintages** NA **drink with** Takeaway food • $7

Penfolds St Henri Cabernet Shiraz

Born of the vigorous rivalry between and radically different winemaking philosophies of Max Schubert and John Davoren. With the exception of emphasis on the highest-quality grape sources, Grange and St Henri are polar opposites. This traditional wine is made from Barossa, Clare and Eden Valleys, McLaren Vale, and Langhorne Creek Shiraz; and Cabernet Sauvignon from Coonawarra and the Barossa Valley. Oak influence is minimal.

🍷🍷🍷🍷 **1995** Medium to full red-purple; the bouquet is intense, with dark blackcurrant and plum fruit supported by subtle oak. The remarkably sweet mid-palate is followed by quite marked, though not harsh, tannins on the finish; subtle oak. Could well evolve further and merit higher points. **rating:** 89

best drinking 2002–2012 **best vintages** '66, '67, '76, '82, '85, '88, '90, '91, '93, '94 **drink with** Steak and kidney pie • $42.50

Penfolds Bin 407 Cabernet Sauvignon

A relatively recent introduction to the Penfolds range, first made in 1990 from a blend of McLaren Vale, Padthaway, Coonawarra, Barossa and Clare Valleys Cabernet Sauvignon. Matured in a mix of 30% new oak (American and French) and 70% hand-me-downs from Bin 707. Initially conceived as a lower-priced wine, the sheer quality of Bin 407 and its already illustrious show record led to its higher price when released.

🍷🍷🍷🍷🍷 **1996** Medium to full red-purple; there is quite evident cedary oak on the bouquet together with gentle berry fruit. The palate has good structure and balance, with fine cassis fruit which is not overripe, and fine tannins on the finish. For all that, not quite up to the other '96 red wines from Penfolds. **rating:** 92

best drinking 2001–2011 **best vintages** '90, '91, '92, '93, '94, '96 **drink with** Venison, kangaroo fillet • $22.50

Penfolds Bin 707 Cabernet Sauvignon

First made in 1964, with a hiatus between 1969 and 1976 as scarce supplies of Cabernet Sauvignon were diverted to Bin 389. Now made from a blend of Coonawarra, Barossa and Eden Valley grapes, vinified in a fashion similar to Grange, with the fermentation finished in American oak barrels. The '94 vintage won three trophies at the 1997 Sydney Wine Show, including the Dr Gilbert Phillips Trophy for Best Red Wine of the Show. Not made in 1995. The '96 is destined for greatness even in the ranks of Bin 707.

🍷🍷🍷🍷🍷 **1996** Full red-purple; concentrated, ripe, cassis and blackberry fruit leaps from the glass, with oak present but trailing behind. The palate is as rich, ripe, succulent and concentrated as the bouquet suggests, redolent of cassis, and long, lingering tannins. **rating:** 97

best drinking 2003–2016 **best vintages** '66, '80, '84, '86, '88, '90, '92, '93, '94, '96 **drink with** Rare scotch fillet • $80

Penfolds Grandfather Port

First released in the 1960s; made from Barossa Valley Shiraz and Mourvedre which is oak-matured in old barrels for decades, and which is finally blended from selection of very old and much younger material. Penfolds has always kept a tight rein on production, thus preserving the all-important old material which forms the base of the wine.

🍷🍷🍷🍷🍷 **NV** Medium to full tawny-red; exceedingly complex, with strong barrel-aged rancio characters and attractively earthy spirit. A long and richly flavoured wine, with the cleansing, drying finish which is so essential to the style. **rating:** 90

best drinking 1999–2009 **best vintages** NA **drink with** Coffee • $79.95

Penfolds Great Grandfather Port

First released in 1994 to celebrate the 150th birthday of Penfolds. In a neat marketing gimmick, 1994 bottles were blended and bottled, utilising the very oldest and best stocks, dating back to the early years of the twentieth century. The concentration and power of the wine is reminiscent of the very old Seppelt Liqueur Tawny Ports.

🍷🍷🍷🍷🍷 **NV** Deep tawny with an olive-green rim; both the bouquet and palate are extraordinarily concentrated and rich, very much into the liqueur style of Port which is unique to Australia. Shows the Brandy spirit which was used in the fortifying process, and which adds yet extra complexity. **rating:** 95

best drinking 1999–2009 **best vintages** NA **drink with** Coffee • $150

penley estate ★★★★☆

McLeans Road, Coonawarra, SA 5263 **region** Coonawarra
ph (08) 8231 2400 **fax** (08) 8231 0589 **open** By appointment
winemaker Kym Tolley **prod.** 20 000 **est.** 1988
prod. range ($18–50 R) Chardonnay, Hyland Shiraz, Shiraz Cabernet, Cabernet Sauvignon, Phoenix Cabernet Sauvignon, Pinot Noir Chardonnay.
summary Owner winemaker Kym Tolley describes himself as a fifth-generation winemaker, the family tree involving both the Penfolds and the Tolleys. He worked 17 years in the industry before establishing Penley Estate and has made every post a winner since, producing a succession of rich, complex, full-bodied red wines and stylish Chardonnays. Now ranks as one of the best wineries in Coonawarra, drawing upon 81 precious hectares of estate plantings.

Penley Estate Shiraz Cabernet

A blend of 50% Shiraz and 50% Cabernet Sauvignon, sourced from Coonawarra (65%), the balance from McLaren Vale (25%) and Barossa Valley (10%). Matured for two years in a mix of 75% American and 25% French (Troncais) oak. Like the Cabernet Sauvignon, a prolific medal winner.

🍷🍷🍷🍷🍷 **1996** Medium red-purple; sweet vanilla oak together with earth, chocolatey fruit of moderate intensity is followed by a palate with attractive red fruit flavours, soft tannins and oak bringing up the rear. **rating:** 90

best drinking 2001–2011 **best vintages** '88, '90, '91, '92, '94, '96 **drink with** Soft ripened cheese • $23

Penley Estate Merlot

Coonawarra is presently synonymous with Cabernet Sauvignon; wine such as this, Pepper Tree and Petaluma make one wonder whether it may ultimately become as famous for its Merlots.

🍷🍷🍷🍷🍷 **1996** Medium purple-red; the bouquet is quite full and very stylish, with some briary Bordeaux notes providing varietal authenticity. The palate is harmonious, with sweet cherry, cedar and briar flavours intermingling, but fruit foremost. A unanimous gold at the 1999 Sydney Royal Wine Show. **rating:** 94

best drinking 2000–2006 **best vintages** '96 **drink with** Fillet of beef • $23

Penley Estate Cabernet Sauvignon

Made from the pick of 38 hectares of estate-grown Cabernet Sauvignon, matured in a cleverly handled mix of American and French oak. Over the years, various vintages of the wine have won a quite extraordinary number of trophies and gold medals.

ΨΨΨΨ **1996** Medium to full red-purple; a powerful, earthy, slightly vegetative bouquet is followed by a palate laden with cassis fruit, soft tannins and subtle oak. Perhaps the points are an over-miserly reaction to a single element of the bouquet; time will tell. **rating:** 89

best drinking 2001–2011 **best vintages** '89, '90, '91, '92, '94 **drink with** Rare beef • $48

penny's hill vineyards NR

Main Road, McLaren Vale, SA 5171 **region** McLaren Vale

ph (08) 8362 1077 **fax** (08) 8362 2766 **open** By appointment

winemaker Ben Riggs (Contract) **prod.** 3200 **est.** 1988

prod. range ($17.25–24.70 R) Chardonnay, Shiraz, Specialized (Shiraz Cabernet Merlot).

summary Penny's Hill is a major new vineyard and winery operation in McLaren Vale owned by Adelaide advertising agency businessman Tony Parkinson and wife Susie. The Penny's Hill vineyard is 25 hectares, and, unusually for McLaren Vale, is close-planted with a thin vertical trellis/thin vertical canopy, the work of consultant viticulturist David Paxton. The innovative red dot packaging was the inspiration of Tony Parkinson, recalling the red dot sold sign on pictures in an art gallery.

pennyweight winery NR

Pennyweight Lane, Beechworth, Vic 3747 **region** Ovens Valley

ph (03) 5728 1747 **fax** (03) 5728 1704 **open** Thur–Tues 10–5

winemaker Stephen Newton Morris **prod.** 1000 **est.** 1982

prod. range ($11–27 CD) A fortified specialist, but also producing limited table wines including Trebbiano Riesling and Shiraz; the primary focus is on Fino and Amontillado Sherries, a range of Ports from Old Tawny to Ruby, Vintage Port, White Port and Muscat.

summary Pennyweight was established by Stephen Morris, great-grandson of G F Morris, founder of Morris Wines. The 3 hectares of vines are not irrigated, and are moving towards organic certification. The business is run by Stephen, together with his wife Elizabeth assisted by their three sons; Elizabeth Morris says, 'It's a perfect world' suggesting Pennyweight is more than happy with its lot in life.

penwortham wine cellars NR

Government Road, Penwortham, SA 5453 **region** Clare Valley

ph (08) 8843 4345 **open** Sat 10–5, Sun, holidays 10–4

winemaker Richard Hughes **prod.** 1500 **est.** 1985

prod. range ($13–16 CD) Riesling, Cabernet Sauvignon, Shiraz.

summary A relatively new arrival on the Clare Valley scene; a '95 unwooded Shiraz tasted in 1996 was fearsomely strong and tannic.

peppers creek NR

Broke Road, Pokolbin, NSW 2321 **region** Lower Hunter Valley

ph (02) 4998 7532 **fax** (02) 4998 7531 **open** Wed–Sun 10–5

winemaker Peter Ireland **prod.** 700 **est.** 1987

prod. range ($20–25 CD) Enzo Bianco, Enzo Rosé, Semillon, Unwooded Chardonnay, Chardonnay, Enzo Rosso, Merlot, Yacht Squadron Port.

summary A combined winery and antique shop which sells all its wine through the cellar door, and runs the Cafe Enzo. The red wines previously tasted were clean and full flavoured, the Merlot coming from the hectare of estate vineyards. No recent tastings.

pepper tree ★★★★

Halls Road, Pokolbin, NSW 2321 **region** Lower Hunter Valley
ph (02) 4998 7539 **fax** (02) 4998 7746 **open** 7 days 9–5
winemaker Chris Cameron, Chris Archer **prod.** 40 000 **est.** 1993
prod. range ($10–30 CD) Sundial White and Red; Chardonnay, Shiraz, Cabernet Franc; Reserve range of Semillon, Chardonnay, Sauvignon Blanc, Traminer, Verdelho, Muscat, Malbec, Coonawarra Merlot and Cabernet Sauvignon.

summary The Pepper Tree winery is situated in the complex which also contains The Convent guesthouse and Roberts Restaurant. The company which now owns Pepper Tree is headed by Chris Cameron, chief winemaker since 1991. It made a decisive move in 1996, formalising the acquisition of a major interest in the Parker (Coonawarra) Estate vineyards, having previously purchased some of the fruit from those vineyards. Pepper Tree has made a determined, and quite successful, effort to establish its reputation as one of Australia's leading producers of Merlot.

Pepper Tree Chardonnay

A wine with a varying geographic base, initially Hunter and Coonawarra, but since changed to King Valley, Padthaway and Hunter Valley.

🍷🍷🍷🍷 **1997** Medium to full yellow; a solid wine with buttery/peachy fruit and some spicy oak on the bouquet. The palate moves more towards melon and citrus, and overall the wine is not especially complex or rich, possibly reflecting the King Valley component. **rating:** 86

best drinking 1999–2002 **best vintages** NA **drink with** Brains • $17

Pepper Tree Shiraz

Well made and clean, at this stage showing pure varietal character; the regional influences will come later.

🍷🍷🍷🍷🍷 **1998** Youthful purple-red; the bouquet is clean and sweet, with rich cherry fruit and subtle oak. The palate continues the play, with generous, clean sweet cherry fruit, soft tannins and similarly subtle oak. **rating:** 90

best drinking 2000–2008 **best vintages** NA **drink with** Loin of lamb • $30

---

pertaringa ★★★★

Cnr Hunt and Rifle Range Roads, McLaren Vale, SA 5171 **region** McLaren Vale
ph (08) 8323 8125 **fax** (08) 8323 7766 **open** Mon–Fri 10–4
winemaker Geoff Hardy, Ben Riggs **prod.** 2000 **est.** 1980
prod. range ($15–25 R) Semillon, Sauvignon Blanc, Shiraz, Liqueur Frontignac and Cabernet Sauvignon.

summary The Pertaringa wines are made from part of the grapes grown by leading viticulturists Geoff Hardy and Ian Leask. The Pertaringa vineyard of 33 hectares was acquired in 1980 and rejuvenated; establishment of the ultra-cool Kuitpo vineyard in the Adelaide Hills began in 1987 and now supplies leading makers such as Southcorp, Petaluma and Shaw & Smith. Retail distribution through South Australia, New South Wales, Victoria and Queensland; exports to UK, US, Canada and Denmark.

Pertaringa Shiraz

A classic example of McLaren Vale shiraz, coming from fully mature vines. It is fruit, rather than oak, which provides the flavour.

🍷🍷🍷🍷🍷 **1997** Medium to full red-purple; solid dark berry fruit with a touch of briar and a hint of oak on the bouquet is followed by a palate with abundant plum and dark cherry fruit, hints of chocolate, fine tannins and a nice touch of oak. **rating:** 90

best drinking 2000–2007 **best vintages** '97 **drink with** Beef casserole • $24

---

petaluma ★★★★★

Spring Gully Road, Piccadilly, SA 5151 **region** Adelaide Hills
ph (08) 8339 4122 **fax** (08) 8339 5253 **open** See Bridgewater Mill
winemaker Brian Croser **prod.** 30 000 **est.** 1976
prod. range ($19–45 R) Riesling, Chardonnay, Coonawarra (Cabernet Blend), Croser (Sparkling); Second label Sharefarmers White and Red. Bridgewater Mill is another second label – see separate entry.
summary The Petaluma empire continues to flourish now taking in both Knappstein Wines and Mitchelton. While running a public-listed group, Brian Croser has never compromised his fierce commitment to quality, and doubtless never will. The Riesling is almost monotonously good; the Chardonnay is the big mover, going from strength to strength; the Merlot another marvellously succulent wine to buy without hesitation.

Petaluma Riesling

A 100% estate-produced wine from Petaluma's Hanlins Hill Vineyard in the Clare Valley, one of the classic Australian Riesling regions. The wine is made with iron discipline, and is a crystal-pure reflection of the interaction of climate, soil and variety. As the notes indicate, it ages with grace.

🍷🍷🍷🍷🍸 **1998** Light to medium yellow-green; as always with the young Petaluma Rieslings, very tight and undemonstrative, with minerally/chalky aromas but not much fruit coming out at this stage. The palate is similarly restrained (in a classic and expected fashion) needing many years to build on the mineral/herb flavours. **rating:** 92

best drinking 2003–2010 **best vintages** '80, '85, '86, '88, '90, '92, '93, '94, '95, '97 **drink with** Sashimi • $20

Petaluma Sharefarmers White

Produced from grapes grown on Petaluma's Sharefarmers Vineyard, the eye in the centre of the storm revolving around the drawing of the borders of Coonawarra. The blend is typically two-thirds Chardonnay, one-third Sauvignon Blanc, and the wine is fermented in stainless steel and early bottled.

🍷🍷🍷🍷 **1997** Light yellow-green; the bouquet is still fresh, with attractive citrus fruit of light to medium intensity. In the mouth, the wine is no less clean and fresh, with a mix of citrus, tropical and passionfruit flavours, and a pleasingly crisp finish. **rating:** 86

best drinking 1998–1999 **best vintages** NA **drink with** Seafood • $20

Petaluma Chardonnay

One of the more elegant and refined Australian Chardonnays which has, however, radically changed its geographic base since it was first made in 1977, starting in Cowra then moving to Coonawarra, then partly to the Clare Valley, and ultimately (since 1990) being made from Piccadilly Valley grapes. The style of the wine has been refined over the period, but has remained remarkably consistent given the quite radically changing regional base.

🍷🍷🍷🍷 **1997** Very light green-yellow; an extremely delicate, unevolved and faintly minerally bouquet is followed by a delicate, crisp apple and melon-accent palate which needs some years to open and build. As it does, will doubtless rate higher points. **rating:** 88

best drinking 2002–2007 **best vintages** '87, '90, '91, '92, '95, '96 **drink with** Slow-roasted Tasmanian salmon • $37

Petaluma Croser

It was the desire to make a great sparkling wine which primarily drew Brian Croser to the Piccadilly Valley in the first instance. The climate is very cool, and ideally suited to the production of the fine base wine for Croser. First made in 1984 from 100% Chardonnay, but quickly moved to its current mix of approximately 50% Chardonnay and 50% Pinot Noir, although the mix between the two varies from 35%–65% (either side) according to the vintage.

🍷🍷🍷🍷🍷 **1996** A mix of minerally, bready, creamy biscuity and fruity characters magically welded together into a coherent and tightly structured palate. Melon, apple and mineral flavours are also to be found. **rating:** 95

best drinking 1999–2000 **best vintages** '86, '88, '90, '92, '94, '96 **drink with** Caviar • $35

Petaluma Sharefarmers Red

The Petaluma Sharefarmers wines now bear the word 'Coonawarra' on their label as the battle over the precise delineation of the boundaries of Coonawarra continues unabated. The wine is a highly variable mix centred on Merlot, Malbec and Cabernet Sauvignon, with the percentages changing markedly from year to year according to the dictates of the vintage.

🍷🍷🍷🍷 **1996** Medium red-purple; the bouquet is of medium intensity, showing a mix of berry, earth and more minerally characters supported by subtle oak. The palate is, as always, light and elegant, driven in this vintage by Merlot (more than 50%) and showing surprisingly little impact from the high level of Malbec (35%). **rating:** 84

best drinking 1998–2002 **best vintages** NA **drink with** Italian, light meat • $20

Petaluma Merlot

Two vintages are typically available at any one time, one on indent for physical delivery 12 months hence, and one representing the balance of the previous year's indent offering sold through normal commercial channels. An interesting way to go, with a substantial cost saving between the indent and normal release wine (the indent price being $30 per bottle). Both the '94 and '95 are at the top end of Australian red winemaking, without definitively answering the question concerning the taste of Merlot.

🍷🍷🍷🍷🍷 **1995** Strong red-purple; a fragrant bouquet showing positive varietal character ranging through berry, cedar and leaf leads into a powerful palate with ripe berry and plum flavours supported by lingering tannins. Classy oak handling adds to the structure of a lovely wine. **rating:** 94

best drinking 2000–2005 **best vintages** '94, '95, '96 **drink with** Veal • $45

Petaluma Coonawarra

A logical counterpart to the Chardonnay in the sense that it is far more elegant and refined than the more typical South Australian (and in particular, Coonawarra) Cabernet. Its regional base has remained the same since 1979, but the varietal composition has changed significantly, moving from Shiraz and Cabernet in '79 through to a Cabernet-dominant blend with around 15% Merlot. As with the Chardonnay, the more recent vintages are best.

🍷🍷🍷🍷 **1996** Medium purple-red; the bouquet is fine, with moderately sweet berry fruit, and the oak, as always, beautifully balanced and integrated. The palate is of medium intensity with clean red berry fruits followed by fine but persistent tannins. Should offer good drinking in the first few years of the next century. **rating:** 89

best drinking 2001–2006 **best vintages** '79, '86, '88, '90, '91, '92, '95 **drink with** Saddle of lamb • $42

peter lehmann ★★★☆

Para Road, Tanunda, SA 5352 **region** Barossa Valley

ph (08) 8563 2500 **fax** (08) 8563 3402 **open** Mon–Fri 9.30–5, weekends, holidays 10.30–4.30

winemaker Peter Lehmann, Andrew Wigan, Peter Scholz, Leonie Lange, Ian Hongell **prod.** 200 000 **est.** 1979

prod. range ($11–50 R) Eden Valley Riesling, Barossa Semillon, Barossa Chenin Blanc, Semillon Chardonnay, Chardonnay, Clancy's Classic Dry White, Botrytis Riesling, Noble Semillon, Grenache, Seven Surveys Dry Red, Barossa Shiraz, Barossa Cabernet Sauvignon,

Clancys Red, and Bin AD 2015 Vintage Port. Premium wines are Reserve Riesling, Reserve Chardonnay, Mentor, Stonewell Shiraz.

summary Public listing on the Stock Exchange has not altered the essential nature of the company, resolutely and single-mindedly focused on Peter Lehmann's beloved Barossa Valley. Some of the top-of-the-range wines are seriously good, the base range reliable rather than inspiring. Exports to the UK through its own subsidiary; also to New Zealand, Asia and the South Pacific and US.

Peter Lehmann Reserve Riesling

Peter Lehmann keeps various wines as Show Reserves, releasing them in whatever order he thinks best. Thus the '94 Show Reserve was released before this wonderfully delicious '93.

🍷🍷🍷🍷🍷 **1993** Glowing yellow-green; the bouquet shows wonderfully aromatic lime juice aromas, the palate being exceptionally harmonious with quite gorgeous fruit and great balance. Richly deserved its trophy success at the 1998 National Wine Show. **rating:** 96

best drinking 1999–2004 **best vintages** '93 **drink with** Prosciutto and melon • $16

Peter Lehmann Shiraz

The blood brother, as it were, to the Peter Lehmann Riesling. A traditional Barossa-style as honest as the day is long, but also reflecting the multitude of vineyard sources available to Peter Lehmann through his long association with the Valley.

🍷🍷🍷🍷 **1997** Medium to full red-purple; there is quite solid dark berry fruit on the bouquet, and the oak influence is appropriately restrained. Pleasant dark berry fruit, soft tannins and gentle oak provide good balance and structure to an attractive commercial red. **rating:** 86

best drinking 1999–2004 **best vintages** '89, '90, '91, '92, '94 **drink with** Spiced Barossa sausage • $17

Peter Lehmann Stonewell Shiraz

The wine is made solely from low-yielding old vineyards of the Stonewell, Ebenezer and Moppa subdistricts of the Barossa Valley. The fermentation is finished in new American oak, in which it is then matured for two years prior to bottling. It is then given three years bottle age before release. The wine has a supremely illustrious show record, with many major trophies to its credit.

🍷🍷🍷🍷½ **1994** Medium to full red-purple; the bouquet is at once complex yet smooth, with dark berry and chocolate fruit whipped through with lots of vanilla oak. The palate has abundant flavour in the dark berry spectrum; fine but evident tannins and good acidity help build structure and length. **rating:** 93

best drinking 2001–2011 **best vintages** '80, '89, '91, '92, '93, '94 **drink with** Kangaroo fillet • $50

Peter Lehmann Clancy's

A blend of Barossa-grown Shiraz, Cabernet Sauvignon, Cabernet Franc and Merlot, spiced with the American oak which is so much part of the Peter Lehmann style. Right from the outset, an unqualified success in the marketplace simply because it represents such good value for money, and because it is ready to drink when released.

🍷🍷🍷🍷 **1997** Medium red-purple; the bouquet is light, with fresh lolly juicy fruit aromas which gain power and amplify on the palate, augmented with soft, ripe tannins and subtle oak. **rating:** 85

best drinking 1999–2003 **best vintages** '91, '92, '94, '96 **drink with** Pasta with tomato or meat sauce • $15

Peter Lehmann Mentor

This wine is based predominantly on Cabernet Sauvignon, blended in varying proportions with Malbec, Merlot and Shiraz which fluctuate from year to year. Fermentation is finished in barrel and then matured for a further two and a half years in French and American oak hogsheads prior to bottling. Incidentally, Mentor is the new label for what was previously styled 'Cellar Collection Cabernet Blend'.

🍷🍷🍷🍷 **1995** Medium to full red-purple; the bouquet offers a mix of rich chocolate and vanilla with a sweet spicy overtone; a quite powerful wine on the palate, with good fruit richness and ripe tannins to close. **rating:** 88

best drinking 2000–2005 **best vintages** '80, '89, '91, '93, '94, '95 **drink with** Spiced beef • $35

Peter Lehmann Cabernet Sauvignon

Made in typical modern McLaren Vale-style, with sweet, ripe cabernet fruit swathed in a gown of American vanilla-accented oak, and the tannins held in restraint.

🍷🍷🍷🍷 **1996** Medium to full red-purple; the bouquet is powerful, with lots of cassis/berry fruit and plenty of American oak flavour. The palate similarly offers a good mix of earthy/berry fruit and vanilla oak; nice length and sweetness. Gold 1998 International Wine Challenge, London. **rating:** 89

best drinking 2000–2005 **best vintages** NA **drink with** Steak and kidney pie • $18

peter rumball wines NR

55 Charles Street, Norwood, SA 5067 **region** Warehouse

ph (08) 8332 2761 **fax** (08) 8364 0188 **open** Mon–Fri 9–5

winemaker Peter Rumball **prod.** 6000 **est.** 1988

prod. range ($12.90–49.90 R) Sparkling Shiraz, Vintage Pinot Noir Chardonnay Brut, The Pink.

summary Peter Rumball has been making and selling sparkling wine for as long as I can remember, but has led a somewhat peripatetic life, starting in the Clare Valley but now operating what I can only describe as a 'warehouse winery' operation, with neither vineyards nor winery of his own. The grapes are purchased and the wines made at various places under the supervision of Peter Rumball. His particular specialty has always been Sparkling Shiraz, and was so long before it became flavour of the month. National retail distribution through Tucker Seabrook, and exports to southeast Asia and US.

Peter Rumball SB11 Sparkling Shiraz

Made from 100% Shiraz, the traditional sparkling red base, half coming from Coonawarra, the remainder from the Barossa Valley and McLaren Vale. The wine spends between eight and 12 months on yeast lees, and is then hand-disgorged, with the addition of a special dosage liqueur at that stage.

🍷🍷🍷🍷 **NV** Dark red; attractive chocolate, red berry and faintly earthy aromas on the bouquet are followed by a lovely chocolatey berry palate, not too sweet, and indeed an excellent wine in its style. **rating:** 88

best drinking 1999–2004 **best vintages** NA **drink with** Roast duck • $20

peterson champagne house NR

Cnr Broke and Branxton Roads, Pokolbin, NSW 2320 **region** Lower Hunter Valley

ph (02) 4998 7881 **fax** (02) 4998 7882 **open** 7 days 9–5

winemaker Gary Reed **prod.** 7000 **est.** 1994

prod. range ($16–28 CD) Sparkling whites include First Creek, Sparkling Ambrosia, Chardonnay Pinot Noir, Semillon Pinot, Chardonnay Blanc de Blanc Millennium, Pinot Noir Chardonnay Pinot Meunier; sparkling reds, Sparkling Shiraz, Sparkling Chambourcin, Rouge Ambrosia and Sparkling Merlot; also table wine Chardonnay and Pinot Noir.

summary Prominently and provocatively situated on the corner of Broke and Branxton Roads, as one enters the main vineyard and winery district in the Lower Hunter Valley. It is an extension of the Peterson family empire, and no doubt very deliberately aimed at the tourist. While the dreaded word 'Champagne' has been retained in the business name, the wine labels now simply say Peterson House, which is a big step in the right direction. Almost all of the wine is sold through cellar door and through the wine club mailing list.

petersons ★★★☆

Mount View Road, Mount View, NSW 2325 **region** Lower Hunter Valley
ph (02) 4990 1704 **fax** (02) 4991 1344 **open** Mon–Sat 9–5, Sun 10–5
winemaker Colin Peterson **prod.** 15 000 **est.** 1971
prod. range ($13–65 CD) Semillon, Chardonnay, Pinot Noir, Shiraz, Cabernet Sauvignon, Sauternes, Vintage Port, Sparkling; Back Block Cabernet Sauvignon, Ian's Selection Cabernet Sauvignon and Glenesk Shiraz are top of the range.
summary After a period in the doldrums, Petersons seems to be resurgent, although there has been no change in the team. Certainly it retains a high reputation in the marketplace, sustained by wines such as the splendidly rich and concentrated Back Block dry reds. Production continues to rise in line with domestic and export markets.

pewsey vale ★★★☆

Brownes Road, Pewsey Vale, SA (vineyard only) **region** Eden Valley
ph (08) 8561 3200 **open** At Yalumba
winemaker Louisa Rose **prod.** 30 000 **est.** 1961
prod. range ($10–15 R) Riesling, Botrytis Riesling, Sauvignon Blanc, Cabernet Sauvignon.
summary Pewsey Vale was a famous vineyard established in 1847 by Joseph Gilbert, and it was appropriate that when S Smith & Son (Yalumba) began the renaissance of the high Adelaide Hills plantings in 1961, they should do so by purchasing Pewsey Vale and establishing 59 hectares of riesling and cabernet sauvignon. Once famous for its Riesling, recent vintages have not been inspiring, tending to be somewhat dilute and unfocused. Exports to all major markets.

Pewsey Vale Riesling

In the late '60s and early '70s one of the great names in Australian Riesling, the '69 vintage still one to bewitch with. In recent times has simply not shone, and perhaps the price is part of the explanation. Quite why Yalumba is content, I quite frankly do not know.

🍷🍷🍷🍸 **1998** Medium yellow-green; while the bouquet is not overly aromatic, it does have quite intense lime and slatey mineral aromas, the palate with quite brisk green apple and lime fruit, rounded off by firm acidity. Not opulent or fleshy, but does have an austere appeal. **rating:** 84

best drinking 2000–2005 **best vintages** NA **drink with** Whiting • $11.95

Pewsey Vale Cabernet Sauvignon

Produced from 13 hectares of estate plantings, and grown in an unequivocally cool climate, particularly when compared to the floor of the Barossa Valley. The site and the climate contribute to an elegant style which can be a little too herbaceous in the cooler vintages, but works well in the warmer years.

🍷🍷🍷🍷 **1997** Medium to full red-purple; quite fragrant mint, leaf and berry fruit on the bouquet is supported by just a hint of oak. The palate shows that slightly herbaceous tannin grip which is often present in the Pewsey Vale Cabernet, but there is also some sweet cassis fruit to provide a counterpoint. **rating:** 86

best drinking 2000–2005 **best vintages** '88, '92, '94 **drink with** Wild mushroom risotto • $15

---

pfeiffer ★★★

Distillery Road, Wahgunyah, Vic 3687 **region** Rutherglen
ph (02) 6033 2805 **fax** (02) 6033 3158 **open** Mon–Sat 9–5, Sun 11–4
winemaker Christopher Pfeiffer **prod.** 12 000 **est.** 1984
prod. range ($9.90–20 CD) Riesling, Auslese Tokay, Chardonnay Semillon, Chardonnay, Spätlese Frontignac, Ensemble (light Rosé-style), Pinot Noir, Shiraz, Shiraz Cabernet, Cabernet Sauvignon, Vintage Port, Old Distillery Tawny, Old Distillery Liqueur Gold (Tokay).

summary Ex-Lindeman fortified winemaker Chris Pfeiffer occupies one of the historic wineries (built 1880) which abound in northeast Victoria, and which is worth a visit on this score alone. The fortified wines are good, and the table wines have improved considerably over recent vintages, drawing upon 21 hectares of estate plantings. The winery offers barbecue facilities, children's playground, gourmet picnic hampers, and dinners (by arrangement).

phillip island vineyard ★★★★

Berrys Beach Road, Phillip Island, Vic **region** Gippsland
ph (03) 5956 8465 **fax** (03) 5956 8465 **open** 7 days 11–7 (Nov–March) 11–5 (April–Oct)
winemaker David Lance **prod.** 2500 **est.** 1993
prod. range ($14–31.60 CD) Sea Spray (Sparkling), Sauvignon Blanc, Cape Woolamai (Semillon Sauvignon Blanc), Summerland (Chardonnay), Newhaven (Riesling Traminer), The Pinnacles (Botrytis Riesling), The Nobbies (Pinot Noir), Berry's Beach (Cabernet Sauvignon), Western Port.
summary A separate operation of Diamond Valley Vineyards, now coming into full flower. 1997 marked the first harvest from the 2.5-hectares of the Phillip Island vineyard, totally enclosed in the permanent silon net which acts both as a windbreak and protection against birds. The quality of the wines across the board must be especially pleasing to the Lance family; this is definitely not a tourist trap cellar door, rather a serious producer of quality wine.

Phillip Island Sauvignon Blanc

Only 100 half-bottles of this wine were made in 1996, the first from Phillip Island. The '97 moved into commercial production scale, and promptly won two trophies (including Best White Wine) at the 1997 Lilydale Wine Show. The '98 is in a much lighter mould.

🍷🍷🍷🍷 **1998** The colour is so pale it is almost water-white; the bouquet is delicate, with light passionfruit aromas, and the palate likewise very much on the delicate side with crisp mineral and slightly fruitier characters. Beautifully made, but not overly concentrated. **rating:** 86

best drinking 1999–2000 **best vintages** '97 **drink with** Fresh crab • $25

Phillip Island Vineyard Chardonnay

This is the first Phillip Island Vineyard estate-grown and produced Chardonnay, previous releases having come from the Yarra Valley. Given the expense of viticulture on the island, which necessitate the entire vineyard being covered by permanent netting, the wine is modestly priced.

🍷🍷🍷🍷 **1997** Medium to full yellow-green; the bouquet exhibits classic melon and fig fruit, a hint of cashew and a nice touch of spicy oak. Toasty/bacony oak is unexpectedly evident on the forepalate, with the delicate fruit streaming out behind. **rating:** 85

best drinking 1999–2003 **best vintages** NA **drink with** Fresh seafood • $25

Phillip Island Vineyard Merlot

Yet another interesting and high-quality release from the estate plantings on Phillip Island, the success of which must be very gratifying to the Lance family.

🍷🍷🍷🍷 **1998** Good red-purple; there are attractive varietal aromas ranging through leaf, earth, mint and berry; these translate into the flavours of the palate, which has excellent texture with soft, fine tannins and restrained oak. There is a touch of greenness in the wine, as much varietal as it is anything else. **rating:** 88

best drinking 2001–2006 **best vintages** NA **drink with** Roast veal • $31.60

Phillip Island Vineyard Cabernet Sauvignon

This is the first Cabernet Sauvignon to be made from grapes grown on Phillip Island's estate vineyard, and is of exceptional interest for this reason alone. Interestingly, the label simply says 'Product of Australia', which might lead one to believe it is a blended wine from many regions – which it is not.

ꝩꝩꝩꝩ **1997** Youthful purple-red; the bouquet offers sweet cassis berry fruit, very subtle oak, and no green characters. The palate runs through blackberry, mulberry and cassis; the only sign of the cool climate comes in the fractionally green tannins on the finish. **rating:** 89

best drinking 2001–2006 **best vintages** '97 **drink with** Roast veal • $25

piano gully ★★☆

Piano Gully Road, Manjimup, WA 6258 **region** Pemberton
ph (08) 9772 3583 **fax** (08) 9771 2886 **open** Weekends, public holidays 10–5
winemaker Haydon White **prod.** 450 **est.** 1987
prod. range ($12–15 CD) Chardonnay, Pinot Noir, Cabernet Sauvignon, Concerto.
summary The 4-hectare vineyard was established in 1987 on rich Karri loam, 10 kilometres south of Manjimup, with the first wine made from the 1991 vintage. Wine quality to date has failed to impress.

pibbin NR

Greenhill Road, Balhannah, SA 5242 **region** Adelaide Hills
ph (08) 8388 4794 **fax** (08) 8398 0015 **open** Weekends 11–5.30
winemaker Roger Salkeld **prod.** 1500 **est.** 1991
prod. range ($15–22 CD) Pinot Noir, Rosé Pinot Noir, White Pinot, Sparkling Pinot.
summary The 7-hectare Pibbin vineyard, near Verdun, is managed on organic principles; owners Roger and Lindy Salkeld explain that the name 'Pibbin' is a corruption of a negro-spiritual word for Heaven, adding that while the wines may not have achieved that lofty status yet, the vineyard has. Pibbin has made a name for itself for producing massive, dense Pinot Noir in a style radically different from that of the rest of the Adelaide Hills.

picardy ★★★★

Cnr Vasse Highway and Eastbrook Road, Manjimup, WA 6260 **region** Pemberton
ph (08) 9779 0036 **fax** (08) 9776 0245 **open** By appointment
winemaker Bill Pannell, Dan Pannell **prod.** 4000 **est.** 1993
prod. range ($28 CD) Chardonnay, Pinot Noir, Shiraz, Merlot Cabernet.
summary Picardy is owned by Dr Bill Pannell and his wife Sandra, who were the founders of Moss Wood winery in the Margaret River region (in 1969). Picardy reflects Bill Pannell's view that the Pemberton area will prove to be one of the best regions in Australia for Pinot Noir and Chardonnay, but it is perhaps significant that the wines to be released include a Shiraz, and a Bordeaux-blend of 50% Merlot, 25% Cabernet Franc and 25% Cabernet Sauvignon. Time will tell whether Pemberton has more Burgundy, Rhône or Bordeaux in its veins. It has lost no time in setting up national distribution, and exports to UK, US and Hong Kong.

Picardy Pinot Noir

1997 marked the second vintage of Pinot Noir from Picardy, and (for the first time) contains a significant proportion of the Burgundian clone 115. As the Pannells themselves comment, the wine has definitely benefited from this clone, showing greater complexity of flavour and more pronounced varietal character than the inaugural 1996 release.

ꝩꝩꝩꝩ½ **1997** Medium red-purple; the fragrant bouquet shows most attractive strawberry and plum fruit, with excellent varietal definition, and appropriately subtle oak. The palate has plenty of presence and concentration, with sweet notes from the French oak rather more evident than on the bouquet, finishing with soft tannins. Absolutely delicious now, but may prove to be fast-developing. **rating:** 92

best drinking 1998–2000 **best vintages** NA **drink with** Braised pheasant • $28

Picardy Shiraz

The first crop of shiraz from Picardy produced only 300 cases of wine, but is an exceptionally promising start.

🍷🍷🍷🍷🍷 **1997** Medium to full red-purple; the bouquet is complex, with some gamey varietal characters similar to those found in the northern Rhône Valley. The palate has good depth and concentration, with ripe black cherry fruit, soft, lingering tannins and very subtle oak. **rating:** 90

best drinking 1999–2003 **best vintages** NA **drink with** Moroccan lamb • $28

Picardy Merlot Cabernet

1997 was also the first release of this wine, a blend of 55% Merlot, 30% Cabernet Franc and 15% Cabernet Sauvignon.

🍷🍷🍷🍷 **1997** Strong red-purple; while not particularly aromatic, the bouquet is powerful, with blackberry/cherry fruit. The palate likewise shows considerable power and concentration, with well-balanced tannins, yet is still very tied up and closed. May well blossom with age. **rating:** 89

best drinking 2002–2007 **best vintages** NA **drink with** Rare beef • $28

piccadilly fields NR

185 Piccadilly Road, Piccadilly, SA 5151 **region** Adelaide Hills
ph (08) 8272 2239 **fax** (08) 8232 5395 **open** Not
winemaker Sam Virgara **prod.** 3000 **est.** 1989
prod. range ($17.95 ML) Chardonnay, Merlot Cabernet Franc Cabernet Sauvignon.
summary Piccadilly Fields draws upon a very substantial vineyard, with much of the production being sold to Petaluma. The plantings include 10 hectares of pinot meunier, 8 hectares of pinot noir, 5 hectares each of chardonnay, merlot and sauvignon blanc, 2 hectares of cabernet franc and 1 hectare of cabernet sauvignon.

pierro ★★★★☆

Caves Road, Willyabrup via Cowaramup, WA 6284 **region** Margaret River
ph (08) 9755 6220 **fax** (08) 9755 6308 **open** 7 days 10–5
winemaker Dr Michael Peterkin **prod.** 7500 **est.** 1979
prod. range ($24.90–49.50 R) Chardonnay, LTC Semillon Sauvignon Blanc, Pinot Noir, Cabernets.
summary Dr Michael Peterkin is another of the legion of Margaret River medical practitioners who, for good measure, married into the Cullen family. Pierro is renowned for its stylish white wines, which often exhibit tremendous complexity. The Chardonnay can be monumental in its weight and complexity. The wines are exported to the UK, US, Japan and Indonesia.

Pierro LTC Semillon Sauvignon Blanc

Made from relatively low-yielding estate-grown grapes, and originally marketed under the 'Les Trois Cuvées' label, now abbreviated to LTC, which (coincidentally) can also stand for 'a little touch of Chardonnay'. The wine tastes as if it may have been wholly or partially barrel-fermented, though no mention of this is made on the label.

🍷🍷🍷🍷 **1998** Medium yellow-green; the bouquet is solid, with pleasantly ripe melon, nectarine and citrus fruit. The palate, too, is generous, softly ripe and concentrated, yet sustained by a touch of mineral on the finish. An interesting style. **rating:** 88

best drinking 1999–2002 **best vintages** '87, '89, '90, '94, '95, '97 **drink with** Veal cutlets • $24.80

Pierro Chardonnay

One of the most distinguished of a band of striking wines from the Margaret River region and which achieved great acclaim during the second half of the 1980s. The style is invariably complex, concentrated and powerful, with the emphasis on secondary rather than primary fruit characters.

ŸŸŸŸ **1997** Excellent full yellow-green; the bouquet is quite rich and solid, with figgy/nutty aromas which are followed by a comparably rich and mouthfilling palate. The 14.5 degrees alcohol is just a little over the top. **rating:** 89

best drinking 2000–2005 **best vintages** '86, '87, '89, '90, '92, '94, '95, '96 **drink with** Seafood pasta • $49.50

Pierro Cabernets

A challenging, high-quality blend of the five red varieties of Bordeaux: Cabernet Sauvignon, Cabernet Franc, Merlot, Petit Verdot and Malbec. The wine spends 40 days and 40 nights in contact with its skins ('making it a wine of biblical proportions' says Michael Peterkin) and spends 18 months in new French oak. A serious wine made in serious style, with relatively little vintage variation.

ŸŸŸŸ **1996** Medium red, with just a touch of purple; both the bouquet and palate are less formidable than previous releases, but all the better for that, with an excellent balance between red berry fruits and the tannins are softer (almost chocolatey) on the finish. **rating:** 88

best drinking 2003–2009 **best vintages** '94, '95 **drink with** Boned leg of lamb • $49.50

piesse brook NR

226 Aldersyde Road, Bickley, WA 6076 **region** Perth Hills

ph (08) 9293 3309 **fax** (08) 9443 2839 **open** Sat 1–5, Sun, public holidays 10–5 and by appointment

winemaker Di Bray, Ray Boyanich (Michael Davies Consultant) **prod.** 1000 **est.** 1974

prod. range ($10–17.50 CD) Chardonnay, Shiraz, Merlot, Cabernet Sauvignon, Cabernet Merlot, Cabernet Shiraz, Cabernova (early-drinking style).

summary Surprisingly good red wines made in tiny quantities, and which have received consistent accolades over the years. The first Chardonnay was made in 1993; a trophy winning Shiraz was produced in 1995. Now has 4 hectares of chardonnay, shiraz, merlot and cabernet sauvignon under vine.

pieter van gent ★★★

Black Springs Road, Mudgee, NSW 2850 **region** Mudgee

ph (02) 6373 3807 **fax** (02) 6373 3910 **open** Mon–Sat 9–5, Sun 11–4

winemaker Pieter van Gent, Philip van Gent **prod.** 10 000 **est.** 1978

prod. range ($9.50–16.90 CD) The only dry wines are the Chardonnay, Müller Thurgau, and Cabernet Sauvignon; the Frontignac, Rivaner, Angelic White and Sundance Soft Red all have varying degrees of sweetness; fortified wines are the specialty including Pipeclay Port, Mudgee White Port, Cornelius Port, Mudgee Oloroso, Pipeclay Muscat, Mudgee Liqueur Frontignac, Pipeclay Vermouth.

summary Many years ago, Pieter van Gent worked for Lindemans, before joining Craigmoor, then moving to his own winery in 1979 where he and his family have forged a strong reputation and following for his fortified wines in particular, although the range extends far wider. The wines are seldom seen outside cellar door.

pikes ★★★★

Polish Hill River Road, Sevenhill, SA 5453 **region** Clare Valley

ph (08) 8843 4370 **fax** (08) 8843 4353 **open** 7 days 10–4

winemaker Neil Pike **prod.** 30 000 **est.** 1984

prod. range ($14–40 R) Riesling, Reserve Riesling, Sauvignon Blanc, Chardonnay, Premio Sangiovese, Shiraz, Reserve Shiraz, Cabernet Sauvignon.

Seppelt Show Reserve Muscat DP63

An extraordinary show reserve wine of great age which invariably scores high gold medal points every time I taste it. The following note (and points) were made in March 1999.

🍷🍷🍷🍷🍷 **NV** Deep mahogany brown, with a mix of gold and green on the rim. Rich, raisined, spicy plum pudding aromas are lifted by perfectly balanced and integrated spirit. The powerful, complex and rich palate – plum pudding and Christmas cake – has a very long, lingering finish.

rating: 96

best drinking 1999–2000 **best vintages** NA **drink with** Dried fruits • $24.95

Seppelt Amontillado Sherry DP116

Just as winemaker James Godfrey is forever seeking ways to make the Show Fino finer and more delicate (without losing character) so he believes in fine Amontillados with sweetness on the mid-palate, but a cleansing, dry finish.

🍷🍷🍷🍷🍷 **NV** Bright mid-gold; the bouquet has a lovely touch of honey over the bite of the rancio; the very elegant, very fresh palate finishes distinctly dry after the mellowness of the bouquet and mid-palate. This is as it should be.

rating: 94

best drinking 1999–2000 **best vintages** NA **drink with** A great winter aperitif • $20.25

Seppelt Show Fino Sherry DP117

DP117 has reigned supreme in the Australian show circuit for decades. It is made using the traditional solera system, resulting in a seamless blend of wines of varying age, protected by the flor yeast which floats on the surface of the wine and is responsible for its unique taste. This driest-of-all Sherries is also the finest, and recent changes in the law have allowed Seppelt winemaker James Godfrey to make the wine even finer by reducing the alcohol content from around 17% to around 15.5%. There is nothing better than a chilled glass of Fino on a hot summer's day. It is essential, incidentally, that the stock be fresh, and that the bottle be kept in the refrigerator after it is opened, and consumed within two or three days of opening.

🍷🍷🍷🍷🍷 **NV** Brilliant green-yellow; the bouquet is strong and stylish, with that faintly nutty, faintly tangy cut which is the hallmark of Fino Sherry. The palate is intense and racy, the flavour lingering in the mouth long after the wine is swallowed, but not so long to stop you taking the next mouthful.

rating: 96

best drinking 1999–2000 **best vintages** NA **drink with** Olives, tapas • $20.25

Seppelt Show Oloroso Sherry DP38

The sweetest of the Seppelt Sherries, yet even here (in the Spanish tradition) sweetness and dryness are delicately balanced.

🍷🍷🍷🍷🍷 **NV** Golden brown; nutty rancio complexity, with just a hint of sweetness, introduce a finely balanced palate with a constant interplay between nutty, honeyed sweetness and drier, rancio characters.

rating: 95

best drinking 1999–2000 **best vintages** NA **drink with** Sweet biscuits • $20.25

Seppelt Show Tawny Port DP90

DP90 has an average age of 21 years, blended from the reserve stocks of very old Tawny Port made from Barossa Valley Shiraz and Grenache, with a little Cabernet Sauvignon. Between 1968 and 1991 alone DP90 won 30 trophies and 106 gold medals at Australian wine shows, making it the most-awarded wine of any style. It is intermittently released in limited quantities. No other Australian Tawny challenges the sheer complexity and finesse of DP90. It is a blend of wines that span more than 100 years, aged in all kinds of barrels (none new, of course) and only a limited quantity is bottled each year to preserve the balance of very old material in the blend.

🍷🍷🍷🍷🍷 **NV** The tawny hues are rimmed with olive-green, immediately proclaiming the age of the wine. The bouquet is fine, fragrant and penetrating, much closer to the Tawny Ports of Portugal than most Australian wines. The palate offers flavours of spice, butterscotch and more

summary Owned by the Pike brothers, one of whom (Andrew) was for many years the senior viticulturist with Southcorp, the other (Neil) a former winemaker at Mitchells. Pikes now has its own winery, with Neil Pike presiding. Generously constructed and flavoured wines are the order of the day. The wines are exported to the UK, US, Canada, Japan, Switzerland and Belgium.

Pikes Riesling

Produced from a little over 3.5 hectares of estate-grown Riesling. In the mainstream of Clare Valley-style, crisp, lively and zesty when young, and developing considerable character with bottle age.

ŸŸŸŸ **1998** Light to medium yellow-green; while the primary fruit aromatics are subdued, with strong mineral, toast and a hint of spice uppermost, the fruit will grow. The palate offers a typically solid, robust Riesling with the strength to go anywhere. **rating:** 89

best drinking 2000–2005 **best vintages** '86, '90, '92, '93, '95 **drink with** Lightly spiced chicken salad • $14

pinelli NR

18 Bennett Street, Caversham, WA 6055 **region** Swan District
ph (08) 9279 6818 **fax** (08) 9377 4259 **open** 7 days 10–6
winemaker Robert Pinelli **prod.** 7000 **est.** 1979

prod. range ($5–16 CD) Limited table wine range centred on Chenin Blanc, Chardonnay, Shiraz and Cabernet Sauvignon, and an extensive range of fortified wines including Cabernet-based Vintage Port. The wines have won a number of medals at the Perth Show in recent years.

summary Dominic Pinelli and son Robert – the latter a Roseworthy Agricultural College graduate – sell 75% of their production in flagons, but are seeking to place more emphasis on bottled-wine sales in the wake of recent show successes with Chenin Blanc.

pipers brook vineyard ★★★★☆

Bridport Road, Pipers Brook, Tas 7254 **region** Northern Tasmania
ph (03) 6382 7527 **fax** (03) 6382 7226 **open** Mon–Fri 10–5, weekends 11–5
winemaker Andrew Pirie **prod.** 60 000 **est.** 1974

prod. range ($16–55 CD) The wine is released in two tiers: Pipers Brook Chardonnay, Riesling, Pinot Gris, Gewurztraminer, Opimium, Pellion (Pinot Noir), and second label Ninth Island Chardonnay, Riesling, Chardonnay, Straits Dry White, Rosé, Tamar Cabernets, Pinot Noir, Pirie Cuvée.

summary The Pipers Brook Tasmanian empire has continued to grow apace. It now has 220 hectares of vineyard supporting the Pipers Brook, Heemskerk and Rochecombe labels, with the major focus, of course, being on Pipers Brook. The wines are exported to the UK, US, Japan, Canada and Singapore, and are distributed throughout Australia by S Smith & Son. However a considerable part of its sales are to its shareholders, who receive a substantial discount, and who have supported the company through to its public status on the Australian Stock Exchange. As ever, fastidious viticulture and winemaking, immaculate packaging and enterprising marketing constitute a potent and effective blend.

Pipers Brook Riesling

First made in 1979. The 3 hectares of riesling at the home Pipers Brook Vineyard are situated on the favourable north- and northeast-facing aspects, and are now approaching 20 years of age. The wine is excellent, and invariably develops well with prolonged cellaring.

ŸŸŸŸŸ **1998** Light to medium yellow-green; there is fine, lime/tropical fruit in abundance on the bouquet; the elegant, tightly structured and long palate presents quite lovely lime-accented flavours, and guaranteeing a long life. One of the best Pipers Brook Rieslings for years. **rating:** 96

best drinking 2002–2012 **best vintages** '82, '91, '92, '93, '94, '98 **drink with** Lemon chicken salad • $23.50

Pipers Brook Chardonnay

First made in 1981. Pipers Brook is now able to draw upon three distinct estate-owned vineyard sources: Pipers Brook Vineyard, Pellion Vineyard and Ninth Island Vineyard. The expanded base has strengthened an already distinguished wine which is made using the full range of techniques including barrel fermentation and partial malolactic fermentation. The style is always elegant, and longer-lived than Australian Chardonnays grown in warmer regions. Incidentally, these wines typically achieve high sugar (and hence alcohol) levels of about 13%.

ŸŸŸŸ **1998** Light to medium yellow-green; the bouquet offers clean, citrus and nectarine fruit matched by subtle oak; there is no obvious mlf influence, nor any botrytis. The crisp, delicate palate proclaims its cool-grown origins, but has the balance and length to mature and grow in bottle. All it needs is time. **rating:** 88

best drinking 2002–2008 **best vintages** '82, '88, '91, '92, '93, '94, '97, '98 **drink with** Veal in white sauce • $29

Pipers Brook Summit Chardonnay

The Summit Chardonnay release is effectively the Pipers Brook reserve, made only in exceptional vintages and in small quantities.

ŸŸŸŸŸ **1997** Light to medium yellow-green; the bouquet is tight and complex, although the oak is quite evident. The palate is, if anything, even more tightly restrained, with mineral, cashew and citrus all running along the base of the wine. Should be exceptionally long-lived. **rating:** 94

best drinking 2004–2014 **best vintages** '97 **drink with** Grilled Tasmanian lobster • $55

Pipers Brook Pirie Cuvée

A blend of 70% Pinot Noir and 30% Chardonnay, the major part of which is fermented in old oak barrels. The wine is taken through malolactic-fermentation and spends more than two years on yeast lees prior to disgorgement.

ŸŸŸŸY **1995** Bright, light green-yellow with fine mousse; the bouquet is extremely complex, with tangy, slightly herbal fruit intermingling with a hint of aldehyde. The powerful palate has intense, tangy citrus and herb flavours, finishing with high acidity. Looks as if it will benefit from further time on cork. **rating:** 93

best drinking 1999–2005 **best vintages** NA **drink with** Tasmanian oysters • $44

Pipers Brook Pellion

First made in 1981 (a bucketful of remarkable wine) but changed its name to Pellion only in 1992. Pellion was an artist on one of the very early voyages of discovery to Tasmania, hence the name; notwithstanding the absence of any varietal claim on the label, the wine is in fact 100% Pinot Noir. Winemaker Andrew Pirie says, 'the expression of this grape in the red soils and climate of the region is so individual that we do not think we should be constrained by names which lead to preconceived ideas as to the taste of the wine'.

ŸŸŸŸY **1997** Light to medium red-purple; a fragrant bouquet with a mix of cherry and more sappy/green characters is followed by a long palate, with cherry fruit, subtle oak and quite firm acid on the finish in true Tasmanian style. **rating:** 90

best drinking 1999–2003 **best vintages** '81, '85, '91, '92, '94, '96, '97 **drink with** Ripe King Island brie • $29

Pipers Brook Opimium

The top Bordeaux-blend in the Pipers Brook stable, now largely sourced from its Tamar Valley vineyard holdings.

ŸŸŸŸ **1997** Medium to full red-purple; the bouquet offers a mix of dark berry, fern and green leaf aromas supported by subtle oak; the palate is powerful but dominated by flavours in the olive/leaf/fern/grass spectrum. **rating:** 85

best drinking 2001–2005 **best vintages** NA **drink with** Roast veal • $33

pirramimma ★★★★

Johnston Road, McLaren Vale, SA 5171 **region** McLaren Vale

ph (08) 8323 8205 **fax** (08) 8323 9224 **open** Mon–Fri 9–5, Sat 11–5, Sun, public holidays 11.30–4

winemaker Geoff Johnston **prod.** 30 000 **est.** 1892

prod. range ($10–20 R) Stock's Hill Semillon Chardonnay, Adelaide Hills Semillon, McLaren Vale Semillon, Stock's Hill Semillon Chardonnay, Hillsview Chardonnay, Stock's Hill Shiraz, Petit Verdot, Hillsview Cabernet Merlot, Cabernet Sauvignon, Ports.

summary An operation with large vineyard holdings of very high quality, and a winery which devotes much of its considerable capacity to contract-processing of fruit for others. In terms of the brand, has been a consistent under-performer during the 1990s. The marketing of the brand does scant justice to the very considerable resources available to it, notably its gold medal Petit Verdot and fine elegant Chardonnay. Exports to the UK and US.

pizzini NR

King Valley Road, Wangaratta, Vic 3768 **region** King Valley

ph (03) 5729 8278 **fax** (03) 5729 8495 **open** 7 days 10–4

winemaker Alfred Pizzini, Joel Pizzini, Mark Walpole **prod.** 4000 **est.** 1980

prod. range ($12–30 CD) Riesling, Sauvingon Blanc, Alfred Pizzini Chardonnay, Sangiovese, Nebbiolo, Alfred Pizzini Shiraz Cabernet, Cabernet.

summary Fred and Katrina Pizzini have been grape growers in the King Valley for over 20 years with over 60 hectares of vineyard. Grape growing (rather than winemaking) still continues to be the major focus of activity, but their move into winemaking has been particularly successful, and I can personally vouch for their Italian cooking skills. It is not surprising, then, that their wines should span both Italian and traditional varieties.

Pizzini Sangiovese

Sangiovese, of course, is the grape of Tuscany, and is one of the Italian varieties which the grape growers of the King Valley of Italian extraction have planted with relish. Just because it has a link back home does not give it any inherent status, but this is a very nice wine.

ΥΥΥΥ **1997** Medium red-purple; there are quite high-toned, slightly sharp-edged fruit aromas on the bouquet with minimal oak influence. The palate opens with similar high-toned cherry pip flavours but then rolls on to softer notes with sweet, dusty tannins on the finish. **rating:** 89

best drinking 2001–2005 **best vintages** NA **drink with** Pizza, of course • $16

plantagenet ★★★★☆

Albany Highway, Mount Barker, WA 6324 **region** Great Southern

ph (08) 9851 2150 **fax** (08) 9851 1839 **open** Mon–Fri 9–5, weekends 10–4

winemaker Gavin Berry, Gordon Parker **prod.** 40 000 **est.** 1974

prod. range ($11–45 CD) Riesling, Omrah Sauvignon Blanc, Omrah Chardonnay (unoaked), Mount Barker Chardonnay, Fronti, Fine White, Fine Red, Pinot Noir, Shiraz, Henry II, Cabernet Sauvignon, Mount Barker Brut.

summary The senior winery in the Mount Barker region which is making superb wines across the full spectrum of variety and style – highly aromatic Riesling, tangy citrus-tinged Chardonnay, glorious Rhône-style Shiraz, ultra-stylish Cabernet Sauvignon and an occasional inspiring Pinot Noir. Exports to the US, UK, Germany, Austria, Singapore, Japan, Switzerland and Hong Kong.

Plantagenet Mount Barker Riesling

Draws upon 6.2 hectares of estate vineyards, all of which are now fully mature, and which (along with a similar amount of Cabernet Sauvignon) constitute the major estate plantings. First made in 1975, one of the flagships not only for Plantagenet but for the region as a whole.

🍷🍷🍷🍷🍸 **1998** Light green-yellow; the bouquet is clean, with lime/mineral fruit in a more restrained mould than usual. The palate is similarly light, crisp and clean, with a mix of delicate, floral lime and passionfruit flavours. **rating:** 90

best drinking 2000–2005 **best vintages** '81, '83, '86, '92, '94, '95, '96, '98 **drink with** Most Asian dishes • $16.50

Plantagenet Mount Barker Chardonnay

Usually draws upon four different vineyards in the Mount Barker region deliberately harvested at varying levels of ripeness to increase the flavour complexity. It is barrel-fermented in a mix of Vosges, Allier and Troncais oak, with a percentage taken through malolactic fermentation. It is invariably a generously proportioned wine, while still retaining some of the tighter, cool-grown characters one would expect.

🍷🍷🍷🍷 **1998** Medium yellow-green; the bouquet is of light to medium intensity, with peach and melon fruit gently supported by oak. The palate is restrained, with citrus/melon fruit, and an attractively crisp finish. Could well develop into something special over the next few years. **rating:** 87

best drinking 2000–2005 **best vintages** '81, '83, '86, '92, '94, '95 **drink with** Breast of chicken • $24

Plantagenet Mount Barker Shiraz

It draws upon 4 hectares of estate plantings, the style varying with the vintage. In cooler years the vibrant pepper/spice characters are to the fore, in warmer years it tends more to liquorice, dark cherry and berry. The one consistent feature is the quality of the wine. Spends 18 months in 100% French oak, the majority first and second use. The '96 is a welcome return to top form after an indifferent '95.

🍷🍷🍷🍷🍸 **1996** Medium to full red-purple; a quite potent bouquet with solid berry and earth varietal character is followed by a palate redolent of sweet, black cherry and blueberry fruit; succulent and satisfying. **rating:** 93

best drinking 2001–2010 **best vintages** '82, '83, '85, '90, '91, '93, '94, '96 **drink with** Hare, squab • $29.50

Plantagenet Cabernet Sauvignon

First made in 1974 (in fact at Sandalford in the Swan Valley) and has established itself as one of the West Australian classics over the intervening years. While primarily based upon estate-grown Cabernet Sauvignon, Malbec, Cabernet Franc and Merlot have all contributed to the wine over the last decade, with the core of the wine coming from the Plantagenet Bouverie Vineyard at Denbarker.

🍷🍷🍷🍷 **1996** Medium red-purple; the bouquet is fairly light and relatively earthy, with not overmuch berry or cassis fruit. The palate lacks flesh, although the soft tannin balance for the light, red berry fruit is on the mark. **rating:** 85

best drinking 1999–2003 **best vintages** '81, '83, '85, '86, '90, '91, '94 **drink with** Rack of lamb • $25.30

---

platt's ★★☆

Mudgee Road, Gulgong, NSW 2852 **region** Mudgee

ph (02) 6374 1700 **fax** (02) 6372 1055 **open** 7 days 9–5

winemaker Barry Platt **prod.** 4000 **est.** 1983

prod. range ($9–12 CD) Chardonnay, Semillon, Gewurztraminer, Cabernet Sauvignon.

summary Inconsistent and often rather unhappy use of oak prevents many of the wines realising their potential.

plunkett ★★★

Lambing Gully Road, Avenell, Vic 3664 **region** Central Victorian High Country
ph (03) 5796 2150 **fax** (03) 5796 2147 **open** 7 days 11–5
winemaker Sam Plunkett **prod.** 10 000 **est.** 1980
prod. range ($15–20 CD) The top-of-the-range wines are released under the Strathbogie Ranges label (Riesling, Chardonnay, Merlot, Cabernet Merlot); standard wines under the Blackwood Ridge brand of Riesling, Gewurztraminer, Unwooded Chardonnay, Sauvignon Blanc Semillon, and Shiraz.
summary The Plunkett family first planted grapes way back in 1968, establishing 3 acres with 25 experimental varieties. Commercial plantings commenced in 1980, with 100 hectares now under vine. While holding a vigneron's licence since 1985, the Plunketts did not commence serious marketing of the wines until 1992, and has now settled down into producing an array of wines which are pleasant and well-priced. The wines are exported to the US, UK and Malaysia.

pokolbin estate ★★☆

McDonalds Road, Pokolbin, NSW 2321 **region** Lower Hunter Valley
ph (02) 4998 7524 **fax** (02) 4998 7765 **open** 7 days 10–6
winemaker Contract **prod.** 3500 **est.** 1980
prod. range ($15–40 CD) Semillon, Riesling, Pokolbin Horse Coaches Verdelho, Unwooded Chardonnay, Show Reserve Chardonnay, Pinot Noir, Shiraz Merlot, Shiraz; Port.
summary An unusual outlet, offering its own-label wines made under contract by Trevor Drayton, together with the wines of Lake's Folly, Peacock Hill and Pothana, and with cheap varietal 'cleanskins'. Wine quality under the Pokolbin Estate label is very modest, although the 1997 Hunter Riesling (perversely, true Riesling, not Semillon) won a silver medal and was the top-pointed wine in its class at the 1997 Hunter Valley Wine Show.

poole's rock ★★★☆

Lot 41 Wollombi Road, Broke, NSW 2330 **region** Lower Hunter Valley
ph (02) 9667 1622 **fax** (02) 9667 1442 **open** Not
winemaker Philip Ryan (Contract) **prod.** 4000 **est.** 1988
prod. range ($20–24 R) Chardonnay.
summary Sydney merchant banker David Clarke has had a long involvement with the wine industry, including a directorship of McGuigan Brothers Limited. The 5-hectare Poole's Rock vineyard, planted purely to chardonnay, is his personal venture. The wine has retail distribution throughout Australia, and is exported to the UK.

Poole's Rock Chardonnay

Right from the outset, this 100% estate-grown wine has been made in full-blown fashion by contract-winemaker Iain Riggs (of Brokenwood). The propensity of Hunter Valley Chardonnay to produce a rich, buttery, peachy wine was given free rein, complexed by partial malolactic fermentation and maturation in French Vosges oak barriques. The '98 broke through with a gold medal at the fiercely contested 1998 Hunter Valley Wine Show. The wine was matured in 27% new French oak barrels, the remainder being between one and five years old. Thirteen per cent was taken through malolactic fermentation, and it spent six months on lees.

🍷🍷🍷🍷🍷 **1998** Glowing yellow-green; a full, complex, peach and honey bouquet with relatively subtle oak. Relative to the bouquet, and to the propensity of Hunter Valley Chardonnays to blow out, is really quite restrained, if a heavyweight prize-fighter can be so described. **rating:** 90

best drinking 1999–2000 **best vintages** '92, '93, '95, '98 **drink with** Creamy pasta • $22

poplar bend NR

RMB 8655 Main Creek Road, Main Ridge, Vic 3928 **region** Mornington Peninsula
ph (03) 5989 6046 **fax** (03) 5989 6460 **open** Weekends and public holidays 10–5, also by appointment
winemaker David Briggs **prod.** 350 **est.** 1988
prod. range ($16–28 ML) Pineau Chloe, Cabernet Chloe, Sparkling Chloe, Pinot Noir, Cellar Reserve Pinot Noir, Cabernet Shiraz.
summary Poplar Bend was the child of Melbourne journalist, author and raconteur Keith Dunstan and wife Marie, who moved into full-scale retirement in 1997, selling Poplar Bend to David Briggs. The changes are few; the label still depicts Chloe in all her glory, which could be calculated to send the worthy inhabitants of the Bureau of Alcohol, Tobacco and Firearms (of the US) into a state of cataleptic shock.

port phillip estate ★★★★★

261 Red Hill Road, Red Hill, Vic 3937 **region** Mornington Peninsula
ph (03) 5989 2708 **fax** (03) 5989 2891 **open** Weekends and public holidays 12–5
winemaker Lindsay McCall (Contract) **prod.** 4000 **est.** 1987
prod. range ($20–32 R) Sauvignon Blanc, Chardonnay, Pinot Noir, Reserve Pinot Noir, Reserve Shiraz.
summary Established by leading Melbourne QC Jeffrey Sher, who, having briefly flirted with the idea of selling Port Phillip, has decided to continue – not surprising, given the quality of the wines. Lindsay McCall proves his success at Paringa Estate is no fluke; the Port Phillip wines are also excellent, hitting the high point in 1997 which is unlikely to ever be exceeded.

Port Phillip Estate Sauvignon Blanc

Yet another good wine from Port Phillip Estate, made without artifice and simply relying on the cool Mornington Peninsula climate. Cold-fermented in stainless steel, and bottled early to retain maximum fruit freshness.

YYYY **1998** Light green-yellow; the bright, fresh and crisp bouquet is grassy rather than tropical, and while the palate shows the same herbal, grassy fruit, it has intensity and length. **rating:** 86

best drinking 1999–2000 **best vintages** NA **drink with** Shellfish • $22

Port Phillip Estate Chardonnay

Produced from a little over 1.8 hectares of chardonnay planted in 1988. High natural acidity often forces Mornington Peninsula winemakers to rely on the malolactic fermentation to a considerable degree, and not always to the benefit of the wines. One of the attractions of this wine is that the malolactic characters are not overdone.

YYYY **1997** Medium yellow-green; the bouquet is moderately intense, with clean citrus and melon fruit and gentle mlf barrel-ferment oak aromas. In surprising contrast to the powerhouse red wines in '97, the palate is elegant, but has length and persistence. **rating:** 88

best drinking 2000–2005 **best vintages** '96, '97 **drink with** Lobster, shellfish • $24

Port Phillip Estate Reserve Pinot Noir

A challenging but majestic wine which is a testament both to the exceptional quality of the '97 vintage in the Mornington Peninsula and to Lindsay McCall's skills as a winemaker. In the space of a few weeks it won two trophies at the 1999 Sydney Royal Wine Show including the Trophy for the Best 1997 Red Table Wine of Show, as well as the Trophy for the Best Table Wine Exhibited by a Small Producer. Had it been nominated directly, it would also have won the trophy for Best Pinot Noir of Show. Only a week earlier it had topped the 1999 *Winewise* Small Makers Competition Pinot Noir Class. Like the Paringa Estate of '97, it is an exceptionally concentrated and powerful wine which pushes the Pinot Noir envelope just about as far as is possible.

ŸŸŸŸŸ **1997** Full red-purple; the bouquet is exceptionally intense and complex with deep, dark plum and briar aromas. Opulent plummy pinot fruit, rich and mouthfilling, is offset by a nice touch of spice, partly from the fruit and partly from oak. The wine also has considerable tannin support which is more likely to aid than harm it as it matures further. This is an all duck or no dinner style. **rating:** 97

best drinking 1999–2004 **best vintages** NA **drink with** Duck • $30

Port Phillip Estate Reserve Shiraz

A twin to the Reserve Pinot Noir of the same vintage, and in precisely the same way attests to the outstanding year and the equally outstanding skills of Lindsay McCall as winemaker.

ŸŸŸŸŸ **1997** Medium to full red-purple; a marvellous tapestry of ripe and rich dark cherry, spice and liquorice fruit aromas on the bouquet logically leads into a massively powerful and complex wine on the palate with abundant tannin. Should be very long-lived. **rating:** 96

best drinking 2002–2012 **best vintages** '97 **drink with** Braised oxtail • $32

portree ★★★☆

Powells Track via Mount William Road, Lancefield, Vic 3455 **region** Macedon

ph (03) 5429 1422 **fax** (03) 5429 2205 **open** Weekends and public holidays at Antique Centre, Lancefield 11–5

winemaker Ken Murchison **prod.** 1500 **est.** 1983

prod. range ($15–26 ML) Chardonnay, Greenstone, Damask (Cabernet Franc Rosé), Quarry Red (Cabernet Franc Merlot).

summary Owner Ken Murchison selected his 4-hectare Macedon vineyard after studying viticulture at Charles Sturt University and being strongly influenced by Dr Andrew Pirie's doctoral thesis. All of the wines show distinct cool-climate characteristics, the Quarry Red having distinct similarities to the wines of Chinon in the Loire Valley. However, it is with Chardonnay that Portree has done best, and which is its principal wine (in terms of volume). As from the 1998 vintage the wines were made at a newly-constructed on-site winery.

Portree Chardonnay

Produced from 3 hectares of estate plantings. An interesting vertical tasting in October 1996 included all of the vintages made between 1991 and 1995, the '91 and '93 being particularly good and demonstrating the ageing capacity of the wine. On this bloodline, the '97 should repay cellaring.

ŸŸŸŸ **1997** Medium to full yellow-green; the bouquet is substantial, showing some barrel-ferment characters, but still tight. The extremely powerful palate shows the 14.2° alcohol, but is also multilayered and multifaceted. **rating:** 89

best drinking 2000–2005 **best vintages** '91, '93, '97 **drink with** Fish terrine • $26

port stephens wines NR

69 Nelson Bay Road, Bobs Farm, NSW 2316 **region** Other Wineries of NSW

ph (02) 4982 6411 **fax** (02) 4982 6411 **open** 7 days 10–5

winemaker Contract **prod.** 3500 **est.** 1984

prod. range ($10–19.50 CD) Chardonnay, Tri-Blend, Tomaree White, Late Harvest, Golden Sands, Shiraz, Cabernet Merlot, Cabernet Sauvignon, Sparkling and Fortifieds.

summary Planting of the quite substantial Port Stephens Wines vineyard began in 1984, and there are now 4 hectares of vines in production. The wines are made under contract by John Baruzzi at Wilderness Estate in the Hunter Valley, but are sold through an attractive, dedicated cellar-door sales outlet on site.

pothana NR

Carramar, Belford, NSW 2335 **region** Lower Hunter Valley
ph (02) 6574 7164 **fax** (02) 6574 7209 **open** By appointment
winemaker David Hook **prod.** 2000 **est.** 1984
prod. range ($NA) Chardonnay, Semillon, Pinot Noir, Adina Pinot Grigio; The Gorge range includes Unwooded Chardonnay, Classic Dry White, Chardonnay, Verdelho, Semillon and Shiraz.
summary Principally sold through Pokolbin Estate and by mailing list; the Chardonnay is a soft, buttery/toasty wine in mainstream Hunter Valley-style.

preston peak NR

Old Wallangarra Road, Wyberba via Ballandean, Qld 4382 **region** Granite Belt
ph (07) 4630 9499 **fax** (07) 4630 9477 **open** At Wyberba Vineyard
winemaker Philippa Hambleton **prod.** 6000 **est.** 1994
prod. range ($18–26 ML) Sauvignon Blanc, Semillon Chardonnay, Sauvignon Blanc Semillon, Shiraz, Cabernets.
summary The spectacular growth plans of dentist owners Ashley Smith and Kym Thumpkin have seemingly slowed to a more realistic level, although production has doubled over the past year. Winemaking continues at Wyberba, with the proposed winery at Toowoomba on hold.

primo estate ★★★★★

Old Port Wakefield Road, Virginia, SA 5120 **region** Adelaide Plains
ph (08) 8380 9442 **fax** (08) 8380 9696 **open** June–Sept Mon–Fri 9–5, Sat, holidays 10–4.30
winemaker Joseph Grilli **prod.** 20 000 **est.** 1979
prod. range ($12–100 R) Colombard, La Magia Botrytis Riesling, Shiraz, Cabernet Merlot, Adelaide Shiraz, Joseph Moda Amarone Cabernet Merlot, Joseph Sparkling Red.
summary Roseworthy dux Joe Grilli has risen way above the constraints of the hot Adelaide Plains to produce an innovative and always excellent range of wines. The biennial release of the Joseph Sparkling Red (in its tall Italian glass bottle) is eagerly awaited, the wine immediately selling out. However, the core lies with the zingy, fresh Colombard, the velvet-smooth Adelaide Shiraz and the distinguished, complex Joseph Cabernet Merlot.

Primo Estate Colombard

Joe Grilli has always been able to conjure something quite magical from the 4.5 hectares of estate plantings of colombard. The variety is known for its capacity to hold its natural acidity in hot climates (and the Adelaide Plains are hot) but no one else seems to be able to invest the wine with the fruit freshness and crispness – almost Sauvignon Blanc-like – achieved by Joe Grilli.

🍷🍷🍷🍷 **1998** Light to medium yellow-green; the bouquet has that unique gooseberry/herbal lift which only Primo Estate seems to achieve with the variety. The palate is lively and fresh, with a touch of $CO2$ giving it a tingle; the natural acidity also adds to the interest and length of the wine. **rating:** 87

best drinking 1999–2000 **best vintages** NA **drink with** Oysters, shellfish • $12

Primo Estate Joseph La Magia Botrytis Riesling

The original Primo Estate late-harvest dessert wines were made using rack-dried, artificially botrytised grapes. Since 1991 the style (and the source) has changed radically, now utilising Eden Valley Riesling which is either naturally botrytised, or naturally raisined. Says Joe Grilli 'This wine is not a study in how concentrated can a sweet wine be ... we are aiming for a wine that is delicious to drink'. In 1998 the wine received a label change, now known as Joseph La Magia Botrytis Riesling.

🍷🍷🍷🍷 **1998** Medium to full yellow-green; the bouquet is quite complex, with lime marmalade fruit, the palate with intensely sweet lime juice flavours well balanced by acidity. **rating:** 87

best drinking 1999–2003 **best vintages** '91, '92, '93, '95 **drink with** Orange soufflé • $23

Primo Estate Joseph Moda Amarone Cabernet Merlot

Although the front label does not make reference to it, the back label says 'moda amarone' – a modest claim which has brought the wrath of the Italians down on the head of Joe Grilli, and his promise to desist from using it. The wine does in fact use the amarone methods of partially drying the red grapes before fermentation. The blend varies between Coonawarra and McLaren Vale according to the vintage.

🍷🍷🍷🍷🍷 **1997** Strong red-purple; rich, chocolate, prune and blackberry fruit aromas give a hint of the concentrated palate, with sweet cherry and bitter chocolate its counterpoints. The unifying force is the excellent tannin which runs right through the palate, yet is in no sense hard. For the record a blend of 90% Coonawarra Cabernet Sauvignon and 10% Merlot from McLaren Vale and Coonawarra. **rating:** 94

best drinking 2002–2012 **best vintages** '81, '84, '86, '90, '91, '93, '94, '95, '96, '97 **drink with** Bistecca Fiorentina • $40

prince albert ★★★★

100 Lemins Road, Waurn Ponds, Vic 3221 **region** Geelong
ph (03) 5241 8091 **fax** (03) 5241 8091 **open** By appointment
winemaker Bruce Hyett **prod.** 500 **est.** 1975
prod. range ($22.90 ML) Pinot Noir.

summary Australia's true Pinot Noir specialist (it has only ever made the one wine) which also made much of the early running with the variety: the wines always show good varietal character, and have rebounded after a dull patch in the second half of the 1980s. In 1998 the vineyard and winery was certified organic by OVAA Inc. Apart from the mailing list, the wine is sold through fine wine retailers in Sydney and Melbourne, with a little finding its way to the UK.

providence vineyards NR

236 Lalla Road, Lalla, Tas 7267 **region** Northern Tasmania
ph (03) 6395 1290 **fax** (03) 6395 1290 **open** 7 days 10–5
winemaker Andrew Hood (Contract) **prod.** 600 **est.** 1956
prod. range ($18.50–24.95 CD) Semillon, Botrytis Semillon, Chardonnay, Pinot Noir in exceptional years may be released under the Miguet label.

summary Providence incorporates the pioneer vineyard of Frenchman Jean Miguet, now owned by the Bryce family which purchased it in 1980. The original 1.3-hectare vineyard has been expanded to a little over 3 hectares, as well as grafting over unsuitable grenache and cabernet (left from the original plantings) to chardonnay and pinot noir and semillon. Miguet in fact called the vineyard 'La Provence', reminding him of the part of France from whence he came, but after 40 years the French authorities forced a name change to Providence.

Providence Vineyards Botrytis Semillon (375 ml)

A truly remarkable effort from this historic (in Tasmanian terms) vineyard, ever so skilfully made by Andrew Hood.

🍷🍷🍷🍷🍸 **1998** Light to medium yellow-green; a very complex bouquet, looking as much like Riesling as Semillon, and with just a hint of toast, presumably from oak. The wonderfully sweet and rich palate is very long and intense; paradoxically for Tasmania, it seems to need half a gram more acidity. **rating:** 90

best drinking 1999–2003 **best vintages** NA **drink with** Rich fruit-based ice cream • $21.50

punters corner ★★★★

Cnr Riddoch Highway and Racecourse Road, Coonawarra, SA 5263 **region** Coonawarra
ph (08) 8737 2007 **fax** (08) 8737 2007 **open** 7 days 10–5
winemaker Balnaves (Contract) **prod.** 10 000 **est.** 1988

prod. range ($19–24 ML) Chardonnay, Shiraz, Cabernet Merlot, Cabernet Sauvignon.
summary The quaintly named Punters Corner started off life in 1975 as James Haselgrove, but in 1992 was acquired by a group of investors who quite evidently had few delusions about the uncertainties of viticulture and winemaking, even in a district as distinguished as Coonawarra. The arrival of Peter Bissell as winemaker at Balnaves has paid immediate dividends, with some quite excellent reds from 1996 and Chardonnay from 1997 the result. Sophisticated packaging and label design add to the appeal of the wines.

Punters Corner Shiraz

Sourced from both the Victoria and Albert Block and the Punters Corner Cellardoor Block, and employing sophisticated winemaking techniques. The V & A component is fermented on skins for four days, followed by barrel fermentation in new American oak, while the Cellardoor component is given extended post-fermentation maceration.
▼▼▼▼▽ **1997** Medium red-purple; the bouquet is clean, with moderately intense and pleasant earthy/berry varietal fruit aromas. The palate is particularly attractive, with sweet cherry/berry fruit supported and surrounded by well-balanced and integrated oak. The more you taste it, the more you like it. **rating:** 90
best drinking 2001–2007 **best vintages** '97 **drink with** Mushroom risotto • $22

Punters Corner Cabernet Sauvignon

The sourcing and making of this wine is similar to the Shiraz. Part, coming from the V & A vineyard, is barrel-fermented; most comes from low-yielding old vines on the high-bank red soil on the Cellardoor Block, which is given extended maceration.
▼▼▼▼ **1997** Medium to full red-purple; the bouquet ranges through berry, briary, mint and leaf; the palate is quite uncompromising in terms of varietal flavour, but has some carbonic maceration-like characters. One gold, two silver medals in national wine shows. **rating:** 86
best drinking 2002–2007 **best vintages** NA **drink with** Yearling beef • $24

queen adelaide ★★

Sturt Highway, Waikerie, SA 5330 **region** Barossa Valley
ph (08) 8541 2588 **fax** (08) 8541 3877 **open** Not
winemaker Nigel Logos **prod.** 700 000 **est.** 1858
prod. range ($6–8 R) Rhine Riesling, Chenin Blanc, Semillon Chardonnay, Chardonnay, Spaetlese Lexia, Sauvignon Blanc, Regency Red.
summary The famous brand established by Woodley Wines, and some years ago subsumed into the Seppelt and now Southcorp group. It is a pure brand, without any particular home, either in terms of winemaking or fruit sources, but is hugely successful; Queen Adelaide Chardonnay is and has for some time been the largest selling bottled white wine in Australia. However, the use of agglomerate corks precludes assessment of the true quality of the wines, because of the near-certainty of lesser or greater degrees of cork taint. The move to synthetic corks cannot come too soon; all the wines show agglomerate cork taint to a lesser or greater degree.

quelltaler ★★★★

Main North Road, Watervale, SA 5452 **region** Clare Valley
ph (08) 8843 0003 **fax** (08) 8843 0096 **open** Mon–Fri 8.30–5, weekends 11–4
winemaker David O'Leary, Allen Hart **prod.** 150 000 **est.** 1856
prod. range ($12–34 CD) The Annie's Lane range includes Riesling, Semillon Chardonnay, Shiraz and Cabernet Merlot; The Clare Essentials range includes Carlsfield Vineyard, Polish Hill River Vineyard and Prospect Vineyard Rieslings; also The Essentials O'Deas Vineyard Coonawarra Cabernet.
summary The wheel has turned full circle, and after some regrettable decisions to progressively change the name of Quelltaler to Eaglehawk Estate and Black Opal, this great legacy of the

nineteenth century once again proudly bears the Quelltaler name. The Eaglehawk brand continues, but is made at Wolf Blass in the Barossa Valley, and has a southeastern Australia origin; Black Opal is made, but only for export; Quelltaler as a brand name is, at least for the time being, in suspense; and the Clare Valley portfolio made at Quelltaler is sold under the Annie's Lane label. The name comes from Annie Weyman, a turn-of-the-century local identity.

Annie's Lane Riesling

Produced from a blend of Watervale and Polish Hill-grown grapes. Notwithstanding the emergence of three Quelltaler single vineyard Rieslings, this wine suffers nothing in comparison.

🍷🍷🍷🍷🍸 **1998** Light to medium yellow-green; the aroma immediately proclaims this as a tight, classic regional style, with crisp, firm, minerally fruit on the bouquet followed by a very lively lime and mineral palate, long and intense, well balanced and with excellent development potential. **rating:** 91

best drinking 2002–2008 **best vintages** '96, '97, '98 **drink with** Asparagus • $13

The Clare Essentials Carlsfield Vineyard Riesling

Carlsfield Vineyard was planted in 1935 on a sweeping hillside of red loam over limestone. The vineyard is situated just to the southeast of the Quelltaler winery at an elevation of 470 metres. The riesling is the celebrated Geisenheim clone.

🍷🍷🍷🍷🍸 **1998** Light to medium yellow-green; the bouquet is firm, with mineral notes touched by hints of citrus blossom. A spotlessly clean and elegant wine with fleeting sweet fruit on the mid-palate which will expand with time; long, clean, crisp finish. **rating:** 92

best drinking 2001–2008 **best vintages** '98 **drink with** Eggplant terrine • $14

The Clare Essentials Polish Hill River Vineyard Riesling

The vineyard was planted in 1979 (by Wolf Blass) on a site of clay loam over the classic Polish Hill River slate. It matures several weeks later than the Watervale vineyards, and in classic Polish Hill River style, produces a tight, more restrained wine which has the capacity to age wonderfully.

🍷🍷🍷🍷🍸 **1998** Light to medium yellow-green; the bouquet is clean and firm, relatively subdued in typical Polish Hill-style, with hints of talc and mineral. The palate has the requisite balance but, as with the bouquet, is very subdued at this stage. The judgment is (at least partially) that of the future. **rating:** 90

best drinking 2004–2010 **best vintages** '97, '98 **drink with** Grilled sardines • $14

Annie's Lane Contour Vineyard Shiraz

Part of the individual range of wines based on the large and now relatively old vineyards of Quelltaler; contour planting was avant garde in the 1940s and '50s.

🍷🍷🍷🍷🍸 **1996** Strong red-purple; the voluptuous bouquet has a mixture of chocolate and sweet vanilla bean oak, followed by blackberry, chocolate and vanilla tumbling around abundantly on the palate. **rating:** 90

best drinking 2001–2010 **best vintages** NA **drink with** Steak and kidney pie • $34

Annie's Lane Cabernet Merlot

A blend of 85% Cabernet Sauvignon and 15% Merlot, the Cabernet Sauvignon coming in part from Annie's Lane Vineyard in Polish Hill and from other Clare Valley growers, the Merlot component from the Quelltaler Estate vineyard. Fermented in small 4-tonne open fermenters, then matured in French oak (20% new) for a period of 15 months.

🍷🍷🍷🍷🍸 **1997** Medium to full red-purple; the bouquet is full and rich, with ripe cassis fruit and chocolate, the oak under control. A very rich and concentrated dark berry palate, with touches of chocolate and persistent ripe tannins. **rating** 93

best drinking 2002–2012 **best vintages** '94, '95, '97 **drink with** Illabo spring lamb • $15

The Essentials O'Deas Vineyard Coonawarra Cabernet

The first release in the so-called 'Essentials' range. The vineyard was one of the first planted by Ron Haselgrove in the 1960s to the old Reynella clone and this is a more than decent budget-priced Coonawarra Cabernet.

ŸŸŸŸ **1997** Medium to full red-purple; the bouquet is rich, with lots of red and dark berry fruit counterbalanced by masses of oak. The palate offers rich, sweet, red berry fruit in a cocoon of soft vanilla oak; soft tannins on the finish. **rating:** 89

best drinking 2000–2005 **best vintages** NA **drink with** Ragout of lamb • $16

raleigh winery NR

Queen Street, Raleigh, NSW 2454 **region** Other Wineries of NSW
ph (02) 6655 4388 **fax** (02) 6655 4265 **open** 7 days 10–5
winemaker Lavinia Dingle **prod.** 1600 **est.** 1982
prod. range ($13–25 CD) Semillon Chardonnay, Traminer Riesling, Rouge (Rosé), Late Harvest, Shiraz Cabernet Merlot, Port.
summary Raleigh Winery lays claim to being Australia's most easterly vineyard. The vineyard was initiated in 1982, and purchased by Lavinia and Neil Dingle in 1989, with the wine produced in part from 1 hectare of vines planted to no less than six varieties. The wines have won bronze medals at the Griffith Wine Show.

ravenswood lane vineyard NR

Ravenswood Lane, Hahndorf, SA 5245 **region** Adelaide Hills
ph (08) 8388 1250 **fax** (08) 8388 7233 **open** Not
winemaker Stephen Pannell, Hardys (Contract) **prod.** 1500 **est.** 1993
prod. range ($22–50 ML) The Gathering Sauvignon Blanc, Beginning Chardonnay, Reunion Shiraz, 19th Meeting Cabernet.
summary With a sales and marketing background, John and Helen Edwards opted for a major lifestyle change when they began the establishment of the first of the present 28.1 hectares of vineyards in 1993. Most of the production is sold to Hardys, which makes a small quantity of high-quality wines for sale under the Ravenswood Lane Label.

Ravenswood Lane Reunion Shiraz

A striking first-up red, with the Hardy winemaking stamp firmly on it, but also with some fruit characters reminiscent of the extreme southern end of the Napa Valley around a little town called Coombsville. The aroma and flavour is quite unique, and I have only encountered it once or twice before.

ŸŸŸŸŸ **1997** Medium red-purple; a very complex Rhône-style with game, liquorice and boot polish aromas, all in the varietal slot. The palate offers yet more liquorice, spice and cherry, complexed by stylish oak. **rating:** 95

best drinking 2001–2011 **best vintages** '97 **drink with** Kangaroo or game • $50

ray-monde NR

250 Dalrymple Road, Sunbury, Vic 3429 **region** Sunbury
ph (03) 5428 2657 **fax** (03) 5428 3390 **open** Sundays or by appointment
winemaker John Lakey **prod.** 770 **est.** 1988
prod. range ($25 CD) Pinot Noir.
summary The Lakey family has established a little under 4 hectares of pinot noir on their 230-hectare grazing property at an altitude of 400 metres. Initially the grapes were sold to Domaine Chandon, but in 1994 son John Lakey (who had gained experience at Tarrawarra, Rochford, Virgin Hills, Coonawarra plus a vintage in Burgundy) commenced making the wine – and very competently.

reads ★★

Evans Lane, Oxley, Vic 3678 **region** King Valley
ph (03) 5727 3386 **fax** (03) 5727 3559 **open** Mon–Sat 9–5, Sun 10–6
winemaker Kenneth Read **prod.** 1900 **est.** 1972
prod. range ($7.50–13 CD) Riesling, Chardonnay, Sauvignon Blanc, Crouchen, Cabernet Shiraz, Cabernet Sauvignon, Port.
summary Limited tastings have not impressed, but there may be a jewel lurking somewhere, such as the medal-winning though long-gone 1990 Sauvignon Blanc.

redbank winery ★★★★

Sunraysia Highway, Redbank, Vic 3467 **region** Pyrenees
ph (03) 5467 7255 **fax** (03) 5467 7248 **open** Mon–Sat 9–5, Sun 10–5
winemaker Neill Robb **prod.** 58 000 **est.** 1973
prod. range ($9.90–66 CD) The range centres on a series of evocatively named red wines, with Sally's Paddock the flagship, followed by Cabernet Sauvignon and Cabernet Franc; then Hard Hill Cabernet Sauvignon, Fighting Flat Shiraz and Spud Gully Pinot; and various specialties available cellar door. Long Paddock Shiraz, Long Paddock Chardonnay and Emily Brut are cheaper, larger-volume second labels.
summary Neill Robb makes very concentrated wines, full of character; the levels of volatile acidity can sometimes be intrusive, but are probably of more concern to technical tasters than to the general public. Sally's Paddock is the star, a single vineyard block with an esoteric mix of Cabernet, Shiraz and Malbec and which over the years has produced many great wines.

red edge NR

Golden Gully Road, Heathcote, Vic 3523 **region** Bendigo
ph (03) 9370 9565 **fax** (03) 9370 9565 **open** Not
winemaker Peter Dredge **prod.** 300 **est.** 1971
prod. range ($23–27 R) Shiraz, Cabernet Sauvignon.
summary Red Edge is a new name on the scene, but the vineyard dates back to 1971, at the renaissance of the Victorian wine industry. In the early 1980s it produced the wonderful wines of Flynn & Williams, and has now been rehabilitated by Peter and Judy Dredge, producing two quite lovely wines in their inaugural 1997 vintage. For the time being, at least, Peter Dredge continues to keep body and soul together by making the wines at Wildwood and at Witchmount Estate, Rockbank. Trying to eke a living out of 300 cases of what in these days are moderately-priced wines is simply not possible.

Red Edge Shiraz

Produced from unirrigated and very low-yielding estate vineyards planted in 1971. The wine is neither fined nor filtered, and is deliberately not overoaked.

🍷🍷🍷🍷 **1997** Medium to full red-purple; the fruit aromas are suitably complex, ranging through blackberry, earth and black cherry. The wine is full, round and soft in the mouth, with gentle tannins. **rating:** 87

best drinking 2000–2007 **best vintages** NA **drink with** Beef stroganoff • $27

Red Edge Cabernet Sauvignon

Like the Shiraz, produced from low-yielding, unirrigated estate vines. A minimalist handling approach has been adopted; perhaps a little more interference would add more polish.

🍷🍷🍷🍷 **1997** Medium to full red-purple; the bouquet is powerful, with strong leafy/olivaceous fruit varietal character. The palate, too, shows potent, primary cabernet varietal character, giving the impression it might have been improved if worked more in barrel. However, the tannins are fine and soft. **rating:** 86

best drinking 2002–2007 **best vintages** NA **drink with** Roast lamb • $23

redgate ★★★☆

Boodjidup Road, Margaret River, WA 6285 **region** Margaret River
ph (08) 9757 6488 **fax** (08) 9757 6308 **open** 7 days 10–5
winemaker Andrew Forsell **prod.** 8000 **est.** 1977
prod. range ($14–24.50 CD) Classic Semillon Sauvignon Blanc, OFS Semillon, Sauvignon Blanc, Sauvignon Blanc Reserve, Chenin, Late Harvest Riesling, Cabernet Sauvignon, Pinot Noir Méthode Champenoise, Port.
summary Twenty hectares of vineyard provide the base for one of the larger wineries of the Margaret River region which probably has a lower profile than it deserves. The wines do have limited distribution in the eastern States, and export markets in Singapore, Japan, Taiwan, Denmark and Switzerland have been established.

red hill estate ★★★☆

53 Redhill–Shoreham Road, Red Hill South, Vic 3937 **region** Mornington Peninsula
ph (03) 5989 2838 **fax** (03) 5989 2855 **open** 7 days 11–5
winemaker Jenny Bright **prod.** 10 000 **est.** 1989
prod. range ($14–36 CD) Particular emphasis on Méthode Champenoise, but also producing Unoaked Chardonnay, Chardonnay, Riesling, Sauvignon Blanc, Hill Block Pinot (Rosé), Pinot Noir and Cabernet Sauvignon; also Muscat (from Rutherglen material) available cellar-door only.
summary Sir Peter Derham and family completed the construction of an on-site winery in time for the 1993 vintage, ending a period in which the wines were made at various wineries under contract arrangements. The 10-hectare vineyard is one of the larger plantings on the Mornington Peninsula, and the tasting room and restaurant have a superb view across the vineyard to Westernport Bay and Phillip Island.

Red Hill Estate Chardonnay

An almost startlingly different style from that normally encountered from the Mornington Peninsula, normally bracingly crisp, clean and direct. Years such as 1996, however, impose their own imperative (in the form of botrytis) balanced by perfect vintages such as 1998.

🍷🍷🍷🍷 **1998** Medium yellow-green; a complex bouquet, with high-quality spicy oak set around tangy fruit. The palate provides more of the same, with plenty of life and brightness; should develop well. **rating:** 89

best drinking 1999–2004 **best vintages** '98 **drink with** Seafood • $25

Red Hill Estate Chardonnay Pinot Noir

A blend of 60% Chardonnay and 40% Pinot given 15 months on lees and six months on cork after disgorgement.

🍷🍷🍷🍷 **1996** Light to medium yellow-green; the bouquet is quite firm with some stony minerally notes together with green apple; I wonder whether there is a deliberate touch of aldehyde. The palate is well balanced, introducing a touch of strawberry in the flavour range, and a long finish. All in all, sophisticated sparkling wine. **rating:** 89

best drinking 1998–1999 **best vintages** NA **drink with** Aperitif • $28

Red Hill Estate Sparkling Pinot Noir Blanc de Noirs

Made entirely from estate-grown pinot noir, given 18 months on yeast lees and a further three months on cork after disgorgement and prior to release.

🍷🍷🍷🍷🍷(half) **1996** The pale bronze-pink colour denotes the Pinot Noir base wine, and leads into an attractive bouquet with faint spice and musk overtones to the more bready yeast autolysis notes. Distinct strawberry Pinot flavours run right through the palate to the finish, which has a most attractive and striking flavour. **rating:** 91

best drinking 1998–1999 **best vintages** NA **drink with** Delicate seafood • $36

Red Hill Estate Pinot Noir

The Red Hill Pinot Noir is sourced from three different vineyards at Merricks, Red Hill and Main Ridge. To add additional complexity, four clones (MV6, D2V5, G5V15 and D5V12, often grown on different canopies such as Geneva Double Curtain or Scott Henry) are used. Each parcel is picked separately and processed separately with both pre- and post-fermentation maceration.

ΨΨΨΨ **1998** Medium red-purple; the bouquet offers the full spectrum of berry through to forest aromas, with nicely controlled oak. The palate is quite stylish in a slippery, sappy mode, with a long finish. Quite Burgundian in its inflection. **rating:** 86

best drinking 1999–2002 **best vintages** NA **drink with** Tea-smoked duck • $25

redman ★★★

Riddoch Highway, Coonawarra, SA 5253 **region** Coonawarra
ph (08) 8736 3331 **fax** (08) 8736 3013 **open** Mon–Fri 9–5, weekends 10–4
winemaker Bruce Redman, Malcolm Redman **prod.** 18 000 **est.** 1966
prod. range ($13–25 R) Shiraz (formerly labelled as Claret), Cabernet Merlot (the first new wine in 26 years), Cabernet Sauvignon.
summary After a prolonged period of mediocrity, the Redman wines are showing sporadic signs of improvement, partly through the introduction of modest amounts of new oak, even if principally American. It would be nice to say the wines now reflect the full potential of the vineyard, but there is still some way to go.

reedy creek vineyard NR

Reedy Creek, via Tenterfield, NSW 2372 **region** Northern Slopes Zone
ph (02) 6737 5221 **fax** (02) 6737 5200 **open** 7 days 9–5
winemaker Bruce Humphery-Smith (Contract) **prod.** 800 **est.** 1971
prod. range ($10–20 CD) Italia Chardonnay, Chardonnay, Merlot, Shiraz Mourvedre, Old Vine Shiraz, Shiraz, Valambrossa Liqueur, Muscat Liqueur.
summary Like so many Italian settlers in the Australian countryside, the De Stefani family has been growing grapes and making wine for its own consumption for over 30 years at its Reedy Creek property near Tenterfield, in the far north of New South Wales. What is more, like their compatriots in the King Valley, the family's principal activity until 1993 was growing tobacco, but the continued rationalisation of the tobacco industry led the De Stefanis to turning a hobby into a commercial exercise. The vineyard has now been expanded to 4.5 hectares, and the first commercial vintage of Shiraz was made in 1995, with Chardonnay following in 1998. The wines are made by the incredibly industrious Bruce Humphery-Smith at Rimfire Vineyards at MacLagan, and are sold cellar door from the maturation cellar opened in 1997.

reg drayton wines ★★★☆

Cnr Pokolbin Mountain and McDonalds Roads, Pokolbin, NSW 2321 **region** Lower Hunter Valley
ph (02) 4998 7523 **fax** (02) 4998 7523 **open** 7 days 10–5
winemaker Tyrrell's (Contract) **prod.** 3000 **est.** 1989
prod. range ($16–25 CD) Lambkin Semillon, Lambkin Verdelho, Pokolbin Hills Chardonnay, Pokolbin Hills Chardonnay Semillon, Three Sons Shiraz, Pokolbin Hills Shiraz, Pokolbin Hills Cabernet Shiraz, Port.
summary Reg and Pam Drayton were among the victims of the Seaview/Lord Howe Island air crash in October 1984, having established Reg Drayton Wines after selling their interest in the long-established Drayton Family Winery. Their daughter Robyn (a fifth-generation Drayton and billed as the Hunter's first female vigneron) and husband Craig continue the business, which draws chiefly upon the Pokolbin Hills Estate, but also takes fruit from the historic Lambkin Estate vineyard. The wines are made for them at Tyrrell's.

Reg Drayton Pokolbin Hills Shiraz

Produced from 5 hectares of estate plantings on the Pokolbin Hills vineyard, and matured in a mix of new (30%) and predominantly one-year-old American oak barrels. Yet another consistent show medal winner.

🍷🍷🍷🍷 **1997** Medium red-purple; the bouquet is pleasantly soft, with cherry/berry fruit and a hint of Hunter earth. The palate follows on logically, with sweet black cherry fruit, a hint of oak and ample but soft tannins. **rating:** 88

best drinking 2000–2007 **best vintages** '91, '94, '97 **drink with** Roast lamb • $20

reilly's creek ★★★

226A Lower White Hills Road, Relbia, Tas 7258 **region** Northern Tasmania
ph (03) 6391 8974 **open** Not
winemaker Andrew Hood (Contract) **prod.** 100 **est.** 1995
prod. range ($16 ML) Riesling, Pinot Noir.
summary Reilly's Creek is the label for Relbia Vineyards, which sells most of its grapes to contract-winemaker Andrew Hood. Small quantities are made for it by Hood and, as one would expect, are of good quality.

reilly's wines NR

Cnr Hill and Burra Streets, Mintaro, SA 5415 **region** Clare Valley
ph (08) 8843 9013 **fax** (08) 8337 4111 **open** 7 days 10–5
winemaker Justin Ardill **prod.** 5000 **est.** 1994
prod. range ($13–35 CD) Watervale Riesling, Late Picked Riesling, Chardonnay, Semillon, Sparkling Grenache, Old Bushvine Grenache, Clare Valley Shiraz, Dry Land Shiraz, Cabernet Sauvignon, Port.
summary Justin and Julie Ardill are among the newest arrivals in the Clare Valley, with just a handful of vintages under their belt. An unusual sideline of Reilly's Cottage is the production of an Extra Virgin Olive Oil; unusual in that it is made from wild olives found in the Mintaro district of the Clare Valley. Retail distribution in South Australia, Victoria and New South Wales.

Reilly's Old Bush Vine Grenache

It is fascinating to see how finely balanced truly ripe grenache has to be, and how quickly it steps over the line into outright jammy confection characters. A little is fine, as this wine attests, but more is too much.

🍷🍷🍷🍷½ **1996** Medium to full red; there is plenty of sweet strawberry/berry fruit on the bouquet, and authentic, sweet slightly jammy varietal fruit on the palate, very typical of Grenache, and well balanced. **rating:** 90

best drinking 1999–2003 **best vintages** NA **drink with** Steak and kidney pie • $15

renmano ★★★★

Sturt Highway, Renmark, SA 5341 **region** Riverland
ph (08) 8586 6771 **fax** (08) 8586 5939 **open** Mon–Sat 9–5
winemaker Glenn James, Tony Ingle **prod.** 1.6 million **est.** 1914
prod. range ($9.95 R) Chairman's Selection Chardonnay; River Breeze is a second label.
summary Part of the BRL Hardy group. A radical change in winemaking technique and philosophy in 1996 has wrought miracles with the Chairman's Selection Chardonnay, now the only premium wine under the Renmano label.

reynell ★★★★☆

Reynell Road, Reynella, SA 5161 **region** McLaren Vale
ph (08) 8392 2222 **fax** (08) 8392 2202 **open** 7 days 10–4.30
winemaker Stephen Pannell **prod.** NFP **est.** 1838

prod. range ($36.90 R) Basket Pressed Shiraz, Basket Pressed Merlot, Basket Pressed Cabernet Sauvignon.

summary Reynell is the name under which all wines from the historic Reynella winery (once called Chateau Reynella) are released. What is more, the range of wines has been compressed, and taken into the super-premium category with the initial release in July 1997 of three multi-award-winning wines, all effectively Show Reserve releases.

Reynell Basket Pressed Shiraz

The initial release of this wine (the '94) had won seven gold medals prior to its release, and is the type of wine which will continue to amass gold medals so long as it is exhibited in wine shows. Hugely powerful and concentrated, it was matured in American oak for 24 months. Initially that oak is married well with the wine but does gradually build-up on retasting, and ultimately threatens the balance of the wine. Those more tolerant of the flavour of American oak will find no problem with it, however.

🍷🍷🍷🍷🍷 **1995** Strong red-purple; a lively and complex bouquet with a mix of sweet leather, spice and cherry fruit spilling over into the palate which is replete with lingering tannins and cleverly-accented oak. Gold medal 1998 National Wine Show. **rating:** 94

best drinking 2000–2010 **best vintages** '94, '95 **drink with** Barbecued rump steak • $36.90

Reynell Basket Pressed Cabernet Sauvignon

A wine which oscillates between being simply powerful and being downright intimidating, successfully occupying the Australian heavyweight wine boxing ring.

🍷🍷🍷🍷🍷 **1996** Full purple-red; a very powerful wine, with lots of cedary/leafy Bordeaux-like aromas on the bouquet, and more of the same on the palate. The only question for the long-term development of the wine lies with its very powerful tannins which may outdistance the fruit. Gold medal 1998 National Wine Show. **rating:** 92

best drinking 2004–2010 **best vintages** '94, '95, '95 **drink with** Leave it in the cellar • $36.90

reynolds yarraman ★★★★

Yarraman Road, Wybong, NSW 2333 **region** Upper Hunter Valley

ph (02) 6547 8127 **fax** (02) 6547 8013 **open** Mon–Sat 10–4, Sun, public holidays 11–4

winemaker Jon Reynolds, Nic Millichip **prod.** 15 000 **est.** 1967

prod. range ($16–25 CD) From the Hunter Valley: Semillon, Chardonnay, Shiraz; from Orange: Chardonnay, Cabernet Sauvignon; Cabernet Merlot is a blend from both the Hunter Valley and Orange.

summary With the Orange region steadily assuming greater importance for Reynolds Yarraman, wine quality (and consistency) likewise continues to increase, although the Semillons will of course remain Hunter Valley-sourced. The skills of Jon Reynolds as a winemaker have never been in doubt, and as the size and maturity of the Orange vineyards grow, it seems certain that even better wines will appear in the future. The wines are exported to the UK, US Europe and Asia.

Reynolds Yarraman Hunter Valley Semillon

Made entirely from estate-grown grapes from the Yarraman Vineyard picked at a relatively high baumé (for Semillon) of 12.5 degrees, tank-fermented and given extended lees contact. Partial malolactic fermentation adds yet further to the structure of the wine.

🍷🍷🍷🍷🍷 **1998** Light green-yellow; classic, fine, crisp herb, nettle and grass aromas set the scene for a wine with an attractive and unusually complex mid-palate, and a long, fine lingering finish. **rating** 90

best drinking 1999–2004 **best vintages** '91, '92, '96, '98 **drink with** Mussels • $16

Reynolds Yarraman Orange Chardonnay

The two most distinguished wines in the Reynolds portfolio now come not from the Upper Hunter but from Orange, whence Jon Reynolds has sourced increasing amounts of fruit over the recent years. The Chardonnay was the first 100% Orange district wine made by Jon Reynolds; barrel-fermented in French oak, with the full range of malolactic fermentation and lees contact. It comes from the Bloodwood and Bantry Grove vineyards.

🍷🍷🍷🍷🍷 **1997** Bright light to medium yellow-green; the bouquet is intense but still quite tight, with subtle oak; the palate has most attractive melon and nectarine fruit flavours supported by well-handled oak, with the fruit running through a long finish. **rating:** 91

best drinking 1999–2003 **best vintages** '97 **drink with** Sautéed veal • $19

ribbon vale estate ★★★☆

Lot 5 Caves Road, Willyabrup via Cowaramup, WA 6284 **region** Margaret River
ph (08) 9755 6272 **fax** (08) 9755 6337 **open** Weekends, holidays 10–5
winemaker Mike Davies **prod.** 4000 **est.** 1977
prod. range ($15–29 CD) Semillon, Semillon Sauvignon Blanc, Sauvignon Blanc, Cabernet Merlot, Merlot, Cabernet Sauvignon.
summary Makes crisp, herbaceous Semillon and Sauvignon Blanc (and blends), ideal seafood wines, and austere, very firm Cabernets all in mainstream regional-style. Retail distribution in Western Australia, New South Wales and Victoria.

richfield vineyard NR

Bruxner Highway, Tenterfield, NSW 2372 **region** Other Wineries of NSW
ph (02) 6737 5588 **fax** (02) 6737 5598 **open** Not
winemaker Contract **prod.** NA **est.** 1997
prod. range Chardonnay, Shiraz, Merlot, Cabernet Sauvignon, with the first release unlikely before the end of the decade.
summary Richfield Vineyard points to the tyranny of State boundaries. Established at the instigation of Denis Parsons of Bald Mountain vineyards in the Granite Belt (Queensland), Richfield is little more than 30 kilometres south of Bald Mountain vineyards as the crow flies. A little over 11 hectares were planted in 1997, with plans to at least double those plantings in 1998. All of the indications are that the Tenterfield–Granite Belt area will become a very significant cross-border wine growing region.

richmond grove ★★★★

Para Road, Tanunda, SA 5352 **region** Barossa Valley
ph (08) 8563 2184 **fax** (08) 8563 2804 **open** Mon–Fri 10–5, weekends 10–4
winemaker John Vickery **prod.** NFP **est.** 1977
prod. range ($9.95–19.95 R) Eden Valley Traminer Riesling, Watervale Riesling, Barossa Riesling, Oak Matured Chablis, Cowra Chardonnay, Cowra Verdelho, French Cask Chardonnay, Hunter Valley Classic Dry White, Marlborough Sauvignon Blanc, Barossa Shiraz, Cabernet Merlot.
summary Richmond Grove now has two homes, including one in the Barossa Valley, where John Vickery presides. It is owned by Orlando Wyndham, and draws its grapes from diverse sources. The Richmond Grove Barossa Valley and Watervale Rieslings made by John Vickery represent the best value for money (for Riesling) year in, year out. If these were the only wines produced by Richmond Grove, it would have five-star rating. Exports to the UK.

Richmond Grove Watervale Riesling

With John Vickery's vast experience and impeccable contacts, it is not surprising that Richmond Grove Rieslings should be as exceptionally good as they are.

▼▼▼▼▽ **1998** Light to medium yellow-green; a clean and crisp bouquet with mineral and lime aromas moving into lime, mineral and toast on a palate with great length and grip. A classic in the making which is quite certain to achieve significantly higher points when mature. **rating:** 90

best drinking 2003–2008 **best vintages** '94, '96, '97, '98 **drink with** Asparagus with hollandaise sauce • $14.95

Richmond Grove Marlborough Sauvignon Blanc

When Richmond Grove decided to expand its varietal range to include Sauvignon Blanc, it had no hesitation in looking to Marlborough, New Zealand. John Vickery supervised the making in a local winery, vintage taking place after the completion of the Barossa season. The wine is made in the fashion of a Riesling, cold-fermented and early-bottled, so it all came naturally to John Vickery. Gold medal winner 1999 Liquorland Royal Easter Wine Show.

▼▼▼▼▽ **1998** Light yellow-green; a moderately intense, clean bouquet with clear but unaggressive varietal character is followed by a very good palate, particularly in the length of its flavour and the pleasing tightness to the fruit. A cut above many of the Sauvignon Blancs of the vintage. **rating:** 92

best drinking 1999–2000 **best vintages** NA **drink with** Shellfish • $14.95

Richmond Grove Barossa Shiraz

The wine is made using a variety of techniques, all adding up to sophistication. Varying temperatures are used during fermentation, and while the majority of the wine is given post-fermentation maceration (or skin contact) part completed its primary fermentation in barrel. The pressings are back-blended, and the partly clarified wine is placed in American oak barrels where it spends the next 18 months.

▼▼▼▼ **1996** Strong red-purple; while it may be in the style mainstream, the bouquet is concentrated, with dark berry fruit and quite subtle vanillin oak. The palate has abundant fruit weight, power and concentration, with black cherry and touches of mint; firm tannins come through on the finish. Excellent oak balance and integration. **rating:** 89

best drinking 2000–2006 **best vintages** NA **drink with** Rib of beef • $19.95

rimfire vineyards ★★★☆

Bismarck Street, MacLagan, Qld 4352 **region** Other Wineries of Qld

ph (07) 4692 1129 **fax** (07) 4692 1260 **open** 7 days 10–5

winemaker Bruce Humphery-Smith (Consultant), Tony Connellan **prod.** 5000 **est.** 1991

prod. range ($10.50–18 CD) Settlers Blend, Estate Chardonnay, Oak Matured Chardonnay, Estate Colombard, Pioneer White, Shiraz, Colonial Cabernet, Light Fruity Red; Fortifieds.

summary The Connellan family (parents Margaret and Tony and children Michelle, Peter and Louise) began planting the 6-hectare Rimfire Vineyards in 1991 as a means of diversification of their very large (1500-hectare) cattle stud in the foothills of the Bunya Mountains, 45 minutes drive northeast of Toowoomba. Rimfire has had one success after another in Queensland wine shows, and on this yardstick has to be regarded as one of the best producers in Queensland.

rivendell ★★★

Lot 328 Wildwood Road, Yallingup, WA 6282 **region** Margaret River

ph (08) 9755 2235 **fax** (08) 9755 2295 **open** 7 days 10–5

winemaker Mike Davies, Jan Davies (Contract) **prod.** 2750 **est.** 1987

prod. range ($12.50–14.50 CD) Semillon Sauvignon Blanc, Honeysuckle Late Harvest Semillon, Verdelho, Shiraz Cabernet.

summary With 13.5 hectares of vineyards coming into bearing, production for Rivendell will increase significantly over the coming years. The cellar-door sales facility is in a garden-setting complete with restaurant. An unusual sideline is the sale of 50 types of preserves, jams and chutneys. No recent tastings.

riverbank estate NR

126 Hamersley Road, Caversham, WA 6055 **region** Swan District
ph (08) 9377 1805 **fax** (08) 9377 2168 **open** Weekends and public holidays 10–5
winemaker Robert James Bond **prod.** 3500 **est.** 1993
prod. range ($12–16 CD) Semillon, Verdelho, Chenin, Chardonnay, Cabernet.
summary Robert Bond, a graduate of Charles Sturt University and Swan Valley viticulturist for 20 years, established RiverBank Estate in 1993. He draws upon 11 hectares of estate plantings, and, in his words, 'the wines are unashamedly full-bodied, produced from ripe grapes in what is recognised as a hot grape growing region'.

riverina wines ★★★★☆

700 Lidman Way, Griffith, NSW 2680 **region** Riverina
ph (02) 6962 4122 **fax** (02) 6962 4628 **open** 7 days 9–5.30
winemaker Sam Trimboli **prod.** 220 000 **est.** 1969
prod. range ($7.50–14.50 CD) An extensive range of varietal wines under the Ballingal Estate, Ridgewood Estate and Warburn Estate labels, the former being slightly higher priced, each including Chardonnay, Semillon, Semillon Chardonnay, Shiraz, Cabernet Sauvignon and Cabernet Merlot, with a few additions under the Ballingal Estate label. There is also a range of sparkling wines and fortifieds in both bottle and cask.
summary One of the large producers of the region drawing upon 1100 hectares of estate plantings. While much of the wine is sold in bulk to other producers, selected parcels of the best of the grapes are made into table wines with quite spectacular success. At the 1997 National Wine Show, Riverina Wines won an astonishing six gold medals, topping no less than four classes. That success has, it seems, given rise to the introduction of the Show Reserve wines, and while the 1997 success has not been equalled since, the Show Reserve wines continue to justify their label. Exports to the UK.

Warburn Estate Show Reserve Verdelho

One thousand cases made; there appear to have been no tricks with the making, but the vineyard fruit certainly had something extra.

🍷🍷🍷🍷 **1998** Medium yellow-green; the moderately intense bouquet is clean with pleasant fruit, but it is the palate which really opens up with well-above-average intensity to tangy, Sauvignon Blanc-like fruit surrounded with touches of honey; well judged acidity adds both balance and length. **rating:** 86

best drinking 1999–2000 **best vintages** NA **drink with** Stir-fried prawns • $12.50

Ballingal Estate Premium Selection Chardonnay

Riverina Wines always seems to shine at the National Wine Show, perhaps because its wines develop so quickly, and also thanks to high-quality, estate-grown grapes.

🍷🍷🍷🍷½ **1998** Medium yellow-green; the bouquet is fragrant, with quite prominent oak; the palate is lively, long and quite intense with citrus and melon fruit, the oak being much less assertive. Gold medal 1998 National Wine Show. **rating:** 92

best drinking 1998–1999 **best vintages** '96, '98 **drink with** Fricassee of veal • $10

Warburn Estate Show Reserve Chardonnay

Two thousand cases made of the wine; it is fermented in a mix of American and French oak barriques, and left on its lees for eight months with the barrels rolled to the 3 o'clock position. An unusual but labour-saving approach which in no way diminished the quality of the wine, which has won five gold medals and a trophy.

1998 Medium to full yellow-green; the bouquet is complex and rich, with full, tangy fruit and the clever but obvious use of oak. The palate is quite intense, and finer than the bouquet suggests, although experience shows it will develop quickly. **rating:** 93

best drinking 1999–2000 **best vintages** '98 **drink with** KFC • $12.50

Ballingal Estate Botrytis Semillon

Given Riverina Wines skill with its dry table wine portoflio, it comes as no surprise that it should also produce an excellent Botrytis Semillon.

1997 Deep gold; super-intense mandarin and cumquat aromas lead into an incredibly powerful, sweet and intense wine. It may lack finesse, but is certainly seductive. **rating:** 90

best drinking 1999–2001 **best vintages** '97 **drink with** Rich, sweet dessert • $13.50

Warburn Estate Show Reserve Shiraz

The distinguishing feature of the Show Reserve wines, and the reason why they have done so well (comparatively speaking) in national wine shows, is the above-average intensity of the fruit. This in turn stems from strictly controlled yields in the vineyard.

1997 Medium red-purple; the bouquet is moderately intense, with smooth, somewhat plain fruit and lots of American oak. The palate offers another dimension, with extra concentration to the dark cherry fruit and good tannins on the finish. **rating:** 86

best drinking 1999–2003 **best vintages** NA **drink with** Rack of lamb • $14.50

Warburn Estate Merlot

Has been a consistent over-achiever in recent years, and in so doing once again demonstrates the degree to which top-quality wine is made in the vineyard.

1997 Medium red-purple; both the bouquet and palate show strong merlot varietal character, with slightly green/herbal, minty characters. Overall, a pretty wine, with the appropriate structure for Merlot. Gold medal 1998 National Wine Show. **rating:** 93

best drinking 1999–2003 **best vintages** '97 **drink with** Osso bucco • $14.50

---

riversands vineyards NR

Whytes Road, St George, Qld 4487 **region** Queensland Zone
ph (07) 4625 3643 **fax** (07) 4625 5043 **open** Mon–Sat 8–6, Sunday 9–4
winemaker Ballandean Estate (Contract) **prod.** 2000 **est.** 1990
prod. range ($9–15 CD) Sauvignon Blanc Semillon, Chardonnay, Major Mitchell White, Three Rivers Red, Dr Seidel's Soft Red, Ellen Meacle Merlot, Golden Liqueur Muscat, Gaolhouse Port.

summary Riversands is situated on the banks of the Balonne River near St George in the southwest corner of Queensland. It is a mixed wine grape and table grape business, acquired by present owners Alison and David Blacket in 1996. The wines are very competently made under contract at Ballandean Estate and have already accumulated a number of silver and bronze medals. The Chardonnay is particularly meritorious.

Riversands Chardonnay

Fermented in new French oak, but removed after a brief period of maturation and bottled in August of the year of vintage.

1998 Light to medium green-yellow; the bouquet is light and clean, with varietal melon fruit and little or no oak evident. The palate is clean and fresh, light-bodied but unforced, and again showing delicate varietal fruit. **rating** 84

best drinking 1999–2001 **best vintages** NA **drink with** Mud crab • $12

---

robinsons family vineyards ★★★

Curtin Road, Ballandean, Qld 4382 **region** Granite Belt
ph (07) 4684 1216 **fax** (07) 4684 1216 **open** 7 days 9–5
winemaker Craig Robinson **prod.** 2000 **est.** 1969
prod. range ($16–19 CD) Sauvignon Blanc Semillon, Chardonnay, Lyra Dry White, Traminer, Late Harvest Traminer, Shiraz, Shiraz Cabernet, Cabernet Sauvignon, Sparkling.
summary The conjunction of a picture of a hibiscus and 'cool climate' in prominent typeface on the labels is a strange one, but then that has always been the nature of Robinsons Family Vineyards. The red wines can be very good, particularly when not overly extracted and tannic.

robinvale ★★☆

Sea Lake Road, Robinvale, Vic 3549 **region** Murray Darling (Vic)
ph (03) 5026 3955 **fax** (03) 5026 1123 **open** Mon–Fri 9–6, Sun 1–6
winemaker Bill Caracatsanoudis **prod.** 15 000 **est.** 1976
prod. range ($8.50–20 CD) A kaleidoscopic array of wines including five preservative-free wines, white wines which run from Retsina through to Auslese Muscat Hamburg, Dry Marsanne, red wines which encompass Lambrusco, Scarlet Bliss, Cabernet Sauvignon/Franc, Kokkineli, Fruity Rosé; and fortified wines ranging from Mavrodaphne to Cream Marsala, with a few Vintage Ports thrown in for good measure.
summary Robinvale claims to be the only winery in Australia to be fully accredited with the Biodynamic Agricultural Association of Australia. Most, but not all, of the wines are produced from organically-grown grapes, with certain of the wines made preservative-free. Production has increased dramatically, no doubt reflecting the interest in organic and biodynamic viticulture and winemaking. Exports to the UK and Japan.

rochecombe vineyard NR

Baxter's Road, Pipers River, Tas 7252 **region** Northern Tasmania
ph (03) 6382 7122 **fax** (03) 6382 7231 **open** 7 days 10–5
winemaker Fiona West **prod.** 14 000 **est.** 1985
prod. range ($18–24.95 R) Riesling, Chardonnay, Sauvignon Blanc, Pinot Noir, Cabernet Sauvignon Cabernet Franc Merlot.
summary Rochecombe, complete with its much-expanded and state-of-the-art winery, became part of the Pipers Brook Group in February 1998. With its excellent location and restaurant, it will continue to be a significant attraction, but the future of the brand is obscure.

rochford ★★★☆

Romsey Park, Rochford, Vic 3442 **region** Macedon
ph (03) 5429 1428 **fax** (03) 5429 1066 **open** By appointment
winemaker David Creed **prod.** 2500 **est.** 1983
prod. range ($19–32 R) Chardonnay, Pinot Noir, Cabernet Sauvignon; Romsey Park is second label.
summary In February 1998 Helmut Konecsny and Yvonne Lodoco-Konecsny acquired Rochford. David Creed continues as winemaker, and it is the Konecsnys' intention to leave wine style unchanged with the emphasis on Chardonnay and Pinot Noir, and gradually increase production.

Rochford Pinot Noir

Pinot has consistently been the outstanding wine from Rochford, typically exhibiting both complexity and richness of flavour, and helping establish the reputation of the Macedon region as yet another area suited to this fickle variety.

🍷🍷🍷🍷 **1997** Dark red-purple; the bouquet is solid, with dark blood plum fruit and a touch of forest floor, yet despite the colour and the fruit, is still slightly one dimensional. The palate is likewise compressed and concentrated; a most unusual Pinot Noir which needs bottle age and which could become exceptionally good. **rating:** 88

best drinking 2002–2007 **best vintages** '91, '92, '93, '95, '96 **drink with** Jugged hare • $32

rockford ★★★★

Krondorf Road, Tanunda, SA 5352 **region** Barossa Valley
ph (08) 8563 2720 **fax** (08) 8563 3787 **open** Mon–Sat 11–5
winemaker Robert O'Callaghan, Chris Ringland **prod.** 19 500 **est.** 1984
prod. range ($9.75–40 CD) Eden Valley Riesling, Local Growers Semillon, Alicante Bouchet, White Frontignac, Basket Press Shiraz, Sparkling Black Shiraz, Dry Country Grenache, Cabernet Sauvignon, Tawny Port.

summary The wines are sold through Adelaide retailers only (and cellar door), and are unknown to most eastern Australian wine-drinkers, which is a great pity, for these are some of the most individual, spectacularly flavoured wines made in the Barossa today, with an emphasis on old low-yielding dry-land vineyards. This South Australian slur on the palates of Victoria and NSW is exacerbated by the fact that the wines are exported to Switzerland, the UK and NZ; it all goes to show we need proper authority to protect our living treasures.

Rockford Black Shiraz

This, quite simply, is a great sparkling Shiraz, inspired by the Sparkling Burgundies of Colin Preece at Great Western, and made using fundamentally the same techniques. The base wine is matured in large old wood for three years before being tiraged, and then left on lees for a year before disgorgement and further cellaring prior to release. This particular bottle was part of the September '96 disgorgement. Almost impossible to procure; sold only through Adelaide retailers and by mailing list and cellar door, and sells out almost overnight with a limit of six bottles per customer. The unique personality of Robert O'Callaghan comes rocketing through the newsletter, which at the end of the day is probably the best way of getting hold of these scarce wines.

🍷🍷🍷🍷🍷 **NV** Dark red, but with some brick hues evident. There is an attractive mix of spice and earth aromas on the bouquet, but the palate is something else, with that fine, faintly spicy, faintly earthy taste of mature Shiraz of the old Great Western style. It is neither heavy nor sweet, and has tremendous balance and length. **rating:** 94

best drinking 1999–2020 **best vintages** NA **drink with** Needs no accompaniment • $42.50

romavilla NR

Northern Road, Roma, Qld 4455 **region** Other Wineries of Qld
ph (07) 4622 1822 **fax** (07) 4622 1822 **open** Mon–Fri 8–5, Sat 9–12, 2–4
winemaker David Wall, Richard Wall **prod.** 3000 **est.** 1863
prod. range ($11–35 CD) An extensive range of varietal and generic table wines and fortified wine styles including Madeira and Tawny Port are on sale at the winery; the Very Old Tawny Port is made from a blend of material ranging in age from ten to 25 years.

summary An amazing, historic relic, seemingly untouched since its nineteenth-century heyday, producing ordinary table wines but still providing some extraordinary fortifieds, including a truly stylish Madeira, made from Riesling and Syrian (the latter variety originating in Persia). David Wall has now been joined by son Richard in the business which will hopefully ensure continuity for this important part of Australian wine history.

rosabrook estate NR

Rosa Brook Road, Margaret River, WA 6285 **region** Margaret River
ph (08) 9757 2286 **fax** (08) 9757 3634 **open** 7 days 10–4 Nov–Apr, Thur–Sun 11–4 May–Oct
winemaker Simon Keall **prod.** 5000 **est.** 1980

prod. range ($12–19 CD) Semillon, Semillon Sauvignon Blanc, Chardonnay, Autumn Harvest Riesling, Botrytis Riesling, Cabernet Merlot.

summary The 7-hectare Rosabrook Estate vineyards have been established progressively since 1980, with no less than nine varieties planted. The cellar-door facility is housed in what was Margaret River's first commercial abattoir, built in the early 1930s, with a new winery constructed in 1993. No recent tastings. Limited retail distribution in NSW and WA.

rosemount estate (hunter valley) ★★★★★

Rosemount Road, Denman, NSW 2328 **region** Upper Hunter Valley
ph (02) 6549 6400 **fax** (02) 6549 6499 **open** Mon–Sat 10–4; Sun summer 10–4, winter 12–4
winemaker Philip Shaw **prod.** 700 000 **est.** 1969

prod. range ($8.99–50.95 R) A very large range of wines which in almost all instances are varietally identified, sometimes with the conjunction of vineyards at the top end of the range, and which in the case of the lower-priced volume varietals increasingly come from all parts of southeast Australia. Names and label designs change regularly but the emphasis remains on the classic varietals. Roxburgh Chardonnay is the white flag-bearer, Mountain Blue Shiraz is the red leader; Chardonnay, Shiraz, Mountain Blue Shiraz Cabernet and Cabernet Sauvignon under the standard labels consistently excellent at the price. In 1997 a Yarra Valley Chardonnay was added to the regional range, which also encompasses Coonawarra, Orange and Mudgee.

summary Rosemount Estate has achieved a miraculous balancing act over the past years maintaining – indeed increasing – wine quality while presiding over an ever-expanding empire and ever-increasing production. The wines are consistently of excellent value; all have real character and individuality; not a few are startlingly good.

Rosemount Estate Diamond Label Sauvignon Blanc

While made by Rosemount for many years, took a new (and better) turn with the 1997 vintage, which incorporates the first harvest from Rosemount's new The Range Vineyard in the Adelaide Hills, first planted in 1994 and dedicated to the production of sauvignon blanc. The '97 vintage was a blend of 60% Hunter Valley and 40% Adelaide Hills fruit, but with the latter component providing the character the wine has. It is a reasonable assumption that the percentage of Adelaide Hills material will increase in future years, although it will be interesting to see whether Rosemount makes a separate, higher-priced release.

🍷🍷🍷🍷 **1997** Light to medium yellow-green; a fragrant and lively bouquet with clear but not aggressive varietal character is followed by a fresh, clean and crisp palate with nicely weighted gooseberry and herb fruit. There is appropriate acidity on the finish. **rating:** 87

best drinking 1998–1999 **best vintages** NA **drink with** Pan-fried prawns • $15

Rosemount Estate Diamond Label Chardonnay

The only wine in this large so-called premium varietal range to continue to have and claim a Hunter Valley origin. The wine is made to throw maximum emphasis on fairly rapid flavour development; part of the wine underwent malolactic fermentation while ageing in French and American oak, while part was tank-fermented and held on lees prior to blending.

🍷🍷🍷 **1998** Medium yellow-green, quite full. The bouquet is clean, with some slightly minerally/green notes, not expected given the vintage, but the palate has good mouthfeel with peach and citrus flavours. Clearly given only the briefest contact with oak, for it has many of the characters of unoaked Chardonnay. **rating:** 79

best drinking 1998–1999 **best vintages** '84, '86, '87, '90, '93, '95 **drink with** Veal parmigiana • $14.95

Rosemount Estate Rose Label Orange Chardonnay

The vineyard was planted in 1989 at an elevation of 900 metres. The planting density is high, and a high-wall vertical trellis is used. The wine undergoes barrel fermentation and usually undergoes

100% malolactic fermentation. Both the '95 and '96 vintages won gold medals at the 1997 Liquorland National Wine Show, an outstanding achievement.

🍷🍷🍷🍷🍷 **1997** Light green-yellow; a fine and elegant bouquet with aromas of citrus, apple, melon and nectarine introduces a well-crafted wine, slow-developing, with unforced fruit and oak, and very much in the style of preceding vintages. **rating:** 90

best drinking 1999–2004 **best vintages** '92, '95, '96, '97 **drink with** Oyster soup • $23.95

Rosemount Estate Show Reserve Chardonnay

The grapes for this wine are sourced from favoured blocks on two of Rosemount's principal Hunter Valley vineyards – Roxburgh and Giants Creek. The winemaking methods are nothing if not complex, with the Giants Creek component being processed in a quite different fashion from that of the Roxburgh fruit. Likewise, the juices were cold-settled for periods ranging between six and 48 hours, deliberately giving different levels of solids in the musts. Most of the wine was transferred straight to barrel for the initiation of fermentation, with 60% French Allier and 40% American oak being used. The malolactic fermentation occurs spontaneously.

🍷🍷🍷🍷🍷 **1997** Medium yellow-green; the bouquet is quite substantial, though avoids phenolic heaviness, with fig and melon fruit and a nice touch of spicy/toasty oak. The palate offers fig and melon varietal fruit running through nicely-handled oak; good acidity on the finish. A winner of a trophy and two gold medals, and which seems to have unusual cellaring potential for a Hunter Chardonnay, and particularly one from a fairly difficult vintage. **rating:** 91

best drinking 1999–2002 **best vintages** NA **drink with** Pan-fried veal • $24.95

Rosemount Estate Diamond Label Shiraz

A wine which has been largely responsible for the phenomenal success of Rosemount Estate in the US, winning consistently high points from the world's most influential wine magazine, *Wine Spectator*. While Rosemount has never made any secret of the fact that it does subtly alter the balance of the wines according to the market destination, the quality of this wine throughout the 1990s has been extraordinarily consistent and extraordinarily good. Like the Cabernet Sauvignon, drawn principally from McLaren Vale, Langhorne Creek and Mudgee; is aged in American oak for ten months prior to bottling.

🍷🍷🍷🍷🍷 **1998** Medium to full red-purple; the bouquet and palate show unusual depth of flavour and extract for a wine at this price point, with a wonderful mix of liquorice, ripe berry and chocolate fruit running through the palate, but then softening off on the finish, which is exactly what one would wish for in a wine destined for early consumption. Gold medal 1999 Sydney Wine Show. **rating:** 93

best drinking 1999–2003 **best vintages** '88, '90, '91, '92, '94, '96, '98 **drink with** Lamb shanks • $14.95

Rosemount Estate Shiraz Cabernet

Rosemount has consistently achieved more with this early-release, early-drinking red wine than any other Australian producer. The wine shows no sign of having been rushed into bottle, but is usually released within six months of vintage.

🍷🍷🍷🍷 **1998** Medium purple-red; the moderately intense but clean bouquet has aromas of berry and a hint of leaf, and the wine has a sweet, softly fruity mid to back palate precisely made for early drinking. Oak is not evident, but the wine does not have the callow, undermade character which sometimes disfigures early-drinking styles from other makers. **rating:** 87

best drinking 1998–1999 **best vintages** NA **drink with** Lasagne • $11

Rosemount Estate Mountain Blue Shiraz Cabernet

The first regional release by Rosemount from Mudgee, but doubtless not the last. The Rosemount jigsaw puzzle continues to spread across the premium wine-growing regions of Australia, but without any compromise on quality, as evidenced by the trophy and seven gold medals which this wine has won. A blend of 90% Shiraz and 10% Cabernet Sauvignon, much of the Shiraz coming from the 40-year-old shiraz planted on the Mountain Blue vineyard, and which is low yielding. Matured in a mix of French and American oak for 18 months, 100% new. The '96 had won four major trophies and nine gold medals by the end of 1998.

🍷🍷🍷🍷🍷 **1996** Medium to full red-purple; the cascading fruit of the bouquet runs through blackberry, blackcurrant, chocolate and a touch of prune; here, as on the palate, the mix of new French and American oak has been skilfully balanced and integrated. The palate is as voluptuous as you would expect, yet very well balanced, with sweet fine tannins. A mere four trophies and five gold medals to its credit. **rating:** 96

best drinking 2002–2012 **best vintages** '95, '96 **drink with** Barbecued beef • $50.95

rosemount estate (mclaren vale) ★★★★★

Ingoldby Road, McLaren Vale, SA 5171 **region** McLaren Vale
ph (08) 8383 0001 **fax** (08) 8383 0456 **open** Mon–Fri 10–5, weekends 11–5
winemaker Charles Whish **prod.** 100 000 **est.** 1888
prod. range ($20–69.95 CD) Ryecroft Unwooded Chardonnay, Balmoral Syrah, Show Reserve Shiraz, GSM (Grenache Shiraz Mourvedre blend), Traditional (Cabernet blend), Ryecroft Cabernet Shiraz.
summary The specialist red wine arm of Rosemount Estate, responsible for its prestigious Balmoral Syrah, Show Reserve Shiraz and GSM.

Rosemount Estate Show Reserve Shiraz

Rosemount has been making a Reserve Shiraz from low-yielding McLaren Vale vines for many years now. It in fact pre-dates the acquisition of the Ryecroft vineyard and winery which now constitutes the Rosemount base in McLaren Vale. It also pre-dates the development of the super-premium Balmoral Syrah, which uses similar (albeit the very top) fruit sources. The majority of the vines are between 50 and 100 years of age. It is matured in a mix of new and one-year-old American oak barrels for 20 months. The wine took on a new complexion with the 1995 vintage, which was a blend of 60% grapes grown at Orange, NSW and 40% from McLaren Vale. Its four gold medals attest to its quality.

🍷🍷🍷🍷🍷 **1995** Medium red, with a tinge of purple; the scented, elegant bouquet with aromas of fruit spice and cedary oak leads into a very elegant palate, with a touch of McLaren Vale chocolate on the mid-palate followed by fine, subtle tannins on the long finish. Very classy. **rating:** 94

best drinking 2000–2010 **best vintages** '90, '91, '93, '94, '95 **drink with** Marinated beef • $25.90

Rosemount Estate Balmoral Syrah

The name comes from the 1852 Hunter Valley homestead of the Oatley family, founders of the Rosemount Estate. Tastings of all the wines so far released ('89 to '94) have shown an incredible consistency of style, with the 100-year-old vines at Rosemount's Ryecroft Vineyard in McLaren Vale at the heart of the wine. Less than 2000 cases are made each year, 50% being exported to the United States.

🍷🍷🍷🍷🍷 **1996** Dense purple-red; the bouquet is enormously rich, with chocolate and prune fruit woven through dense oak. The palate is exceptionally rich, dense and succulent; somehow or other, the tannin and oak levels have been kept under control. A great example of the Henry VIII red wine style. **rating:** 96

best drinking 2006–2016 **best vintages** '86, '87, '88, '89, '91, '93, '95, '96 **drink with** Char-grilled rump • $69.95

Rosemount Estate GSM

Grenache (typically 50%), Shiraz (typically 40%) and Mourvedre (typically 10%) grapes are selected from old, low-yielding vines drawn from vineyards across McLaren Vale, and are separately fermented in a mix of traditional open fermenters and vinomatics (rotary fermenters). The wines are matured for 18 months in a mix of new and used American oak, and a final blend decision taken shortly prior to bottling, the blend components varying slightly according to the outcome of vintage. A wine which adds lustre to the cornucopia of Rhône-style wines from McLaren Vale. The '95 received the Chairman's Trophy at the 1998 Sydney International Wine Competition.

YYYY **1996** Strong purple-red; a rich and powerful array of dark berry and more savoury fruit on the bouquet is followed by a densely structured palate with a mix of dark berry and liquorice flavours. Just a fraction muscle-bound early in its life, and should improve markedly given time to mature. **rating:** 88

best drinking 2001–2007 **best vintages** '94, '95, '96 **drink with** Beef with olives • $24.95

Rosemount Estate Traditional

When Rosemount acquired the Ryecroft winery in 1991, Ryecroft had a brand simply called 'Traditional'. Rosemount has continued using the name, applying it to, what in Bordeaux, France would be described as a 'traditional blend of Cabernet Sauvignon, Merlot and Petit Verdot'. As with the GSM, the percentage will vary from one year to the next, but is typically 70% Cabernet Sauvignon, 20% Merlot and 10% Petit Verdot. Likewise as with the GSM, the components are separately fermented and matured, blending taking place after oak maturation.

YYYYY **1996** Very good purple-red; the bouquet is complex, with the skilful use of high-quality oak very evident. A most interesting wine on the palate, powerfully structured, with a mix of ripe and more cedary characters, then a touch of chocolate (regional) and well-balanced tannins. **rating:** 94

best drinking 2001–2007 **best vintages** '91, '94, '95, '96 **drink with** Char-grilled beef • $23.95

Ryecroft Cabernet Shiraz

The Ryecroft label was reintroduced to the market by Rosemount early in 1999. The Cabernet Shiraz has 50% of each variety sourced from a wide variety of vineyards in South Australia and Victoria. It is matured for nine months, predominantly in a combination of stainless steel and Vinomatics to emphasise the fruit while reducing tannins, and given light oak maturation to add structure without detracting from the fruit flavour. Yet another masterly exercise by Rosemount.

YYYY **1998** Youthful purple-red; bright, fresh, sweet cherry/berry fruit with minimal oak evident on the bouquet leads into an attractive, light to medium weight wine, with the same fruit flavours as those of the bouquet, and nicely gauged tannins. Fresh and breezy. **rating:** 89

best drinking 1999–2001 **best vintages** NA **drink with** Steak and kidney pie • $13

rosewhite vineyards NR

Happy Valley Road, Rosewhite via Myrtleford, Vic 3737 **region** Ovens Valley
ph (03) 5752 1077 **open** Weekends and public holidays 10–5, 7 days January
winemaker Joan Mullett **prod.** 700 **est.** 1983
prod. range ($10 CD) Traminer, Chardonnay, Pinot Noir, Shiraz, Cabernet Sauvignon, Tawny Port.
summary After a career with the Victorian Department of Agriculture, agricultural scientists Ron and Joan Mullett began the establishment of Rosewhite in 1983, and have since established a little over 2 hectares of vineyards at an altitude of 300 metres.

rossetto ★★☆

Farm 576 Rossetto Road, Beelbangera, NSW 2686 **region** Riverina
ph (02) 6963 5214 **fax** (02) 6963 5542 **open** Mon–Sat 8.30–5.30
winemaker Eddy Rossi **prod.** 500 000 **est.** 1930
prod. range ($5.50–22 R) Several ranges, the commercial Wattle Glen series ($5.50), the Silky Oak range ($8.50–$9.50), the Rossetto premium varietals ($9–$16) and the Promenade Range of Riverina Chardonnay, Watervale Riesling and Riverina Cabernet Merlot ($12.50–$20); also fortifieds.
summary Another family-owned and run Riverina winery endeavouring to lift the profile of its wines, although not having the same spectacular success as Riverina Wines. Rossetto does have distributors in each State, but of course much of the total production is sold in bulk to other makers.

rothbury estate ★★★★

Broke Road, Pokolbin, NSW 2321 **region** Lower Hunter Valley
ph (02) 4998 7555 **fax** (02) 4998 7553 **open** 7 days 9.30–4.30
winemaker Adam Eggins, Rob Cuadagnini **prod.** 30 000 **est.** 1968
prod. range ($12–22 R) At the top comes the Individual Vineyard range of Hunter Valley Semillon, Chardonnay and Shiraz; next the Hunter Valley range of varietals; and finally varietals from Mudgee and Cowra.
summary Rothbury celebrated its 30th birthday in 1998, albeit not quite in the fashion that founder and previous chief executive Len Evans would have wished. After a protracted and at times bitter takeover battle, it became part of the Fosters/Mildara empire. Exports and distribution via Mildara Blass.

Rothbury Estate Brokenback Semillon

One of three wines at the top of the quality tree for Rothbury, and, fittingly in my view, the best of the three (the other two being Chardonnay and Shiraz respectively). It seems almost superfluous to say it is made in the traditional fashion, cold-fermented in stainless steel and early-bottled.

🍷🍷🍷🍷🍷 **1998** Light green-yellow; crisp herb and grass aromas, more intense than the Rothbury Hunter Valley Semillon, lead into a palate which is surprisingly intense, with a touch of passionfruit/tropical fruit to add zest and life. Promises to become something special given time. **rating:** 92

best drinking 2002–2008 **best vintages** '97, '98 **drink with** Sushi • $18

Rothbury Estate Hunter Valley Semillon

Has appeared under various guises since the first experimental vintage in 1971 (first fully commercial 1972). Winemakers have come and gone; labels have changed, recently with increasing rapidity; and, inevitably, quality has wandered all over the place, at times great, at times not. The '97 and '98 vintages suggest there is hope for the wine yet.

🍷🍷🍷🍷 **1998** Light green-yellow; light, crisp, mineral and herb aromas lead into a very correct varietal palate with good length, and the certainty of developing well. **rating:** 88

best drinking 2001–2007 **best vintages** '72, '73, '74, '75, '76, '79, '84, '89, '93, '94, '96, '97, '98 **drink with** Smoked eel • $15

---

rotherhythe ★★★★☆

Hendersons Lane, Gravelly Beach, Exeter, Tas 7251 **region** Northern Tasmania
ph (03) 6394 4869 **open** By appointment
winemaker Steven Hyde **prod.** 1600 **est.** 1976
prod. range ($16–26.95 CD) Chardonnay, Pinot Noir, Cabernet Sauvignon, Pinot Chardonnay.

summary At the 1996 Tasmanian Wines Show Rotherhythe swept all before it, winning trophies for Most Successful Exhibitor, Best Light to Medium Bodied Red Wine, Best Full Bodied Red Wine and Best Wine of Show. Ironically, two days later, Dr Steven Hyde sold the vineyard, although he has retained all of the existing wine stocks and will remain involved in the winemaking for some time to come. In both 1997 and again in 1998 Rotherhythe was awarded the trophy for Most Successful Exhibitor at the Tasmanian Wines Show. Retail distribution through Sutherland Cellars, Melbourne.

Rotherhythe Cabernet Sauvignon

Dr Steven Hyde made a series of great Cabernets between 1990 and 1994 which earned Rotherhythe a cascade of trophies and medals between 1992 and 1999 at the Tasmanian Wines Show.

ƳƳƳƳƳ **1994** Full red-purple; the bouquet is rich with some welcome dark chocolate aromas, the palate big, powerful and complex with blackberry and chocolate fruit offering the generosity lacking in many Tasmanian Cabernets. **rating:** 90

best drinking 1999–2005 **best vintages** '90, '91, '92, '93, '94 **drink with** Braised lamb shanks • $26.95

rouge homme ★★★☆

Riddoch Highway, Coonawarra, SA 5263 **region** Coonawarra
ph (08) 8736 3205 **fax** (08) 8736 3250 **open** 7 days 10–5
winemaker Paul Gordon **prod.** 64 000 **est.** 1954
prod. range ($10–40 R) Semillon, Chardonnay, Unoaked Chardonnay, Pinot Noir, Reserve Pinot Noir, Shiraz Cabernet, Cabernet Merlot, Cabernet Sauvignon.

summary From time to time I have described Rouge Homme as the warrior brand of the Lindeman Group Coonawarra operations. In recent times it has proved a formidable warrior, most surprisingly with its Pinot Noir, but also with its Cabernet and Cabernet blend wines.

Rouge Homme Cabernet Sauvignon

While in price terms the flagship of the range, does not seem to have the quality edge over the other wines that one might expect. Certainly it is well made, coming as it does from 100% Coonawarra cabernet sauvignon, and being matured in French oak hogsheads for 15 months. The weight and concentration of the wine has increased significantly in recent vintages. The about-to-be-released '96 was a gold medal winner at the 1997 Liquorland National Wine Show.

ƳƳƳƳ **1995** Medium red-purple; the moderately intense bouquet is predominantly in the earthy herbaceous spectrum of cabernet varietal character but sweetens up on the palate with a mix of chocolate, mint, leaf and earth – but the sweet notes predominant – supported by subtle oak. **rating:** 86

best drinking 1999–2004 **best vintages** '88, '90, '91, '94 **drink with** Roast leg of lamb • $16.30

ruker wines NR

Barton Highway, Dickson, ACT 2602 **region** Canberra District
ph (02) 6230 2310 **fax** (02) 6230 2818 **open** Weekends, public holidays 10–5
winemaker Richard Ruker **prod.** 500 **est.** 1991
prod. range ($15 CD) Riesling, Gewurztraminer.

summary Barbara and Richard Ruker, with the assistance of eldest daughter Niki, planted 2 hectares of riesling and traminer in 1984. The cellar door-cum-winery is a farmshed, subsequently converted to an office and then to its present function of winery and restaurant; it is finished with heavy wooden beams salvaged from a railway bridge near Tarago and clad with the remains of an old slab hut, while the tables are made from huge red and yellow box trees cut down when the vineyard was planted.

rumbalara ★★☆

Fletcher Road, Fletcher, Qld 4381 **region** Granite Belt
ph (07) 4684 1206 **fax** (07) 4684 1299 **open** 7 days 9–5
winemaker Bob Gray **prod.** 2000 **est.** 1974
prod. range ($9–19.50 CD) Barrel Fermented Semillon, Granitegolde, Light Shiraz, Cabernet Sauvignon, Pinot Noir, Cabernet Shiraz and a range of fortified wines, Cider and Vermouth.
summary Has produced some of the Granite Belt's finest, honeyed Semillon and silky, red berry Cabernet Sauvignon, but quality does vary. The winery incorporates a spacious restaurant, and there are also barbecue and picnic facilities.

ryland river NR

RMB 8945 Main Creek Road, Main Ridge, Vic 3928 **region** Mornington Peninsula
ph (03) 5989 6098 **fax** (03) 9899 0184 **open** Weekends and public holidays 10–5 or by appointment
winemaker John W Bray **prod.** 2000 **est.** 1986
prod. range ($15–30 CD) Semillon Sauvignon Blanc, Chardonnay, Cabernet Sauvignon, Jack's Delight Tawny Port and Muscat.
summary John Bray has been operating Ryland River at Main Ridge on the Mornington Peninsula for a number of years, but not without a degree of controversy over the distinction between Ryland River wines produced from Mornington Peninsula grapes and those produced from grapes purchased from other regions. A large lake with catch-your-own trout and a cheese house are general tourist attractions.

rymill ★★★☆

The Riddoch Run Vineyards, Coonawarra, SA 5263 **region** Coonawarra
ph (08) 8736 5001 **fax** (08) 8736 5040 **open** 7 days 10–5
winemaker John Innes **prod.** 50 000 **est.** 1970
prod. range ($12–25 R) Sauvignon Blanc, Chardonnay, March Traminer, June Traminer, Shiraz, Merlot Cabernets, Cabernet Sauvignon and Sparkling.
summary The Rymills are descendants of John Riddoch, and have long owned some of the finest Coonawarra soil upon which they have grown grapes since 1970, with present plantings of 140 hectares. Peter Rymill made a small amount of Cabernet Sauvignon in 1987, but has long since plunged headlong into commercial production, with winemaker John Innes presiding over the striking winery portrayed on the label. Australian distribution is through Negociants Australia; exports go to all of the major markets in Europe, North America and Asia.

Rymill Coonawarra Sauvignon Blanc

Rymill has been producing a Sauvignon Blanc for almost a decade, but is now able to use grapes from both its younger Mount Stones Vineyard as well as the older Riddoch Run Vineyard. The wine is made from up to seven separate components, picked at varying degrees of ripeness, and with varying contributions to the flavour.

🍷🍷🍷🍸 **1996** Light straw-yellow; the bouquet is, with a mix of relatively austere mineral and herb aromas, but with nuances of ripe fruit. The palate shows a similar mix of flavours and characters, slightly sweeter than the bouquet, and with a faintly hot finish. **rating:** 83

best drinking 1998–1999 **best vintages** NA **drink with** Calamari • $14

saddlers creek ★★★☆

Marrowbone Road, Pokolbin, NSW 2321 **region** Lower Hunter Valley
ph (02) 4991 1770 **fax** (02) 4991 1778 **open** 7 days 9–5
winemaker John Johnstone **prod.** 10 000 **est.** 1989

prod. range ($17–28 CD) The Estate selection of Marrowbone Chardonnay, Sauternes, Marrowbone Pinot Noir, Cabernet Merlot, Equus Shiraz, Equus Hunter Shiraz, Equus McLaren Shiraz, Bluegrass Cabernet Merlot, Bluegrass Cabernet Sauvignon (magnum); Reserve selection Malolactic Semillon, Unwooded Chardonnay, Sauvignon Blanc, Sauvignon Blanc Semillon.

summary Made an impressive entrance to the district with consistently full-flavoured and rich wines. Marrowbone Chardonnay and Equus Hunter Shiraz are its best wines. Limited retail distribution in New South Wales, Queensland and Victoria.

Saddlers Creek Marrowbone Chardonnay

The top-end of the Saddlers Creek range, with all of the rich fruit one expects from good Hunter Chardonnay, allied with some good French oak. The seemingly unusual name is not so unusual once you find that Saddlers Creek is situated on Marrowbone Road.

🍷🍷🍷🍷🍷 **1997** Medium yellow-green; the bouquet has slightly honeyed overtones to the peachy fruit and subtle oak. A stylish and (for the Hunter Valley) unusual palate with texture and richness, good oak integration, nice acidity, all in all adding up to a stylish wine which is ageing slowly and with grace. **rating:** 90

best drinking 1999–2005 **best vintages** '95, '97 **drink with** Bone marrow in brioche • $25

st gregory's NR

Bringalbert South Road, Bringalbert South via Apsley, Vic 3319 **region** Far South West Victoria

ph (03) 5586 5225 **open** By appointment

winemaker Gregory Flynn **prod.** NFP **est.** 1983

prod. range ($14 ML) Port.

summary Unique Port-only operation selling its limited production direct to enthusiasts (by mailing list).

st hallett ★★★★☆

St Hallett's Road, Tanunda, SA 5352 **region** Barossa Valley

ph (08) 8563 2319 **fax** (08) 8563 2901 **open** 7 days 10–4

winemaker Stuart Blackwell, Cathy Spratt **prod.** 65 000 **est.** 1944

prod. range ($9.50–35 CD) Poacher's Blend (White), Eden Valley Riesling, Semillon Sauvignon Blanc, Semillon Select, Sweet Meredith, Chardonnay, Gamekeeper's Reserve (Red), Faith Shiraz, Blackwell Shiraz, Grenache, Cabernet Merlot, Old Block Shiraz.

summary Nothing succeeds like success, and St Hallett continues to grow, significantly expanding the range of its Shiraz-based wines, but also coming up with wines such as the multiple gold medal and trophy winning 1997 Eden Valley Riesling. One has to say that some of the wines (even Old Block Shiraz) don't seem to have quite the intensity (or is it excitement?) they once had. Elegant and smooth, yes, but with a certain sameness about the wines. It has established its own distribution network in the UK, and actively exports to Europe, North America and Asia.

St Hallett Eden Valley Riesling

A wine which burst from the ruck in 1997, winning a trophy at the 1997 Barossa Valley Wine Show, and then going on to win a gold medal in Class 1 at the 1997 Liquorland National Wine Show. Eden Valley Riesling at its best.

🍷🍷🍷🍷🍷 **1998** Medium yellow-green; the bouquet is quite powerful, with hints of the herb and mineral which seem to have characterised many of the '98 vintage Rieslings. The palate has good structure, with lime, herb and mineral flavours; crisp, clean and good acidity. Not quite up to the stellar '97 vintage. **rating:** 90

best drinking 2001–2006 **best vintages** '97 **drink with** Crab • $15

St Hallett Poacher's Blend

A somewhat improbable blend of Chenin Blanc, Semillon, Sauvignon Blanc and Riesling, with the components coalescing rather than warring. The wine is stainless steel-fermented and early-bottled.

🍷🍷🍷🍷 **1998** Light to medium green-yellow; the fresh, lively and tangy herbal/citrus aromas are followed by an equally fresh and zippy palate, in which the hint of sweetness is balanced by acidity. **rating:** 87

best drinking 1999–2000 **best vintages** NA **drink with** Mussel and saffron soup • $9.50

st huberts ★★★★

Maroondah Highway, Coldstream, Vic 3770 **region** Yarra Valley
ph (03) 9739 1118 **fax** (03) 9739 1096 **open** Mon–Fri 9–5, weekends 10.30–5.30
winemaker Fiona Purnell **prod.** 5000 **est.** 1966
prod. range ($20–30 R) Roussanne, Chardonnay, Pinot Noir, Cabernet Sauvignon, Cabernet Merlot; under the second label Rowan Sauvignon Blanc, Chardonnay, Shiraz, Pinot Noir, Cabernet Merlot.

summary The changes have come thick and fast at St Huberts, which is now part of the Mildara Blass (Rothbury Estate) Group. It has produced some quite lovely wines, notably Chardonnay and Cabernet Sauvignon. Plans are afoot for the rebuilding of the ornate nineteenth-century winery on its original (recently repurchased) site.

St Huberts Roussanne

The Yarra Valley was famous for its Marsanne in the nineteenth century, and St Hubert's neighbour Yeringberg reintroduced that grape to the Valley in the 1970s, following with small plantings of roussanne, paralleled by plantings at St Huberts. This is one of the rarest wines in Australia, and 1994 marked the first release from St Huberts; cellar door is the best bet to find the wine.

🍷🍷🍷🍷 **1998** Deep yellow-gold; there are quite powerful aromas of dried flowers and dried orange peel on the bouquet, followed by equally generous palate flavour with notes of honey, honeysuckle and hay to go with the sweet fruit of the bouquet. **rating:** 89

best drinking 1999–2001 **best vintages** NA **drink with** Pork with apple or peach • $19

St Huberts Cabernet Sauvignon

Almost 50% of the 20 hectares of estate plantings at St Huberts are cabernet sauvignon, some of it original plantings, and more grafted over from other varieties. The '77 St Huberts Cabernet caused a sensation at the time, both for its quality and its then astronomically high price of $17 a bottle. Clever winemaking continues to produce rich, smooth wines 20 years later, selling at much the same price.

🍷🍷🍷🍷🍷 **1997** Strong red-purple; the bouquet has attractive sweet cassis berry fruit and cedary oak. The palate opens up with sweet blackcurrant fruit, carried along by the same sweet oak of the bouquet which marries well without dominating. **rating:** 92

best drinking 2002–2007 **best vintages** '77, '88, '90, '91, '92, '96, '97 **drink with** Rich casserole dishes • $22

st leonards ★★★☆

Wahgunyah, Vic 3687 **region** Rutherglen
ph (02) 6033 1004 **fax** (02) 6033 3636 **open** Thur–Sun and holidays 11–3
winemaker Peter Brown **prod.** NFP **est.** 1860
prod. range ($12.50–35 CD) Kalara Red and White; Carlyle Chardonnay, Wahgunyah Shiraz, Carlyle Ruby Cabernet.

summary An old favourite, relaunched in late 1997 with a range of three premium wines cleverly marketed through a singularly attractive cellar door and bistro at the historic winery on the banks of the Murray. All Saints and St Leonards are now wholly owned by Peter Brown; the vast majority of the wines are sold through cellar door and by mailing list.

St Leonards Carlyle Chardonnay

This is the top-of-the-range release for St Leonards; some of the previous vintages have been excessively oaky, but no such problem exists with the '97.

🍷🍷🍷🍷 **1997** Medium to full yellow-green; the bouquet is quite toasty and rich, but there is a nice tangy cut from the fruit. A well-made wine, with lots of mid-palate buttery, peachy fruit, but not overoaked or overblown. **rating:** 87

best drinking 1999–2002 **best vintages** '94 **drink with** Creamy pasta • $12.50

st mary's NR

V & A Lane, via Coonawarra, SA 5277 **region** Other Wineries of SA

ph (08) 8736 6070 **fax** (08) 8736 6045 **open** 7 days 10–4

winemaker Barry Mulligan **prod.** 4000 **est.** 1986

prod. range ($12–20 CD) Riesling, Chardonnay, Shiraz, House Block Cabernet Sauvignon.

summary Established by the Mulligan and Hooper families in 1986, but with Tyrrell's Vineyards purchasing the Hooper interest in the vineyards (though not the brand name St Mary's) in 1995. The winemaking operation continues as a separate entity, now wholly owned by the Mulligans.

st matthias NR

113 Rosevears Drive, Rosevears, Tas 7277 **region** Northern Tasmania

ph (03) 6330 1700 **fax** (03) 6330 1975 **open** 7 days 10–5

winemaker Alain Rousseau **prod.** 4000 **est.** 1983

prod. range ($13–21 CD) Riesling, Chardonnay, Pinot Noir, Cabernet Sauvignon Merlot, Brut, Cuvee Printemps Dry White, Cuvee Printemps Dry Red.

summary Acquired by Moorilla Estate in 1995 and, after a quiet period, now fully back in business. The wines are made at Moorilla Estate by Alain Rousseau, but are sold through the cellar door (along with those of Moorilla itself).

St Matthias Chardonnay

The St Matthias label is now effectively a second label of Moorilla, with the grapes coming from the St Matthias Vineyard on the Tamar River.

🍷🍷🍷🍷 **1998** Light to medium yellow-green; the bouquet is clean and smooth, with solidly ripe fruit and subtle oak. The palate is equally clean and gentle, with citrus melon fruit, a touch of oak, and lingering acidity. **rating:** 87

best drinking 1999–2002 **best vintages** NA **drink with** Pan-fried veal • $16.95

st peter's edenhope wines NR

Whitton Stock Route, Yenda, NSW 2681 **region** Riverina

ph (0418) 421 000 **fax** (02) 4285 3180 **open** Not

winemaker Wilton Estate (Contract) **prod.** 17 000 tonnes **est.** 1978

prod. range ($7.99–18.99 CD) A wide variety of wines under the Edenhope, St Peter's and Sydney labels.

summary St Peter's Edenhope Wines is not only the physical home of the perhaps better-known Wilton Estate (as tenant) but also produces three ranges of wine for itself, Edenhope being conventionally distributed through the wholesale/retail chain, and St Peter's and Sydney sold direct ex-winery to selected outlets. Managing Director Stephen Chatterton's other claim to fame is as a fanatical fly-fisher and fly-tyer which necessarily means he is made of the right stuff.

salem bridge wines NR

Salem Bridge Road, Lower Hermitage, SA 5131 **region** Adelaide Hills
ph (08) 8380 5240 **fax** (08) 8380 5240 **open** Not
winemaker Barry Miller **prod.** NFP **est.** 1989
prod. range ($16–18 R) Cabernet Franc, Shiraz, Cabernet Sauvignon.
summary Barry Miller acquired the 45-hectare Salem Bridge property in the Adelaide Hills of South Australia in 1988. A little under 2 hectares of cabernet franc were planted in 1989, and cabernet franc has been the only commercial release prior to 1999. However, a further 13 hectares have been planted to cabernet sauvignon, shiraz and merlot, with a Shiraz and Cabernet Sauvignon release in the pipeline. The wine is made offsite by contract-winemaking with input from Barry Miller.

salisbury estate ★★★

Campbell Avenue, Irymple, Vic 3498 **region** Murray Darling (Vic)
ph (03) 5024 6800 **fax** (03) 5024 6605 **open** Mon–Sat 10–4.30
winemaker Bob Shields, David Martin **prod.** 385 000 **est.** 1977
prod. range ($5–13 R) Top-end wines under the Milburn Park label are Chardonnay and Cabernet Sauvignon and under the Salisbury Estate label Show Reserve Chardonnay and Show Reserve Cabernet Sauvignon. Then comes the standard Salisbury Estate range of Rhine Riesling, Chardonnay Semillon, Sauvignon Blanc Semillon, Chardonnay, Cabernet Sauvignon, Cabernet Merlot; the Castle Crossing range is even cheaper, consisting of Fumé Blanc, Colombard Chardonnay, Spätlese Rhine Riesling, Chenin Blanc, Chambourcin, Shiraz Malbec Mourvedre, Claret; Acacia Ridge non-vintage generics bring up the rear, with two wines in the Tennyson Vineyard off to one side.
summary Salisbury Estate, Milburn Park and Castle Crossing are three of the principal brands of Australian Premium Wines Limited, a public company which moved onto the lists of the Australian Stock Exchange in 1998, having acquired the 157-hectare Kasbah Vineyard near Loxton and a 40-hectare property at Koppamurra in South Australia (both in 1995) and Haselgrove Wines (in mid-1997). Also produces the equivalent of 4.9 million litres of bulk wine.

Salisbury Estate Milburn Park Chardonnay

A very limited-production wine, with only 1500 cases being made. The wine was barrel-fermented in new Seguin Moreau French and American oak hogsheads, left on lees for six months and then bottled. It won the Trophy for Best Victorian Dry White Table Wine at the 1998 Royal Melbourne Wine Show, and also won a gold medal at the 1998 Rutherglen Wine Show. The label, incidentally, does not claim Victorian appellation, but it necessarily has to be for entry into the various Victorian-based wine shows.

🍷🍷🍷🍷 **1998** Medium yellow-green; the barrel-ferment characters drive the bouquet and the palate. The wine has some elegance, particularly on the palate, but you really do have to accept oak as the primary flavour. **rating:** 88

best drinking 1999–2000 **best vintages** NA **drink with** Brains in black butter • $15

salitage ★★★★

Vasse Highway, Pemberton, WA 6260 **region** Pemberton
ph (08) 9776 1771 **fax** (08) 9776 1772 **open** 7 days 10–4
winemaker Patrick Coutts, Ross Pamment **prod.** 12 000 **est.** 1989
prod. range ($18–34.50 R) Chardonnay, Unwooded Chardonnay, Sauvignon Blanc, Pinot Noir, Cabernet Merlot; Treehouse range Chardonnay Verdelho, Pinot Noir, Shiraz and Cabernets.
summary Salitage is the showpiece of Pemberton. If it had failed to live up to expectations, it is a fair bet the same fate would have befallen the whole of the Pemberton region. The quality and

style of Salitage has varied substantially, presumably in response to vintage conditions and yields. It still remains the key producer in the Pemberton region, but the variability is a little unsettling as is the propensity to age quickly. Key retail distribution in all States, and exports to the UK, Germany, Switzerland, US, Canada, Hong Kong, Japan, Singapore, Taiwan and NZ.

Salitage Chardonnay

Estate-grown and produced, always with strong barrel-ferment and oak influences. As with the Pinot Noir, has tended to develop fairly quickly.

YYYY **1997** Medium to full yellow-green; the bouquet is full, with nutty/figgy fruit and no extreme oak character as in previous vintages. The palate follows down the same track, with figgy fruit, although there are some toasty/herbal notes which catch slightly on the finish. **rating:** 87

best drinking 2000–2004 **best vintages** NA **drink with** Turkey breast • $30.50

saltram ★★★★

Angaston Road, Angaston, SA 5353 **region** Barossa Valley
ph (08) 8564 3355 **fax** (08) 8564 2209 **open** 7 days 10–5
winemaker Nigel Dolan **prod.** 50 000 **est.** 1859
prod. range ($10–49.95 R) At the top is No. 1 Shiraz; then Mamre Brook, now 100% Barossa and comprising Chardonnay, Shiraz and Cabernet Sauvignon; Metala Black Label and White Label; and the Saltram Classic range sourced from southeast Australia.
summary There is no doubt that Saltram has taken giant strides towards regaining the reputation it held 30 or so years ago. Under Nigel Dolan's stewardship, grape sourcing has come back to the Barossa Valley for the flagship wines, a fact of which he is rightly proud. The red wines, in particular, have enjoyed great show success over the past few years, with No. 1 Shiraz, Mamre Brook and Metala leading the charge. Exports to the UK, US, southeast Asia and NZ.

Saltram Classic Chardonnay

A commercial wine with an abundance of honest flavour, always best drunk when young.

YYYY **1998** Medium to full yellow-green; a big, solid, buttery bouquet with some peach and subtle oak. A resounding rich, full Barossa style, but better than average thanks to the subtle oak handling and lack of phenolic extraction. A trophy winner at the 1998 Royal Perth Wine Show. **rating:** 86

best drinking 1999–2000 **best vintages** NA **drink with** Roast turkey • $10

Saltram Mamre Brook Chardonnay

On the face of it, bringing the source for a top-flight Chardonnay back to the Barossa adds a degree of difficulty which might easily be avoided by looking either to the Adelaide Hills, Eden Valley or McLaren Vale. However, this wine shows just what can be done.

YYYYY **1998** Medium yellow-green; the bouquet is complex and tangy, far removed from the normal Teutonic heaviness of Barossa Chardonnay. The palate does offer generous peach and nectarine fruit coupled with nice oak; good length. Despite all of its virtues, will develop quickly and should be drunk sooner rather than later. **rating:** 90

best drinking 1999–2001 **best vintages** '98 **drink with** Cold Barossa mettwurst • $17

Saltram Mamre Brook Shiraz

As from the 1996 vintage, the Mamre Brook Red has split itself from a Cabernet Shiraz blend into its two component parts, with a Mamre Brook Shiraz and a Mamre Brook Cabernet Sauvignon. It has also reverted to a 100% Barossa origin. The wine is made in the old, open fermenters of the Saltram winery with heading down boards used to encourage extraction. The wine spends 18 months in a mix of American (60%) and French (40%) oak.

YYYY **1996** Full red-purple; quite dense dark berry, dark chocolate and mint aromas are supported by plenty of oak. The palate is in classic Barossa Shiraz-style, with abundant sweet fruit, fleeting hints of leaf and mint, offset by charry oak. Unrecognisable from the wine first tasted in March 1998. **rating:** 88

best drinking 1999–2004 **best vintages** NA **drink with** Grilled calf's liver • $18

Saltram No. 1 Shiraz

Produced from old, low-yielding, dry-grown Barossa shiraz. Made in an extremely bold style, with lashings of oak. The '96 was first tasted when it was not long in bottle; a year later the wine had improved out of all recognition, a common feature of the marvellous vintage much of South Australia encountered in 1997.

YYYYY **1996** Medium to full red-purple; voluminous berry/briary fruit is matched by tangy oak on the bouquet. The palate brings chocolate and vanilla to the berry fruit adding complexity to the flavour, while tannins running right through the palate provide excellent structure. **rating:** 92

best drinking 2001–2010 **best vintages** NA **drink with** Richly sauced casserole • $29

Saltram Metala White Label Shiraz Cabernet

Draws upon a century-old block of vines at Metala Creek, and first made by present winemaker Nigel Dolan's father, Brian, in 1959. The 1961 vintage won the inaugural Jimmy Watson Trophy, adding to the lineage of one of Australia's enduring brands. A blend of 65% Shiraz and 35% Cabernet Sauvignon; the standard release under the famous Metala label with its striking script-like printing and numbered bottles.

YYYY **1996** Medium red-purple; the aromas of the bouquet run through red berry, with touches of spice, vanilla and mineral, and that attractive sweet berry fruit runs through the palate, finishing with a gentle tannin and mineral grip. **rating:** 87

best drinking 1999–2003 **best vintages** '96 **drink with** Italian • $16

Saltram Mamre Brook Cabernet Sauvignon

Like its brother, Mamre Brook Shiraz, as from the 1996 vintage 100% Barossa-grown. It is sourced from vineyards adjacent to the Angaston winery and from the eastern part of the Barossa Valley floor. It is matured in predominantly new Nevers oak for 18 months before bottling.

YYYY **1997** Medium to full red-purple; nicely balanced and integrated cedary oak combines with olivaceous fruit on the bouquet. The same sweet oak helps shape and balance the palate; I am not entirely convinced about the fruit character of this particular wine. **rating:** 87

best drinking 2001–2007 **best vintages** '96 **drink with** Smoked kangaroo • $18

Saltram Mr Pickwick Tawny Port

One of Australia's best-known Tawny Ports made from Barossa Shiraz and Grenache, and judiciously using a blend of very old and much younger material.

YYYYY **NV** Light to medium tawny, with just a hint of red; there is pronounced rancio with a hint of spice and clean, penetrating spirit on the bouquet. The palate is well balanced, showing a range of nutty/nutmeg/spice flavours with a clean, lingering finish, showing the skilled blending of younger and older material. Good acidity. **rating:** 91

best drinking 1999–2000 **best vintages** NA **drink with** After dinner • $49.95

sandalford ★★★☆

West Swan Road, Caversham, WA 6055 **region** Swan District

ph (08) 9874 9374 **fax** (08) 9274 2154 **open** 7 days 10–5

winemaker Bill Crappsley, Severine Dombret **prod.** 60 000 **est.** 1840

prod. range ($12.95–36 R) At the bottom end under the Caversham label Chenin Verdelho, Late Harvest Cabernet Shiraz; then the 1840 Collection of Semillon Sauvignon Blanc, Chardonnay and Cabernet Merlot; under the premium range Margaret River Mount Barker Riesling, Margaret River Verdelho, Mount Barker Margaret River Chardonnay, Mount Barker

Margaret River Shiraz, Mount Barker Margaret River Cabernet Sauvignon; also excellent fortifieds, notably Sandalera; also the new Element brand including Verdelho, Chardonnay and Cabernet Shiraz.

summary The arrival of Bill Crappsley as winemaker coupled with the refurbishment of the winery, has heralded major changes at Sandalford. Wine quality has improved year by year, with Chenin Blanc and Chardonnay leading the way, but not alone. The quality of the labelling and packaging has also taken a giant leap forwards. Exports to the UK, Switzerland, US, Japan, Singapore and Hong Kong.

Sandalford Mount Barker Margaret River Shiraz

First appeared on the scene with the '93 vintage, and has proved the synergies of the relatively uncommon regional blend.

ŸŸŸŸ **1997** Medium red-purple; solid earthy fruit blends with spicy vanilla oak on the bouquet; a quite elegant palate with attractive spice and gentle berry flavours, although the tannins are ever so slightly persistent. **rating:** 89

best drinking 2002–2007 **best vintages** '94, '95, '97 **drink with** Lamb chops • $22.50

sandalyn wilderness estate NR

Wilderness Road, Rothbury, NSW 2321 **region** Lower Hunter Valley
ph (02) 4930 7611 **fax** (02) 4930 7611 **open** 7 days 10–5
winemaker Adrian Sheridan (Contract) **prod.** 3500 **est.** 1988
prod. range ($14–20 CD) Semillon, Verdelho, Chardonnay, Pinot Noir.

summary Sandra and Lindsay Whaling preside over the picturesque cellar-door building of Sandalyn on the evocatively named Wilderness Road, where you will find a one-hole golf range and views to the Wattagan, Brokenback and Molly Morgan ranges. The estate has 8.85 hectares of vineyards.

sand hills vineyard ★★☆

Sandhills Road, Forbes, NSW 2871 **region** Other Wineries of NSW
ph (02) 6852 1437 **fax** (02) 6852 4401 **open** Mon–Sat 9–5, Sun 12–5
winemaker Jill Lindsay, John Saleh (Contract) **prod.** 900 **est.** 1920
prod. range ($10–13 CD) Classic Dry White, Chardonnay, Shiraz, Shiraz Cabernet, Tawny Port.

summary Having purchased Sand Hills from long-term owner Jacques Genet, the Saleh family have replanted the vineyard to appropriate varieties, with over 3 hectares of premium varieties having been established between 1989 and 1995. Winemaking duties are split between John Saleh at Sand Hills (Pinot Noir and fortifieds), Charles Sturt University (Cabernet Shiraz) and Jill Lindsay of Woodonga Hill (white wines).

sandstone ★★★

CMB Carbunup River, WA 6280 **region** Margaret River
ph (08) 9755 6271 **fax** (08) 9755 6292 **open** Not
winemaker Mike Davies, Jan Davies **prod.** 750 **est.** 1988
prod. range ($18–22 R) Semillon, Cabernet Sauvignon.

summary The family operation of consultant-winemakers Mike and Jan Davies, who also operate very successful mobile bottling plants. The wines are made at Ribbonvale, where the Davies work as consultants and contract-winemakers for others.

scarborough ★★★★

Gillards Road, Pokolbin, NSW 2321 **region** Lower Hunter Valley
ph (02) 4998 7563 **fax** (02) 4998 7786 **open** 7 days 9–5
winemaker Ian Scarborough **prod.** 10 000 **est.** 1985

prod. range ($19.50–25 CD) A Chardonnay specialist, making token quantities of Pinot Noir.

summary Ian Scarborough put his white winemaking skills beyond doubt during his years as a consultant, and his exceptionally complex and stylish Chardonnay is no disappointment. Vintage conditions permitting, Ian Scarborough makes two styles: a rich, traditional buttery White Burgundy version for the Australian market (exemplified by the 1995 with its mustard-gold label) and a lighter, more elegant Chablis-style (under a blue-silver label) for the export market. Exports to the UK and US.

scarpantoni estate ★★★☆

Scarpantoni Drive, McLaren Flat, SA 5171 **region** McLaren Vale
ph (08) 8383 0186 **fax** (08) 8383 0490 **open** Mon–Fri 10–5, weekends 11–5
winemaker Michael Scarpantoni, Filippo Scarpantoni **prod.** 15 000 **est.** 1979

prod. range ($8–25 CD) Block 1 Riesling, Sauvignon Blanc, Unwooded Chardonnay, Chardonnay, Fleurieu Brut, Gamay, Fiori, School Block (Cabernet Shiraz Merlot), Block 3 Shiraz, Cabernet Sauvignon, Botrytis Riesling, Tawny Port, Vintage Port.

summary While an erratic producer at times, and not helped by the earlier use of agglomerate corks, has made some excellent wines in recent years which – if repeated – would earn the winery an even higher rating.

Scarpantoni Estate Block 3 Shiraz

Yet another McLaren Vale Shiraz of high quality made from old McLaren Vale vines.

🍷🍷🍷🍷🍸 **1997** Dense, impenetrable purple-red; complex oak, chocolate, earth and berry aromas on the bouquet are followed by a ripe, rich palate with luscious, sweet berry and chocolate flavours. **rating:** 91

best drinking 2001–2010 **best vintages** '96, '97 **drink with** Seared kangaroo fillet • $17

scarp valley ★★★★

6 Robertson Road, Gooseberry Hill, WA 6076 **region** Perth Hills
ph (08) 9454 5748 **open** By appointment
winemaker Hainault (Contract) **prod.** 25 **est.** 1978

prod. range ($15 ML) Hermitage.

summary Owner Robert Duncan presides over what has to be one of the smallest producers in Australia, with one-quarter acre of shiraz and 30 cabernet sauvignon vines producing a single cask of wine each year if the birds do not get the grapes first.

Scarp Valley Darling Range Hermitage

The only wine produced by Scarp Valley, and a single barrel at that.

🍷🍷🍷🍷 **1998** Bright medium purple-red; the youthful bouquet shows fresh, earthy berry varietal fruit, the palate sweet, ripe cherry, a hint of mint and supple, ripe tannins. In a different league from the prior releases. **rating:** 88

best drinking 2002–2008 **best vintages** '98 **drink with** Fillet steak • $15

schmidts tarchalice NR

Research Road, Vine Vale via Tanunda, SA 5352 **region** Barossa Valley
ph (08) 8563 3005 **fax** (08) 8563 0667 **open** Mon–Sat 10–5, Sun 12–5
winemaker Christopher Schmidt **prod.** 1500 **est.** 1984

prod. range ($8.50–19.75 CD) Barossa Riesling, Eden Valley Riesling, Barossa Chardonnay, Barossa Semillon, Auslese Riesling, Shiraz Cabernet/Cabernet Franc, four different Ports and Old Liqueur Frontignac.

summary Typically has a range of fully mature wines at low prices available at cellar door.

scotchmans hill ★★★★

190 Scotchmans Road, Drysdale, Vic 3222 **region** Geelong
ph (03) 5251 3176 **fax** (03) 5253 1743 **open** 7 days 10.30–4.30
winemaker Robin Brockett **prod.** 30 000 **est.** 1982
prod. range ($18.50–25 R) Riesling, Sauvignon Blanc, Chardonnay, Pinot Noir, Cabernet Merlot; Spray Farm is the second label.

summary Situated on the Bellarine Peninsula, southeast of Geelong, with a very well-equipped winery and first class vineyards. It is a consistent performer with its Pinot Noir, and has a strong following in both Melbourne and Sydney for its well-priced, well-made wines. A doubling in production has seen the establishment of export markets to the UK and Holland. The second label of Spray Farm takes its name from a National Trust property with panoramic views of Port Phillip Bay and Melbourne, which has also been planted to vines by the Brown family, and is run as a distinct vineyard and brand operation. The same four varieties are produced, but at a lower price-point across the range.

Scotchmans Hill Sauvignon Blanc

Estate-grown, and with moderately high yields of 10 to 12.5 tonnes per hectare. Cold-fermented in stainless steel and early bottled, it is a pleasant, direct no-frills style.

🍷🍷🍷🍷 **1998** Light straw-green; the aromas of passionfruit and gooseberry are varietally correct and of light to medium intensity. The light, delicate but pleasant palate precisely replicates the characters of the bouquet. **rating:** 85

best drinking 1999–2000 **best vintages** '96 **drink with** Light seafood • $20

Scotchmans Hill Chardonnay

Estate-grown with yields of between 7.5 and 10 tonnes per hectare. The wine is oaked, but in a very low-key fashion.

🍷🍷🍷🍷 **1998** Light to medium green-yellow; the bouquet is light, with clear melon fruit and just the barest whiff of oak. The palate adds a hint of cashew, a touch of toasty oak but is essentially driven by the melon fruit and crisp acidity. **rating:** 85

best drinking 1999–2003 **best vintages** '91, '92, '94, '97 **drink with** Smoked salmon • $25

Spray Farm Chardonnay

Produced from slightly higher-yielding vineyards than Scotchmans Hill, and a wine which spends a briefer time in oak. Indeed, the oak influence is barely, if at all, perceptible.

🍷🍷🍷🍷 **1998** Light to medium green-yellow; the fresh melon fruit of the bouquet is very typical of cool-grown chardonnay, as is the relatively delicate but quite long palate. A minimalist approach to the winemaking has been adopted. **rating:** 85

best drinking 1999–2002 **best vintages** NA **drink with** Light seafood • $17

Scotchmans Hill Pinot Noir

Estate-grown; the most distinguished of the Scotchmans Hill wines in most vintages, but the southernmost parts of Victoria did not have an easy time of it in 1995 or 1996, rebounding in 1997 and 1998.

🍷🍷🍷🍷 **1998** Medium purple-red; not particularly aromatic, but does have solid dark plum varietal fruit. The palate, likewise, offers quite solid plummy fruit, but with little texture or expression. Tasted only a few months after bottling, and very likely suffering a degree of bottle shock, as so many Pinot Noirs do. **rating:** 86

best drinking 2000–2004 **best vintages** '91, '92, '94, '97 **drink with** Squab • $25

Spray Farm Pinot Noir

Completes the trio of wines released under both the Scotchmans Hill and Spray Farm label. This wine has particular appeal at its price point.

ŦŦŦŦ **1998** Medium red-purple; slightly earthy/stemmy notes actually add a greater degree of complexity to the wine when compared to Scotchmans Hill. It also has plenty of early and mid-palate flavour and complexity, before falling away markedly on the finish. **rating:** 85

best drinking 1999–2001 **best vintages** NA **drink with** Smoked meats • $17.50

scotts brook NR

Scotts Brook Road, Boyup Brook, WA 6244 **region** Other Wineries of WA

ph (08) 9765 3014 **fax** (08) 9765 3015 **open** Weekends, school holidays 10–5 or by appointment

winemaker Aquila Estate (Contract) **prod.** 1000 **est.** 1987

prod. range ($11–19 CD) Riesling, Autumn Harvest White, Chardonnay, Cabernet Sauvignon.

summary The Scotts Brook winery at Boyup Brook (equidistant between the Margaret River and Great Southern regions) has been developed by local schoolteachers Brian Walker and wife Kerry – hence the opening hours during school holidays. There are 17.5 hectares of vineyards, but the majority of the production is sold to other winemakers, with limited quantities being made by contract.

seaview ★★★★☆

Chaffey's Road, McLaren Vale, SA 5171 **region** McLaren Vale

ph (08) 8323 8250 **fax** (08) 8323 9308 **open** Mon–Fri 9–4.30, Sat 10–5, Sun 11–4

winemaker Steve Chapman **prod.** 500 000 **est.** 1850

prod. range ($10–32 R) Increasingly tied to McLaren Vale, with only the Riesling and the sparkling wines using fruit from outside the region. Riesling, Chardonnay, Semillon Sauvignon Blanc, Verdelho, Shiraz, Grenache and Cabernet Sauvignon make up the basic range. Recently introduced super-luxury Edwards & Chaffey range of Chardonnay, Shiraz, Cabernet Sauvignon and Pinot Noir Chardonnay, the latter replacing Edmond Mazure.

summary A maker of table wines which are frequently absurdly underpriced and perhaps suffer in consequence and, of course, of some of the country's best-known sparkling wines, which have gone from strength to strength over recent years. Moreover, the addition of the super-premium Edwards & Chaffey range has done much to change perceptions and lift the profile of the brand. Exports to the UK and US.

Seaview Riesling

The one thing certain about Seaview Riesling is that it is not made from McLaren Vale grapes, riesling being a very poor performer in that part of the world. Instead, grapes are sourced from the classic riesling regions of the Eden Valley and Clare Valley, with the precise composition varying from year to year.

ŦŦŦŦ **1998** Light to medium yellow-green; the bouquet is crisp and quite fragrant, with lemon zest and lime aromas. The palate is quite full and rich as befits the 1997 vintage, with passionfruit overtones. Slightly grippy finish. Outstanding value at the price and volume. **rating:** 85

best drinking 1999–2004 **best vintages** NA **drink with** Marinated octopus • $10

Seaview Vintage Reserve Chardonnay Blanc de Blanc

First made in 1991, and immediately caught the eye. It is made from 100% chardonnay grown in the Adelaide Hills and Eden Valley, and spends 18 months on yeast lees before disgorgement. The '92, '93, '94 and '95 have all been major trophy and gold medal winners, the '95 collecting two trophies in 1998.

🍷🍷🍷🍷🍷 **1995** Glowing light to medium yellow-green; the bouquet is unexpectedly complex and tangy, with some of the secondary aromas one finds in (true) Champagne. The palate then comes with almost surprisingly feathery elegance, light and soft, yet has length. Very appealing. **rating:** 94

best drinking 1999–2002 **best vintages** '92, '93, '94, '95 **drink with** Crustacea • $16.95

Seaview Pinot Chardonnay Brut

A very complex blend of Pinot Noir and Chardonnay sourced predominantly from Coonawarra, Padthaway and the Eden Valley in South Australia, but with small components from the New South Wales alps, Strathbogie Ranges of Victoria and Adelaide Hills. It spends two years on yeast lees before disgorgement.

🍷🍷🍷🍷🍷 **1996** Light green-yellow, with good mousse. The bouquet starts with a range of lemony, bready and creamy characters, then a touch of aldehyde appears. The palate is fresh, clean and quite delicate, with a long, well-balanced finish. **rating:** 91

best drinking 1999–2001 **best vintages** '92, '93, '94, '95 **drink with** Oysters • $16.95

Seaview Edwards & Chaffey Pinot Chardonnay

The premium sparkling wine from Seaview, first made in 1991. A blend of about 70% Pinot Noir and 30% Chardonnay, sourced predominantly from the Adelaide Hills and Yarra Valley. The wine spends three years on yeast lees.

🍷🍷🍷🍷🍷 **1995** Medium yellow-green; the bouquet is distinctly aldehydic, with complex bready aromas running through to a rich but nicely balanced palate with nutty/toasty autolysis characters, and good balance. **rating:** 93

best drinking 1999–2001 **best vintages** '93, '95 **drink with** Salmon roulade • $22.95

Seaview Shiraz

Introduced in 1992 as part of the repositioning and refocusing of the Seaview range. Prior to 1992 there was a Shiraz Cabernet which was sourced from many South Australian regions. This wine effectively replaces it, and is a distinct improvement. The wine spends 11 months in previously used American and French oak barrels.

🍷🍷🍷🍷 **1997** Medium to full red-purple; a solid wine on the bouquet with regional chocolate and earth varietal fruit together with a touch of American oak. The palate is in classic regional style, with lots of flavour and excellent structure for a wine of this price. **rating:** 85

best drinking 1999–2004 **best vintages** NA **drink with** Lamb shashlik • $13

Seaview Edwards & Chaffey Shiraz

First made in 1992, with no release in 1993, returning with the 1994 and 1995 vintages. Sourced entirely from premium McLaren Vale vineyards, the wine spends 17 months in 100% new French oak – an interesting departure from the normal pattern of American oak for big, rich South Australian Shiraz. Right from the first vintage, a wine of exceptional quality.

🍷🍷🍷🍷🍷 **1996** Dense red-purple; the aromas are powerful and concentrated, with blackberry fruit, hints of spice and mulberry married with abundant sweet oak. The palate is rich, thick and concentrated, with ripe raspberry fruit, and chewy tannins running throughout. There is plenty of night here; a bit of daylight might come as a welcome relief. **rating:** 92

best drinking 2006–2016 **best vintages** '92, '94, '95, '96 **drink with** Leave it in the cellar • $32

Seaview Edwards & Chaffey Cabernet Sauvignon

The sister wine to the Edwards & Chaffey Shiraz, made from old, low-yielding McLaren Vale cabernet and matured in new French oak for 18 months. Over the relatively few years the wine has been in production, it has won numerous trophies and gold medals.

🍷🍷🍷🍷🍷 **1996** Dark red-purple; the bouquet is flooded with ripe blackberry and cassis fruit together with charry/toasty/smoky oak. The palate is currently at war with itself, at once powerful, potent and concentrated, yet somehow lacking the structure one expects. Time may well sort it out. **rating:** 90

best drinking 2004–2010 **best vintages** '92, '94, '96 **drink with** Char-grilled rump • $32

seldom seen vineyard NR

Craigmoor Road, Mudgee, NSW 2850 **region** Mudgee
ph (02) 6372 4482 **fax** (02) 6372 1055 **open** 7 days 9.30–5
winemaker Barry Platt **prod.** 4500 **est.** 1987
prod. range ($11–16 CD) Semillon (wooded and unwooded), Chardonnay Semillon, Chardonnay, Traminer.
summary A substantial grape grower (with 18 hectares of vineyards) which reserves a proportion of its crop for making and release under its own label. No recent tastings.

seppelt ★★★★★

Seppeltsfield via Nuriootpa, SA 5355 **region** Barossa Valley
ph (08) 8568 6200 **fax** (08) 8562 8333 **open** Mon–Fri 10–5, Sat 10.30–4.30, Sun 11–4
winemaker James Godfrey, Jonathan Ketley **prod.** NFP **est.** 1851
prod. range ($8–3000 R) The great wines of Seppeltsfield are first and foremost Para Liqueur Port, Show Tawny Port DP90, Seppeltsfield Fino Sherry and Dorrien Cabernet Sauvignon. The other wines in the Seppelt portfolio are handled at Great Western. The 100 Year Old Para Liqueur Port is the $3000 a bottle jewel in the crown.
summary A multi-million-dollar expansion and renovation programme has seen the historic Seppeltsfield winery become the production centre for the Seppelt wines, adding another dimension to what was already the most historic and beautiful major winery in Australia. It is now home to some of the unique fortified wines in the world, nurtured and protected by the passionate James Godfrey. Worldwide distribution.

Seppelt Dorrien Cabernet Sauvignon

Based upon the Seppelt Dorrien Vineyard in the Barossa Valley from which it takes its name, sometimes including small components from Coonawarra and Langhorne Creek, and a dab of Merlot here and there. A prolific trophy and gold medal show winner in every vintage between 1988 and 1994 inclusive. Two years maturation in French Nevers oak always gives the wine tremendous impact. Only released when the quality is considered sufficiently high.

🍷🍷🍷🍷🍷 **1994** Medium red-purple; the smooth bouquet offers an attractive mix of dark berry, cassis and mint fruit together with subtle oak. Has a quite lovely, sweet mid-palate, rounded and supple, powerful yet elegant. **rating:** 95

🍷🍷🍷🍷🍷 **1993** Medium to full red, still with touches of purple. A wonderfully mature wine with sweet berry and chocolate fruit flavours balanced by quite firm tannins on the finish. Trophy winner 1998 National Wine Show. **rating:** 94

best drinking 2000–2008 **best vintages** '88, '89, '91, '92, '93, '94 **drink with** The richest possible red meat dish • $31.60

Seppelt Rutherglen Show Tokay DP57

A very old Tokay from Rutherglen; here the winemaking trick is to keep the wine fresh without taking away from the predominantly aged characteristics of the wine.

🍷🍷🍷🍷🍷 **NV** Mahogany gold; complex, sweet and rich tea-leaf, raisin and plum pudding aromas are followed by an equally complex, multiflavoured palate which leaves the mouth fresh thanks to its perfect balance. **rating:** 94

best drinking 1999–2019 **best vintages** NA **drink with** Fine, dark chocolate • $24.95

nutty characters, but it is the length of flavour and finish which is absolutely remarkable. Given its age, arguably the most undervalued wine on the Australian market today. **rating:** 97

best drinking 1999–2019 **best vintages** NA **drink with** Dried fruits and nuts • $88.50

Seppelt Para Liqueur Port

Produced from old, dry-grown shiraz, grenache and mourvedre, principally on the floor of the Barossa Valley. It is a blend of many vintages in a solera system which was established in 1973, and using wines dating back to 1940. Each successive blend is given a bottling number; that current in 1999 was Bottling 118.

🍷🍷🍷🍷🍷 **NV** Bottling 118. Dark mahogany tinged with green; the bouquet is complex, and both richer and sweeter than DP90 with malt, butterscotch and strong rancio characters. The palate has complex structure and great power, yet paradoxically has an almost dry finish, and no biscuity aftertaste. **rating:** 94

best drinking 1999–2019 **best vintages** NA **drink with** Coffee, chocolate • $19.50

seppelt great western ★★★★★

Moyston Road, Great Western, Vic 3377 **region** Grampians
ph (03) 5361 2239 **fax** (03) 5361 2200 **open** 7 days 10–5
winemaker Ian McKenzie (Chief) **prod.** NFP **est.** 1865
prod. range ($5.95–51 R) Méthode Champenoise comprising (from the bottom up) Brut Reserve, Imperial Reserve, Rosé Reserve, Grande Reserve, Sunday Creek Pinot Noir Chardonnay, Fleur de Lys, Harpers Range, Rhymney Sparkling Sauvignon Blanc, Original Sparkling Shiraz and Salinger; table wines include Moyston Unoaked Chardonnay and Cabernet Shiraz; Sheoak Riesling; Terrain Series Chardonnay and Cabernet Sauvignon; Eden Valley Botrytis Gewurztraminer, Corella Ridge Chardonnay, Harpers Range Cabernet Sauvignon, Chalambar Shiraz, Sunday Creek Pinot Noir, Drumborg Riesling, Partalunga Vineyard Chardonnay, Great Western Shiraz, and Drumborg Cabernet Sauvignon. Great Western Hermitage and Show Reserve Sparkling Burgundy are the flag-bearers alongside Salinger.
summary Australia's best-known producer of sparkling wine, always immaculate in its given price range, but also producing excellent Great Western-sourced table wines, especially long-lived Shiraz and Australia's best Sparkling Shirazes. Now the production centre for many Southcorp Group brands, with a vast new bottling plant and attendant warehouse facilities. Worldwide distribution.

Seppelt Drumborg Riesling

An interesting wine, not made every year, and by no means released in vintage sequence. The time of release depends entirely on winemaker Ian McKenzie's judgment, and thus the '93 was not released until late 1998.

🍷🍷🍷🍷🍷 **1993** Developed yellow colour; there are lime, toast and kerosene characters on the bouquet, the palate a mix of soft, toasty lime marmalade flavours. One wonders whether the wine has developed very rapidly over the last year or so. **rating:** 90

best drinking 1999–2003 **best vintages** NA **drink with** Chinese prawns with cashews • $22

Seppelt Sheoak Riesling

First released in 1994, and made entirely from Great Western Riesling. Chronically under-appreciated and underpriced in the market, and deserves far more recognition. A gold and silver medal winner at national wine shows in 1996 and 1997.

🍷🍷🍷🍷 **1998** Light to medium yellow-green; a clean and crisp bouquet showing classic dry riesling characters. The youthful palate is no less crisp, clean and lively, finishing bone-dry, and with excellent development potential. **rating:** 89

best drinking 2002–2008 **best vintages** '96 **drink with** Asian cuisine • $13.90

Seppelt Drumborg Sauvignon Blanc

Seppelt's Drumborg vineyard has always been on the climatic and viticultural edge, but is slowing releasing its secrets; the wines are always interesting, sometimes challenging. 1998 was the first commercial release after a cellar-door trial in 1997.

🍷🍷🍷🍷 **1998** Light green-yellow; the bouquet is clean, fresh and crisp, with mineral and herb aromas of light to medium intensity. The palate follows precisely the same track, but unlike many of the Sauvignon Blancs from 1998, has some real presence and weight. **rating:** 86

best drinking 1999–2000 **best vintages** NA **drink with** Crab salad • $22.50

Seppelt Drumborg Pinot Gris

The first Pinot Gris from Seppelt's Drumborg vineyard, providing further proof that this variety will do very well in the coolest parts of Australia. 1998 was the first commercial release.

🍷🍷🍷🍷 **1998** Light green-yellow; the highly aromatic bouquet offers lime blossom, spice and honeysuckle aromas, the palate crisp, fresh and elegant flavours which run more in a lime/citrus spectrum than elsewhere. Nicely balanced acidity. **rating:** 87

best drinking 1999–2000 **best vintages** NA **drink with** Prosciutto and melon • $22.50

Seppelt Corella Ridge Chardonnay

First made in 1991, utilising grapes grown in premium cool-climate Victorian regions, and 100% barrel-fermented in new and one-year-old French oak barriques. Matured on its yeast lees for nine months. In 1996 the grapes came from Great Western, Strathbogie Ranges and Drumborg. A much underrated and under-appreciated wine.

🍷🍷🍷🍷🍷 **1997** Medium yellow-green; the bouquet is extremely complex for a wine at this price, with strong barrel-ferment characters supporting the tangy melon/citrus fruit, which attests to the cool-climate regions from which the wine comes. The palate is quite rich with positive (even slightly assertive) oak, but the fruit is not hidden. **rating:** 90

best drinking 1999–2002 **best vintages** '97 **drink with** Fish mornay • $13.50

Seppelt Fleur de Lys Méthode Champenoise

A blend of Chardonnay, Pinot Noir and Pinot Meunier sourced from cool-climate vineyard sites in the southern New South Wales alps, Victoria and South Australia. The 1993 is a particularly distinguished example of Fleur de Lys.

🍷🍷🍷🍷🍷 **1993** Medium yellow-green; a very complex bouquet with attractive bready overtones to the rich fruit. A totally delicious wine on the palate, with gently sweet mid-palate flavours, good balance and perfectly judged dosage. **rating:** 94

best drinking 1999–2000 **best vintages** '93 **drink with** Oysters or shellfish • $12

Seppelt Salinger Méthode Champenoise

Seppelt's flagship Méthode Champenoise. Both the varietal composition and the regional base has varied substantially over the years; but typically is a complex blend of Pinot Noir and Chardonnay from Tumbarumba in the Snowy Mountains of New South Wales, the Yarra Valley and Drumborg in Victoria, and Adelaide Hills in South Australia. Salinger is disgorged progressively during its 12-month release span.

🍷🍷🍷🍷🍷 **1993** Light green-yellow; the bouquet is firm, with a complex amalgam of tangy fruit and more bready characters emanating from the time the wine spent on lees. It has a wonderful palate, with gently sweet fruit on the mid-palate; delicate, yet long and lingering, with a dry finish. **rating:** 93

best drinking 1999–2000 **best vintages** '83, '89, '90, '91, '93 **drink with** Aperitif, oysters, shellfish • $26.50

Seppelt Sunday Creek Pinot Chardonnay

A non-vintage blend of base wines of varying ages sourced from Coonawarra, Barooga, Padthaway, Great Western and Drumborg.

NV Medium yellow-green; the bouquet is quite firm, with an almost minerally cut to the fruit. The palate, likewise, is crisp, clean, bright, fresh and lively with the mineral characters predominant. A quite distinctive style. **rating:** 84

best drinking 1999–2000 **best vintages** NA **drink with** Aperitif • $9.95

Seppelt Great Western Show Sparkling Shiraz

Made from old vine shiraz grown at Seppelt Great Western, matured in large oak casks for one year before tiraging, and then on yeast lees for nine to ten years before disgorgement. Always an exceptionally complex wine, which will live for decades, as the classic wine notes demonstrate. (The wine was formerly labelled Sparkling Burgundy.)

1987 Medium to full red; an exceptionally complex and fragrant bouquet with a mix of spice, leaf and liquorice, proclaiming its Great Western shiraz origins. An utterly unmistakable wine on the palate, with berry, leaf, spice and liquorice flavours running through a long finish which is neither sweet nor dry. **rating:** 96

best drinking 2002–2022 **best vintages** '44, '46, '54, '61, '64, '67, '84, '85, '86, '87, '90, '91 **drink with** Borscht • $58.50

Seppelt Sunday Creek Pinot Noir

The Sunday Creek name comes from Western Victoria, where the creek in question flows near the Seppelt Drumborg Vineyard. As with all of the Sunday Creek releases, Drumborg is always a key component (together with fruit from Tumbarumba and the Strathbogies) in the limited production of this underrated wine. The '96 and '97 have both been prolific trophy and gold medal winners, leaving no doubt about the proposition that this has been the best value Pinot Noir on the Australian market over the past few years.

1997 Strong red-purple; a bouquet of rich, ripe dark plum and cherry fruit, and abundant sweet plummy fruit, supple oak and soft tannins on the palate all add up to a Pinot Noir with the widest possible market appeal. **rating:** 92

best drinking 1999–2001 **best vintages** '94, '96, '97 **drink with** Quail or duck • $16.90

Seppelt Chalambar Shiraz

Part of the Seppelt Victorian portfolio, using grapes grown predominantly at Great Western, but including components from the Ovens Valley, Strathbogie Ranges and Bendigo, all premium areas. It is matured in a mix of new and one-year-old French and American oak casks for 12 months, but is fruit- rather than oak-driven. It has been a prolific gold medal and trophy winner over the past four or five years; the '96 vintage added to the tally with three trophies and three gold medals, and the '97 looks set to follows in its tracks.

1997 Strong purple-red. A wonderfully scented, opulent and ripe bouquet is followed by a rich, almost lush palate with abundant fruit, finishing with soft tannins. Gold medal 1998 National Wine Show. **rating:** 95

best drinking 2002–2007 **best vintages** '53–'63, '91, '93, '94, '95, '96, '97 **drink with** Braised game dishes • $18.50

Seppelt Drumborg Cabernet Sauvignon

A limited release, and, what is more, only made in particularly good vintages – which, in this neck of the woods, means the dryer, warmer years.

1994 Medium red-purple; the bouquet proclaims the cool climate, with a mix of berry, leaf and chocolate aromas. The palate is intense and potent, starting with briar and cedar flavours followed by bitter chocolate and a touch of leaf. Austerely imperious. **rating:** 90

best drinking 2000–2010 **best vintages** NA **drink with** Marinated lamb • $39.50

serventy ★★☆

Valley Home Vineyard, Rocky Road, Forest Grove via Margaret River, WA 6286 **region** Margaret River

ph (08) 9757 7534 **fax** (08) 9757 7534 **open** Fri–Sun, holidays 10–4

winemaker Peter Serventy **prod.** 1500 **est.** 1984

prod. range ($15 CD) Chardonnay, Pinot Noir, Shiraz.

summary Peter Serventy is nephew of the famous naturalist Vincent Serventy and son of ornithologist Dominic Serventy. It is hardly surprising, then, that Serventy should practise strict organic viticulture, using neither herbicides nor pesticides. The wines, too, are made with a minimum of sulphur dioxide, added late in the piece and never exceeding 30 parts per million.

sevenhill cellars ★★★★

College Road, Sevenhill via Clare, SA 5453 **region** Clare Valley

ph (08) 8843 4222 **fax** (08) 8843 4382 **open** Mon–Fri 8.30–4.30, Sat, public holidays 9–4

winemaker Brother John May, John Monten **prod.** 24 000 **est.** 1851

prod. range ($9–20 CD) St Aloysius (Chenin Blanc, Chardonnay, Verdelho blend), Semillon, Riesling, College White, Traminer Frontignac, St Ignatius (Cabernet Sauvignon, Malbec, Franc and Merlot blend), Shiraz, Cabernet Sauvignon, Fortifieds, Sacramental Wine.

summary One of the historical treasures of Australia; the oft-photographed stone wine cellars are the oldest in the Clare Valley, and winemaking is still carried out under the direction of the Jesuitical Manresea Society and in particular, Brother John May. Quality is very good, particularly that of the powerful Shiraz, all the wines reflecting the estate-grown grapes from old vines. Extensive retail distribution throughout all States; exports to NZ, Switzerland and the US.

severn brae estate NR

Lot 2 Back Creek Road (Mount Tully Road), Severnlea, Qld 4352 **region** Granite Belt

ph (07) 4683 5292 **fax** (07) 3391 3821 **open** Weekends 9–5 or by appointment

winemaker Bruce Humphery-Smith **prod.** 300 **est.** 1990

prod. range ($14–16 ML) Chardonnay, Shiraz, Liqueur Muscat.

summary Patrick and Bruce Humphery-Smith have established 5.5 hectares of chardonnay with relatively close spacing and trained on a high two-tier trellis. Winery and cellar-door facilities were completed in time for the 1995 vintage. Prior to that time, the Chardonnay was made at Sundown Valley winery.

seville estate ★★★★☆

Linwood Road, Seville, Vic 3139 **region** Yarra Valley

ph (03) 5964 2622 **fax** (03) 5964 2633 **open** Not

winemaker Iain Riggs **prod.** 7000 **est.** 1970

prod. range ($13–45 ML) Chardonnay, Chardonnay Sauvignon Blanc, Pinot Noir, Shiraz, Cabernet Sauvignon; new GP label of Chardonnay Semillon and Cabernet Sauvignon in honour of winery founder Dr Peter McMahon.

summary In February 1997 a controlling interest in Seville Estate was acquired by Brokenwood (of the Hunter Valley), and interests associated with Brokenwood. I was one of the founding partners of Brokenwood, and the acquisition meant that the wheel had turned full circle. This apart, Seville Estate will add significantly to the top-end of the Brokenwood portfolio, without in any way competing with the existing styles.

Seville Estate Chardonnay

Reflects both the skills of the winemaking team headed by Iain Riggs and the 25-year-old estate vines.

▼▼▼▼▽ **1998** Excellent medium yellow-green colour. The bouquet is clean and smooth, with ripe melon fruit which spills over into a generously flavoured palate with sweet melon and fig fruit, all supported by subtle oak. **rating:** 92

best drinking 1999–2003 **best vintages** NA **drink with** Yabbies or marron • $24.95

Seville Estate Shiraz

Consistently the best of the Seville Estate wines but little has been known outside a select circle simply because production was so limited. It is matured in a mixture of French and American oak, but it is the fruit which really drives the wine.

▼▼▼▼▼ **1997** Full red-purple; the bouquet is rich and powerful, with abundant dark berry/cherry fruit, with oak in support. The high-flavoured palate shows rather more oak extract, but the fruit is there as are chewy tannins. A striking example from a warm, low-yielding Yarra vintage. Gold medal 1999 Sydney Royal Wine Show. **rating:** 94

best drinking 2002–2007 **best vintages** '88, '90, '91, '92, '93, '94, '97 **drink with** Pot-au-feu • $40.95

shantell ★★★★

1974 Melba Highway, Dixons Creek, Vic 3775 **region** Yarra Valley
ph (03) 5965 2264 **fax** (03) 5965 2331 **open** Thur–Mon 10.30–5
winemaker Shan Shanmugam, Turid Shanmugam **prod.** 1500 **est.** 1980
prod. range ($15–32 CD) Semillon, Chardonnay, Pinot Noir, Cabernet Sauvignon, Sparkling.
summary The substantial and now fully mature Shantell vineyards provide the winery with a high-quality fruit source; part is sold to other Yarra Valley makers, the remainder vinified at Shantell. In January 1998 Shantell opened a new cellar door situated at 1974 Melba Highway, 50 metres along a service road from the highway proper. Chardonnay, Semillon and Cabernet Sauvignon are its benchmark wines, sturdily reliable, sometimes outstanding (witness the 1997 Chardonnay).

Shantell Chardonnay

The Shantell vineyard has been producing high-quality chardonnay grapes for well over a decade producing elegant fruit-driven wines which age with grace.

▼▼▼▼▼ **1997** Light to medium green-yellow; the bouquet is complex, with powerful fruit and well-handled oak, but the wine comes into its own even more on the palate. It has excellent mouthfeel, weight and intensity, richly tangy, with a long finish. Multiple trophy winner 1998 Southern Victorian Wines Show. **rating:** 95

best drinking 1998–2002 **best vintages** '90, '92, '94 **drink with** Yarra Valley smoked trout • $20

sharmans ★★★☆

Glenbothy, RSD 175 Glenwood Road, Relbia, Tas 7258 **region** Northern Tasmania
ph (03) 6343 0773 **fax** (03) 6343 0773 **open** By appointment
winemaker James Chatto **prod.** 700 **est.** 1987
prod. range ($16.50–18 ML) Riesling, Sauvignon Blanc, Chardonnay, Pinot Noir.
summary Mike Sharman has very probably pioneered one of the most promising wine regions of Tasmania, not far south of Launceston but with a distinctly warmer climate than (say) Pipers Brook. Ideal north-facing slopes are home to a vineyard now approaching 3 hectares, most still to come into bearing. The few wines produced in sufficient quantity to be sold promise much for the future.

shaw & smith ★★★★★

PO Box 172, Stirling, SA 5152 **region** Adelaide Hills
ph (08) 8370 9911 **fax** (08) 8370 9339 **open** Not
winemaker Martin Shaw **prod.** 25 000 **est.** 1989
prod. range ($16–25 R) Sauvignon Blanc, Unoaked Chardonnay, Reserve Chardonnay.
summary Has progressively moved from a contract grape growing base to estate production, with the development of a 40-hectare vineyard at Woodside in the Adelaide Hills. Wine quality has been exemplary throughout, and the wines have wide international distribution including the UK, Italy, Japan, US, Canada, Hong Kong and Malaysia.

Shaw & Smith Sauvignon Blanc

The first vintages were produced on grapes grown on Geoff Hardy's Range Vineyard at Kuitpo in the Adelaide Hills; the '96 contained 20% from Shaw & Smith's new Woodside Vineyard in the Hills, and since that time the Woodside Vineyard has become the predominant source.

🍷🍷🍷🍷 **1998** Light green-yellow; the bouquet offers light, crisp mineral and grassy aromas but no sweet fruit. While the wine is very competently crafted, it struggles to do more than soothe the palate. A difficult vintage. **rating:** 86

best drinking 1999–2000 **best vintages** '92, '93, '95 **drink with** Grilled whiting • $17

Shaw & Smith Reserve Chardonnay

First made in 1992 and undoubtedly merits the Reserve designation. Using fruit from Geoff Hardy's vineyard, winemaker Martin Shaw applies the full gamut of Burgundian techniques of barrel fermentation, extended time on yeast lees and partial malolactic fermentation, using only finest French oak. A mini vertical tasting in January 1997 of all four vintages then released underlined how well these wines mature in bottle.

🍷🍷🍷🍷🍷 **1997** Medium to full yellow-green; one of those wines which is subtle yet complex, with the seamless balance and integration of all of the components, and in particular fruit and oak. On the palate, elegant nectarine and citrus fruit with a substrate of more creamy/nutty flavour is once again complemented by the especially sensitive oak handling. **rating:** 93

best drinking 2000–2007 **best vintages** '92, '94, '95, '96, '97 **drink with** Baked schnapper • $25

shottesbrooke ★★★★

Bagshaws Road, McLaren Flat, SA 5171 **region** McLaren Vale
ph (08) 8383 0002 **fax** (08) 8383 0222 **open** Mon–Fri 10–4.30, weekends and public holidays 11–5
winemaker Nick Holmes **prod.** 8000 **est.** 1984
prod. range ($16–20 CD) Fleurieu Sauvignon Blanc, Chardonnay, Eliza Shiraz, Shiraz, Merlot, Cabernet Merlot Malbec.
summary Now the full-time business of former Ryecroft winemaker Nick Holmes, made from grapes grown on his vineyard at Myoponga, at their best showing clear berry fruit, subtle oak and a touch of elegance. A compact, handsome new winery was erected prior to the 1997 vintage. Exports to the UK and US supplement distribution through all Australian States.

Shottesbrooke Fleurieu Sauvignon Blanc

Unwooded and with masses of fruit flavour.

🍷🍷🍷🍷🍷 **1998** Light green-yellow; the bouquet is quite intense with tangy herb and mineral aromas. The palate has good flavour, balance and length; no exotic fruit characters or frills, but does have some depth. **rating:** 90

best drinking 1999–2001 **best vintages** '95, '98 **drink with** Baby octopus • $16

Shottesbrooke Shiraz

There is a remarkable consistency to the style of all of the Shottesbrooke wines, white and red, an elegance which is in some ways at odds with the usually very rich McLaren Vale style. This is a good example of the Shottesbrooke approach.

🍷🍷🍷🍷 **1997** Medium to full red-purple; smooth, dark berry fruit of moderate intensity is supported by subtle oak on the bouquet; the palate is lifted and held together by lingering, soft, chewy tannins. **rating:** 87

best drinking 2000–2005 **best vintages** NA **drink with** Osso bucco • $18

Shottesbrooke Cabernet Sauvignon Merlot Malbec

A classic Bordeaux-blend, usually made in an elegant, reserved style, breaking free every now and then into a more striking mode, as it did in 1997.

🍷🍷🍷🍷🍷 **1997** Bright red-purple; quite powerful, youthful varietal characters on the bouquet with slightly earthy overtones. The palate is far less aggressive, with gentle cassis fruit and well-balanced, fine tannins. Gold medal 1998 National Wine Show; Topped the Cabernet class at the 1999 *Winewise* Competiton. **rating:** 94

best drinking 2001–2006 **best vintages** '97 **drink with** Yearling steak • $20

silvan winery NR

Lilydale–Silvan Road, Silvan, Vic 3795 **region** Yarra Valley
ph (03) 9737 9392 **open** Weekends, public holidays 11–6
winemaker John Vigliaroni **prod.** 500 **est.** 1993
prod. range ($8 CD) Chardonnay, Pinot Noir, Cabernet, Cabernet Shiraz Merlot, Merlot.
summary One of the newest and smallest of the Yarra Valley wineries; tastings are held in the Vigliaronis' spacious Italian villa.

simon hackett ★★★

PO Box 166, Walkerville, SA 5081 **region** Other Wineries of SA
ph (08) 8232 4305 **fax** (08) 8223 3714 **open** Not
winemaker Simon Hackett **prod.** 14 000 **est.** 1981
prod. range ($12–28 R) Barossa Valley Semillon, Barossa Valley Chardonnay, McLaren Vale Anthony's Reserve Shiraz, McLaren Vale Old Vine Grenache, McLaren Vale Foggo Road Cabernet Sauvignon.
summary Simon Hackett runs a very interesting operation, owning neither vineyards nor winery, but purchasing grapes and then making the wines at various establishments on a lend-lease basis. With considerable industry experience, he is thus able to produce a solid range of wines at competitive prices. The red wines are distinctly better than the whites.

sinclair wines NR

Graphite Road, Glenoran, WA 6258 **region** Pemberton
ph (08) 9421 1399 **fax** (08) 9421 1191 **open** By appointment
winemaker Brenden Smith **prod.** 1300 **est.** 1993
prod. range ($16–20 CD) Sauvignon Blanc, Chardonnay, Merlot, Cabernet Sauvignon.
summary Sinclair Wines is the child of Darelle Sinclair, a science teacher, wine educator and graduate viticulturist from Charles Sturt University and John Healy, a lawyer, traditional jazz musician and graduand wine marketing student of Adelaide University, Roseworthy Campus. Five hectares of estate plantings are coming into production, with the first wines released in August 1998.

sittella wines NR

100 Barrett Road, Herne Hill, WA 6056 **region** Swan District
ph (08) 9296 2600 **fax** (08) 9296 2600 **open** Weekends and public holidays 11–4
winemaker Candy Johnsson (Contract) **prod.** NA **est.** 1998
prod. range ($11.95–14.95 CD) Chenin Blanc, Verdelho, Chardonnay, Tawny Port.
summary Perth couple Simon and Maaike Berns acquired a 7-hectare block at Herne Hill, making the first wine in February 1998 and opening the most attractive cellar-door facility later in the year. They also own the Wildberry Springs Estate vineyard in the Margaret River region, which commenced to provide grapes from the 1999 vintage.

s kidman wines NR

Riddoch Highway, Coonawarra, SA 5263 **region** Coonawarra
ph (08) 8736 5071 **fax** (08) 8736 5070 **open** 7 days 9–5
winemaker John Innes (Contract) **prod.** 7000 **est.** 1984
prod. range ($11–18 CD) Riesling, Sauvignon Blanc, Cabernet Sauvignon, Shiraz.
summary One of the district pioneers, with a 16-hectare estate vineyard which is now fully mature. No recent tastings; limited retail distribution in Melbourne and Adelaide.

skillogalee ★★★★

Off Hughes Park Road, Sevenhill via Clare, SA 5453 **region** Clare Valley
ph (08) 8843 4311 **fax** (08) 8843 4343 **open** 7 days 10–5
winemaker Dave Palmer **prod.** 7000 **est.** 1970
prod. range ($14–25 R) Riesling, Late Picked Riesling, Gewurztraminer, Chardonnay, Shiraz, The Cabernets, Fortifieds.
summary David and Diana Palmer purchased the small hillside stone winery from the George family at the end of the 1980s, and have capitalised to the full on the exceptional fruit quality of the Skillogalee vineyards. The winery also has a well-patronised lunchtime restaurant. All of the wines are generous and full flavoured, particularly the reds.

smithbrook ★★★☆

Smith Brook Road, Middlesex via Manjimup, WA 6258 **region** Pemberton
ph (08) 9772 3557 **fax** (08) 9772 3579 **open** 7 days 10–4
winemaker Matt Steel **prod.** 14 000 **est.** 1988
prod. range ($22.50–28 R) Sauvignon Blanc, Chardonnay, Early Release Chardonnay, Merlot, Cabernet Merlot.
summary Smithbrook is a major player in the Pemberton region with 60 hectares of vines in production. A majority interest was acquired by Petaluma in 1997, but will continue its role as a contract grower for other companies, as well as supplying Petaluma's needs and making relatively small amounts of wine under its own label. Perhaps the most significant change has been the removal of Pinot Noir from the current range of products, and the introduction of Merlot.

Smithbrook Sauvignon Blanc

The first Sauvignon Blanc release from the Smithbrook vineyard in Pemberton, and a promising start. Stainless steel-fermented, and early-bottled, it nonetheless has plenty of weight.
🍷🍷🍷🍷 **1998** Medium yellow-green; the aromas are very rich and ripe, running through tropical fruit, gooseberry and passionfruit. The palate offers more of the same ripe, tropical gooseberry flavours, and a relatively soft finish. **rating:** 87
best drinking 1998–1999 **best vintages** NA **drink with** Sashimi • $16

snowy river winery NR

Rockwell Road, Berridale, NSW 2628 **region** Other Wineries of NSW
ph (02) 6456 5041 **fax** (02) 6456 5005 **open** 7 days 10–5
winemaker Contract **prod.** 2500 **est.** 1984
prod. range ($10–20 CD) Alpine Dry White, Semillon, Sauvignon Blanc, Semillon Chardonnay, Müller Thurgau Sylvaner, Sieger Rebe [sic], Rhine Riesling Auslese, Snow Bruska, Port.
summary Claims the only Eiswein to have been made in Australia, picked on 8 June 1990 after a frost of –8 degrees Celsius. Also makes a Trocken Beeren Auslese [sic] picked mid-May from the vineyard situated on the banks of the Snowy River, one hour from Mount Kosciusko. One suspects many of the wines are purchased from other makers.

somerset hill wines NR

891 McLeod Road, Denmark, WA 6333 **region** Great Southern
ph (08) 9840 9388 **fax** (08) 9840 9394 **open** By appointment
winemaker Brenden Smith (Contract) **prod.** 600 **est.** 1995
prod. range Sauvignon Blanc, Pinot Noir.
summary Graham Lipson commenced planting 9 hectares of pinot noir, chardonnay, semillon and sauvignon blanc in 1995, and plans to open for cellar-door sales in newly built but rustic-style stone building in late 1999, offering the first commercial vintage (also 1999). The full name of the venture is Somerset Hills Wines and Lavender, and lavender crafts will also be on sale.

sorrenberg ★★★

Alma Road, Beechworth, Vic 3747 **region** Ovens Valley
ph (03) 5728 2278 **fax** (03) 5728 2278 **open** Mon–Fri by appointment, most weekends 1–5 (phone first)
winemaker Barry Morey **prod.** 1200 **est.** 1986
prod. range ($19–25 CD) Sauvignon Blanc Semillon, Chardonnay, Gamay, Cabernet Sauvignon, Havelock Hills Shiraz.
summary Barry and Jan Morey made their first wines in 1989 from the 2.5-hectare vineyard situated on the outskirts of Beechworth. Wine quality has steadily improved since the early days, with a particular following for its Gamay.

southern grand estate NR

c/o 111 Goulburn Street, Sydney, NSW 2333 **region** Upper Hunter Valley
ph (02) 9282 0987 **fax** (02) 9211 8130 **open** Not
winemaker Jon Reynolds (Contract) **prod.** 1600 **est.** 1997
prod. range ($16.50–19.95 R) Semillon, Chardonnay, Cabernet Sauvignon.
summary Southern Grand Estate is the successor of the long-lost Hollydene Estate, the label of which (though not the vineyards) disappeared almost 20 years ago. The Kho family purchased the Hollydene vineyard from Orlando Wyndham in 1997, and, using their long involvement in the Australian hospitality and hotel industry, are planning to build a holiday and convention resort on the property. In the meantime, winemaking is in the capable hands of Jon Reynolds.

Southern Grand Estate Chardonnay

The estate-grown grapes are picked very ripe (typically around 13 to 13.5° baumé) and fermented in new French oak barriques. Thirty per cent of the total undergoes malolactic fermentation. The style of the wine suggests that the wine is given only brief contact with oak.

🍷🍷🍷🍸 **1997** Medium yellow-green; the bouquet is quite solid, initially a little closed but unfolding to show nectarine fruit but no obvious oak. The palate is quite complex with some faintly green/mineral edges to the mid-palate fruit, but finishing very well, with a gently creamy/cashew flavour and texture from the mlf component. Once again, oak is very much in the background. **rating:** 84

best drinking 1999–2001 **best vintages** NA **drink with** Grilled chicken breast • $17.50

Southern Grand Estate Cabernet Sauvignon

The grapes are, of course, estate-grown from vines which are now old. After fermentation, the wine is matured in new American oak barriques for a period of ten months prior to bottling in the February following vintage.

🍷🍷🍷🍷 **1997** Bright red-purple of medium depth. The bouquet is clean and soft, with gentle but attractive cedar, mulberry and blackberry aromas. The palate breaks into two parts: quite succulent fruit on the mid-palate, and then a slightly metallic tweak on the finish. Vanillin American oak is not overdone. **rating:** 83

best drinking 2000–2005 **best vintages** NA **drink with** Lamb in red wine sauce • $19.95

spring vale vineyards ★★★☆

Spring Vale, Swansea, Tas 7190 **region** Southern Tasmania

ph (03) 6257 8208 **fax** (03) 6257 8598 **open** Weekends, holidays 10–5

winemaker Andrew Hood (Contract) **prod.** 3000 **est.** 1986

prod. range ($19.50–29.50 CD) Gewurztraminer, Chardonnay, Pinot Gris, Pinot Noir.

summary Rodney Lyne has progressively established 1.5 hectares each of pinot noir and chardonnay and then added half a hectare each of gewurztraminer and pinot gris; the latter produced a first crop in 1998. After frost problems, Spring Vale is now enjoying the fruits of two excellent vintages in 1997 and 1998.

Spring Vale Chardonnay

Produced from a little over 1 hectare of estate plantings which have recovered from the frosts of 1996, but which, in common with much of Tasmania, experienced low yields in 1997, with tiny, concentrated berries.

🍷🍷🍷🍷🍷 **1997** Medium yellow-green; very potent Burgundian characters with intense, tangy grapefruit and smoky barrel-ferment characters. The palate is very intense and concentrated, with long, persistent flavour and beautifully balanced acidity. No part of the wine underwent malolactic fermentation. **rating:** 94

best drinking 1999–2007 **best vintages** '97 **drink with** Veal or pork • $22.50

Spring Vale Pinot Noir

Through both conviction and necessity, Andrew Hood (the Spring Vale contract-winemaker) destems and conventionally ferments the Pinot Noirs he makes, thereby excluding whole bunches. One can hardly argue with the approach, particularly when the grapes come from Tasmania's east coast, one of the most favoured spots for pinot noir in Australia.

🍷🍷🍷🍷 **1997** Good red-purple; the fragrant bouquet first offered a mix of spice and more sappy/leafy characters, but plum and cherry progressively evolved in the glass. The palate is clean and smooth, with soft cherry/plum fruit and a silky, lingering finish. **rating:** 89

best drinking 1999–2003 **best vintages** NA **drink with** Roast duck • $29.50

springviews wine NR

Woodlands Road, Porongurup, WA 6324 **region** Great Southern

ph (08) 9853 2088 **fax** (08) 9853 2098 **open** 7 days 10–5

winemaker Howard Park (Contract) **prod.** NA **est.** 1994

prod. range ($15–20 CD) Riesling, Chardonnay, Cabernet Sauvignon.

summary Andy and Alice Colquhoun planted their 5-hectare vineyard (2 hectares each of chardonnay and cabernet sauvignon an 1 hectare of riesling) in 1994. The wine is made at Howard Park by John Wade, and is sold through the cellar door and mailing list.

stanley brothers ★★★

Barossa Valley Way, Tanunda, SA 5352 **region** Barossa Valley
ph (08) 8563 3375 **fax** (08) 8563 3758 **open** 7 days 9–5
winemaker Lindsay Stanley **prod.** 6000 **est.** 1994
prod. range ($11–22 CD) Sylvaner, Full Sister Semillon, Chardonnay Pristine, John Hancock Shiraz, Thoroughbred Cabernet, Cabernet Shiraz, Late Harvest Sylvaner; NV Black Sheep; Fortifieds.
summary Former Anglesey winemaker and industry veteran Lindsay Stanley established his own business in the Barossa Valley when he purchased (and renamed) the former Kroemer Estate in late 1994. As one would expect, the wines are competently made, the '97 red wines and '98 whites all receiving between 80 and 85 points, the most interest being the 1998 Full Sister Semillon. Twenty-one hectares of estate plantings have provided virtually all of the grapes for the business.

stanton & killeen wines ★★★★☆

Jacks Road, Murray Valley Highway, Rutherglen, Vic 3685 **region** Rutherglen
ph (02) 6032 9457 **fax** (02) 6032 8018 **open** Mon–Sat 9–5, Sun 10–5
winemaker Chris Killeen **prod.** 20 000 **est.** 1875
prod. range ($10–37.50 CD) A red wine and fortified wine specialist, though offering Chardonnay, Riesling, White Frontignac and Dry White as well as Cabernet Sauvignon, Shiraz, Durif, Old Tawny Port, Old Rum Port, Vintage Port, Liqueur Port, Liqueur Tokay, Liqueur Muscat and top-of-the-range Special Old Liqueur Muscat.
summary Chris Killeen has skilfully expanded the portfolio of Stanton & Killeen, but without in any way compromising its reputation as a traditional maker of smooth, rich reds, some very good Vintage Ports, and attractive, fruity Muscats and Tokays.

Stanton & Killeen Riesling

A wine which did extremely well at the 1999 *Winewise* Small Makers Competition, both on my score sheet and on those of my fellow judges. A major surprise for a winery which has been producing a string of excellent red wines and even better Muscats. I am not certain, but suspect the origin of the fruit is from one of the nearby Alpine valleys.

🍷🍷🍷🍷🍸 **1998** Light green-yellow; the crisp bouquet has considerable presence with a blend of aromatic herb, nettle and citrus fruit. The palate is elegant, fresh and delicate, moving more to lime and passionfruit flavours; well-balanced and dry finish. **rating:** 93

best drinking 1999–2003 **best vintages** '98 **drink with** Calamari • $12

Stanton & Killeen Jack's Block Shiraz

In 1921 fourth generation winemaker Jack Stanton planted the block of shiraz from which this wine is made. He lived an extraordinarily active life until he died, aged 95 in 1990; almost to the very end he made a daily visit to the winery. He would doubtless have approved greatly of this wine, and be more than happy that Chris Killeen is such a gifted winemaker.

🍷🍷🍷🍷🍸 **1997** Dense red-purple; wondrously sweet, ripe liquorice, earth and berry fruit aromas are reflected in the opulently ripe and concentrated palate, awash with dark berry and spice fruit, finishing with just a hint of vanilla oak. **rating:** 92

best drinking 2000–2012 **best vintages** NA **drink with** Rich red meat, game • $17.50

Stanton & Killeen Durif

A rare grape variety brought to northeast Victoria about the turn of the century, and which may or may not be the same as California's petite syrah. Wherever grown, it makes a massively dense and potent wine, characteristics enhanced by the climate of northeast Victoria, to which it is ideally suited. Yet another example of the all-round winemaking skills of Chris Killeen, topping the Other Red Class at the 1999 *Winewise* Small Makers Competition, and doing so by a wide margin.

🍷🍷🍷🍷🍷 **1997** Deep red-purple; the bouquet is as robust and powerful as one should expect from the variety, with briary/earthy fruit and a touch of spice. The palate is extractive and powerful, with ripe, big berry, big tannin flavours. Not for the faint-hearted nor the impatient. **rating:** 90

best drinking 2002–2012 **best vintages** '97 **drink with** Two-inch-thick rump steak • $20

Stanton & Killeen Collectors Rutherglen Muscat

Came top of the Fortified Class at the 1999 *Winewise* Small Makers Competition. A particularly good example of a wine which uses both old and younger material.

🍷🍷🍷🍷🍷 **NV** Brown, with touches of olive; the bouquet is complex and concentrated, with plum pudding/raisin fruit. On the palate the younger component adds life without detracting from the depth of the raisiny fruit. Clean spirit and well balanced. **rating:** 95

best drinking 1999–2000 **best vintages** NA **drink with** Friut cake • $39.95

Stanton & Killeen Tawny Port

Yet another excellent wine style from Stanton & Killeen, showing considerable skill in the selection and blending of the base wines.

🍷🍷🍷🍷🍷 **NV** Medium red-tawny; the bouquet is clean with nice butterscotch overtones and, like the palate, shows positive but not aggressive rancio. **rating:** 93

best drinking 1999–2000 **best vintages** NA **drink with** Aperitif or coffee • $22.50

Stanton & Killeen Vintage Port

A blend of 80% Shiraz, 15% Touriga and 5% Durif, principally drawn from 75-year-old vines. Threatens to displace the Hardys Vintage Ports as the most successful on the Australian show circuit. Both the '92 and '93 vintages are literally festooned with trophies and gold medals. The '93 accumulated five trophies and 13 gold medals between 1994 and 1998, with the number of awards greatest in 1998, including the trophy for Best Port at the Royal Adelaide Wine Show of that year. Ludicrously underpriced.

🍷🍷🍷🍷🍷 **1993** Dark red, with the first hints of colour change starting to appear. Fragrant spice, liquorice and cedar aromas are married with almost sweet, caressingly smooth, spirit. Chocolate, liquorice, cinnamon and blackberry fruits run through the near-dry palate, far less sweet than most Australian Vintage Ports, and unambiguously better for that. A wine of the highest quality. **rating:** 97

best drinking 2000–2020 **best vintages** '92, '93 **drink with** Nuts • $22.50

steels creek estate NR

1 Sewell Road, Steels Creek, Vic 3775 **region** Yarra Valley
ph (03) 5965 2448 **fax** (03) 5965 2448 **open** Weekends and public holidays 10–6
winemaker Simon Peirce **prod.** 400 **est.** 1981
prod. range ($12–25 CD) Colombard, Chardonnay, Shiraz, Cabernet Sauvignon.
summary While only a tiny operation, with 1.7 hectares of vineyard planted at various times between 1981 and 1994, Steels Creek Estate has an on-site winery where the wines are made with assistance from consultants, but increasingly by Simon Peirce, who has recently completed his associate diploma in Applied Science (Winegrowing) at Charles Sturt University. Steels Creek 1996 Cabernet Sauvignon won a silver medal at the Southern Victorian Wine Show in 1998, an excellent outcome for what was an extremely difficult vintage with the late-ripening varieties.

stefano lubiana ★★★★

60 Rowbottoms Road, Granton, Tas 7030 **region** Southern Tasmania
ph (03) 6263 7457 **fax** (03) 6263 7430 **open** Fri–Mon 10–3 or by appointment
winemaker Steve Lubiana **prod.** 2000 **est.** 1990
prod. range ($16–40 CD) Primavera Riesling, Chardonnay, Pinot Noir, Primavera Pinot, NV Brut.

summary Steve Lubiana has moved from one extreme to the other, having run Lubiana Wines at Moorook in the South Australian Riverland for many years before moving to the Tamar Valley region of Tasmania to set up a substantial winery which acts as both contract-maker and maker for its own label wines, and which was also known as Granton Vineyard. He has progressively grown his contract-winemaking business, and in particular the sparkling wine side. For the past three years, has made the trophy-winning sparkling wine at the Tasmanian Wines Show, first for Barrington Estate and (in 1998) for Elsewhere Vineyard. The estate-produced Stefano Lubiano wines come from 5.5 hectares of beautifully located vineyards sloping down to the Derwent River.

Stefano Lubiana Primavera Riesling

Produced from estate-grown grapes on the banks of the Derwent River. The 1998 vintage scored well in both the 1999 *Winewise* Small Makers Competition and the 1999 Tasmanian Wines Show.

🍷🍷🍷🍷🍷 **1998** Light yellow-green; the bouquet is initially closed, showing the effects of bottling sulphur dioxide early in its life, an effect which will completely dissipate with further time in bottle. The wine exhibits its class on the powerful and long palate, with an intense mix of lime and more minerally/savoury characters. **rating:** 90

best drinking 2000–2008 **best vintages** '98 **drink with** Marinated scallops • $16

Stefano Lubiana Vintage Brut

Produced from pinot noir and chardonnay grown on the banks of the Derwent River, spending a minimum of three years on yeast lees prior to disgorgement, and thereafter disgorged progressively. Gold medal and trophy winner at the 1999 Tasmanian Wines Show.

🍷🍷🍷🍷🍷 **1995** Light straw-gold; the bouquet is complex and tangy, with notes of bread and brioche attesting to the prolonged time on lees. The palate is exceptionally harmonious and rich, with perfectly balanced acidity on a long, clean, lingering finish. Gold medal and trophy 1999 Tasmanian Wines Show. **rating:** 94

best drinking 1999–2000 **best vintages** '95 **drink with** Shellfish • $40

steins ★★★

Pipeclay Lane, Mudgee, NSW 2850 **region** Mudgee
ph (02) 6373 3991 **fax** (02) 6373 3709 **open** 7 days 10–4
winemaker Robert Stein, Greg Barnes **prod.** 5000 **est.** 1976
prod. range ($10–16.50 CD) Semillon, Semillon Riesling, Traminer, Chardonnay, Late Harvest Riesling, Rosé, Mt Buckaroo Dry Red, Shiraz, Cabernet Sauvignon and a range of Muscats and Ports; Robert Stein range of Semillon, Chardonnay, Shiraz, Cabernet Shiraz, Cabernet Sauvignon.

summary The sweeping panorama from the winery is its own reward for cellar-door visitors. Right from the outset Steins has been a substantial operation, but has managed to sell the greater part of its production direct from the winery by mail order and cellar door, with limited retail distribution in Sydney, but nowhere else. Wine quality has been very good from time to time, with the Shiraz and Chardonnay variously coming out on top.

Robert Stein Chardonnay

The '95 was an extremely good wine which deservedly won a gold medal at the 1996 Australian Small Makers Wine Show, and I must say came as a considerable – albeit pleasant – surprise. The '96 is good, but not quite in the same class.

🍷🍷🍷🍷 **1996** Medium to full yellow-green; there are slightly baggy overtones to the bouquet, but the palate largely puts matters to right, still fresh and young with tangy melon fruit, finishing with good acidity. **rating:** 85

best drinking 1999–2001 **best vintages** '95 **drink with** Rack of veal • $15

stephen john wines ★★★

Government Road, Watervale, SA 5452 **region** Clare Valley
ph (08) 8843 0105 **fax** (08) 8843 0105 **open** 7 days 11–5
winemaker Stephen John **prod.** 4000 **est.** 1994
prod. range ($10–50 CD) Watervale Riesling, Watervale Pedro Ximinez, Chardonnay, Clare Valley Shiraz, Clare Valley Cabernet Sauvignon, Traugott Cuvée Sparkling Burgundy.
summary The John family is one of the best-known names in the Barossa Valley, with branches running Australia's best cooperage (AP John & Sons) and providing the chief winemaker of Lindemans (Philip John) and the former chief winemaker of Quelltaler (Stephen John). Stephen and Rita John have now formed their own family business in the Clare Valley, based on a 6-hectare vineyard overlooking the town of Watervale, supplemented by modest intake from a few local growers. The cellar-door sales area is housed in an 80-year-old stable which has been renovated, and is full of rustic charm. The significantly increased production has led to the appointment of distributors in each of the eastern States, and to limited exports to the US.

Stephen John Clare Valley Shiraz

Produced from dry-grown grapes, and fermented in an open vat with hand-plunging. Matured in a mix of 70% American and 30% French oak barrels for 18 months.
🍷🍷🍷🍷 **1997** Strong red-purple; the bouquet is rich, with mint overtones to the abundant ripe fruit. The palate has plenty of the same minty berry fruit, a nice touch of vanilla oak, and finishes with soft tannins. **rating:** 87
best drinking 2001–2007 **best vintages** NA **drink with** Marinated beef • $19.50

sterling heights ★★★★

Faulkners Road, Winkleigh, Tas 7275 **region** Northern Tasmania
ph (03) 6396 3214 **fax** (03) 6396 3214 **open** By appointment
winemaker Moorilla Estate (Contract) **prod.** 350 **est.** 1988
prod. range ($11–17 CD) Riesling, Chardonnay, Breton Rose, Pinot Noir.
summary With just over 1.5 hectares of vines, Sterling Heights will always be a small fish in a small pond. However, the early releases had considerable success in wine shows, and the quality is all one could expect.

Sterling Heights Chardonnay

The wine is made in tiny quantities by the enormously talented Alain Rousseau of Moorilla Estate. It is barrel-fermented, but, interestingly, is not taken through malolactic fermentation, which is no doubt part of the reason why the wine has such intense fruit flavour. The 1997 vintage won the trophy for Best Chardonnay at the 1999 Tasmanian Wines Show. Unusually for Tasmania, the vineyard is established on relatively poor sandy soils, similar to those of Elsewhere Vineyard.
🍷🍷🍷🍷🍷 **1997** Medium yellow-green; the complexity and power of the wine are immediately evident on the bouquet, with tangy stone fruit and citrus aromas supported by subtle barrel-ferment oak. The palate is likewise rich, complex and concentrated, with ripe stone fruit flavours and subtle oak playing a pure support role. **rating:** 94
best drinking 1999–2003 **best vintages** NA **drink with** Sweetbreads • $15.50

Sterling Heights Pinot Noir

Sterling Heights, however small, has an exceptional vineyard site; the Pinot completed a completely outstanding 1999 Tasmanian Wines Show for the brand, which won silver or gold medals with the majority of its entries.

🍷🍷🍷🍷🍷 **1997** Medium red-purple; the bouquet has deep plummy fruit mixed with hints of leaf and stalk. The substantial palate offers the same complex mix of plum, stem and briar characters, still developing. **rating:** 90

best drinking 1999–2003 **best vintages** '94, '97 **drink with** Kangaroo • $17

stone ridge ★★★

Limberlost Road, Glen Aplin, Qld 4381 **region** Granite Belt
ph (07) 4683 4211 **open** 7 days 10–5
winemaker Jim Lawrie, Anne Kennedy **prod.** 1450 **est.** 1981
prod. range ($10–40 CD) Under the Stone Ridge label Semillon, Chardonnay, Shiraz, Cabernet Malbec; under the Mount Sterling label Dry Red Shiraz.
summary Spicy Shiraz is the specialty of the doll's house-sized winery, but the portfolio has progressively expanded over recent years to include two whites and the only Stanthorpe region Cabernet Malbec (and occasionally a straight varietal Malbec). No recent tastings.

stoney vineyard NR

Campania, Tas 7026 **region** Southern Tasmania
ph (03) 6260 4174 **fax** (03) 6260 4390 **open** Mon–Fri 9–4, weekends by appointment
winemaker Peter Althaus **prod.** 6000 **est.** 1973
prod. range ($18–50 CD) Domaine A is the top label with Cabernet Sauvignon and Pinot Noir; second label is Stoney Vineyard with Aurora (wood-matured Sylvaner), Sauvignon Blanc, Pinot Noir, Cabernet Sauvignon.
summary The striking black label of the premium Stoney Vineyard wine, dominated by the single, multicoloured 'A', signified the change of ownership from George Park to Swiss businessman Peter Althaus. The NR rating for the winery is given in deference to Peter Althaus, who has no faith whatsoever in Australian wine judges or critics, and profoundly disagrees with their ratings. If I were bold enough to give a rating, it would not be less than four and a half stars. Retail distribution in Melbourne, Sydney and Perth; exports to Switzerland, China and Singapore.

Stoney Vineyard Sauvignon Blanc

Produced, as are all the Stoney Vineyard wines, from estate-grown grapes picked very ripe. Destemmed, then crushed, and fermented in stainless steel, all with the aim of the lowest possible phenolic level.

🍷🍷🍷🍷 **1998** Light green-yellow; a crisp and clean bouquet with stony/minerally aromas is followed by a palate with interesting flavours headed towards tropical/banana, and not overtly varietal. Something of a surprise after the bouquet. **rating:** 87

best drinking 1999–2001 **best vintages** NA **drink with** Gravlax • $25

Stoney Vineyard Cabernet Sauvignon

A first crop made from three-year-old vines, and – like many first crop wines – produced a brilliant outcome. As at February 1999 still available from cellar door.

🍷🍷🍷🍷🍷 **1994** Bright, strong red-purple; the bouquet is powerful and concentrated, with some briary notes running through dark berry fruit. The wine comes alive on the palate, with lovely sweet berry fruit cosseted by soft, fine tannins. **rating:** 93

best drinking 1999–2009 **best vintages** '94 **drink with** Braised beef • $25

Stoney Vineyard Domaine A Cabernet Sauvignon

The core of this wine comes from a small planting of 26-year-old cabernet sauvignon planted by Stoney Vineyard's founder George Park in the early 1970s. Includes token quantities of Merlot, Cabernet Franc and Petit Verdot, usually comprising about 10% of the total blend. Spends over two years in 100% new French oak, and given further time on cork before release. Four hundred cases and 200 magnums of the '95 were made, no '96 will be released; the '97 will be released at the end of 1999 and, tasted from barrel, should be every bit as good as the '95.

🍷🍷🍷🍷🍷 **1995** Medium red-purple; the bouquet is fine and elegant with secondary briary/cedary, rather than primary fruit, aromas. On the palate a genuinely sophisticated wine in an elegant European mould, with very fine tannins and structure, and excellent oak balance and integration.

rating: 94

best drinking 1999–2009 **best vintages** '86, '88, '90, '92, '94, '95, '97 **drink with** Tasmanian venison • $50

---

stonier wines ★★★★★

362 Frankston–Flinders Road, Merricks, Vic 3916 **region** Mornington Peninsula
ph (03) 5989 8300 **fax** (03) 5989 8709 **open** 7 days 12–5
winemaker Tod Dexter **prod.** 20 000 **est.** 1978
prod. range ($14–37 CD) Chardonnay, Pinot Noir and Cabernet under the Stonier and Stonier Reserve labels.

summary Looked at across the range, Stonier is now the pre-eminent winery in the Mornington Peninsula; its standing is in turn based more or less equally on its Chardonnay and Pinot Noir under the Reserve label. The acquisition of a 70% interest by Petaluma in 1998 has given both Stonier's (and the Mornington Peninsula as a whole) even greater credibility than hitherto. It has already led to the appointment of Menzendorf & Co as UK distributor; the wines are also sold throughout Asia.

Stonier Chardonnay

This is the standard release from the winery, made in far greater quantities than the Reserve, although the label design is not greatly different. Curiously, relatively cheap bottles, corks and capsules are used; one might argue the wine deserves better. It is produced from grapes sourced largely from 10 hectares of estate plantings. A total style contrast to the Reserve Chardonnay, which is exactly as it should be.

🍷🍷🍷🍷🍷 **1998** Light to medium green-yellow; very fragrant passionfruit and grapefruit aromas; despite those high aromatics the palate is quite delicate, with a rather light finish, and carries its 13.5 degrees alcohol with grace.

rating: 90

best drinking 1999–2004 **best vintages** '91, '93, '94, '97, '98 **drink with** Pasta, rich seafood • $19

Stonier Reserve Chardonnay

Produced from 3.5 hectares of estate-grown grapes, and made using the full gamut of Burgundian techniques including barrel fermentation, lees contact and malolactic fermentation. The wines are exceptionally complex and stylish; and successive vintages have all received accolades.

🍷🍷🍷🍷🍷 **1997** Medium yellow-green; the bouquet is composed of contrasting characters from quite evident barrel-fermentation and oak maturation on the one hand, and some quite attractive minerally undertones to the fruit. The palate opens with citrus and melon fruit, then to cashew and finally with spicy French oak.

rating: 92

best drinking 2001–2006 **best vintages** '86, '88, '91, '93, '94, '95, '96, '97 **drink with** Milk-fed veal • $37

Stonier Reserve Pinot Noir

If it were possible, the Reserve Pinot Noir has an even more distinguished track record than the Reserve Chardonnay. The 1992 vintage was nominated as the Best Pinot Noir in the 1994/95 Penguin *Good Australian Wine Guide*, by Huon Hooke and Mark Shield, while the 1993 vintage won two major trophies at the 1994 Royal Adelaide Wine Show, including the trophy for Best

Varietal Dry Red Table Wine – Any Variety. The wine is exceedingly complex and stylish. None was made in 1996, but 1997 saw an emphatic return to top quality.

🍷🍷🍷🍷🍷 **1997** Medium red-purple; the bouquet is complex, with powerful oak and carbonic maceration characters. The outstanding feature of the wine is the tremendous length of flavour, with fruit, oak and fine-grained tannins all running through on the finish. 1998 National Wine Show gold medal winner. **rating:** 94

best drinking 1999–2003 **best vintages** '90, '91, '92, '93, '94, '95, '97 **drink with** Coq au vin • $35

strathérne vale estate NR

Campbell Street, Caballing, WA 6312 **region** Other Wineries of WA
ph (08) 9881 2148 **fax** (08) 9881 3129 **open** Not
winemaker James Pennington (Contract) **prod.** 600 **est.** 1980
prod. range A single red wine made from a blend of Cabernet Sauvignon, Zinfandel, Merlot and Shiraz.
summary Stratherne Vale Estate stretches the viticultural map of Australia yet further. It is situated near Narrogin, which is north of the Great Southern region and south of the most generous extension of the Darling Ranges. The closest viticultural region of note is at Wandering, to the northeast.

strathkellar NR

Murray Valley Highway, Cobram, Vic 3644 **region** Goulburn Valley
ph (03) 5873 5274 **fax** (03) 5873 5270 **open** 7 days 10–6
winemaker Chateau Tahbilk (Contract) **prod.** 2500 **est.** 1990
prod. range ($9–15 CD) Chenin Blanc, Chardonnay, Late Picked Chenin Blanc, Shiraz, Muscat, Tokay, Putters Port.
summary Dick Parkes planted his 5.5-hectare vineyard to chardonnay, shiraz and chenin blanc in 1990, and has the wine contract-made at Chateau Tahbilk by Alister Purbrick. No tastings, but the fact that the wines are made at Chateau Tahbilk is a sure guarantee of quality, and the prices are modest.

Strathkellar Chardonnay

An unpretentious, no-frills approach to winemaking from warm-grown, irrigated chardonnay grapes, and which offers flavour and value for money.

🍷🍷🍷🍷 **1996** Medium to full yellow-green; the bouquet is clean, with solid buttery peachy fruit, the palate following on in the same mould with plenty of honest Chardonnay flavour, nice balancing acidity, and avoiding excessive phenolics (oak or otherwise). **rating:** 85

best drinking 1999–2000 **best vintages** NA **drink with** Takeaway chicken • $12

straws lane NR

Cnr Mount Macedon Road and Straws Lane, Hesket, Vic 3442 **region** Macedon Ranges
ph (03) 9654 9380 **fax** (03) 9663 6300 **open** Not
winemaker Stuart Anderson, Hanging Rock, Cope-Williams (Contract) **prod.** 1000 **est.** 1987
prod. range ($18–35 ML) Gewurztraminer, Macedon Blanc de Noirs, Pinot Noir.
summary The Straws Lane vineyard was planted 12 years ago, but the Straws Lane label is a relatively new arrival on the scene; after a highly successful 1995 vintage, adverse weather in 1996 and 1997 meant that little or no wine was made in those years, but the pace has picked up again with the excellent 1998 vintage. Stuart Anderson guides the making of the Pinot Noir, Hanging Rock Winery handles the Gewurztraminer and the sparkling wine base, and Cope-Williams looks after the tiraging and maturation of the sparkling wine. It's good to have co-operative neighbours.

stringy brae NR

Sawmill Road, Sevenhill, SA 5453 **region** Clare Valley
ph (08) 8843 4313 **fax** (08) 8843 4313 **open** Weekends and public holidays 10–5 or by appointment
winemaker Contract (Mitchell) **prod.** 1800 **est.** 1991
prod. range ($11.50–17.50 CD) Riesling, Shiraz, Sir Lancelot, Mote Hill, Cabernet Sauvignon.
summary Donald and Sally Willson have established over 8 hectares of vineyards that since the 1996 vintage have produced all the grapes for their wines. (Previously grapes from Langhorne Creek were used.) The Australian domestic market is serviced direct by mail order, but the wines are exported to the UK and the US.

Stringy Brae Riesling

Produced from grapes grown on 0.8 hectare of vines at the Stringy Brae vineyard supplemented by grapes grown on the Slate Creek vineyard, Watervale. An extremely powerful and concentrated wine, made from very ripe grapes.

🍷🍷🍷🍷🍸 **1998** Bright but strong green-yellow; the bouquet is very rich, with an attractive mix of lime and more tropical fruit. The palate likewise is crammed full of fruit flavour, making it a very attractive, early-drinking style at an equally attractive price. **rating:** 92

best drinking 1999–2003 **best vintages** NA **drink with** Asparagus • $12

stuart range estates NR

67 William Street, Kingaroy, Qld 4610 **region** Other Wineries of Qld
ph (07) 4162 3711 **fax** (07) 4162 4811 **open** 7 days 9–5
winemaker Adam Chapman **prod.** 11 000 **est.** 1997
prod. range ($9.50–15 CD) Chardonnay, Goodger Vineyard Chardonnay, Range White and Range Red (both semi-sweet), Shiraz (unwooded).
summary Stuart Range Estates is a prime example of the extent and pace of change in the Queensland wine industry, coming from nowhere in 1997 to crushing just under 120 tonnes of grapes in its inaugural vintage in 1998, producing 7000 cases. The anticipated production for 1999 is around 11 000 cases, the grapes being supplied by seven growers in the South Burnett Valley with 52 hectares planted by 1996. A state-of-the-art winery has been established within an old butter factory building, and the immensely experienced and talented Adam Chapman installed as winemaker. The barrel-fermented Chardonnay won the trophy for the Best Queensland White Wine in the annual *Courier-Mail* Top 100 Wine Competition. Very much a case of watch this space.

Stuart Range Estates Chardonnay

A blend of wooded and unwooded Chardonnay, the latter prevailing, which is no bad thing. The barrel-fermented component was kept in barrel (American oak) for three months with lees stirring.

🍷🍷🍷🍸 **1998** Light to medium yellow-green; pleasant melon fruit on the bouquet is supported by barely perceptible oak, while on the palate all of the action comes from the gentle melon fruit. **rating:** 84

best drinking 1999–2000 **best vintages** NA **drink with** Spicy Thai • $13

Stuart Range Estates Goodger Vineyard Chardonnay

Barrel-fermented in a mix of American and French barriques to good effect. There is very competent winemaking at work here.

🍷🍷🍷🍷 **1998** Light to medium yellow-green; a well-made wine, quite tangy with citrus and melon fruit and a nice touch of barrel-ferment. The palate is elegant, fresh and nicely balanced with lively fruit and well-integrated oak. **rating:** 88

best drinking 1999–2001 **best vintages** NA **drink with** Char-grilled salmon • $15

stumpy gully ★★★☆

1247 Stumpy Gully Road, Moorooduc, Vic 3933 **region** Mornington Peninsula
ph (03) 5978 8429 **fax** (03) 5978 8429 **open** First weekend of each month 12–5
winemaker Frank Zantvoort, Wendy Zantvoort **prod.** 2000 **est.** 1988
prod. range ($15–22 CD) Sauvignon Blanc, Chardonnay, Marsanne, Pinot Grigio, Pinot Noir, Merlot Cabernet, Merlot Shiraz.
summary Frank and Wendy Zantvoort have progressively established 9 hectares of vineyard planted to chardonnay, marsanne, sauvignon blanc, pinot noir, cabernet sauvignon and merlot, electing to sell 80% of the production to local winemakers, and vinifying the remainder, with impressive results. The '97 Cabernet Sauvignon (84 points) neatly reflected the warm, dry vintage.

summerfield ★★★★

Main Road, Moonambel, Vic 3478 **region** Pyrenees
ph (03) 5467 2264 **fax** (03) 5467 2380 **open** 7 days 9–6
winemaker Ian Summerfield, Mark Summerfield **prod.** 2300 **est.** 1979
prod. range ($12–25 CD) Sauvignon Blanc Chardonnay, Trebbiano, Shiraz, Cabernet Shiraz, Cabernet Sauvignon.
summary A specialist red wine producer, the particular forte of which is Shiraz. The wines since 1988 have been consistently excellent, luscious and full bodied and fruit-driven but with a slice of vanillin oak to top them off.

Summerfield Shiraz

Produced from 3.56 hectares of estate plantings, and now just a single release without differentiation of what was previously a varietal Shiraz (regionally sourced) and an estate-grown wine styled 'Estate'. Incidentally, Summerfield has moved straight from the '94 to the '96 vintage. With the 1997 vintage, Summerfield released two wines: the Premium Shiraz, and, under a different label, the Second Release Shiraz.

🍷🍷🍷🍷 **1997** Second Release. Medium to full red-purple; the bouquet is clean, with moderately intense earthy/berry shiraz fruit; the palate ranges through plum, spice and a touch of liquorice; nicely balanced and flavoured, with minimal oak input. ($19.95) **rating:** 89

🍷🍷🍷🍷½ **1997** Premium. Dense red-purple, there is very concentrated, rich dark cherry plum fruit on both the bouquet and palate. Lingering tannins and subtle oak close a wine with exceptional development potential. **rating:** 90

best drinking 2001–2006 **best vintages** '88, '90, '91, '92, '93, '94, '96, '97 **drink with** Beef stew • $25

Summerfield Cabernet Sauvignon

Produced from 2.1 hectares of estate plantings. The early releases employed a great deal of new American oak, but in recent years the touch has become much more subtle, and the wines have not suffered in consequence. As with the Shiraz, bypassed the '95 vintage.

🍷🍷🍷🍷½ **1997** Dense red-purple; an extraordinarily concentrated, rich and ripe wine from start to finish, with a mix of essencey berry, crushed ant, cinnamon and spice aromas on the bouquet, and a towering palate, with incredible untamed depth of fruit. The points given are an article of faith in the future. **rating:** 90

best drinking 2007–2017 **best vintages** '88, '90, '91, '92, '93, '94, '97 **drink with** Rare beef • $25

---

surveyor's hill winery NR

215 Brooklands Road, Wallaroo, NSW 2618 **region** Canberra District
ph (02) 6230 2046 **fax** (02) 6230 2048 **open** Weekends and public holidays or by appointment

winemaker Hardys (Contract) **prod.** 1000 **est.** 1986

prod. range ($13–25 R) Riesling, Estate Dry White, Alexandra's Block Chardonnay, Touriga Dry Red, Shiraz, Cabernet Sauvignon.

summary Surveyor's Hill has 7.5 hectares of vineyard, but most of the grapes are sold to Hardys, which vinifies the remainder for Surveyor's Hill, which should guarantee the quality of the wines sold.

sutherland ★★☆

Deasey's Road, Pokolbin, NSW 2321 **region** Lower Hunter Valley

ph (02) 4998 7650 **fax** (02) 4998 7603 **open** 6 days 10–4.30

winemaker Neil Sutherland, Nicholas Sutherland **prod.** 7000 **est.** 1979

prod. range ($15–40 CD) Chardonnay, Chenin Blanc, Chenin Cremant, Shiraz, Cabernet Shiraz, Cabernet Sauvignon.

summary With substantial and now fully mature vineyards to draw upon, Sutherland is a more or less consistent producer of generous, mainstream Hunter whites and reds, albeit with a high level of phenolic extraction.

sutherland smith wines NR

Cnr Falkners Road and Murray Valley Highway, Rutherglen, Vic 3685 **region** Rutherglen

ph (03) 6032 8177 **fax** (03) 6032 8177 **open** Weekends 10–5 or by appointment

winemaker George Sutherland Smith **prod.** 1000 **est.** 1993

prod. range ($10.90–16.50 CD) Riesling, Josephine (Riesling Traminer), Chardonnay, Merlot, Cabernet Shiraz, Port.

summary George Sutherland Smith, for decades managing director and winemaker at All Saints, has opened up his own small business at Rutherglen, making wine in the refurbished Emu Plains winery, originally constructed in the 1950s. He draws upon fruit grown in a leased vineyard at Glenrowan and also from grapes grown in the King Valley.

tait wines NR

Yaldara Drive, Lyndoch, SA 5351 **region** Barossa Valley

ph (08) 8524 5000 **fax** (08) 8524 5220 **open** Weekends 10–5 and by appointment

winemaker Contract **prod.** 500 **est.** 1994

prod. range ($12–18 CD) Chardonnay, Bush Vine Grenache, Shiraz, Cabernet Sauvignon.

summary The Tait family has been involved in the wine industry in the Barossa for over 100 years, making not wine, but barrels. Their recent venture into winemaking has been extremely successful, with four of the five wines on release in early 1998 having won either a silver or gold medal. Retail distribution through single outlets in Melbourne, Adelaide and Sydney.

talijancich NR

26 Hyem Road, Herne Hill, WA 6056 **region** Swan District

ph (08) 9296 4289 **fax** (08) 9296 1762 **open** Sun–Fri 11–5

winemaker James Talijancich **prod.** 10 000 **est.** 1932

prod. range ($17–85 CD) Verdelho, Voices Dry White, Grenache, Shiraz, Julian James White Liqueur, Julian James Red Liqueur, Liqueur Tokay.

summary A former fortified wine specialist (a 1969 Liqueur Tokay was released in July 1999) now making a broad range of table wines, with particular emphasis on Verdelho: on the third Saturday of August each year there is a tasting of fine three-year-old Verdelho table wines from both Australia and overseas. Also runs an active wine club, and exports to China, Japan and Hong Kong. An extensive building programme is currently underway, including the family homestead being converted to Voices Café, due to be opened in September 1999.

tallara NR

Cassilis Road, Mudgee, NSW 2850 **region** Mudgee
ph (02) 6372 2408 **fax** (02) 6372 6924 **open** Not
winemaker Simon Gilbert (Contract) **prod.** 1800 **est.** 1973
prod. range ($15 ML) Chardonnay, Cabernet Sauvignon.
summary Tallara's large, 54-hectare vineyard, was first established in 1973 by KPMG partner Rick Turner. Only a fraction of the production has ever been made into wine, currently by Simon Gilbert. It is sold principally through a mailing list, with a little local distribution.

taltarni ★★★☆

Taltarni Road, Moonambel, Vic 3478 **region** Pyrenees
ph (03) 5467 2218 **fax** (03) 5467 2306 **open** 7 days 10–5, winter school holidays only
winemaker Shane Clohesy, Chris Markell, Phillipe Bru **prod.** 70 000 **est.** 1972
prod. range ($10–75 R) Sauvignon Blanc, Blanc des Pyrenees, Rosé, Reserve des Pyrenees, Merlot Cabernet, Shiraz, Cabernet Sauvignon, Merlot; Sparkling Blanc de Blanc Tete de Cuvee, Cuvee Brut, Brut Tache, Clover Hill.
summary In the shadow of the departure (on the best of terms) of long-serving winemaker and chief executive Dominique Portet, Taltarni seems to have backed off the high levels of tannin and extract evident in the older vintage wines, which ought really to be a step in the right direction, but (perversely) seems to have robbed the red wines of some of their (formidable) character. A rock and a hard place, it seems. Exports to all the major markets including the UK, US, Canada, Japan, Hong Kong, Switzerland, Sweden and extensively thoughout southeast Asia and Western Europe.

Taltarni Sauvignon Blanc

Made in an uncomplicated and direct style, and in good years (such as 1996) produces a very attractive wine. In the warmer vintages it seems to lose varietal definition. The 1996 won a major trophy in an international competition in the US, while the '97 and '98 are, if anything, even better.

🍷🍷🍷🍷🍸 **1998** Very pale green-yellow; the bouquet is elegant and delicate, yet also offering quite intense gooseberry and passionfruit aromas. There is similar fruit expression on the palate; an outstanding example from this region not known for white wines. **rating:** 91

best drinking 1999–2000 **best vintages** '96, '97, '98 **drink with** Calamari • $15

Taltarni Cuvee Brut

A blend of Chardonnay and Pinot Noir sourced from Taltarni's vineyard in the Pyrenees but, even more significantly, also from the Clover Hill vineyard in Tasmania. It spends a minimum of 18 months on lees, and the Tasmanian component has significantly improved the wine.

🍷🍷🍷🍷 **NV** Light green-yellow; the bouquet is moderately complex, with some strawberry overtones. The palate is fine and elegant, the Tasmanian component sharpening the focus and the length, the Taltarni component softening the sometimes assertive acidity of Tasmanian sparkling wine. **rating:** 86

best drinking 1999–2000 **best vintages** NA **drink with** Hors d'oeuvres • $14

Taltarni Shiraz

Produced from a little over 12 hectares of estate-grown grapes, and has established a track record for its strength and longevity. The wines are never fruity or fragrant; durability and honesty are their hallmarks.

🍷🍷🍷🍷 **1996** Medium red-purple; the clean bouquet offers red berry and chocolate fruit of medium intensity and is followed by a palate with the fruit supported by soft tannins and a gentle touch of sweet oak. Much the best for some time from Taltarni. Developed very well over the past year or so. **rating:** 86

best drinking 2001–2007 **best vintages** '84, '88, '90, '91, '92, '96 **drink with** Gippsland blue cheese • $22

talunga NR

Adelaide to Mannum Road, (PO Box 134) Gumeracha, SA 5233 **region** Adelaide Hills
ph (08) 8389 1222 **fax** (08) 8389 1233 **open** Wed–Sun and public holidays 10.30–5
winemaker Vince Scaffidi **prod.** 2000 **est.** 1994
prod. range ($8–22.50 CD) Sauvignon Blanc, Semillon, Chardonnay, Yearling Blend, Shiraz, Sangiovese Franc, Sangiovese Merlot, Pinot Noir, Cabernet Sauvignon.
summary Talunga owners Vince and Tina Scaffidi are among the largest contract grape growers in the Adelaide Hills, operating Gumeracha Vineyards with 165 hectares planted. A small part of their Gumeracha Vineyards output is vinified under the Talunga label.

tamar ridge NR

PO Box 1513, Launceston, Tas 7250 **region** Northern Tasmania
ph (03) 6334 6044 **fax** (03) 6334 6050 **open** Not yet
winemaker Julian Alcorso **prod.** 7000 **est.** 1994
prod. range ($NA) Sauvignon Blanc.
summary Tamar Ridge is the most recent venture of grass-fed-beef magnate Joe Chromy into wine production. When he sold Heemskerk and Rochecombe to Pipers Brook in 1998, he retained a substantial vineyard in the Tamar Valley, which presently has 28 hectares in bearing and which is to be increased to 70 hectares over the next two to three years. The new winery, situated on the edge of a large dam (or lake, for that is what it looks like) is a striking piece of architecture, and a large restaurant is in the plans for the future.

Tamar Ridge Sauvignon Blanc

An extremely promising wine for a first crop wine made under unusual and difficult circumstances.
🍷🍷🍷🍷 **1998** Light green-yellow; the bouquet has ultra-potent aromas of grass, herb and gooseberry. A wine with lots of flavour on the palate including some riper notes which give some flesh, but the acidity is there to provide balance. Very good style. **rating:** 90

best drinking 1999–2001 **best vintages** NA **drink with** Shellfish • $17.95

tamburlaine ★★★☆

McDonalds Road, Pokolbin, NSW 2321 **region** Lower Hunter Valley
ph (02) 4998 7570 **fax** (02) 4998 7763 **open** 7 days 9.30–5
winemaker Mark Davidson **prod.** 40 000 **est.** 1966
prod. range ($14–22 CD) Semillon, Verdelho, Sauvignon Blanc, 3 Parishes Chardonnay, The Chapel Reserve Chardonnay, Botrytis Semillon, Petite Fleur (Sparkling), Cabernet Merlot Malbec, The Chapel Reserve Red, Barrique Blend Petite Sirah, Old Muscat.
summary A thriving business which, notwithstanding the fact that it has doubled its already substantial production in recent years, sells over 90% of its wine through cellar door and by mailing list (with an active tasting club members' cellar programme offering wines which are held and matured at Tamburlaine). Unashamedly and deliberately focused on the tourist trade, (and of course, its club members).

tanglewood downs NR

Bulldog Creek Road, Merricks North, Vic 3926 **region** Mornington Peninsula
ph (03) 5974 3325 **fax** (03) 5974 4170 **open** Sun–Mon 12–5
winemaker Ken Bilham, Wendy Bilham **prod.** 1000 **est.** 1984

prod. range ($25 CD) Riesling, Gewurztraminer, Chardonnay, Pinot Noir, Cabernet Sauvignon, Cabernet Franc Merlot.

summary One of the smaller and lower-profile wineries on the Mornington Peninsula, with Ken Bilham quietly doing his own thing on 2.5 hectares of estate plantings. Winery lunches and dinners are available by arrangement.

tannery lane vineyard NR

174 Tannery Lane, Mandurang, Vic 3551 **region** Bendigo
ph (03) 5439 5011 **fax** (03) 5439 5891 **open** By appointment
winemaker David Anderson (Contract) **prod.** 200 **est.** 1990
prod. range ($15 CD) Sangiovese, Shiraz, Cabernet Merlot.

summary Ray and Anne Moore established their tiny vineyard in 1990, gradually establishing a total of 2 hectares of shiraz, cabernet sauvignon, cabernet franc, sangiovese, merlot and nebbiolo. Their Sangiovese is the only such wine coming from the Bendigo region at the present time. The micro-production is sold through cellar door only and then only while stocks last, which typically is not for very long. A phone call will quickly establish whether or not wine is available.

tantemaggie NR

Kemp Road, Pemberton, WA 6260 **region** Pemberton
ph (08) 9776 1164 **fax** (08) 9776 1164 **open** Weekends 9–5
winemaker Contract **prod.** 110 **est.** 1987
prod. range ($21 CD) Cabernet Sauvignon.

summary Tantemaggie was established by the Pottinger family with the help of a bequest from a deceased aunt named Maggie. It is part of a mixed farming operation, and by far the greatest part of the 20 hectares is under long-term contract to Houghton. The bulk of the plantings are cabernet sauvignon and verdelho, the former producing the light-bodied style favoured by the Pottingers.

tarrawarra estate ★★★★☆

Healesville Road, Yarra Glen, Vic 3775 **region** Yarra Valley
ph (03) 5962 3311 **fax** (03) 5962 3887 **open** 7 days 10.30–4.30
winemaker Clare Halloran **prod.** 16 000 **est.** 1983
prod. range ($19–38 R) Chardonnay and Pinot Noir each released under the Tarrawarra Estate and second label Tunnel Hill.

summary Slowly evolving Chardonnay of great structure and complexity is the winery specialty; robust Pinot Noir also needs time and evolves impressively if given it. The second label Tunnel Hill wines are more accessible when young, and better value for those who do not wish to wait for the Tarrawarra wines to evolve.

Tarrawarra Pinot Noir

Produced from 6.5 hectares of estate plantings of several different clones. As with the Chardonnay, much attention is paid to style and structure: the grapes are picked when very ripe, and the winemaking techniques are designed to gain maximum extraction of colour and flavour from the grapes. The wines are often awkward when young, powerfully impressive with age and have won a significant number of trophies in Australian wine shows in recent years.

🍷🍷🍷🍷 **1997** Light to medium red-purple; although the fruit is relatively light, the bouquet is complex, assisted by charry oak. The high toast oak is a fraction assertive on the palate, threatening the core of sweet cherry fruit, but one has to expect the wine will come together with time. **rating:** 88

best drinking 2000–2004 **best vintages** '88, '90, '91, '92, '94, '95, '96 **drink with** Squab • $38

tarwin ridge NR

Wintles Road, Leongatha South, Vic 3953 **region** Gippsland
ph (03) 5664 3211 **fax** (03) 5664 3211 **open** Weekends and holidays 10–5
winemaker Brian Anstee **prod.** 700 **est.** 1983
prod. range ($16–27 CD) Sauvignon Blanc, White Merlot, Pinot Noir, Pinot Noir Premium, Cabernet Merlot.
summary For the time being Brian Anstee is making his wines at Nicholson River under the gaze of fellow social worker Ken Eckersley; the wines come from 2 hectares of estate pinot and half a hectare each of cabernet and sauvignon blanc.

tatachilla ★★★★☆

151 Main Road, McLaren Vale, SA 5171 **region** McLaren Vale
ph (08) 8323 8656 **fax** (08) 8323 9096 **open** Mon–Sat 10–5, Sundays and public holidays 11–5
winemaker Michael Fragos, Justin McNamee **prod.** 150 000 **est.** 1901
prod. range ($12.95–36 CD) Clarendon Vineyard Riesling, Late Harvest Riesling, Adelaide Hills, Sauvignon Blanc, Growers (Chenin Blanc Semillon Sauvignon Blanc), Padthaway Chardonnay, Bluestone Brut, Sparkling Malbec, Merlot, Foundation Shiraz, Keystone (Grenache Shiraz), Padthaway Cabernet Sauvignon, Partners (Cabernet Sauvignon Shiraz), Tawny Port.
summary Tatachilla was reborn in 1995, but with an at-times tumultuous history going back to 1901. For most of the time between 1901 and 1961 the winery was owned by Penfolds, but was closed in that year before being reopened in 1965 as the Southern Vales Co-operative. In the late 1980s it was purchased and renamed The Vales, but did not flourish, and in 1993 was purchased by local grower Vic Zerella and former Kaiser Stuhl chief executive Keith Smith. After extensive renovations, the winery was officially reopened in 1995, and has won a number of tourist awards and accolades. The star turns are Keystone (Grenache Shiraz) and Foundation Shiraz, bursting with vibrant fruit. It has 480 hectares of vineyards, and the expanding production services markets throughout Europe, Hong Kong and Taiwan.

Tatachilla Adelaide Hills Sauvignon Blanc

No-frills, straightforward winemaking; it remains to be seen how much downwards pressure New Zealand will place on the price of wines such as this.

🍷🍷🍷🍷 **1998** Bright, light green-yellow; the bouquet is clean and smooth with fair fruit weight, and it is not a surprise to find some structured, minerally flavours akin to those of white Bordeaux on the palate. Good length. **rating:** 86

best drinking 1999–2001 **best vintages** NA **drink with** Calamari • $19.95

Tatachilla Foundation Shiraz

Foundation Shiraz is Tatachilla's flag-bearer, taking its name from the foundation stone of the winery which was laid in place in 1901 by founder Cyril Pridmore. Open fermentation, using a mixture of heading down boards and plunging, is followed by partial barrel fermentation, before the wine is matured in a mix of predominantly American (70% new) oak for 18 months. Has had considerable show success, particularly with the '95.

🍷🍷🍷🍷 **1996** Medium to full purple-red; there is abundant, bright cherry and plum fruit on both the bouquet and palate, but the oak is slightly raw, and firm tannins swoop on the finish. May well improve significantly with time in bottle, but not in the same class as the '95. **rating:** 87

best drinking 2003–2010 **best vintages** '95 **drink with** Kangaroo fillet • $36

Tatachilla McLaren Vale Shiraz

Overall, Tatachilla has a record second only to Rosemount in producing high-flavoured red wines which can be released when very young, and which are usually ready for immediate consumption. By contrast, this wine undoubtedly needs time.

🍷🍷🍷🍷 **1997** Strong red-purple; the bouquet is powerful, with dense fruit and slightly raw oak. There is similarly lots of sweet dark berry/cherry/chocolate fruit on the palate, but it is somewhat callow. May well improve with further time in bottle. **rating:** 87

best drinking 2002–2007 **best vintages** NA **drink with** Steak and kidney pie • $22.95

Tatachilla Keystone

A blend of Grenache and Shiraz, the '94 vintage won the trophy for Best Lighter Bodied Dry Red Table Wine at the 1996 Sydney International Wine Competition, receiving top points from all nine judges. It is made from dry-grown 60-year-old bush-pruned McLaren Vale grenache and shiraz vines, open fermented, and matured for 12 months in a mix of new American and used French hogsheads. The '96 and '97 were also good wines, the '98 a little disappointing.

🍷🍷🍷🍷🍸 **1997** Bright purple; there are lively cherry pepper and spice aromas on the bouquet, with a tangy repetition of those characters in the mouth. Lovely early-drinking style. **rating:** 92

best drinking 1999–2000 **best vintages** '94, '96, '97 **drink with** Roast pigeon breast • $17.95

Tatachilla Clarendon Vineyard Merlot

1995 was the first vintage under this label, produced from grapes grown at an altitude of 215 metres above the floor of McLaren Vale. The wine is made in open fermenters, pressed before dryness, and matured in a mix of predominantly French (80%) and American oak, half-new and half-used. The '97 has won a trophy and four gold medals in national wine shows.

🍷🍷🍷🍷🍸 **1997** Very deep, dense red-purple; the bouquet is powerful, with ripe dark chocolate and earthy fruit, power which carries through on the palate. Here dark fruit and dark chocolate flavours are supported by persistent tannins. All in all, a classic show style though not necessarily a classic Merlot. **rating:** 91

best drinking 2002–2007 **best vintages** '96, '97 **drink with** Rack of veal • $31

Tatachilla Partners Cabernet Sauvignon Shiraz

The inaugural release (1996) of Partners won a gold medal at the 1997 International Wine Challenge in London, an auspicious start. But this blend of Cabernet Sauvignon and Shiraz, sourced from the relatively unfashionable South Australian Riverland and Adelaide Plains areas hit even greater heights with the '97 release, winning a gold medal at the Liquorland National Wine Show in Canberra in November 1997, before receiving the Arthur Kelman Trophy at the 1998 Royal Sydney Wine Show. The '98 continues the line, although one can ask whether such a big wine is really so suited to early release.

🍷🍷🍷🍷 **1998** Deep red-purple; powerful, youthful dark berry fruit floods the bouquet and crams the palate. There is no obvious oak influence, but the wine avoids the metallic/tanky/callow characters of lesser early-released red wines. **rating:** 89

best drinking 2002–2005 **best vintages** '96, '97 **drink with** Marinated beef • $14.95

Tatachilla McLaren Vale Cabernet Sauvignon

Tatachilla has obtained a first class network of high-quality grape suppliers; cabernet sauvignon is not always McLaren Vale's long suit, but this wine shows what can be done.

🍷🍷🍷🍷🍸 **1997** Medium to full red-purple; solid, ripe blackcurrant fruit and some regional chocolate aromas introduce a rich, full-bodied wine with lots of depth to the fruit and to the ripe tannins; subtle oak. **rating:** 92

best drinking 2002–2010 **best vintages** '97 **drink with** Grilled beef • $24.50

taylors ★★★

Taylors Road, Auburn, SA 5451 **region** Clare Valley

ph (08) 8849 2008 **fax** (08) 8849 2240 **open** Mon–Fri 9–5, Sat and public holidays 10–5, Sun 10–4

winemaker Susan Mickan, Kelvin Budarick **prod.** 250 000 **est.** 1972

prod. range ($7.45–14.50 R) The premium range consists of Chardonnay, Clare Riesling, White Clare, Promised Land Unwooded Chardonnay, Pinot Noir, Shiraz and Cabernet Sauvignon; the lower-priced Clare Valley range consists of Riesling, Dry White, Sweet White and Dry Red.

summary Taylors continues to flourish and expand, with yet further extensions to its vineyards. Now totalling almost 500 hectares, by far the largest holding in Clare Valley. There have also been substantial changes on the winemaking front, both in terms of the winemaking team and in terms of the wine style, the latter moving to fresher, earlier-release wines. Change always brings a measure of pain, but I wonder whether Taylors has not gone from one extreme to the other. Widespread national distribution, with exports to NZ, Germany, Northern Ireland, Malaysia and Singapore.

temple bruer ★★★☆

Milang Road, Strathalbyn, SA 5255 **region** Langhorne Creek
ph (08) 8537 0203 **fax** (08) 8537 0131 **open** Mon–Fri 9.30–4.30
winemaker Nick Bruer **prod.** 12 000 **est.** 1980

prod. range ($13.95–24.95 R) Riesling, Verdelho, Chenin Blanc, Botrytis Riesling, Cornucopia Grenache, Cabernet Merlot, Reserve Merlot, Shiraz Malbec, Sparkling Cabernet Merlot.

summary Always known for its eclectic range of wines, Temple Bruer (which also carries on a substantial business as a vine propagation nursery) has seen a sharp lift in wine quality. Clean, modern, redesigned labels add to the appeal of a stimulatingly different range of red wines. Part of the production from the 24 hectares of estate vineyards is sold to others, the remainder being made under the Temple Bruer label; in a sign of the times, it has found distribution in the US, Europe and Asia.

Temple Bruer Shiraz Malbec

A blend of 70% Shiraz and 30% Malbec from the 1.5 hectares of shiraz and half a hectare of malbec, both estate plantings. A traditional Langhorne Creek blend which works very well. The wine is made in old-style open fermenters, and matured in American oak, but without that oak making more than a subliminal contribution, which is all to the good. Engaged itself in a contest with a wine of radically different style (the '97 Port Phillip Reserve Shiraz) at the 1999 *Winewise* Small Makers Competition, and by a close vote, elegance prevailed over power.

🍷🍷🍷🍷🍷 **1996** Medium red-purple; the bouquet is pristinely clean and fruit-driven; a marvellously lively and elegant wine on the palate with cherry/berry fruit, a touch of spice, and only minimal oak influence. **rating:** 94

best drinking 1999–2003 **best vintages** '81, '83, '87, '89, '90, '91, '96 **drink with** Grilled steak • $15.95

Temple Bruer Reserve Merlot

The Langhorne Creek area should provide excellent merlot because of its (the region's) propensity towards soft, silky fruity red wines – which is what Merlot should be.

🍷🍷🍷🍷 **1996** Medium red-purple; quite sweet berry fruit aromas, moderately intense, are supported by subtle oak on the bouquet; similar berry fruits together with a touch of mint run through the palate, finishing with quite firm tannins. **rating:** 85

best drinking 2001–2006 **best vintages** NA **drink with** Escalope of veal • $24.95

---

templer's mill NR

Orange Agricultural College, Leeds Parade, Orange, NSW 2800 **region** Orange
ph (02) 6360 5509 **fax** (02) 6360 5698 **open** Friday 2–5 or by appointment
winemaker Jon Reynolds **prod.** 4000 **est.** 1997
prod. range ($16–18 CD) Chardonnay

summary Templer's Mill is a most interesting alliance of winemaker (Jon Reynolds), businessman-entrepreneur-cum-vigneron (Gary Blom of Barrington Estate in the Upper Hunter Valley) and the Orange Agricultural College, which is part of the University of Sydney. Together they are erecting a substantial winery which will cater not only for the needs of the joint venture partners, but also other growers in the district, as well as providing a teaching facility for the College. It looks a very promising venture.

Templer's Mill Chardonnay

The first release from the Sydney University's Orange campus vineyard, skilfully made by Jon Reynolds.

🍷🍷🍷🍷 **1997** Glowing yellow-green; the bouquet is clean, smooth and fresh with pure nectarine fruit and a bare whisper of oak. The palate, likewise, is fresh, lively and crisp, with a firm backbone of acidity which will ensure longevity though making the wine a touch formidable right now. Patience will be rewarded. **rating:** 85

best drinking 2000–2004 **best vintages** NA **drink with** Marinated octopus • $18

tempus two wines ★★★☆

Hermitage Road, Pokolbin, NSW 2321 **region** Lower Hunter Valley
ph (02) 9818 7222 **fax** (02) 9818 7333 **open** 7 days 10–5
winemaker Peter Hall **prod.** 20 000 **est.** 1997
prod. range ($14–22 R) Cowra Chardonnay, Hunter Merlot; followed by the Reserve range of Clare Riesling, Semillon, Vine Vale Shiraz, Hollydene Cabernet Sauvignon, Botrytis Semillon.

summary Tempus Two is the new name for Hermitage Road Wines, a piece of doggerel akin to that of Rouge Homme, except that it is not Franglais, but a mix of Latin (Tempus means time) and English. I should not be too critical, however; the change was forced on the winery by the EU Wine Agreement and the prohibition of the use of the word 'hermitage' on Australian wine labels. Nor should the fracas over the labels disguise the fact that some very attractive wines have appeared so far, and will do so in the future, no doubt.

Tempus Two Reserve Clare Riesling

The precise origin of the wine is not stated, other than the Clare Valley, which is probably sufficient in itself.

🍷🍷🍷🍷 **1998** Light to medium green-yellow; the bouquet shows fine toast, mineral and citrus-lime aromas; the palate lives up to the bouquet with intense citrus-lime fruit and a long finish. **rating:** 88

best drinking 1999–2008 **best vintages** NA **drink with** Asparagus salad • $18

Hermitage Road Reserve Semillon

Produced from grapes grown on the Somerset Vineyard in Oakey Creek Road, with a yield of between 2 and 3 tonnes per acre. Cold-fermented in stainless steel at 14 degrees for two weeks.

🍷🍷🍷🍷 **1998** Light green-yellow; while the bouquet is light and clean, as befits a young Semillon, has underlying weight. The palate, too, has considerable length and structure, certain to become a very good wine with time. **rating:** 87

best drinking 2003–2010 **best vintages** '98 **drink with** Flathead fillets • $18

Tempus Two Reserve Vine Vale Shiraz

With its Barossa Valley background, this wine emphasises the reach of the Tempus Two Wines portfolio, but leads one to wonder how it could ever be given a coherent identity or brand focus, no matter how good the wine quality.

🍷🍷🍷🍷 **1997** Bright but full purple-red; the moderately intense bouquet is clean, with smooth dark berry fruit and subtle oak. The palate picks up the pace with rich dark cherry fruit flavours, and balanced, lingering tannins. A fruit-driven style, and all the better for that. **rating:** 87

best drinking 2000–2006 **best vintages** NA **drink with** Braised steak • $22

terrace vale ★★★

Deasey's Lane, Pokolbin, NSW 2321 **region** Lower Hunter Valley
ph (02) 4998 7517 **fax** (02) 4998 7814 **open** 7 days 9–5
winemaker Alain Leprince **prod.** 8000 **est.** 1971
prod. range ($10–17.50 CD) Semillon, Chardonnay, Semillon Chardonnay, Gewurztraminer, Elizabeth Sauvignon Blanc, Fine Hunter White, Pinot Noir, Shiraz, Fine Hunter Red, Cabernet Sauvignon, Sparkling.
summary Long-established but relatively low-profile winery heavily dependent on its cellar-door and local (including Newcastle) trade. Smart new packaging has lifted the presentation, and, as always, there are one or two very good wines among the portfolio.

Terrace Vale Semillon Chardonnay

A blend which seems far more fashionable in the US and the UK than it is here, but which (as this wine shows) has a legitimate place in the scheme of things.

🍷🍷🍷🍷 **1997** Medium yellow-green; has acquired bottle-developed toasty characters on the bouquet a little more quickly than one might expect, but has good length and style on the palate, with an attractive mix of lime/herb and peachy fruit. **rating:** 87

best drinking 1999–2002 **best vintages** '97 **drink with** Blanquette of veal • $10

t'gallant ★★★★☆

Mornington Road, Red Hill, Vic 3937 **region** Mornington Peninsula
ph (03) 5989 6565 **fax** (03) 5989 6577 **open** 7 days 10–5
winemaker Kathleen Quealy, Kevin McCarthy **prod.** 12 000 **est.** 1990
prod. range ($19–33 R) An ever-changing list of names (and avant-garde label designs) but with Unwooded Chardonnay and Pinot Gris at the centre. Labels include Chardonnay, Lot 2 Chardonnay, Flag Pinot Grigio, Tribute Pinot Gris, Celia's White Pinot,Cape Schanck Pinot Grigio, IO McCabe Late Harvest Pinot Gris, Holystone, Cyrano Pinot Noir, Juno Pinot Noir and a range of wines under the Lyncroft label.
summary Husband and wife consultant-winemakers Kathleen Quealy and Kevin McCarthy are starting to carve out an important niche market for the T'Gallant label, noted for its innovative label designs and names. The acquisition of a 15-hectare property, and the planting of 10 hectares of pinot gris gives the business a firm geographic base, as well as providing increased resources for its signature wine. The yearly parade of new (usually beautiful and striking, it is true) labels designed by Ken Cato do not make my life at all easy. No sooner is the database built up than it is discarded for next year's rash of labels. La Baracca Trattoria is open seven days for lunch and for specially booked evening events.

T'Gallant Cape Schanck Pinot Grigio Chardonnay

A new and interesting direction for T'Gallant, very obviously aimed at the smart café and brasserie market, designed to go at the top of the list and into the best value category. One can only speculate the particular blend is marriage of convenience, but that hardly matters.

🍷🍷🍷🍷 **1998** Full straw-yellow; the bouquet is rich, with ripe papaya and exotic fruit aromas. Those flavours are positively decadent on the palate, which I like; inevitably, though, the wine falls away somewhat on the finish. **rating:** 84

best drinking 1999–2000 **best vintages** '98 **drink with** Virtually anything • $15

T'Gallant Pinot Grigio

A wine, the name of which seems to change with the direction of the wind; in 1997 the word 'Flag' was part of the label, but in '98 not. The key is the use of the name 'Pinot Grigio' (the Italian version) as opposed to Pinot Gris (the French). The Pinot Grigio is always lighter, crisper and less opulent than the Pinot Gris. It has come under various guises over the years, at one time called Flag, but in 1998 Hickson Vineyard.

🍷🍷🍷🍷 **1998** Light green-yellow; the bouquet is crisp, quite firm and minerally. The palate is firm and long, with an aftertaste of hay and mineral; a bone-dry food style, with Italianate overtones.

rating: 89

best drinking 1999–2000 **best vintages** NA **drink with** Smoked salmon • $20

T'Gallant Tribute Pinot Gris

An extraordinarily powerful and complex wine at its best, sometimes nearing 15 degrees alcohol. The grapes come from the vineyard of Madeleine and Patrick McCabe. The Tribute, incidentally, is to the distinguished viticulturist and consultant Max Loader.

🍷🍷🍷🍷 **1998** Light green-yellow; the bouquet is clean, quite fragrant, with those elusive hints of spicy fruit. The palate is likewise intense, with hints of spice, mineral and herbs; it carries its awesome 14.5 degrees alcohol with amazing nonchalance. **rating:** 88

best drinking 1999–2003 **best vintages** '94, '95, '97 **drink with** Gravlax • $26

T'Gallant Triumph Pinot Gris

Yes, another new label from T'Gallant, but justified by the wine in the bottle, made in the fashion of a Vendage Tardive from Alsace. It was picked between 4 May and 16 May, with considerable botrytis infection. It finished with 14.3 degrees alcohol and 24 grams per litre of residual sugar.

🍷🍷🍷🍷🍷 **1998** Light to medium yellow-green; the bouquet is fragrant with floral and honey aromas which are quite delicious. The palate has rich flavour, lots of flesh and balanced acidity. The only problem with a wine such as this is when to serve it. **rating:** 90

best drinking 1999–2004 **best vintages** NA **drink with** Fresh fruit • $19

T'Gallant Unwooded Chardonnay

Quealy and McCarthy are firm believers in the virtues of Unwooded Chardonnay, but recognise the paramount necessity of fully ripe grapes which are largely free from botrytis. Made from selected contract-growers across the length and breadth of the Mornington Peninsula, the wine typically has sufficient alcohol to give it the requisite flesh and character. Almost inevitably, yet another label design.

🍷🍷🍷🍷🍷 **1997** Light green-yellow; the bouquet is crisp and clean, slow-developing and belying its 13.8 degrees alcohol. The palate is necessarily a little simple (no oak and not overly modified by malolactic influences) but with nice melon fruit which carries the alcohol well. **rating:** 90

best drinking 1999–2001 **best vintages** NA **drink with** Sweet white-fleshed fish • $20

T'Gallant IO McCabe Late Harvest Pinot Gris

A freakish wine produced from a small parcel of vines on the McCabe Vineyard. The grapes reached normal maturity on 1 May, and were only then attacked by botrytis, being picked at the end of the third week in May. This is precisely the sequence needed to produce the best botrytis wines, permitting the development of full varietal character before the concentrating effect of botrytis takes hold. For all that, and like the Vendage Tardive wines of Alsace, they are often easier to respect than to love. Since a grand total of 43 cases were made, this hardly matters.

🍷🍷🍷🍷 **1998** Light green-yellow; the bouquet is clean and smooth, yet it is almost impossible to point to any single fruit character or aroma. The palate, too, has much more to do with texture, oozing as it does with alcohol and glycerol, than it does with flavour. It is in every way the vinous equivalent to goose, which is the recommended food match of T'Gallant, and with which I have no quarrel. **rating:** 86

best drinking 1999–2005 **best vintages** '98 **drink with** Goose • $33

T'Gallant Lyncroft Pinot Noir

The 5-hectare Lyncroft vineyard is owned by Geoff Slade and partner Anita Zierner, but is managed by T'Gallant and, of course, the wines are made by T'Gallant. A massive 200 cases of this wine were made, not quite in the class of Lindemans Bin 65, but perhaps giving you a fighting chance of finding it somewhere outside cellar door. It is certainly worth the search, coming as it does from an excellent Mornington vintage.

🍷🍷🍷🍷 **1998** Medium red-purple; the bouquet is clean and fresh, with direct cherry fruit and just a whisper of more stemmy characters. The attractive, moderately-weighted palate has cherry and plum fruit supported by subtle oak. The '97 vintage was always going to be a hard act to follow, but this wine makes a good showing of so doing. **rating:** 88

best drinking 1999–2002 **best vintages** NA **drink with** Breast of duck • $26

thalgara estate NR

De Beyers Road, Pokolbin, NSW 2321 **region** Lower Hunter Valley

ph (02) 4998 7717 **fax** (02) 4998 7774 **open** 7 days 10–5

winemaker Steve Lamb **prod.** 3000 **est.** 1985

prod. range ($15–30 CD) Chardonnay, Show Reserve Chardonnay, Semillon Chardonnay, Shiraz, Show Reserve Shiraz, Shiraz Cabernet.

summary A low-profile winery but given to surprising show success, never more so than at the 1997 Hunter Valley Wine Show when it won the Doug Seabrook Memorial Trophy for Best Dry Red of Show with its 1995 Show Reserve Shiraz.

the green vineyards ★★★★

1 Alber Road, Upper Beaconsfield, Vic 3808 **region** Yarra Valley

ph (03) 5944 4599 **fax** (03) 5944 4599 **open** Weekends by appointment

winemaker Sergio Carlei **prod.** 2000 **est.** 1994

prod. range ($12–25 CD) Yarra Valley Riesling, Mornington Sauvignon Blanc, Yarra Valley Chardonnay, Sunbury Pinot Noir, Heathcote Shiraz, Yarra Valley Cabernets.

summary The Green Vineyards has come a long way in a little time, with Sergio Carlei graduating from home winemaking in a suburban garage to his own (real) winery in Upper Beaconsfield, which happens to fall just within the boundaries of the Yarra Valley. As the product range attests, many of the wines come from grapes grown by others for Green Vineyards, but Carlei does have 2.25 hectares of pinot noir, and his preferred source is the Yarra Valley. He has already produced a number of remarkably stylish wines, with more in the pipeline.

the gurdies NR

St Helier Road, The Gurdies, Vic 3984 **region** Gippsland

ph (03) 5997 6208 **fax** (03) 5997 6511 **open** 7 days 10–5

winemaker Peter Kozik **prod.** 1000 **est.** 1991

prod. range ($15–18 R) Riesling, Pinot Noir, Shiraz, Cabernet Sauvignon.

summary The only winery in the southwest Gippsland region, established on the slopes of The Gurdies hills overlooking Westernport Bay and French Island. Plantings of the 3.5-hectare vineyard commenced in 1981, but no fruit was harvested until 1991 owing to bird attack. A winery has been partially completed, and it is intended to increase the vineyards to 25 hectares and ultimately build a restaurant on site.

the melbourne wine company ★★★☆

PO Box 74, Bannockburn, Vic 3331 **region** Warehouse
ph (03) 5281 7477 **fax** (03) 5281 7477 **open** Not
winemaker Scott Ireland, Martin Williams **prod.** NA **est.** 1998
prod. range ($15–19 R) Riesling, Pinot Gris, Chardonnay, Pinot Noir; premium range under the Provenance label.
summary The Melbourne Wine Company was originally conceived by Donleavy Fitzpatrick, but is now owned by five partners headed by winemakers Scott Ireland and Martin Williams. The concept is to source grapes from the four major grape-growing regions encircling Melbourne: the Yarra Valley, Macedon, Geelong and the Mornington Peninsula. The principal range of wine places the emphasis on light, fresh styles which do not need cellaring. The most distinctive feature of distinctive packaging is the use of crown seals, yet another solution to cork taint. The initial releases precisely fulfilled the partners' aims, offering simple enjoyment.

The Melbourne Wine Company Pinot Gris

Produced from grapes grown at the evocatively-named (but almost unpronounceable) Bunjil Moolaghurk vineyard at Geelong.
▼▼▼▽ **1998** The colour has distinct pink tinges, quite legitimate in the context of the variety. The bouquet is crisp, with faint blossom aromas and some more minerally notes. The fresh, clean and crisp palate has nice mouthfeel, moving from Gris to Grigio in style towards the finish as it dries slightly and those minerally characters become more pronounced. Overall, works well. **rating:** 84
best drinking 1999–2000 **best vintages** NA **drink with** Fish or white meat • $19

The Melbourne Wine Company Chardonnay

The winemaking team of Scott Ireland and Martin Williams took grapes from the Geelong and Alpine regions of Victoria (a reasonable assumption is that the King Valley is included in the term 'Alpine regions', which does not have an official registration) to make this rich but seemingly unoaked Chardonnay.
▼▼▼▽ **1998** Bright, light green-yellow; the bouquet is light, with fresh melon and apple aromas which track precisely on the palate where you find identical characters. Although it doesn't say so on the label, almost certainly unoaked. **rating:** 84
best drinking 1999–2000 **best vintages** NA **drink with** White-fleshed fish • $16

The Melbourne Wine Company Pinot Noir

Produced from a blend of grapes sourced from Seymour, Geelong and Ballarat.
▼▼▼▼ **1998** Medium to full red-purple; the initial impression on the bouquet is of solid plummy fruit, which likewise opens up the palate. However, both the bouquet and palate have a slightly edgy character, difficult to pin down and with uncertain ramifications for the future. **rating:** 86
best drinking 1999–2001 **best vintages** NA **drink with** Tea-smoked duck • $19

the minya winery NR

Minya Lane, Connewarre, Vic 3227 **region** Geelong
ph (03) 5264 1397 **open** Public holidays and by appointment
winemaker Susan Dans **prod.** 1330 **est.** 1974
prod. range ($12–18 CD) Gewurztraminer, Chardonnay, Unwooded Chardonnay, Grenache, Cabernet Shiraz Merlot.
summary Geoff Dans first planted vines in 1974 on his family's dairy farm, followed by further plantings in 1982 and 1988. I have not tasted any of the wines.

the silos estate NR

Princes Highway, Jaspers Brush, NSW 2535 **region** Shoalhaven
ph (02) 4448 6082 **fax** (02) 4448 6246 **open** Wed–Sun 10–5
winemaker Gaynor Sims, Kate Khoury **prod.** 1000 **est.** 1985
prod. range ($7–16 CD) Traminer Riesling, Semillon, Chardonnay, Sauvignon Blanc, Wileys Creek Brut, Shiraz Cabernet, Tawny Port, Liqueur Muscat.
summary Since 1995, Gaynor Sims and Kate Khoury, together with viticulturist Jovica Zecevic, have worked hard to improve the quality of the wine, starting with the 5 hectares of estate vineyards but also in the winery. The winery continues to rely on the tourist trade, however, and the wines do not appear in normal retail channels.

the warren vineyard NR

Conte Road, Pemberton, WA 6260 **region** Pemberton
ph (08) 9776 1115 **fax** (08) 9776 1115 **open** 7 days 11–5
winemaker Andrew Forsell **prod.** 600 **est.** 1985
prod. range ($17–25 CD) Chardonnay, Riesling, Merlot, Cabernet Merlot, Cabernet Sauvignon, Vintage Port.
summary The 1.4-hectare vineyard was established in 1985, and is one of the smallest in the Pemberton region, coming to public notice when its 1991 Cabernet Sauvignon won the award for the Best Red Table Wine from the Pemberton Region at the 1992 SGIO Western Australia Winemakers Exhibition.

the willows vineyard ★★★

Light Pass Road, Light Pass, Barossa Valley, SA 5355 **region** Barossa Valley
ph (08) 8562 1080 **fax** (08) 8562 3447 **open** 7 days 10.30–4.30
winemaker Peter Scholz, Michael Scholz **prod.** 4000 **est.** 1989
prod. range ($10.50–19 R) Riesling, Semillon, Shiraz, Cabernet Sauvignon.
summary The Scholz family has been grape growers for generations, and have a little over 37 hectares of vineyards, selling part, and retaining part of the crop. Current generation winemakers Peter and Michael Scholz could not resist the temptation to make smooth, well-balanced and flavoursome wines under their own label. These are all marketed with some years bottle age.

thistle hill ★★★☆

McDonalds Road, Mudgee, NSW 2850 **region** Mudgee
ph (02) 6373 3546 **fax** (02) 6373 3540 **open** Thur–Mon 10–4
winemaker David Robertson **prod.** 4000 **est.** 1976
prod. range ($11–22 CD) Riesling, Semillon, Chardonnay, Pinot Noir, Merlot, Cabernet Sauvignon.
summary David and Leslie Robertson produce supremely honest wines, always full of flavour and appropriately reflecting the climate and terroir. Some may be a little short on finesse, but never on character. Chardonnay and Cabernet Sauvignon lead the way, and age well.

thomas NR

23-24 Crowd Road, Gelorup, WA 6230 **region** Geographe
ph (08) 9795 7925 **open** By appointment
winemaker Gill Thomas **prod.** 600 **est.** 1976
prod. range ($4.50–25 CD) Pinot Noir, Cabernet Sauvignon.
summary I have not tasted the elegant wines of Bunbury pharmacist Gill Thomas for several years; they are only sold to a local clientele.

thornhill/the berry farm NR

Bessel Road, Rosa Glen, WA 6285 **region** Margaret River
ph (08) 9757 5054 **fax** (08) 9757 5116 **open** 7 days 10–4.30
winemaker Eion Lindsay **prod.** NFP **est.** 1990
prod. range ($11.50–25 CD) Under the Thornhill label Classic Dry Semillon, Sauvignon Blanc, Cabernet Sauvignon, Tickled Pink (Sparkling Cabernet Sauvignon), Still Tickled Pink (Light Cabernet Sauvignon). Under The Berry Farm label a range of fruit-based wines including Sparkling Strawberry and Plum Port.
summary Although I have not enjoyed the Thornhill table wines, the fruit wines under The Berry Farm label are extraordinarily good. The sparkling strawberry wine has intense strawberry flavour; the plum port likewise, carrying its 16% alcohol with remarkable ease.

tilba valley NR

Glen Eden Vineyard, Corunna Lake via Tilba, NSW 2546 **region** Other Wineries of NSW
ph (02) 4473 7308 **open** Mon–Sat 10–5, Sun 11–5
winemaker Barry Field **prod.** 1400 **est.** 1978
prod. range ($10–14 CD) Traminer Riesling, Semillon, Chardonnay, Cabernet Hermitage.
summary A strongly tourist-oriented operation, serving a ploughman's lunch daily from noon to 2 pm. Has 5 hectares of estate vineyards; no recent tastings.

tim adams ★★★★

Warenda Road, Clare, SA 5453 **region** Clare Valley
ph (08) 8842 2429 **fax** (08) 8842 3550 **open** Mon–Fri 10.30–5, weekends 11–5
winemaker Tim Adams **prod.** 15 000 **est.** 1986
prod. range ($11.50–30 CD) Riesling, Semillon, Botrytis Semillon, Botrytis Reisling, The Fergus (Grenache), Shiraz, Aberfeldy Shiraz, Cabernet.
summary Tim and Pam Adams have built a first class business since Tim Adams left his position as winemaker at Leasingham in 1985. Nine local growers provide the grapes for the enterprise, which has consistently produced wines of exceptional depth of flavour, and which also makes significant quantities of wine under contract for others in the district. Extensive distribution through all Australian States; exports to the UK and US.

Tim Adams Semillon

It is made to develop early, with substantial oak input (75% American, 25% French) and 12 hours skin contact used prior to fermentation. The wine is barrel-fermented and spends five months in oak.

🍷🍷🍷🍷 **1998** Medium to full yellow-green; the usual full-frontal style with lots of American oak and some fruit refusing to capitulate on the bouquet; the palate is oak-driven as always, very faithful to the style. **rating:** 86

best drinking 1999–2002 **best vintages** '89, '90, '92, '94, '95, '96 **drink with** Rich white meat dishes • $14

Tim Adams Aberfeldy Shiraz

The top-of-the-tree release for Tim Adams, which is given four weeks post-fermentation maceration on skins, and then taken to new American oak where it spends the next 22 months. Bottled February two years after vintage.

🍷🍷🍷🍷½ **1997** Medium to full red-purple; the bouquet is flooded with sweet, rich spicy berry fruit and the very clever use of oak. The opulently rich palate consummates the marriage between oak (the dominant partner) and dark fruit flavours, all supported by soft but ample tannins. **rating:** 92

best drinking 2005–2015 **best vintages** '97 **drink with** Char-grilled rump • $30

Tim Adams The Fergus

Predominantly made from Grenache, with small amounts of Cabernet Sauvignon, Cabernet Franc and Shiraz (quite why the Cabernets are used I do not know). The grapes come from a vineyard of Ferg and Vyv Mahon and are picked ripe to give a finished alcohol of 14.5 degrees. Like The Aberfeldy, the wine spends four weeks macerating on skins post-fermentation.

🍷🍷🍷🍷 **1998** Medium red-purple; there is smooth berry fruit, with chocolate overtones and subtle oak on the bouquet. The palate unexpectedly diverts into uncompromising minty flavours, which some will like more than others. Whatever one's view on the mint, the structure of the wine is good. **rating:** 86

best drinking 1999–2002 **best vintages** NA **drink with** Designer hamburgers • $16

tim gramp ★★★★

Mintaro Road, Watervale, SA 5452 **region** Clare Valley
ph (08) 8431 3338 **fax** (08) 8431 3229 **open** Weekends and holidays 10.30–4.30
winemaker Tim Gramp **prod.** 5000 **est.** 1990
prod. range ($14–25 R) Watervale Riesling, McLaren Vale Shiraz, McLaren Vale Grenache, Watervale Prism Cabernet Sauvignon.

summary Tim Gramp has quietly built up a very successful business with a limited product range, and – by keeping overheads to a minimum – provides good wines at modest prices. The operation is supported by 2 hectares of cabernet sauvignon around the cellar door.

Tim Gramp Watervale Riesling

With the underlying philosophy of matching variety and region, it is not surprising Tim Gramp's Riesling comes from Watervale in the Clare Valley. Made in a bold, full-frontal style, but which can improve dramatically with time in bottle.

🍷🍷🍷🍷🍷 **1998** Medium yellow-green; the bouquet is moderately intense and clean, with lime backed by touches of mineral and toast. The palate is elegant and clean, with the long finish augmented by crisp acidity. **rating:** 90

best drinking 2000–2008 **best vintages** '97, '98 **drink with** Bouillabaisse • $14

Tim Gramp McLaren Vale Shiraz

The inaugural 1991 Shiraz won a trophy, six gold and two silver medals, including a gold medal at the 1993 Intervin International Wine Show in New York. The wine is made in a full-throated, full-blooded style, using low-yielding dry-grown grapes and lots of American oak, and works to perfection. The wine is sourced from Willunga (in McLaren Vale) and spends 14 months in oak.

🍷🍷🍷🍷🍷 **1997** Bright red-purple; a super-elegant bouquet with cherry and spice fruit which carries through into the palate. The tannins are present, but are not harsh, and the oak subtle. Gold medal 1998 National Wine Show. **rating** 93

best drinking 2002–2007 **best vintages** '91, '92, '94, '96, '97 **drink with** Barbecued, marinated steak • $25

Tim Gramp Watervale Prism Cabernet Sauvignon

A big, robust Clare Valley-style with the near-certainty of ageing well.

🍷🍷🍷🍷 **1997** Medium red-purple; ripe cassis and mint fruit dominates the bouquet, with similar ripe but not jammy fruit on the forepalate, finishing with quite persistent tannins. **rating:** 86

best drinking 2001–2007 **best vintages** NA **drink with** Barbecued beef • $16

tingle-wood ★★★★★

Glenrowan Road, Denmark, WA 6333 **region** Great Southern
ph (08) 9840 9218 **open** 7 days 9–5
winemaker Brenden Smith (Contract) **prod.** 1000 **est.** 1976

prod. range ($16–20 CD) Yellow Tingle (Riesling), Late Harvest Yellow Tingle, Red Tingle (Cabernet Shiraz), Ruby Tingle (Port-style).

summary An intermittent producer of Riesling of extraordinary quality, although birds and other disasters do intervene and prevent production in some years.

tinlins NR

Kangarilla Road, McLaren Flat, SA 5171 **region** McLaren Vale
ph (08) 8323 8649 **fax** (08) 8323 9747 **open** 7 days 9–5
winemaker Warren Randall **prod.** 30 000 **est.** 1977

prod. range ($1.50–3.40 CD) Generic table, fortified and flavoured wines sold for $1.50 for table wines and $3.40 for fortified wines.

summary A very interesting operation, run by former Seppelt sparkling winemaker Warren Randall, drawing upon 100 hectares of estate vineyards which specialises in bulk-wine sales to the major Australian wine companies. A small proportion of the production is sold direct through the cellar door at mouthwateringly low prices to customers who provide their own containers and purchase by the litre. McLaren Vale's only bulk-wine specialist.

tinonee vineyard NR

Milbrodale Road, Broke, NSW 2330 **region** Lower Hunter Valley
ph (02) 6579 1308 **fax** (02) 6579 1146 **open** Weekends and public holidays 10–5
winemaker Adrian Sheridan (Contract) **prod.** 3000 **est.** 1997
prod. range ($15–18 CD) Chardonnay, Verdelho, Merlot, Shiraz.

summary Ian Craig has established 14 hectares of vineyards on a mix of red volcanic and river flat soils at Broke. Part are in production, and the remainder will come into bearing by the end of the decade, ultimately producing 5000 cases of wine per year.

tintilla wines NR

725 Hermitage Road, Pokolbin, NSW 2335 **region** Lower Hunter Valley
ph 0411 214 478 **fax** (02) 9767 6894 **open** By appointment
winemaker Jon Reynolds (Contract) **prod.** 1000 **est.** 1993
prod. range ($15–30 CD) Semillon, Sangiovese, Shiraz, Merlot.

summary The Lusby family has established a 7.5-hectare vineyard (including 1 hectare of sangiovese) on their northeast-facing vineyard, with its red clay and limestone soil. They have also established an olive grove producing four different types of olives which are cured and sold on the estate. The first wine release promises much for the future.

Tintilla Shiraz

Ten per cent of the wine is fermented using the carbonic maceration technique, the remaining part a 'natural' fermentation. The wine spends a little under 12 months in American oak. The 1997 was a tentative start, the '98 in a different league altogether.

ΨΨΨΨ **1998** Bright purple-red; the complex bouquet ranges through chocolate, earth and briar, but with rich, dark cherry the lasting impression. The palate picks up on that impressive cherry fruit, providing plenty of total flavour; the American oak has not been allowed to more than subtly influence what is a very good wine. **rating:** 89

best drinking 2001–2008 **best vintages** NA **drink with** Rare scotch fillet • $25

tipperary hill estate NR

Alma–Bowendale Road, Alma via Maryborough, Vic 3465 **region** Bendigo
ph (03) 5461 3312 **fax** (03) 5461 3312 **open** Weekends 10–5, or by appointment
winemaker Paul Flowers **prod.** 300 **est.** 1986
prod. range ($16–24 CD) Shiraz, Pinot Noir, Cabernets.

summary All of the wine is sold through the cellar door and on-site restaurant, open on Sundays. Says Paul Flowers, production depends 'on the frost, wind and birds', which perhaps explains why this is very much a part-time venture. Situated 7 kilometres west of the city of Maryborough, Tipperary Hill Estate is the only winery operating in the Central Goldfields Shire. Winemaker Paul Flowers built the rough-cut pine winery and the bluestone residential cottage next-door with the help of friends. Together with wife Margaret he also operates a restaurant.

tizzana NR

518 Tizzana Road, Ebenezer, NSW 2756 **region** Other Wineries of NSW
ph (02) 4579 1150 **fax** (02) 4579 1216 **open** Weekends, holidays 12–6
winemaker Peter Auld **prod.** 200 **est.** 1887
prod. range ($5.50–16.50 CD) Estate-grown and made Shiraz, Cabernet Sauvignon, Port; cleanskin wines under Tizzana Selection label.
summary The only estate wines tasted (several years ago) were not good, but the historic stone winery is most certainly worth a visit, and a wide selection of Tizzana Selection wines from other makers gives a broad choice.

tollana ★★★★

Tanunda Road, Nuriootpa, SA 5355 **region** Barossa Valley
ph (08) 8560 9389 **fax** (08) 8562 2494 **open** Mon–Sat 10–5, Sun 1–5
winemaker Neville Falkenberg **prod.** 19 000 **est.** 1888
prod. range ($10–21 R) Eden Valley Riesling, Eden Valley Chardonnay, Coonawarra Botrytis Riesling, Hermitage, Show Reserve Shiraz, Eden Valley Adelaide Hills Cabernet Sauvignon Bin TR222.
summary As the Southcorp Wine Group moves to establish regional identity for its wines, Tollana is emphasising its Eden Valley base. Seemingly as a by-product of Penfolds development of Yattarna and related wines, the Tollana Chardonnay style has become more elegant, now standing comfortably alongside the flavoursome Riesling and Shiraz.

Tollana Eden Valley Riesling

For the better part of two decades, the Tollana Riesling has been one of Australia's better-kept secrets, particularly under its prior ownerships. It ages well, and has not infrequently collected trophies at national wine shows as a mature wine.

🍷🍷🍷🍷 **1998** Light to medium yellow-green; as always, plenty of weight with lime and tropical fruit. The palate is rich, relatively full-bodied, with masses of lime-flavoured fruit, giving the appearance it will develop quite quickly. **rating:** 85

best drinking 1999–2003 **best vintages** '86, '87, '90, '92, '93, '97 **drink with** Seafood salad • $11.90

Tollana Sauvignon Blanc

Stainless steel-fermented, the wine utilises grapes primarily grown in the Adelaide Hills.

🍷🍷🍷🍷 **1998** Light green-yellow; the bouquet is firm and discreet, but with notes of both mineral and more tropical fruit. A cleverly made wine with a touch of residual sugar to fill out the palate. **rating:** 86

best drinking 1999–2000 **best vintages** NA **drink with** Vegetable terrine • $11.90

Tollana Eden Valley Chardonnay

The wine is barrel-fermented in a temperature-controlled fermentation area and matured in new and used French oak for seven months. The majority undergoes malolactic fermentation prior to final blending and bottling.

🍷🍷🍷🍷 **1996** Glowing yellow-green; the bouquet is quite solid and smooth with good oak balance and integration, the palate likewise showing smooth melon and fig fruit. One of those wines which has neither highlights nor lowlights. **rating:** 86

best drinking 1998–2001 **best vintages** '88, '90, '92, '93 **drink with** Chicken poached in white wine • $10

Tollana Eden Valley Adelaide Hills Chardonnay

A blend of Eden Valley and Adelaide Hills material, attesting to the changes both in fruit source and wine style at Tollana.

🍷🍷🍷🍷 **1997** Medium yellow-green; the bouquet is quite fresh, with well-integrated and balanced fruit and oak; there is elegant melon and citrus fruit on the forepalate, with gentle oak picking up towards the finish. **rating:** 88

best drinking 1999–2002 **best vintages** NA **drink with** Breast of chicken • $12.95

Tollana Coonawarra Botrytis Riesling

Made from heavily botrytised Coonawarra material, and sometimes containing a little Gewurztraminer. A much-underrated wine outside of the show ring; it has won an awesome number of medals since the 1992 vintage in wine shows.

🍷🍷🍷🍷🍷 **1997** The golden-yellow colour precedes a wine with the aromas of a tropical cocktail, so complex it almost seems as if the wine has had a touch of oak, which it hasn't. The super-rich and concentrated palate has an exceptionally long finish perfectly balanced by acidity. **rating:** 95

best drinking 1998–2000 **best vintages** '97 **drink with** Poached pears and ice cream • $11.50

Tollana Show Reserve Shiraz

Tollana's Show Reserve wine is only made in exceptional vintages, thus on average it is made once every two years. It is 100% Eden Valley Shiraz, matured for 12 months in new French and American oak barrels.

🍷🍷🍷🍷 **1995** Medium to full red-purple; the bouquet is rich, with dense, dark fruit and strong vanilla oak, but on the palate sweet berry and black cherry fruit flavours, supported by fine tannins are not overwhelmed by the oak. **rating:** 87

best drinking 1999–2006 **best vintages** '91, '93, '95 **drink with** Hearty red meat dishes • $21

Tollana Cabernet Sauvignon Bin TR222

Another Tollana classic with a proud history. Made entirely from Eden Valley Cabernet Sauvignon, it is matured in new and one-year-old American and French oak for 15 months. Like the Riesling, it has been a prolific and consistent gold medal winner. The wine comes from the Woodbury Vineyard, which has 60 hectares of fully mature cabernet sauvignon, the best parcels going to produce this wine which has been a prolific gold medal winner over the years.

🍷🍷🍷🍷 **1996** Medium to full red-purple; the bouquet is quite complex, with an interplay between savoury, leafy cabernet and American oak providing both depth and texture. There is plenty of sweet, dark berry fruit on the palate, and again a vanilla coating of American oak. In its style, very good. **rating:** 86

best drinking 2001–2010 **best vintages** '86, '88, '90, '91, '92, '93, '95, '96 **drink with** Braised lamb • $18.80

toorak estate NR

Toorak Road, Leeton, NSW 2705 **region** Riverina

ph (02) 6953 2333 **fax** (02) 6953 4454 **open** Mon–Sat 9–5

winemaker Robert Bruno **prod.** 10 000 **est.** 1965

prod. range ($7–14 CD) Rhine Riesling, Traminer Riesling, Colombard Chardonnay, Semillon, Chardonnay, Autumn Harvest, Lambrusco Red and White, Leeton Shiraz, Shiraz Cabernet, Ruby Cabernet, Cabernet Sauvignon, Sparkling, Fortifieds.

summary A traditional, long-established Riverina producer with a strong Italian-based clientele around Australia. Production has been increasing significantly, utilising 65 hectares of estate plantings and grapes purchased from other growers.

torbreck vintners ★★★★☆

Roenfeldt Road, Marananga, SA 5352 **region** Barossa Valley
ph (08) 8562 4155 **fax** (08) 8562 4195 **open** Not
winemaker David Powell **prod.** 400 **est.** 1994
prod. range ($24–41 ML) The Steading (Grenache Mataro Shiraz), RunRig (Shiraz Viognier).
summary Torbreck has made a major impact since its first releases in 1997 of a 1995 RunRig and 1996 The Steading. David Powell's family has assembled small patches of old vine shiraz, grenache and mourvedre in the Barossa Valley which they are sharefarming, complementing a small vineyard of their own. The two vintages of each of the red wines so far released have been of very high quality, opulently concentrated and rich.

Torbreck RunRig Shiraz Viognier

We are told the highland clans used a 'Runrig' system to distribute land among their clansmen in a series of widely dispersed holdings, the emphasis not on any one farm, but rather the communal elements of the whole. Hence the wine, in which Shiraz from old dry-grown vineyards is blended with 3% to 5% Viognier, in the words of Torbreck 'complementing the strength and complexities of these individual parcels of fruit, whilst giving the resulting wine a further dimension'.

🍷🍷🍷🍷🍷 **1996** Excellent, deep red-purple. There is voluminous rich, concentrated and lusciously ripe, chocolate-accented fruit on both bouquet and palate. The latter is crammed with luscious fruit and oak, and will certainly repay those who lock the wine away for ten years or more.

rating: 94

best drinking 2006–2016 **best vintages** '96 **drink with** Game • $41

Torbreck The Steading Grenache Shiraz

The inspiration for the name, and the wine, is explained thus by Torbreck 'on a highland farm the collection of barns, stables and outbuildings is known as a steading'. A blend of Grenache, Mourvedre and Shiraz, which in 1997 moved to 60%/20%/20% compared to the 55%/35%/10% in 1996, and which is likely to be the approximate future blend. One thousand cases made in 1997.

🍷🍷🍷🍷🍷 (4½) **1997** Medium red, with some purple hues; the bouquet shows quite pronounced jammy/berry Grenache varietal character, which is as it should be, together with a nicely judged touch of vanilla oak. The wine has tremendous character on the palate, with that very sweet ripe grenache fruit coming through on the faintly, but far from unpleasantly jammy mid-palate. Very seductive.

rating: 91

best drinking 2002–2007 **best vintages** '97 **drink with** Lamb Provençale • $24

trafford hill vineyard NR

Lot 1 Bower Road, Normanville, SA 5204 **region** Fleurieu Zone
ph (08) 8558 3595 **open** By appointment
winemaker John Sanderson, Allan Dyson (Consultant) **prod.** 300 **est.** 1996
prod. range ($9.80–16.50 ML) Riesling, Monique Family Reserve Blend.
summary Irene and John Sanderson have established one and a quarter hectares of vineyard at Normanville, on the coast of the Fleurieu Peninsula near to its southern extremity. Irene carries out all the viticulture, and John Sanderson makes the wine with help from district veteran Allan Dyson. Distribution is through local restaurants, the remainder through mail order and cellar door.

treehouse vineyard NR

257 Richmond Road, Cambridge, Tas 7170 **region** Southern Tasmania
ph (03) 6248 5367 **fax** (03) 6248 5367 **open** Wed–Sun 10–5
winemaker Andrew Hood (Contract) **prod.** NA **est.** 1991
prod. range ($18 CD) Riesling, Pinot Noir.
summary Gradon and Margaret Johnstone established Treehouse in 1991 with the planting of their first grapes. Restaurant meals and casual food is available at the Treehouse Wine Centre, and all of the wines are sold through the cellar door and by mail order.

Treehouse Vineyard Riesling

A typical Tasmanian Riesling, with good acidity and the promise of cellaring well. Produced in tiny quantities, normally a difficult challenge for delicate white wines, but no problem for the experienced contract-winemaker Andrew Hood.

YYYY **1998** Light green-yellow; the bouquet has distinctive chalky, almost dusty aromas but the palate is more conventional in a well-balanced no-frills style with good length, although not so much depth. **rating:** 86

best drinking 1999–2004 **best vintages** NA **drink with** Caesar salad • $18

treen ridge estate NR

Parker Road, Pemberton, WA 6260 **region** Pemberton
ph (08) 9776 1131 **fax** (08) 9776 1176 **open** Weekends 10–5
winemaker Andrew Mountford, Donnelly River Wines (Contract) **prod.** NFP **est.** 1992
prod. range ($NA) Riesling, Sauvignon Blanc, Shiraz, Cabernet Sauvignon.
summary As the details indicate, the venture is in its infancy, with the first wine sales in 1997. Bed and breakfast accommodation is under construction. Draws upon 2 hectares of vines.

treeton estate ★★★

North Treeton Road, Cowaramup, WA 6284 **region** Margaret River
ph (08) 9755 5481 **fax** (08) 9755 5051 **open** 7 days 10–6
winemaker David McGowan **prod.** 3000 **est.** 1984
prod. range ($15–17 R) Chardonnay, Riesling, Estate White, Petit Rouge, Shiraz, Liqueur Muscat.
summary David McGowan and wife Corinne purchased the 30-hectare property upon which Treeton Estate is established in 1982, beginning to plant the vines two years later. David has done just about everything in his life, and in the early years was working in Perth, which led to various setbacks for the vineyard. The wines are light and fresh, sometimes rather too much so.

trentham estate ★★★☆

Sturt Highway, Trentham Cliffs, NSW 2738 **region** Murray Darling and Swan Hill (NSW)
ph (03) 5024 8888 **fax** (03) 5024 8800 **open** Mon–Fri 8.30–5, weekends 9.30–5
winemaker Anthony Murphy, Shane Kerr **prod.** 30 000 **est.** 1988
prod. range ($8.50–14 R) Riesling, Sauvignon Blanc, Colombard Chardonnay, Chardonnay, Noble Taminga, Pinot Noir, Merlot, Grenache Shiraz, Shiraz, Burke & Wills Tawny Port; Tresoli White and Red, Falling Leaf Autumn White and Red.
summary Remarkably consistent tasting notes across all wine styles from all vintages since 1989 attest to the expertise of ex-Mildara winemaker Tony Murphy, now making the Trentham wines from his family vineyards. Indeed, Trentham seems to be going from strength to strength with each succeeding vintage, exemplified by the Tresoli White and Red wines. The winery restaurant is also recommended.

Trentham Estate Merlot

Since the early 1990s Tony Murphy has often managed to conjure up something special with Merlot; the Mildura region is not the most likely site for a variety such as this.

🍷🍷🍷 **1997** Light to medium red with just a touch of purple; the bouquet does have varietal character in a somewhat sappy/vegetal mode, the palate with a touch more sweet fruit. Has some of the slightly silky texture one is entitled to expect from Merlot. **rating:** 84

best drinking 2000–2003 **best vintages** '92, '93, '94 **drink with** Marinated rabbit • $12.50

trevor jones ★★★★☆

Barossa Valley Highway, Lyndoch, SA 5351 **region** Barossa Valley
ph (08) 8524 4303 **fax** (08) 8524 4880 **open** 7 days 9–6
winemaker Trevor Jones **prod.** 2000 **est.** 1996
prod. range ($15–25 CD) Riesling, Virgin Chardonnay, Dry Grown Shiraz, Cabernet Merlot.
summary Trevor Jones is an industry veteran, with vast experience in handling fruit from the Barossa Valley, Eden Valley and Adelaide Hills. He has finally taken the step of introducing his own strikingly-designed label, using grapes purchased from various contract growers, with the first wines going on sale in 1996, although including earlier vintages.

Trevor Jones Riesling

Produced from grapes grown at high altitude from the Adelaide Hills, the Wilton Ridge of the Eden Valley, and the Upper Hermitage Hills area.

🍷🍷🍷🍷 **1997** Medium yellow-green; the bouquet is clean, with gently floral lime and passionfruit aromas. The same fruit flavours run through the palate surrounding a minerally core. **rating:** 86

best drinking 1999–2004 **best vintages** NA **drink with** Grilled scampi • $15

Trevor Jones Dry Grown Shiraz

Produced from what are said to be three of the oldest vineyards in the southern Barossa Valley, dry-grown and yielding less than 1.5 tonnes per acre. The fermentation is allowed to reach higher temperatures at the onset, while the middle and final phases are controlled at colder temperatures to capture flavour and finesse. Extended maturation in both new and older American oak and new French oak then follows.

🍷🍷🍷🍷🍷 **1995** Full red-purple, still bright and deep; the bouquet is redolent of sweet berry, chocolate and vanilla, the palate offering luscious cherry and mint fruit, soft but persistent tannins, and well-integrated and balanced oak. All in all, a beautiful wine. **rating:** 94

best drinking 2000–2015 **best vintages** '95 **drink with** Beef Wellington • $25

Trevor Jones Cabernet Merlot

A mix of high altitude cabernet sauvignon, Barossa cabernet sauvignon and merlot, all picked fully ripe, as the 13.5 degrees alcohol attests.

🍷🍷🍷🍷 **1994** Medium red-purple; sweet chocolate/earthy/berry aromas, with no hint of vegetal green characters lead into a wine with delicate sweet mid-palate fruit, and an attractive soft tannin finish. **rating:** 87

best drinking 1999–2004 **best vintages** '94 **drink with** Beef carpaccio • $23

trio station NR

17–21 Piper Street, Kyneton, Vic 3444 **region** Macedon
ph (03) 5423 2755 **fax** (03) 5423 2756 **open** 7 days 10–6
winemaker Mark Sheppard **prod.** NA **est.** 1996
prod. range ($12.95–17.95 R) Trio Station Dry Riesling, Cabernet Rosé, Shiraz Cabernet.
summary Trio Station is one of the businesses of Vincorp Wineries Limited, a listed company (the other, and very important, winery is Virgin Hills). It is situated in an historic building in the

heart of the township of Kyneton, Victoria, and was established in co-operation with the Macedon Ranges shire council. It draws upon 28 hectares of vineyard in the Geographe region of Western Australia, and a further 40 hectares at the nearby Glenhope Vineyard in the Macedon Ranges. Various marketing and labelling issues were still being resolved in early 1999.

tuck's ridge ★★★☆

37 Red Hill–Shoreham Road, Red Hill South, Vic 3937 **region** Mornington Peninsula
ph (03) 5989 8660 **fax** (03) 5989 8579 **open** 7 days 12–5
winemaker Daniel Greene **prod.** 10 000 **est.** 1988
prod. range ($14–25 CD) Semillon, Riesling, Chardonnay, Pinot Noir, Altera Pinot Noir, Cabernet Sauvignon Merlot, Vues Méthode Champenoise.
summary After an initial burst of frenetic activity following its launch in July 1993, Tuck's Ridge has slowed down a little. Nonetheless, plantings have been increased to a little over 25 hectares, making it one of the largest vineyards in production on the Mornington Peninsula.

Tuck's Ridge Riesling

The warm and dry conditions of the 1997 and 1998 vintages were ideal for the Mornington Peninsula, and in particular for the late-ripening riesling.
🍷🍷🍷🍷 **1998** Light yellow-green; the bouquet is fresh and lively with fragrant nettle/herb aromas; the palate is similarly fresh and delicate, yet has the length and balance to develop. **rating:** 89
best drinking 2000–2005 **best vintages** NA **drink with** Gravlax • $16

Tuck's Ridge Chardonnay

Made using all of the politically correct techniques, including 'wild' (naturally occurring) malolactic fermentation, lees contact and minimal filtration.
🍷🍷🍷🍷 **1998** Medium yellow-green; the bouquet is clean, with attractive nectarine fruit; any oak present cannot be readily detected. There is plenty of ripe nectarine fruit on the palate which helps invest the wine with good texture; once again, no apparent oak influence. **rating:** 86
best drinking 2000–2004 **best vintages** NA **drink with** Thai cuisine • $18

Tuck's Ridge Pinot Noir

Produced from 10 hectares of estate plantings and released in two modes: the cheaper and simpler Altera, and the more complex varietal wine.
🍷🍷🍷🍷 **1998** Light to medium red-purple; the bouquet is soft, moderately intense but with quite complex plum and forest aromas. Tasted in March 1999, the palate was still evolving and on the light side, though having good flavour. **rating:** 85
best drinking 1999–2003 **best vintages** '94, '97 **drink with** Grilled Atlantic salmon • $25

tulloch NR

'Glen Elgin', De Beyers Road, Pokolbin, NSW 2321 **region** Lower Hunter Valley
ph (02) 4998 7580 **fax** (02) 4998 7682 **open** Mon–Fri 9–4.30, weekends 10–4.30
winemaker Patrick Auld **prod.** 26 000 **est.** 1895
prod. range ($11.20–13.80 R) Unoaked Chardonnay, Classic Hunter White, Semillon Chardonnay, Verdelho, Justina (Fruity White), Classic Rich Red, Cabernets, Hector of Glen Elgin (Hermitage), Fortifieds.
summary A once-great name and reputation which suffered enormously under multiple ownership changes, with a loss of identity and direction. In production terms at least, it has found its feet, for it is now the centre of winemaking activities in the Hunter Valley for the Lindeman, Hungerford Hill and Tulloch brands, the last suckling at the breast of the Verdelho, and doing very nicely. Once again, a disclosure of interest, as for Lindemans and Hungerford Hill. Neither the winery nor the wines (in the CD version) are rated because of my (perhaps nominal) role as Group Winemaker (see Lindemans and Hungerford Hill).

tumbarumba wine cellars NR

Sunnyside, Albury Close, Tumbarumba, NSW 2653 **region** Tumbarumba
ph (02) 6948 3055 **fax** (02) 6948 3055 **open** Weekends and public holidays, or by appointment
winemaker Charles Sturt University (Contract) **prod.** 600 **est.** 1990
prod. range ($15–25 CD) Chardonnay, Pinot Noir and Pinot Chardonnay sparkling wines under the Black Range label, with further individual labels likely for the future.
summary Tumbarumba Cellars has taken over the former George Martins Winery (itself established in 1990) to provide an outlet for wines made from Tumbarumba region grapes. It is essentially a co-operative venture, involving local growers and businessmen, and with modest aspirations to growth.

tumbarumba wine estates NR

Maragle Valley via Tumbarumba, NSW 2653 **region** Tumbarumba
ph (02) 6948 4457 **fax** (02) 6948 4457 **open** Not
winemaker Charles Sturt University (Contract) **prod.** NA **est.** 1995
prod. range ($NA) Chardonnay, Pinot Chardonnay Sparkling.
summary Having established his vineyards progressively since 1982, Frank Minutello decided to seek to add value (and interest) to the enterprise by having a small proportion of his production vinified at Charles Sturt University, commencing with the 1995 vintage. The wines are sold by mail order and from The Elms Restaurant in Tumbarumba.

turkey flat ★★★★☆

Bethany Road, Tanunda, SA 5352 **region** Barossa Valley
ph (08) 8563 2851 **fax** (08) 8563 3610 **open** 7 days 11–5
winemaker Peter Schulz **prod.** 10 000 **est.** 1990
prod. range ($11–24 CD) Semillon, Rosé, Grenache Noir, Butchers Block, Shiraz, Cabernet Sauvignon.
summary The establishment date of Turkey Flat is given as 1990, but it might equally well have been 1870 (or thereabouts) when the Schulz family purchased the Turkey Flat vineyard, or 1847 when the vineyard was first planted to the very shiraz which still grows today. In addition there are 6 hectares of very old grenache, and 3 hectares of much younger semillon and cabernet sauvignon. A significant part of the output is sold to some of the best-known Barossa Valley makers, not the least being Charles Melton, St Hallett, and Rockford. Retail distribution in Adelaide, Melbourne and Sydney; exports to US, UK, Belgium and Switzerland.

Turkey Flat Semillon

Produced from 45-year-old vines, and fermented solely in stainless steel. The absence of oak is a joy to behold in Barossa Valley terms.

YYYY **1998** Light to medium yellow-green; the wine shows attractive young semillon varietal character, with minerally citrus aromas. The palate, likewise, has plenty of varietal character, power and length. Should develop very well in bottle. **rating:** 86

best drinking 1999–2005 **best vintages** NA **drink with** St George whiting • $12

Turkey Flat Rosé

A quite delicious wine, made from a blend of 80% Grenache, 10% Cabernet Sauvignon and 10% Shiraz. The addition of the Cabernet Sauvignon and Shiraz components in fact takes the wine from a conventional Rosé-style towards a light-bodied dry red, an elusive goal for most winemakers.

YYYY **1998** Bright pink-red; the bouquet is clean, with hints of caramel and strawberry; the palate seems a little sweeter than normal, but has attractive herb, citrus and spice flavours. **rating:** 85

best drinking 1999–2000 **best vintages** NA **drink with** Nothing or anything • $11.50

Turkey Flat Shiraz

Based upon a precious patch of 145-year-old vines at the heart of the Turkey Flat Vineyard. It was made at Rockford by Chris Ringland, and has had the Adelaide wine press in a paroxysm of delight since the day the first vintage was released.

YYYY **1997** Medium red, with just a touch of purple. The bouquet is clean, moderately intense, with mint and red berry aromas. The palate offers chocolate, berry and a touch of mint, with soft tannins and minimal oak influence. All in all, a lovely, fleshy wine. **rating:** 88

best drinking 2000–2007 **best vintages** '90, '92, '93, '94 **drink with** Smoked kangaroo • $24

Turkey Flat Cabernet Sauvignon

Joins the very good Turkey Flat Shiraz, and indeed in 1997 outpointed its better-known stablemate. Made in the generously proportioned and fruity style which has gained Turkey Flat such a loyal band of followers.

YYYYY **1997** Medium red-purple; the bouquet is quite powerful, with clean berry, earthy fruit and subtle oak. Attractive cassis berry fruit supported by sweet, ripe tannins and good oak combine to produce an extremely attractive wine. **rating:** 90

best drinking 2000–2010 **best vintages** '97 **drink with** Roast leg of lamb • $22.50

turramurra estate NR

RMB 4327 Wallaces Road, Dromana Vic 3926 **region** Mornington Peninsula

ph (03) 5987 1146 **fax** (03) 5987 1286 **open** 12–5 first weekend of the month or by appointment

winemaker David Leslie **prod.** 750 **est.** 1989

prod. range ($14–25 CD) Sauvignon Blanc, Chardonnay, Pinot Noir, Shiraz, Cabernet Sauvignon.

summary Dr David Leslie gave up his job as a medical practitioner after completing the Bachelor of Applied Science (Wine Science) degree at Charles Sturt University to concentrate on developing the family's 10-hectare estate at Dromana. Wife Paula is the viticulturist. Limited retail distribution in Melbourne and Sydney.

Turramurra Estate Chardonnay

The second vintage of this wine from Turramurra Estate and which benefited from the excellent and very warm growing season. The alcohol is just a touch under 14 degrees (13.9) which testifies to the ripeness of the grapes. The usual full complement of winemaking techniques were used to produce the wine.

YYYY **1997** Medium to full yellow-green; the bouquet is full, with nutty, melon and fig aromas complemented by subtle oak. A big wine in the mouth, with pronounced mid-palate sweetness, though it does dry off pleasantly on the finish notwithstanding its alcohol. **rating:** 84

best drinking 1999–2002 **best vintages** NA **drink with** Brains in black butter sauce • $25

Turramurra Estate Pinot Noir

The 1997 vintage was a good one on the Mornington Peninsula, and Dr David Leslie has learned his winemaking well.

YYYY **1997** Medium purple-red; quite powerful plummy fruit is married with a fair amount of oak on the bouquet. The palate is big and rich, a no-holds-barred style with plummy fruit, plenty of oak and extract, and tannins too. A full-frontal style, but good. **rating:** 89

best drinking 1999–2002 **best vintages** '97 **drink with** Duck breast • $20

Turramurra Estate Shiraz

A singularly impressive achievement for what was a difficult vintage for the late-ripening varieties. The wine spent 22 months in a mix of French and American oak barriques, but it is the fruit which provides the essential character.

🍷🍷🍷🍷🍷 **1996** Medium red-purple; the bouquet is of medium intensity and clean, with black cherry fruit dominant, and showing neither green nor particularly spicy characters. On the palate strong Rhône Valley characters come rocketing through; the wine has lovely texture and feel, fine yet imposing. **rating:** 92

best drinking 1999–2004 **best vintages** NA **drink with** Roast squab • $20

twin bays NR

Lot 1 Martin Road, Yankalilla, SA 5203 **region** Other Wineries of SA
ph (08) 8267 2844 **fax** (08) 8239 0877 **open** Weekends and holidays
winemaker Bruno Giorgio, Alan Dyson **prod.** 1000 **est.** 1989
prod. range ($10–19 CD) Riesling, Light Red, Rosado, Wild Grenache, Shiraz, Cabernet Sauvignon, Fortifieds.

summary Twin Bays operates the first winery in the Yankalilla district, one hour's drive south of Adelaide on the Fleurieu Peninsula. Two hectares of estate plantings have been established, but until these and future plantings come into bearing, the wines are made by district veteran Alan Dyson and Adelaide doctor and specialist Bruno Giorgio from grapes partly estate-grown and partly purchased from McLaren Vale. Retail distribution in Sydney, Melbourne and Adelaide.

twin valley estate NR

Hoffnungsthal Road, Lyndoch, SA 5351 **region** Barossa Valley
ph (08) 8524 4584 **fax** (08) 8524 4978 **open** Weekends 10–5
winemaker Fernando Martin, Kay Martin **prod.** 3500 **est.** 1990
prod. range ($9–18 CD) Traminer, Frontignac Spätlese, Eden Valley Rhine Riesling, Semillon Chardonnay, Cabernet Sauvignon Franc, Classic Burgundy, Pinot Cabernet, White Port, Martin's Mead.

summary While Fernando Martin has always had his sights set firmly on the tourist trade, the Twin Valley Estate wines are more than acceptable, the spicy, limey Frontignac Spätlese being a particularly good example of its kind.

tyrrell's ★★★★★

Broke Road, Pokolbin, NSW 2321 **region** Lower Hunter Valley
ph (02) 4993 7000 **fax** (02) 4998 7723 **open** Mon–Sat 8–5
winemaker Andrew Spinaze **prod.** 620 000 **est.** 1858
prod. range ($7–40 R) At the bottom end the large-volume Long Flat White and Red; next in price is Old Winery Chardonnay, Semillon, Chardonnay Semillon, Semillon Sauvignon Blanc, Riesling, Pinot Noir, Shiraz, Cabernet Merlot; next the Individual Vineyard range of wines including Shee-Oak Chardonnay, Stevens Semillon, Lost Block Semillon, Brookdale Semillon, Fordwich Verdelho, Brokenback Shiraz, Stevens Shiraz; at the very top Vat 1 Semillon, Vat 6 Pinot Noir, Vat 9 Shiraz, Vat 47 Chardonnay.

summary A quite extraordinary family winery which has grown up from an insignificant base in 1960 to become one of the most influential mid-sized companies, successfully competing with wines running all the way from cheap, volume-driven Long Flat White up to the super-premium Vat 47 Chardonnay, which is one of Australia's best. There is a similar range of price and style with the red wines, and in recent years Tyrrell's has simply never faltered within the parameters of price and style. Exports to all of the major markets throughout North America, Europe and southeast Asia.

Tyrrell's Lost Block Semillon

The Lost Block is in fact part of the famous HVD Vineyard, and was so named after the 1993 vintage when the grapes were mistakenly allowed to ripen longer than normal, i.e. someone forgot about them.

🍷🍷🍷🍷🍷 **1997** Medium yellow-green; honey characters are starting to appear matched by light citrus and toast on the bouquet. A lively mix of honey and crisp, grassy semillon provides mid-palate sweetness; good length and finish. **rating:** 91

best drinking 2002–2007 **best vintages** '93, '94, '96 **drink with** Calamari • $18

Tyrrell's Vat 1 Semillon

One of the great, classic Hunter Valley Semillons, produced from unirrigated vines which, because of their superior soils, do in fact yield well. Released as a young wine through the Tyrrell's mailing list and cellar door, but re-released through the retail trade at various intervals according to the vintage. Vat 1 Semillons have won 20 trophies, 101 gold, 110 silver and 109 bronze medals since 1971. Never looked better than at the 1998 National Wine Show where the '91 won three trophies including the Len Evans Trophy for Best Wine of Show.

🍷🍷🍷🍷🍷 **1994** Glowing yellow-green; the bouquet is fine, with lemony/citrussy fruit and no errant aromas. The palate is still very crisp, with a youthful mix of herb and lemon, and a long finish. **rating:** 95

🍷🍷🍷🍷🍷 **1991** Full, brilliant yellow-green; the bouquet is wonderfully smooth and rich, and the palate is quite exceptional given the hot, dry vintage: it has retained its freshness, acidity and length throughout that magic combination of honey and brioche flavours. (A re-release). **rating:** 95

best drinking 2000–2010 **best vintages** '75, '76, '77, '86, '87, '89, '90, '91, '92, '93, '94 **drink with** Pan-fried veal • $35

Tyrrell's Moon Mountain Chardonnay

Produced from Tyrrell's Moon Mountain Vineyard and other sandy soil vineyards along the flats. The wine is barrel-fermented in 50% new and 50% one-year-old barriques, but is in no sense overoaked. A consistent performer over recent years.

🍷🍷🍷🍷 **1998** Medium yellow-green; the bouquet is quite complex, with good integration of melon fruit and oak. The palate has quite lively, citrussy fruit and good length; once again, the oak input is subtle. **rating:** 89

best drinking 1999–2001 **best vintages** '97, '98 **drink with** Light pasta • $22

Tyrrell's Old Winery Chardonnay

A wine which is as much a testament to the production skills of the Tyrrell's winemaking team as to the vast storehouse of accumulated knowledge in handling this variety. It is deliberately made in a richer, earlier-maturing style than Vat 47, and perhaps for this reason is often very nearly as well-treated in wine shows as a young wine, being a prolific medal winner. The message is, drink it, don't cellar it. The wine is a blend of Hunter Valley, Liverpool Plains (Camden, NSW) and McLaren Vale fruit, 30% barrel-fermented.

🍷🍷🍷🍷 **1997** Medium to full yellow-green; the bouquet is complex, with clever use of oak allied with solid peach and melon fruit. The palate is equally powerful and complex, with lots going on, anchored by the fruit weight. **rating:** 88

best drinking 1999–2000 **best vintages** NA **drink with** Turkey breast • $11.50

Tyrrell's Shee-Oak Chardonnay

The Shee-Oak Vineyard, with its sandy flats, was purchased in 1981, and the vines planted in 1982. This is classic Hunter white country, and it shows through in the wine. Unoaked Chardonnay has quickly gained a dreadful reputation as vapid and lollyish, but as Cole Porter said, it ain't necessarily so.

▼▼▼▼▽ **1997** Medium yellow-green; a smooth and supple bouquet with peach and melon aromas is followed by a wine with great structure and grip on the palate. Only 12.5 degrees alcohol, and in traditional Chablis (French) style. **rating:** 93

best drinking 1999–2003 **best vintages** NA **drink with** Steamed fish • $18

Tyrrell's Vat 47 Pinot Chardonnay

First made in the appalling vintage of 1971, sharing with the Craigmoor wine of the same year the honour of being the first Chardonnay labelled as such and sold this century. Has since unequivocally stamped itself as one of the great marques, with recent vintages going from strength to strength, the '95 winning the Bert Bear Trophy for Best One Year Old White Wine at the 1996 Royal Sydney Wine Show, and topping its class in the same show in 1997. The '97 followed in the footsteps of the '96 wine, winning the very significant Bert Bear Memorial Trophy for Best Current Vintage White Wine at the 1998 Royal Sydney Wine Show.

▼▼▼▼ **1998** Light to medium green-yellow; the bouquet is admittedly complex, but there are some errant aromas there which do not impress. The palate is lively and crisp enough, but generally seems to miss the mark. Very difficult to know what is going on here. **rating:** 86

▼▼▼▼▼ **1997** Glowing yellow-green; melon and nectarine fruit is balanced by the very clever use of barrel-ferment and toasty oak on the bouquet. The palate emphasis is on the nectarine and melon fruit, running through to a long finish. The oak is ever present but never dominant. **rating:** 95

best drinking 1999–2001 **best vintages** '82, '84, '85, '89, '91, '94, '95, '96, '97 **drink with** Fresh, slow-cooked salmon • $40

Tyrrell's Brokenback Shiraz

This wine comes from the Brokenback Vineyard established by Rothbury Estate almost 30 years ago, part of which was acquired by Tyrrell's some years ago. The fully mature vines are producing great fruit; the '97 fared exceptionally well at the 1998 Liquorland National Wine Show in Canberra, winning several trophies. There it seemed to have greater depth and complexity, but whichever way you look at it, it is a nice wine.

▼▼▼▼▽ **1997** Medium red-purple; the bouquet is clean, with fresh berry, spice and liquorice aromas. There is genuine fruit and varietal complexity on the palate, underlined by the delicate oak handling. Just a little light-on. **rating:** 90

▼▼▼▼▼ **1997** Full purple-red; the bouquet is redolent of classic black cherry, faintly spicy varietal fruit. The palate is, quite simply, glorious, with great varietal character, elegance and structure. **rating:** 94

best drinking 2000–2005 **best vintages** NA **drink with** Rack of lamb • $22

Tyrrell's Rufus Stone Shiraz

Part of a new and imaginatively-packaged range from Tyrrell's featuring regions outside the Hunter Valley (Coonawarra, Heathcote and McLaren Vale), with various vintages introduced onto the market in 1998.

▼▼▼▼▽ **1997** Heathcote. Dense red-purple; an almost extravagantly rich, ripe and voluptuous bouquet with cascades of cherry and spice leads into a highly seductive, powerful palate with dark cherry fruit and persistent tannins. **rating:** 93

best drinking 2002–2012 **best vintages** '96, '97 **drink with** Lasagne • $19.50

Tyrrell's Stevens Shiraz

One of the Individual Vineyard wines of Tyrrell's, sourced from one of the great old shiraz vineyards in the Pokolbin region, owned by George and Neil Stevens. 1993 was the first vintage made by Tyrrell's from this prime source.

🍷🍷🍷🍷🍷 **1995** Medium red-purple; the bouquet is clean, of medium intensity, with a mix of sweet berry, spice and chocolate fruit aromas. The delicious palate has a range of spice, cedar, berry and vanilla flavours, with soft tannins on a long finish supported by good acidity. **rating:** 90

best drinking 1999–2009 **best vintages** '94, '95 **drink with** Moroccan lamb • $30

undercliff NR

Yango Creek Road, Wollombi, NSW 2325 **region** Lower Hunter Valley
ph (02) 4998 3322 **fax** (02) 4998 3322 **open** Weekends 10–4 or by appointment
winemaker James Luxton, Janet Luxton **prod.** 1000 **est.** 1990
prod. range ($12–18 CD) Semillon, Shiraz, Sparkling Shiraz.
summary A combined winery and studio owned and run by James and Janet Luxton situated at Wollombi, on the edge of the Hunter Valley. Janet Luxton is the artist, having studied etching and printing at the Jerusalem Print Workshop, and James Luxton is the printer. The wines, produced from 2.5 hectares of estate vineyards, have won a number of awards in recent years at the Hunter Valley Wine Show and the Hunter Valley Small Winemakers Show. All of the wine is sold through cellar door.

van de scheur NR

O'Connors Lane, Pokolbin, NSW 2321 **region** Lower Hunter Valley
ph (02) 4998 7789 **fax** (02) 4998 7789 **open** Weekends 10–5
winemaker Kees Van De Scheur **prod.** 1500 **est.** 1995
prod. range ($16.50 CD) Semillon, Chardonnay, Shiraz.
summary Kees Van De Scheur is a Hunter Valley veteran, having spent the last 25 years in the Hunter Valley, first with the Robson Vineyard, and then Briar Ridge, before leaving in November 1993 to establish his own winery and label. He has purchased part of the historic Ingleside property, established by vigneron Frederick Ingle in 1872. After a hiatus of 60 years, vines have returned, with an initial planting of a little over 1 hectare (semillon, chardonnay and shiraz), increasing to 4 hectares by 1997. In the meantime, grapes purchased within the Hunter Valley provide the base for the Van De Scheur Wines.

vasse felix ★★★★★

Cnr Caves Road and Harmans Road South, Willyabrup, WA 6284 **region** Margaret River
ph (08) 9755 5242 **fax** (08) 9755 5425 **open** 7 days 10–5
winemaker Clive Otto, Will Shields **prod.** 60 000 **est.** 1967
prod. range ($12–50 R) Classic Dry White, Theatre White and Red, Forest Hill Riesling, Semillon, Chardonnay, Noble Riesling, Classic Dry Red, Shiraz, Flinders Bay Cabernet Sauvignon, Sparkling Brut, Heytesbury.
summary The 1999 vintage will see Vasse Felix wines produced in a brand-new 2000-tonne winery situated 200 metres to the west of the current facility; the old winery will be dedicated entirely to the restaurant and tasting rooms. A new 140-hectare vineyard at Jindong in the north of the Margaret River will supply a large part of the increased fruit intake. National Australian distribution; exports to the UK, US, Switzerland, Singapore and Hong Kong.

Vasse Felix Forest Hill Riesling

The Forest Hill Vineyard is the oldest planting in the southern part of Western Australia, with 1 hectare of riesling (and a hectare of cabernet sauvignon) established back in 1966. The vineyard remained in the Pearce family for almost 30 years until purchased by Vasse Felix.

🍷🍷🍷🍷 **1998** Light to medium yellow-green; the very rich bouquet has potent lime/pastille aromas which come through strongly on the palate. Overall, generous almost to the point of a fault. **rating:** 87

best drinking 1999–2004 **best vintages** NA **drink with** Vegetable terrine • $16

Vasse Felix Semillon

Drawn primarily from 4 hectares of estate plantings. A very richly structured wine with an unusual depth of flavour, enhanced by 100% barrel fermentation in mainly new French oak, and five months in barrel thereafter.

▼▼▼▼ **1998** Light green-yellow; obvious barrel-fermentation oak characters are immediately apparent, wrapping the fruit in a silky cocoon of spice and vanilla. The palate is likewise strongly influenced by high-quality oak; it will appeal greatly to some for this very reason, but for me it is on the intrusive side. Semillon being the variety it is, time may well provide better balance. **rating:** 88

best drinking 2002–2006 **best vintages** '92, '93, '95, '96 **drink with** Coquilles St Jacques • $22.50

Vasse Felix Heytesbury Chardonnay

One of the super-premium wine from Vasse Felix, made entirely from grapes grown on the Forest Hill Vineyard planted in 1978.

▼▼▼▼ **1997** Medium to full yellow-green; a powerful bouquet with a mix of fruit, oak and malolactic influences is followed by similarly high-flavoured but slightly heavy-handed palate. Good but not great. **rating:** 87

best drinking 1999–2003 **best vintages** NA **drink with** Rich seafood • $35

Vasse Felix Shiraz

Draws upon a little under 7 hectares of estate plantings, and over the years has produced some outstanding wines. The 1983 vintage, tasted in 1993, was wonderfully complex, rich and full with a touch of Rhône Valley gaminess. Since 1991 the quality of the wine has often been outstanding, and it has enjoyed great success in wine shows, the 1996 winning trophies for Best Shiraz and Best Varietal Dry Red Table Wine (Any Variety) at the 1997 Royal Adelaide Wine Show. The '97 is disappointing.

▼▼▼▼ **1997** Medium purple-red; sweet oak is the first impression on the bouquet, and likewise on the palate, with sweet milkshake oak flavours. Shiraz fruit is hiding there somewhere, but really the oak is severely overdone. **rating:** 85

best drinking 2000–2005 **best vintages** '83, '85, '88, '90, '91, '92, '94, '96 **drink with** Rich casseroles • $33

Vasse Felix Heytesbury

Heytesbury is the flagship red of Vasse Felix, taking its name from Heytesbury Holdings, the company led by Janet Holmes à Court since the death of her husband, and which owns Vasse Felix. It is a blend of Cabernet Sauvignon, Merlot and Cabernet Franc which is generously oaked. The '96 won the trophy for Best Wine of Show at the 1997 Western Australia SGIO Wine Awards.

▼▼▼▼▽ **1997** Medium to full purple-red; while the oak is quite powerful, it is well integrated and balanced with earthy cassis cabernet fruit. The message is continued on the palate, where the clever use of high-quality oak throughout does not submerge the cassis and berry flavours; finishing with fine tannins. **rating:** 93

▼▼▼▼▽ **1996** Dense red-purple; a very concentrated bouquet with blackberry, mint, briar and some spicy oak. The palate is no less concentrated and inky with powerful blackberry/briary fruit and oodles of cedary oak. Other judges may disagree, but this is hardly an exercise in restraint. **rating:** 90

best drinking 2002–2012 **best vintages** '95, '96, '97 **drink with** Beef Bordelaise • $50

Vasse Felix Cabernet Sauvignon

For decades the flag-bearer of Vasse Felix, and traditionally made in a lighter, more elegant style than most of the big names in the Margaret River, and none the worse for that. In recent years, the style has become more powerful and hence more typical of the region as a whole.

ΨΨΨΨ **1997** Medium red, with some purple hues remaining but showing the first signs of development. The bouquet has earthy berry fruit offset by cleverly weighted vanilla oak. The same play occurs on the palate, with light to medium weight tannins and well-handled soft oak.

rating: 87

best drinking 1999–2004 **best vintages** '85, '88, '90, '91, '94, '95, '96 **drink with** Lamb cutlets • $30

veritas ★★★☆

94 Langmeil Road, Tanunda, SA 5352 **region** Barossa Valley
ph (08) 8563 2330 **fax** (08) 8563 3158 **open** Mon–Fri 9–5, weekends 11–5
winemaker Rolf Binder, Christa Deans **prod.** 14 000 **est.** 1955
prod. range ($10–30 CD) Riesling, Semillon, Chardonnay, Cabernet Shiraz, Cabernet Merlot, Binder's Bull's Blood, Shiraz Mourvedre Pressings, Fortifieds.
summary The Hungarian influence is obvious in the naming of some of the wines, but Australian technology is paramount in shaping the generally very good quality. Veritas has 28 hectares of estate vineyards to draw on. A near-doubling of production has coincided with the establishment of export markets to UK, Germany, Switzerland, Belgium, Holland and US.

vicarys ★★☆

Northern Road, Luddenham, NSW 2745 **region** Other Wineries of NSW
ph (02) 4773 4161 **fax** (02) 4773 4411 **open** Mon–Fri 9–5, weekends 10–5
winemaker Chris Niccol **prod.** 1700 **est.** 1923
prod. range ($12–36 CD) Chardonnay, Semillon, Riesling, Gewurztraminer, Fumé Blanc, Cabernet Sauvignon, Shiraz Cabernet Merlot, Sparkling, Fortifieds.
summary Vicarys justifiably claims to be the Sydney region's oldest continuously operating winery, having been established in a very attractive, large, stone shearing shed built about 1890. Most of the wines come from other parts of Australia, but the winery does draw upon 1 hectare of estate traminer and 3 hectares of chardonnay for those wines, and has produced some good wines of all styles over the years.

vico NR

Farm 1687 Beelbangera Road, Griffith, NSW 2680 **region** Riverina
ph (02) 6962 2849 **open** Mon–Fri 9–5
winemaker Ray Vico **prod.** 1200 **est.** 1973
prod. range ($5–15 CD) Semillon, Late Harvest Semillon, Barbera, Cabernet Sauvignon, Liqueur Muscat.
summary Ray Vico has been growing grapes for many years, with 9 hectares of vines, more recently deciding to bottle and sell part of the production under the Vico label. At $60 to $70 a dozen for the table wines, and $15 per bottle for the 1984 Liqueur Muscat, the prices are positively mouthwatering – the more so given that 1997 Barbera, 1996 Cabernet Sauvignon and 1984 Liqueur Muscat all won silver medals at the Australian Small Winemakers Show, the '97 Semillon collecting a bronze medal.

victorian alps wine co NR

Great Alpine Road, Gapsted, Vic 3737 **region** King Valley
ph (03) 5751 1992 **fax** (03) 5751 1368 **open** Due to be opened in September 1999
winemaker Shayne Cunningham **prod.** 15 000 **est.** 1996
prod. range ($10–25 R) Merlot, Cabernet Sauvignon, Petit Verdot, Cabernet Franc.

summary Victorian Alps Wine Co is primarily a contract grape crushing and fermenting facility, processing grapes grown in the King and Ovens Valleys and adjacent regions for wineries spread across various parts of Australia. Its total crush in 1998 was 2400 tonnes, of which only a small part will find its way onto the market as branded, bottled wine. Indeed, the first commercial release (and unveiling of the brand name and packaging) is scheduled for September 1999. By that time the cellar door will be open, and exports will be underway.

villa primavera NR

Mornington–Flinders Road, Red Hill, Vic 3937 **region** Mornington Peninsula
ph (03) 5989 2129 **open** Weekends, public holidays 10–5, January 7 days 10–5
winemaker Gennaro Mazzella **prod.** 300 **est.** 1984
prod. range ($18–30 CD) Chardonnay, Pinot Noir, Limoncello, Méthode Champenoise.
summary A most unusual operation, which is in reality a family Italian-style restaurant at which the wine is principally sold and served, and which offers something totally different on the Mornington Peninsula. A consistent winner of tourism and food awards, it is praised by all who go there, particularly for the concerts staged throughout January each year.

vincorp NR

17-21 Piper Street, Kyneton, Vic 3444 **region** Macedon
ph (03) 5423 2755 **fax** (03) 5423 2756 **open** 7 days 10–6
winemaker Mark Sheppard **prod.** NA **est.** 1996
prod. range ($8.95–27.95 R) The 'One' range of White One (Sauvignon Blanc from Western Australia); Red One (unoaked Shiraz from Victoria and New South Wales); and Grand One (premium Cabernet Merlot from Central Victoria).
summary A smartly-packaged but curiously named trio of wines from Vincorp, which joined the lists of the Stock Exchange in 1998, and also owns – but runs as separate ventures – Virgin Hills and Trio Station. Where the range will head in the future remains to be seen; it had a somewhat uncertain start.

vintina estate NR

1282 Nepean Highway, Mount Eliza, Vic 3930 **region** Mornington Peninsula
ph (03) 9787 8166 **fax** (03) 9775 2035 **open** 7 days 9–5
winemaker Jim Filippone, Kevin McCarthy (Consultant) **prod.** 400 **est.** 1985
prod. range ($12–14 CD) Chardonnay, Semillon, Pinot Gris, Pinot Noir, Cabernet Sauvignon.
summary The initial releases of Vintina (the only wines tasted to date), were mediocre. With competent contract-winemaking, improvement can be expected. However, no recent tastings.

violet cane vineyard NR

13 Wallace Court, Glen Aplin, Qld 4381 **region** Granite Belt
ph 0418 739 257 **fax** (07) 4162 5328 **open** Not
winemaker Adam Chapman **prod.** 60 **est.** 1994
prod. range ($22–40 ML) Semillon, Merlot, Sparkling.
summary The intriguingly named Violet Cane Vineyard, and no less startlingly labelled wine, is the tiny personal business of Ballandean winemaker Adam Chapman, who manages to fit in three winemaking lives: one at Ballandean, one for Violet Cane, and the last as a Flying Winemaker travelling to Europe each Australian spring. It is perhaps fitting that Adam Chapman should have married Elle Sigurdardottir in her native Iceland in June 1998.

Violet Cane Merlot

Produced from 1 hectare of estate merlot grown on two planting densities and trellises: the first on 1.8-metre row spacing with 1 metre between vines (high density) and the second on a one by one spacing and a very low trellis. The latter is the traditional spacing of Bordeaux and Burgundy and is the highest density planting of merlot in Australia. The wine is matured in new American oak.

🍷🍷🍷🍷🍷 **1997** Medium red-purple; the bouquet is quite exotic, with a mix of spicy/savoury/gamey fruit with tinges of fresh earth and quite subtle oak. The wine has the delicate texture that I believe Merlot should have, with fresh, bright cherry fruit. The structure is good, although does fall away a little on the palate, where the virtual absence of tannin detracts somewhat. Nonetheless, a wine of genuine quality. **rating:** 90

best drinking 1999–2004 **best vintages** '97 **drink with** Char-grilled kangaroo • $22

virage ★★★

13B Georgette Road, Gracetown, WA 6284 **region** Margaret River
ph (08) 9755 5318 **fax** (08) 9755 5318 **open** Not
winemaker Bernard Abbott **prod.** 1000 **est.** 1990
prod. range ($13–20 R) Sauvignon Blanc, Semillon Chardonnay, Traminer Riesling, Cabernet Shiraz Zinfandel, Cabernet Merlot.
summary Former Vasse Felix winemaker Bernard Abbott, together with wife Pascale, acquired (under long-term lease) the former government research station vineyard at Bramley Estate in 1990. Bernard Abbott makes the wines at a local Margaret River winery, and sells them by mailing list and direct to retailer and restaurants in Perth, Melbourne and Sydney.

virgin hills ★★★★★

Salisbury Road, Lauriston West via Kyneton, Vic 3444 **region** Macedon
ph (03) 5422 7444 **fax** (03) 5422 7400 **open** By appointment
winemaker Mark Sheppard, David Watson **prod.** 4000 **est.** 1968
prod. range ($43.95 R) A single Cabernet Sauvignon Shiraz Merlot Blend called Virgin Hills; occasional limited Reserve release.
summary Virgin Hills has now become part of the public-listed Vincorp Group, a turn of events which has seen the return of Mark Sheppard (assisted by David Watson) to the role he held for so many years as winemaker. All eyes will be focused on what is in many ways an icon winery which constantly faces the climatic odds with its late-ripening varieties in an unequivocally cool region.

Virgin Hills

An entirely estate-grown blend (in descending order) of Cabernet Sauvignon, Shiraz, Merlot, Malbec and occasionally a touch of Pinot Noir. Not only are the grapes organically grown, but since 1988 the wine has been made without the use of added sulphur dioxide. It is a brave move which makes world's best preservative-free red; only time will tell how these wines age over a decade.

🍷🍷🍷🍷 **1996** Medium red, already showing some development. A wine which shows some of the sweet characters of the '93 mixed with softer leafy/cedary/tobacco aromas. The palate is a mix of sweet berry/minty characters and more cedary leafy notes. Looks as if it will mature reasonably quickly. **rating:** 85

best drinking 1999–2004 **best vintages** '74, '75, '76, '80, '82, '85, '88, '90, '91, '92, '95 **drink with** Duck • $43.95

voyager estate ★★★★☆

Lot 1 Stevens Road, Margaret River, WA 6285 **region** Margaret River
ph (08) 9757 6354 **fax** (08) 9757 6494 **open** 7 days 10–4
winemaker Stuart Pym **prod.** 20 000 **est.** 1978
prod. range ($16–30 R) Chenin Blanc, Sauvignon Blanc Semillon, Semillon, Chardonnay, Cabernet Merlot, Shiraz Grenache.

summary Formerly Freycinet Estate, renamed after its purchase (in May 1991) from Western Australian viticulturist Peter Gherardi. Much money has been spent on the property by new owner, the mining magnate Michael Wright (including the expansion of the vineyards to over 51 hectares), although the winery itself remains in strictly utilitarian form. The wines are rich and opulent, particularly the white wines.

Voyager Estate Sauvignon Blanc Semillon

Voyager Estate regularly produces high class Sauvignon Blanc and Semillon (the latter also offered as a straight varietal) from its now fully mature vineyards.

🍷🍷🍷🍷🍷 **1998** Light green-yellow; there is excellent fruit weight and intensity to the bouquet, with pleasantly grassy (but not too green) aromas. The well-weighted and balanced palate displays a mix of nectarine, gooseberry and more herbaceous flavours rippling through a long finish. **rating:** 93

best drinking 1999–2002 **best vintages** NA **drink with** Marinated octopus • $19

Voyager Estate Chardonnay

In the majority of vintages Voyager Estate produces a Chardonnay of complexity, style and verve typical of the Margaret River at its very best.

🍷🍷🍷🍷🍷 **1997** Medium yellow-green; there is intense, ripe melon and fig fruit smoothly married with quality oak on the bouquet. The tight structure of the palate promises longevity; nectarine and a touch of grapefruit are beautifully balanced with subtle oak. **rating:** 94

best drinking 2000–2007 **best vintages** '92, '93, '95, '96, '97 **drink with** Braised pork neck • $28

Voyager Estate Shiraz Grenache

A wine way out of left field, but there is in fact an old planting of grenache at Voyager Estate left over from the days of Peter Gherardi and Freycinet (as Voyager was then called).

🍷🍷🍷🍷🍷 **1997** Strong red-purple; the bouquet is powerful and concentrated, with very good dark berry aromas. The palate is similarly power-packed with dark berry fruits and soft tannins. I suppose it is exactly what you might expect from a Shiraz Grenache blend grown in Margaret River and made by Stuart Pym. **rating:** 90

best drinking 1999–2004 **best vintages** NA **drink with** Game stew • $20

wa-de-lock ★★★

Stratford Road, Maffra, Vic **region** Gippsland

ph (03) 5147 3244 **fax** (03) 5147 3132 **open** Thur–Tues 10–5

winemaker Graeme Little **prod.** 2500 **est.** 1987

prod. range ($9.50–24.50 R) Chameleon White Pinot, Chardonnay, Reserve Chardonnay, Sauvignon Blanc, Noble Sauvignon Blanc, Pinot Noir, Cabernet Merlot, Tawny Port.

summary The initial plantings of pinot noir, cabernet sauvignon and sauvignon blanc in 1987 have been progressively expanded by increases in those varieties and the addition of chardonnay, nebbiolo, shiraz, merlot and durif, with 12 hectares under vine, some of it still coming into production. The quality of the wines has improved steadily as Graeme Little's handling of oak has become more assured, and the range of wines has increased. The wines have distributors in Melbourne, Sydney, Tasmania and Perth.

Wa-De-Lock Sauvignon Blanc

The 1995 vintage was released under a new label, which will in future be used on all Wa-De-Lock wines. The label change coincided with improvement in the quality of the Sauvignon Blanc thanks to the exclusion of the indifferent oak which had been used between 1991 and 1994. The '98 follows in the footsteps of the '95.

🍷🍷🍷🍷 **1998** Medium yellow-green; a relatively light and subdued bouquet with slight matchbox aromas that did, however, blow off as the wine aired in the glass. A pleasant Sauvignon Blanc on the palate with some depth and mineral characters, though lacking the exotic bloom of the very best Sauvignon Blancs. **rating:** 82

best drinking 1999–2000 **best vintages** '95 **drink with** Seafood pasta • $15.95

Wa-De-Lock Pinot Noir

There is no question that the Gippsland region can and does produce pinot noir with good varietal character and considerable weight and substance in the drier vintages. Over the years, Wa-De-Lock has gradually come to terms with this most temperamental of grape varieties.

🍷🍷🍷🍷 **1998** Light to medium red; the bouquet is light, with a mix of strawberry and mint aromas; the palate has surprising length and strength; you feel it on the tip and middle of the tongue in particular. One of those Pinots which sneaks up on you. **rating:** 85

best drinking 1999–2000 **best vintages** '93, '95, '97 **drink with** Risotto • $17.50

wallington wines NR

Nyrang Creek Vineyard, Canowindra, NSW 2904 **region** Cowra

ph (02) 6344 7153 **fax** (02) 6344 7153 **open** By appointment

winemaker Don Buchanan, Murray Smith (Contract) **prod.** 800 **est.** 1992

prod. range ($14–22 CD) Chardonnay, Cabernet Sauvignon.

summary Anthony and Margaret Wallington commenced the development of their Nyrang Creek Vineyard with a little over 2 hectares of cabernet sauvignon in 1992, followed by 7 hectares of chardonnay in 1994, then shiraz (2.5 hectares) and semillon (0.75 hectare) in 1995 and ultimately 0.75 hectare each of cabernet franc and pinot noir in 1998. Most of the production is sold, but Don Buchanan at Arrowfield makes the Wallington Chardonnay and Murray Smith of Canobolas-Smith Wines at Orange makes the Cabernet Sauvignon and Shiraz. The quality of the wines is such that exports to the US have already commenced.

walsh family wines NR

90 Walnut Road, Bickley, WA 6076 **region** Perth Hills

ph (08) 9291 7341 **fax** (08) 9291 7341 **open** 7 days 9–5

winemaker Celine Rousseau (Contract) **prod.** 95 **est.** 1995

prod. range ($14 CD) Shiraz.

summary Walsh Family Wines is aptly named: it is a partnership of the Walshes and their eight children. One of those children is establishing a vineyard near Bridgetown in the Great Southern, the grapes from which will ultimately form part of the Walsh Family winery intake.

wandering brook estate NR

PO Box 32, Wandering, WA 6308 **region** Other Wineries of WA

ph (08) 9884 1064 **fax** (08) 9884 1064 **open** Weekends 9.30–6

winemaker Steve Radikovich **prod.** 2000 **est.** 1989

prod. range ($10–14.95 CD) Verdelho, Chardonnay, Unwooded Chardonnay, Soft Red and White, Cabernet Sauvignon, Sparkling Verdelho, Port.

summary Laurie and Margaret White have planted 10 hectares of vines on their 130-year-old family property in a move to diversify. Up to 1994 the wines were made at Goundrey, since 1994 at the nearby Hotham Valley. Renamed Wandering Brook Estate late 1994; up till then known as Redhill Estate.

wandin valley estate ★★★★

Wilderness Road, Rothbury, NSW 2321 **region** Lower Hunter Valley

ph (02) 4930 7317 **fax** (02) 4930 7317 **open** 7 days 10–5

winemaker Geoff Broadfield **prod.** 10 000 **est.** 1973

prod. range ($13–25 R) Pavilion Range Dry White, Classic White and Dry Red; Estate Range of Semillon, Chardonnay, Cabernet Sauvignon, Shiraz, Ruby Cabernet and Muscat; top of the range WVE Cabernet Brut Champagne, Reserve Chardonnay, Bridie's Shiraz and Reserve Cabernet Sauvignon.

summary The former Millstone vineyard now owned by the producer of Australian televisions classic 'A Country Practice' who has acquired the services of Allanmere winemaker Geoff Broadfield. Rapidly developing Chardonnays have been the focal point of Wandin Valley's considerable show success. The estate also boasts a Cope-Williams-type village cricket oval and extensive cottage accommodation.

Wandin Valley Estate WVE Reserve Cabernet Sauvignon

Like the Reserve Chardonnay, made in restricted quantities, typically 500 cases. A consistent show medal winner, but the '96 was absolutely the best wine to come from Wandin Valley since 1991, and the '97 is another excellent wine.

YYYY **1997** Bright red-purple; there is exemplary, firm varietal cabernet fruit on the bouquet; the palate, likewise, is firm and youthful, with cassis berry fruit on the mid-palate immediately encased in firm tannins on the finish. While not a lush style, has considerable development potential. **rating:** 89

best drinking 2002–2012 **best vintages** '91, '96, '97 **drink with** Braised duck • $25

waninga ★★★★

Hughes Park Road, Sevenhill via Clare, SA 5453 **region** Clare Valley

ph (08) 8843 4395 **fax** (08) 8843 4395 **open** 7 days 10–5

winemaker Tim Adams, Jeffrey Grosset (Contract) **prod.** 1500 **est.** 1989

prod. range ($12–18 CD) Skilly Hills Riesling, Late Picked Riesling, Chenin Blanc, Chardonnay, Shiraz, Reserve Shiraz, Cabernet Sauvignon, Port.

summary The large (37.5 hectares) vineyards owned by Waninga were established in 1974, but it was not until 1989 that a portion of the grapes was withheld from sale and vinified for the owners. Since that time, Waninga has produced some quite lovely wines, having wisely opted for very competent contract-winemaking.

wansbrough wines NR

Richards Road, Ferguson, WA 6236 **region** Geographe

ph (08) 9728 3091 **fax** (08) 9728 3091 **open** Weekends 10-5

winemaker Willespie Wines (Contract) **prod.** 250 **est.** 1986

prod. range ($12-18 CD) Riesling, Semillon, Sauvignon Blanc, Constantia (late-picked Semillon), Shiraz Cabernet, Port.

summary Situated east of Dardanup in the picturesque Ferguson Valley, Wansbrough enjoys views of the distant Geographe Bay and the nearer state forest, with the Bibblemun Track running along its northern and eastern borders. To taste the wine you need either to order by mail or visit the Wansbrough restaurant on weekends.

wantirna estate NR

Bushy Park Lane, Wantirna South, Vic 3152 **region** Yarra Valley

ph (03) 9801 2367 **fax** (03) 9887 0225 **open** By appointment

winemaker Reg Egan, Maryann Egan **prod.** 900 **est.** 1963

prod. range ($40 CD) Isabella Chardonnay, Lily Pinot Noir, Cabernet Merlot.

summary Situated well within the boundaries of the Melbourne metropolitan area Wantirna Estate is part of a nature reserve. The only retail outlet for the wine is Richmond Hill Cellars; all the remainder is sold through mail order, to selected restaurants and to a few overseas customers.

In deference to Reg Egan's very firmly held views on the subject, neither the winery nor the wines are rated.

Wantirna Estate Lily Pinot Noir

A new label for a wine which has been made for almost 20 years at Wantirna Estate, adorned with a Michael Leunig drawing, as far removed from the original Wantirna Estate label as the earth is from the sun. The wine is just great.

1997 Light to medium red; some cherry fruit on the bouquet with light, briary undertones isn't the most promising of starts, but the wine really comes to life on the mid to back palate. What I describe as a 'slippery' style, finishing with fine tannins.

best drinking 1999–2003 **best vintages** '91, '92, '94, '96 **drink with** Peking duck • $40

wards gateway ★★☆

Barossa Valley Highway, Lyndoch, SA 5351 **region** Barossa Valley
ph (08) 8524 4138 **open** 7 days 9–5.30
winemaker Ray Ward (plus Contract) **prod.** 600 **est.** 1979
prod. range ($7.50–15.50 CD) Riesling, Frontignac, Chardonnay, Frontigna Spatlese, Barossa Shiraz, Cabernet Sauvignon, Port.
summary The very old vines surrounding the winery produce the best wines, which are made without frills or new oak and sold without ostentation.

warrabilla NR

Murray Valley Highway, Rutherglen, Vic 3685 **region** Rutherglen
ph (02) 6035 7242 **fax** (02) 6035 7242 **open** 7 days 10–5
winemaker Andrew Sutherland-Smith **prod.** 4500 **est.** 1986
prod. range ($12–19 CD) Chardonnay, Brimin Series Cabernet Shiraz, Glenrowan Cabernet Sauvignon, Glenrowan Shiraz, Vintage Port, Liqueur Muscat.
summary Former All Saints winemaker Andrew Sutherland-Smith has leased a small winery at Corowa to make the Warrabilla wines from a 4-hectare vineyard developed by himself and Carol Smith in the Indigo Valley.

warramate ★★★

27 Maddens Lane, Gruyere, Vic 3770 **region** Yarra Valley
ph (03) 5964 9219 **fax** (03) 5964 9219 **open** 7 days 10–6
winemaker Jack Church, David Church **prod.** 900 **est.** 1970
prod. range ($16–26 CD) Riesling, Shiraz, Cabernet Sauvignon.
summary Wine quality has been variable in recent years; it would seem that the oak in some of the older barrels questionable. At their best, reflect the distinguished site on which the vineyard sits.

Warramate Shiraz

Produced from just under half a hectare of estate-grown plantings, typically providing about 200 cases of wine. The plantings are not irrigated, and enjoy a prime north-facing slope.

YYYY 1995 Medium to full red-purple; the bouquet is quite dense, with liquorice, spice and blackberry varietal fruit verging into a smooth palate with nuances of black cherry joining the other characters. A considerable success for Warramate, echoing that of its '95 Cabernet Sauvignon. **rating:** 85

best drinking 2000–2005 **best vintages** '88, '92, '93 **drink with** Spiced lamb • $25

warraroong estate NR

Wilderness Road, Lovedale, NSW 2321 **region** Lower Hunter Valley
ph (02) 4930 7594 **fax** (02) 4930 7199 **open** 7 days 10–5
winemaker Adam Rees, Greg Silkman (Consultant) **prod.** NA **est.** 1988
prod. range ($15–18 CD) Semillon Chardonnay, Chardonnay, Shiraz, Malbec.
summary Warraroong Estate was formerly Fraser Vineyard, and adopted its new name after it changed hands in 1997. The name 'Warraroong' is an Aboriginal word for hillside, reflecting the southwesterly aspect of the property looking back towards the Brokenback Range and Watagan Mountains. The label design is from a painting by local Aboriginal artist Kia Kiro who, while coming from the Northern Territory, is living and working in the Hunter Valley.

Warraroong Estate Semillon

Estate-grown from ten-year-old vines, and, once again, made with a minimum of artifice.
1998 Light straw-green; the bouquet is, of course, youthful, with powdery herbal aromas, followed by a palate with distinct lemony/grassy grip. Has presence and length, and will mature well. **rating:** 88
best drinking 2001–2005 **best vintages** NA **drink with** Shellfish • $18

Warraroong Estate Sauvignon Blanc

Quite clearly, a wine cold-fermented in stainless steel and finished with a little sweetness. I am uncertain whether this is estate-grown, or even from the Hunter Valley; if it is, it is an outstanding achievement.
1998 Light green-yellow; the bouquet is light and fresh, with crisp mineral aromas, while the palate neatly balances crisp, clean mineral herb fruit against a touch of residual sugar. **rating:** 84
best drinking 1999–2000 **best vintages** NA **drink with** Light seafood • $15

warrenmang vineyard ★★★☆

Mountain Creek Road, Moonambel, Vic 3478 **region** Pyrenees
ph (03) 5467 2233 **fax** (03) 5467 2309 **open** 7 days 9–5
winemaker Luigi Bazzani **prod.** 4000 **est.** 1974
prod. range ($11–50 CD) Bazzani Chardonnay Chanin Blanc, Cabernet Shiraz Dolcetto, Saluté (Méthode Champenoise), Vintage Port; Valley of the Kings Chardonnay Chenin Blanc, Cabernet Shiraz; Warrenmang Late Harvest Traminer, Vintage Port; flagship wine is a limited release of a once-only bottling of 1991–1997 Luigi Riserva (Cabernet blend).
summary Warrenmang is now the focus of a superb accommodation and restaurant complex created by former restaurateur Luigi Bazzani and wife Athalie, which is in much demand as a conference centre as well as for weekend tourism. The striking black Bazzani label is gradually overtaking the Warrenmang label in importance, and is responsible for the growth in the volume of production. It is partially sourced from contract-growers; the estate wines are, as their name suggests, estate-grown.

Warrenmang Estate Shiraz

Produced from 3 hectares of very low-yielding estate plantings, vines which have tended to produce wines of awesome power and extract, needing a light touch in the winery. Recent releases have increasingly headed in the right direction.
1997 Dense red-purple; powerful, dark chocolate fruit is matched by attractive cigar box oak on the bouquet; concentrated, powerhouse plum and chocolate-flavoured fruit is matched by abundant oak and long, persistent tannins on the palate. **rating:** 92
best drinking 2002–2012 **best vintages** '82, '85, '92, '94, '97 **drink with** Strong cheese or red meat • $28

Warrenmang Luigi Riserva

Without doubt, this must be one of the most unusual wines ever released onto the Australian market. Luigi Bazzani says 'Its not an epitaph, but an announcement, and is a blend of a parcel of 1991 wine kept in French oak for six years, and refermented with a portion of 1997 Warrenmang Shiraz'. Such techniques are not unknown in Italy, but a span of six years between the two parcels is most unusual. The wine is available only from the cellar door, and only 400 cases were made. The exact blend components are 46% 1997 Shiraz, 16% 1991 Cabernet Franc, 13% 1991 Cabernet Sauvignon, 19% 1991 Merlot and 6% 1991 Shiraz.

🍷🍷🍷🍷🍷 **1991–1997** Bright medium to full purple-red; the bouquet is clean, with plenty of berry cassis fruit and subtle oak. The palate is really interesting, offering a seductive mix of red small fruits and Swiss chocolate, denying any particular age. **rating:** 97

best drinking 2000–2007 **best vintages** NA **drink with** Venison pie • $50

water wheel ★★★☆

Bridgewater-on-Loddon, Bridgewater, Vic 3516 **region** Bendigo
ph (03) 5437 3060 **fax** (03) 5437 3082 **open** Mon–Sat 9–5, Sun 12–5
winemaker Peter Cumming, Bill Trevaskis **prod.** 25 000 **est.** 1972
prod. range ($13–16 R) Chardonnay, Riesling, Sauvignon Blanc, Pinot Noir, Shiraz, Cabernet Sauvignon; grapes from other districts under premium Wing Fields label.
summary Peter Cumming gained great respect as a winemaker during his four-year stint with Hickinbotham Winemakers, and his 1989 purchase of Water Wheel was greeted with enthusiasm by followers of his wines. Recent releases have been of consistent quality and modest price, being distributed throughout Australia, and with export markets in NZ, Asia, the UK and Europe, and the US and Canada.

Water Wheel Shiraz

The Shiraz is always presented in a way which encourages early consumption, for the wines are never tannic or extractive, and put the emphasis on fresh fruit. However, in many vintages there is the substance for medium-term cellaring.

🍷🍷🍷🍷½ **1997** Medium red-purple; there is abundant varietal character on the bouquet, with liquorice and spice of medium to full intensity, backed by some cedary oak. An elegant palate with those liquorice, cedar and spice flavours is nicely rounded off with fine tannins. Thoroughly deserved its gold medal at the Melbourne Wine Show in 1998. **rating:** 90

best drinking 2000–2004 **best vintages** '94, '96, '97 **drink with** Cold meats • $16

Water Wheel Cabernet Sauvignon

Produced predominantly from plantings of cabernet sauvignon established when the Cumming family took over control of Water Wheel. Right from the outset these plantings produced wines with intense fruit flavours. A small percentage (15%) of Cabernet Franc and Merlot adds complexity.

🍷🍷🍷🍷 **1997** Medium red-purple; gentle briar berry aromas do not reflect the palate, which has considerably more weight, with dark berry, chocolate and persistent tannins. **rating:** 85

best drinking 2001–2007 **best vintages** '90, '91, '92, '97 **drink with** Braised beef • $16

watson NR

75 Monbulk-Seville Road, Wandin East, Vic 3139 **region** Yarra Valley
ph (03) 5964 3059 **fax** (03) 5964 3059 **open** Weekends and public holidays 11–5
winemaker Oakridge (Contract) **prod.** 700 **est.** 1986
prod. range ($13–16.50 CD) Chardonnay, Cabernet Sauvignon.

summary Production from the 4-hectare vineyard has been sold to other makers in some years, and in others part has been contract-made for owner Geoffrey Watson. It has made a somewhat uncertain entry onto the marketplace.

waybourne NR

60 Lemins Road, Waurn Ponds, Vic 3221 **region** Geelong
ph (03) 5241 8477 **fax** (03) 5241 8477 **open** By appointment
winemaker David Cowburn (Contract) **prod.** 730 **est.** 1980
prod. range ($12–16 ML) Riesling, Trebbiano, Pinot Gris, Cabernet Sauvignon.
summary Owned by Tony and Kay Volpato, who have relied upon external consultants to assist with the winemaking. No recent tastings.

wayne thomas wines NR

26 Kangarilla Road, McLaren Vale, SA 5171 **region** McLaren Vale
ph (08) 8323 9737 **fax** (08) 8323 9737 **open** 7 days 12–5
winemaker Wayne Thomas **prod.** 4000 **est.** 1994
prod. range ($12.50–18 ML) Sauvignon Blanc, Chardonnay, Shiraz, Cabernet Sauvignon, Premium Brut.
summary Wayne Thomas is a McLaren Vale veteran, having commenced his winemaking career in 1961, working for Stonyfell, Ryecroft and Saltram before establishing Fern Hill with his wife Pat in 1975. When they sold Fern Hill in April 1994 they started again, launching the Wayne Thomas Wines label, using contract-grown grapes sourced from throughout McLaren Vale. The wines are exported to the UK and to California, as well as enjoying limited retail distribution through all Australian States except WA.

wellington ★★★★☆

Cnr Richmond and Denholms Roads, Cambridge, Tas 7170 **region** Southern Tasmania
ph (03) 6248 5844 **fax** (03) 6243 0226 **open** By appointment
winemaker Andrew Hood **prod.** 2500 **est.** 1990
prod. range ($20–24 ML) Riesling, Iced Riesling, Chardonnay, Pinot Noir.
summary Consultant winemaker Andrew Hood (ex-Charles Sturt University) and wife Jenny have constructed a state-of-the-art winery on land leased from the University of Tasmania. The 2000-case production of Wellington is dwarfed by the 4500 cases contract-made for others, but the wines are always flawlessly crafted, particularly the Chardonnay.

Wellington Riesling

Another wine from the Wellington stable to repay cellaring, and, as with the other wines, immaculately made. Sourced not from southern Tasmania, but from a vineyard at Relbia, on the outskirts of Launceston. The 1996 was held back in the cellar for further ageing and released after the '97, a strategy which resulted not only in a gold medal but three trophies at the 1999 Tasmanian Wines Show for the '96.

🍷🍷🍷🍷🍷 **1996** Bright green-yellow; the influence of botrytis is evident on both the bouquet and palate, but is very much to the advantage of the wine, rather than to its disadvantage. The high-toned fruit aromas of the bouquet are followed by a remarkably intense palate, with sweet lime juice flavours rippling through from the very start to the very finish. **rating:** 94

best drinking 1999–2004 **best vintages** '96, '97, '98 **drink with** Sautéed prawns • $20

Wellington Chardonnay

Andrew Hood is nothing if not economical in his back label description of this wine as 'a dry, lightly wooded wine made from fruit grown mainly in northern Tasmania'. The '95 has matured superbly, winning a gold medal at the 1997 Tasmanian Wines Show, the very personification of elegance. The '97 followed suit at the 1998 Tasmanian Wines Show. The '92, tasted socially (in February 1998) is as fresh as a daisy, with years in front of it yet.

🍷🍷🍷🍷🍷 **1998** Bright yellow-green; a typically refined bouquet with gentle melon fruit, a touch of citrus and subtle oak. The same tune is played by the palate, fine and elegant, ultra-cool-grown yet not green, simply stylish. **rating:** 93

best drinking 1999–2004 **best vintages** '92, '95, '96, '97, '98 **drink with** Gravlax • $24

Wellington Iced Riesling

Produced using freeze concentration of the unfermented juice in a stainless steel tank. Water freezes first, and is removed in solid form, lifting the sugar, acidity (and flavour) of the remaining juice which is then conventionally cold-fermented. The '97 was awarded the trophy for Best Unwooded Table Wine (Dry or Sweet) at the 1998 Tasmanian Wines Show.

🍷🍷🍷🍷 **1998** Glowing yellow-green; there are quite intense lime and herb aromas, with lime the dominant flavour on the palate, and just a hint of herb. Well balanced, with a long, lingering finish thanks to Tasmanian acidity. **rating:** 89

best drinking 2001–2008 **best vintages** '97, '98 **drink with** Sorbet • $20

Wellington Pinot Noir

Produced from grapes grown near Cambridge in the Coal River Valley close to Andrew Hood's winery, with both the 1994 and 1995 vintages in the top echelon of Tasmanian Pinots. It consistently shows strong, dark plum fruit reminiscent of the east coast Pinot Noirs, and quite different from those of northern Tasmania. Winemaker Andrew Hood takes a deliberately simple approach to the fermentation of the Wellington Pinot Noir, in part due to the constraints of a very busy winery, and in part to a belief that the Pinot Noir he works with does not need more complexity.

🍷🍷🍷🍷 **1997** Medium red-purple; a fragrant, light and elegant spicy cherry bouquet is followed by a similarly light and elegant palate, with pronounced cool-climate fruit manifesting itself in a hint of leaf and mint. **rating:** 86

best drinking 1999–2001 **best vintages** '94, '95, '97 **drink with** Tasmanian salmon • $24

wendouree ★★★★★

Wendouree Road, Clare, SA 5453 **region** Clare Valley

ph (08) 8842 2896 **open** By appointment

winemaker Tony Brady **prod.** 2500 **est.** 1895

prod. range ($26–30 ML) Shiraz Malbec, Shiraz Mataro, Cabernet Malbec, Cabernet Sauvignon, Muscat of Alexandria.

summary The iron fist in a velvet glove best describes these extraordinary wines. They are fashioned with passion and yet precision from the very old vineyard with its unique terroir by Tony and Lita Brady, who rightly see themselves as custodians of a priceless treasure. The 100-year-old stone winery is virtually unchanged from the day it was built; this is in every sense a treasure beyond price.

west cape howe wines NR

PO Box 548, Denmark, WA 6333 **region** Great Southern

ph (08) 9848 2959 **fax** (08) 9848 2903 **open** Not

winemaker Brenden Smith, Dave Cleary **prod.** 300 **est.** 1997

prod. range ($18–22 R) Sauvignon Blanc, Shiraz.

summary Brenden Smith was senior winemaker at Goundrey Wines for many years, and has branched into business on his own with a contract-winemaking facility for growers throughout the Great Southern region. The 1998 vintage will see West Cape Howe wines crush 250 tonnes for 11 different producers; a tiny amount of wine will be made under the West Cape Howe label from the 1999 vintage.

westend ★★★

1283 Brayne Road, Griffith, NSW 2680 **region** Riverina
ph (02) 6964 1506 **fax** (02) 6962 1673 **open** Mon–Fri 9–4.30
winemaker William Calabria, James Cecato **prod.** 25 000 **est.** 1945
prod. range ($5.90–19.95 CD) Outback Traminer Riesling, Semillon Colombard, Sauvignon Blanc, Shiraz, Richland Semillon Chardonnay, Chardonnay, Cabernet Sauvignon, Cabernet Merlot, Shiraz, Merlot; Port and Liqueur Muscat; are followed by 3 Bridges range of Chardonnay, Cabernet Sauvignon and Golden Mist Botrytis Semillon.
summary Along with a number of Riverina producers, West End is making a concerted move to lift both the quality and the packaging of its wines, spearheaded by the 3 Bridges range which has an impressive array of gold medals to its credit since being first released in April 1997. It has also ventured into the export market, with distribution in the UK, US and Switzerland.

Westend Richland Chardonnay

A very lightly oaked Chardonnay (the subtlety of the oak being uncommon in the Riverina and wholly commendable) which has a dash of Sauvignon Blanc fruit. The 1998 vintage was selected in the 1999 Sydney International Wine Competition Top 100 and given the Best Value Award in the Fuller Bodied White Wine Class.

🍷🍷🍷🍷 **1998** The colour is an extraordinarily brilliant green-yellow; the bouquet is clean, the oak ever so subtle and very well integrated into the gentle tropical fruit aromas. The palate is clean, gently ripe and smooth, avoiding the coarse phenolics which often disfigure wines such as this. **rating:** 87

best drinking 1999–2000 **best vintages** NA **drink with** Pasta marinara • $9.90

Westend 3 Bridges Golden Mist Botrytis Semillon

Yet another botrytised Semillon from Griffith. It is 100% estate-grown at Hanwood, and has been made each year since 1995. As one would expect, a prolific winner of trophies and gold medals, even in the highly competitive botrytised Semillon wine show classes of Australia.

🍷🍷🍷🍷🍷 **1996** Glowing yellow-green; the bouquet is magnificently rich and intense with cumquat and mandarin aromas; the similarly rich and intense palate introduces pear and peach flavours to go with the mandarin; excellent acidity on the finish keeps the wine fresh and prevents it from cloying. Two trophies and five gold medals to its credit. **rating:** 94

best drinking 1999–2002 **best vintages** NA **drink with** Fruit tart • $19.95

Westend 3 Bridges Cabernet Sauvignon

A wine which had considerable show success over the years, the '97 continuing that record with a gold medal at Perth in 1998 and a very highly commended at Rutherglen in 1997.

🍷🍷🍷🍷½ **1997** Medium to full red-purple; the bouquet offers unusually concentrated blackberry fruit and subtle oak. The palate has plenty of depth with attractive, sweet cassis/berry fruit, soft and ripe tannins, and appropriate sweet oak. **rating:** 90

best drinking 1999–2003 **best vintages** NA **drink with** Roast beef • $17.95

westfield ★★★☆

Cnr Memorial Avenue and Great Northern Highway, Baskerville, WA 6056 **region** Swan District
ph (08) 9296 4356 **fax** (08) 9296 4356 **open** 7 days 10–5.30
winemaker John Kosovich **prod.** 11 000 **est.** 1922
prod. range ($17–22 CD) Verdelho, Sauvignon Blanc, Unwooded Chardonnay, Chardonnay, Bronze Wing Chardonnay, Chenin Blanc, Semillon, Riesling, Verdelho, Bronze Wing Verdelho, Bronze Wing Merlot, Shiraz, Cabernet Sauvignon, Vintage Port, Liqueur Muscat, Sparkling.

summary Consistent producer of a surprisingly elegant and complex Chardonnay; the other wines are more variable, but from time to time has made attractive Verdelho and excellent Cabernet Sauvignon. 1998 has seen the first release of wines partly or wholly coming from the family's new planting at Pemberton, those being Swan/Pemberton blends released under the Bronze Wing label. Limited retail distribution in Perth, Melbourne and Sydney.

Westfield Verdelho

Seventy per cent of the wine comes from the Kosovich family's Bronze Wing Vineyard at Pemberton, the remaining 30% from the Swan Valley. John Kosovich believes it will repay cellaring for up to eight years.

ΨΨΨΨ **1998** Light green-yellow; the bouquet is clean, fresh and bright, with a nice balance between fruit salad aromas and more grassy overtones. The palate is bright, fresh and lively, finishing with lingering acidity. Certainly does cry out for cellaring. **rating:** 86

best drinking 2001–2007 **best vintages** NA **drink with** Grilled fish • $17

Westfield Chardonnay

John Kosovich will soon have to open a trophy room at his winery to accommodate all of the trophies and gold medals which his Chardonnays have accumulated over recent years; virtually every vintage since 1988 has been showered with gold. Still sourced entirely from the Swan Valley; the unwooded version contains 40% Pemberton fruit.

ΨΨΨΨ **1997** Medium yellow-green; smooth nectarine and a touch of honey drive the clean, moderately intense bouquet. The palate is brisk and clean, with surprising acidity; I would never pick it as a pure Swan Valley wine. Like the Verdelho, will richly repay cellaring. **rating:** 85

best drinking 2001–2005 **best vintages** Virtually all **drink with** Marron or crayfish • $17

wetherall NR

Naracoorte Road, Coonawarra, SA 5263 **region** Coonawarra
ph (08) 8737 2104 **fax** (08) 8737 2105 **open** 7 days 10–4
winemaker Michael Wetherall **prod.** 3000 **est.** 1991
prod. range ($19–25 CD) Chardonnay, Cabernet Sauvignon, Shiraz.
summary The Wetherall family has been growing grapes in Coonawarra for more than 30 years, and Michael Wetherall (a Roseworthy graduate), has been responsible for overseeing wine production since Wetherall extended its operations into winemaking in 1991. Most of the grapes are still sold; no recent tastings.

wharncliffe NR

Summerleas Road, Kingston, Tas 7050 **region** Southern Tasmania
ph (03) 6229 7147 **fax** (03) 6229 2298 **open** Not
winemaker Andrew Hood (Contract) **prod.** 32 **est.** 1990
prod. range ($20 ML) Chardonnay.
summary With total plantings of 0.75 hectare, Wharncliffe could not exist without the type of contract-winemaking service offered by Andrew Hood, which would be a pity, because the vineyard is beautifully situated on the doorstep of Mount Wellington, the Huon Valley and the Channel regions of southern Tasmania.

whispering hills NR

54 Gibbs Road, Majorca, Vic 3465 **region** Bendigo
ph (03) 5964 6070 **fax** (03) 5964 6231 **open** Not
winemaker Murray Lyons, Ron Snep **prod.** 200 **est.** 1994
prod. range ($16 ML) Cabernet Sauvignon.
summary The minuscule production of Whispering Hills is limited to one wine (a Cabernet Sauvignon) which is sold by mail order and word of mouth.

whisson lake NR

PO Box 91, Uraidla, SA 5142 **region** Adelaide Hills
ph (08) 8390 1303 **fax** (08) 8390 3822 **open** By appointment
winemaker Roman Bratasiuk (Contract) **prod.** 300 **est.** 1985
prod. range ($25–35 CD) Pinot Noir.
summary Mark Whisson is primarily a grape grower, with 4.5 hectares of close-planted, steep-sloped north-facing vineyard. A small quantity of the production is made for the Whisson Lake label by Roman Bratasiuk, best known as the owner/winemaker of Clarendon Hills.

whitehorse wines NR

4 Reid Park Road, Mount Clear, Vic 3350 **region** Ballarat
ph (03) 5330 1719 **fax** (03) 5330 1288 **open** Weekends 11–5
winemaker Noel Myers **prod.** 900 **est.** 1981
prod. range ($10–18 CD) Riesling, Riesling Müller Thurgau, Chardonnay, Pinot Noir, Cabernet Shiraz.
summary The Myers family has moved from grape growing to winemaking, utilising the attractive site on its sloping hillside south of Ballarat. Four hectares of vines are in production, with pinot noir and chardonnay the principal varieties.

white rock vineyard NR

1171 Railton Road, Kimberley, Tas 7304 **region** Northern Tasmania
ph (03) 6497 2156 **fax** (03) 6497 2156 **open** By appointment
winemaker Richard Richardson **prod.** 500 **est.** 1991
prod. range ($17 CD) Chardonnay, Pinot Noir.
summary With a 1-hectare vineyard, more or less equally split between chardonnay and pinot noir, production remains tiny, but the White Rock wines may be found in a number of outlets in the nearby town of Deloraine and at Tasmanian Flavours in Launceston. The style is light and delicate.

wignalls wines ★★★★

Chester Pass Road (Highway 1), Albany, WA 6330 **region** Great Southern
ph (08) 9841 2848 **fax** (08) 9842 9003 **open** 7 days 12–4
winemaker Bill Wignall **prod.** 5000 **est.** 1982
prod. range ($12–32 R) Chardonnay, Sauvignon Blanc, Late Harvest Frontignac, Pinot Noir, Reserve Pinot Noir, Cabernet Sauvignon, Tawny Port, White Port.
summary A noted producer of Pinot Noir which has extended the map for the variety in Australia. The Pinots have tremendous style and flair, but do age fairly quickly. The white wines are elegant, and show the cool climate to good advantage. A new winery was constructed and opened for the 1998 vintage, utilising the production from the 16 hectares of estate plantings.

Wignalls Chardonnay

Produced from 2 hectares of estate plantings. It is usually fairly discreet when first released, but builds flavour and character with a year or two in the bottle. In September 1998 the '96 won the trophy for Best Australian White Wine at the Japan International Wine Challenge.

🍷🍷🍷🍷 **1997** Medium yellow-green; the floral, honey and citrus bouquet is quite different from the norm; the palate shows the impact of the softening effect of malolactic fermentation on the melon fruit, introducing more creamy/nutty flavours. **rating:** 85

🍷🍷🍷🍷🍷 **1996** Brilliant, light green-yellow. A marvellously elegant wine on the bouquet with subtle, figgy fruit, and a fine yet intense palate running through fig, melon and nectarine flavours, supported by subtle oak. **rating:** 94

best drinking 1999–2002 **best vintages** '85, '88, '91, '92, '96 **drink with** Milk-fed veal • $22

Wignalls Pinot Noir

An enigmatic wine, even by the standards of the perversely unpredictable Pinot Noir variety. By and large, has tended to look excellent when young but to develop with disconcerting rapidity, sometimes problematically. Note also there are tiny quantities of a Reserve Pinot released from time to time.

🍷🍷🍷🍷🍷 **1997** Light to medium red; the bouquet is light with a mix of sappy, strawberry and earthy aromas. An elegant wine on the palate with good structure and length, the flavours in the foresty/sappy/strawberry spectrum. **rating:** 91

best drinking 1998–1999 **best vintages** '85, '86, '88, '91, '93, '95, '97 **drink with** Seared Tasmanian salmon • $26

Wignalls Reserve Pinot Noir

A relatively new arrival on the scene for Wignalls, which does show an extra degree of weight and concentration. However, like the standard Pinot Noir, it does tend to develop relatively quickly, with cellaring a doubtful proposition.

🍷🍷🍷🍷 **1998** Light red; the bouquet has fragrance and lift, with strawberry fruit, the palate with better acidity than the varietal release, with clever oak handling also lingering on the finish. **rating:** 88

best drinking 1999–2001 **best vintages** NA **drink with** Seared Tasmanian salmon • $32

wild dog NR

South Road, Warragul, Vic 3820 **region** Gippsland

ph (03) 5623 1117 **fax** (03) 5623 6402 **open** 7 days 10–5

winemaker John Farrington **prod.** 2500 **est.** 1982

prod. range ($12–18 CD) Riesling, Chardonnay, Rosé, Pinot Noir, Shiraz, Cabernet Sauvignon.

summary An aptly named winery, which produces somewhat rustic wines from the 11.5 hectares of estate vineyards; even the Farringtons say that the Shiraz comes 'with a bite', also pointing out that there is minimal handling, fining and filtration. Be warned.

wild duck creek estate NR

Spring Flat Road, Heathcote, Vic 3523 **region** Bendigo

ph (03) 5433 3133 **fax** (03) 5433 3133 **open** By appointment

winemaker David Anderson **prod.** 500 **est.** 1980

prod. range ($15–25 CD) Springflat Shiraz, Alan's Cabernets, The Blend, Duck Muck (which rivals Demondrille's Purgatory in the contest for the worst wine name).

summary The first release of Wild Duck Creek Estate from the 1991 vintage marks the end of 12 years of effort by David and Diana Anderson, who commenced planting the 4.5-hectare vineyard in 1980, made their first tiny quantities of wine in 1986, the first commercial quantities of wine in 1991, and built their winery and cellar-door facility in 1993. The 1996 Shiraz was the public choice as the Best Red Wine at the 1997 Victorian Winemakers Exhibition. All of the wines are sold by mail order and cellar door, and no wine was available at the time of going to press (all sold).

wilderness estate NR

Branxton Road, Pokolbin, NSW 2321 **region** Lower Hunter Valley

ph (02) 4998 7755 **fax** (02) 4998 7750 **open** 7 days 9–5

winemaker John Baruzzi, Josef Lesnik **prod.** 25 000 **est.** 1986

prod. range ($16.30–17.80 CD) The premium varietal range of Unwooded Chardonnay, Reserve Chardonnay, Unwooded Semillon, Individual Block Semillon, Shiraz, Cabernet Merlot and Merlot is under the Wilderness Estate label; the second Black Creek label encompasses a similar range of lower-priced varietals.

summary Long-term Wyndham Estate winemaker John Baruzzi has formed a 50-50 joint venture with Joe Lesnik, resulting in the former Lesnik Family Winery now renamed Wilderness Estate. The Lesnik label will be phased out, with all wines from the '95 vintage and onwards being released either under the Wilderness Estate label or under the Black Creek label. The '97 Reserve Chardonnay is a good wine with plenty of fruit and balanced use of what appears to be American oak. National distribution through Normans is supplemented by exports to the US, Canada and UK.

wildwood ★★★

St John's Lane, Wildwood, Bulla, Vic 3428 **region** Sunbury
ph (03) 9307 1118 **fax** (03) 9331 1590 **open** 7 days 10–6
winemaker Dr Wayne Stott, Peter Dredge **prod.** 2000 **est.** 1983
prod. range ($20–32 CD) Chardonnay, Viognier, Cabernets, Merlot Cabernet Franc, Pinot Noir, Shiraz.

summary Wildwood is situated just 4 kilometres past Melbourne airport. The vineyard and cellar door are situated at an altitude of 130 metres in the Oaklands Valley, which provides unexpected views back to Port Phillip Bay and the Melbourne skyline. Plastic surgeon Wayne Stott has taken what is very much a part-time activity rather more seriously than most by undertaking (and completing) the Wine Science degree at Charles Sturt University.

Wildwood Shiraz

Given the reverence with which Craiglee Shiraz is greeted, it should be no surprise to find Wildwood coming up with a lovely wine in a vintage made to measure for the variety in the Sunbury region.

🍷🍷🍷🍷🍷 **1997** Extraordinarily dense purple-red colour; the bouquet is rich and concentrated, with masses of dark berry fruit together with hints of liquorice and leather. An exceptional wine from a micro-yield, with the black cherry and blackberry fruit soaking up whatever oak was accorded to it. **rating:** 94

best drinking 2005–2015 **best vintages** '97 **drink with** Coq au vin • $25

Wildwood Cabernets

The '93 was the winner of a gold medal at the Royal Melbourne Wine Show in 1993 as an unbottled wine, but made the grade through bottling and bottle maturation. The '97 is the best vintage since.

🍷🍷🍷🍷 **1997** Dense red-purple; the bouquet is fully ripe, with dark berry fruit supported by more briary/cedary notes. The palate is impressive, bordering on oppressive, with abundant fruit and even more abundant tannins. By all means buy it, but do not open it other than with the prescription of a medical practitioner. Dr Wayne Stott will suffice. **rating:** 88

best drinking 2007–2017 **best vintages** '93, '97 **drink with** Oxtail • $20

willespie ★★★☆

Harmans Mill Road, Willyabrup via Cowaramup, WA 6284 **region** Margaret River
ph (08) 9755 6248 **fax** (08) 9755 6210 **open** 7 days 10.30–5
winemaker Michael Lemmes **prod.** 4000 **est.** 1976
prod. range ($15–35 R) Sauvignon Blanc, Semillon Sauvignon Blanc, Verdelho, Riesling, Shiraz, Cabernet Sauvignon, Merlot; Harmans Mill White and Harmans Mill Red are cheaper second-label wines.

summary Willespie has produced many attractive white wines over the years, typically in brisk, herbaceous Margaret River-style. All are fruit- rather than oak-driven; the newer Merlot also shows promise. The wines have had such success that the Squance family (which founded and owns Willespie) has announced plans to substantially increase winery capacity, drawing upon an additional 25 hectares of estate vineyards in the course of establishment.

Willespie Sauvignon Blanc

In many ways, the flag-bearer for Willespie. The typical Margaret River herbaceous characters are usually accompanied by a riper gooseberry spectrum, particularly in better vintages. Twenty per cent of the wine is briefly fermented in high toast American oak, but has little obvious impact. Perhaps this is why it was grouped in the unwooded class at the 1999 *Winewise* Small Makers Competition (where it scored very well indeed) showing a touch of oak, which in fact improves a very good wine.

🍷🍷🍷🍷🍷 **1998** Light green-yellow; the bouquet is complex with abundant passionfruit and gooseberry aromas, and the palate is similarly in full-on tropical mode without being flabby or sweet. Instead it has real presence, texture and flavour. **rating:** 91

best drinking 1999–2002 **best vintages** '87, '90, '91, '93, '95, '98 **drink with** Lighter Chinese dishes • $24

Willespie Semillon Sauvignon Blanc

Estate-grown, and produced from a total of over 6 hectares of plantings of the two varieties. Made in straightforward, unwooded style, it is a blend which works particularly well in the Margaret River region, in no small measure due to the essentially similar flavour profile of both components. The Semillon simply provides a little more structure for the blend.

🍷🍷🍷🍷🍷 **1997** Light straw-green; the bouquet is moderately fragrant with grass and citrus aromas, and no oak evident. The palate has considerable length and persistence, tight, and finishing with well-balanced acidity. Has developed very well over the past year, but is in no particular hurry. **rating:** 90

best drinking 1999–2004 **best vintages** '97 **drink with** Grilled fish • $23

Willespie Cabernet Sauvignon

As with all the Willespie wines, estate-grown, produced from 4 hectares of plantings. Made in the austere end of the Margaret River spectrum, but certainly very typical.

🍷🍷🍷🍷 **1995** Medium to full red-purple; the bouquet offers relatively austere but crystal clear Cabernet varietal fruit. The palate comes as a pleasant surprise, for it is not the least bit austere, helped in part by the clever use of sweet oak, but also by the lovely berry and bitter chocolate flavours of the fruit. The tannins, too, are soft. **rating:** 88

best drinking 2001–2008 **best vintages** NA **drink with** Butterfly leg of lamb • $34

williams rest NR

Lot 195 Albany Highway, Mount Barker, WA 6324 **region** Great Southern
ph (08) 9367 3277 **fax** (08) 9367 3328 **open** Not
winemaker Contract **prod.** NA **est.** 1972
prod. range ($NA) Granite Flats White and Red.
summary A long-established vineyard, planted way back in 1972, which is now part of the Selwyn Wine Group. The vineyard is named after Benjamin Williams, an eight-year-old boy who was accidentally killed by a mail coach in 1890, hence the name Williams Rest.

willow bend ★★★☆

Lyndoch Valley Road (PO Box 107), Lyndoch, SA 5351 **region** Barossa Valley
ph (08) 8524 4169 **fax** (08) 8524 4169 **open** Not
winemaker Wayne Dutschke **prod.** 600 **est.** 1990
prod. range ($16–22 R) Chardonnay, Shiraz, Shiraz Merlot Cabernet, Shiraz Cabernet Merlot.
summary Wayne Dutschke has had ten years of winemaking experience with major wine companies in South Australia, Victoria and New South Wales, but has returned to South Australia to

join his uncle, Ken Semmler, a leading grape grower in the Barossa Valley and now in the Adelaide Hills. No recent tastings, simply because Willow Bend sells out of wine in less than six months each year. Annual release in August, there is limited retail distribution in Sydney and Melbourne.

willow creek ★★★☆

166 Balnarring Road, Merricks North, Vic 3926 **region** Mornington Peninsula
ph (03) 5989 7448 **fax** (03) 5989 7584 **open** 7 days 10–5
winemaker Simon Black **prod.** 8000 **est.** 1989
prod. range ($16–25 CD) Unoaked Chardonnay, Tulum Chardonnay, Pinot Noir, Cabernet Sauvignon, Sparkling Cuvee.
summary Yet another significant entrant in the fast-expanding Mornington Peninsula area, with 15 hectares of vines planted to cabernet sauvignon, chardonnay and pinot noir. The cellar-door sales area boasts picnic areas, barbecue facilities, trout fishing and bocce; lunches are served every day, and dinners by appointment. Expansion of the cellar door was completed by January 1998, with a winery constructed for the 1998 vintage.

Willow Creek Pinot Noir

The wine is made using a variety of techniques, with selected parcels undergoing barrel fermentation, and others extended skin maceration. It can spend up to 18 months in new and used French oak barriques, as it did in 1997, a response to the low-yielding and intense vintage.
🍷🍷🍷🍷🍷 **1997** Medium red-purple; the bouquet is still to evolve secondary complexity, but does have very good fruit in a somewhat linear fashion at the moment. The palate, too, has all the ingredients, with cherry and plum fruit running through a long finish, aided by fine tannins which give grip but which do not threaten the integrity of the wine. The fruit has swallowed up the oak treatment given. **rating:** 90
best drinking 1999–2004 **best vintages** '91, '94, '95, '97 **drink with** Gently spiced Asian food • $25

will taylor wines NR

1 Simpson Parade, Goodwood, SA 5034 **region** Other Wineries of SA
ph (08) 8271 6122 **fax** (08) 8271 6122 **open** Not
winemaker Various contract **prod.** 1300 **est.** 1997
prod. range ($20 R) Clare Valley Riesling, Adelaide Hills Sauvignon Blanc, Hunter Valley Semillon.
summary Will Taylor is a partner in the leading Adelaide law firm Finlaysons specialising in wine law. Together with Suzanne Taylor, he has established a classic negociant wine business, having wines contract-made to his specification. Moreover, he chooses what he considers to be the best regions for each variety, and will be adding a Coonawarra Cabernet Sauvignon to the roster in vintage 2000. A Victorian Pinot Noir is also planned. Limited retail distribution, and small exports to the US; the wines are principally sold via mailing list.

Will Taylor Clare Valley Riesling

A traditional Clare Riesling, albeit with a little more character and flavour than some. Very competently made.
🍷🍷🍷🍷 **1998** Medium yellow-green; a spotlessly clean and quite aromatic bouquet, starting with citrus and running almost into stone fruit and blossom characters. The palate is harmonious and well-balanced; should develop very well, but needs time for its full personality to evolve. **rating:** 89
best drinking 2002–2008 **best vintages** NA **drink with** Whiting • $19.50

Will Taylor Hunter Valley Semillon

Produced from Pokolbin semillon grown in one of the rare dry Hunter vintages. Stainless steel-fermented and relatively early-bottled (in September of the year of vintage, along with the Riesling and Sauvignon Blanc).

🍷🍷🍷🍷 **1998** Medium to full yellow-green; the bouquet reflects the vintage, with a mix of buttery and more minerally aromas, perhaps fractionally advanced. The palate has abundant soft, ripe semillon, with hints of butter and banana; good back palate and finish. **rating:** 87

best drinking 2000–2004 **best vintages** NA **drink with** Light pasta • $19.50

wilmot hills vineyard NR

407 Back Road, Wilmot, Tas 7310 **region** Northern Tasmania
ph (03) 6492 1193 **fax** (03) 6492 1193 **open** 7 days 9–7
winemaker John Cole, Ruth Cole **prod.** NA **est.** 1991
prod. range ($15–20 CD) Muller Thürgau, Pinot Noir, El Nino Pinot Noir, fruit wines and ciders.

summary The beautiful Wilmot Hills Vineyard is situated on the western side of Lake Barrington, not far from the Cradle Mountain road, with marvellous views to Mount Roland and the adjacent peaks. It is very much a family affair, established by John and Ruth Cole, and produces both wine and cider. John Cole spent 18 years in Melbourne participating in engineering design and some graphic art, Ruth working in the hospitality industry for ten years and making fruit wines for 20 years. The neat on-site winery was both designed and built by the Coles, as was much of the wine and cider-making equipment.

wilson vineyard ★★★★

Polish Hill River, Sevenhill via Clare, SA 5453 **region** Clare Valley
ph (08) 8843 4310 **open** Weekends 10–4 May–Oct
winemaker Daniel Wilson **prod.** 4500 **est.** 1974
prod. range ($14.50–24 CD) Gallery Series Riesling, Cabernet Sauvignon, Hippocrene Sparkling Burgundy, Chardonnay, Zinfandel, Liqueur Gewurztraminer.

summary Dr John Wilson is a tireless ambassador for the Clare Valley and for wine (and its beneficial effect on health) in general. His wines were made using techniques and philosophies garnered early in his wine career, and can occasionally be idiosyncratic, but in recent years have been most impressive. The winemaking mantle has now passed to his son, Daniel. The wines are sold through cellar door and retail in Sydney, Melbourne, Brisbane and Adelaide; no mailing list.

Wilson Gallery Series Riesling

This powerful Riesling from the Polish Hill River vineyards is almost invariably at the upper end of the Clare Valley hierarchy. A bottle of 1991 tasted in January 1997 was magnificent, a great Riesling at the height of its power and complexity. The '98 should go down the same path; these are Rieslings with the capacity to age exceptionally well.

🍷🍷🍷🍷 **1998** Excellent, light green-yellow; the bouquet is crisp, with slightly hard mineral and herb aromas which will soften and fill out as the wine ages. The palate is light, fresh and well-balanced, with a nice dry finish. Once again, there is not a super-abundance of fruit, but the wine has definite potential to improve significantly given time. **rating:** 86

best drinking 2000–2010 **best vintages** '85, '90, '91, '92, '94, '96, '97, '98 **drink with** Japanese cuisine • $16.50

wilton estate ★★★

Whitton Stock Route, Yenda, NSW 2681 **region** Riverina
ph (02) 6968 1303 **fax** (02) 6968 1328 **open** Mon–Fri 9–5
winemaker Ralph Graham **prod.** 250 000 **est.** 1977

prod. range ($6.99–15.95 CD) Chardonnay Reserve, Botrytis Semillon, NV Brut, Shiraz Cabernet, Cabernet Merlot; also wines under Hidden Valley and Yenda Vineyards labels.

summary Wilton Estate draws grapes and wine from various parts of southern Australia and New South Wales for its dry table wines, but having outstanding success with its Botrytis Semillon from locally grown fruit. It shares the winemaking facilities at St Peters. The wines are sold in the UK by separate distributors for the Yenda Vineyards label and for the Wilton Estate label.

windowrie estate NR

Windowrie, Canowindra, NSW 2804 **region** Cowra

ph (02) 6344 3234 **fax** (02) 6344 3227 **open** 7 days 10–6

winemaker Rodney Hooper, Tobias Ansted **prod.** 42 000 **est.** 1988

prod. range ($10–19 CD) Chardonnay, Cabernet Sauvignon; the Mill range of Unwooded Chardonnay, Semillon Sauvignon Blanc, Traminer Riesling and Shiraz Cabernet.

summary Windowrie Estate was established in 1988 on a substantial grazing property at Canowindra, 30 kilometres north of Cowra, and in the same viticultural region. Most of the grapes from the 240-hectare vineyard are sold to other makers, with increasing quantities being made for the Windowrie Estate and The Mill labels, the Chardonnays enjoying show success.

windy ridge vineyard NR

Foster–Fish Creek Road, Foster, Vic 3960 **region** Gippsland

ph (03) 5682 2035 **open** Holiday weekends 10–5

winemaker Graeme Wilson **prod.** 350 **est.** 1978

prod. range ($19–24 CD) Traminer, Pinot Noir, Cabernet Sauvignon Malbec, Vintage Port, Georgia's Liqueur Pinot Noir, Graeme's Late Bottled Vintage Port.

summary The Windy Ridge Vineyard was planted between 1978 and 1986, with the first vintage not taking place until 1988. Winemaker Graeme Wilson favours prolonged maturation, part in stainless steel and part in oak, before bottling his wines, typically giving the Pinot Noir two and a half years and the Cabernet four years before bottling. Robin Bradley accorded the 1995 Pinot Noir (released Easter 1998) five stars.

winewood NR

Sundown Road, Ballandean, Qld 4382 **region** Granite Belt

ph (07) 4684 1187 **fax** (07) 4684 1187 **open** Weekends, public holidays 9–5

winemaker Ian Davis **prod.** 650 **est.** 1984

prod. range ($12–15 CD) Chardonnay, Chardonnay Marsanne, Shiraz Marsanne, MacKenzies Run (Cabernet blend), Muscat.

summary A weekend and holiday activity for schoolteacher Ian Davis and town-planner wife Jeanette; the tiny winery is a model of neatness and precision planning. The use of Marsanne with Chardonnay and Semillon shows an interesting change in direction. Has a little over 3 hectares of estate plantings. All wine sold through cellar door.

winstead ★★★★☆

Lot 7 Winstead Road, Bagdad, Tas 7030 **region** Southern Tasmania

ph (03) 6268 6417 **fax** (03) 6268 6417 **open** Wed–Sun 11–5

winemaker Andrew Hood (Contract) **prod.** 400 **est.** 1989

prod. range ($14.50–17.50 CD) Riesling, Pinot Noir.

summary The good news about Winstead is the outstanding quality of its extremely generous and rich Pinot Noirs, rivalling those of Freycinet for the abundance of their fruit flavour without any sacrifice of varietal character. The bad news is that production is so limited, with only half a hectare each of riesling and pinot noir being tended by fly-fishing devotee Neil Snare and wife Julieanne. Retail distribution through Sutherland Cellars, Melbourne.

Winstead Riesling

Produced from a little over half a hectare of estate plantings. One of the most consistent Tasmanian Rieslings, and which richly repays cellaring, not to mention service with trout caught by Winstead owner Neil Snare.

🍷🍷🍷🍷🍷 **1998** Pale yellow-green; a high-toned bouquet with lots of toasty spicy fruit leads into a similarly powerful, rich and flavoursome palate with an array of lime, toast and spice fruit flavours. **rating:** 91

best drinking 2001–2006 **best vintages** NA **drink with** Pan-fried trout • $15.85

winters vineyard NR

Clarke Road, O.B. Flat via Mount Gambier, SA 5290 **region** Mount Gambier
ph (08) 8726 8255 **fax** (08) 8726 8255 **open** 7 days 10–5
winemaker Bruce Gregory **prod.** 500 **est.** 1988
prod. range ($10–14 CD) Chardonnay, Cabernet Sauvignon.
summary Former restaurateurs Martin and Merrilee Winter have established 8 hectares of vineyards 6 kilometres south of Mount Gambier and about 60 kilometres south of Coonawarra proper. The wines are contract-made, with an ultimate production target of 2500 cases. Light, leafy but pleasant Cabernet Sauvignon shows the cool climate, but is well made with a nice touch of cedary vanillin oak.

wirilda creek ★★★☆

RSD 90 McMurtrie Road, McLaren Vale, SA 5171 **region** McLaren Vale
ph (08) 8323 9688 **fax** (08) 8323 9688 **open** 7 days 10–5
winemaker Kerry Flanagan **prod.** 1500 **est.** 1993
prod. range ($11–22 CD) Oak Matured Semillon, Sauvignon Blanc, Trinity – The Blend (Cabernet Malbec Shiraz), Cabernet Merlot, Shiraz, Rare Shiraz, Rare Liqueur, Port.
summary Wirilda Creek may be one of the newer arrivals in McLaren Vale, but it offers the lot: wine, lunch every day (Pickers Platters reflecting local produce) and accommodation (four rooms opening onto a private garden courtyard). Co-owner Kerry Flanagan (with partner Karen Shertock) has had great experience in the wine and hospitality industries: a Roseworthy graduate (1980) he has inter alia worked at Penfolds, Coriole and Wirra Wirra, and also owned the famous Old Salopian Inn for a period of time. The red wines are the best bet, the white wines a less safe haven.

wirra wirra ★★★★☆

McMurtie Road, McLaren Vale, SA 5171 **region** McLaren Vale
ph (08) 8323 8414 **fax** (08) 8323 8596 **open** Mon–Sat 10–5, Sun 11–5
winemaker Benn Riggs, Dr Tony Jordan **prod.** 75 000 **est.** 1969
prod. range ($13–30 R) The Cousins (Sparkling), Hand Picked Riesling, Late Picked Riesling, Semillon Sauvignon Blanc, Scrubby Rise Semillon, Sauvignon Blanc, Scrubby Rise Semillon Sauvignon Blanc Chardonnay, Chardonnay, Wood Matured Semillon Sauvignon Blanc, The Angelus Cabernet Sauvignon, Pinot Noir, Church Block (Cabernet Shiraz Merlot), RSW Shiraz, Original Blend (Grenache Shiraz), Fortifieds.
summary Long-respected for the consistency of its white wines, Wirra Wirra has now established an equally formidable reputation for its reds. Right across the board, the wines are of exemplary character, quality and style, The Angelus Cabernet Sauvignon and RSW Shiraz battling with each other for supremacy. Long may the battle continue. The arrival in late 1998 of Dr Tony Jordan, formerly managing director of Domaine Chandon, signals the start of a new

phase in the development of what has been a singularly successful winery. It also represents the wheel turning full circle, for Wirra Wirra was one of Tony Jordan's first clients when he and Brian Croser set up their Oenotech consultancy business in the late 1970s.

Wirra Wirra Chardonnay

A blend of McLaren Vale and Adelaide Hills fruit, barrel-fermented in a mix of new, one-year-old and two-year-old barrels coming respectively from the Vosges, Allier and Troncais forests. If this were not enough, a portion was stainless steel-fermented and back-blended. The wine was given partial lees contact for eight months. In other words, it had the full book thrown at it in the winery.

🍷🍷🍷🍷🍸 **1997** Medium yellow-green; the bouquet has good richness and weight, with nectarine and melon fruit supported by subtle oak. A full-flavoured and weighted wine on the palate, with melon and apple fruit flavours, subtle oak and nicely judged acidity on the finish. **rating:** 90

best drinking 1999–2002 **best vintages** '82, '89, '91, '92, '94, '96, '97 **drink with** Wiener schnitzel • $18

Wirra Wirra RSW Shiraz

RSW Shiraz is named after Robert Strangways Wigley, who founded Wirra Wirra in 1894. While The Angelus has brought much recognition to Wirra Wirra, it is arguable that, viewed since 1994, RSW Shiraz is its best wine. In both vintages it is a beautifully elegant, constructed and structured wine, showing a particular sensitive use of French (80%) and American (20%) oak.

🍷🍷🍷🍷 **1996** Medium red-purple; a clean, smooth bouquet with red berry and subtle oak is followed by a fresh, fruit-driven palate, with red berry/cherry flavours and soft tannins. **rating:** 89

best drinking 2001–2006 **best vintages** '94, '95 **drink with** Smoked beef • $30

Wirra Wirra Vineyard Series Shiraz

Part of a special, very limited production Vineyard Series. The '94 was released in December 1998, and only 168 dozen were made. The fruit was grown by David Paxton, and the wine was matured in a mix of French and American oak.

🍷🍷🍷🍷🍸 **1994** Medium to full red; there is a complex array of ripe plum and mulberry fruit contrasting with secondary/earthy aromas on the bouquet. The palate has that classic mix of McLaren Vale chocolate and dark berry fruit, augmented by sweet tannins and positive vanilla oak. **rating:** 90

best drinking 1999–2010 **best vintages** NA **drink with** Kangaroo fillet • $58

Wirra Wirra The Angelus Cabernet Sauvignon

Named after a one-tonne bell which used to ring at St Ignatius Church, Norwood; a Trott whimsy for The Angelus is only made in 'ring the bell' vintages – four in total up to 1992, including the '91 which received national acclaim as the top wine at the 1992 Sydney International Wine Competition. The precise source varies from year to year, typically 65% McLaren Vale and 35% Coonawarra material. It spends 20 months in French oak.

🍷🍷🍷🍷🍸 **1996** Medium purple-red; a wonderfully fine and elegant bouquet with aromas of berry, earth and cedar is followed by an equally fine and elegant palate. Here sweet berry/cherry fruit come to the fore, balanced by subtle oak and fine tannins. **rating:** 93

best drinking 2001–2010 **best vintages** '86, '90, '91, '92, '95, '96 **drink with** Fillet of beef • $30

wise wines NR

Lot 4 Eagle Bay Road, Dunsborough, WA 6281 **region** Margaret River
ph (08) 9756 8627 **fax** (08) 9756 8770 **open** 7 days 10.30–4.30
winemaker Siobhan Lynch **prod.** 20 000 **est.** 1986

prod. range ($16–34.50 R) Sauvignon Blanc Semillon, Aquercus Chardonnay (Unwooded), Chardonnay, Late Harvest (Chenin Blanc, Semillon, Muscat), Pinot Noir, Classic Soft Red, Shiraz Merlot, Cabernet Sauvignon, Tawny Port.

summary Wise Wines, headed by Perth entrepreneur Ron Wines, brings together the 20.5-hectare Eagle Bay Vineyard at Meelup, the 10.3-hectare Donnybrook Valley Vineyard at Donnybrook, and the 4-hectare Bramley Estate Vineyard at Margaret River. The appointment of Siobhan Lynch, formerly winemaker at Chatsfield, has coincided with a sharp increase in production, and exports to the US.

wolf blass ★★★★

Bilyara Vineyards, Sturt Highway, Nuriootpa, SA 5355 **region** Barossa Valley
ph (08) 8562 1955 **fax** (08) 8562 4127 **open** Mon–Fri 9.15–4.30, weekends 10–4.30
winemaker John Glaetzer (Red), Wendy Stuckey (White) **prod.** 150 000 **est.** 1966

prod. range ($11–100 R) White wines under White, Yellow, Green and Gold labels, with emphasis on Riesling and blended Classic Dry White; red wines under Red, Yellow, Brown, Grey and Black labels with emphasis on Cabernet Sauvignon, Shiraz and blends of these. Also sparkling and fortified wines. The Eaglehawk now roosts here too.

summary Although merged with Mildara and now under the giant umbrella of the Fosters Brewing Group, the brands (as expected) have been left largely intact, and – so far at least – the style of the wines has changed little. The red wines continue to be very oaky and to my palate, at least, increasingly old-fashioned. The white wines (made by Wendy Stuckey) are particularly impressive, none more so than the Gold Label Riesling. Worldwide distribution via Mildara Blass International.

Eaglehawk Riesling

Now made at Wolf Blass by Wendy Stuckey, and uses grapes grown in various parts of southeast Australia. In 1996 produced an outstanding wine, a gold medal winner at the 1997 Sydney Wine Show, the subsequent vintages simply underlining the skills of the winemaking team headed by Stuckey.

🍷🍷🍷🍷🍸 **1998** Bright straw-green; a quite powerful bouquet with some spicy talc aromas. A similarly powerful, intense palate with lime and spice flavours running through to a lingering finish. Gold medal 1998 National Wine Show, and 1999 Sydney Royal Wine Show. **rating:** 92

best drinking 1998–2001 **best vintages** '90, '92, '93, '96 **drink with** Terrine of scallops • $11

Wolf Blass Gold Label Riesling

Has been a prolific gold medal and trophy winner in national wine shows over the past four or five years, every vintage winning at least one trophy and multiple gold medals. A blend of Eden Valley and Clare Valley material.

🍷🍷🍷🍷🍷 **1998** Light to medium yellow-green; the bouquet is clean, with tropical lime fruit of medium intensity, which however builds significantly on the palate, where intense lime-accented fruit has considerable length. **rating:** 94

best drinking 1999–2004 **best vintages** '90, '92, '95, '96, '97 **drink with** Salad Niçoise • $15

Wolf Blass Traminer Riesling

Year in, year out, Wolf Blass produces this blend, always of immaculate quality, and specifically aimed at the Asian restaurant market. Seldom, however, is it as fine as it was in 1998.

🍷🍷🍷🍷🍸 **1998** Light straw-green; very elegant, fine and lightly spicy aromas are followed by rose petal and spice flavours on the palate. Has good acidity, and is not the least bit over-sweet. Gold medal winner 1998 National Wine Show. **rating:** 93

best drinking 1998–2000 **best vintages** '98 **drink with** Asian seafoods • $11.95

Eaglehawk Semillon Sauvignon Blanc

Sourced (I presume) from the Quelltaler Vineyards in the Clare Valley, and also made by the enormously talented Wendy Stuckey. Gold medal winner Class 2 1999 Sydney Royal Wine Show.

🍷🍷🍷🍷🍷 **1998** Light to medium yellow-green; the bouquet is lifted and tangy, with Semillon very much to the fore. The palate has the intensity and length of flavour required of a gold medal, untrammelled by oak. **rating:** 92

best drinking 1999–2000 **best vintages** '98 **drink with** Pasta marinara • $11

Wolf Blass Pinot Chardonnay Brut

An often underestimated wine which has a habit of bobbing up with gold medals and trophies in wine shows, and which doubtless draws upon the group experience in sparkling winemaking through the medium of Yellowglen.

🍷🍷🍷🍷 **1996** Light straw-green; a clean, very fine and delicate bouquet is followed by a similarly delicate citrus and nectarine-flavoured palate. Very fine, perhaps a little too much so. **rating:** 86

best drinking 1999–2000 **best vintages** NA **drink with** A sunny spring morning • $17

Wolf Blass Brown Label Classic Shiraz

Sourced from McLaren Vale, the Barossa Valley and Langhorne Creek areas and, as one would expect, given lengthy maturation in new American oak barrels.

🍷🍷🍷🍷 **1997** Medium red-purple; the bouquet is of freshly-turned earth but in a strongly varietal mould, with obvious American oak. The palate has berry, chocolate and vanilla flavours coming together in a sweetly tasty finish supported by soft tannins. Much the best wine I can remember from this label. **rating:** 88

best drinking 1999–2004 **best vintages** '87, '88, '90, '91, '93, '97 **drink with** Spaghetti Bolognese • $25

Wolf Blass Black Label Cabernet Shiraz

Only limited quantities are made, which at the price seems a pretty smart decision by those in charge of such matters. In a sign of the times, Black Label is now a Cabernet Merlot blend, not the Cabernet Shiraz blend of the prior 20 years. The style, however, is unchanged.

🍷🍷🍷🍷 **1995** Medium to full red-purple; the bouquet is complex, with cedar, vanilla, berry and earth aromas. A powerful wine on the palate with a mix of dark berry and vanilla, and quite pronounced tannins. **rating:** 89

best drinking 2000–2005 **best vintages** '86, '88, '90, '91, '95 **drink with** Steak with wild mushrooms • $95

---

woodend winery NR

82 Mahoneys Road, Woodend, Vic 3442 **region** Macedon
ph (03) 5427 2183 **fax** (03) 5427 4007 **open** Weekends 10–6
winemaker Howard Bradfield **prod.** 600 **est.** 1983
prod. range ($15–30 CD) Unwooded Chardonnay, Pinot Noir, Cabernet Franc.
summary Woodend Winery (for a while known as Bluestone Bridge, a name which had to be relinquished due to trademark problems) draws upon 2.5 hectares of vineyard established way back in 1983, although the winemaking is of much more recent origin. The wines are available at cellar door, but are also distributed through wholesaler Australian Prestige Wines.

woodlands NR

Cnr Caves and Metricup Roads, Willyabrup via Cowaramup, WA 6284 **region** Margaret River
ph (08) 9755 6226 **fax** (08) 9321 6385 **open** Weekends by appointment
winemaker David Watson, Dorham Mann (Consultant) **prod.** 1000 **est.** 1973
prod. range ($20–30 CD) Chardonnay, Pinot Noir, James Cabernet, Emily Cabernets.

summary Burst on the scene with some superlative Cabernet Sauvignons early on, but did not manage to maintain the momentum; and indeed made no red wine in 1988, 1989 or 1991. The 1992 red wines marked a return to form, but no recent tastings.

woodonga hill NR

Cowra Road, Young, NSW 2594 **region** Hilltops
ph (02) 6382 2972 **fax** (02) 6382 2972 **open** 7 days 9–5
winemaker Jill Lindsay **prod.** 4000 **est.** 1986
prod. range ($12.50–21 CD) Dry Rhine Riesling, Sauvignon, Chardonnay, Botrytis Semillon, Auslese Gewurztraminer, Meunier, Shiraz, Vintage Port, Cherry Liqueur Port.
summary Early problems with white wine quality appear to have been surmounted. The majority of the wines on release in 1998 had won bronze or silver medals at regional wine shows in NSW and Canberra, and Jill Lindsay is also a successful contract-winemaker for other small producers.

wood park NR

RMB 1139 Bobinawarrah–Whorouly Road, Milawa, Vic 3678 **region** King Valley
ph (03) 5727 3367 **fax** (03) 5727 3682 **open** By appointment
winemaker John Stokes, Rick Kinzbrunner **prod.** 1800 **est.** 1989
prod. range ($15–30 CD) Meadow Creek Chardonnay, Shiraz Cabernet.
summary The first vines were planted at Wood Park in 1989 by John Stokes as part of a diversification programme for his property at Bobinawarrah in the hills of the Lower King Valley to the east of Milawa. The bulk of the 8-hectare production is sold to Brown Brothers, with a further 8 hectares of vineyard being established for Southcorp. In an unusual twist, Stokes acquires his chardonnay from cousin John Leviny, one of the King Valley pioneers with his vineyard at Meadow Creek. To complicate matters further, all four vintages of Chardonnay ('95 to '98) were made by Rick Kinzbrunner. 1995 Cabernet Shiraz won a silver medal at the Victorian Wines Show in 1996, the '96 and '97 Shiraz following suit in 1997.

woodstock ★★★★☆

Douglas Gully Road, McLaren Flat, SA 5171 **region** McLaren Vale
ph (08) 8383 0156 **fax** (08) 8383 0437 **open** Mon–Fri 9–5, weekends, holidays 12–5
winemaker Scott Collett **prod.** 15 000 **est.** 1974
prod. range ($12–30 CD) Riesling, Semillon, Chardonnay, Douglas Gully Semillon, Semillon Sauvignon Blanc, Botrytis Sweet White, Grenache, Cabernet Sauvignon, Shiraz, Vintage Port, Tawny Port and Muscat. The Stocks Shiraz is a recently introduced flagship.
summary One of the stalwarts of McLaren Vale, producing archetypal, invariably reliable, full-bodied red wines and showing versatility with spectacular botrytis sweet whites and high-quality (14-year-old) Tawny Port. Also offers a totally charming reception-cum-restaurant which understandably does a roaring trade with wedding receptions. It has also branched out by establishing 22.5 hectares of vineyards in the Limestone Coast Zone, supplementing its 15.7 hectares in McLaren Vale. The wines are exported to the UK, Canada, New Zealand, Philippines and Taiwan.

Woodstock The Stocks Shiraz

First made in the 1991 vintage, and released in late 1994. It is made from century-old vines, and matured in new American oak hogsheads. The subsequent vintages continue the line: very ripe, concentrated fruit with positive use of new American oak.

🍷🍷🍷🍷🍷 **1996** Medium to full red-purple; concentrated sweet berry fruit is woven through equally sweet, vanilla-accented American oak. The palate is wonderful, with luscious dark cherry/berry fruit, appropriate oak, and lingering tannins. The best for years, a tribute to the '96 vintage and to its ability to soak up oak. **rating:** 95

best drinking 2001–2011 **best vintages** '91, '94, '95, '96 **drink with** Barbecued steak • $30

Woodstock Cabernet Sauvignon

Produced from a little over 6.5 hectares of estate plantings. The vines are old, and the fruit concentration substantial. Like the Shiraz, has been incredibly consistent over the years, and represents great value.

1997 Medium red-purple; the moderately intense bouquet is quite complex, but brings together cassis, earth and subtle oak in a convincing fashion. Attractive, ripe cassis berry fruit, supported by good tannins, drives the palate. **rating:** 90

best drinking 2002–2007 **best vintages** '82, '84, '91, '92, '94, '97 **drink with** Game pie • $20

woody nook ★★★★

Metricup Road, Busselton, WA 6280 **region** Margaret River
ph (08) 9755 7547 **fax** (08) 9755 7547 **open** 7 days 10–4.30
winemaker Neil Gallagher **prod.** 3000 **est.** 1982
prod. range ($14–22 CD) Chenin Blanc, Sauvignon Blanc, Classic Dry White, Late Picked Chenin Blanc, Late Harvest Semillon, Merlot, Cabernet Sauvignon; Gallagher's Choice Cabernet Sauvignon is top of the range.
summary This improbably named and not terribly fashionable winery has produced some truly excellent wines in recent years, with its Classic Dry White and Cabernet Sauvignon both starring at the 1999 *Winewise* Small Makers Competition, having put in a similar performance at prior *Winewise* competitions, and likewise at the Mount Barker Wine Show.

Woody Nook Gallagher's Choice Cabernet Sauvignon

As the name suggests, the pick of the vintage, and winemaker Neil Gallagher certainly got it right in 1993, for this wine is far superior to the standard Cabernet Sauvignon of the same year. The '95 was placed first in the Cabernet Class at the 1998 *Winewise* Small Makers Competition, and (amazingly) the 1996 performed precisely the same feat at the 1999 competition.

1996 Full purple-red; the bouquet is powerful, with that assertive gravelly Margaret River regional character, but the wine comes together marvellously well on the stylish, well-balanced palate. This is Cabernet at its masculine best. **rating:** 94

best drinking 2000–2007 **best vintages** '93, '95, '96 **drink with** Herbed rack of lamb • $22

wrights NR

Harmans South Road, Cowaramup, WA 6284 **region** Margaret River
ph (08) 9755 5314 **fax** (08) 9755 5459 **open** 7 days 10–4.30
winemaker Henry Wright **prod.** 2500 **est.** 1973
prod. range ($12–25 CD) Premium Estate (Semillon Riesling), Semillon, Hermitage, Henry Wright's Chardonnay White Port.
summary Continues to go about its business and selling its wines locally.

wyanga park ★★★

Baades Road, Lakes Entrance, Vic 3909 **region** Gippsland
ph (03) 5155 1508 **fax** (03) 5155 1443 **open** 7 days 9–5
winemaker Andrew Smith **prod.** 6000 **est.** 1970
prod. range ($12–22 CD) Riesling Traminer, Colombard, Estate Grown Chardonnay, Miriam's Fancy Chardonnay, Rosé, Boobialla (medium sweet white), Shiraz, Shiraz Cabernet Sauvignon, Fortifieds.
summary Offers a broad range of wines of diverse provenance directed at the tourist trade one of the Chardonnays and the Cabernet Sauvignon are estate-grown. Winery cruises up the north arm of the Gippsland Lake to Wyanga Park are scheduled four days a week throughout the entire year. The best of the current releases are, without question, the two Chardonnays.

Wyanga Park Estate Grown Chardonnay

There are two releases of Chardonnay, one estate-grown, the other Miriam's Fancy, which does not carry any estate or regional appellation claim.

ΥΥΥΥ **1998** Light to medium yellow-green; ripe, sweet peachy/buttery fruit has the odd green note running through it. The palate has plenty of fruit flavour, intensity and length; one of those wines in which it is difficult to tell whether oak has been used or not. **rating:** 86

best drinking 2001–2005 **best vintages** NA **drink with** Creamy pasta • $20

wyldcroft estates NR

98 Stanleys Road, Red Hill South, Vic 3937 **region** Mornington Peninsula

ph (03) 5989 2646 **fax** (03) 5989 2646 **open** Weekends and public holidays 10–5

winemaker Kevin McCarthy (Contract) **prod.** 800 **est.** 1987

prod. range ($18–21 CD) Chardonnay, Unwooded Chardonnay, Pinot Noir, Cabernet Sauvignon.

summary Richard Condon and Sharon Stone commenced planting Wyldcroft Estates in 1987, extending the plantings in 1993 and 1996 to the present total of just under 3 hectares, constructing a mudbrick winery and cellar door in 1995.

wyndham estate ★★★

Dalwood Road, Dalwood, NSW 2335 **region** Lower Hunter Valley

ph (02) 4938 3444 **fax** (02) 4938 3422 **open** Mon–Fri 9.30–5, weekends 10–4

winemaker Robert Paul **prod.** NFP **est.** 1828

prod. range ($6.95–24.95 R) In ascending order: Bin TR2 Classic White and Classic Red; Chablis Superior Semillon Sauvignon Blanc, Bin 777 Semillon Chardonnay; Bin 222 Chardonnay, Bin 111 Verdelho, Bin 333 Pinot Noir, Bin 555 Shiraz; Oak Cask Chardonnay, Bin 444 Cabernet Sauvignon, Bin 888 Cabernet Merlot; Show Reserve Semillon and Shiraz.

summary An absolutely reliable producer of keenly priced mid-range table wines which are smoothly and precisely aimed at those who enjoy wine but don't wish to become over-involved in its mystery and intrigue. Every now and then it comes up with a wine of surprising quality, although there does seem to be some variation between different batch bottlings.

wynns coonawarra estate ★★★★★

Memorial Drive, Coonawarra, SA 5263 **region** Coonawarra

ph (08) 8736 3266 **fax** (08) 8736 3202 **open** 7 days 10–5

winemaker Sue Hodder **prod.** NFP **est.** 1891

prod. range ($8–80 R) Riesling, Chardonnay, Shiraz, Cabernet Shiraz Merlot, Black Label Cabernet Sauvignon, Michael Shiraz, John Riddoch Cabernet Sauvignon; also Ovens Valley Shiraz (not sourced from Coonawarra).

summary The large-scale production has in no way prevented Wynns from producing excellent wines covering the full price spectrum from the bargain basement Riesling and Shiraz through to the deluxe John Riddoch Cabernet Sauvignon and the more-recently introduced Michael Shiraz. Even with recent price increases, Wynns offers extraordinary value for money.

Wynns Coonawarra Estate Riesling

Arguably the best value Riesling in the country – a label revamp has slightly lifted the cachet of the wine, however much I personally disapprove of the new label. Extensive vertical tastings of the wines show that the better vintages can live for 20 years or more, becoming great classics in the course of so doing.

ΥΥΥΥ **1998** Light to medium yellow-green; an extremely correct, light, crisp lime and passionfruit-accented bouquet leads into an elegant, fine and correct palate. Note that there is some significant bottle variation with this wine. **rating:** 89

best drinking 2001–2010 **best vintages** '90, '91, '93, '95, '96 **drink with** Tiger prawns • $11

Wynns Coonawarra Estate Chardonnay

A wine which has evolved dramatically over the years. 1985 was the first vintage to utilise barrel fermentation; in 1992 winemaking was moved back to Coonawarra from McLaren Vale, and French oak was introduced (previously German and American was used). By 1995 only French oak was being employed, and in 1996 the barrel size had started to change from puncheon to barrique, with tight-grained French oak. All of these changes have progressively tightened and refined a wine which deserves greater recognition from critics and consumers – it has, in fact, done very well in wine shows in recent years.

🍷🍷🍷🍷🍷 **1998** Medium yellow-green; the bouquet is an exercise in restrained complexity, with subtle cashew, creamy overtones to the integrated fruit and oak. The palate, likewise, is at once complex yet restrained, with the flavours welded together, and good acidity on the finish. A poor man's Giaconda. **rating:** 91

best drinking 1999–2004 **best vintages** '92, '93, '94, '96, '97, '98 **drink with** Fillet of pork • $14

Wynns Coonawarra Estate Shiraz

Estate means what it says; this is 100% estate-grown Coonawarra Shiraz, a wine which vies with Penfolds Koonunga Hill dry red for the title of best-value red wine in Australia. A vertical tasting in March 1997 climaxed with the magnificent '53, '54 and '55 vintages. While the new generation wines may not last for 40 years, those made in the 1990s are the best since the 1950s, and the recommended drinking range should be regarded as strictly nominal. Well-corked and well-cellared, the wines have almost indefinite life.

🍷🍷🍷🍷 **1997** Medium red-purple, but showing a little more development than one would expect. The bouquet has a distinct touch of cinnamon spice, together with subtle oak. The palate is pleasant, with gently spicy fruit, but lacks the concentration of the '96 and '95 vintage wines. **rating:** 86

best drinking 2000–2005 **best vintages** '54, '55, '62, '65, '70, '85, '86, '89, '90, '91, '92, '93, 94, '95, '96 **drink with** Spiced lamb • $16

Wynns Coonawarra Estate Michael Shiraz

First made in the outstanding Coonawarra vintage of 1990, to stand alongside the John Riddoch Cabernet Sauvignon. It takes its name from the most famous of all of the Wynns wines, the glorious 1955 Michael Hermitage, which still rates as one of the top half-dozen wines made in Australia since the Second World War. A prolific trophy and gold medal winner, every bit as powerful as the Riddoch. No '95 was made or released.

🍷🍷🍷🍷🍷 **1996** Dense purple-red; the bouquet is typically super-concentrated, oozing ripe black cherry fruit and heaps of oak. The palate is at once profound, yet smooth, almost glossy, with black cherry fruit and cedar/vanilla oak. The tannins are ample but round and soft. **rating:** 95

best drinking 2006–2016 **best vintages** '90, '91, '93, '94, '96 **drink with** Leave it in the cellar • $80

Wynns Coonawarra Estate Cabernet Shiraz Merlot

A blend of Cabernet and Shiraz, with a little Merlot added since the end of the 1980s. It is aged in a mixture of new and used American and French barrels for 15 months, and the aim is to produce a more elegant and slightly lighter style of red wine than that offered by the other red wines under the Wynns label.

🍷🍷🍷🍷 **1996** Medium to full red-purple; the bouquet is moderately rich and ripe, particularly in the usual context of this wine; dark berry, chocolate and faintly earthy fruit is supported by a hint of oak. The palate is well-balanced, with pleasant black cherry and mulberry fruit; once again, the oak is subtle. **rating:** 87

best drinking 1999–2004 **best vintages** '86, '88, '90, '91, '92, '96 **drink with** Yearling beef • $16

Wynns Coonawarra Estate Black Label Cabernet Sauvignon

Given the volume in which this wine is made (said to be over 40 000 cases) it has to be the most important Cabernet in Australia, a powerful testament to the synergy between Coonawarra and Cabernet Sauvignon. Another dyed-in-the-wool classic with a magnificent history.

🍷🍷🍷🍷🍷 **1996** Medium to full red-purple; the bouquet is extremely attractive, with an array of cassis, blackberry and mint fruit characters, the palate elegant and stylish. Instead of succumbing to the temptation of extracting every bit of character from the wine, it has been allowed to express itself in an unforced fashion. **rating:** 94

best drinking 2001–2011 **best vintages** '53, '57, '58, '62, '82, '86, '88, '90, '91, '94, '95, '96 **drink with** Roast beef • $22

Wynns Coonawarra Estate John Riddoch Cabernet Sauvignon

First made in 1982, and only vintaged in the best years, and then from the finest material available. It is matured in new French and American oak hogsheads for 12–15 months, and is a wine of enormous concentration and power. Vertical tastings over the past few years have shown that at somewhere about seven years of age the wine typically undergoes a remarkable transformation, opening up almost overnight. We are still to see how long the ensuing plateau will last, for none of the wines back to 1982 have started to decline. Not made in 1995.

🍷🍷🍷🍷🍷 **1996** Dark, deep red-purple; the bouquet is as concentrated and rich as one would expect, with a mix of blackberry and blackcurrant fruit; ample but not excessive oak. The palate is immensely powerful and concentrated, precisely as one would expect from Riddoch in a vintage such as '96, with a full panoply of dark, black fruit flavours, tannin and oak. **rating:** 96

best drinking 2006–2016 **best vintages** '82, '85, '86, '88, '90, '91, '94, '96 **drink with** Leave it in the cellar • $80

xanadu wines ★★★★

Terry Road, Margaret River, WA 6285 **region** Margaret River
ph (08) 9757 2581 **fax** (08) 9757 3389 **open** 7 days 10–5
winemaker Jürg Muggli **prod.** 16 000 **est.** 1977
prod. range ($13.50–165 R) Semillon, Chenin Blanc, Chardonnay, Unwooded Chardonnay, Secession (Semillon Sauvignon Blanc Chenin Blanc), Semillon, Late Harvest Riesling, Featherwhite (Rosé), Shiraz, Merlot, Cabernet Franc, Cabernet Sauvignon, Cabernet Reserve.
summary Samuel Taylor Coleridge would thoroughly approve of the labels on the Xanadu wines, and one imagines would be equally pleased with wine quality – quality which can be excitingly variable, but is more often good than not.

Xanadu Semillon

Yet another wine from Margaret River which emphasises first what a great region for the variety the Margaret River is, and secondly, how different the style is from that of the Hunter River, and in particular, how well the wine lends itself to sophisticated handling techniques (including the use of oak). This wine is whole-bunch pressed, is taken through a full malolactic fermentation, and then barrel-aged – techniques one normally associates with Chardonnay. Yet it is in no sense overworked or overblown. Produced from 20-year-old vines.

🍷🍷🍷🍷🍷 **1998** Medium yellow-green; a very clean, powerful and smooth bouquet with herb and honey aromas is followed by a very long palate and clean finish with excellent acidity. The oak influence throughout is perfectly judged in its restraint. **rating:** 93

best drinking 1999–2005 **best vintages** '97, '98 **drink with** Richer fish dishes • $20

Xanadu Secession

A very typical Margaret River blend of Semillon, Sauvignon Blanc and Chenin Blanc, more commonly called 'classic white'; the Secession title is pure Xanadu, or rather Xanadu and Lenton Brae.

🍷🍷🍷🍷 **1998** Light green-yellow; the bouquet is crisp and clean, with quite tangy citrus fruit, and also some more tropical notes. The palate is light to medium-bodied, crisp and lively, with pleasant citrus and mineral flavours. **rating:** 86

best drinking 1999–2001 **best vintages** '85, '87, '90, '92, '93 **drink with** Marron • $17

Xanadu Chardonnay

A stylish, restrained style using many of the same techniques employed by Xanadu to produce its Semillon. The wine is whole-bunch pressed, barrel-fermented and given lees contact, and gives the impression of having at least some malolactic fermentation, although none is mentioned in the background material furnished by the winery.

🍷🍷🍷🍷 **1997** Light to medium green-yellow; the bouquet is bright and fresh, with faintly citrussy fruit and minimal oak influence. The palate, too, is surprisingly light (for the Margaret River region) and crisp, repeating those citrussy notes. The lack of power and complexity may be seen by some as a welcome relief. **rating:** 88

best drinking 2000–2004 **best vintages** NA **drink with** White-fleshed fish • $29.50

yaldara wines ★★★

Gomersal Road, Lyndoch, SA 5351 **region** Barossa Valley
ph (08) 8524 4200 **fax** (08) 8524 4678 **open** 7 days 9–5
winemaker Robert Thumm, Jim Irvine **prod.** 650 000 **est.** 1947

prod. range ($4–90 R) A kaleidoscopic array of wines under (in ascending order) the Ducks Flat, Lyndoch Valley, Acacia Hill, Lakewood, Julians, and the super-premium The Farms labels. The Lakewood range is the largest, covering all major wine styles and varietals. There is also a substantial range of sparkling, non-alcoholic and fortified wines.

summary The second-largest family-owned winery in the Barossa Valley, with an ornate Chateau-like building housing a collection of porcelain and art; there is also a motor inn and restaurant, and a reception and convention centre set among lakes and tree-lined creeks. Has 75 hectares of established vineyards, and has for long been an active exporter to all of the major markets.

yalumba ★★★★☆

Eden Valley Road, Angaston, SA 5353 **region** Barossa Valley
ph (08) 8561 3200 **fax** (08) 8561 3393 **open** Mon–Fri 8.30–5, Sat 10–5, Sun 12–5
winemaker Simon Adams **prod.** 55 000 **est.** 1849

prod. range ($7–55 R) Under the Yalumba label (in ascending order) Oxford Landing range, Galway Hermitage, and Christobels Dry White, Family Selection range, The Menzies Cabernet Sauvignon, The Signature Collection and Octavius Shiraz. Separate brand identities for Hill-Smith Estate, Pewsey Vale and Heggies, with strong emphasis on key varietals Riesling, Chardonnay, Semillon and Cabernet Sauvignon. Angas Brut is a leader in the sparkling wine market, with Yalumba D at the top end of the quality tree.

summary Family-owned and run by Robert Hill-Smith; much of its prosperity in the late 1980s and early 1990s turned on the great success of Angas Brut in export markets, but the company has always had a commitment to quality and shown great vision in its selection of vineyard sites and brands. In particular, it has always been a serious player at the top end of full-bodied (and full-blooded) Australian reds. Exports to all major markets.

Yalumba Barossa Growers Semillon

Made in uncompromising Barossa-style from fruit grown by the Hahn, Rozenzweig, Koch, Grossman and Johns families. The inaugural release won a gold medal at the 1997 Barossa Valley

Wine Show, underlining how typical (and Teutonic) the style is, although it was also selected in the Top 100 1998 Sydney International Wine Competition. The '98 emphasises the inherent quality of the grapes.

🍷🍷🍷🍷🍷 **1998** Light to medium yellow-green; the bouquet has excellent varietal character and intensity, with lemon and herb aromas leading into a positively flavoured palate which has plenty of fruit and flesh, particularly for a young Semillon. **rating:** 90

best drinking 2000–2005 **best vintages** '97, '98 **drink with** Smoked chicken • $14.95

Yalumba Antipodean

A strikingly packaged blend of Semillon, Sauvignon Blanc and Viognier coming from Yalumba's network of vineyards in the Adelaide Hills, and fashioned without the use of oak.

🍷🍷🍷🍷 **1998** Light yellow-green; the aromas are fresh and attractive, running through faintly grassy to more tropical passionfruit; on the palate the Sauvignon Blanc component comes to the fore on a lively and fresh wine. **rating:** 89

best drinking 1999–2000 **best vintages** '96 **drink with** Blue-lipped mussels • $16

Yalumba D

Originally deliberately and consistently made at the fuller end of the Australian sparkling wine spectrum. A complex, rich, mouthfilling-style, although the levels of aldehyde in the older vintages were somewhat controversial. Was typically a blend of Pinot Noir from Eden Valley, Coonawarra, Adelaide Hills; Chardonnay from Coonawarra and Eden Valley; and Pinot Meunier from Eden Valley.

🍷🍷🍷🍷🍷 **1996** Light to medium yellow-green; clean, with very lively, fresh citrussy fruit aromas quite different from the normal D style. Citrus and ripe apple flavours on the palate mirror the bouquet continuing the impression of a style change, possibly due to the sourcing of some base material from Tasmania. **rating:** 90

best drinking 1999–2002 **best vintages** '90, '91, '93, '95, '96 **drink with** Richer seafood dishes • $31

Yalumba Octavius Shiraz

Octavius is the super-premium Yalumba red. The first vintage was 1988 (Coonawarra Cabernet) but the two subsequent vintages have been of old Barossa Shiraz, and this is where the future of the wine will lie. The distinguishing feature of the wine is its opulent oak treatment: it is matured in barrels made at Yalumba's own cooperage from American oak from Missouri which was seasoned for eight years before being made into barrels. More is not always best, but in this instance it is. Incidentally, the barrels are unusually small (octaves) which increases the impact of the oak.

🍷🍷🍷🍷🍷 **1995** VI. Medium red-purple; the bouquet is quite fragrant, but, as always, cedary oak is very evident, with fruit of medium intensity. The palate does have red berry, blackberry and raspberry fruit, plus oodles of high-quality American oak. **rating:** 90

best drinking 2005–2015 **best vintages** '88, '90, '92, '93, '95 **drink with** The biggest steak imaginable • $55

Yalumba Bush Vine Grenache

Produced from 70-year-old vines grown by the Anderson, Burgmeister, Habermann and Wachter families. Previous releases have tended to be jammy yet dilute, which may seem to be a contradiction in terms; perhaps I am suggesting they lack structure. Whatever, the '97 seems to me to be a vast improvement.

🍷🍷🍷🍷 **1997** Medium to full red-purple; a full, ripe and concentrated bouquet with distinct juicy berry varietal character, but not jammy. The palate provides more of the same, with good juicy berry flavour running through to a softly tannic finish. **rating:** 88

best drinking 1999–2003 **best vintages** NA **drink with** Ravioli • $16

Yalumba Signature Cabernet Shiraz

An Australian Classic, dating back to 1962, but deriving from Sir Robert Menzies' declaration at a lunch in Adelaide that the '61 Special Vintage Galway Claret was 'the finest Australian wine I have ever tasted'. A blend of 65% Coonawarra Cabernet Sauvignon and 35% old-vine Barossa Shiraz, it spends 24 months in American oak prior to bottling, but never seems to be overwhelmed by it.

🍷🍷🍷🍷 **1995** Medium red-purple; a fragrant bouquet with dark berry and plum fruit married with scented oak; the flavours of the palate run through berry, plum, cedar and earth, with the sweet oak evident but not overdone. **rating:** 87

best drinking 2000–2007 **best vintages** '62, '66, '75, '81, '85, '88, '90, '91, '92, '93 **drink with** Rare roast beef • $30

Yalumba The Menzies Cabernet Sauvignon

The Menzies is a 20-hectare vineyard established by Yalumba in Coonawarra, planted on terra rossa soil within the heart of the Coonawarra region. Its name derives from the particular affection Sir Robert Menzies had for the wines of Yalumba.

🍷🍷🍷🍷 **1996** Medium red-purple; the lively bouquet features blackcurrant, spice and leaf aromas with minimal oak input. The palate opens with firm redcurrant and cassis fruit flavours followed by no less firm tannins. A wine in its infancy with considerable development potential. **rating:** 88

best drinking 2003–2010 **best vintages** '90, '91, '94, '96 **drink with** Topside steak • $20

yanwirra ★★★

Redman Road, Denmark, WA 6333 **region** Great Southern

ph (08) 9386 3577 **fax** (08) 9386 3578 **open** Not

winemaker John Wade (Contract) **prod.** 500 **est.** 1989

prod. range ($12.50–19.50 CD) Riesling, Sauvignon Blanc, Semillon Sauvignon Blanc, Cabernet Merlot.

summary Perth anaesthetist Ian McGlew and wife Liz have a liquorice allsorts 4-hectare vineyard, with contract-winemaking by John Wade. The white wines are particularly modestly priced, being sold by word of mouth and mail order.

yarrabank ★★★★★

42 Melba Highway, Yarra Glen, Vic 3775 **region** Yarra Valley

ph (03) 9730 2188 **fax** (03) 9730 2189 **open** 7 days 10–5

winemaker Claude Thibaut, Tom Carson, Darren Rathbone **prod.** 2000 **est.** 1993

prod. range ($30 R) Thibaut & Gillet Cuvée Brut, Cuvée Rosée

summary The 1997 vintage saw the opening of the majestic new winery established as part of a joint venture between the French Champagne House Devaux and Yering Station, and which adds another major dimension to the Yarra Valley. Until 1997 the Yarrabank Cuvée Brut was made under Claude Thibaut's direction at Domaine Chandon, but henceforth the entire operation will be conducted at Yarrabank. Four hectares of dedicated 'estate' vineyards have been established at Yering Station; the balance of the intake comes from other growers in the Yarra Valley and southern Victoria. Wine quality has been quite outstanding, the wines having a delicacy unmatched by any other Australian sparkling wines.

Yarrabank Brut Cuvée

The first release of Yarrabank is a blend of 50% Pinot Noir and 50% Chardonnay, half of each variety coming, respectively, from the Yarra Valley and Mornington Peninsula. It spent three years on yeast lees prior to disgorgement. The rating of the '93 previously appearing was erroneous; it should have been five glasses (or 94 points), the same as for the 1994 (and 1995) vintages.

1995 Light straw-yellow; a fine, delicate, crisply elegant bouquet is followed by a palate as fine and elegant as the bouquet promises. A feature of the wine is the way the components of aroma, flavour and texture are seamlessly welded together. Long carry and finish; be patient. **rating:** 94

best drinking 2002–2007 **best vintages** '93, '94, '95 **drink with** Aperitif, shellfish • $30

yarra burn ★★★★

Settlement Road, Yarra Junction, Vic 3797 **region** Yarra Valley
ph (03) 5967 1428 **fax** (03) 5967 1146 **open** 7 days 10–5
winemaker Tom Newton, Ed Carr, Stephen Pannell **prod.** 4500 **est.** 1975
prod. range ($18–41 R) Sauvignon Blanc Semillon, Chardonnay, Pinot Noir, Shiraz, Cabernet Sauvignon, Sparkling Pinot, Chardonnay Pinot; Bastard Hill Chardonnay, Bastard Hill Pinot Noir.
summary Acquired by BRL Hardy in 1995, and destined to become the headquarters of Hardy's very substantial Yarra Valley operations, the latter centring on the 1000-tonne production from its Hoddles Creek vineyards. The new brand direction is slowly taking shape, though not helped by the very difficult 1995 and 1996 Yarra Valley vintages.

Yarra Burn Sauvignon Blanc Semillon

Since 1995, a well-put-together blend of Yarra Valley and King Valley fruit; comparisons with the Sauvignon Blanc of Yarra Ridge are inevitable, and Yarra Burn comes off better in that comparison.

1998 Light green-yellow; the bouquet is clean and fresh, but does not display particularly strong varietal fruit character. The palate is fresh and crisp, with some mineral and herb flavours, finishing with good acidity. **rating:** 84

best drinking 1999–2000 **best vintages** NA **drink with** Light seafood • $18

Yarra Burn Bastard Hill Chardonnay

The Bastard Hill duo of Chardonnay and Pinot Noir were launched in November 1996. The provocative – and evocative – name comes from the breathtakingly steep hillsides of the Hoddles Creek vineyard acquired by Hardys in 1993. According to the vineyard workers, the slopes are not only a bastard to walk up, but also a bastard to walk down. The '95 was a gold medal winner at the 1996 Sydney Wine Show.

1995 Glowing yellow-green; the bouquet offers very particular intense, lime/citrus/grapefruit aromas which are undeniably attractive, but the palate seems out of balance due to the high acidity which runs through to the finish. **rating:** 87

best drinking 2000–2004 **best vintages** '95 **drink with** Richly sauced fish • $41

Yarra Burn Chardonnay Pinot Noir

Sourced from Hardys Hoddles Creek vineyard, high in the upper Yarra Valley. The percentage of Chardonnay and Pinot varies from vintage to vintage; sometimes the Chardonnay is a little over 50%, in other years it can be the other way around. Part of the wine goes through malolactic fermentation, and spends 24 months on yeast lees.

1996 Light green-yellow; a complex, multilayered bread, citrus and faintly aldehydic aromas lead into a palate in which sweet strawberry and citrus fruit dominates on entry in quite a striking fashion, followed by more creamy/bready notes on the finish. **rating:** 90

best drinking 2000–2004 **best vintages** NA **drink with** Salmon caviar • $18

Yarra Burn Pinot Noir

After two particularly troublesome vintages in 1995 and 1996, the Upper Yarra Valley came into its own in 1997, and again in 1998. Produced from grapes grown in BRL Hardy's very extensive

Upper Yarra Valley vineyards. The '97 Yarra Burn and Seppelt Sunday Creek were regarded as the best value Pinot Noirs on the Australian market.

🍷🍷🍷🍷 **1998** Medium to full purple-red; rich plummy fruit is supported by charry oak on the bouquet, with a mix of plummy and more minty fruit on the generously-sized palate. Perhaps the finish is just the faintest bit hard. **rating:** 87

best drinking 2000–2004 **best vintages** '97 **drink with** Ragout of venison • $19.80

Yarra Burn Bastard Hill Pinot Noir

The Bastard Hill duo of Chardonnay and Pinot Noir were launched in November 1996. Much of the production goes to sparkling wine, but in excellent vintages such as 1994 and 1997, limited quantities of high-quality table wine are produced. Five hundred cases made.

🍷🍷🍷🍷 **1997** Dense red-purple; a very complex bouquet with outstandingly rich plum and briar fruits augmented by clever oak. The palate is hugely powerful and, to my taste at least, distressingly tannic. There are so many good things in the wine I don't understand why so much tannin was left in it. **rating:** 89

best drinking 2007–2012 **best vintages** '94, '97 **drink with** Game pie • $41

yarra edge ★★★☆

PO Box 390, Yarra Glen, Vic 3775 **region** Yarra Valley
ph (03) 9730 1107 **fax** (03) 9739 0135 **open** At Yering Station
winemaker Tom Carson, Darren Rathbone **prod.** 2000 **est.** 1984
prod. range ($27.50–28.50 CD) Chardonnay, Cabernets.

summary Now leased to Yering Station, which makes the wines but continues to use the Yarra Edge brand for grapes from this estate. Tom Carson, the Yering Station winemaker, was briefly winemaker/manager at Yarra Edge, and knows the property intimately, so the rich style can be expected to continue.

Yarra Edge Chardonnay

Made entirely from estate-grown fruit, and now produced at Yering Station. The style is usually rich and concentrated.

🍷🍷🍷🍷½ **1997** Light to medium yellow-green; the bouquet is full, with solid peach and melon fruit supported by subtle oak. The palate is sustained by balanced acidity on the finish providing a framework for that ever-so-typical Yarra Valley melon fruit flavour. **rating:** 90

best drinking 2002–2007 **best vintages** '92, '93, '94, '97 **drink with** Smoked salmon pasta • $27.50

yarra ridge ★★★★

Glenview Road, Yarra Glen, Vic 3755 **region** Yarra Valley
ph (03) 9730 1022 **fax** (03) 9730 1131 **open** 7 days 10–5
winemaker Rob Dolan **prod.** 40 000 **est.** 1983
prod. range ($18–40 R) Chardonnay, Sauvignon Blanc, Botrytis Semillon, Pinot Noir, Reserve Pinot Noir, Merlot, Shiraz, Cabernet Sauvignon.

summary Now under the sole ownership and control of Mildara Blass, but with the ever-affable Rob Dolan continuing to work winemaking and production miracles at a winery which is strained to its limits. Recent vineyard plantings in the Yarra Valley, and continued purchasing of Yarra Valley grapes, means that the majority of the wines will continue to be Yarra Valley sourced. Sometimes it is not easy to tell which are, and which aren't, even if one has a master's degree in label-reading and interpretation.

Yarra Ridge Sauvignon Blanc

This is the wine which launched the Yarra Ridge ship, but both time and volume of production seem to have caught up with it. The back label now tells us that it is 'grown in the renowned Yarra Valley with small parcels of Sauvignon Blanc selected from other cooler regions of Victoria'. I don't think the author means regions which are cooler than the Yarra Valley, and does mean to include the King Valley.

🍷🍷🍷🍷 **1998** Light green-yellow; the bouquet is clean, of moderate intensity with no obvious varietal character. The palate is well-balanced, with faint mineral and herb flavours, but is essentially a captive of the year and perhaps the regions and crop levels. **rating:** 84

best drinking 1999–2000 **best vintages** NA **drink with** Fish and chips • $20

Yarra Ridge Reserve Pinot Noir

A genuine reserve wine, made in limited quantities and built from the time the grapes are picked. Hit new heights of opulence in the 1996 vintage, although the style (and the degree of oak) is bound to polarise opinions. The '97 is a worthy follow-on wine from a great pinot vintage.

🍷🍷🍷🍷🍷 **1997** Medium to full red-purple; the bouquet is powerful and concentrated with aromas of plum and earth, the palate rich, powerful and concentrated, with massive fruit extract (cherry and plum) together with masses of tannins and oak. For some it will be too ripe and too aggressive, others will enjoy its full-frontal approach. Regardless, time should help. Gold medal winner 1998 National Wine Show. **rating:** 92

best drinking 2000–2005 **best vintages** '94, '96, '97 **drink with** Game, jugged hare • $39

Yarra Ridge Merlot

Many observers believe that merlot (rather than cabernet sauvignon) will be the Bordeaux grape most suited to many of the newer sites in the Yarra Valley, and certainly to the less-warm sites. The '95 won a gold medal and trophy at the Adelaide Wine Show in 1996, giving credence to that view.

🍷🍷🍷🍷 **1997** Medium red-purple; the bouquet offers cedary, foresty notes with the fruit aromas seemingly in decline, but the palate is far more attractive with sweet fruit, hints of spice and the appropriate structure and gentle tannins. **rating:** 88

best drinking 1999–2003 **best vintages** '95, '97 **drink with** Rack of veal • $20.50

yarra valley hills ★★★★☆

Delaneys Road, Warranwood, Vic 3134 **region** Yarra Valley

ph (03) 5962 4173 **fax** (03) 5962 4059 **open** Weekends, public holidays 11–5

winemaker Martin Williams (Consultant) **prod.** 15 000 **est.** 1989

prod. range ($16–27 CD) Warranwood Riesling, Log Creek Sauvignon Blanc, Kiah Yallambee Chardonnay, Log Creek Pinot Noir, Log Creek Cabernet Sauvignon.

summary Former schoolteacher Terry Hill has built-up a very successful empire in a short period of time through leasing two substantial vineyards and principally acting as a grape supplier to others, with a small proportion of the grapes being contract-made by a range of Yarra Valley winemakers. No 1997 vintage tastings.

Yarra Valley Hills Warranwood Riesling

In most previous vintages, Yarra Valley Hills Chardonnays and Pinot Noirs have been the best in the portfolio. In 1998 they were quite disappointing, with odd fermentation characters, but the Riesling came into its own.

🍷🍷🍷🍷 **1998** Light yellow-green; there is a dusting of cinnamon spice over more conventional, if light, riesling varietal fruit, perhaps coming from a touch of Gewurztraminer. The palate is pleasant, faintly spicy with good balance and length. Not a heavyweight, but has flavour. **rating:** 89

best drinking 1999–2003 **best vintages** '98 **drink with** Sashimi • $17.95

yarra yarra ★★★★☆

239 Hunts Lane, Steels Creek, Vic 3775 **region** Yarra Valley
ph (03) 5965 2380 **fax** (03) 9830 4180 **open** By appointment
winemaker Ian Maclean **prod.** NFP **est.** 1979
prod. range ($25–40 CD) Semillon Sauvignon Blanc, Merlot, Cabernets.
summary Notwithstanding its tiny production, the wines of Yarra Yarra have found their way onto a veritable who's who listing of Melbourne's best restaurants. This has encouraged Ian Maclean to increase the estate plantings from 2 hectares to over 7 hectares during the 1996 and 1997 seasons. The demand for the wines will only be intensified by the quality of the current releases.

Yarra Yarra Merlot

Produced in very limited quantities; estate-grown.

🍷🍷🍷🍷 **1995** Medium red-purple; the moderately intense bouquet is clean and smooth, with hints of cedar, briary and mint. The palate opens solidly, with briar, mint and plum flavours, although perhaps needing a touch more back palate structure. **rating:** 87

best drinking 2000–2005 **best vintages** NA **drink with** Risotto • $40

---

yarra yering ★★★★★

Briarty Road, Coldstream, Vic 3770 **region** Yarra Valley
ph (03) 5964 9267 **fax** (03) 5964 9239 **open** Sat, public holidays 10–5, Sun 12–5
winemaker Bailey Carrodus **prod.** 6000 **est.** 1969
prod. range ($30–100 CD) Dry White No 1 (Sauvignon Blanc Semillon), Chardonnay, Pinot Noir, Dry Red No 1 (Bordeaux-blend), Dry Red No 2 (Rhône-blend), Merlot (tiny quantities at $100 a bottle), Underhill Shiraz, Portsorts.
summary Dr Bailey Carrodus makes extremely powerful, occasionally idiosyncratic wines from his 30-year-old, low-yielding unirrigated vineyards. Both red and white wines have an exceptional depth of flavour and richness, although my preference for what I believe to be his great red wines is well known.

Yarra Yering Pinot Noir

Made in a no-holds-barred style from low yielding, unirrigated vines. Long on power but not necessarily finesse.

🍷🍷🍷🍷½ **1997** Medium red-purple; sweet, spicy plummy fruit and well-handled oak on the bouquet introduce a wine with powerful, plummy fruit, which glides across the tongue; still building texture. **rating:** 90

best drinking 2002–2010 **best vintages** NA **drink with** Squab • $38

Yarra Yering Underhill Shiraz

Made entirely from the former Prigorje Vineyard, which adjoins that of Yarra Yering and is now, indeed, part of the Yarra Yering estate – and has been so for some years. Here, too, the vines are old, low-yielding and unirrigated.

🍷🍷🍷🍷🍷 **1997** Medium purple-red; attractive if slightly light spicy berry fruit is accompanied by a nice touch of charry oak on the bouquet. An elegant, supple palate with attractive spice and black cherry flavours; fine tannins run throughout the palate. **rating:** 94

best drinking 2002–2012 **best vintages** '91, '92, '93, '97 **drink with** Victorian parmesan cheese • $38

Yarra Yering Dry Red No 2

Predominantly Shiraz, with a little Viognier and a few scraps of other things from time to time. Entirely estate-grown, of course, and produced from vines which are now over 25 years old.

🍷🍷🍷🍷🍷 **1997** Medium to full purple-red; excellent, ripe liquorice and cherry varietal fruit is complexed by oak on the bouquet. As expected, dark berry, liquorice and plum fruit flavours make the palate what it is, although ripe tannins do ripple through from start to finish. **rating:** 94

best drinking 2006–2016 **best vintages** '80, '81, '86, '89, '90, '91, '92, '93, '96, '97 **drink with** Beef bourguignon • $38

Yarra Yering Dry Red No 1

Predominantly Cabernet Sauvignon, with small quantities of Merlot, Cabernet Franc and Malbec, and a tiny contribution of Petit Verdot. Entirely estate-grown from low-yielding, unirrigated vines, and matured in high-quality 100% new French oak. It is a sign of truly great wine produced anywhere in the world if it is matured in new oak without showing any obvious signs of having been so made.

🍷🍷🍷🍷🍷 **1997** Medium to full red-purple; red berry fruits, a touch of leaf and nicely judged, slightly charry oak make up the bouquet; the wine has the depth of structure needed for a wine built to live. Dark berry fruits and tannins run through the palate from start to finish. **rating:** 94

best drinking 2006–2016 **best vintages** '80, '81, '86, '89, '90, '91, '93, '94, '96, '97 **drink with** Roast leg of lamb • $38

Yarra Yering Portsorts

Only Bailey Carrodus could come up with a name such as this for his Vintage Port, made from the classic Portuguese varieties planted on a single, dedicated vineyard immediately abutting his original plantings. When I first came to the Yarra Valley and planted Coldstream Hills in 1985, Bailey suggested to me I should plant the port varieties. I thought he was joking. Now I know he was not.

🍷🍷🍷🍷🍷 **1997** Medium to full purple-red; a fragrant, complex bouquet with lifted spirit and spicy fruit is followed by a youthful palate with pleasantly dry, spicy characters, the spirit fractionally intrusive. **rating:** 90

best drinking 2000–2010 **best vintages** NA **drink with** As many friends as possible • $38

yass valley wines NR

9 Crisps Lane, Murrumbateman, NSW 2582 **region** Canberra District

ph (02) 6227 5592 **fax** (02) 6227 5592 **open** Weekends, public holidays 11–5 or by appointment

winemaker Michael Withers **prod.** 800 **est.** 1979

prod. range ($9–14 CD) Riesling, Traminer, Chardonnay, Chardonnay Semillon, Shiraz, Cabernet Sauvignon.

summary Michael Withers and Anne Hillier purchased Yass Valley in January 1991, and have subsequently rehabilitated the existing run-down vineyards and extended the plantings. Mick Withers is a chemist by profession and has completed a Wine Science degree at Charles Sturt University; Anne is a registered psychologist and has completed a Viticulture diploma at Charles Sturt. No recent tastings.

yaxley estate NR

31 Draws Field Road, Copping, Tas 7174 **region** Southern Tasmania

ph (03) 6253 5222 **fax** (03) 6253 5222 **open** 7 days 10–6.30

winemaker Andrew Hood (Contract) **prod.** 300 **est.** 1991

prod. range ($16.50–18 CD) Pinot Gris, Sauvignon Blanc, Chardonnay, Pinot Noir.

summary While Yaxley Estate was established back in 1991, it was not until 1998 that it offered each of the four wines from its vineyard plantings, which total 1.7 hectares. Once again, the small batch handling skills (and patience) of contract-winemaker Andrew Hood have made the venture possible.

Yaxley Estate Pinot Gris

Produced from half a hectare of estate plantings, and made in conventional style, fermented in stainless steel and early-bottled. The 1998 won a silver medal at the 1999 Tasmanian Wines Show.

ΨΨΨΨ **1998** Light green-yellow; the bouquet is clean and firm, with good fruit intensity and a touch of spice. The palate has both length and balance, with citrus blossom and honeysuckle flavours well in the mainstream of Pinot Gris, and offering much promise for the future.

rating: 89

best drinking 1999–2002 **best vintages** NA **drink with** Summer salads • $16.50

yellowglen ★★★★

Whites Road, Smythesdale, Vic 3551 **region** Ballarat
ph (03) 5342 8617 **fax** (03) 5333 7102 **open** Mon–Fri 10–5, weekends 11–5
winemaker Nick Walker **prod.** 100 000 **est.** 1975
prod. range ($10–25 R) Brut Cremant, Brut Pinot Chardonnay, Brut Rosé, Cuvée Victoria, Vintage Pinot Chardonnay, Y, Yellow.

summary Just as the overall quality of Australian sparkling wine has improved out of all recognition over the past five or so years, so has that of Yellowglen. Initially the quality lift was apparent at the top end of the range, but now extends right to the non-vintage commercial releases.

Yellowglen Vintage Cuvée Victoria

It would seem that Cuvée Victoria is regarded by Yellowglen as a proprietary brand name which no longer has any geographical base or meaning, for it is a blend of 55% Chardonnay and 45% Pinot Noir sourced not from Victoria, but from the Adelaide Hills and Eden Valley – or so the literature says. Whatever, it is usually a totally delicious wine. The 1998 release of the 1995 vintage featured snappy new packaging and a new bottle shape, carried through into the 1996.

ΨΨΨΨ **1996** Pale straw-green; there are interesting notes to the bouquet, initially of spice, followed by more bready, yeasty characters. The palate is crisp, clean, and quite long, with particularly well-balanced acidity and dosage.

rating: 87

best drinking 1999–2002 **best vintages** '90, '91, '92, '95, '96 **drink with** Aperitif • $25

yeringberg ★★★★★

Maroondah Highway, Coldstream, Vic 3770 **region** Yarra Valley
ph (03) 9739 1453 **fax** (03) 9739 0048 **open** By appointment
winemaker Guill de Pury **prod.** 1000 **est.** 1863
prod. range ($30–40 CD) Chardonnay, Yeringberg White (Marsanne/Roussanne), Pinot Noir, Yeringberg Red (Cabernet-blend).

summary Makes wines for the next millennium from the low-yielding vines re-established on the heart of what was one of the most famous (and infinitely larger) vineyards of the nineteenth century. In the riper years, the red wines have a velvety generosity of flavour which is rarely encountered, yet never lose varietal character, while the Yeringberg White takes students of history back to Yeringberg's fame in the nineteenth century. The wines are exported to the UK, US, Switzerland and Germany.

Yeringberg Marsanne/Roussanne

In fact simply labelled 'Yeringberg' (as is the dry red) in the tradition of the nineteenth-century label, which is faithfully reproduced. The wine is predominantly Marsanne, with just a touch of Roussanne, and in in the warmer years ('94, '97, '98) achieved a level of flavour and richness which one imagines the great wines of the nineteenth century possessed.

🍷🍷🍷🍷🍷 **1998** Light green-yellow; typically, a very clean bouquet with slightly honeyed overtones to the fruit salad base. Notwithstanding its freakish 15 degrees alcohol, the tangy flavours of the palate, ranging from fruit salad through to citrus, leave the mouth fresh. Very impressive. **rating:** 90

best drinking 2000–2010 **best vintages** '94, '97, '98 **drink with** Snowy Mountains trout • $30

Yeringberg Chardonnay

Produced from the half-hectare of estate plantings, and made in necessarily very limited quantities. This restricts its opportunity for show entries, but it has been a consistent trophy and gold medal winner at the Lilydale Wine Show.

🍷🍷🍷🍷🍷 **1998** Light to medium yellow-green; restrained citrus, melon and grapefruit aromas, with the usual subtle oak, lead into a classic, youthful Yarra Valley Chardonnay palate: melon fruit with a substrate of mineral giving it a crisp, punchy effect. The oak on the palate is less convincing, but that is of small moment; overall, the remarkable feature of the wine is its elegance, which belies its 14 degrees alcohol. **rating:** 91

best drinking 2000–2008 **best vintages** '88, '90, '91, '92, '93, '94, '97, '98 **drink with** Sweetbreads • $30

Yeringberg Pinot Noir

Estate-grown, and usually at the fuller end of the Pinot Noir spectrum, with the capacity to age better than most of its counterparts.

🍷🍷🍷🍷 **1997** Good, bright red-purple; the bouquet has a haunting overtone of cut grass or crushed nettle when first approached, with more conventional foresty/sappy/plum coming through as the wine sits in the glass. The palate is altogether more conventional than the bouquet, of medium weight, with nicely accented varietal plum and cherry fruit; the usual restrained use of oak. Would have been higher-pointed were it not for that slightly distracting bouquet. **rating:** 89

best drinking 1999–2005 **best vintages** '97 **drink with** Squab • $40

Yeringberg Dry Red

Produced from an estate-grown blend of Cabernet Sauvignon, Merlot, Cabernet Franc and Malbec (with Cabernet Sauvignon dominant, and the other components in descending order as listed). The vineyard is established on the precise site of the great nineteenth-century plantings, albeit but a fraction of the size of those vineyards, enjoying a prime north-facing slope.

🍷🍷🍷🍷🍷 **1997** Medium red; the bouquet opens quietly, with earthy/berry fruit and subtle oak. It is on the palate that the elegant blackcurrant/cassis fruit-driven palate comes to the fore, with a long, lingering finish woven through with fine tannins. The grapes were fully ripe, but the wine does not show its 14 degrees alcohol. The best Dry Red for some years, requiring time to flower. **rating:** 93

best drinking 2002–2012 **best vintages** '85, '86, '88, '90, '91, '93, '94, '97 **drink with** Yarra Valley venison • $40

yering farm NR

St Huberts Road, Yering, Vic 3770 **region** Yarra Valley

ph (03) 9735 4161 **fax** (03) 9735 4012 **open** Not

winemaker Alan Johns, Rob Dolan (Contract) **prod.** 850 **est.** 1989

prod. range ($18–25 ML) Chardonnay, Cabernet.

summary Alan and Louise Johns established their 12-hectare vineyard in 1989 on the site of the original Yeringa winery built by the Deschamps family in the last century. Between 1992 and 1998 the wines were made by Alan Johns, and since then by Rob Dolan at Yarra Ridge, which purchases most of the production from the vineyard.

yering grange vineyard NR

14 McIntyre Lane, Coldstream, Vic 3770 **region** Yarra Valley
ph (03) 9739 1172 **fax** (03) 9739 1172 **open** By appointment
winemaker John Ellis (Contract) **prod.** 300 **est.** 1989
prod. range ($15 CD) Cabernet Sauvignon.

summary Yering Grange has 2 hectares of cabernet sauvignon under vine, part being sold and part made under the Yering Grange label by John Ellis at Hanging Rock. The tiny production is sold through a mailing list.

yering station ★★★★☆

Melba Highway, Yering, Vic 3770 **region** Yarra Valley
ph (03) 9730 1107 **fax** (03) 9739 0135 **open** Thur–Sun 10–5
winemaker Tom Carson, Darren Rathbone, Paul Kaan, Dan Buckle **prod.** 25 000 **est.** 1988
prod. range ($12.50–42 CD) Baraks Bridge Chardonnay, Pinot Noir, Shiraz, Botrytis Semillon; Yering Station Sauvignon Blanc, Chardonnay, Pinot Noir Rosé ED, Pinot Noir, Cabernet Merlot; Reserve range of Chardonnay, Pinot Noir and Shiraz; also Verjuice in 375 ml bottles.

summary The historic Yering Station (or at least the portion of the property on which the cellar-door sales and vineyard are established) was purchased by the Rathbone family in January 1996, and is now the site of a joint venture with the French Champagne House Devaux. A spectacular and very large winery has been erected which handles the Yarrabank sparkling wines, and the Yering Station and Yarra Edge table wines. Has immediately become one of the focal points of the Yarra Valley, particularly with the historic Chateau Yering next door. Here, luxury accommodation and the finest dining in the Yarra Valley is available.

Yering Station Chardonnay

1996 marked the transition from the old regime at Yering Station to the new, following its acquisition in the middle of 1996 by the Rathbone family. Tom Carson was installed as winemaker and is responsible for taking the wine through to bottle. As from 1997, all winemaking has been and will be carried out in the large, state-of-the-art winery part-completed in time for the 1997 vintage.

🍷🍷🍷🍷🍷 **1997** Medium yellow-green; the bouquet is rich, with fig and nectarine fruit, a touch of cream from malolactic fermentation, and subtle spicy oak all contributing. On the palate, a fine, intense and elegant wine with nice grip (mineral and acid) to the finish. Will be slow-developing. **rating** 91

best drinking 2000–2005 **best vintages** NA **drink with** Poached Tasmanian salmon • $19.50

Yering Station Reserve Chardonnay

Produced in very limited quantities, and is in every sense a Reserve wine. A Chardonnay to cellar.

🍷🍷🍷🍷🍷 **1997** Light to medium yellow-green; a quite complex bouquet with creamy/nutty overtones attesting to some malolactic fermentation influence. The palate is in similar style, elegant, with nutty/creamy overtones, and well balanced. **rating:** 97

best drinking 2000–2005 **best vintages** NA **drink with** Pan-fried veal • $42

Yering Station Pinot Noir Rosé ED

The juice is given 12 to 18 hours maceration or skin contact, and then barrel-fermented, and left in oak for four months before bottling. Either way, it works well, and is the very model of a modern-day brasserie wine, catching the wave (how long will it last?) of enthusiasm for all things Rosé.

ŸŸŸŸ **1998** Slightly blackish aspects don't give immediate visual appeal, but the bouquet strikes back, with interesting earthy/strawberry notes which are quite Champagne-like. The palate, too, has some stylish, tangy characters and real bite to the finish. **rating:** 88

best drinking 1999–2000 **best vintages** NA **drink with** Seafood, Singapore noodles – you name it • $17.50

Yering Station Pinot Noir

The standard release of Yering Station, and in 1998 the only Pinot Noir made (no Reserve).

ŸŸŸŸ **1998** Medium red; the bouquet offers a range of spicy, earthy and more foresty aromas, the style of the wine coming through on a long, lingering sappy finish together with a faint hint of mint. **rating:** 86

best drinking 1999–2002 **best vintages** '91, '94, '96, '97 **drink with** Smoked quail • $21

Yering Station Reserve Pinot Noir

The newly-introduced Reserve wines from Yering Station fully deserve the title, and the substantially higher price they command. The '97 was a trophy winner at the 1998 Southern Victorian Wines Show.

ŸŸŸŸŸ **1997** Intense, dark purple-red; there is concentrated plum and briar fruit on the bouquet with some foresty underlay. An extremely powerful and concentrated wine on the palate with plummy fruit and reasonably substantial tannins. Will be very long-lived for a Pinot Noir. **rating:** 94

best drinking 2000–2007 **best vintages** '97 **drink with** Jugged hare • $42

Yering Station Reserve Shiraz

The initial vintage ('97) of Reserve Shiraz was made in minuscule quantities, but is in many ways the benchmark for Yarra Shiraz.

ŸŸŸŸŸ **1997** Medium red-purple; the bouquet is smooth, with red berries predominant, but spice also present. The palate shows perfectly ripe black cherry/berry fruit with a touch of liquorice, surrounded by soft tannins and subtle oak. **rating:** 94

best drinking 2001–2010 **best vintages** '97 **drink with** Venison • $42

yungarra estate NR

Yungarra Drive, Dunsborough, WA 6281 **region** Margaret River

ph (08) 9755 2153 **fax** (08) 9755 2310 **open** 7 days 10–5

winemaker Erland Happ (Contract) **prod.** 1450 **est.** 1988

prod. range ($10.50–12 CD) Semillon, Sauvignon Blanc, Quartet (Semillon, Sauvignon Blanc, Chenin Blanc, Verdelho), Chenin Blanc Verdelho, Pink Opal (sweet red table wine made from Cabernet and Merlot), Springtime (sweet Sauvignon Blanc, Verdelho), Cabernet Sauvignon, Cabernet Merlot, Royale.

summary Yungarra Estate is a combined tourist lodge and cellar-door facility set on a 40-hectare property overlooking Geographe Bay. The 9-hectare vineyard was first planted in 1988, producing its first wines in 1992, contract-made by Erland Happ. Cellar-door sales commenced in 1993, and there are five bed and breakfast cottages on the Yungarra Estate property.

zappacosta estate NR

Farm 161 Hanwood Road, Hanwood, NSW 2680 **region** Riverina

ph (02) 6963 0278 **fax** (02) 6963 0278 **open** By appointment

winemaker Judy Zappacosta, Dino Zappacosta **prod.** 22 000 **est.** 1996

prod. range ($12 CD) Riesling, Semillon, Dry White, Shiraz.

summary Zappacosta Estate, briefly known as Hanwood Village Wines, is a relatively new business, with the first release from the 1996 vintage.

zarephath wines NR

Moorialup Road, East Porongurup, WA 6324 **region** Great Southern
ph (08) 9853 1152 **fax** (08) 9841 8124 **open** 7 days 9–4
winemaker Brenden Smith **prod.** 300 **est.** 1994
prod. range ($15–20 CD) Riesling, Chardonnay, Pinot Noir, Shiraz, Cabernet Sauvignon.
summary The Zarephath vineyard is owned and operated by Brothers and Sisters of The Christ Circle, a Benedictine community. They say the most outstanding feature of the location is the feeling of peace and tranquility which permeates the site, something I can well believe on the basis of numerous visits to the Porongurups.

zema estate ★★★★☆

Riddoch Highway, Coonawarra, SA 5263 **region** Coonawarra
ph (08) 8736 3219 **fax** (08) 8736 3280 **open** 7 days 9–5
winemaker Matt Zema, Nick Zema **prod.** 10 000 **est.** 1982
prod. range ($17–40 CD) Shiraz, Cabernet Sauvignon, Family Selection Cabernet Sauvignon, Cluny (Cabernet-blend).
summary Zema is one of the last outposts of hand-pruning and hand-picking in Coonawarra, the various members of the Zema family tending a 40-hectare vineyard progressively planted between 1982 and 1994 in the heart of Coonawarra's terra rossa soil. Winemaking practices are straightforward; if ever there was an example of great wines being made in the vineyard, this is it.

Zema Estate Shiraz

Zema Estate's very conservative approach to viticulture has paid big dividends, with outstanding wines produced consistently over the past 15 years. Matured for 16 months in a mix of French and American oak, mostly used.

🍷🍷🍷🍷🍷 **1997** Medium to full red-purple; full, dark cherry and plum fruit aromas are followed by a powerful, concentrated, fruit-driven palate with sweet fruit on the mid-palate then persistent tannins on the finish. These need time to soften. **rating:** 93

best drinking 2002–2012 **best vintages** '84, '86, '88, '92, '94, '96, '97 **drink with** Bistecca Fiorentina • $17

Zema Estate Cluny

First made in 1993. An estate-grown blend of Cabernet Sauvignon (65%), Merlot (25%) and the remainder Cabernet Franc and Malbec. The '96 is the best to this point of time, the '97 not far behind.

🍷🍷🍷🍷 **1997** Full red-purple, bright but deep; the bouquet is quite powerful with berry and leaf aromas, the palate with a mix of cassis, a touch of sweet mint, and other dark berry fruits; finishes with soft tannins and subtle oak. **rating:** 89

best drinking 2002–2007 **best vintages** '96, '97 **drink with** Braised lamb • $17

Zema Estate Cabernet Sauvignon

Just as is the case with the Shiraz, produced from dryland, hand-pruned, hand-picked vines. A 100% Cabernet Sauvignon wine, matured in small French and American wood, but with the fruit – rather than the oak – driving the wine.

🍷🍷🍷🍷🍷 **1997** Medium red-purple; the bouquet is clean and smooth with pleasant red berry fruit and a touch of herbaceousness. The palate is classically restrained and modulated in true Cabernet style. Good fruit, no frills and fine tannins to close. **rating:** 90

best drinking 2002–2012 **best vintages** '84, '86, '88, '92, '93, '96, '97 **drink with** Barbecued lamb • $18

Zema Estate Family Selection Cabernet Sauvignon

A reserve selection of the best grapes; unlike many such wines does not rely on an extra shipload of oak to give it status.

🍷🍷🍷🍷🍷 **1997** Dense red-purple; powerful cassis and blackcurrant fruit on the bouquet, together with a touch of oak, is followed by a very tight and powerful palate offering pure Cabernet flavour and structure. A wine for the purist, the tannins guaranteeing longevity. **rating:** 94

best drinking 2005–2015 **best vintages** NA **drink with** Game • $40

zuber estate NR

Northern Highway, Heathcote, Vic 3523 **region** Bendigo

ph (03) 5433 2142 **open** 7 days 9–6

winemaker A Zuber **prod.** 450 **est.** 1971

prod. range ($10–12 CD) Chardonnay, Pinot Noir, Shiraz, Cabernet Sauvignon.

summary A somewhat erratic winery which is capable of producing the style of Shiraz for which Bendigo is famous, but does not always do so. No recent tastings.

new zealand

wineries and wines

wine regions of new zealand

key to regions

1 Northland and Matakana
2 Auckland Area
3 Waiheke Island
4 Waikato and Bay of Plenty
5 Gisborne/Poverty Bay
6 Hawke's Bay
7 Wairarapa/Martinborough
8 Nelson
9 Marlborough
10 Canterbury
11 Otago

akarangi NR

River Road, Havelock North, Hawke's Bay **region** Hawke's Bay
ph (06) 877 8228 **fax** (06) 877 7947 **open** Weekends, public holidays and summer 9–5
winemaker Morton Osborne **prod.** 780 **est.** 1988
prod. range ($9.90–15 CD) Sauvignon Blanc, Müller Thurgau, Chenin Blanc, Chardonnay, Cabernet Sauvignon.
summary Former contract grape growers now making and selling tiny quantities cellar door and through one or two local shops. Morton and Vivien Osborne have 5 hectares of vineyards, and operate the cellar-door sales through a century-old Presbyterian church moved onto the property.

alana estate NR

Puruatanga Road, Martinborough **region** Wairarapa
ph (06) 306 9784 **fax** (06) 306 9784 **open** 7 days 10–6
winemaker John Kavanagh **prod.** 1150 **est.** 1995
prod. range ($15–25 CD) Riesling, Sauvignon Blanc, Chardonnay, Pinot Noir.
summary Ian and Alana Smart acquired their prime vineyard site, situated between Dry River and Te Kairanga in 1995, no mean feat given the scarcity of such sites on the Martinborough terraces. It came after a decade of living in London and travelling incessantly, and is now their permanent home. Twelve hectares of vineyard were planted in 1995, a further 4 hectares in 1999, with the remaining plantable 6 hectares to be planted in 2000. A completed stage one of a gravity-fed winery has been built into the side of a hill, and production is projected to increase to about 12 000 to 15 000 cases by 2002. Winemaker John Kavanagh has joined Alana Estate from Palliser Estate, bringing both industry and local knowledge with him.

alexander vineyard NR

Dublin Street Extension, Martinborough (PO Box 87) **region** Wairarapa
ph (06) 306 9389 **open** Not
winemaker Kingsley Alexander, Deborah Alexander **prod.** 650 **est.** 1991
prod. range ($26.50 CD) Bordeaux-blend.
summary The Alexanders share with Benfield & Delamere the conviction that Martinborough is best-suited to the Bordeaux varieties of cabernet sauvignon, cabernet franc and merlot which they have planted on a high-density, low-trellis, guyot-pruned configuration. The first small vintage was in 1994; an on-site winery was built for the 1996 vintage.

allan scott wines ★★★★

Jacksons Road, RD3, Blenheim **region** Marlborough
ph (03) 572 9054 **fax** (03) 572 9053 **open** 7 days 9.30–4.30
winemaker Paddy Borthwick **prod.** 30 000 **est.** 1990
prod. range ($15–20 R) Marlborough Sauvignon Blanc, Marlborough Riesling, Autumn Riesling, Chardonnay.
summary The collapse of Vintech in 1995 accelerated former Corbans' chief viticulturist Allan Scott's plans for his own winery and full-time winemaker (previously the wines were contract-made at Vintech). Thus from 1996 a winery joined the attractive cellar-door sales and restaurant open seven days a week from noon to 4 pm, utilising 63 hectares of estate vineyards. Wine quality is all one could ask for.

Allan Scott Marlborough Riesling

Drawn from about 8 hectares of estate plantings. A wine with a distinguished show history going back to the '92 vintage.

ŸŸŸŸ 1998 Light yellow-green; the bouquet is clean, but relatively closed; the palate has citrus and lime flavours braced by a touch of mineral; good length supported by a carefully judged touch (8.5 grams) of residual sugar which is all but undetectable. **rating:** 87

best drinking 1999–2003 **best vintages** '92, '93, '97 **drink with** Fresh asparagus • $15

Allan Scott Marlborough Sauvignon Blanc

Given normal vintage conditions this is a classic example of Marlborough Sauvignon Blanc, reflecting the viticultural skills and experience of Allan Scott and the maturity of the estate vineyards. The 1998 is a particularly good example; this wine succeeds where so many failed.

ŸŸŸŸY 1998 Light to medium yellow-green; the bouquet is fragrant and clean, with gooseberry fruit aromas; the palate has excellent pungency with more gooseberry flavour; the crisp acidity gives a classic Sauvignon Blanc bite to the finish. **rating:** 93

best drinking 1999–2000 **best vintages** '92, '94, '97, '98 **drink with** Sugar-cured tuna • $18

Allan Scott Chardonnay

Made in full-frontal New Zealand style, fermented in new French oak and taken through malolactic fermentation, with 13.5 degrees alcohol.

ŸŸŸŸ 1998 Light to medium green-yellow; somewhat surprisingly, the bouquet is clean and tight, and the oak in balance with the fruit. The palate opens in similar fashion, but then the oak and mlf characters come through forcefully on the finish. **rating:** 85

best drinking 1999–2000 **best vintages** NA **drink with** Pork chops • $20

alpha domus NR

1829 Maraekakaho Road, RD1, Bridge Pa, Hastings **region** Hawke's Bay

ph (06) 879 6752 **fax** (06) 879 6952 **open** Mon–Fri 10–4, summer 7 days 10–5

winemaker Evert Nijzink **prod.** 10 000 **est.** 1996

prod. range ($14–38 R) Chardonnay, Semillon, Semillon Sauvignon Blanc, Sauvignon Blanc, Leonarda Late Harvest Semillon, Rosé, Pinot Noir, Cabernet Merlot, The Navigator (Merlot blend); AD Selection Barrel Fermented Semillon, Barrel Fermented Chardonnay, Noble Selection.

summary An estate-based operation drawing upon 16 hectares of vineyards, and enjoying significant growth in production. The wines have attracted considerable favourable comment in NZ, with export markets in the Netherlands and the UK presently under development.

arahura vineyard NR

Ness Valley Road, Clevedon **region** Auckland Area

ph (09) 292 8749 **fax** (09) 292 8749 **open** 10–4 summer

winemaker Ken Mason, Tim Mason **prod.** 700 **est.** 1991

prod. range ($20–25 CD) Merlot, Merlot Cabernet Sauvignon, Franc Malbec Cabernet Sauvignon.

summary Retired judge Ken Mason and wife Dianne are following in the footsteps of Tony Molloy QC by venturing into a new viticultural area and specialising in a single Bordeaux-style red (the plantings also include a little cabernet franc and merlot). A micro-winery was built in 1997 to handle the production from the 2 hectares of vineyards. If the Masons are as successful as Tony Molloy (of St Nesbit) they will have done well.

ashwell vineyards NR

Kitchener Street, Martinborough **region** Wairarapa

ph (04) 472 0519 **fax** (04) 389 8748 **open** Not

winemaker John Phipps **prod.** 250 **est.** 1989

prod. range ($17.50–28 ML) Sauvignon Blanc, Rosé, Pinot Noir, Cabernet Merlot.

summary Vivienne and John Phipps planted 2 hectares of vines in 1989, only to suffer severe frost damage in 1991. Undaunted, they doubled plantings in 1994, and were appropriately rewarded with silver medals for both their 1997 Pinot Noir and Sauvignon Blanc at the 1998 Air New Zealand Wine Awards. The Pinot Noir also received a five-star rating from Australia's *Winestate* magazine in March 1999.

askerne NR

267 Te Mata–Magateretere Road, Havelock North **region** Hawke's Bay
ph (06) 877 6085 **fax** (06) 877 2089 **open** Sat 10–5, Sun 10.30–4.30, weekends summer 10–5
winemaker Sorrelle Pearson, Jenny Dobson (Consultant) **prod.** 1500 **est.** 1993
prod. range ($15.50–22 CD) Riesling, Semillon, Semillon Sauvignon Blanc, Sauvignon Blanc, Chardonnay.
summary Askerne is the venture of John Loughlin, son of Dr John Loughlin of Waimarama Estate, who has named his vineyard after the Yorkshire town which was his wife's birthplace. It runs counter to conventional Hawke's Bay wisdom by having its 5.6-hectare vineyard planted entirely to white varieties: in descending order sauvignon blanc, riesling, semillon, chardonnay, gewurztraminer and optima.

ata rangi ★★★★★

Puruatanga Road, Martinborough **region** Wairarapa
ph (06) 306 9570 **fax** (06) 306 9523 **open** 7 days 11–5 Sept–Mar
winemaker Clive Paton, Oliver Masters **prod.** 6000 **est.** 1980
prod. range ($16–35 CD) Craighall Chardonnay, Petrie Chardonnay, Dalnagairn Chardonnay (Hawke's Bay), Summer Rosé, Pinot Noir, Celebre (Cabernet, Syrah, Merlot blend).
summary Consistently ranks among the best wineries in New Zealand, let alone Martinborough. Both the Pinot Noir and Celebre are remarkable for their depth of colour and sweetness of fruit, showing the impact of full physiological ripeness. A splendid new winery was commissioned for the 1996 vintage, handling the grapes from the 19 hectares of estate plantings as well as the grapes purchased from other regions. The wines are exported to Australia, the UK and US. Murphy's Law has meant no tastings in the past 12 months.

babich ★★★★

Babich Road, Henderson **region** Auckland Area
ph (09) 833 7859 **fax** (09) 833 9929 **open** Mon–Fri 9–5, Sat 9–6, Sun 11–5
winemaker Neil Culley **prod.** 85 000 **est.** 1916
prod. range ($7.55–25 R) The Patriarch Chardonnay, Cabernet Sauvignon; Irongate Chardonnay, Cabernet Merlot; Mara Estate Chardonnay, Sauvignon, Merlot, Cabernet Sauvignon, Syrah; also varietal/regional wines such as Marlborough Sauvignon Blanc, Hawke's Bay Sauvignon Blanc, East Coast Chardonnay.
summary Continues to uphold the reputation it gained in the 1960s, but has moved with the times in radically changing its fruit sources and wine styles. Particularly given the volume of production, quality is admirably consistent, with the expanded Mara Estate range leading the way, and strong support from Irongate Chardonnay. It now has 114 hectares of estate vineyards, with 57 hectares in the Gimblett Road region of Hawke's Bay, 42 hectares in the Awatere Valley of Marlborough and the remainder around the Henderson winery. Wines are exported throughout Europe, Asia and North America.

Babich Marlborough Riesling

Produced from grapes grown in the Pigou Vineyard in the Rapaura district of Marlborough, from vines which were crop-thinned during the growing season to enhance the flavour development. Only free-run and light-press juice is used in the wine, with a very cold fermentation to capture the maximum riesling flavour. The residual sweetness of 15 grams per litre is on the generous side.

YYYY 1998 Light green-yellow; herb, mineral and lime aromas; some of the sweetness does show through on the finish, suggesting less might have been better. On the other hand, should age attractively. **rating:** 85

best drinking 1999–2002 **best vintages** NA **drink with** Chinese • $15

Babich Hawke's Bay Sauvignon Blanc

The wine is made at Babich's winery at Henderson, Auckland, using grapes grown in the Hawke's Bay region, and is 100% Sauvignon Blanc, made without the use of oak.

YYYY 1998 Light to medium yellow-green; the bouquet is clean, of medium intensity, with a touch of gooseberry and some herbal notes. The palate again has some minerally notes which at least stop it from being flabby, and provides some structure. **rating:** 85

best drinking 1999–2000 **best vintages** NA **drink with** Pasta marinara • $12

Babich Mara Estate Sauvignon

Babich says the wine is modelled on the classic dry white wines of Graves in Bordeaux: barrel-fermented, ripe, complex and subtle. Actually, that is not a bad description. As the wine name indicates, the grapes are grown on Babich's Mara Estate on Gimblett Road, Hawke's Bay, generally acknowledged to be one of the foremost subregions of Hawke's Bay. It is barrel-fermented and aged on yeast lees for nine months.

YYYY 1998 Light to medium yellow-green; ripe gooseberry fruit rather than subtle oak is the driving force of the bouquet, and once again the pleasantly flavoured and balanced palate shows the carefully controlled use of oak which does help provide both structure and length to the finish. Whether the wine benefited from the 14 degrees alcohol is another matter. **rating:** 87

best drinking 1999–2000 **best vintages** '93 **drink with** Grilled eggplant and zucchini • $16

Babich Marlborough Sauvignon Blanc

At the opposite end of the spectrum to Babich's opulent Mara Estate Sauvignon (from Hawke's Bay). This Marlborough version is made in the traditional, no-frills approach, but succeeds brilliantly thanks to the quality of the fruit. Has won a number of gold medals at NZ wine shows, and it is not hard to see why.

YYYY 1998 Light to medium yellow-green; the fresh and crisp bouquet has gently grassy/nettle aromas; the strongly-flavoured palate runs through those grassy/nettle characters into more tropical passionfruit and gooseberry characters. Good balance and mouthfeel. **rating:** 88

best drinking 1998–1999 **best vintages** '91, '92, '94, '96, '97 **drink with** Sautéed prawns • $14

Babich Marlborough Pinot Gris

The second vintage of Pinot Gris for Babich, grown on the Pigou family vineyard. Tasted twice, once with a fair degree of sulphur dioxide showing.

YYYY 1998 Pale straw-green; a crisp, stony, faintly spicy bouquet with what appears to be some sulphur dioxide holding back the fruit. The flavour picks up noticeably on the back palate and finish, as does that inimitable, slightly velvety, texture of Pinot Gris. **rating:** 86

best drinking 1999–2001 **best vintages** NA **drink with** Sweet salad • $15

Babich Irongate Chardonnay

Produced from relatively low-yielding vineyards on the shingle soils to the west of Hastings, and rated as a classic by both Bob Campbell and Michael Cooper. The wine receives the full winemaking treatment, being barrel-fermented (50% new) and held on lees for nine months, with evident malolactic fermentation.

🍷🍷🍷🍷🍷 **1996** Medium yellow-green; complex, bottle-developed nutty overtones to the fruit; subtle oak. The palate has good length and grip, with nice peach, melon and cashew fruit retention in a soft, generous mould. However, do not delay in drinking it. **rating:** 90

best drinking 1999–2000 **best vintages** '96 **drink with** Tortellini • $25

bazzard estate NR

Awa Road, RD1, Kumeu **region** Auckland Area

ph (09) 412 8486 **fax** (09) 412 8486 **open** By appointment

winemaker Sarah Hennessy **prod.** 500 **est.** 1991

prod. range ($18–28 CD) Chardonnay, Merlot, Pinot Noir, Huapai Reserve Pinot Noir.

summary Charles Bazzard is a former Buckinghamshire solicitor (and transient waterfront worker) who, together with wife Kay, has developed an organically grown hillside vineyard in the Awa Valley, near Huapai. Pinot Noir is the current specialty with chardonnay and cabernet merlot planted and earmarked for future releases. Various tastings of the Pinot Noirs in 1997 and 1998 simply confirmed that Auckland is several hundred kilometres too far north to produce Pinot Noir with varietal character.

benfield & delamere NR

Cambridge Road, Martinborough **region** Wairarapa

ph (06) 306 9926 **fax** (06) 306 9926 **open** By appointment

winemaker Bill Benfield, Sue Delamere **prod.** 350 **est.** 1987

prod. range ($35–38 ML) 'Martinborough', a single Cabernet Sauvignon Merlot Cabernet Franc blend. A second label is in the offing.

summary Wellington architect Bill Benfield and partner librarian Sue Delamere have single-mindedly set about recreating Bordeaux, with an ultra-high density, very low-trellised vineyard and utilising 'conservative' techniques of the kind favoured by the Bordelaise. All of the tiny production is sold by mailing list.

bentwood wines NR

Akaroa Highway, Tai Tapu, Canterbury **region** Canterbury

ph (03) 329 6191 **fax** (03) 329 6192 **open** By appointment

winemaker Grant Whelan **prod.** 550 **est.** 1991

prod. range ($12.50–15.95 CD) Riesling, Pinot Blanc, Pinot Noir.

summary Ray Watson has established a 2-hectare vineyard on the Banks Peninsula, his interest in wine fired after a 12-month sojourn living on a vineyard in France. The first wines were released from the 1995 vintage; the Pinot Blanc is already a silver medal winner.

black ridge ★★★★

Conroys Road, Earnscleugh, Alexandra **region** Otago

ph (03) 449 2059 **fax** (03) 449 2597 **open** 7 days 10–5

winemaker Dean Shaw **prod.** 3000 **est.** 1981

prod. range ($12.50–29 R) Riesling, Chardonnay, Gewurztraminer, Pinot Noir.

summary The formidable, rocky vineyard site at Black Ridge is legendary even in New Zealand, where toughness is taken for granted. The 8-hectare vineyard will always be low-producing, but the wines produced to date have all had clear and bracing varietal character. The outstanding Pinot Noir is a particularly good example of what the site can produce. Exports to the UK and US.

Black Ridge Pinot Noir

A bitter-sweet reward to Black Ridge, the '98 following on as it does to the outstanding 1997 vintage: Mike Wolter, who made the '97, died a few months later in a winemaking accident. The

'98 was a gold medal winner at the 1999 Liquorland Royal Easter Wine Show. Produced from 3 hectares of estate plantings.

YYYYY 1998 Medium to full red-purple; a generous, sturdy and solid bouquet with excellent plummy variety fruit is followed by an abundance of sweet plummy fruit on the rich, ripe full-bodied palate. **rating:** 94

best drinking 1999–2003 **best vintages** '97, '98 **drink with** Braised duck • $29

bladen NR

1346 Conders Bend Road, Renwick, Marlborough **region** Marlborough
ph (03) 572 9417 **fax** (03) 572 9217 **open** 7 days 12–5
winemaker Simon Waghorn (Contract) **prod.** 7500 **est.** 1997
prod. range ($13.50–18 CD) Riesling, Gewurztraminer, Pinot Gris, Sauvignon Blanc, Sauvignon Blanc Oak Aged.

summary The McDonald family began establishing the Bladen vineyard in 1989, inspired by travels through Europe in the mid-1980s. The name has two derivations: the burial place of Sir Winston Churchill, but also a combination of the McDonald children, Blair and Deni. Initially all of the grapes were sold to Grove Mill, but since 1997 part of 6-hectare vineyard production has been vinified by contract-winemaker Simon Waghorn.

blue rock vineyard ★★☆

Dry River Road, Martinborough **region** Wairarapa
ph (06) 306 9353 **fax** (06) 306 9353 **open** 7 days 11–6
winemaker Jenny Clark **prod.** 3500 **est.** 1986
prod. range ($14–30 CD) Chardonnay, Sauvignon Blanc, Riesling Bone Dry, Pinot Noir, Cabernet Sauvignon, Magenta Méthode Traditionelle.

summary Blue Rock is a partnership run by the Clark family in a bid to diversify the activities carried out on its 200-hectare sheep and cattle farm. There are now 16 hectares of windswept vineyards servicing a winery which was built in 1992, and a winery tasting room-cum-café-style restaurant completed in 1994, overlooking a 4-hectare park complete with lake, wildlife habitat and barbecue facilities.

bradshaw estate NR

291 Te Mata Road, Havelock North **region** Hawke's Bay
ph (06) 877 8017 **fax** (06) 876 5494 **open** 7 days 10–5
winemaker Hans Peet **prod.** 1700 **est.** 1994
prod. range Dry White, Medium White, Chardonnay, Cabernet Merlot.

summary Wayne and Judy Bradshaw have established their operation on the historic Vidal's No. 1 Vineyard and Homestead, and in 1996 opened a new winery with an attendant restaurant on the vineyard. The winery restaurant is open Friday to Sunday and every day throughout January.

briar vale estate NR

Kelliher Lane, Alexandra, Central Otago **region** Otago
ph (03) 448 8221 **fax** (03) 448 8221 **open** Weekends and public holidays 11–4
winemaker John Currie **prod.** 150 **est.** 1990
prod. range ($14–23 CD) Riesling, Chardonnay, Pinot Blanc, Pinot Noir.

summary Alsace was the inspiration for the varieties chosen by John and Judy Currie when they established their 1.75-hectare vineyard on a steep north-facing slope above their cherry orchard. The cool climate carries the threat of spring frosts, however, and until the installation of frost protection (via overhead sprinklers) the crops were significantly reduced; better things are now on the way.

brick bay NR

c/o PO Box 28270, Remuera, Auckland **region** Auckland Area
ph (09) 524 2831 **fax** (09) 524 2831 **open** Not
winemaker Contract **prod.** NA **est.** NA
prod. range ($NA) Pinot Gris.
summary A brand-new winery which made a dramatic entrance with its gold medal-winning Pinot Gris.

Brick Bay Pinot Gris

Gold medal winner at the 1999 Liquorland Royal Easter Wine Show.
YYYY **1998** Pale straw-green; an interesting wine with lots of character and flavour, ever so slightly obscured by free sulphur dioxide. It has good spicy fruit on the bouquet and palate, and attractive mid-palate richness, even if (on my score sheet) the finish was slightly sugary and slightly bland.
rating: 89
best drinking 1999–2003 **best vintages** NA **drink with** Cheese soufflé • NA

brookfields vineyards ★★★★☆

Brookfields Road, Meeanee, Napier **region** Hawke's Bay
ph (06) 834 4615 **fax** (06) 834 4622 **open** Mon–Sat 9–5, Sun 12–4.
winemaker Peter Robertson **prod.** 8000 **est.** 1937
prod. range ($16–46 R) Chardonnay, Gewurztraminer, Sauvignon Blanc, Fumé Blanc, Pinot Gris, Cabernet, Reserve Cabernet Sauvignon, Gold Label Cabernet Merlot.
summary Peter Robertson has worked hard since acquiring Brookfields in 1977, producing grassy Sauvignon Blanc, lightly-oaked, understated Chardonnay and – best of all – the powerful, structured Gold Label Cabernet Merlot, now his highly regarded top-of-the-range release. A particular feature of his wines is their ability to age with grace.

brownlie brothers NR

6 Franklin Road, Bayview **region** Hawke's Bay
ph (06) 836 6250 **open** 7 days 9–6
winemaker Chris Brownlie **prod.** 180 **est.** 1991
prod. range ($NA) Chardonnay, Sauvignon Blanc, Gewurztraminer, Pinot Noir.
summary Chris and Jim Brownlie have progressively established 15 hectares of vineyards, selling most of the grapes to other wineries, but recently taking the plunge and vinifying a small part of the production for mail order and cellar-door sales.

bullrush wines NR

Main Road, Waimauku, Kumeu (PO Box 132) **region** Auckland Area
ph (09) 358 2952 **fax** (09) 366 7112 **open** Not
winemaker Mark Robertson **prod.** NA **est.** 1995
prod. range ($NA) Sauvignon Blanc, Chardonnay.
summary Bullrush Wines is based on a 16-hectare vineyard owned by Matua Valley winemaker Mark Robertson and wine judge Jane Osborne. So far, the wines have only been marketed through Cardmember Wines.

cairnbrae wines ★★★★

Jacksons Road, RD3, Blenheim **region** Marlborough
ph (03) 572 8048 **fax** (03) 572 7018 **open** 7 days 9–5
winemaker Matt Thomson, Kim Crawford (Consultant) **prod.** 14 000 **est.** 1981
prod. range ($17–23 R) Riesling, Reserve Riesling, Semillon, Sauvignon Blanc, Barrel Fermented Sauvignon Blanc, Chardonnay, Reserve Chardonnay.

summary The Brown family (Daphne, Murray and Dion) established 18 hectares of vineyard progressively from 1981, selling the grapes to Corbans until 1992, when part of the production was made for them by Kim Crawford, and the label was launched with immediate success. A fast-growing label to watch, with Pinot Noir and Pinot Gris in the pipeline, likewise exports to Australia, US, Hong Kong, Holland, Denmark and Canada.

Cairnbrae Riesling

Produced from fruit grown on the Cairnbrae Jacksons Road Vineyard, complemented by grapes from the Ellin Vineyard at the confluence of the Omaka and Wairau Valley.

▼▼▼▼ **1998** Light to medium yellow-green; the bouquet has elevated tropical fruit aromas with a mix of pineapple, apricot and citrus. The palate is rich and soft in an easy-drinking style, but is well balanced. **rating:** 87

best drinking 1999–2003 **best vintages** '98 **drink with** Vegetarian • $17

Cairnbrae Reserve Chardonnay

As is appropriate, made in diametrically opposed style to the varietal Chardonnay. The Reserve uses the full gamut of barrel fermentation, malolactic fermentation and opulent American oak.

▼▼▼▼▽ **1998** Medium yellow-green; a very complex bouquet with full-on malolactic fermentation and barrel-ferment American oak characters. The palate is big, rich, high-flavoured and uncompromising; it does raise the question whether less would have been better still. **rating:** 90

best drinking 1999–2000 **best vintages** '90 **drink with** Chinese chicken dishes • $22

canadoro NR

New York Street, Martinborough **region** Wairarapa
ph (04) 387 9761 **fax** (04) 387 9761 **open** Via The Grape Vine
winemaker Chris Lintz, Greg Robins **prod.** 400 **est.** 1993
prod. range ($23–28 CD) Chardonnay, Cabernet Sauvignon.
summary A weekend operation for Wellington residents Greg and Lesley Robins. The 1.25-hectare vineyard is planted to cabernet sauvignon and chardonnay, but is due to be expanded over the next few years; Greg Robins makes the wine with assistance from Chris Lintz.

canterbury house vineyards NR

780 Glasnevin Road, RD3, Amberley **region** Canterbury
ph (03) 314 6700 **fax** (03) 314 6905 **open** 7 days 10–4
winemaker Mark Rattray **prod.** 15 000 **est.** 1994
prod. range ($16.90–23.90 CD) Riesling, Sauvignon Blanc, Chardonnay, Pinot Gris, Viognier, Muscat, Pinot Noir, Merlot, Cabernet Sauvignon.
summary Californian Michael Reid (and his wife) came to New Zealand for a vacation, and are now engaged in a most ambitious vineyard and winery development in the Waipara region. The first of five planned stages of winery construction was completed in time for the 1998 vintage; when the final phase is completed it will have a production capacity of 150 000 cases. The first 40 hectares of vineyard have been established, with a further 110 hectares due to be planted over the coming years. If that is not the most expensive vacation ever taken, I don't know what is.

Canterbury House Sauvignon Blanc

First made in 1997 from estate-grown grapes, the vines then being three years old. The second vintage ('98) won a gold medal at the 1999 Liquorland Royal Easter Wine Show. Another wine to demonstrate that 1998 was a far better vintage for Sauvignon Blanc in Waipara/North Canterbury than it was further north in New Zealand.

🍷🍷🍷🍷🍷 **1998** Light green-yellow; the bouquet is fine, with relatively neutral mineral and herb-accented fruit. The palate has excellent length, intensity and balance, even though the oak is not splashy. **rating:** 94

best drinking 1999–2000 **best vintages** NA **drink with** Oysters, shellfish • $17

cellier le brun ★★★★

Terrace Road, Renwick **region** Marlborough
ph (03) 572 8859 **fax** (03) 572 8814 **open** 7 days 9–5
winemaker Allan McWilliams **prod.** 18 000 **est.** 1985
prod. range ($15–40 CD) Méthode Champenoise specialist with a large range of both vintage and non-vintage wines, including the Daniel Le Brun range of Brut, Brut Taché, Vintage, and Blanc de Blancs. Small quantities of Sauvignon Blanc, Chardonnay and Pinot Noir table wine also made and sold under the Terrace Road label.
summary For almost a decade has produced some of New Zealand's highly rated sparkling wines, initially somewhat erratic, but now much more consistent in style – a style which tends to the baroque, but which seems to be exactly what New Zealanders like and want. The Courtyard Café restaurant is open 7 days from 9 am to 5 pm. A doubling of production over the past few years bears eloquent testimony to the quality of the wines which are now also exported to Canada.

Cellier Le Brun Terrace Road Sauvignon Blanc

Cellier Le Brun branched out into table (i.e. non-sparkling) wines in 1995, and made a fairly inauspicious start. However, the follow-on vintages marked a much surer touch. Produced from estate-grown grapes and cold-settled and cold-fermented in stainless steel in the traditional fashion.

🍷🍷🍷🍷 **1998** Light to medium yellow-green; the bouquet is clean and firm, with minerally notes to the fore but also some stone fruit characters. The palate then shows a touch of stone fruit slightly away from the usual Sauvignon Blanc flavour, but has good grip and excellent length. Nice wine. **rating:** 88

best drinking 1999–2000 **best vintages** NA **drink with** Salmon terrine • $15

chancellor wines of waipara ★★★

133 Mount Cass Road, Waipara **region** Canterbury
ph (03) 314 6834 **fax** (03) 314 6894 **open** By appointment
winemaker Kym Rayner **prod.** 4300 **est.** 1982
prod. range ($18.95–24.95 CD) Mount Cass Road Waipara Riesling, Sauvignon, Chardonnay and Cabernet Sauvignon; Nor'Wester Marlborough Chardonnay, Marlborough Cabernet Merlot.
summary Having been grape growers for 15 years Anthony and Helen Willy took the plunge of establishing the Chancellor Wines brand in 1995, and also almost doubling the estate plantings to 11 hectares, which will see Pinot Noir added to the portfolio in the near future. A winery, cellar door and restaurant are all planned for the future. Wines are exported to Australia, the UK and US.

chard farm ★★★★☆

Chard Road, RD1, Gibbston **region** Otago
ph (03) 442 6110 **fax** (03) 441 8400 **open** 7 days 11–5
winemaker Duncan Forsyth, Rob Hay **prod.** 15 000 **est.** 1987
prod. range ($12–33 R) Riesling, Gewurztraminer, Sauvignon Blanc, Pinot Gris Surlie, Judge and Jury Chardonnay, Closeburn Chardonnay, Southern Lakes Chardonnay, Pinot Noir, Bragato Reserve Pinot Noir.

summary Perched precariously between sheer cliffs and the fast-flowing waters of the Kawarau River, Chard Farm is a tribute to the vision and courage of Rob and Gregory Hay. At a latitude of 45°S, viticulture will never be easy, but Chard Farm has made every post a winner to date, supplementing production from the 20-hectare vineyard with grapes purchased from Marlborough. The Chardonnay and Pinot Noir are superb, especially the intermittent prestige releases.

chifney wines ★★★

Huangarua Road, Martinborough **region** Wairarapa
ph (06) 306 9495 **fax** (06) 306 9493 **open** 7 days 11–5 (usually)
winemaker Sue Chifney **prod.** 1800 **est.** 1980
prod. range ($18–27 CD) Bottle Fermented Chiffonnay (Sparkling Chenin Blanc), Chenin Blanc, Chardonnay, Chenin Chardonnay Reserve, Enigma (Cabernet Merlot blend), Cabernet Sauvignon, Tawny Port.
summary After a long reign (he opened Wairarapa's first winery) Stan Chifney died in 1996; the business is being carried on by Rosemary Chifney together with daughter Sue, who has undertaken the winemaking responsibilities.

c j pask winery ★★★★☆

1133 Omahu Road, Hastings **region** Hawke's Bay
ph (06) 879 7906 **fax** (06) 879 6428 **open** Mon–Fri 9–5, Sat and public holidays 10–5, Sun 11–4
winemaker Kate Radburnd **prod.** 32 000 **est.** 1985
prod. range ($9.50–25 CD) Sauvignon Blanc, Chenin Blanc, Chardonnay, Reserve Chardonnay, Pinot Noir, Merlot, Cabernet Merlot, Cabernet Sauvignon; Roy's Hill White and Red are second label.
summary Ex-cropduster pilot Chris Pask became one of the most highly regarded grape growers in Hawke's Bay; his coup in securing former Vidal winemaker Kate Radburnd (née Marris) has paid the expected dividends. Production has increased rapidly, and the wines have had significant and consistent success in NZ and international wine shows thanks to the complexity of the Chardonnays and the supple, sweet fruit of its Cabernet Merlots and Reserve Cabernet Sauvignons. The wines find their way to Australia, Canada, US, Japan, Thailand, UK and the Netherlands.

C J Pask Reserve Chardonnay

As with all the C J Pask wines, estate-grown. The wine is entirely barrel-fermented in Francois Freres French oak, and is given five months lees contact before being racked and cleaned up and returned to barrel for a further five months. Interestingly, and consistently with all of the Pask Chardonnays, does not undergo malolactic fermentation. The '97 was tasted twice, once with adverse notes about the excessive and slightly raw oak. The note which follows is a compromise.

🍷🍷🍷🍷 **1997** Light to medium yellow-green; powerful, spicy oak is the first impression on the bouquet, but there is attractive fruit too. The palate has plenty of flavour and length, again with oak playing a major role. **rating:** 88

best drinking 1999–2000 **best vintages** '95, '96 **drink with** Blue-lipped mussels in sauce • $25

C J Pask Reserve Merlot

Gold medal winner at the 1999 Liquorland Royal Easter Wine Show where, for various reasons, I tasted the wine on several occasions throughout the judging. On the first tasting I thought it was excessively oak-dominated; on other tastings I had no such problem, although the oak undoubtedly makes a major contribution to the quality and style of the wine.

🍷🍷🍷🍷🍷 **1997** Medium red-purple; soft, sweet red plum fruit is surrounded by gently cedary, spicy oak on the bouquet. The palate has excellent Merlot structure, the tannins being fine, the fruit sweet and the level of extract not overly high. Once again, spicy cedary oak provides another dimension. **rating:** 94

best drinking 1999–2004 **best vintages** NA **drink with** Grilled calf's liver • $25

claddagh vineyards NR

Puruatanga Road, Martinborough **region** Wairarapa
ph (06) 306 9264 **fax** (06) 306 9264 **open** By appointment
winemaker Russell Pearless **prod.** 300 **est.** 1991
prod. range ($14.95–25 CD) Pinot Noir, Cabernet Sauvignon.
summary Presently a weekend and holiday occupation for computer industry executives Russell and Suzanne Pearless, but when the 5-hectare vineyard (also planted to chardonnay and sauvignon blanc) comes into full bearing, the level of involvement will doubtless increase.

clearview estate ★★★★★

Clifton Road, RD2, Te Awanga **region** Hawke's Bay
ph (06) 875 0150 **fax** (06) 875 1258 **open** 7 days 10–6 Dec–Feb, winter Fri–Sun 10–5
winemaker Tim Turvey **prod.** 4000 **est.** 1989
prod. range ($14–30 CD) Beach Head Chardonnay, Reserve Chardonnay, Black Reef Riesling, Te Awanga Sauvignon Blanc, Fumé Blanc, Reserve Fumé Blanc, Reserve Cabernet Franc, Reserve Merlot, Reserve The Old Olive Block (Cabernet blend), Cape Kidnappers Cabernet, Basket Press Cabernet, Blush, Sea Red (dessert red wine), Noble 51 (botrytised Chardonnay).
summary Clearview Estate is situated on a shingly site first planted by Anthony Vidal in 1916 on the coast of Te Awanga; it has been replanted since 1988 with chardonnay, cabernet sauvignon, cabernet franc and merlot, with grapes also coming from a neighbouring vineyard. All of the wines to date have been of exceptional quality, especially the magically concentrated and complex Chardonnay. The icing on the cake is an outstanding restaurant, rated by Bob Campbell as Hawke's Bay's best.

clifford bay estate ★★★★

PO Box 1088, Blenheim **region** Marlborough
ph (03) 578 4617 **fax** (03) 578 4619 **open** Not
winemaker Glen Thomas (Vavasour Wines – Contract) **prod.** 11 000 **est.** 1994
prod. range ($14.95–18.95 R) Sauvignon Blanc, Chardonnay.
summary Clifford Bay made the most spectacular imaginable entry onto the scene, winning two gold medals (Air New Zealand Wine Awards and Christchurch Show) with its first wine, a 1997 Sauvignon Blanc. It is the venture of Eric and Beverley Bowers, Graham and Thelma Ca ns, and Chris Wilson. Viticultural advice has come from Richard Bowling, and the wine is made by the masterful Glen Thomas. Twenty hectares of sauvignon blanc, chardonnay and riesling are under vine. The wines are exported to the UK.

Clifford Bay Sauvignon Blanc

Grown on river terraces adjoining the Awatere River, and late-picked. Fifteen per cent of the wine is barrel-fermented in new French oak barriques, the remainder in stainless steel.

🍷🍷🍷🍷 **1998** Light to medium yellow-green; the bouquet is quite aromatic with sweet gooseberry and passionfruit aromas; that passionfruit comes through quite strongly on the palate, which has moderate length and good balance. **rating:** 89

best drinking 1999–2000 **best vintages** '97 **drink with** Sugar-cured tuna • $18.95

clifton road NR

Clifton Road, Te Awanga **region** Hawke's Bay
ph (06) 875 0748 **fax** (06) 876 6211 **open** Not
winemaker Preston Group (White), Tim Turvey (Red) – both Contract **prod.** 900 **est.** 1992
prod. range Sauvignon Blanc, Cabernet Sauvignon.

summary The 2.5-hectare vineyard owned by Wayne Harrison and Terri Coats is established close to the sea on the same shingle soils as Clearview Estate. Not surprisingly, there is a similarity in both the quality and style of the wines. A label to watch, even if production will always be small.

cloudy bay ★★★★★

Jacksons Road, Blenheim **region** Marlborough
ph (03) 572 8914 **fax** (03) 572 8065 **open** 7 days 10–4.30
winemaker Kevin Judd **prod.** 80 000 **est.** 1985
prod. range ($24.50–41 R) Sauvignon Blanc, Chardonnay, Pinot Noir, Cabernet Merlot, Pelorus (sparkling).

summary The other arm of Cape Mentelle, masterminded by David Hohnen and realised by Kevin Judd, his trusted lieutenant from day one. A marketing tour de force, it became a world-recognised brand in only a few years, but the wine quality and style should not be underestimated: Hohnen and Judd may share a great sense of humour, but they are perfectionists in every way, the wines consistently great. A warped New World view, perhaps, but I rate the Sauvignon Blanc the best in the world, all vintages taken into account. Kevin Judd, incidentally, could as easily earn a living as a photographer, he has a rare talent. The wines are available in Australia, Japan, US, Singapore, Canada, Thailand, Hong Kong, Indonesia and South Africa and most parts of Europe.

Cloudy Bay Sauvignon Blanc

The most famous New World Sauvignon Blanc with an international reputation second to none. The creative team of David Hohnen and Kevin Judd are disarmingly modest about the wine, however correctly they may point to the perfect marriage between the variety and the climate and soil of Marlborough. There is also the attention to detail, the discipline and the creative intelligence required to make a wine of such distinction with such consistency.

🍷🍷🍷🍷🍸 **1998** Light to medium yellow-green; a crisp, clean bouquet with a slightly subdued mix of lime, herb and gooseberry aromas. The palate is more delicate than the usual Cloudy Bay-style, but, as ever, is beautifully balanced with a long, cleansing finish. **rating:** 92

best drinking 1998–1999 **best vintages** '92, '94, '96, '98 **drink with** Virtually any seafood dish • $24.50

Cloudy Bay Chardonnay

The extensive chardonnay plantings of Cloudy Bay (almost 35 hectares in total) are increasingly directed to the production of Pelorus, with the newer plantings dedicated to this purpose. However, Cloudy Bay produces a Chardonnay of real stature and complexity, made using prolonged yeast lees contact, malolactic fermentation and ageing in French oak. About 70–80% of the wine is barrel-fermented, 20% in new French oak.

🍷🍷🍷🍷🍷 **1997** Quite full straw-yellow colour, although reassuringly pale by normal NZ standards. The bouquet is concentrated and complex, with distinctly Burgundian overtones, and typical of the striking fruit character of the best Chardonnays from the low-yielding vintage. The palate is exceptionally tight and intense, fruit-driven, even though malolactic fermentation (60%) and oak (25% new) have contributed to the complexity of the wine. **rating:** 96

best drinking 1999–2001 **best vintages** '87, '91, '93, '94, '95, '97 **drink with** Sweetbreads • $32

Cloudy Bay Pelorus

A blend of Pinot Noir and Chardonnay, made with consultancy advice and direction from Californian-born and trained Harold Osborne. The wine is aged for three years on yeast lees, and is given further bottle age prior to release. The most positively flavoured and structured sparkling wines from Australasia, which is quite impossible to give a point-rating to.

1994 Deep gold-bronze; the bouquet is quite massive, with orange peel and madeirised aromas followed by the biggest, boldest and brassiest palate imaginable. A total renegade in the conventional scheme of things.

best drinking 1999–2001 **best vintages** NA **drink with** Medium-weight shellfish or fish dishes • $41

Cloudy Bay Pinot Noir

The fully commercial release of Pinot Noir from Cloudy Bay has been an even more drawn-out affair than Penfolds' creation of Yattarna. Cloudy Bay made a Pinot Noir back in 1989 which I thought was of excellent quality. The wine is made in open-topped stainless steel fermenters, typically with five days of pre-fermentation cold soaking. Once commenced, fermentation is encouraged to reach 32°C, and following a short primary fermentation the wine is taken directly to French oak barrels, 45% being new. The wine spends an unusually long 16 months in oak, and is eggwhite-fined before bottling.

🍷🍷🍷🍷🍷 **1997** Very deep, strong colour for a Pinot Noir. Dark briary, plum and forest aromas run through a complex bouquet; the palate is interesting, with plenty of length and flavour in the plum/briar spectrum but with acidity quite obvious and making the wine somewhat linear.

rating: 91

best drinking 1999–2003 **best vintages** '97 **drink with** Venison • $32

collards ★★★★☆

303 Lincoln Road, Henderson, Auckland **region** Auckland Area

ph (09) 838 8341 **fax** (09) 837 5840 **open** Mon–Sat 9–5, Sun 11–5

winemaker Bruce Collard, Geoff Collard **prod.** 20 000 **est.** 1910

prod. range ($6–25 CD) Riesling, Queen Charlotte Riesling, Chardonnay (Rothesay, Hawke's Bay, Marlborough, Blakes Mill), Chenin Blanc, Sauvignon Blanc (Rothesay, Marlborough), Barrique Fermented Semillon, Queen Charlotte Marlborough Pinot Noir, Marlborough Syrah, Rothesay Cabernet Sauvignon, Cabernet Merlot, Tawny Port.

summary At the same time a family-owned and run business which is a bastion of conservatism, adopting a low promotional profile, but which consistently produces fastidiously crafted wines of excellent quality, and which has moved with the times in developing new wines and labels. Exports to Singapore, Japan, UK, Germany, Holland, Denmark and Hong Kong.

coopers creek NR

State Highway 16, Huapai **region** Auckland Area

ph (09) 412 8560 **fax** (09) 412 8375 **open** Mon–Fri 9–5.30, weekends 10.30–5.30

winemaker Simon Nunns **prod.** 60 000 **est.** 1980

prod. range ($10–25 CD) Hawke's Bay Riesling, Chardonnay, Pinot Noir, Merlot, Cabernet Sauvignon Franc, Semillon Chardonnay, First Edition, Marlborough Sauvignon Blanc, Gisborne Chardonnay and Huapai Cabernet Merlot; the Reserve Range of Hawke's Bay Reserve Riesling, Reserve Oak Aged Sauvignon Blanc, Swamp Reserve Chardonnay, Wild Ferment Chardonnay, Late Harvest Riesling, Coopers Gold, Hawke's Bay Reserve Pinot Noir, Hawke's Bay Reserve Cabernet Sauvignon and Huapai Reserve Cabernet Merlot.

summary A long-term producer of stylish white wines sourced from Gisborne, Hawke's Bay and Marlborough, respectively. They are full of character and flavour, but avoid the heavy, coarse phenolics which were once so much part of the white wine scene in New Zealand. The mislabelling scandal of 1998 was still making television news in March 1999, and seems unlikely to go away in a hurry, but Coopers Creek seems to have largely shrugged it off, continuing to export to all corners of the globe. Perhaps the most telling sign of hidden turmoil has been the recent departure of Kim Crawford as chief winemaker.

Coopers Creek Hawke's Bay Riesling

A consistently excellent wine from a region not known for the quality of its Riesling; grown by Jim Scotland. A remarkable fact is that every vintage since 1989 has won a gold medal.

🍷🍷🍷🍷 **1998** Light yellow-green; the bouquet has hints of spice together with tropical lime fruit; there are similar flavours on the palate, again with a touch of spice, and a pleasant finish. **rating:** 88

best drinking 1999–2001 **best vintages** '90, '91, '96 **drink with** Smoked salmon salad • $14.95

Coopers Creek Marlborough Sauvignon Blanc

This is the engine-room of the Coopers Creek winery, with 10 000 cases made in a no-frills, unwooded style. However, some complexity comes from the two differently sited vineyards providing the grapes.

🍷🍷🍷🍷 **1998** Light yellow-green; the bouquet is quite lively and zesty, ranging from herbal to hints of gooseberry. The flavour carries well onto the finish, which is well balanced and has a touch of elegance. **rating:** 88

best drinking 1999–2001 **best vintages** '91, '94, '96, '97 **drink with** Crustacea • $14.95

Coopers Creek Hawke's Bay Chardonnay

Former winemaker Kim Crawford believed in ripe grapes, something fairly readily achieved in the Hawke's Bay climate. Barrel fermentation in American oak, lees contact and 100% malolactic fermentation all add to the impact of a powerful wine style.

🍷🍷🍷🍷½ **1997** Medium yellow-green; a tangy bouquet with a mix of mineral and citrus aromas, and a touch of cashew which comes through on the palate. A stylish and restrained wine which will hurry slowly. Retasted March 1999, has developed a little more quickly than anticipated, but offers plenty of complexity and richness. **rating:** 90

best drinking 1998–1999 **best vintages** '86, '89, '91, '92, '95, '96, '97 **drink with** Sweet and sour pork • $19.95

Coopers Creek Hawke's Bay Merlot

While the very warm 1998 vintage caused problems for Sauvignon Blanc in New Zealand, it gave rise to what many regard as the best yet vintage in Hawke's Bay for the cabernet varieties. The wine was fermented and macerated on skins for three weeks, and aged for three months in new and used American oak before relatively early bottling. A wine which takes on Chilean Merlot in every way.

🍷🍷🍷🍷 **1998** Vivid, youthful purple-red; the bouquet offers a mix of redcurrant and earthy fruit, with only the barest hint of oak. The palate is fresh and vibrant, with attractive small red berry fruits. Simple perhaps, but enjoyable nonetheless. **rating:** 86

best drinking 1999–2000 **best vintages** NA **drink with** Lamb chops • $16

corbans (auckland) ★★★★☆

426–448 Great North Road, Auckland **region** Auckland Area

ph (09) 837 3390 **fax** (09) 836 0005 **open** Mon–Sat 9–6

winemaker Michael Kluczko **prod.** In excess of 20 million litres (approx 750 000 plus 3 million casks) **est.** 1902

prod. range ($9–36 R) Headed by newly introduced flagship range Cottage Block; then the premium varietal ranges of Corbans Private Bin and Cooks Winemaker's Reserve; the negociant-type label Robard & Butler; and the low-priced Corbans White Label Collection. The regional brands are listed under the other Corbans entries.

summary New Zealand's second largest wine group with a turnover exceeding $NZ100 million, 500 hectares of estate vineyards and another 500 hectares of contracted vineyards. Wine quality is exemplary, setting the pace for others to follow. The wines are exported to Europe, Asia, US, Canada and Iceland.

Corbans White Label Johannesburg Riesling

The cheapest of the Corbans' Rieslings; the Johannesburg tag does not denote a special clone, but is simply the name often used for Riesling in the US. The '98 was a gold medal winner at the 1999 Liquorland Royal Easter Wine Show, Auckland. Outstanding value for money.

🍷🍷🍷🍷🍷 **1998** Light green-yellow; there is a mix of soft lime and tighter, more minerally, fruit on the bouquet; the palate has crisp apple and lime flavours, with well-balanced acidity and a long finish. **rating:** 93

best drinking 1999–2003 **best vintages** '98 **drink with** Thai take-away • $9

Corbans (Auckland) White Label Sauvignon Blanc Semillon

Produced from a blend of 50% Sauvignon Blanc and 50% Semillon, although the region of origin is not stated.

🍷🍷🍷🍷 **1998** Light yellow-green; the bouquet is moderately aromatic, with an attractive touch of herbaceousness yet not excessive. The palate is light and crisp, with herb and mineral hints, and nicely balanced acidity. Happily, fermented dry. **rating:** 84

best drinking 1999–2000 **best vintages** NA **drink with** Cold seafood • $8.95

corbans (gisborne) ★★★★☆

Solander Street, Gisborne **region** Gisborne

ph (06) 867 1269 **fax** (06) 867 8467 **open** By appointment

winemaker Daniel Alcorso **prod.** NA **est.** 1972

prod. range ($8–30 R) The winery supplies Chardonnay, Riesling, Sauvignon Blanc and Müller Thurgau variously used in the Cottage Block, Cooks, Huntaway and Corbans Private Bin ranges.

summary Gisborne unfairly labours under the reputation of simply being a bulk-wine producer, no doubt because the major wine companies have such large holdings there, but do not seek publicity for their wineries, and also no doubt because of the dearth of small, high-profile makers, The Millton Vineyard and Matawhero being the exceptions which prove the rule.

corbans (hawke's bay) ★★★★☆

91 Thames Street, Napier **region** Hawke's Bay

ph (06) 835 2011 **fax** (06) 835 9791 **open** Not

winemaker Kirsty Walton **prod.** NA **est.** 1944

prod. range ($11–36 R) The principal regionally identified brand is Longridge of Hawke's Bay; Chardonnay, Sauvignon Blanc, Gewurztraminer, Chenin Blanc, Pinot Noir, Cabernet Sauvignon, Cabernet Franc and Merlot are produced for the Longridge, Cooks and Corbans Private Bin ranges. Verde Méthode Traditionelle.

summary Established on what was the original McWilliam's winery, although no longer recognisable as such. Arguably the most important red wine production facility for Corbans.

corbans (marlborough) ★★★★☆

Jacksons Road, RD3, Blenheim **region** Marlborough

ph (03) 572 8198 **fax** (03) 572 8199 **open** 7 days 10–4

winemaker David Freschi (previous) **prod.** NA **est.** 1989

prod. range ($9.50–36 R) The Marlborough winery flag-bearers are Stoneleigh Vineyard Riesling, Cottage Block Noble Riesling, Cottage Block Marlborough Sauvignon Blanc, Private Bin Marlborough Chardonnay, Cottage Block Pinot Noir, Cottage Block Merlot and Private Bin Amberly Riesling; also the full Stoneleigh range and many others, including the new (1998) varietal Estate range.

summary A much expanded and upgraded winery, some outstanding vineyards, notably the Stoneleigh Vineyard, and some outstanding winemaking skills (notably from former winemaker

Alan McCorkindale) have all contributed to the quality of the impressive range of wines coming from Corbans Marlborough winery.

Corbans Stoneleigh Riesling

As the name attests, produced from Corbans Stoneleigh Vineyard in Marlborough, and coming (in price terms) near the top of the Corbans range. It is, indeed, virtually a separate brand, and promoted in its own right. As prior vintages (notably the multi-trophy winning '91) attest, ages superbly if vintage conditions are good.

🍷🍷🍷🍷 **1998** Light green-yellow; there is abundant, ripe lime fruit on the bouquet, with touches of passionfruit. The palate is soft, with gentle citrus flavours, and just a touch of perceptible sweetness on the finish. **rating:** 85

best drinking 1999–2003 **best vintages** '86, '89, '90, '91, '94 **drink with** Pan-fried sole • $13.95

Corbans Estate Marlborough Sauvignon Blanc

An important part of a new budget-priced range developed by Corbans, aimed particularly at the export market. The corresponding wine in New Zealand is in fact sourced from Hawke's Bay and is so labelled.

🍷🍷🍷🍷 **1998** Light green-yellow; the bouquet is very fragrant, bursting with passionfruit and gooseberry. The palate has lots of soft, ripe flavour, and is inevitably slightly broad, but at the price one can hardly complain. **rating:** 87

best drinking 1999–2000 **best vintages** '97 **drink with** Roasted red capsicum • $9.50

Corbans Stoneleigh Sauvignon Blanc

A no-frills but ever-reliable unwooded Marlborough Sauvignon Blanc that normally contains about 15% Semillon. Produced from Corbans' Stoneleigh Vineyard, which gives the wine its brand name, and treated to traditional, cold stainless steel fermentation. Typically has lower-than-threshold residual sugar.

🍷🍷🍷🍷 **1998** Light to medium yellow-green; the bouquet is clean, with some fragrance and hints of passionfruit and gooseberry, all excellent for the year. As the bouquet promises, a wine which has life and lift on the palate, with fresh fruit and a crisp, lively finish. **rating:** 89

best drinking 1999–2000 **best vintages** '92, '94, '96, '98 **drink with** Shellfish • $13.95

Corbans Stoneleigh Chardonnay

Produced from the Stoneleigh Vineyard in Marlborough. Barrel-fermented in French oak barriques for six months, with 75% taken through malolactic fermentation.

🍷🍷🍷🍷 **1998** Light to medium yellow-green; the bouquet is quite sophisticated, with a mix of melon, citrus and fig fruit; the palate is textured, with restrained but pleasantly evident mlf characters; the fruit has been retained, and the finish is quite dry. **rating:** 87

best drinking 1999–2001 **best vintages** NA **drink with** Pan-fried veal • $14.95

Corbans Private Bin Marlborough Noble Riesling (375 ml)

Ranks well below the Cottage Block Marlborough Noble Riesling, but nonetheless was the top wine in its class at the 1999 Liquorland Royal Easter Wine Show, winning a gold medal.

🍷🍷🍷🍷🍷(half) **1997** Medium to full yellow-orange; a complex bouquet ranges through peach, lime and marmalade, followed by a very rich, very long palate with lingering acidity. **rating:** 93

best drinking 1999–2003 **best vintages** NA **drink with** Home-made fruit ice-cream • $12

cottle hill winery NR

Cnr State Highway 10 and Cottle Hill Drive, Kerikeri **region** Northland and Matakana
ph (09) 407 5203 **fax** (09) 407 6808 **open** 7 days 10–5
winemaker Michael Bendit **prod.** 1000 **est.** 1997

prod. range ($14–22 CD) Sauvignon Blanc, Chardonnay, Bay Breeze (Sauvignon Blanc Chardonnay), Cabernet Sauvignon, Cabernet Sauvignon Reserve.

summary Michael and Barbara Webb are fugitives from 'the southern California rat-race'. They first arrived in the Bay of Islands on their yacht in 1992, and have now returned to establish Cottle Hill Winery.

covell estate NR

Troutbeck Road, Galatea, RD1, Murupara **region** Waikato and Bay of Plenty
ph (07) 366 4827 **fax** (07) 366 4071 **open** 7 days 10–4 by appointment
winemaker Bob Covell, Robert Covell Jnr, Norm Iles **prod.** 1000 **est.** NA
prod. range ($12–20 CD) Riesling, Chardonnay, Pinot Noir, Rata (Cabernet Merlot).
summary Owners Bob and Desarei Covell have established this vineyard using strict biodynamic organic standards, and mature their wines for extended periods in oak, and give them further time in bottle before release.

crab farm NR

125 Main Road, Bay View, Hawke's Bay **region** Hawke's Bay
ph (06) 836 6678 **open** 7 days 10–5
winemaker Hamish Jardine **prod.** 2600 **est.** 1989
prod. range ($10–17 CD) Gewurztraminer, Sauvignon Blanc, Chardonnay, Pinot Noir, Merlot, Cabernet Sauvignon.
summary Hamish Jardine has worked at both Chateau Reynella and Matawhero; the family vineyards were planted in 1980 and are now mature, so given the equable Hawke's Bay climate there is no reason why the wines should not succeed. A seafood restaurant has recently been added, open from the end of October to Easter.

cross roads winery ★★★★

State Highway 50, Korokipo Road, Fernhill, Napier **region** Hawke's Bay
ph (06) 879 9737 **fax** (06) 879 6068 **open** 7 days 10–5
winemaker Malcolm Reeves, Ken Sanderson **prod.** 10 000 **est.** 1990
prod. range ($10–33 CD) Gewurztraminer, Dry Riesling, Late Harvest Riesling, Chardonnay, Reserve Chardonnay, Oak Aged Sauvignon, Sauvignon, Pinot Noir, Cabernet Merlot, Reserve Cabernet Merlot, Talisman, Stormy Ports.
summary A very successful partnership between Malcolm Reeves, wine journalist and Massey University lecturer, and computer entrepreneur Lester O'Brien. Right from vintage, the wines have received widespread critical acclaim, and have enjoyed great success in wine shows. Production has grown as projected, and a new winery facility has been completed. Draws in part on 5 hectares of various of red vinifera varieties which go to produce the super-premium Talisman red. The identity of those varieties (and hence the blend) is deliberately kept confidential.

daniel schuster ★★★★

192 Reeces Road, Omihi Valley, RD3, Amberley **region** Canterbury
ph (03) 314 5901 **fax** (03) 314 5902 **open** By appointment
winemaker Danny Schuster, Mark Neville **prod.** 5000 **est.** 1986
prod. range ($19–30 R) Canterbury Chardonnay, Petrie Vineyard Selection Chardonnay, Canterbury Pinot Noir, Reserve Pinot Noir.
summary Austrian-born, German-trained Danny Schuster must now rank as one of the leading consultant viticulturists in the world, his clients ranging from Stag's Leap, Neibaum Coppola, Moraga and Spotswoode in the Napa Valley to Antinori in Tuscany. The mix is all the more fascinating when one considers that the Napa Valley makers are all producing powerful and dense

Cabernet-based red wines; that the climate of Tuscany is as far removed from that of Canterbury as one could imagine; and that at home Danny Schuster is known for his pioneering work in the production of Pinot Noir from the Canterbury/Waipara region. Truly, a man for all seasons, producing wines of equally variable (seasonal) quality, at times exhilarating, at times depressing. Exports to Australia, UK and US.

Daniel Schuster Canterbury Chardonnay

Much less winemaker inputs in this wine than with the Petrie Vineyard Selection Chardonnay but perhaps helped by the concentration and small crops of 1997; whichever way, a most attractive wine.

🍷🍷🍷🍷 **1997** Light green-yellow; restrained melon and citrus fruit on the bouquet is supported by subtle oak; the harmonious palate proceeds down precisely the same pathway, with no obvious mlf character to obscure the elegant citrus/melon fruit. Here, too, the oak has not been overplayed. **rating:** 88

best drinking 1999–2003 **best vintages** NA **drink with** Trout mousse • $20

Daniel Schuster Petrie Vineyard Selection Chardonnay

The Petrie Vineyard Selection is drawn from the best blocks on the estate Petrie Vineyard at Rakaia. Hand-harvested grapes are barrel-fermented and lees-aged in a mix of new and used Vosges and Troncais barriques for 15 months with a spontaneous malolactic fermentation. Only 250 cases produced.

🍷🍷🍷🍷 **1998** Medium yellow-green; the subtle use of barrel-ferment and malolactic-ferment influences produce a gently complex and stylish bouquet, which swells out on a powerful, rich and sweet palate, that sweetness coming from fruit, glycerine and alcohol rather than residual sugar. **rating:** 87

best drinking 1999–2000 **best vintages** '96, '97, '98 **drink with** Fresh abalone • $26.95

Daniel Schuster Canterbury Pinot Noir

Typically a blend of grapes grown on the Petrie Vineyard at Rakaia and the Omihi Hill Vineyard at Waipara. An early-drinking style, it seems to have been made very simply, but knowing Schuster's philosophy, there is probably more to it than meets the eye.

🍷🍷🍷🍷 **1998** Medium red-purple; bright, fresh cherry aromas, with the barest whiff of oak are followed by a direct, fresh, firm berry-flavoured palate. All in all, a little simple. **rating:** 86

best drinking 1999–2002 **best vintages** '89, '92, '93, '95, '96, '98 **drink with** Tea-smoked duck • $20

darjon vineyards NR

North Eyre Road, Swannanoa, North Canterbury **region** Canterbury

ph (03) 312 6054 **fax** (03) 312 6544 **open** Fri–Sun and public holidays 12–5

winemaker John Baker **prod.** 500 **est.** 1992

prod. range ($20 CD) The minute estate-produced Swannanoa range of Pinot Noir and Riesling is supplemented by wines from Marlborough, and from other New Zealand and Australian boutique producers.

summary A new arrival on the Christchurch scene, run by former amateur winemaker John Baker and his wife Michelle. The restaurant was opened on site in 1994, coinciding with the first production from the estate plantings (earlier wines were made from purchased grapes) and all of the Darjon wine will be sold through the restaurant, mail list and cellar-door sales.

de gyffarde NR

Gifford Road, Rapaura, RD3, Blenheim **region** Marlborough

ph (03) 572 8189 **fax** (03) 572 8189 **open** At Marlborough Vintners, Rapaura Road, Blenheim

winemaker Graeme Paul **prod.** 4000 **est.** 1995

prod. range ($15–16 CD) Sauvignon Blanc, Chardonnay under both de Gyffarde and Lofthouse labels.

summary English-born owners Di and Rod Lofthouse were 20-year veterans of the film and television industry before establishing their 6-hectare vineyard in 1989, and moving into winemaking from 1995. They have now taken the process one step further by becoming part-owners of Marlborough Vintners Limited. This operates the new winery, commissioned for the 1998 vintage, which makes the wine for de Gyffarde and three other similar-sized Marlborough wineries. Over 90% of the production is exported to the UK, and a little to Australia.

de Gyffarde Single Vineyard Sauvignon Blanc

First made in 1995, with the third vintage moving closer to Rod Lofthouse's stated aim of winning a gold medal, the 1997 receiving a silver medal at the 1997 Air New Zealand Wine Awards. The 1998 is a nice wine, but seems unlikely to attain the gold medal status sought. The Single Vineyard, incidentally, which now forms part of the wine name is the de Gyffarde Vineyard in Rapaura, Marlborough.

ΥΥΥΥ **1998** Light to medium yellow-green; the bouquet has some crisp minerally edges together with the tropical fruit so characteristic of the year. There is a hint of gooseberry on the mid to back palate to accompany the more mineral and herbal notes; the finish is crisp and clean. **rating:** 86

best drinking 1999–2000 **best vintages** NA **drink with** Thai prawns • $16

delegat's wine estate ★★★★☆

Hepburn Road, Henderson **region** Auckland Area

ph (09) 836 0129 **fax** (09) 836 3282 **open** Mon–Fri 10–5, weekends 10–6

winemaker Michael Ivicevich **prod.** NFP **est.** 1947

prod. range ($19.95–25 R) Estate label of Chardonnay, Sauvignon Blanc and Cabernet Merlot; top-of-the-range Proprietors Reserve label of Chardonnay, Fumé Blanc, Cabernet Sauvignon and Merlot. Also vineyard-designated Chardonnay from Hawke's Bay, Oyster Bay Chardonnay and Sauvignon Blanc, and Sauvignon Blanc from Marlborough.

summary Delegat's now sources most of its grapes from Hawke's Bay, utilising its own vineyards there and contract growers, but has added the Oyster Bay range from Marlborough to its repertoire. The quality of the wines is seldom less than good, with a number of excellent wines under the Proprietors Reserve label, conspicuously the Chardonnay.

Delegat's Proprietors Reserve Chardonnay

Made from hand-harvested, estate-grown grapes; 50% is barrel-fermented in a mixture of new and one-year-old French oak barriques, and a sensibly restrained 25% undergoes malolactic fermentation. Consistently one of the highest-rated New Zealand Chardonnays, although in some years I prefer the more direct appeal of the Oyster Bay-style.

ΥΥΥΥΥ **1998** Medium to full yellow-green; a complex and rich bouquet showing the sophisticated use of oak, with the same impact coming through on the palate. However, there is also plenty of tangy fruit, and good acidity on the finish; gold medal Auckland Wine Show. **rating:** 90

best drinking 1999–2000 **best vintages** '86, '89, '91, '92, '94, '96 **drink with** Smoked salmon • $25

de redcliffe estates ★★★☆

Lyons Road, Mangatawhiri Valley, Bombay Hills, near Auckland **region** Waikato and Bay of Plenty
ph (09) 233 6314 **fax** (09) 233 6215 **open** 7 days 9.30–5
winemaker Mark Compton, Daryl Solijan **prod.** 28 000 **est.** 1976
prod. range ($14.95–27.95 R) Marlborough Riesling, Marlborough Sauvignon Blanc, Mangatawhiri Chardonnay, Hawke's Bay Pinot Noir, Hawke's Bay Cabernet Merlot, Hawke's Bay Cabernet Merlot Franc; 'The Dedication Series' is top-end label, along with Mangatawhiri Chardonnay.
summary The Waikato's answer to the Napa Valley, with the $7 million Hotel du Vin, luxury restaurant, wine tours, lectures, the lot; briefly listed on the Stock Exchange, but now Japanese-owned. Production continues to increase, and De Redcliffe now has vineyards in both Hawke's Bay and Marlborough. Increasing production has not prevented winemaker Mark Compton from making some very good wines, although it must be said a few missed the mark.

dry river ★★★★★

Puruatanga Road, Martinborough **region** Wairarapa
ph (06) 306 9388 **fax** (06) 306 9275 **open** Not
winemaker Neil McCallum **prod.** 3500 **est.** 1979
prod. range ($27–36 CD) Craighall Amaranth Riesling, Botrytis Selection Riesling, Chardonnay, Gewurztraminer, Sauvignon Blanc, Pinot Gris, Pinot Noir, Syrah.
summary Winemaker/owner Neil McCallum is a research scientist with a Doctorate from Oxford University, with winemaking very much a part-time occupation. He has justifiably gained an international reputation for the exceptional quality of his wines, which he jealously guards and protects. Each is made in tiny quantities, and sells out immediately on release, but minuscule quantities are now making their way to Australia. Some rate Dry River as New Zealand's best winery, and I'm not sure I would disagree.

Dry River Craighall Amaranth Riesling

Produced from the Craighall Vineyard, part-owned by Dry River and in every respect a remarkable wine. Bob Campbell MW has bestowed on the '98 the highest points he has ever given to a New Zealand wine (of any variety). I am slightly less enthusiastic, if only because the wine shows far more development (admittedly bringing with it amazing complexity) than I would expect from a one-year-old Riesling with an indefinite cellaring future.

🍷🍷🍷🍷🍷 **1998** Full yellow-green; the bouquet is overflowing with classic lime and toast aromas, with the complexity and depth one would expect from a five- to ten-year-old wine. The palate, likewise, has tremendous depth and length, with awesomely rich lime and toast flavours; immaculately balanced, with a seemingly dry finish. **rating:** 95

best drinking 1999–2004 **best vintages** '93, '96, '97, '98 **drink with** Cold-smoked salmon • $21

Dry River Martinborough Chardonnay

Like the Riesling, one of the great classics of New Zealand. Made using the full bag of tricks, including barrel fermentation (but with restrained use of new oak), prolonged lees contact and neatly-handled mlf. In radically different style from so many of New Zealand Chardonnays, relying on finesse, elegance and understatement, rather than brassy blonde.

🍷🍷🍷🍷🍷 **1997** Medium yellow-green; discreet cashew and melon fruit is seamlessly interwoven with oak on both the bouquet and palate. Individual flavours do not stand out at any point in a superbly stylish and intense wine, with excellent acidity, and a long, lingering finish. **rating:** 96

best drinking 1999–2006 **best vintages** '90, '92, '94, '95, '96, '97 **drink with** Milk-fed veal • $30

Dry River Martinborough Pinot Noir

For many observers, New Zealand's finest Pinot Noir and certainly its most sought after. Immaculately crafted, as are all of Neil McCallum's wines, but with all of the robust fruit of Martinborough at its best.

🍷🍷🍷🍷🍷 **1997** Medium red-purple; the initial impression on the bouquet is of quite evident oak, but from this point onwards the fruit takes over, relegating the oak to the back row of the stalls. Plum, forest and briar aromas are followed by a palate which initially seems plush and opulent, but which becomes finer and finer, and ever more interesting, with each sip or mouthful. One of those wines which only reveals all it has to say with the food, and when tasted over a period of an hour or so. There is a silky fineness to the dark plum and sweet oak which is rare in world Pinot Noir terms. Simply glorious; it is hard to see how anyone could dream of doing better than this. **rating:** 97

best drinking 1999–2007 **best vintages** '89, '90, '91, '93, '94, '96, '97 **drink with** Coq au vin • $36

---

eskdale winegrowers NR

Main Road, Eskdale **region** Hawke's Bay
ph (06) 836 6302 **open** Mon–Sat 9–5
winemaker Kim Salonius **prod.** Under 1000 **est.** 1973
prod. range ($NA) Gewurztraminer, Chardonnay, Cabernet.
summary Having gained winemaking experience at McWilliam's, Canadian-born Kim Salonius and family have established a small 4-hectare estate operation, making wines in very small quantities sold cellar door which have gained a strong reputation for consistency of style.

esk valley estate ★★★★☆

745 Main Road, Bay View, Napier **region** Hawke's Bay
ph (06) 836 6411 **fax** (06) 836 6413 **open** 7 days 9–5.30 summer, 9–5 winter
winemaker Gordon Russell **prod.** 30 000 **est.** 1933
prod. range ($14.50–39.95 CD) Black Label Chenin Blanc, Chenin Blanc, Riesling, Chardonnay, Merlot Rose, Merlot; Reserve Chardonnay, Late Harvest Chenin Blanc; also The Terraces, a super-premium single estate vineyard Bordeaux-blend sold by mail order only, when two years old.
summary The little brother in the Villa Maria-Vidal family, but with the ultra-premium The Terraces standing boldly in the top echelon of New Zealand reds, and making the winery rating difficult. Which is not to say that some of the other wines in the portfolio aren't impressive; they are.

Esk Valley Estate Reserve Merlot Malbec Cabernet Sauvignon

The wine is fermented in open fermenters with hand plunging, and is pressed to barrel where it undergoes malolactic fermentation and matures for 17 months before final blending and bottling. One suspects most of this wine would have been incorporated in Esk Valley's classic The Terraces were it not for the fact that no The Terraces was made in 1996. That may have been a tough call for judgment, for this is a truly delicious wine.

🍷🍷🍷🍷🍷 **1996** Medium to full red-purple; the bouquet is quite remarkable, with wonderful sweet, cassis berry fruit cascading through to the palate. Here true ripeness and sweetness run through from start to finish, yet the wine is not the least bit jammy or over-extracted. A great wine by any standard. Gold medal and trophy winner 1998 and 1999 Liquorland Royal Easter Show and gold medal 1998 Air New Zealand Wine Awards. **rating:** 96

best drinking 2000–2010 **best vintages** NA **drink with** Aged beef • $40

fairhall downs estate ★★★★☆

814 Wrekin Road, RD2, Brancott Valley, Marlborough **region** Marlborough
ph (03) 572 8356 **fax** (03) 572 8357 **open** Not
winemaker John Forrest (Contract) **prod.** 7200 **est.** 1996
prod. range ($14–19 ML) Sauvignon Blanc, Chardonnay.
summary Ken Small and Stuart Smith have been grape growers in Marlborough since 1982, supplying Montana and Villa Maria from their 20-hectare vineyard at the top of the Brancott Valley Road. In 1996 they launched their own label, with John Forrest as contract-winemaker, and using the Forrest Estate winery facility. Instant success followed with both the 1996 and 1998 Sauvignon Blancs winning trophies. The wines are exported to Australia and US.

Fairhall Downs Estate Sauvignon Blanc

Draws upon 10 hectares of estate sauvignon blanc and 2.25 hectares of estate semillon; it represents only a small part of the vineyard output. Cold-fermented in stainless steel, the wine has 4–9% (depending on the vintage – 9% in 1998) Semillon included to add structure. A 20-day spread in harvest dates is also used to maximise fruit complexity. The 1998 won the trophy at the 1998 Liquorland Top 100 Wine Show.

🍷🍷🍷🍷🍷 **1998** Light to medium yellow-green; a highly aromatic bouquet exudes tropical passionfruit and gooseberry fruit. The palate is full of fruit, yet retains acidity and balance. An exceptional achievement for a vintage that did not readily give results such as this. **rating:** 94

best drinking 1999–2000 **best vintages** '98 **drink with** Tempura • $17

Fairhall Downs Estate Chardonnay

Produced from grapes grown at the head of the Brancott Valley, and 20% barrel-fermented.

🍷🍷🍷🍷 **1997** Light to medium yellow-green; the bouquet is complex and tangy, with fruit to the fore, but oak evident. A concentrated and powerful palate with nutty flavours ranging through to grapefruit and citrus. Significantly more concentrated than the '98. **rating:** 90

best drinking 1999–2000 **best vintages** '97 **drink with** Wok-fried king prawns • $18.95

fairmont estate NR

Gladstone Road, RD2, Gladstone, Wairarapa **region** Wairarapa
ph (06) 379 8498 **fax** (06) 379 5498 **open** 7 days 9–5
winemaker Jon McNab **prod.** 700 **est.** 1996
prod. range ($12–24 CD) Riesling, Sauvignon Blanc, Chardonnay, Pinot Noir.
summary Jon McNab started his wine career at Martinborough Vineyard 'as a general dogsbody', working for Larry McKenna for two years before taking up an assistant winemaker position in Germany. Thereafter he commuted between Germany and Martinborough Vineyard for several years before coming back to Fairmont Estate and its first on-site vintage in 1997. (The initial vintage was made off site by Chris Lintz at Lintz Estate.) Fairmont is in the Gladstone subregion of Wairarapa, situated on the free-draining alluvial Ruamahanga River terrace.

felton road wines ★★★★★

Bannockburn, RD, Central Otago **region** Otago
ph (03) 445 0885 **fax** (03) 445 0881 **open** 7 days 11–5 Nov–Apr, Mon–Fri 11–5 May–Oct
winemaker Blair Walter **prod.** 6000 **est.** 1991
prod. range ($16.50–34 CD) Dry Riesling, Riesling, Chardonnay, Barrel Fermented Chardonnay, Pinot Noir, Pinot Noir Block 3.
summary Overnight Felton Road has become a major player in the Central Otago wine scene. Twelve hectares of vines were established between 1992 and 1994, with a further 2 hectares to be planted in 1999. The grapes from the first two vintages (1995 and 1996) were sold to Gibbston Valley, the first Felton Road wines being produced from the '97 vintage. Winemaker Blair Walter

has had an impeccable apprenticeship for making Pinot Noir and Chardonnay, including a lengthy stint at Giesen, followed by Sokol Blosser (Oregon), Tarrawarra (Yarra Valley), Rippon Vineyard (Central Otago) and Domaine de L'Arlot (Nuits St Georges, France). He has designed and built a 200-tonne winery drawing upon this varied yet specialised experience; not surprisingly, Felton Road wines have made a major impact in both New Zealand and Australia with the complex, concentrated Pinot Noir leading the way, and even winning a gold medal for the 1998 Riesling at the 1999 Liquorland Royal Easter Show. Small quantities are already being exported to Australia and the UK.

fiddler's green NR

Georges Road, Waipara **region** Canterbury
ph (03) 314 6979 **fax** (03) 314 6978 **open** Mon–Sat 11.30–5
winemaker Petter Evans (Contract) **prod.** 1400 **est.** 1994
prod. range ($15–17 CD) Riesling, Sauvignon Blanc.
summary Fiddler's Green has been established by Christchurch lawyer Barry Johns and his wife Jenny; 20 hectares of vines (5 hectares each of sauvignon blanc and semillon, 4 hectares each of riesling and pinot noir, and 2 hectares of chardonnay) have been established on the 30-hectare site. New Burgundian clones of both chardonnay and pinot noir have been chosen, and the expectation is that Pinot Noir will be the flagship wine.

Fiddler's Green Waipara Riesling

The 11 tonnes of riesling which went to make this wine were hand-picked over a four-day period from 31 March to 3 April 1998, and the wine was cold-fermented in stainless steel. 752 cases were made.

YYYY **1998** Light yellow-green; the bouquet is clean, moderately intense with delicate lime blossom aromas replicated on the palate, although the wine then does soften and finish slightly short. **rating:** 86

best drinking 1999–2002 **best vintages** NA **drink with** Seafood antipasto • $12.50

Fiddler's Green Waipara Sauvignon Blanc

Like the Fiddler's Green Riesling, hand-picked (25-26 March) and cold-fermented in stainless steel. 587 cases made.

YYYY **1998** Light to medium yellow-green; the bouquet has an attractive aromatic fruit lift, with ripe apple notes and a touch of spice. The palate has some delicacy, with neatly counterbalanced sweet fruit and good acidity on the finish. **rating:** 89

best drinking 1999–2000 **best vintages** NA **drink with** Blue swimmer crab • $13.50

forrest estate ★★★★☆

Blicks Road, Renwick, Marlborough **region** Marlborough
ph (03) 572 9084 **fax** (03) 572 9084 **open** Mon–Sun 10–5 Sept–May
winemaker John Forrest **prod.** 10 000 **est.** 1989
prod. range ($16.95–32.95 R) Marlborough Chardonnay, Sauvignon Blanc, Semillon, Riesling, Gibsons Creek (Cabernet blend), Indian Summer Late Harvest, Botrytis Riesling, Newton Forrest Cornerstone, Merlot.
summary Former biochemist and genetic engineer John Forrest has had considerable success since his first vintage in 1990, relying initially on purchased grapes but with a 30-hectare vineyard now planted. Wine quality has been remarkably consistent right across the range, perhaps reflecting John Forrest's strong grounding in chemistry.

Forrest Estate Sauvignon Blanc

An interesting wine in which a small or greater portion of Semillon is used in the blend, a small proportion in the cooler years and a higher proportion in the warmer years; overall ranges between 4% and 15%. Part of the wine is taken through malolactic fermentation. John Forrest is forever looking for ways to add weight and complexity to his Sauvignon Blanc, including multiple pickings over varying degrees of ripeness.

🍷🍷🍷🍷🍷 **1998** Light to medium yellow-green; a clean, elegant passionfruit-accented bouquet introduces a lovely wine with similarly elegant passionfruit, gooseberry and tropical fruit flavours. Is neither heavy nor sweet. **rating:** 94

best drinking 1999–2000 **best vintages** '91, '92, '94, '95, '97, '98 **drink with** Fresh snapper • $15

Forrest Estate Marlborough Chardonnay

Intensive viticulture (trimming, leaf-plucking and bunch-thinning) to limit the yield to 10 tonnes per hectare. Between 15% and 30% of the wine is barrel-fermented and taken through malolactic fermentation, and all of the wine will spend between six and 13 months on lees. In a strange double, the barrels are from Dargaud & Jaegle and Demptos.

🍷🍷🍷🍷 **1997** Medium yellow-green; the oak comes rocketing through on the bouquet and also on the palate; however, the wine is greatly helped by good acidity, which picks up the fruit and which gives length. **rating:** 85

best drinking 1999–2001 **best vintages** NA **drink with** Pan-fried veal • $20

Forrest Estate Botrytis Riesling (375 ml)

Depending on the season, John Forrest will make two late harvest styles, one or none at all. For example in 1998 he made an Indian Summer Riesling, from raisined rather than botrytised fruit, but also made a Botrytis Riesling. An exceptionally good wine, it has 8.5 grams of acidity and 144 grams per litre of residual sugar.

🍷🍷🍷🍷🍷 **1998** Medium to full yellow-green; the bouquet offers intense botrytis aromas yet magically retains the pure lime juice varietal character of riesling. This is not an easy double to achieve. The palate fulfils the promise of the bouquet, gloriously full and balanced by lingering acid on the finish. **rating:** 94

best drinking 1999–2004 **best vintages** '98 **drink with** Baked apples • $22

foxes island wines ★★★★

PO Box 1039, Blenheim **region** Marlborough
ph (03) 578 6221 **fax** (03) 578 4482 **open** 7 days 10–3 at Wairau River Wine Shop
winemaker John Belsham **prod.** 2380 **est.** 1992
prod. range ($28–30 R) Chardonnay, Pinot Noir.
summary Former Hunter's winemaker John Belsham runs Rapaura Vintners contract-winemaking business (formerly Vintech), but since 1992 has made small quantities of wine under the Foxes Island label, the name of the pinot noir vineyard he is establishing. The Chardonnay is – and will in the future be – made from purchased grapes.

Foxes Island Chardonnay

John Belsham deliberately adopts a very low profile for Foxes Island, emphasising that his main business is the contract-making of wine for others through Rapaura Vintners (formerly Vintech) and emphatically not Foxes Island. The grapes are purchased from the Rose and Jenkins families. The wine spends 12 months in oak.

🍷🍷🍷🍷🍷(half) **1997** Medium yellow-green; the bouquet is complex, but has relatively restrained barrel-ferment and malolactic-ferment inputs. Suave, smooth melon fruit, with just a hint of creamy/nutty mlf influence, runs through a long, well-balanced palate. **rating:** 90

best drinking 1999–2001 **best vintages** '94, '96, '97 **drink with** Blanquette of veal • $30

framingham ★★★★

Conders Bend Road, Marlborough **region** Marlborough
ph (03) 572 8884 **fax** (03) 572 9884 **open** 7 days 11–5
winemaker Ant Mackenzie **prod.** 16 000 **est.** 1982
prod. range ($15–28 CD) Classic Riesling, Dry Riesling, Medium Riesling, Reserve Late Harvest Riesling, Sauvignon Blanc, Chardonnay, Merlot.
summary Rex and Paula Brooke-Taylor established their 13-hectare vineyard in 1981, being content to sell the grapes in the intervening years to various makers, most conspicuously Grove Mill and Corbans. Since 1994 a rapidly increasing proportion of their grape production has been vinified under the Framingham label, with exports to Australia, UK, Denmark, Holland and US.

Framingham Sauvignon Blanc

Made in a no-frills style, but with some quite excellent fruit in the majority of vintages.

🍷🍷🍷🍷 **1998** Light green-yellow; the bouquet is clean, of moderate intensity, offering a mix of herb/grass and more minerally notes. The palate is direct, no frills, with fair length and underlying fruit sweetness carrying through to the finish. All in all, a better-than-average outcome for '98. **rating:** 87

best drinking 1999–2000 **best vintages** NA **drink with** Fried oysters • $15

Framingham Reserve Late Harvest Riesling (375 ml)

Produced from grapes grown in the Renwick district of Marlborough, ultra-late-picked, and with consequently high sugar levels.

🍷🍷🍷🍷 **1998** Glowing yellow-green; there are strong peach overtones to the fruit, together with lots of lime on the bouquet. An impressive, powerful palate in big, bold style. **rating:** 89

best drinking 1999–2004 **best vintages** NA **drink with** Fruit tart • $28

fromm winery ★★★★☆

Godfrey Road, RD2, Blenheim **region** Marlborough
ph (03) 572 9355 **fax** (03) 572 9366 **open** Sat 11–5, summer holidays Tues–Sat 11–5
winemaker Hatsch Kalberer, George Fromm **prod.** 6000 **est.** 1992
prod. range ($15–38 R) Released under the La Strada label are Reserve Chardonnay, Pinot Noir, Clayvin Vineyard Pinot Noir, Syrah, Merlot, Merlot Cabernet Syrah, Reserve Malbec, Clayvin Vineyard Cabernet Sauvignon.
summary Swiss-born and resident George Fromm, wife Ruth, and former Matawhero winemaker Hatsch Kalberer have formed a dynamic team to produce exceptionally full-flavoured wines, with the emphasis on red wines. Fromm has 21 hectares of estate vineyards coming into maturity, and, while the emphasis remains on Pinot Noir, produces an utterly eclectic range of wines, with Sangiovese in the pipeline. Exports to Australia, UK, Hong Kong, Austria and (of course) Switzerland.

fullers NR

86 Candia Road, Swanson, Auckland **region** Auckland Area
ph (09) 833 7026 **fax** (09) 832 1778 **open** Mon–Sat 9–6, Sun 11–6
winemaker Ray Allen **prod.** 300 **est.** NA
prod. range ($6–18 CD) Sauvignon Blanc, Cabernet Merlot and a range of sparkling and fortified wines.
summary A west Auckland landmark specialising in functions for up to 240 people at a time, year round, with all of the estate-produced wine sold through this outlet.

gatehouse wines NR

Jowers Road, RD6, Christchurch **region** Canterbury
ph (03) 342 9682 **open** Mon–Sat 10–5 Nov–Feb, Sat 10–5 Mar–Oct
winemaker Peter Gatehouse **prod.** 1300 **est.** 1989
prod. range ($NA) Chardonnay, Gewurztraminer, Riesling, Pinot Noir, Merlot, Cabernet Sauvignon.
summary The Gatehouse family made its first wines in 1989 from estate plantings commenced in the early 1980s. The initial release was under the Makariri label, but the wines will henceforth be released under the Gatehouse label, and it is planned to increase production through the purchase of additional grapes from contract growers.

gibbston valley ★★★★

State Highway 6, Gibbston, RD1, Queenstown **region** Otago
ph (03) 442 6910 **fax** (03) 442 6909 **open** 7 days 10–5.30
winemaker Grant Taylor **prod.** 15 000 **est.** 1989
prod. range ($15–40 R) Riesling, Gewurztraminer, Central Otago Sauvignon Blanc, Marlborough Sauvignon Blanc, Pinot Gris, Greenstone (unoaked Chardonnay), Chardonnay, Pinot Noir, Gold River (Pinot Noir), Reserve Pinot Noir.
summary A highly professional and attractive winery, restaurant and cellar-door sales facility situated near Queenstown which has been an outstanding success since the day it opened. Viticulture poses special problems, and both varietal selection and determining style will inevitably take time. However, a neat, modern production facility, New Zealand's largest cellar caves (opened December 1995). The arrival of Grant Taylor as winemaker, with significant international experience, has lifted wine quality further, particularly with the elusive Pinot Noir. The wines are distributed in Australia through Negociants Australia.

Gibbston Valley Marlborough Sauvignon Blanc

Right from the outset, Gibbston Valley has not hesitated to venture outside the confines of Central Otago to supplement its wine intake. In four vintages since 1992 it has made a Sauvignon Blanc from Marlborough grapes, wines which have won many medals (including gold) in New Zealand wine shows, the most recent success going to the '97 which won a gold medal at the 1997 Liquorland Top 100 Wine Competition. The '98 scored well at the 1999 *Winewise* Small Makers Competition because it offered greater fruit intensity and varietal character than the majority.

🍷🍷🍷🍷🍷 (4 full, 1 empty) **1998** Light green-yellow; the bouquet is quite intense and clean albeit with lower aromatic qualities than the palate suggests it should have. Once you have the wine in the mouth, there is plenty of ripe fruit without in any way compromising the varietal character; well-balanced acidity and length. A good wine. **rating:** 92

best drinking 1999–2000 **best vintages** '92, '94, '96, '97, '98 **drink with** Whitebait • $20

Gibbston Valley Pinot Noir

Produced in limited quantities from the relatively small estate plantings of the variety. The short growing season carries with it threats at both the start and finish of the season; the yields are low, and the wine is very much subject to vintage conditions. The '97 keeps up the record.

🍷🍷🍷🍷🍷 **1997** Youthful purple-red; a very stylish and complex bouquet offering an amalgam of fruit, oak and carbonic maceration characters. The palate has abundant flavour and structure, complex and rich, and will age well. **rating:** 94

best drinking 1999–2005 **best vintages** '94, '96, '97 **drink with** Spiced quail • $29

---

giesen estate ★★★★☆

Burnham School Road, Burnham **region** Canterbury
ph (03) 347 6729 **fax** (03) 347 6450 **open** Mon–Sat 10–5
winemaker Andrew Blake **prod.** 50 000 **est.** 1981
prod. range ($9–35 CD) Müller Thurgau, Riesling, Dry Riesling, Extra Dry Riesling, Late Harvest Riesling, Marlborough Sauvignon Blanc, Canterbury Burnham School Road Chardonnay, Reserve Chardonnay, Pinot Noir and Reserve Pinot Noir (in each case both from Canterbury and Marlborough); Merlot; Selwyn River Chenin Chardonnay and Pinot Noir.
summary Determination, skill and marketing flair have seen Giesen grow from obscurity to one of the largest family-owned and run wineries in New Zealand. Given the Giesens' Rhine Valley origins it is not surprising that they have done so well with aromatic, non-wooded white wines, but have also gained acclaim for impressive Chardonnay and Pinot Noir. The ever-increasing production has allowed distribution throughout Europe, the US and Canada, and the Pacific including Australia through Negociants.

Giesen Estate Marlborough Sauvignon Blanc

In recent years Giesen has established its own vineyards in Marlborough, and this wine now comes from its Dillon's Point Vineyard. In typical Giesen fashion, it is also fermented dry, using partial skin contact and extended lees contact to add both complexity and balance.

🍷🍷🍷🍷🍸 **1998** Light green-yellow; lively citrus and gooseberry aromas announce a fresh, crisp wine with touches of herb along with the citrus and gooseberry, and a long palate aided by perfectly judged acidity. An extremely good outcome for a very difficult vintage. **rating:** 90

best drinking 1999–2000 **best vintages** NA **drink with** Full-flavoured Asian seafood • $14.95

Giesen Reserve Canterbury Burnham School Road Chardonnay

Estate-grown, and given the full winemaking treatment before being held back and given an extra year's bottle age to release. The wine spends 18 months on lees in French oak barriques; 1000 cases were made in 1997.

🍷🍷🍷🍷🍸 **1997** Medium to full yellow-green; the bouquet is complex with charry, high toast barrel-ferment characters which carry through onto the powerful, spicy, nutty forepalate; intense citrus and melon fruit then takes over and carries through to the finish. **rating:** 93

best drinking 1999–2003 **best vintages** '97 **drink with** Veal parmigiana • $35

Giesen Estate Reserve Marlborough Chardonnay

As one would expect, given the full treatment with 100% barrel fermentation, 50% of the barrels new. Made only in good vintages; none made in 1995 or 1996.

🍷🍷🍷🍷🍷 **1997** Medium yellow-green; the bouquet is rich with nutty, malty, honeyed aromas, showing less high toast oak than the Canterbury Reserve. The palate is smooth, with ripe fig and peach fruit seamlessly married with high-quality oak. A feature of the wine is its great length. **rating:** 94

best drinking 1999–2003 **best vintages** '94, '97 **drink with** Fresh Tasmanian salmon • $35

gillan wines ★★★★

454B Rapaura Road, Blenheim, Marlborough **region** Marlborough
ph (03) 572 9979 **fax** (03) 572 9980 **open** 7 days 10.30–5 spring, summer and autumn, and by appointment in winter
winemaker Sam Weaver **prod.** 7000 **est.** 1992
prod. range ($15.95–27.95 CD) Eastfields Sauvignon Blanc, Chardonnay, Merlot, Brut Reserve.

summary Gillan Wines is a partnership between English-born Toni and Terry Gillan and local vignerons Hamish and Anne Young. A white, Mediterranean-style wine cellar and restaurant (serving tapas-style food) opened in December 1996. While being Mediterranean-style, architect Neil Charles-Jones believes it is also a building 'which belongs in the Marlborough landscape while quietly alluding to the great Champagne cellars of France'. That is quite an achievement.

gladstone vineyard ★★★★☆

Gladstone Road, RD2, Carterton, Wairarapa **region** Wairarapa
ph (06) 379 8563 **fax** (06) 379 8563 **open** Tues–Sun 11–5
winemaker Christine Kernohan **prod.** 2000 **est.** 1987
prod. range ($11–25 CD) Riesling, Sauvignon Blanc, Fumé Blanc, Chardonnay, Cabernet Sauvignon Merlot, Red Label Cabernet Merlot, Cabernet Sauvignon, Cafe Red.
summary Gladstone Vineyard was acquired from founder Dennis Roberts by Christine and David Kernahan in February 1996, with Christine now in charge of winemaking. That the transition has been without pain is handsomely demonstrated by the quality of the Sauvignon Blanc.

glenmark wines ★★★

Mackenzies Road, Waipara **region** Canterbury
ph (03) 314 6828 **fax** (03) 314 6828 **open** 7 days 11–5
winemaker Kym Rayner **prod.** 2700 **est.** 1981
prod. range ($11–25 CD) Waipara Riesling, Weka Plains Riesling, Waipara White, Chardonnay, Gewurztraminer, Pinot Noir.
summary Kym Rayner is a very experienced winemaker, and Glenmark is an important part of the Canterbury scene, notwithstanding its relatively small size. Much of the wine is sold cellar door, with the Weka Plains Wine Garden offering a full restaurant service and wine by the glass from October through to April. Bookings are essential.

glover's vineyard ★★☆

Gardner Valley Road, Upper Moutere **region** Nelson
ph (03) 543 2698 **open** 7 days 10–6
winemaker David Glover **prod.** 1900 **est.** 1984
prod. range ($11–18 CD) Sauvignon Blanc, Riesling, Late Harvest Riesling, Pinot Noir, Cabernet Sauvignon.
summary David Glover studied winemaking and viticulture at Charles Sturt University in southern New South Wales during a 17-year stay in Australia. He returned with wife Penny to establish their own vineyard in 1984, struggling with birds and other predators before producing their first wines in 1989. The quality of the white wines has been good, although the muscular, brawny Pinot Noir has pleased others more than it has me.

goldwater estate ★★★★★

18 Causeway Road, Putiki Bay, Waiheke Island **region** Waiheke Island
ph (09) 372 7493 **fax** (09) 372 6827 **open** 7 days 11–4 summer
winemaker Kim Goldwater, Martin Pickering **prod.** 25 000 **est.** 1978
prod. range ($18–90 CD) Marlborough Roseland Chardonnay, Delamore Chardonnay, Dog Point Marlborough Sauvignon Blanc, Waiheke Island Esslin Merlot, Waiheke Island Cabernet Merlot Franc.
summary Goldwater Estate goes from strength to strength. Having initially forged a reputation for its Waiheke Island Cabernet Merlot Franc, it has built on that with its superb Waiheke Island Esslin Merlot and a range of beautifully crafted wines made from Marlborough grapes, with the limited volume Waiheke Island-sourced Delamore Chardonnay providing additional support. The wines come from 11.5 precious hectares on Waiheke Island, and 36 hectares in Marlborough. The wines are exported to Australia (DWS), US, UK, Europe and Asia.

Goldwater Estate Dog Point Marlborough Sauvignon Blanc

Produced from the 11-hectare Dog Point Vineyard in Marlborough, established by Goldwater Estate some years ago. A luscious and striking wine; Kim Goldwater may be without technical training, but certainly knows how to make wine. A small portion of the wine is barrel-fermented to add complexity.

🍷🍷🍷🍷🍷 **1998** Medium yellow-green; the bouquet is clean with abundant ripe, tropical fruit aromas, the palate with near identical tropical gooseberry flavours. It has very considerable length, appropriate acidity and, improbably perhaps, carries its 14.3 degrees alcohol. **rating:** 90

best drinking 1999–2001 **best vintages** '96 **drink with** New Zealand whitebait • $18

Goldwater Estate Marlborough Roseland Chardonnay

Produced from Goldwater's Marlborough vineyard holdings, fermented in French oak barriques and which spends ten months on lees in that oak. Typically has a towering 14 degrees plus alcohol.

🍷🍷🍷🍷 **1997** Medium yellow-green; the bouquet is complex and rich with spicy barrel-ferment aromas. Yellow peach and melon fruit come through on the palate, with the same potent but high-quality oak of the bouquet. Given the benefit of the doubt on the level of oak; the intensity of the fruit should come through with age. **rating:** 85

best drinking 1999–2003 **best vintages** '96, '97 **drink with** Smoked chicken salad • $23

Goldwater Estate Waiheke Island Esslin Merlot

Waiheke Island continues to unveil extremely impressive red wines, and although it cannot challenge Hawke's Bay in terms of volume, is certainly laying down the gauntlet on quality. The wine may be expensive, but it is of world class. The Goldwaters ascribe part of its success to the crumbly clay soils on which it is grown.

🍷🍷🍷🍷🍷 **1997** Medium red-purple; the bouquet is wonderfully fragrant, with a mix of spicy oak and sweet cherry fruit. A glorious wine in the mouth, with soft, silky, supple fine tannins and what can only be described as sexy oak. **rating:** 97

best drinking 2000–2010 **best vintages** '95, '97 **drink with** Venison • $90

Goldwater Estate Waiheke Island Cabernet Merlot Franc

Another extremely distinguished red from Goldwater Estate, acknowledged by Bob Campbell as a New Zealand classic, and it is not hard to see why. The wine spends 17 months in French oak, 50% new, and is typically a blend of about 50% Cabernet Sauvignon, 40% Merlot and 10% Cabernet Franc.

🍷🍷🍷🍷🍷 **1996** Light to medium red-purple; the bouquet is of light to medium intensity, with a mix of savoury, foresty and berry aromas supported by subtle oak. Typically enough, not a heavyweight on the palate, but has truly excellent length, sustained by gentle, savoury tannins and deft use of oak. **rating:** 94

best drinking 2001–2008 **best vintages** '95, '96 **drink with** Lamb fillets • $50

greenhough vineyard ★★★★

Patons Road, RD1, Richmond, Nelson **region** Nelson

ph (03) 542 3868 **fax** (03) 542 3462 **open** Mon–Sat 10–5 Dec–Mar

winemaker Andrew Greenhough **prod.** 2000 **est.** 1991

prod. range ($14–20 CD) Riesling, Sauvignon Blanc, Chardonnay, Pinot Noir, Hope Vineyard Pinot Noir.

summary Yet another name change for what was initially Ranzau, then Pelorus and now Greenhough – the last change a sensible one, dictated by the confusion with the Pelorus Méthode Champenoise of Cloudy Bay. Under whatever name, Andrew Greenhough makes appealing wines, notably the Sauvignon Blanc.

Greenhough Vineyard Hope Vineyard Pinot Noir

Made using similar sophisticated techniques to those employed with the Nelson Pinot Noir.

🍷🍷🍷🍷 **1997** Bright and clear purple-red; the bouquet is fresh, clean, of light to medium intensity showing plummy fruit and subtle oak. The palate is pleasant, offering direct plum fruit flavours in the no-frills style the vineyard seems to produce. **rating:** 86

best drinking 1999–2002 **best vintages** '97 **drink with** Char-grilled salmon • $20

Greenhough Vineyard Nelson Pinot Noir

Four clones of pinot noir (under New Zealand numbering system 22, 5, 6 and 10/5) are utilised and fermented separately prior to blending. Pre-fermentation cold maceration for three days is followed by a warm fermentation, with the total time on skins between ten and 12 days. Matured in a mix of new to three-year-old French oak barriques; coarse earth filtration prior to bottling.

🍷🍷🍷🍷 **1997** Light to medium red; a bright and fresh bouquet with strawberry and cherry aromas together with a touch of oak. The palate is similarly bright and fresh, with red cherry fruit flavours, and slightly dusty/grippy tannins on the finish which may well soften with a little more bottle age. **rating:** 86

best drinking 1999–2002 **best vintages** '97 **drink with** Mushroom risotto • $20

grove mill ★★★★★

Waihopai Valley Road, Marlborough **region** Marlborough

ph (03) 572 8200 **fax** (03) 572 8211 **open** 7 days 11–5

winemaker David Pearce, Sarah Hennesy **prod.** 53 000 **est.** 1988

prod. range ($12.95–30 R) Marlborough Sauvignon Blanc, Chardonnay, Riesling and Pinot Noir, Winemakers Reserve Pinot Noir, Merlot; Lansdowne Chardonnay; also lower-priced Sanctuary Sauvignon Blanc, Chardonnay and Pinotage.

summary Has firmly established itself as a producer of wines of consistently high quality in substantial volumes. Its success in wine shows both in NZ and elsewhere (particularly Australia) underlines the continuing achievements of the winemaking team headed by David Pearce. The wines are now distributed through NZ by Eurowine, Australia by Fesq & Co, and are also found in Hong Kong, US, Canada, Singapore, Japan, UK and the Netherlands.

Grove Mill Marlborough Sauvignon Blanc

It would seem I rate this wine a little higher than most New Zealand judges and writers, although it certainly receives strong support across the board from the critics. To a lesser degree than the Lansdowne Chardonnay, it is a 'love it or hate it' style, and one certainly needs to know what has gone on in the making of the wine. This is a long way from the straightforward no-frills cutthroat Marlborough Sauvignon Blanc-style, utilising partial oak ageing, partial malolactic fermentation, some Semillon if the year is right and, generally speaking, of a lot of thought. If the '96 was outstanding, and it was, the '97 is even more brilliant – a wonderful wine. The extraordinary success of the '98 leaves me no choice but to give it 94 points.

🍷🍷🍷🍷🍷 **1998** Light green-yellow; the bouquet is clean, restrained, with moderately ripe tropical fruit, but still lacking focus in the manner of virtually all '98s. However, the wine is exceptionally easy drinking, with enough varietal character to convince yet not so much to confront. Passionfruit and gooseberry flavours are there, and it is no doubt these which carried the wine to its gold medals at the 1998 Air New Zealand Wine Show, the 1999 Sydney International Wine Competition and the 1999 Liquorland Royal Easter Wine Show. **rating:** 94

best drinking 1999–2000 **best vintages** '92, '94, '96, '97, '98 **drink with** Wairau River trout • $17

Grove Mill Marlborough Pinot Noir

I may stand to be corrected, but I believe this to be the first Pinot Noir made by Grove Mill. A fairly rapid and straightforward fermentation is followed by a little under ten months in barrel.

🍷🍷🍷🍷 **1997** The wine has good colour; the aromas have a slightly briary/foresty/stemmy edge, but are far from unpleasant. The palate, too, has some character and style, but the echoes of the bouquet come on a slightly sappy/green edge to the finish. Almost very good. **rating:** 85

best drinking 1999–2000 **best vintages** NA **drink with** Pastrami • $23

gunn estate NR

85 Ohiti Road, RD9, Hastings **region** Hawke's Bay

ph (06) 874 3250 **fax** (06) 874 3256 **open** By appointment

winemaker Denis Gunn **prod.** 2000 **est.** 1994

prod. range ($16–30 R) Sauvignon Blanc, Chardonnay, Reserve Chardonnay, Dessert Chardonnay, Merlot Cabernet, Reserve Merlot.

summary Denis and Alan Gunn have been contract grape growers since 1982, with 15 hectares of vines providing grapes for many of the best-known names in the Hawke's Bay region. In the interim, Denis Gunn graduated from Roseworthy College, Australia, and became assistant winemaker at Villa Maria in 1993, moving to Kemblefield in 1995, where the 1995 Gunn Estate wines were made. In 1996 production moved to a newly constructed on-site winery.

harrier rise vineyard ★★★★

748 Waitakere Road, RD1, Kumeu **region** Auckland Area

ph (09) 412 7256 **fax** (09) 412 7256 **open** Weekends 12–6

winemaker Tim Harris **prod.** 3600 **est.** 1986

prod. range ($16–30 CD) Cabernet Sauvignon Cabernet Franc, Merlot Cabernets, Cabernet Franc, Uppercase Merlot.

summary The project of Auckland lawyer and wine-writer Tim Harris and wife Alix. The resolution of some complicated vineyard ownership arrangements in 1996 has led to the change of name from Waitakere Road to Harrier Rise, and to the Harrises acquiring full ownership of the 4-hectare Harrier Rise Vineyard, replete with 15-year-old cabernet sauvignon, merlot and cabernet franc. Wine quality is impressive, with ripe flavours and tannins.

hau ariki wines NR

Regent Street, Martinborough **region** Wairarapa

ph (06) 306 9360 **fax** (06) 306 9360 **open** Weekends 9–5

winemaker Chris Lintz (Contract) **prod.** NA **est.** 1994

prod. range ($18–28 CD) Sauvignon Blanc, Rosé, Pinot Noir, Cabernet Sauvignon.

summary Hau Ariki is housed in New Zealand's first marae to be making and selling wine on a commercial basis. The 3.5-hectare vineyard was developed by the marae kaumatua (elder) and managing director George Hawkins together with former Mission Vineyard viticulturist Mike Eden who had retired to Martinborough from Hawke's Bay.

hawkesbridge wines ★★★★

Hawkesbury Road, Renwick, Marlborough **region** Marlborough

ph (03) 572 8024 **fax** (03) 572 9489 **open** 7 days 10.30–4.30

winemaker Contract **prod.** 2200 **est.** 1991

prod. range ($16–24 R) Willowbank Vineyard Sauvignon Blanc, Sophie's Vineyard Chardonnay, Sophie's Vineyard Chardonnay Reserve, Pinot Noir, Cabernet Merlot, Merlot.

summary Hawkesbridge Wines and Estates (to give it its full name) is presently chiefly a contract grape grower, but export demand for its wines is likely to see half the production from its 16 hectares of vines vinified under the Hawkesbridge label. Exports to Australia, UK and Holland.

heron's flight ★★★★☆

Sharp Road, Matakana **region** Northland and Matakana
ph (09) 422 7915 **fax** (09) 422 7915 **open** 7 days 10–6
winemaker David Hoskins **prod.** 1500 **est.** 1987
prod. range ($17–50 CD) La Volee (Chardonnay), Barrique Fermented Matakana Chardonnay, La Cerise (Merlot), Sangiovese, Montepulciano, Matakana Cabernet Sauvignon Merlot, Cabernet Merlot.
summary Having established a small vineyard in 1987, David Hoskins and Mary Evans leased the defunct Antipodean Winery which was the scene of so much marketing hype and excitement in the mid-1980s. (The Antipodean has built a new winery since.) The first Heron's Flight wine (a densely-coloured and flavoured Cabernet Sauvignon) was produced from the 1991 vintage, and was a gold medal winner. The '94 Cabernet Merlot was, if anything, even better, but no wines made in either 1995 or 1996. A small quantity of Sangiovese (New Zealand's first) was made in 1997. No recent tastings.

highfield estate ★★★☆

Brookby Road, RD2, Blenheim **region** Marlborough
ph (03) 577 7133 **fax** (03) 577 7339 **open** 7 days 10–5
winemaker Alistair Soper **prod.** 20 000 **est.** 1990
prod. range ($14.50–50 CD) Riesling, Late Harvest Riesling, Sauvignon Blanc, Chardonnay, Merlot, Sparkling Merlot; Elstree range of Brut Reserve Riesling (a table wine, not sparkling), Reserve Sauvignon Blanc, Chardonnay, Optima (Chardonnay), Botrytised Semillon Sauvignon Blanc, Czar (dessert Riesling), Cuvée Sparkling.
summary Highfield Estate was purchased by an international partnership in late 1991, the English and Japanese limbs of which are associated with the French Champagne House Drappier. The ornate Tuscan-style winery which has since been built is Marlborough's answer to some of the more bizarre edifices of the Napa Valley. Wine quality has been erratic; the flamboyantly packaged Elstree range was introduced onto the market in late 1997 in the super-premium sector, some justifying their price, others not.

Highfield Estate Waipara Riesling

Contract-grown in the Waipara district of Canterbury and an impressive wine, in which the 15 grams per litre of residual sugar is barely noticeable.
🍷🍷🍷🍷 **1998** Light yellow-green; the aromas run through from mineral to lime, and the powerful, concentrated palate has good length and balance. **rating:** 89
best drinking 1999–2004 **best vintages** NA **drink with** Smoked ham hock • $14.50

Highfield Estate Sauvignon Blanc

Produced from estate-grown Sauvignon Blanc and conventionally fermented in stainless steel. Quality has wobbled around a bit in recent years; the '98 is a partial return to form.
🍷🍷🍷🍷 **1998** Light to medium yellow-green; the bouquet is clean, quite crisp and herbal, and without the volatile acidity problems which have appeared from time to time in previous years. The palate has some crisp, slightly stony characters to offset the tropical ripeness of the fruit, although there is a lack of intensity. **rating:** 85
best drinking 1999–2000 **best vintages** NA **drink with** Marinated octopus • $16

Highfield Estate Elstree Optima Chardonnay

The quality flagship for the Highfield Elstree wines, confusingly named because Optima is also a grape variety. Here it is simply a play on the word. It is only made in exceptional vintages (three years in the last seven) and is hand-picked and whole-bunch pressed; it spends 18 months in barrel on lees, 35% new from Seguin Moreau.

🍷🍷🍷🍷🍷 **1997** Full yellow-green; the bouquet is very smooth, showing an assured touch with oak. The fully ripe palate has plenty of powerful, peachy fruit, and the oak is integrated though quite obvious. The wine finishes with soft acidity. **rating:** 93

best drinking 1999–2003 **best vintages** NA **drink with** Grilled spatchcock • $50

Highfield Estate Elstree Cuvée

A blend of 50% Chardonnay and 50% Pinot Noir which spends three years on yeast lees prior to disgorgement. Another wine in the Elstree range which deserves the premium rating. The '95 vintage has won a string of trophies and medals, and topped the sparkling wine class at the 1999 Auckland Wine Show.

🍷🍷🍷🍷 **1995** Light to medium straw-yellow; a wine which goes to the extreme end of style, with pronounced aldehydic characters which are clearly interpreted by the majority of judges as desirable (and which are no doubt intentional), characters which certainly add complexity but are not for the faint-hearted. It is a wine which you must taste for yourself and decide whether or not you like the style; my points and rating are a nominal, indeed cowardly, compromise. **rating:** 88

best drinking 1999–2000 **best vintages** NA **drink with** Fresh shellfish • $36

Highfield Estate Elstree Reserve Pinot Noir

Ideal ripening conditions in the 1997 vintage, coupled with very low yields, has resulted in the first Elstree Reserve Pinot Noir. An excellent wine which thoroughly deserves its status as a Reserve.

🍷🍷🍷🍷🍷 **1997** Medium to full red-purple; the bouquet is complex and savoury, barely hinting at the rich and concentrated dark plum fruit that fills the mouth. Finishes with long, fine, silky tannins and will develop well. **rating:** 90

best drinking 1999–2003 **best vintages** '97 **drink with** Ragout of venison • NA

holmes brothers ★★★

McShane Road, Richmond, Nelson **region** Nelson

ph (03) 544 4230 **fax** (03) 544 4230 **open** 7 days 10–6

winemaker Jane Cooper **prod.** 1600 **est.** 1991

prod. range ($14.95–20 CD) Sauvignon Blanc, Chardonnay, Rosé and Pinot Noir; Richmond Plains is second label red wine.

summary The 4-hectare estate vineyard is claimed by David Holmes to be the most southerly in the world to be certified fully organic, having been run on fully organic principles right from the outset. The tasting-room facilities are shared with Te Mania Estate, and are supported by a range of other local handicraft shops.

huia NR

Rapaura Road, RD3, Blenheim **region** Marlborough

ph (03) 572 8326 **fax** (03) 572 8326 **open** 7 days 10–4.30

winemaker Claire Allan, Mike Allan **prod.** 2600 **est.** 1996

prod. range ($17–23 CD) Gewurztraminer, Sauvignon Blanc, Chardonnay.

summary Owners Claire and Mike Allan bring a wealth of experience to Huia. Both are winemakers, and both have had outstanding careers in Marlborough, variously working at Cloudy Bay, Corbans Marlborough, Rapaura Vintners, Lawsons Dry Hills and Vavasour Wines – as well as working an 'extended stage' in Champagne, France. They acquired their vineyard in Rapaura Road in late 1990, and have now planted 12 hectares of vines, with new Dijon (Burgundy) clones of pinot noir and chardonnay due to come into full production in 1999.

Huia Riesling

A lovely wine picked on 14 April at 23.4° brix and with a low pH.

🍷🍷🍷🍷🍷 **1998** Light yellow-green; the bouquet is tight, with mineral, lime and herb aromas; the palate is only light-bodied, but is supremely elegant, with perfectly balanced acidity and residual sugar. **rating:** 92

best drinking 1999–2003 **best vintages** '98 **drink with** Shellfish • $17

Huia Gewurztraminer

Huia's gewurztraminer vineyard, named Traminer View, is located on the clay run-off of the Wither Hills. The variety and the vineyard are usually very low-yielding, but in 1998 did require bunch thinning. The crop is whole-bunch pressed to avoid the excessive collection of phenolics.

🍷🍷🍷🍷🍷 **1998** Pale straw; the delicate bouquet runs through rose water, lychee and spice, and the flavour is commensurately delicate with the flavours promised by the bouquet appearing again. It is spotlessly clean and the makers have resisted the temptation to flesh it out with sugar, leaving the finish crisp and dry. **rating:** 90

best drinking 1999–2002 **best vintages** NA **drink with** Gently-spiced Asian dishes • $19

Huia Sauvignon Blanc

Both the viticultural and winemaking approach for this wine vary from year to year according to the conditions. Typically, the vineyard will be picked at different times according to the state of ripeness of the grapes, part undergoing indigenous yeast fermentation in a large French oak vat, the remainder using cultured yeasts for stainless steel tank fermentation in traditional Marlborough style. The care taken pays substantial dividends.

🍷🍷🍷🍷 **1998** Light to medium yellow-green; the bouquet is clean, fresh with a mix of herb/mineral and riper passionfruit/gooseberry aromas. The palate has plenty of punch and flavour, with a far better-than-average finish, and can only be seen as a total success in the difficult 1998 vintage. **rating:** 89

best drinking 1999–2000 **best vintages** NA **drink with** Shellfish • $18

Huia Chardonnay

Another very elegant wine from Huia, utilising wild yeast fermentation and a special malolactic fermentation innoculum which does not leave a major mark on the wine.

🍷🍷🍷🍷🍷 **1997** Light to medium yellow-green; the bouquet is elegant and fine, with fresh citrus fruit supported by beautifully balanced and integrated oak. The palate is lively, fresh and crisp in a direct, fruit-driven fashion, a long way apart from the usual NZ style. **rating:** 94

best drinking 1999–2002 **best vintages** '97 **drink with** Full-flavoured fish • $23

hunter's wines ★★★★★

Rapaura Road, Blenheim **region** Marlborough

ph (03) 572 8489 **fax** (03) 572 8489 **open** 7 days 9.30–4.30

winemaker Gary Duke **prod.** 40 000 **est.** 1980

prod. range ($15.95–29.50 R) Chardonnay, Gewurztraminer, Sauvignon Blanc, Oak Aged Sauvignon Blanc, Riesling, Pinot Noir, Cabernet Merlot, Brut; Spring Creek is a newly-introduced second label.

summary Hunter's goes from strength to strength, consistently producing flawless wines with tremendous varietal character. Given the quantity and quality of its production, it is a winery of world standing, and certainly among the top dozen in Australasia: it is hard to choose between its long-lived Riesling, Sauvignon Blanc, Oak Aged Sauvignon Blanc (a tour de force) and subtly complex Chardonnay.

Hunter's Oak Aged Sauvignon Blanc

Having decided to put Sauvignon Blanc in oak, Jane Hunter and her winemaking/consulting team have not done anything by half measures. It is barrel-fermented with a substantial percentage of new wood making a pronounced impact on the wine. Whether it is a better wine than the unoaked version really depends on one's personal view of oak and, for that matter, of Sauvignon Blanc. Universally accepted as a five-star/classic wine.

🍷🍷🍷🍷🍷 **1997** Light to medium yellow-green; the bouquet has plenty of ripe, rich fruit, the oak playing the pure support role which it also does on the palate. Tropical passionfruit flavours fill the mid-palate and finish; maturing with unusual grace. **rating:** 91

best drinking 1999–2001 **best vintages** '88, '89, '91, '93, '94, '96, '97 **drink with** Grilled spatchcock • $23.50

huthlee estate NR

Montana Road, RD5, Hastings, Hawke's Bay **region** Hawke's Bay
ph (06) 879 6234 **fax** (06) 879 6234 **open** Mon–Sat 10–5, Sun 11–4
winemaker Devon Lee **prod.** 1000 **est.** 1991
prod. range ($12–25 R) Pinot Gris, Rosé, Kaweka Red, Cabernet Franc, Cabernet Sauvignon Merlot, Merlot, Reserve Merlot, Cabernet Sauvignon.
summary Devon and Estelle Lee commenced planting their 6-hectare vineyard in 1984, and established an on-site cellar door in 1992. The majority of the grapes are sold to other producers; the best is reserved for their own label.

isabel estate ★★★★☆

Hawkesbury Road, Renwick, Marlborough **region** Marlborough
ph (03) 572 8300 **fax** (03) 572 8383 **open** By appointment
winemaker Jeff Sinnott **prod.** 10 000 **est.** 1982
prod. range ($18.95–28.95 R) Sauvignon Blanc, Chardonnay, Pinot Noir.
summary The 54-hectare Isabel Estate Vineyard was planted in 1982, and until 1994 was purely and simply a grape growing enterprise (and the largest external supplier to Cloudy Bay). In that time it built up a considerable reputation for the quality of its grapes, and in 1994 introduced the Isabel Estate label. It has now taken a further critical step, constructing a 300-tonne winery which was commissioned for the 1998 vintage, and employing a full-time winemaker, of course. Stage 2 will see the construction of an underground cellar for 500 barrels. The wines are exported to Australia (FD&C Wine), US, UK and the Netherlands.

Isabel Estate Marlborough Sauvignon Blanc

Unusually complex winemaking techniques are used to produce this wine, which is made up of a number of different components. The majority is cold-fermented with a neutral yeast and aged on lees for three months. Fifteen per cent of the wine is fermented in French oak barriques and underwent a partial malolactic fermentation. A small portion was whole-bunch pressed directly to barrel and fermented warm. The finished wine has over 7 grams of acid and very low (1.7 grams) residual sugar. A classy act.

🍷🍷🍷🍷🍷 **1998** Light to medium yellow-green; the bouquet exudes passionfruit, white peach and gooseberry aromas; the palate has the same panoply of rich flavours, balanced by a nice touch of acid on the finish. A wine that does have the intensity and focus absent from the majority of the '98 Sauvignon Blancs. **rating:** 91

best drinking 1999–2000 **best vintages** '94 **drink with** Sugar-cured tuna • $19

Isabel Estate Marlborough Chardonnay

Thirty per cent of the wine is fermented in new and used French oak barriques, then aged on lees with stirring for 14 months. The balance is cool-fermented in tank, aged on lees, and allowed to go through a natural malolactic fermentation.

🍷🍷🍷🍷🍷 **1997** Medium yellow-green; the complex bouquet has abundant, soft creamy/nutty aromas running through the melon fruit, supported by subtle oak. Amidst this wealth of plenty, the palate has a touch of finesse and delicacy, aided by well-balanced acidity. **rating:** 90

best drinking 1999–2001 **best vintages** NA **drink with** Crumbed brains • $25

Isabel Estate Marlborough Pinot Noir

Hand-picked fruit is hand-sorted and destemmed, before being fermented in an open fermenter with hand-plunging and natural (wild) yeast activity. Pressed after dryness, and matured in oak (20% new) for 15 months.

🍷🍷🍷🍷 **1997** Medium red-purple; the bouquet is light, but quite aromatic with a nice marriage of strawberry, sappy and oaky aromas. The palate, while of moderate weight and richness, is likewise quite textured, with strawberry and plum flavours following in the track of the bouquet. **rating:** 88

best drinking 1999–2001 **best vintages** '97 **drink with** Smoked quail • $29

jackson estate ★★★★☆

Jacksons Road, Blenheim **region** Marlborough

ph (03) 572 8287 **fax** (03) 572 9500 **open** At Jackson Estate shop, Blenheim airport 9–7.30

winemaker Martin Shaw (Consultant) **prod.** 22 000 **est.** 1988

prod. range ($10.55–29 ML) Riesling, Dry Riesling, Sauvignon Blanc, Chardonnay, Botrytis Riesling, Methode Traditionelle, Pinot Noir. A top-line maker of Sauvignon Blanc and Chardonnay.

summary Long-term major grape growers John and Warwick Stichbury, with leading viticulturist Richard Bowling in charge, own substantial vineyards in the Marlborough area, and have now established their own winery and brand. The wines are exported to Australia, UK, US and France.

Jackson Estate Riesling

As with the Sauvignon Blanc, estate-grown (from 4 hectares of vines) and, again as with the Sauvignon Blanc, showing an unusual degree of ripeness and softness. Perhaps some of these fruit characters derive from the fact that the vines are not irrigated.

🍷🍷🍷🍷 **1998** Light yellow-green; the lime, herb, mineral and spice aromas are varietally correct but slightly reserved. Herb and mineral flavours follow on the palate, which has a tight, crisp and very long finish. A wine for the future. **rating:** 89

best drinking 1999–2005 **best vintages** '92, '94 **drink with** Steamed mussels • $10.55

Jackson Estate Sauvignon Blanc

Produced from 15 hectares of estate plantings established in 1988, and producing the first vintage in 1991. The vines are not irrigated, are hand-pruned and utilise the Scott Henry trellis. The quality of the fruit is beyond dispute, and the winemaking skills of international flying winemaker Martin Shaw add the final touch. The most striking feature of this wine is that in all except the most extreme years it is perfectly weighted and proportioned, However, it did not escape the clutches of the 1998 vintage.

🍷🍷🍷🍷 **1998** Light to medium yellow-green; the bouquet is solid, with ripe tropical fruit which is, however, slightly unfocused. The palate follows on the same track; no one could take offence at a wine such as this, but it does look as if it will mature very quickly, so soft and ripe is it. **rating:** 88

best drinking 1999–2000 **best vintages** '91, '92, '93, '94, '97 **drink with** Calamari • $17

Jackson Estate Reserve Chardonnay

A single vineyard wine, first made in 1994, and thereafter in 1996 and 1997. One hundred per cent barrel-fermented in 70% new French oak and 100% taken through mlf, it is in take-no-hostages-style.

🍷🍷🍷🍷 **1997** Medium yellow-green; the bouquet is typically nutty/malty in high-powered New Zealand style; the palate is similarly impressive, starting with hazelnut characters and then finishing with a twist of acidity. **rating:** 88

best drinking 1999–2001 **best vintages** '94, '96, '97 **drink with** Smoked chicken • $29

johanneshof cellars NR

State Highway 1, Koromiko, RD3, Blenheim **region** Marlborough
ph (03) 573 7035 **fax** (03) 573 7034 **open** Tues–Sun 10–4
winemaker Edel Everling, Warwick Foley **prod.** 2500 **est.** 1991
prod. range ($15–32 R) Riesling, Gewürztraminer, Sauvignon Blanc, Botrytised Sauvignon Blanc Chardonnay, Müller Thurgau, Emmi Méthode Champenoise, Pinot Noir.

summary Marlborough district winemaker Warwick Foley met his wife-to-be Edel Everling in New Zealand and followed her back to Germany (where her family has a winemaking history) to spend five years studying and working, inter alia at Geisenheim. The couple have returned to New Zealand to make European-style wines in an underground cellar blasted into a hillside between Blenheim and Picton.

Johanneshof Cellars Riesling

Normally made in an off-dry or sweet mode, but in 1998 taken to an almost dry style, and succeeds marvellously in this mode.

🍷🍷🍷🍷🍷 **1998** Light yellow-green; the intense bouquet ranges through mineral and citrus aromas, the promise being fulfilled on the palate where intense lime fruit flavours sit over a mineral substrate, giving a European feel to an excellent wine. **rating:** 93

best drinking 2000–2005 **best vintages** '98 **drink with** Asparagus salad • $15.95

john mellars of great barrier island NR

Okupu Beach, Great Barrier Island **region** Auckland Area
ph (09) 429 0361 **fax** (09) 429 0370 **open** By appointment
winemaker John Mellars **prod.** 100 **est.** 1990
prod. range ($35 R) Great Barrier Cabernet.

summary The winery's full name is John Mellars of Great Barrier Island, and indeed the hectare of vines planted on a steep, stony slope facing the nearby sea is the only planting on the island. Output is tiny, and likely to remain so, perhaps fortunate given that access to the cellar door is either a ten-minute beach and track walk or by dinghy from a boat. Judging by the newsletter, those who make the effort will be rewarded by a delightfully eccentric and humorous John Mellars in person.

kaituna valley NR

150 Old Tai Tapu Road, Halswell, Christchurch **region** Canterbury
ph (03) 325 2094 **fax** (03) 322 9272 **open** Not
winemaker Grant Whelan **prod.** 150 **est.** 1993
prod. range ($28–35) Pinot Noir.

summary Grant and Helen Whelan bring considerable skills to this tiny venture; Grant Whelan was a tutor in Wine Science and Viticulture at Lincoln University before becoming winemaker for Rossendale Wines in Christchurch, while Helen has a PhD in plant pathology. The vineyard is established on the Banks Peninsula on a north-facing, non-irrigated slope; extensive canopy work paid dramatic dividends with the first vintage (1993) which won the gold medal and trophy for Champion Pinot Noir at the 1995 Liquorland Royal Easter Wine Show.

Kaituna Valley Pinot Noir

This is not the first time this wine has won a medal at the Liquorland Royal Easter Wine Show and is yet another demonstration of the ability of the Canterbury region to produce top-class pinot noir. The fruit comes from a vineyard on Banks Peninsula planted back in 1979.

🍷🍷🍷🍷🍷 **1997** Medium red-purple; a highly scented, stylish sappy/spicy bouquet is followed by a palate with cascades of bright cherry and strawberry fruit, the barest hint of mint, and good length. **rating:** 93

best drinking 1999–2002 **best vintages** '93, '97 **drink with** Grilled spatchcock • $28

kanuka forest wines NR

Moore Road, Thornton, RD2, Whakatane **region** Waikato and Bay of Plenty
ph (07) 304 9963 **fax** (07) 304 9963 **open** Weekends 10–6, Tues–Fri 3–6
winemaker Tony Hassall **prod.** 500 **est.** 1992
prod. range ($17.95–24.95 CD) Fumé Blanc, Chardonnay, Cabernet Sauvignon Merlot.

summary Tony and Julia Hassall commenced the establishment of their 3-hectare vineyard in 1989, offering their first wine for sale at the end of 1994. The winery and vineyard enjoy spectacular views of the eastern Bay of Plenty. It is the eastern Bay of Plenty's only commercial wine producer; the wines are made exclusively from grapes grown at Thornton.

kawarau estate NR

Cromwell-Wanaka Highway, SH6, Cromwell **region** Otago
ph (03) 445 1315 **fax** (03) 218 7657 **open** By appointment
winemaker Dean Shaw (Contract) **prod.** 1100 **est.** 1992
prod. range ($13–28 CD) Sauvignon Blanc, Chardonnay, Reserve Chardonnay, Pinot Noir, Reserve Pinot Noir.

summary Kawarau Estate is owned by four partners: Wendy Hinton and Charles Finney, and Geoff Hinton and Nicola Sharp-Hinton; Geoff is the vineyard manager and Nicola the marketing and sales manager. The 7-hectare Dunstan Vineyard at Lowburn, 10 kilometres north of Cromwell, which is managed according to strict organic principles, has been the principal source of grapes. All of the wines on the 1998 price list have been medal winners at New Zealand wine shows.

kemblefield estate ★★★☆

Aorangi Road, Hastings **region** Hawke's Bay
ph (06) 874 9649 **fax** (06) 874 9457 **open** Mon–Fri 9–5, weekends 11–3 Labour weekend to Easter, rest of year by appointment
winemaker John Kemble **prod.** 20 000 **est.** 1993
prod. range ($14.95–24.95 CD) Gewürztraminer, Sauvignon Blanc, Reserve Sauvignon Blanc, Chardonnay, Cabernet Merlot, Merlot; also Terrace View range of Sauvignon Blanc, Chardonnay and Cabernet Merlot.

summary With 8.5 hectares of sauvignon blanc, 7.3 hectares of chardonnay, 4.5 hectares of merlot and 2.9 hectares of cabernet sauvignon, Kemblefield Estate has accelerated out of the blocks since it graduated from grape growing to winemaking in 1994. John Kemble, incidentally, is a graduate of UC Davis, and worked in California for 15 years before moving to Hawke's Bay. Exports to the UK, US, Switzerland, Canada and Germany.

Kemblefield Estate Chardonnay

Produced from the small-berried Mendoza clone of chardonnay grown near the Tutaekuri River. It is 100% barrel-fermented in French oak, 25% new. Seventy-five per cent undergoes malolactic fermentation, and is kept on lees for nine months prior to bottling, with fortnightly lees stirring. The '97 won a gold medal at the 1999 Liquorland Royal Easter Wine Show.

🍷🍷🍷🍷🍷 **1997** Medium green-yellow; a very interesting, intense, complex bouquet with strong Burgundian overtones is followed by a long, intense palate with some grapefruit, good oak and a long, positive finish. **rating:** 94

best drinking 1999–2001 **best vintages** NA **drink with** Sautéed veal • $24.95

kenley vineyard NR

Earnscleugh Road, RD, Alexandra **region** Otago
ph (03) 449 2674 **fax** (03) 440 2064 **open** Not
winemaker Mike Woltner, Rudi Bauer (Contract) **prod.** 300 **est.** 1989
prod. range ($15–17 R) Gewurztraminer, Pinot Noir.
summary Ken and Bev Boddy have taken the slow boat in establishing their 1-hectare Kenley Vineyard. Ken Boddy became interested in the possibility of growing grapes in the Central Otago region in the mid-1960s during his time as a staff bacteriologist at the Oamaru Hospital. He corresponded with institutions around the world as well as New Zealand's Te Kauwhata Research Station, the latter giving him scant encouragement, but eventually supplying him with grape cuttings which formed the nucleus of a back garden nursery vineyard. Another 20 years were to pass before the Boddys acquired their present vineyard site (in 1989), planting half a hectare of pinot noir and half a hectare of gewurztraminer. They have 10 hectares available for planting, and Ken Boddy is currently evaluating the potential for scheurebe and viognier through trial plantings of each. Even after this long time, they appear to be in no great hurry.

kerr farm vineyard NR

48 Dysart Lane, Kumeu, Auckland **region** Auckland Area
ph (09) 412 7575 **fax** (09) 412 7575 **open** By appointment
winemaker Contract **prod.** 1000 **est.** 1989
prod. range ($14.95–18.95 CD) Semillon, Sauvignon Blanc, Chardonnay, Pinotage, Cabernet Sauvignon.
summary Jason and Wendy Kerr have established 5.5 hectares of vines on the site of an old Corbans vineyard, their first wines being made in 1995.

kim crawford wines ★★★★☆

236–238 Dominion Road, Mount Eden, Auckland **region** Auckland Area
ph (09) 630 6263 **fax** (09) 630 6293 **open** Not
winemaker Kim Crawford **prod.** 22 000 **est.** 1996
prod. range ($16.95–26.95 R) Marlborough Riesling, Marlborough Sauvignon Blanc, Awatere Sauvignon, Marlbourgh Unoaked Chardonnay, Tietjen Gisborne Chardonnay.
summary Kim Crawford first made his reputation as winemaker at Coopers Creek, then as consultant winemaker for a number of vineyards, and now as the producer of a number of very good wines under his own label (as well as wines made for others), sourced from vineyards in Marlborough, Hawke's Bay and Gisborne. Exports to Australia, UK, US, Canada and Hong Kong.

Kim Crawford Marlborough Dry Riesling

The 1997 vintage of this wine swept all before it in Class 1 at the 1997 Liquorland National Wine Show, exhibiting tremendous power and complexity. The 1998 is a wine in similar mould, which stood out as being radically different from most of the wines in the 1999 *Winewise* Small Makers Wine Competition, whence this note comes.

🍷🍷🍷🍷🍷 **1998** Strong yellow-green; the bouquet is highly aromatic and pungent with floral spice notes introducing characters reminiscent either of a touch of Gewurztraminer and/or a touch of botrytis. Many things contributed to the high-toned flavours of the palate most, but not all, good. Ever so slightly spongy finish. **rating:** 90

best drinking 1999–2002 **best vintages** '97, '98 **drink with** Calamari • $19.95

Kim Crawford Awatere Sauvignon

An extremely rich and complex oaked style of Sauvignon Blanc. Fifty per cent of the wine is barrel-fermented, lees-aged and undergoes malolactic fermentation, with ten months maturation in new American oak. The remaining 50% is tank-fermented and held in stainless steel until blending and bottling.

🍷🍷🍷🍷🍷 **1998** Medium yellow-green; spicy nutmeg oak comes to the fore on both bouquet and palate, but can only be described as sexy, and certainly adds to the enjoyment of the wine if consumed with food. The spicy clove flavours can, however, burn in the absence of a food background. **rating:** 91

best drinking 1999–2000 **best vintages** '96, '97 **drink with** Seafood chowder • $26.95

Kim Crawford Marlborough Sauvignon Blanc

Produced from grapes grown in Marlborough's Awatere Valley, emanating from low-yielding vines. Straightforward cold fermentation in stainless steel and early-bottling followed, the wine quality coming principally from the vineyard selection, and partly from the choice of yeast strain.

🍷🍷🍷🍷 **1998** Light to medium yellow-green; the bouquet is clean, but only of moderate intensity, with gooseberry herbal fruit. The palate is well balanced and harmonious with gentle gooseberry and herb flavours; quite long, and noticeably soft acidity. Not a great year for Sauvignon Blanc. **rating:** 89

best drinking 1999–2000 **best vintages** '97 **drink with** Salmon terrine • $16.95

Kim Crawford Tietjen Gisborne Chardonnay

Made from grapes grown by Paul Tietjen, one of Gisborne's most respected viticulturists. His vineyard sits below the Ormond Hills, and is one of Gisborne's coolest. Oxidatively handled juice was barrel-fermented in new American oak, and kept in barrel for six months with weekly lees stirring. Malolactic fermentation occurred spontaneously.

🍷🍷🍷🍷 **1998** Medium yellow-green; the bouquet is complex with good barrel-ferment characters and tangy fruit. The oak on the palate is poised on the brink of being excessive; much will depend on the circumstances under which the wine is drunk. It certainly has length, and was very highly rated at the 1999 *Winewise* Small Makers Competition. **rating:** 88

best drinking 1999–2000 **best vintages** NA **drink with** New Zealand whitebait • $24.95

kindale wines NR

Falveys Road, Omaka Valley, Blenheim, Marlborough **region** Marlborough

ph (03) 572 8272 **fax** (03) 572 8414 **open** By appointment

winemaker Various **prod.** 250 **est.** 1978

prod. range ($NA) Müller Thurgau, Chardonnay, Pinot Noir.

summary The Hadfield family have been grape growers since 1978, selling their production to Montana Wines. The first wine was made in 1993 for a special occasion; it has now graduated to slightly more commercial levels, with the wine sold through local retail outlets and through the mailing list.

kingsley estate NR

PO Box 1100, Hastings **region** Hawke's Bay

ph (025) 454 780 **fax** (08) 326 9463 **open** Not

winemaker Kingsley Tobin **prod.** 400 **est.** 1991

prod. range ($29 ML) Cabernet Sauvignon.

summary Kingsley Tobin has established a 6-hectare vineyard at Gimblett Road, with certified Bio-Gro™ status. Most of the grapes are sold; a small portion is made by C J Pask winemaker Kate Radburnd, but Tobin does have plans to establish his own storage facility at the vineyard and to expand production. The current tiny production is sold by mail order, with a few cases being exported to Boutique Vineyards, Sydney.

koura bay wines NR

10 Nursery Road, Seddon **region** Marlborough
ph (03) 578 3882 **fax** (03) 578 3771 **open** By appointment
winemaker Simon Waghorn **prod.** 700 **est.** 1992
prod. range ($NA) Whalesback Sauvignon Blanc, Mount Fyffe Chardonnay, Barney's Rock Riesling.

summary Geoff and Dianne Smith have established 21 hectares of vineyard, initially selling their grapes to Nobilo, and gaining recognition for their quality when the 1996 Nobilo Grand Reserve Marlborough Sauvignon Blanc won a gold medal at the Braggato Awards (the medal going to the vineyard, not the winery). This no doubt prompted the Smiths to have part of their production vinified for sale under the Koura Bay label; in 1998 Whitehaven Wines fulfilled that task.

Koura Bay Wines Whalesback Sauvignon Blanc

Koura is a Maori word for crayfish, and the vineyard is on the banks of the Awatere River, which does not number whales among its population, nor even crayfish. Name quibbles to one side, Koura Bay produces classy Sauvignon Blanc.

YYYY **1998** Light green-yellow; the aromas are crisp, and more in the grassy/herbal spectrum, the palate providing a surprising counterpoint with hints of passionfruit and gooseberry. The components all work well together. **rating:** 87

best drinking 1999–2000 **best vintages** NA **drink with** Lightly poached oysters • NA

kumeu river wines ★★★★★

550 Highway 16, Kumeu **region** Auckland Area
ph (09) 412 8415 **fax** (09) 412 7627 **open** Mon–Fri 9–5.30, Sat 11–5.30
winemaker Michael Brajkovich **prod.** 20 000 **est.** 1944
prod. range ($16–32 CD) At the top end come the limited production Maté Vineyard Chardonnay; then follow the Kumeu River range of Sauvignon Semillon, Chardonnay and Merlot Cabernet; the less expensive wines come under the Brajkovich Signature range, with Chardonnay, Sauvignon Semillon, Cabernet Merlot, Merlot Malbec Cabernet Franc and Cabernet Franc.

summary The wines of Michael Brajkovich defy conventional classification, simply because the highly trained, highly skilled and highly intelligent Brajkovich does not observe convention in crafting them, preferring instead to follow his own French-influenced instincts and preferences. The wines enjoy strong export markets in the US, UK, Hong Kong, Singapore, Switzerland, Australia and Brazil.

Kumeu River Sauvignon Semillon

A blend of 90% Sauvignon Blanc and 10% Semillon, barrel-fermented, lees-aged and taken through a full malolactic fermentation.

YYYY **1997** Medium yellow-green; the bouquet is opulent, almost buttery, the palate ripe, rich, textured and buttery, augmented further by malolactic-induced softness. Really on another planet in terms of varietal character. **rating:** 86

best drinking 1999–2002 **best vintages** NA **drink with** Braised pork • $20

Kumeu River Chardonnay

Produced from five different vineyard sites around Kumeu. Whole-bunch pressed direct to barrel, and relies upon indigenous yeasts. Twenty to 25% new French oak is used, and the wine undergoes malolactic fermentation and lees contact in barrel before being bottled just prior to the following vintage.

🍷🍷🍷🍷🍷 **1997** Medium yellow-green; the bouquet is in the usual full style, but the honeyed/nutty components are nicely married together. A seamlessly welded and moulded wine in the mouth which stands up extremely well in major line-ups of New Zealand Chardonnays, and benefits from the crisp, clean finish. **rating:** 90

best drinking 1999–2000 **best vintages** '96, '97 **drink with** Trout • $32

Kumeu River Maté Vineyard Chardonnay

The top-of-the-range Chardonnay from Kumeu River, named by Michael Brajkovich in honour of his late father, Maté Brajkovich – surely one of the all-time great gentlemen of the wine industry anywhere. A single-vineyard wine fermented in French oak, and an outstanding example of the genre. The vineyard, incidentally, was first planted in 1944, but entirely replanted in 1990, with the first wine from the new plantings made in 1993.

🍷🍷🍷🍷🍷 **1997** Light to medium yellow-green; the bouquet is spotlessly clean, showing no sign of ambient temperature fermentation. As ever, fruit, oak and malolactic-fermentation characters coalesce on the bouquet and the exceptionally long palate providing a seamless, flawless wine. **rating:** 94

best drinking 2000–2005 **best vintages** '96, '97 **drink with** Milk-fed veal • $36

lake chalice wines NR

Vintage Lane (Box 66), Renwick **region** Marlborough

ph (03) 572 9327 **fax** (03) 572 9327 **open** Mon–Sat 10.30–4.30

winemaker Chris Gambitsis, Matt Thomson **prod.** 6000 **est.** 1989

prod. range ($13.50–25 CD) Riesling, Sauvignon Blanc, Platinum Oak Aged Sauvignon Blanc, Chardonnay, Botrytised Riesling, Platinum Merlot Cabernet.

summary Lake Chalice Wines is a partnership of three long-time friends, Chris Gambitsis, Ron Wichman and Phil Binning. In 1989 they purchased the 11.5-hectare Falcon Vineyard; the name of the winery comes from a wilderness lake situated in the Richmond Range which borders the northern side of Marlborough's Wairau Plain. The first wine release was in 1993; the first red wine was released in 1997.

Lake Chalice Marlborough Sauvignon Blanc

The Falcon Vineyard was selected for its potential to produce outstanding fruit, say the partners. It is, even by Marlborough standards, extremely stony and free-draining. The '97 was the only gold medal in the 1997 and Older Sauvignon Blanc class at the 1998 National Wine Show.

🍷🍷🍷🍷🍷 **1997** Light to medium yellow-green; the bouquet has potent, herbal gooseberry fruit, the palate all of the power and intensity the vast majority of the '98 wines lack. Voluminous herb and gooseberry flavours run right through to the finish. **rating:** 92

best drinking 1998–2000 **best vintages** NA **drink with** Eggplant terrine • $15

langdale estate NR

Langdale Road, West Melton, Christchurch **region** Canterbury

ph (03) 342 6266 **fax** (03) 342 4059 **open** Tues–Thur 11–5, Fri–Sat 11–late, Sun 11–6

winemaker Carol Bunn **prod.** 2000 **est.** 1989

prod. range ($12.95–18.95 CD) Riesling, Marlborough Sauvignon Blanc, Pinot Gris, Chardonnay, Breidecker, Pinot Noir, Melton Hills Pinot Noir.

summary Based upon 4.5 hectares of estate vineyards planted to pinot noir, riesling, breidecker and pinot gris, with plantings commencing in 1989 and expanded since. Most of the wine is sold through the cellar door and restaurant, and wedding function centre on site.

larcomb vineyard ★★★☆

Larcombs Road, RD5, Christchurch **region** Canterbury
ph (03) 347 8909 **open** Tues–Sun 11–5 Nov–March, Fri–Sun 11–5 Apr–Oct
winemaker Contract **prod.** 2000 **est.** 1985
prod. range ($14–16 CD) Riesling, Breidecker, Gewurztraminer, Pinot Gris, Chardonnay, Pinot Noir.
summary Following its acquisition by Michelle and Warren Barnes in 1995, the winery has apparently obtained a reputation for itself as 'home of Rattle the Rafters Barn Dance', which – if nothing else – is something different.

lawson's dry hills ★★★★

Alabama Road, Blenheim **region** Marlborough
ph (03) 578 7674 **fax** (03) 578 7603 **open** 7 days 10–5
winemaker Mike Just **prod.** 20 000 **est.** 1992
prod. range ($16–25 R) Gewurztraminer, Sauvignon Blanc, Chardonnay, Riesling, Late Harvest Riesling, Pinot Noir.
summary Lawson's Dry Hills is situated on the Wither Hills, which in turn take their name from their parched mid-summer look. It is part-owned by Barbara and Ross Lawson, recently joined by three shareholders who have contributed vineyards giving a total of 22 hectares. The partners have all graduated from being grape growers to winemakers, with conspicuous success and steadily rising production. There is significant distribution of the wine in Australia.

Lawson's Dry Hills Gewurztraminer

The Gewurztraminer was initially the only estate-grown wine under the Lawson label. Always well crafted, sometimes in a delicate mode and in others on a full-blown, rich style reflecting the vintage.

🍷🍷🍷🍷🍷 **1998** Light to medium yellow-green; the bouquet exudes lychee, spice and rose petal aromas; a very big wine on the palate but the very low residual sugar (under 3 grams per litre) helps prevent the wine from cloying, notwithstanding its 14.5 degrees alcohol. **rating:** 90

best drinking 1999–2001 **best vintages** NA **drink with** Delicate Asian seafood • $18

Lawson's Dry Hills Sauvignon Blanc

Another classic Marlborough wine made with skill and discipline. A portion of the wine is fermented in French oak barriques and undergoes malolactic fermentation, and is then back-blended with the major component which is conventionally fermented in stainless steel.

🍷🍷🍷🍷 **1998** Light to medium yellow-green; an aromatic, rich bouquet with some passionfruit and gooseberry is an enticing start, but on the palate more tropical fruit flavours take over, and the alcohol intrudes somewhat. **rating:** 89

best drinking 1999–2000 **best vintages** '94, '97 **drink with** Crayfish • $19

leaning rock vineyard NR

Hillview Road, Alexandra **region** Otago
ph (03) 448 9169 **fax** (03) 448 9169 **open** By appointment
winemaker Mark Hesson, Dhana Pillai **prod.** 1200 **est.** 1991
prod. range ($18–25 CD) Riesling, Gewurztraminer, Chardonnay, Pinot Noir.
summary Notwithstanding bare gravel soils and a northerly slope, spring frosts proved a major problem for geologist owners Mark Hesson and Dhana Pillai, curtailing production until sprinklers were installed prior to the 1996 growing season. Small quantities of strongly flavoured wines were then produced with gradually increasing quantities (particularly Pinot Noir) expected over coming vintages.

limeburners bay NR

112 Hobsonville Road, Hobsonville **region** Auckland Area
ph (09) 416 8844 **open** Mon–Sat 9–6
winemaker Alan Laurenson **prod.** 3500 **est.** 1978
prod. range ($7–19.95 CD) Müller Thurgau, Semillon Chardonnay, Sauvignon Blanc, Chardonnay, Cabernet Merlot, Cabernet Sauvignon.
summary Initially established a reputation for itself in the 1980s with some good Cabernet Sauvignon, but with more variable outcomes in the 1990s. No recent tastings.

lincoln vineyards ★★★

130 Lincoln Road, Henderson **region** Auckland Area
ph (09) 838 6944 **fax** (09) 838 6984 **open** Mon–Sat 9–6, Sun 11–5
winemaker Joseph Papesch **prod.** 35 000 **est.** 1937
prod. range ($9–45 CD) Chardonnay (under a series of labels including Vintage Selection, Gisborne, Show Reserve and Parklands Vineyard), Sauvignon Blanc, Chenin Blanc, Riesling, Müller Thurgau, Cabernet Sauvignon, Merlot. Presidents Selection is newly introduced flagship.
summary A substantial family-owned operation drawing its grapes from Auckland, Gisborne and Hawke's Bay. The labels are avant garde, but the wines have been variable, good at best but sometimes disappointing, even if the prices are competitive. Exports to the UK, US, Australia, Pacific Islands, Japan, Germany and Canada.

linden estate NR

Napier–Taupo Road, SH5, Eskdale **region** Hawke's Bay
ph (06) 836 6806 **fax** (06) 836 6586 **open** 7 days 9–5
winemaker Nick Chan **prod.** 10 000 **est.** 1971
prod. range ($14–32 CD) Sauvignon Blanc, Oak Aged Sauvignon Blanc, Estate White, Chardonnay, Reserve Chardonnay, Merlot, Cabernet Sauvignon, Cabernet Merlot, Reserve Cabernet Merlot.
summary This is the project of retired civil engineer and long-term grape grower Wim van der Linden and family, son John being a tutor in viticulture at the Polytechnic in Hawke's Bay. The estate vineyard was replanted in 1989 to 25 hectares of premium varieties, including a 2.5-hectare hillside planting producing a Reserve wine first released in 1996. The wines are distributed in New Zealand (Eurowine), UK, Canada, US and Taiwan.

lintz estate NR

Kitchener Street, Martinborough **region** Wairarapa
ph (06) 306 9174 **fax** (06) 306 9175 **open** By appointment while stocks last
winemaker Chris Lintz **prod.** 3000 **est.** 1989
prod. range ($16–65 CD) Reserve Riesling, Spicy Traminer, Sauvignon Blanc, Optima Noble Selection, Rosé, Pinot Noir, Moy Hall (Pinot), Cabernet Merlot, Vitesse Cabernet Sauvignon, Bottle Fermented Riesling Brut.
summary New Zealand-born Chris Lintz comes from a German winemaking family, and graduated from Geisenheim. The first stage of the Lintz winery, drawing grapes from the 9-hectare vineyard, was completed in 1991. Since 1996 Lintz Estate has enjoyed much show success, winning numerous gold medals across the full range of wines, both white and red, success which briefly underpinned a public issue of shares in 1998. Lintz was then caught up in a Coopers Creek scandal which made television and newspaper headlines; Lintz admitted that certain of the gold medal-winning wines at the 1998 National Wine Show were different from the commercially released versions.

lombardi wines NR

298 Te Mata Road, Havelock North **region** Hawke's Bay
ph (06) 877 7985 **fax** (06) 877 7816 **open** 7 days 10–5
winemaker Tracy Haslam **prod.** 1000 **est.** 1948
prod. range ($9.95–14.95 CD) Riesling Sylvaner, Sauternes, Sherry, Vermouth, Marsala and flavoured fortifieds.
summary The Australian Riverland transported to the unlikely environment of Hawke's Bay, with a half-Italian, half-English family concentrating on a kaleidoscopic array of Vermouths and sweet, flavoured fortified wines. A change of ownership at the end of 1994 has not signalled any fundamental change in direction – except for a desire to have fun while making better wines.

longbush wines NR

State Highway 2, Manutuke, Gisborne **region** Gisborne
ph (06) 862 8577 **fax** (06) 867 8012 **open** Tues–Sun 10–5 Oct–Easter
winemaker John Thorpe **prod.** 15 000 **est.** 1992
prod. range ($9–20 CD) Woodlands Chardonnay; the Longbush range of Rhine Riesling, Chardonnay, Sauvignon Blanc, Botrytis Riesling, Pinot Noir, Merlot, Kahurangi; Nicks Head Sauvignon Blanc, Muller Muscat, Sea Breeze Classic Dry, Chardonnay, and Merlot.
summary Part of the ever-changing circus of winemaking and brand ventures of the Thorpe Brothers Group. Woodlands is the premium label; Longbush the principal (and mid-range) label; Nicks Head is the third and lowest priced.

longview estate NR

State Highway 1, Whangarei **region** Northland and Matakana
ph (09) 438 7227 **fax** (09) 438 7224 **open** summer, Mon–Sat 8.30–6; winter Sat 8.30–5.30, Sun 9–5
winemaker Mario Vuletich **prod.** 2500 **est.** 1969
prod. range ($11–17 CD) Chardonnay, Gewurztraminer, White Diamond (sweet), Scarecrow Cabernet Sauvignon, Mario's Merlot, Cabernet Merlot, Gumdigger's Port, Golden Sherry, Dry Sherry.
summary Mario and Barbara Vuletich have been involved in viticulture and winemaking since 1969, but have replanted the 6-hectare vineyard on elevated slopes overlooking Whangarei harbour, with the four principal Bordeaux varieties (and shiraz) with the intention of making full-bodied dry reds. That they have succeeded handsomely in so doing is evident by the fact that both the Cabernet Sauvignon and the Merlot have received four stars in *Cuisine* magazine for the '93, '94 and '96 vintages.

loopline vineyard NR

Loopline Road, RD1, Masterton **region** Wairarapa
ph (06) 377 3353 **fax** (06) 378 8338 **open** 7 days 10.30–6
winemaker Frank Parker **prod.** 600 **est.** 1994
prod. range ($16.50–21.50 CD) Chasselas, Riesling, Chardonnay, Sauvignon Blanc, Chasselas Chenin Blanc, Waipipi Red, Joseph's (Red).
summary Frank and Bernice Parker are pioneer viticulturists on the Opaki Plains, 5 kilometres north of Masterton. They have established 1 hectare of riesling, and a quarter of a hectare of chasselas, supplementing their intake with limited quantities of grapes grown by other producers in the region.

lynskeys wairau peaks vineyard NR

765e Godfrey Road, RD2, Blenheim **region** Marlborough
ph (03) 572 7180 **fax** (03) 572 7181 **open** Mon–Sat 9.30–5 while stocks last
winemaker Graeme Paul (Contract) **prod.** 750 **est.** 1998
prod. range ($16.50–29.50 CD) Sauvignon Blanc, Gisborne Mendoza Chardonnay, Pinot Noir.
summary Ray and Kathy Lynskey bring diverse backgrounds to their newly established business. Kathy Lynskey has moved between Sydney and New Zealand at various times, and established her first vineyard in 1989, providing contract-grown chardonnay. She now has a distribution business in Sydney (Lynskeys New Zealand Wine Cellars) selling not only the Lynskey wines but those of Lake Chalice, Le Grys and De Gyffarde. Ray Lynskey is an airline pilot based in Blenheim, flying gliders for relaxation so successfully that he won the world gliding championships at Omarama in 1995. Their 7-hectare vineyard is still coming into production; in the interim contract-grown grapes are being used and vinified at Marlborough Vintners. The first harvest of estate-grown chardonnay, gewurztraminer and merlot is scheduled for the year 2000.

Lynskeys Wairau Peaks Sauvignon Blanc

The first vintage from Wairau Peak, and competently made using the traditional stainless steel fermentation techniques.

🍷🍷🍷🍷 **1998** Light to medium yellow-green; the bouquet is soft, clean, with that mix of slightly blousey, tropical fruit so typical of the year. The palate is equally conformist, pleasantly ripe, and dipping towards the soft finish. **rating:** 84

best drinking 1999–2000 **best vintages** NA **drink with** Crab cakes • $16.50

macmillan wines NR

c/o 19 Gladstone Road, Richmond, Nelson (Contract facilities), NZ **region** Nelson
ph (03) 544 5853 **fax** (03) 544 5853 **open** 7 days 10–5 summer, winter by appointment
winemaker Saralinda MacMillan **prod.** 400 **est.** 1993
prod. range ($16–20 CD) Chardonnay.
summary Former Seifried winemaker Saralinda MacMillan left Seifried in 1992 to have her first child. She has since commenced winemaking on her own account in a tiny way using contract facilities, making Chardonnay and Sauvignon Blanc.

margrain vineyard ★★★

Ponatahi Road, (PO Box 97), Martinborough **region** Wairarapa
ph (06) 306 9292 **fax** (04) 569 2698 **open** Weekends and holidays 11-5
winemaker Strat Canning **prod.** 500 **est.** 1992
prod. range ($14-26 CD) Riesling, Chardonnay, Pinot Noir, Merlot.
summary Graham and Daryl Margrain planted their first vines in 1992, and produced the first wine (Chardonnay) in 1994. The vineyard is now planted to a total of 24 hectares of chardonnay, pinot noir, merlot and pinot gris; eight luxury accommodation villas have been built on an adjoining ridge; and a woolshed has been converted into a conference facility and tasting room. An underground cellar was constructed in 1994, and a winery (including a restaurant) was commissioned for the 1996 vintage. And what did the Margrains do before they established Margrain Vineyard? They spent 25 years in the building industry, of course.

mark rattray vineyards NR

State Highway 1, Waipara **region** Canterbury
ph (03) 314 6710 **fax** (03) 314 6710 **open** 7 days 10–5
winemaker Mark Rattray **prod.** 3200 **est.** 1992
prod. range ($22–27 CD) French Farm Chardonnay, Waipara Chardonnay, Marlborough Sauvignon Blanc, Waipara Pinot Noir, Aquilon Pinot.

summary Mark Rattray is a high-profile wine consultant in the Canterbury district, who has enjoyed much success. Initially based at Waipara Springs, with wife Michelle he has now established the Mark Rattray Vineyards label, while continuing to consult to a number of other makers in the region (including Waipara Springs). Has enjoyed particular success in the UK.

martina vineyard NR

Kopuku Road, Te Kauwhata **region** Waikato and Bay of Plenty
ph (07) 826 7790 **fax** (07) 826 7790 **open** By appointment
winemaker Tony Martin **prod.** 1000 **est.** 1995
prod. range ($9–20 ML) Rhine Riesling, Sauvignon Blanc, Chardonnay, Pinot Noir, Syrah, Martined (Light Red), Malbec Cabernet Sauvignon, Dolce Vita (dessert wine).
summary Italian-born and trained Tony Martin was a winemaker at Cooks Te Kauwhata winery for five years before launching his Martina range of wines in 1995, using grapes sourced from both The Waikato and Hawke's Bay, but with changes in the wind.

martinborough vineyard ★★★★★

Princess Street, Martinborough **region** Wairarapa
ph (06) 306 9955 **fax** (06) 306 9217 **open** 7 days 11–5
winemaker Larry McKenna **prod.** 13 000 **est.** 1980
prod. range ($16.50–60 CD) Riesling, Riesling Late Harvest, Gewurztraminer, Pinot Gris, Sauvignon Blanc, Chardonnay, Chardonnay Late Harvest, Pinot Noir, Pinot Noir Reserve.
summary Australian-born and trained Larry McKenna has established a firm reputation as New Zealand's most skilled producer of Pinot Noir, and with an ability to produce Chardonnay and Riesling of similarly impressive ilk. It is on these wines that the reputation of Martinborough Vineyard rests, although McKenna also makes classy Sauvignon Blanc. The Pinot Noir Reserve, only made in good vintages, is in a class of its own. The wines are distributed in Australia by Negociants, and the UK, Hong Kong, Belgium, Germany and the US.

Martinborough Vineyard Reserve Pinot Noir

The combined efforts of Martinborough, Dry River and Ata Rangi leave the Martinborough/Wairarapa region without any serious challengers as New Zealand's best location for Pinot Chardonnay. This magnificent wine was not released until March 1997, having amassed a string of trophies and other accolades, including the Chairman's Trophy (in fact mine) at the 1997 Sydney International Wine Competition. They don't come much better than this from anywhere in the world.

🍷🍷🍷🍷🍷 **1996** Medium purple-red; the bouquet is quite sweet (in the context of the Martinborough-style) with cherry and plum fruit supported by subtle oak. The palate is silky and long in the mouth with red cherry and plum fruit supported by fine, soft tannins. **rating:** 96

🍷🍷🍷🍷🍷 **1995** Medium purple-red; the bouquet is complex, and distinctly richer and bigger than the varietal release of the same year. The palate is outstanding, silky, long and lingering, with the almost piercing linear flavour expanding into the peacock's tail on the finish. **rating:** 93

best drinking 2000–2006 **best vintages** '94, '96 **drink with** New Zealand venison • $45

matariki wines ★★★★

Gimblett Road, Havelock North **region** Hawke's Bay
ph (06) 877 8002 **fax** (06) 877 8004 **open** Not
winemaker John O'Connor **prod.** 6000 **est.** 1981
prod. range ($19.50–27 CD) Sauvignon Blanc, Chardonnay, Anthology (Bordeaux-blend).
summary John and Rosemary O'Connor purchased their Gimblett Road property in 1981, and are now the owners of the largest individual vineyard in that area, with 30 hectares (of a total 60 hectares) under vine, with syrah and sauvignon blanc planted on pure shingle, and cabernet franc,

cabernet sauvignon, malbec, merlot, semillon and chardonnay on terraces with a greater amount of soil. They have also more recently purchased a limestone terrace property below Te Mata Peak, which is cooler and which has been planted to chardonnay. Currently they share the Trinity Hill winery with John Hancock, which houses their own winemaking equipment, but with plans for a large two-storey winery already drawn up. The quality of the first releases (from 1997) was immaculate; no subsequent tastings.

matawhero wines ★★☆

Riverpoint Road, Matawhero **region** Gisborne
ph (06) 868 8366 **fax** (06) 867 9856 **open** Mon–Sat 9–5
winemaker Denis Irwin **prod.** 6000 **est.** 1975
prod. range ($16–25 CD) Gewurztraminer, Riesling, Chardonnay, Reserve Chardonnay, Sauvignon Blanc, Chenin Blanc, Pinot Noir, Syrah, Cabernet Merlot.
summary The wines have always been cast in the mould of Matawhero's unpredictable founder and owner Denis Irwin: at their best, in the guise of the Gewurztraminer from a good vintage, they are quite superb, racy and powerful; at their worst, they are poor and exhibit marked fermentation problems.

matua valley ★★★★

Waikoukou Road, Waimauku **region** Auckland Area
ph (09) 411 8301 **fax** (09) 411 7982 **open** Mon–Sat 8.30–5, Sun 11–4.30
winemaker Ross Spence, Mark Robertson **prod.** 150 000 **est.** 1974
prod. range ($10.45–37.95 CD) A very large range running from generic and varietal white and red table wines at the bottom end of the price scale, then to the Shingle Peak Marlborough range of Sauvignon Blanc, Riesling, Pinot Gris, Chardonnay, Cabernet Sauvignon; the Hawke's Bay varietal wines including Gewurztraminer and Sauvignon Blanc; then to the premium white and red table wines under the Reserve, Ararimu or Judd Estate labels. Waimauku and Smith Cabernet Sauvignons recent illustrious additions.
summary One of the stalwarts of the New Zealand wine industry, producing a wide range of wines of good quality. The Shingle Peak label has been particularly successful, while the presentation of the Ararimu Chardonnay and Cabernet Sauvignon (not to mention the quality of the wines) set new standards of excellence for New Zealand. There are substantial exports to the UK, US, Western Europe, Canada, Hong Kong, Japan and Australia.

Matua Valley Shingle Peak Sauvignon Blanc

Shingle Peak is the Marlborough label of Matua Valley, almost a separate brand, so far is it distanced from Matua Valley. Like the Hawke's Bay Sauvignon Blanc, made without artifice or intervention, and, like the Hawke's Bay wine, 100% Sauvignon Blanc. The two wines provide a perfect contrast between the style of Hawke's Bay Sauvignon and that of Marlborough.

YYYY **1998** Light green-yellow; the bouquet is light, crisp and herbal, characters that open up on the palate, followed by attractive, zesty lemony acidity on the finish. **rating:** 86

best drinking 1999–2000 **best vintages** NA **drink with** New Zealand whitebait • $17.95

Matua Valley Shingle Peak Marlborough Chardonnay

Forty per cent of the wine is fermented in stainless steel, and 60% in oak barriques of varying age before being aged on lees for eight months. Fifty per cent of the wine is taken through malolactic fermentation. It is a fruit-driven style, with the oak barely evident at the end of the day.

YYY **1997** Medium yellow-green; the bouquet is big, toasty and buttery, with a slight burnt caramel overtone. Peach, citrus and buttery fruit flavours run through a pleasantly weighted palate, finishing with subtle oak. **rating:** 84

best drinking 1999–2000 **best vintages** '97 **drink with** Smoked chicken and pasta • $14

Matua Valley Marlborough Pinot Noir

A wine which challenges many of my beliefs about Pinot Noir, for it was made without any significant difference from handling of other red wines; crushed and fermented in a closed tank with regular pumping over. There must be some well-above-average grapes here.

TTTT 1998 Medium red-purple; the bouquet has fresh cherry/cherry pip fruit supported by subtle oak. There is nice texture and mouthfeel on the palate, with more of those cherry pip flavours. A delicious wine for early drinking. **rating:** 87

best drinking 1999–2001 **best vintages** NA **drink with** Duck risotto • $17.95

mazuran's vineyard NR

255 Lincoln Road, Henderson **region** Auckland Area
ph (09) 838 6945 **open** Mon–Sat 9–6
winemaker Rado Hladilo **prod.** 1000 **est.** 1938
prod. range ($12–350) Sherries and Ports.
summary A Sherry and Port specialist, still surviving on the reputation built for its wines by George Mazuran, who died in 1980. The business is continued by his son and son-in-law.

mcdonald winery ★★★★☆

150 Church Road, Taradale **region** Hawke's Bay
ph (06) 844 2053 **fax** (06) 844 3378 **open** 7 days 9–5
winemaker Tony Prichard **prod.** NFP **est.** 1897
prod. range ($18.35–35 R) Church Road Sauvignon Blanc, Chardonnay, Noble Semillon, Cabernet Sauvignon Merlot; Church Road Reserve Chardonnay, Merlot, Cabernet Sauvignon Merlot; Twin Rivers Cuvée Brut; super delux Cabernet Cuvée released 1997.
summary Montana's acquisition of the historic McDonald Winery in 1989 and its investment of $2 million on refurbishment followed by the announcement of the Cordier joint venture, together with the acquisition of premium Hawke's Bay vineyards, signalled Montana's determination to enter the top end of the market with high-quality Chardonnay and Cabernet Sauvignon. Legal squabbles have forced the adoption of the Church Road name for the wine label.

melness wines NR

1816 Cust Road, Cust, North Canterbury **region** Canterbury
ph (03) 312 5402 **fax** (03) 512 5466 **open** Wed–Sun 10.30–5
winemaker Mathew Donaldson, Lynette Hudson **prod.** 1000 **est.** NA
prod. range ($10–18 CD) Gewurztraminer, Riesling, Chardonnay (spray free), Floral (spray free), Rosé, Pinot Noir.
summary Melness' vineyards are run organically, which makes the utilisation of what is claimed to be the only Lyre Trellis system on the South Island all the more understandable, as it maximises sunlight and wind penetration. Owners Colin and Norma Marshall have established the winery and café in a garden setting. The 1996 Pinot Noir won a number of awards.

merlen wines NR

Vintage Lane, Rapaura Road, Marlborough **region** Marlborough
ph (03) 572 9151 **fax** (03) 572 9751 **open** 7 days 9–5
winemaker Almuth Lorenz **prod.** 6500 **est.** 1987
prod. range ($13–22 CD) Riesling, Sauvignon Blanc, Chardonnay, Semillon, Gewurztraminer.
summary In mid-1997 the long-standing winemaking and administration team headed by the colourful and substantial figure of Almuth Lorenz suddenly parted company with the winery which was publicly offered for sale. Almuth Lorenz is the owner of the brand, and has taken it (and herself) to Rapaura Vintners (a major contract crush winery) where she is making the Merlen wines. No information yet on the direction in which the old Merlen winery will head.

mills reef winery ★★★★☆

Moffat Road, Bethlehem, Tauranga **region** Waikato and Bay of Plenty
ph (07) 576 8800 **fax** (07) 576 8824 **open** 7 days 8–5
winemaker Paddy Preston, Tom Preston **prod.** 35 000 **est.** 1989
prod. range ($8–28 CD) Under the Moffat Road/Mere Road labels Riesling, Sauvignon Blanc, Chardonnay, Pinot Blush and Cabernet Sauvignon; then comes the Reserve range of Riesling, Sauvignon Blanc, Chenin Blanc, Chardonnay, Pinot Noir and Ice Wine Riesling; the top of the range is the Elspeth range of Riesling, Sauvignon Blanc, Chardonnay, Cabernet Sauvignon and Cabernet Merlot; Méthode Champenoise comprises Charisma NV and Vintage.
summary Mills Reef has recently completed a new winery, situated on an 8-hectare chardonnay vineyard within five minutes of Tauranga, incorporating wine tasting and display rooms, a restaurant and a conference/meeting room, together with usual winemaking facilities. The initial releases from Mills Reef were impressive, and after a wobbly period the flagship Elspeth range is among New Zealand's best. With its Hawke's Bay base, it is able to span Chardonnay, Cabernet-based reds and Méthode Champenoise with equal ease. Exports to the UK, US, Japan, Hong Kong, Holland and Germany.

Mills Reef Elspeth Cabernet Merlot

A blend of 70% Cabernet Sauvignon and 30% Merlot, given extensive ageing in French oak. Gold medal 1999 Liquorland Royal Easter Wine Show.
🍷🍷🍷🍷🍷 **1997** Strong red-purple; the bouquet is fine, elegant and cedary with strong overtones of Bordeaux; the palate offers harmonious blackcurrant and cedar flavours supported by ripe tannins and subtle oak. As good as they come (from New Zealand). **rating:** 95
best drinking 2000–2005 **best vintages** NA **drink with** Rack of lamb • $28

Mills Reef Elspeth Cabernet Sauvignon

Completed an impressive double for Mills Reef; like the Elspeth Cabernet Merlot of the same year, winning a gold medal at the 1999 Liquorland Royal Easter Wine Show.
🍷🍷🍷🍷🍷 **1997** Full, deep red-purple; the bouquet has impeccably ripe cassis/blackcurrant fruit supported by a deft touch of oak; the palate continues on logically, with considerable depth to the structure; ripe tannins and well-handled oak do the rest. **rating:** 94
best drinking 2002–2012 **best vintages** '97 **drink with** Aged rump • $28

millton vineyard ★★★★☆

Papatu Road, Manutuke, Gisborne **region** Gisborne
ph (06) 862 8680 **fax** (06) 862 8869 **open** By appointment
winemaker James Millton **prod.** 10 000 **est.** 1984
prod. range ($15–35 CD) Barrique Fermented Chardonnay, Clos de Ste Anne Chardonnay, Chenin Blanc, Tete du Cuvée (botrytised Chenin Blanc), Te Arai River Sauvignon Blanc, Riesling Opou Vineyard, Clos de Ste Anne Pinot Noir, Te Arai River Cabernet Merlot.
summary The only registered organic vineyards in New Zealand using biodynamic methods and banning insecticides and herbicides; winemaking methods are conventional, but seek to limit the use of chemical additives wherever possible. The white wines, particularly botrytised, can be of the highest quality; the Germanic, lime-flavoured Riesling Opou Vineyard is almost always outstanding, while James Millton is doing some of the most exciting things with Chenin Blanc anywhere in the world outside the Loire Valley. Exports to Australia, Belgium, Germany, Hong Kong and the UK.

Millton Riesling Opou Vineyard

As with all of the Millton wines, estate-grown according to strict organic standards. The composition of the finished wine reads like a German Riesling: 10% alcohol, 8.6 grams per litre of acid (natural) and 18.3 grams per litre residual sugar. Just as James Millton endeavours to allow nature to run its course in the vineyard, so he also takes a non-interventionist role in the winery. In fact, the result is (usually but not always) a beautifully balanced wine. The '98 won a gold medal at the 1999 Liquorland Royal Easter Wine Show.

🍷🍷🍷🍷🍷 **1998** Medium green-yellow; intense tropical lime and pineapple aromas are followed by an extremely rich and fruity palate in classic botrytis spätlese-style. **rating:** 93

best drinking 2000–2007 **best vintages** '96, '98 **drink with** Shellfish in creamy sauce • $21

Millton Chenin Blanc

The wine is made from grapes hand-picked over a number of days, barrel-fermented and aged for six months in oak. The '94 was the trophy winner for the Champion Dry White Wine Other Varieties at the 1995 Air New Zealand Wine Awards, and a remarkable wine in all respects. Follow-on vintages (when made) have been good, rather than sensational.

🍷🍷🍷🍷 **1997** Medium yellow-green; apricot and peach fruit aromas supported by subtle oak lead into an extremely concentrated and rich palate with obvious botrytis influence producing apricot and peach fruit flavours, with a soft finish. **rating:** 88

best drinking 1999–2002 **best vintages** '92, '93, '94 **drink with** Seafood bisque • $19

Millton Noble Chenin Blanc (375 ml)

Known in New Zealand as Tete du Cuvée, and made in this fashion for the first and so far only time in 1994. An extraordinary wine, strongly reminiscent of the wines of Bonnezeaux in the Loire Valley in a high botrytis year.

🍷🍷🍷🍷🍷 **1994** Deep gold; the bouquet is intense with apricot, cumquat and marzipan aromas flowing into a gloriously rich palate with quite exceptional flavour and balance, offering honey, marzipan and apricot flavours girdled with tight acidity. **rating:** 97

best drinking 1999–2009 **best vintages** '94 **drink with** Rich dessert • $35

miro NR

Browns Road, Waiheke Island **region** Waiheke Island

ph (09) 372 7854 **fax** (09) 372 7056 **open** By appointment

winemaker Stephen White, Barnett Bond **prod.** 700 **est.** 1994

prod. range ($55 R) A single Bordeaux-blend of Cabernet Sauvignon, Merlot, Cabernet Franc and Malbec.

summary Dr Barnett Bond and wife Cate Vosper are the most recent arrivals on the beautiful Waiheke Island scene, planting their first vines in 1994 and extending the vineyard to 3 hectares in 1996. A luxury holiday cottage on site is available overlooking the steep north-facing slopes of the vines looking out to the Onetangi Bay.

mission estate winery ★★★☆

Church Road, Taradale **region** Hawke's Bay

ph (06) 844 2259 **fax** (06) 844 6023 **open** Mon–Sat 8.30–5.30, Sun 11–4

winemaker Paul Mooney **prod.** 74 000 **est.** 1851

prod. range ($12–33 CD) Mission Estate range of Gewurztraminer, Riesling, Pinot Gris, Sauvignon Blanc, Chardonnay, Merlot, Cabernet Merlot, Cabernet Sauvignon, Botrytised Semillon and Iced Wine; Mission Reserve range of Riesling, Semillon, Pinot Gris, Sauvignon Blanc, Chardonnay and Cabernet Merlot; and the premium Jewelstone Selection range of Chardonnay, Noble Riesling, Gimblett Road Syrah and Gimblett Road Cabernet Merlot.

summary New Zealand's oldest winemaker, owned by the Society of Mary. Once content to make honest, basically unpretentious wines at modest prices, it has developed some top-end wines since 1992, notably the Jewelstone range. Production has been increased by 50% in recent years, partly reflecting the buoyancy of the New Zealand wine industry, but also the quality of the Mission Estate wines.

Mission Estate Jewelstone Gimblett Road Cabernet Merlot

A deserving gold medal at the 1999 Liquorland Royal Easter Wine Show.

🍷🍷🍷🍷🍷 **1997** Deep, dense purple-red; powerful, concentrated cassis/berry fruit aromas are supported by quite subtle oak. The palate is extremely concentrated and quite extractive; the tannins are fractionally hard, and the wine needs time, but should reward patience. **rating:** 92

best drinking 2002–2012 **best vintages** '97 **drink with** Venison • $33

montana wines ★★★★

171 Pilkington Road, Glen Innes, Auckland **region** Auckland Area

ph (09) 570 5549 **fax** (09) 527 1113 **open** 7 days 9.30–5.30

winemaker Jeff Clarke (Chief) **prod.** 35 000 tonnes (2.25 million case equivalent) **est.** 1977

prod. range ($10.95–30 CD) A vast range headed by Marlborough Sauvignon Blanc, Riesling, Chardonnay and Cabernet Sauvignon; Renwick Estate Chardonnay, Timara Riesling, Brancott Estate Sauvignon Blanc, Fairhall Estate Cabernet Sauvignon, Saints Sauvignon Blanc (also all from Marlborough); Ormond Estate Chardonnay (Gisborne); important sparkling wines headed by Deutz Marlborough Cuvée (Brut and Blanc de Blanc) and Lindauer (Special Reserve Brut de Brut, Brut, Sec and Rosé); large volume Wohnsiedler Müller Thurgau, Blenheimer and Chablisse; Church Road Chardonnay and Cabernet Sauvignon are top-of-the-range. The bright-blue bottle of Azure Bay was a colourful (and commercially significant) addition in 1996.

summary Has a more dominant position than does Southcorp through Seppelts-Penfolds-Lindemans in Australia, as it produces 50% of New Zealand's wine. Having parted company with Seagram many years ago, it formed joint ventures with Deutz for sparkling winemaking and Cordier with its Church Road winery offshoot. As one might expect, the wines are invariably well crafted right across the range, even if most attention falls on its Marlborough Sauvignon Blanc. It has recently decided to tackle the difficult Australian market in earnest, and deserves to succeed.

Montana Wines Timara Riesling

A budget-priced wine, 100% Riesling and 100% Marlborough. Did particularly well in the warm 1998 vintage.

🍷🍷🍷🍷 **1998** Light yellow-green; the bouquet is quite floral and scented, with gentle lime and toast varietal characters. A soft and easy wine on the palate with some tropical fruit flavours and a soft acid finish. **rating:** 88

best drinking 1999–2001 **best vintages** NA **drink with** Roast capsicum terrine • $13

morton estate ★★★★

State Highway 2, RD2, Kati Kati **region** Waikato and Bay of Plenty

ph (07) 552 0795 **fax** (07) 552 0651 **open** 7 days 10.30–5

winemaker Evan Ward **prod.** 90 000 **est.** 1978

prod. range ($11.95–33 CD) At the top end is the Black Label range of Chardonnay, Pinot Noir, Merlot, Late Harvest Chardonnay and Méthode Champenoise; next is the White Label range of Chardonnay, Sauvignon Blanc and Cabernet Merlot (all Hawke's Bay); then follows the Stone Creek range of Riesling, Sauvignon Blanc and Chardonnay; and the budget-priced Mill Road range of Chardonnay, Sauvignon Blanc, Dry White, Müller Thurgau, Dry Red and Cabernet Merlot, with Boars Leap Dry White bringing up the rear.

summary Now owned by John Coney, with Evan Ward in charge of winemaking, long-term winemaker John Hancock having left to head up his new Hawke's Bay winery. It will be interesting to watch the development of wine style; it seems probable that the more restrained approach of recent years will continue.

morworth estate NR

Shands Road, Christchurch **region** Canterbury
ph (03) 349 5014 **fax** (03) 349 4419 **open** By appointment
winemaker Dayne Sherwood **prod.** 4000 **est.** 1995
prod. range ($9–17 CD) Riesling, Breidecker, Pinot Noir.
summary Leonie and Chris Morkane have established 13 hectares of vines on the outskirts of Christchurch, predominantly planted to pinot noir (6 hectares) and riesling (3 hectares), with lesser quantities of breidecker, pinot gris and gewurztraminer.

Morworth Estate Riesling

Produced from the first estate plantings dating back to 1993. Conventionally fermented in stainless steel and early-bottled.

🍷🍷🍷🍷 **1997** Light yellow-green; the bouquet is clean, of medium intensity with a mix of lime, citrus and passionfruit aromas, bordering on the tropical. A crisp, clean and long palate is balanced by tingling acidity on the finish, which masks the sweetness. Attractive wine. **rating:** 86

best drinking 1999–2003 **best vintages** NA **drink with** Fresh asparagus • $14

Morworth Estate Breidecker

The indispensable Jancis Robinson discloses that breidecker is a hybrid of müller thurgau and siebel 7053 specially developed for New Zealand at Geisenheim, and has been grown in New Zealand since 1962. She says 'it has the advantages of wild vine resistance to rot and downy mildew without too aggressively savage a flavour'.

🍷🍷🍷 **1997** Light yellow-green; the bouquet is clean, relatively neutral with faint stone fruit aromas. On the palate green apple joins the stone fruit; while not intense, is well balanced. A good outcome for a utilitarian variety. **rating:** 82

best drinking 1999–2000 **best vintages** NA **drink with** Summer salads • $14

mount edward NR

Coalpit Road, Gibbston, Queenstown **region** Otago
ph (03) 442 6113 **fax** (03) 442 9119 **open** By appointment
winemaker Alan Brady, Grant Taylor **prod.** 300 **est.** 1995
prod. range ($18–30 CD) Riesling, Pinot Noir.
summary Alan Brady was the driving force behind the establishment and success of Gibbston Valley winery. In 1998 he retired as general manager of Gibbston Valley and commissioned Queenstown architect Michael Wyatt to design a new small winery on Brady's private vineyard in Coalpit Road. Here 1 hectare of riesling is planted, which he supplements with pinot noir (his continuing passion) purchased from growers in the Otago region. Tastings – and tutored tastings – are by appointment only, and are always conducted by Alan Brady in person.

mountford vineyard NR

434 Omihi Road, Waipara, North Canterbury **region** Canterbury
ph (03) 314 6819 **fax** (03) 314 6819 **open** By appointment
winemaker Daniel Schuster (Consultant) **prod.** 300 **est.** 1990
prod. range ($20–26 CD) Chardonnay, Pinot Noir.
summary Michael and Buffy Eaton have established 4 hectares of vines on an east-facing slope of the Waipara Valley, producing the first tiny vintage in 1996 and scheduled to come into full production by 1999. They already offer bed and breakfast accommodation and a restaurant, and have plans for a winery. In the meantime, local Pinot Noir doyen Danny Schuster is making the wine for the Eatons.

mount linton wine company NR

Hammerichs Road, Rapaura, Blenheim **region** Marlborough
ph (03) 572 9911 **fax** (03) 572 9486 **open** Not
winemaker Tim Macfarlane **prod.** 2000 **est.** 1992
prod. range ($16–19 R) Sauvignon Blanc, Chardonnay.
summary A small and new entrant in Marlborough, employing contract-winemaking of a portion of estate-grown grapes (10 hectares chardonnay and 5 hectares sauvignon blanc) produced on a tiny part of a very large and long-established pastoral business owned by the Macfarlane family.

mount riley NR

Old Renwick Road, Blenheim **region** Marlborough
ph (09) 486 0286 **fax** (09) 486 0643 **open** By appointment
winemaker Bill 'Digger' Hennessy **prod.** 12 000 **est.** 1995
prod. range ($13.95–16.95 R) Sauvignon Blanc, Chardonnay, Cabernet Merlot.
summary Mount Riley is a joint venture between Auckland-based businessmen and wine enthusiasts Steve Hotchin and John Buchanan, and Marlborough vigneron and winemaker Allan Scott. Mount Riley owns three vineyards in the Wairau Valley, and has developed a fourth vineyard in Seventeen Valley, 10 kilometres south of Blenheim planted to clonally selected pinot noir. In all, Mount Riley owns 42 hectares of vineyards, and is in no sense a second label of Allan Scott, even though it shares winemakers and wine facilities with Allan Scott Wines.

Mount Riley Semillon

An interesting wine, entirely barrel-fermented in four- to five-year old barrels, but with the malolactic fermentation suppressed, and picked at 23° brix.

🍷🍷🍷🍷 **1998** Light yellow-green; the bouquet is light and crisp with grassy herbal fruit and little oak evident. The palate has nice flavour and feel, with a mix of honey, mead and more citrussy flavours. Unconventional, perhaps, but worth the effort. **rating:** 86

best drinking 1999–2001 **best vintages** NA **drink with** Smoked eel • $16.95

Mount Riley Sauvignon Blanc

Made in a classically uncomplicated fashion, night-harvested by machine and cold-fermented in stainless steel at low temperatures for three weeks. Both at juice and wine stage minimal intervention and maximum fruit protection pays handsome dividends.

🍷🍷🍷🍷 **1998** Light to medium yellow-green; the bouquet offers gentle tropical/gooseberry/peachy fruit; the nicely balanced palate is not intense, but the fruit does carry through to the end. Has won two silver medals. **rating:** 86

best drinking 1999–2000 **best vintages** NA **drink with** Clams • $15.95

moutere hills vineyard ★★★☆

Sunrise Valley, RD1, Upper Moutere, Nelson **region** Nelson
ph (03) 543 2288 **fax** (03) 543 2288 **open** October–Easter 11–5
winemaker Simon Thomas **prod.** 1200 **est.** 1996
prod. range ($13.50–20 CD) Nelson Riesling, Barrique Fermented Sauvignon Blanc, Chardonnay, Sunrise Valley Red (Rosé), Merlot, Cabernet Merlot.
summary Moutere Hills winery was established in an old shearing shed by owners Simon and Alison Thomas. Overlooking the Moutere Valley, it draws upon 1.5 hectares of estate vineyards supplemented by small quantities of grapes purchased from local growers. Wines are available by the glass matched by light meals.

mudbrick vineyard NR

Church Bay Road, Oneroa, Waiheke Island **region** Waiheke Island
ph (09) 372 9050 **fax** (09) 372 9052 **open** 7 days 10–6 summer, Fri–Sun 10–6 winter
winemaker Aran Knight **prod.** 200 **est.** 1996
prod. range ($25–35 CD) Chardonnay, Cabernet Sauvignon Merlot Malbec Cabernet Franc, Croll Vineyard Cabernet Syrah.
summary The metamorphosis from accountancy to winemaking and restaurateurs/hoteliers is about as radical as they come, but Nick and Robyn Jones have accomplished it. Their 4-hectare vineyard, planted to cabernet sauvignon, merlot, malbec, cabernet franc, shiraz and a little chardonnay, is still coming into production, but the restaurant and accommodation are fully operational.

muddy water NR

434 Omihi Road, Waipara **region** Canterbury
ph (03) 377 7123 **fax** (03) 377 7130 **open** Not
winemaker Kym Rayner (Contract) **prod.** 1400 **est.** 1992
prod. range ($NA) Sauvignon Blanc, Pinot Noir.
summary I must say, I don't think I could ever be persuaded to call my winery (least of all my wines) Muddy Water, even if it does happen to be the literal translation of the Maori word 'Waipara'. However, Michael East, a Christchurch doctor, and his wife Jane, who studied viticulture at Lincoln University, had different ideas when they established Muddy Water in 1992. They already have 11 hectares of vineyard planted, with another 4 hectares on the drawing board, and room for a total of 25 hectares. At present, all of the wines are sold by retail and mail order

Muddy Water Pinot Noir

In my view the wine was decidedly unlucky not to win more than a bronze medal at the 1999 Liquorland Royal Easter Wine Show, although I have to acknowledge that the style is far less opulently fruity than that of many of the top New Zealand Pinot Noirs, veering, as it does, more towards Burgundy.

▼▼▼▼▽ **1997** Light to medium red-purple; a very stylish, elegant, sappy/foresty bouquet; the palate has a touch of plum but sappy Burgundian flavours drive through the above-average length and long, lingering finish of the palate. **rating:** 93

best drinking 1999–2003 **best vintages** NA **drink with** Char-grilled salmon • NA

mud house ★★★☆

Conders Bend Road, Renwick, Marlborough **region** Marlborough
ph (03) 572 9490 **fax** (03) 572 9491 **open** By appointment
winemaker Graeme Paul, Matt Thomson (Consultant) **prod.** 11 000 **est.** 1993
prod. range ($16–31 CD) Sauvignon Blanc, Chardonnay, Pinot Noir offered both under the Le Grys and Mud House labels.
summary Mud House marks the end of an odyssey dating back to 1066; Jennifer Joslin's family name (Le Grys) dates back to that time, and the Marlborough vineyard was purchased by John and Jennifer Joslin at the end of a six-year sailing trip around the world. A mudbrick guesthouse built in 1995 is ultimately to be followed by a mudbrick tasting room and cellar. Production has almost trebled over the past few years, and the wines are distributed (through separate agencies for Mudhouse and Le Grys respectively) in the UK and Australia.

Mud House Sauvignon Blanc

Typically a blend of 86% Sauvignon Blanc and 14% Semillon, a percentage chosen with care to allow the wine to be labelled simply as Sauvignon Blanc in any part of the wine world. The fermentation follows the classic approach of cold settling and low temperature.

🍷🍷🍷🍷 **1998** Light green-yellow; the bouquet is clean, moderately ripe with gooseberry notes. The palate provides the flavours promised by the bouquet, having both depth and some grip on the finish which helps the wine structurally. **rating:** 88

best drinking 1999–2000 **best vintages** NA **drink with** Summer salad • $20

Le Grys Sauvignon Blanc

The Le Grys label is simply an alternative label to Mud House, initially created for distribution reasons, but now having a genuinely independent existence. That difference stems from the different fermentation techniques, with 10% fermented in new French oak, and 30% undergoing malolactic fermentation. The varietal base of 86% Sauvignon Blanc and 14% Semillon is the same as the Mud House Sauvignon Blanc. The '97 won a gold medal at the 1997 Air New Zealand Wine Awards, and thoroughly deserved that success.

🍷🍷🍷 **1998** Medium yellow-green; the bouquet is clean, with some lift to the aromas but not entirely convincing. While the complex winemaking techniques would normally be expected to give a superior result, they seem to have done nothing in the 1998 vintage which naturally tends to softness and broadness. In the end, the wine simply lacks crispness and focus, and as the note indicates, I prefer the Mud House version. **rating:** 84

best drinking 1999–2000 **best vintages** NA **drink with** Tortellini • $16

Le Grys Chardonnay

Predominantly sourced from the Le Grys estate vineyard, but with 10% coming from the Beatrice estate. Made in a fairly direct style, with some lees contact but not a great deal of oak.

🍷🍷🍷🍷 **1997** Medium to full yellow-green; the bouquet is quite developed, with toasty/buttery aromas, the palate similarly forward, with pleasant, sweet buttery/peachy fruit and oak under restraint. **rating:** 85

best drinking 1999–2000 **best vintages** NA **drink with** Brains in black butter • NA

muirlea rise ★★☆

50 Princess Street, Martinborough **region** Wairarapa

ph (06) 306 9332 **open** Not

winemaker Willy Brown **prod.** 700 **est.** 1991

prod. range ($NA) Pinot Noir, Justa Red, Apres Wine Liqueur, Sibbald (Rhône-style Cabernet Sauvignon).

summary Former Auckland wine distributor Willy Brown has established a 1.9-hectare vineyard. Since the first wine release of a 1991 Pinot Noir, the accent has remained on that variety, but with an extraordinarily eclectic gaggle of other wines which are decidedly left-of-centre in style, and for that matter, quality.

murdoch james estate ★★★★

c/o Barbara Turner, 15 Cologne Street, Martinborough **region** Wairarapa

ph (06) 306 9193 **fax** (06) 306 9120 **open** Not

winemaker Oliver Masters (Contract) **prod.** 500 **est.** 1986

prod. range ($22–33 CD) Pinot Noir, Shiraz.

summary The origin of Murdoch James Estate goes back to 1986 when Roger and Jill Fraser planted 2.5 hectares of shiraz and pinot noir. The Frasers' plans were interrupted in 1989 when Roger was transferred to Melbourne by his employer, and the decision was made to sell the grapes rather than make wine. In 1993 a limited quantity was contract-made at Ata Rangi, and each year since that time about 180 cases of Pinot Noir have been made annually, and will be made until 2003. Tiny quantities of Shiraz are also made each year.

nautilus wines ★★★★

Blicks Road, Renwick, Marlborough **region** Marlborough
ph (09) 572 9529 **fax** (09) 572 9529 **open** 7 days 10.30–4.30
winemaker Clive Jones **prod.** 28 000 **est.** 1985
prod. range ($12.50–29.90 R) Chardonnay, Reserve Chardonnay, Sauvignon, Pinot Noir, Cabernet Sauvignon Merlot, Cuvée Marlborough Brut; Twin Islands is the second label, sold only in NZ.
summary Nautilus is ultimately owned by Yalumba of Australia. Until 1996 the wines were made by Yalumba winemaker Alan Hoey at Matua Valley, but from that vintage most were made at Rapaura Vintners (of which Nautilus is now a part-owner) under the direction of former Brokenwood (Australia) winemaker Matt Harrop, now replaced by Clive Jones. Draws both upon estate vineyards (7 hectares at Renwick, 10 hectares at the Awatere River), and contract-grown grapes. Given its parentage, with Yalumba itself having a major agency export business around the world, it is not surprising that Nautilus is exported through much of Europe, Asia, the Pacific Islands and Canada.

Nautilus Marlborough Sauvignon

Sauvignon Blanc was the first wine to be released under the striking and beautiful Nautilus label, and remains the most important wine in terms of quantity; invariably clearly defined. Principally made from vineyards situated in the Awatere Valley, the remainder from two vineyards in the Wairau Valley.

🍷🍷🍷🍷 **1998** Light to medium yellow-green; the bouquet is clean, soft and with ample tropical fruit, the palate pleasantly innocuous and typically ripe. **rating:** 83

best drinking 1999–2000 **best vintages** '89, '90, '91, '94 **drink with** Crayfish • $16.80

---

neudorf vineyards ★★★★★

Neudorf Road, Upper Moutere, Nelson **region** Nelson
ph (03) 543 2643 **fax** (03) 543 2955 **open** 7 days 10–5 Nov–Easter
winemaker Tim Finn **prod.** 6000 **est.** 1978
prod. range ($16–35 CD) The Moutere label is reserved for estate-grown wines, notably Chardonnay and Pinot Noir; Nelson typically indicates a mix of estate-grown and locally purchased grapes from the Nelson region; the second label is Neudorf Village. Wines encompass Chardonnay, Sauvignon Blanc, Semillon, Riesling, Pinot Noir and Cabernet Sauvignon.
summary Tim Finn has produced some of Australasia's most stunningly complex and rich Chardonnays, outstanding in any class. But his skills do not stop there, spanning all varieties, which are consistently of show medal standard. Exports to Australia, UK, US, France, Denmark.

Neudorf Moutere Pinot Noir

If Chardonnay presents a challenge in the Nelson climate, Pinot Noir presents an even greater one – the relatively high summer rainfall poses particular difficulties. The contrast between the '91, '92, '93 and '94 vintages shows how big an impact vintage plays, for these are all quite different wines in style, the '91 muscular, the '92 fragrant and cherry-accented, the '93 spicy, the '94 powerful and plummy. The '96 has echoes of the '92. 1997 saw the first release of a Reserve Pinot Noir or, at least, the first I have tasted.

🍷🍷🍷🍷🍷 **1997** Reserve. Medium to full red-purple; a solid but complex bouquet redolent of dark plum fruit and pleasingly discreet oak is followed by an exceptionally complex palate, with fine tannins running through to the wonderfully long finish; plum is once again the dominant flavour. **rating:** 96

best drinking 2000–2007 **best vintages** '83, '90, '91, '92, '94, '97 **drink with** New Zealand venison • $28

nevis bluff NR

c/o Central Otago Wine Company, McMaltry Road, Cromwell **region** Otago
ph (03) 477 8784 **fax** (03) 477 1186 **open** Not
winemaker Contract **prod.** NA **est.** 1995
prod. range ($NA) Pinot Gris.
summary Yet another impressive Central Otago label, owned by Dunedin chartered accountant Bill Dawson.

Nevis Bluff Pinot Gris

A star performer at the 1999 Liquorland Royal Easter Wine Show, where it received a gold medal, causing much comment (largely favourable) because of the unusual but very appealing fruit aromas and flavours.

🍷🍷🍷🍷🍷 **1998** Very intense, fragrant, flowery passionfruit, spice and citrus aromas lead into a striking wine on the palate with similarly intense citrus, lime, herb and passionfruit flavours. Crisp acidity adds to the length of the finish. **rating:** 95

best drinking 1999–2004 **best vintages** '98 **drink with** Fish with beurre blanc sauce • NA

newton forrest estate NR

Cnr State Highway 50 and Gimblett Road, Hawke's Bay **region** Hawke's Bay
ph (06) 879 4416 **fax** (06) 876 6020 **open** Not
winemaker John Forrest (Contract) **prod.** 1200 **est.** 1988
prod. range ($28 ML) Cornerstone Vineyard Cabernet Merlot.
summary Newton Forrest is a joint venture between Hawke's Bay grape grower Bob Newton and Marlborough winemaker and vigneron John Forrest of Forrest Estate. It has produced a single Cornerstone Vineyard Cabernet Merlot each year since 1994, the '95 vintage winning a gold medal at the 1997 Auckland Royal Easter Wine Show.

ngatarawa wines ★★★★

Ngatarawa Road, Bridge Pa, RD5, Hastings **region** Hawke's Bay
ph (06) 879 7603 **fax** (06) 879 6675 **open** 7 days 11–5
winemaker Alwyn Corban, Peter Gough **prod.** 30 000 **est.** 1981
prod. range ($12.95–60 R) The top wines under the Glazebrook label include Chardonnay and Cabernet Merlot; the lesser-priced Stables range comprises Chardonnay, Sauvignon Blanc, Classic White, and Cabernet Merlot. Also Alwyn Noble Botrytis.
summary Alwyn Corban is a highly qualified and highly intelligent winemaker from a famous New Zealand wine family, who has elected to grow vines organically and make wines which sometimes (but certainly not always) fall outside the mainstream. Challenging and interesting, and not to be taken lightly. The wines are exported to the UK, Europe, Asia, US and Canada.

Ngatarawa Stables Sauvignon Blanc

Since 1997 Alwyn Corban has aged the Stables Sauvignon Blanc in seasoned oak puncheons for two months to add textural and structural complexity to the wine. For the prior decade it had simply been fermented in stainless steel and early-bottled. Like all of Corban's wines, restrained, with the accent on secondary characters.

🍷🍷🍷🍷 **1998** Light yellow-green; the bouquet is very much in the Ngatarawa style, minerally and not fruity. The palate is fresh, crisp and again minerally, the wine scoring points for the balance and feel of the finish. **rating:** 85

best drinking 1999–2001 **best vintages** NA **drink with** Avocado salad • $12.95

Ngatarawa Glazebrook Chardonnay

Estate-grown, hand-picked grapes are barrel-fermented in 100% new French oak, with extended time on yeast lees. Eighty per cent of the wine is taken through malolactic fermentation. The aim of Alwyn Corban is to take the wine into secondary flavours and textures, and he succeeds admirably in his aim.

ŸŸŸŸ **1997** Medium to full yellow; the bouquet is complex, although the oak is subtle, the complexity coming from the mlf influences which run through the wine. The palate is clean, of medium weight, not at all lush or phenolic, with nicely balanced oak. **rating:** 86

best drinking 1999–2003 **best vintages** NA **drink with** Smoked salmon • $20.95

Ngatarawa Alwyn Noble Harvest

Not made every year, but when it is made and when conditions are right, scales the ultimate heights. The grapes are allowed to hang on the vines for long after normal harvest, with a mix of botrytis and raisining lifting the sugar levels over 40 degrees brix. Unusually, the wine is barrel-fermented and held in oak for 18 months, yet shows no sign either of oak or oxidation. Some of the great German botrytis wines are made this way, but relatively few in the New World.

ŸŸŸŸŸ **1996** Glowing yellow-green; the intensely perfumed bouquet has aromas of cumquat, peach, lime and honey, the palate offering some orange and lime peel flavours as well. Balanced acidity in a classy wine. Retasted March 1999 and has developed superbly, assuming distinctly Germanic overtones. **rating:** 94

best drinking 2000–2004 **best vintages** '92, '94, '96 **drink with** Dappled sunlight • $25.95

Ngatarawa Glazebrook Cabernet Merlot

A blend of 65% Cabernet Sauvignon and 35% Merlot, which spends 15 months in oak, 80% new. The fruit power absolutely swallows up that oak, which is not the least intrusive.

ŸŸŸŸY **1996** Bright red-purple; the bouquet is clean, with earthy Cabernet varietal fruit aromas to the fore. The palate is typically restrained, with earthy Cabernet, a touch of briar and forest; the oak is just there, giving some vanilla cedar hints; finishes with balanced tannins. Retasted March 1999 and has developed excellently, thanks to the sweet fruit and ripe tannins. Re-rated accordingly. **rating:** 90

best drinking 2000–2006 **best vintages** '90, '91, '95, '96 **drink with** Rib of veal • $20.95

nga waka vineyard ★★★★

Kitchener Street, Martinborough **region** Wairarapa

ph (06) 306 9832 **fax** (06) 306 9832 **open** Weekends 1–5 while stocks last

winemaker Roger Parkinson **prod.** 3000 **est.** 1988

prod. range ($23–30 R) Riesling, Sauvignon Blanc, Chardonnay.

summary Roseworthy-trained Roger Parkinson produces the Nga Waka wines from 4 hectares of estate plantings in the heart of the Martinborough Terraces. The early promise of the vineyard came into full flower in the unlikely environment of the 1995 vintage with the performance of the 1995 Sauvignon Blanc at the Air New Zealand Wine Awards of that year. Subsequent vintages have been good, but not quite so exciting as the '95.

nobilo ★★★★

Station Road, Huapai **region** Auckland Area

ph (09) 412 9148 **fax** (09) 412 7124 **open** Mon–Fri 9–5, Sat 10-5, Sun 11-4

winemaker Darryl Woolley, Greg Foster **prod.** 375 000 **est.** 1943

prod. range ($7.95-29 R) Marlborough Sauvignon Blanc, Gewurztraminer, Chardonnay and Cabernet Sauvignon; Poverty Bay Chardonnay; Müller Thurgau, White Cloud, Huapai Pinotage and Hawke's Bay Cabernet. More recently (in 1996 and 1997) the Fall Harvest (low price), Icon (mid-price) and Grand Reserve (high price) varietal ranges have been introduced.

summary One of the more energetic and effective wine marketers, with production heavily focused on white wines sourced from Gisborne, Hawke's Bay and Marlborough. A long-time exponent of flamboyant label and packaging redesign, but the quality of the top-of-the-range Grand Reserve wines leaves nothing to be desired. Exports to the UK, much of Europe, North America and Singapore.

Nobilo Tietjen Chardonnay

From the highly rated Tietjen Vineyard in Gisborne, which over the years has provided a number of outstanding wines to various purchasers. The wine was not made in either 1996 or 1997.

🍷🍷🍷🍷 **1998** Medium yellow-green; the bouquet shows strong, high toast barrel-ferment characters, but comes into balance on the palate, where citrussy fruit takes over and runs through to a long finish. **rating:** 88

best drinking 1999–2001 **best vintages** '94, '98 **drink with** Salmon risotto • $18

obsidian NR

Te Makiri Road, Waiheke Island **region** Waiheke Island
ph (09) 372 6100 **fax** (09) 372 6100 **open** Not
winemaker Kim Crawford **prod.** 1500 **est.** 1993
prod. range ($45 R) A single red wine made from about 70% Cabernet Sauvignon and 30% Merlot, with just a dash of Malbec and Cabernet Franc.
summary Although still in its infancy (the first vintage was 1997) destined to be an important part of the Waiheke Island scene. A partnership between Andrew Hendy, owner of Coopers Creek, winemaker Kim Crawford and businessman Lindsay Spilman, it brings together the formidable winemaking skills of Crawford and the largest (7.3-hectare) vineyard on Waiheke Island, established in a natural amphitheatre. The wine is initially offered en primeur by mailing list, and then to selected on-premise outlets on Waiheke Island.

odyssey wines NR

32 Henderson Valley Road, Henderson **region** Auckland Area
ph (09) 309 1969 **fax** (09) 309 1969 **open** Not
winemaker Rebecca Salmond **prod.** 1500 **est.** 1994
prod. range ($13.95–16.95 R) Hawke's Bay Sauvignon Blanc, Gisborne Chardonnay, Reserve Gisborne Chardonnay, Cabernet Merlot, Reserve Kumeu Cabernet Sauvignon.
summary Rebecca Salmond is winemaker for Pleasant Valley, and makes the wines for her Odyssey brand at that facility. There are no cellar-door sales; the wines are sold via retail and mail order.

ohinemuri estate NR

Moresby Street, Karangahake **region** Waikato and Bay of Plenty
ph (07) 862 8874 **fax** (07) 862 8847 **open** 7 days 10–6
winemaker Horst Hillerich **prod.** 1000 **est.** 1989
prod. range ($15–20 CD) Chardonnay, Classique (Chenin Blanc Chardonnay blend), Gewurztraminer, Riesling, Sauvignon Blanc, Pinotage.
summary German-born, trained and qualified winemaker Horst Hillerich came to NZ in 1987, first working at Totara before establishing Ohinemuri Estate. An atmospheric restaurant was duly opened at the newly constructed winery in the Karangahake Gorge in 1993; Hillerich has produced some highly regarded Gewurztraminer and Sauvignon Blanc. All of the grapes, incidentally, are purchased from growers in Hawke's Bay, The Waikato and Gisborne.

okahu estate NR

Okahu Road, Kaitaia **region** Northland and Matakana
ph (09) 408 2066 **fax** (09) 408 2686 **open** 7 days 10–6 Oct–June, closed Jul–Aug
winemaker Michael Bendit **prod.** 5000 **est.** 1984
prod. range ($12.95–22.95 CD) 90 Mile Semillon Chardonnay and Cabernet Merlot; Clifton Riesling and Chardonnay; Montgomery Chardonnay, Beverly's Blush, Te Hana Pinot Noir, Kaz Semillon and Shiraz; Chardonnay Desserté and Riesling Desserté; Old Brother John's Tawny Port, Don de Monte Oloroso Sherry.
summary The 90 Mile wines (respectively blends of Chardonnay, Semillon and Arnsburger, and Cabernet Merlot, Pinotage and Pinot Noir) signal the location of Okahu Estate at the bottom end of the 90 Mile Beach. Recently the focus of the estate plantings of 2.5 hectares has switched to semillon and shiraz, which are said to show considerable promise; the plethora of other wines are made from grapes purchased from other regions.

olssen's of bannockburn NR

306 Felton Road, Bannockburn, Central Otago **region** Otago
ph (03) 445 1716 **fax** (03) 445 0050 **open** 7 days 10–4
winemaker Duncan Forsythe **prod.** 1900 **est.** 1989
prod. range ($17.95–26.50 CD) Riesling, Gewurztraminer, Sauvignon Blanc, Chardonnay, Pinot Noir.
summary Heather McPherson and John Olssen began the establishment of their 10-hectare vineyard (and now 5 hectares of rural garden and outdoor eating sites) in 1989. For the first three years after the vineyard came into bearing the grapes were sold to Chard Farm, but in 1997 the first wines were made under the Olssen's of Bannockburn label. A new but substantial enterprise in the Central Otago scene.

omaka springs estate NR

Kennedys Road, RD2, Blenheim, Marlborough **region** Marlborough
ph (03) 572 9933 **fax** (03) 572 9934 **open** 7 days 12–5
winemaker Ian Marchant **prod.** 18 000 **est.** 1992
prod. range ($7.99–15.99 R) Riesling, Semillon, Sauvignon Blanc, Chardonnay, Cabernet Merlot.
summary Omaka Springs is a substantial operation, with a state-of-the-art winery built at the end of 1994 now reaching capacity (20 000 cases) and drawing upon 38 hectares of estate vineyards. The wines are marketed in New Zealand, the US, UK, Sweden.

Omaka Springs Marlborough Riesling

Nothing fancy going on here, just Marlborough-grown riesling, cold-fermentation in stainless steel and early-bottling.

▼▼▼▼ **1998** Light yellow-green; the bouquet is quite full, with lime, tropical aromas and the palate provides a quite soft, pleasant, easy-drinking wine with nice balance. **rating:** 85

best drinking 1999–2002 **best vintages** NA **drink with** Asian dishes • $11

Omaka Springs Sauvignon Blanc

Made in mainstream Marlborough-style, unoaked but with up to 12% Semillon incorporated to give a little more complexity and depth. The '97 is more concentrated and stylish than either the '96 or '98.

▼▼▼▽ **1998** Light straw-green; light, minerally and citrus/gooseberry aromas are followed by a soft palate which makes for easy drinking, but doesn't escape the clutches of the vintage. **rating** 84

best drinking 1999–2000 **best vintages** '94, '97 **drink with** Sautéed scallops • $11

omarama vineyard NR

Omarama Avenue, Omarama, North Otago **region** Otago
ph (03) 438 9708 **fax** (03) 438 9708 **open** Summer 11–9, winter 11–6
winemaker Mike Wolter **prod.** NA **est.** 1991
prod. range ($12–16 CD) Pinot Gris, Muscat, Müller Thurgau.
summary All of the output of Omarama Vineyard is sold through the cellar door and, more particularly, through the vineyard's wine bar and café.

opihi vineyard NR

Gould's Road, Opihi, Pleasant Point, South Canterbury **region** Canterbury
ph (03) 614 7232 **fax** (03) 614 7234 **open** By appointment
winemaker Tony Coakley (Contract) **prod.** 700 **est.** 1991
prod. range ($10–18 CD) Riesling, Müller Thurgau, Chardonnay, Pinot Gris, Pinot Noir.
summary Plantings at Opihi Vineyard commenced in 1991 with half a hectare each of müller thurgau, riesling, pinot noir and chardonnay, followed the next year by 2.5 hectares of pinot gris. The vineyard is established on a north-facing slope of Timaru clay loam with superb views across to the snow-clad Two Thumb Range. The tiny production is chiefly sold by mail order, with limited South Canterbury regional distribution.

Opihi Vineyard Pinot Gris

Like the Nevis Bluff Pinot Gris, a gold medal winner 1999 Liquorland Royal Easter Wine Show. Unlike the Nevis Bluff wine, however, far more conventional in terms of aroma and flavour.

🍷🍷🍷🍷🍸 **1998** Light green-yellow; a spotlessly clean wine on both bouquet and palate, showing very pure but restrained pinot gris fruit which makes any particular descriptor elusive. Neatly balanced by a light touch of residual sugar on the finish. **rating:** 91

best drinking 1999–2003 **best vintages** NA **drink with** Trout mousse • $18

pacific vineyards ★★★

90 McLeod Road, Henderson **region** Auckland Area
ph (09) 838 9578 **fax** (09) 838 9578 **open** Mon–Sat 9–6
winemaker Steve Tubic **prod.** 10 000 **est.** 1936
prod. range ($15–19 CD) Under the Phoenix label comes Gisborne Gewurztraminer, Marlborough Riesling, Marlborough Sauvignon Blanc, Gisborne Chardonnay, Merlot, Cabernet Sauvignon; also Quail Farm Gisborne Chardonnay and Cabernet Sauvignon, and Millie's Merlot.
summary One of the more interesting New Zealand wineries, notwithstanding its low profile, which has at various times produced very large quantities of wine (sold in cask and bulk) but is now refocusing on its bottled wine production (in more limited quantities) utilising grapes from Marlborough, Hawke's Bay and Gisborne, and on the other side of the fence has ventured into beer brewing.

palliser estate ★★★★☆

Kitchener Street, Martinborough **region** Wairarapa
ph (06) 306 9019 **fax** (06) 306 9946 **open** 7 days 10–6
winemaker Allan Johnson **prod.** 20 000 **est.** 1989
prod. range ($16–30 CD) Chardonnay, Sauvignon Blanc, Riesling, Late Harvest Riesling, Methode Champenoise, Pinot Noir; Pencarrow Sauvignon Blanc, Chardonnay and Pinot Noir.
summary Palliser Estate has produced a series of highly regarded and highly awarded wines from its state-of-the-art winery right from its first vintage in 1989, and which has grown rapidly in recent years. My tasting notes indicate high scores across the full range of the wines produced with the perfumed silky Pinot Noir to the fore. The wines are distributed in Australia through Negociants, and have a wide range of active markets in Europe, Asia and North America.

Palliser Estate Riesling

Produced both as a dry Riesling (as in this case) with occasional delectable late-harvest, botrytised versions. The '93 was the top gold medal winner in its class at the 1994 National Wine Show of Australia in Canberra, outpointing all of the leading Australian Rieslings (as well as numerous NZ contenders). The '96 won a gold medal at the Air New Zealand Wine Awards, the '98 following suit at the 1999 Liquorland Royal Easter Wine Show.

▼▼▼▼▼ **1998** Bright, medium to full green-yellow; a resounding, big and rich tropical-accented bouquet is followed logically and inevitably by an exceptionally generous wine in the mouth, with soft, lime pastille flavours. Interesting vintage variation at work with a wine which is always in the top class, but which varies significantly in its flavour profile from year to year. **rating:** 94

best drinking 2000–2007 **best vintages** '93, '96, '97, '98 **drink with** Lightly poached asparagus • $18

Palliser Estate Sauvignon Blanc

The Wellington/Wairarapa region seems to produce excellent Sauvignon Blanc year in, year out, a capacity underlined by the success the region has had in the annual Air New Zealand Wine Awards. Palliser Estate has been one of the most consistent and impressive performers.

▼▼▼▼▽ **1998** Light straw-green; the bouquet is quite powerful, although the aromatics are slightly subdued. An elegant wine on the palate with minerally herb characters more than tropical fruit. **rating** 91

best drinking 1999–2000 **best vintages** '91, '92, '94, '95, '96, '97, '98 **drink with** Grilled fish • $18

---

park estate winery NR

2087 Pakowhai Road, RD3, Napier **region** Hawke's Bay

ph (06) 844 8137 **fax** (06) 844 6800 **open** 7 days 10–5.30

winemaker Owen Park **prod.** 5000 **est.** 1992

prod. range ($10–20 CD) Riesling, Gewurztraminer, Sauvignon Blanc, Chardonnay, Merlot, Cabernet Sauvignon, Sparkling Sauvignon Cuvee, Late Harvest Muscat; Bell Tower is cheaper second label.

summary Owen and Dianne Park run a thriving and varied enterprise offering both fruit and grape-based wines (and 35 different types of fudge) from a large mission-style winery and restaurant. Wine production (from grapes, that is) constitutes a modest part of the business.

parker méthode champenoise NR

91 Banks Street, Gisborne **region** Gisborne

ph (06) 867 6967 **fax** (06) 867 6967 **open** 7 days 9.30–6

winemaker Phil Parker **prod.** 1000 **est.** 1987

prod. range ($NA) Dry Flint, Classical Brut, Rosé Brut, Light Red.

summary A Méthode Champenoise specialist which has caused much interest and comment. Has not entered the show ring and I have not tasted the wines. The winery also has a restaurant open for lunch and dinner every day of the week.

pegasus bay ★★★★☆

Stockgrove Road, Waipara, RD2, Amberley **region** Canterbury

ph (03) 314 6869 **fax** (03) 314 6869 **open** 7 days 10–5

winemaker Matthew Donaldson, Lynnette Hudson **prod.** 16 000 **est.** 1986

prod. range ($18–31 CD) Chardonnay, Sauvignon Blanc Semillon, Riesling, Aria, Pinot Noir, Maestro (Bordeaux-blend); Main Divide is second label range of Aged Riesling, Riesling, Marlborough Sauvignon Blanc, Merlot Cabernet, Pinot Noir.

summary Leading wine-writer and wine judge Professor Ivan Donaldson (a neurologist) has, together with his wife and family, established the largest winery in Waipara, with 20 hectares of vineyards in bearing and a large and striking cathedral-like winery. Son Matthew is a Roseworthy graduate, and in every respect this is a serious operation. A winery restaurant adds to the attraction for visitors. Wine quality is consistently good, the wines with style and verve. Exports to Australia, US and Japan.

Pegasus Bay Riesling

Made in the typically adventurous Pegasus Bay fashion, given eight months on lees, and incorporating 10% of botrytised grapes. The 11 grams per litre of residual sugar are well balanced.

🍷🍷🍷🍷▽ **1998** Light to medium yellow-green; the complex bouquet offers a mix of lime, tropical and more spicy notes, the palate stacked with flavour running through lime to cumquat marmalade. Striking but successful in a mildly off-dry style. **rating:** 90

best drinking 1999–2004 **best vintages** '98 **drink with** Prosciutto and melon • $19

Pegasus Bay Sauvignon Blanc Semillon

The Sauvignon Blanc component (70%) is cold-fermented in stainless steel, the Semillon (30%) component barrel-fermented and taken through malolactic fermentation. Both components are aged separately on lees for approximately eight months before blending and bottling.

🍷🍷🍷🍷 **1997** The colour is very advanced for the age of the wine; the powerful, potent bouquet has strong herbal/grass aromas, and potent clove/spice oak comes through on the finish. As always, boldly sculptured, a little short on finesse, but with abundant character; reflects the microsopic yield of the vineyard. **rating:** 89

best drinking 1999–2000 **best vintages** NA **drink with** Fettuccini • $26

Pegasus Bay Chardonnay

The hand-picked Mendoza clone grapes are whole-bunch pressed, and the wine is fermented in a mix of new (30%) and used French oak barriques. Part is not inoculated with yeast, but utilises indigenous yeast, and the wine undergoes a partial but natural malolactic fermentation on lees. This wine certainly gets the full book thrown at it in terms of winemaking, and is unashamedly at the baroque end of the spectrum.

🍷🍷🍷🍷 **1997** Medium yellow-green; complex nutty, creamy aromas on the bouquet are followed by a palate with almost untamed power and concentration, and considerable complexity. A slightly bitter finish detracts marginally from the wine. **rating:** 88

best drinking 1999–2002 **best vintages** '96, '97 **drink with** Fricassee of veal • $28

Pegasus Bay Canterbury Pinot Noir

Pegasus Bay has produced astonishing rich, deeply coloured and full-bodied Pinot Noirs right from the outset. An extremely impressive wine which proclaims its unfiltered origins. The Donaldson family has also worked very hard in evaluating fermentation and maturation techniques for the variety, all of which have paid off.

🍷🍷🍷🍷🍷 **1997** Medium to full red-purple; a powerful, complex bouquet with lots of high-toast oak to accompany the fruit is followed by flavours of plum, briar, forest, cherry and of course the oak promised by the bouquet. Very complex, and reflecting strong winemaker inputs. **rating:** 94

best drinking 1999–2003 **best vintages** '93, '94, '95, '97 **drink with** Rare breast of squab • $28

Main Divide Pinot Noir

This second label of Pegasus Bay is produced from Canterbury grapes, not necessarily estate. The slightly cloudy colour attests to the fact that the wine is not filtered, and thus should have great appeal to Robert Parker et al.

🍷🍷🍷🍷 **1997** Slightly cloudy; a complex bouquet with faintly gamey, green canopy characters, together with a touch of pepper. The usual full-on winemaking techniques of Pegasus Bay are evident in the quite stylish, long, sappy palate. Better (possibly riper) grapes would have produced an even better wine.

rating: 86

best drinking 1999–2001 **best vintages** NA **drink with** Corned beef • $17

peninsula estate ★★★★

52A Korora Road, Oneroa, Waiheke Island **region** Waiheke Island
ph (09) 372 7866 **fax** (09) 372 7840 **open** Weekends 1–4
winemaker Christopher Lush **prod.** 1400 **est.** 1986

prod. range ($23–33 CD) The top-of-the-line release is Peninsula Estate Cabernet Merlot; the intermittent second label is Oneroa Bay Cabernet Merlot. Both in fact include a small percentage of Cabernet Franc and Malbec.

summary The Peninsula Estate Cabernet Merlot comes from a 5.5-hectare estate vineyard situated on a peninsula overlooking Oneroa Bay. The spectacular vineyard has produced some equally spectacular wines.

pierre estate NR

Elizabeth Street, Waikanae **region** Wairarapa
ph (04) 293 4604 **open** Not
winemaker Peter Heginbotham **prod.** NFP **est.** 1969

prod. range ($13.95–19 ML) Blanc du Noir (Pinot Noir), Cabernet Sauvignon.

summary Waikanae is situated on the coast north of Wellington; the wines are estate-grown, made and produced in the 'Chateau' and underground cellars completed in 1991 by Wellington optometrist Peter Heginbotham. I have not tasted the wines.

pleasant valley wines ★★★

322 Henderson Valley Road, Waitakere **region** Auckland Area
ph (09) 838 8857 **fax** (09) 838 8456 **open** Mon–Sat 9–6, Sun 11–6
winemaker Rebecca Salmond **prod.** 10 000 **est.** 1902

prod. range ($5.95–19 CD) Gewurztraminer, Sauvignon Blanc, Chenin Chardonnay, Chardonnay, Riesling, Pinotage, together with a range of fortified wines, chiefly Sherries but also Port.

summary A former moribund fortified winemaker, revitalised since 1984 and now complementing its stocks of old fortified wines with well-made table wines sourced from Hawke's Bay, Gisborne and Marlborough, supplementing a 7-hectare estate vineyard at Henderson.

pomona ridge vineyard NR

Pomona Road, Ruby Bay, Nelson **region** Nelson
ph (03) 540 2769 **fax** (03) 540 2769 **open** Weekends and holidays 10–5
winemaker Peter Hancock **prod.** 330 **est.** 1993

prod. range ($15–20 CD) Pinot Noir.

summary Pomona Ridge claims to be New Zealand's only Pinot Noir specialist, but I am not sure that is a correct claim. Peter Hancock is a self-taught winemaker; the tiny production is handled in a converted garage with eminently satisfactory results.

ponder estate NR

New Renwick Road, Blenheim **region** Marlborough
ph (03) 572 8642 **fax** (03) 572 9034 **open** Mon–Sat 10.30–4.30, Sun 11–4
winemaker Graham Paul **prod.** 12 000 **est.** 1987

prod. range ($18–26 CD) Marlborough Sauvignon Blanc, Chardonnay.

summary With 25 hectares of vineyard and olive grove, Ponder Estate is primarily a grape grower (supplying chardonnay, sauvignon blanc and riesling to Matua Valley for its Shingle Peak label) but is rapidly increasing the amount vinified for its own label. It also has a press house for its own olive oil production. Exports to Australia, UK and US.

pouparae park NR

Bushmere Road, Gisborne **region** Gisborne
ph (06) 867 7931 **fax** (06) 867 7909 **open** 7 days 10–6
winemaker Alec Cameron **prod.** 800 **est.** 1994
prod. range ($7.50–15 R) Riesling, Chardonnay, Solstice Blanc (Müller Thurgau, Dr Hogg Muscat blend), Pinotage, First Light Red (Pinot Noir).
summary Pouparae Park was established on the family's property by Alec and Rachel Cameron in 1994. 'Pouparae' means 'high vantage point', and First Light Red (made from Pinot Noir) is said to be made from vines which are the first to see the light in the world each day.

purple heights estate NR

Main West Coast Road, RD6 (PO Box 31110), Christchurch, NZ **region** Canterbury
ph (03) 358 2080 **fax** (03) 325 3843 **open** Not
winemaker Dayne Sherwood (Contract) **prod.** NA **est.** 1996
prod. range Riesling, Noble Riesling.
summary Partners Delwyn and John Mathieson and David and Diana Jackson named their vineyard after the colour of the nearby foothills, planting 2 hectares to riesling. Having sold the grapes to other makers for many years, the partners ventured into winemaking (via contract at Sherwood Estate) in 1996. There are no sales to the public; all wine is sold wholesale to retailers and restaurants.

quarry road estate NR

Waerenga Road, RD1, Te Kauwhata **region** Waikato and Bay of Plenty
ph (07) 826 3595 **fax** (07) 826 3595 **open** 7 days 8–6
winemaker Toby Cooper, Jenny Gander, Nikki Cooper **prod.** 4000 **est.** 1996
prod. range ($8.50–20 CD) Grape juices, table wines, sparkling wines, fortified wines and liqueurs.
summary 1996 is a nominal year of establishment, for this is the former Aspen Ridge, acquired by the Cooper family in that year. With the aid of consultancy advice, they intend to move more towards the production of premium table wines and to expand the cellar-door facilities.

quartz reef NR

PO Box 63, Cromwell, Central Otago **region** Otago
ph (03) 445 1135 **fax** (03) 445 1180 **open** By appointment
winemaker Rudi Bauer **prod.** 2000 **est.** 1996
prod. range ($21–23 R) Pinot Noir, Méthode Champenoise.
summary Quartz Reef is a joint venture between Rudi Bauer and Clotilde Chauvet, both of whom are Rippon winemakers. Three hectares have been planted to pinot noir and chardonnay, with plantings intended to extend to 10 hectares.

rabbit ridge NR

407 Taylor Road, Waimauku **region** Auckland Area
ph (09) 411 8556 **fax** (09) 411 8556 **open** By appointment
winemaker Matua Valley (Contract) **prod.** 300 **est.** 1990
prod. range ($19.75 R) Chardonnay, Merlot.
summary Former architect and now leading NZ fine wine wholesaler and importer Paul Mitchell (of Wine Direct Imports) is indulging in a little poaching at Rabbit Ridge. He had established a 2-hectare vineyard 30 minutes from the centre of Auckland, with his house situated

in the middle of the vineyard; it is sufficiently elevated for him to see rabbits feeding in the vineyard from the lounge. The wines are contract-made at Matua Valley, and Paul Mitchell has no intention of hopping into a fermenter at vintage time. However, he does concede there are plans to build a small cellar-door sales and barrel storage area; in the meantime the wines are available from Wine Direct or by mail order.

redmetal vineyards NR

2006 Maraekakaho Road, RD1, Bridge Pa, Hastings, NZ **region** Hawke's Bay
ph (06) 879 8768 **fax** (06) 879 7187 **open** Not
winemaker Grant Edmonds **prod.** 1500 **est.** 1992
prod. range ($22.95–29.95 R) Rosé, Merlot, Basket Press Merlot Cabernet.
summary A joint venture between the vastly experienced winemaker Grant Edmonds, wife Sue and Diane and Gary Simpson, with initial plantings of 4.5 hectares (on an 8-hectare block) on an alluvial silt over gravel soil known locally as red metal. The wines are exported to the UK.

rippon vineyard ★★★★☆

Mount Aspiring Road, Lake Wanaka **region** Otago
ph (03) 443 8084 **fax** (03) 443 8084 **open** 7 days 11.30–5 Aug–May, or by appointment
winemaker Russell Lake **prod.** 5000 **est.** 1975
prod. range ($13.50–54 CD) Gewurztraminer, Riesling, Osteiner, Hotere White, Chardonnay, Gamay Rosé, Pinot Noir, Pinot Noir Selection.
summary Claimed, with some justification, to be the most beautifully sited vineyard in the world, situated on the edge of Lake Wanaka (which is responsible for the remarkable site climate), with the snow-clad New Zealand Alps painting a striking backdrop. Right across the range, Rippon has produced some outstanding wines, none more so than the Pinot Noir. In recent years, Rippon has moved to Bio-Gro™ certified organic status. The wines are exported to both Australia and the UK in restricted quantities.

Rippon Vineyard Riesling

Made in tiny quantities, and considered as expensive as it is rare in NZ. In 1998 picked very ripe (13.5° alcohol) and fermented down to a threshold level of 4.5 grams per litre of residual sugar.

🍷🍷🍷🍷🍷 **1998** Light to medium yellow-green; the bouquet is very tight, showing pure riesling characters of mineral, toast and herb. The palate is extremely long, intense and striking, with herb and lemon zest flavours running through to the finish. **rating:** 92

best drinking 2000–2010 **best vintages** NA **drink with** Vegetable terrine • $21.50

Rippon Vineyard Pinot Noir

Eighty per cent of the fruit is destemmed, 20% fermented as whole bunches, with foot stamping (pigeage). The wine spends 11 months in French oak and is racked once, with no filtration prior to bottling.

🍷🍷🍷🍷🍷 **1996** Medium red; the bouquet is distinctive, with predominantly plum and some cherry fruit; no excess carbonic maceration characters. The palate has crystal clear Pinot varietal flavour in a similar plum/cherry spectrum, and is not jammy. Subtle oak; good length. Retasted March 1999 and seems to be fading somewhat, although a '91 Rippon tasted at the same time was amazingly fresh and fragrant. **rating:** 90

best drinking 1999–2000 **best vintages** '90, '92, '93 **drink with** Coq au vin • $31

---

riverside wines ★★☆

Dartmoor Road, Puketapu, Napier **region** Hawke's Bay
ph (06) 844 4942 **fax** (06) 844 4671 **open** Summer 7 days 11–5, winter by appointment
winemaker Russell Wiggins **prod.** 10 000 **est.** 1989

prod. range ($14–26 CD) Dartmoor Sauvignon Blanc, Barrel Fermented Sauvignon Blanc, Dartmoor Chardonnay, Stirling Chardonnay, Reserve Chardonnay, Rosé, Merlot, Cabernet Merlot.

summary Ian and Rachel Cadwallader have established 14 hectares of vines on their farm which are coming progressively into production. The wine is made on site in the small winery above the Dartmoor Valley. It has to be said that overall wine quality was far from exciting, but the arrival of Russell Wiggins as winemaker in late 1998 is almost certain to rectify matters. In the meantime, exports to US and Thailand are underway.

rockwood cellars NR

James Rochford Place, RD5, Hastings **region** Hawke's Bay

ph (06) 879 8760 **fax** (06) 879 4158 **open** By appointment

winemaker Tony Bish **prod.** 10 000 **est.** 1995

prod. range (under $15 R) Sauvignon Blanc, Chardonnay, Cabernet Merlot.

summary Rockwood Cellars aims primarily at the export market, with a varied range of modestly priced wines all selling for less than $NZ15 and is in fact part of the Sacred Hill wine group, albeit with its separate brand identity.

rongopai wines ★★★★

Te Kauwhata Road, Te Kauwhata **region** Waikato and Bay of Plenty

ph (07) 826 3981 **fax** (07) 826 3462 **open** Mon–Fri 9–5, Sat 10–5, Sun 11–5

winemaker Tom van Dam **prod.** 15 000 **est.** 1985

prod. range ($11–32 CD) Riesling, Sauvignon Blanc, Oak Aged Sauvignon Blanc, Winemakers Selection Sauvignon Blanc, Chenin Blanc, Chardonnay, TK Reserve Chardonnay, Reserve Botrytis Chardonnay, Botrytis Reserve, Pinot Noir, Merlot Malbec, Waerenga (Cabernet blend), TK Reserve Merlot.

summary Now owned solely by Tom van Dam and wife Faith, but going from strength to strength, it would seem. The reputation of Rongopai rests fairly and squarely upon its spectacular botrytised wines which have enjoyed equal quantities of show success and critical acclaim throughout most of the 1990s. Both Chardonnay and the more conventional Riesling are used in these wines; the minimal use of chemicals and herbicides in the vineyards promotes late-season botrytis, countered by vine-trimming, leaf-plucking and bunch-thinning for the conventional table wines. All grapes are hand-picked.

rosebank estate NR

Cnr Johns and Groynes Drive, Belfast, Christchurch **region** Canterbury

ph (03) 323 8539 **fax** (03) 323 8538 **open** 7 days 10–5

winemaker Mark Lennard **prod.** 3000 **est.** 1993

prod. range ($8–15 CD) Riesling, Sauvignon Blanc, Chardonnay, Marlborough Chardonnay, Canterbury Chardonnay, Reserve Canterbury Chardonnay, Müller Thurgau, Directors' White, Sparkling Sekt, Pinot Noir, Cabernet Shiraz.

summary Situated only minutes from the city centre and six minutes from Christchurch airport, this is as much an entertainment centre as it is a winery, with a beautiful garden setting containing hundreds of roses, rhododendrons and camellias and lunch served from the restaurant each day, and à la carte dinner from Wednesday to Sunday from 6 pm. The Waipara vineyard will ultimately provide 50% of the production; in the meantime most of the grapes are being sourced from Marlborough. In 1996 a cricket ground was established at Rosebank in village-green style.

rossendale wines NR

150 Old Tai Tapu Road, Christchurch **region** Canterbury
ph (03) 322 7780 **fax** (03) 332 9272 **open** 7 days 10–5
winemaker Grant Whelan **prod.** 5000 **est.** 1987
prod. range ($12–18 CD) Chardonnay, Barrel Selection Chardonnay, Riesling, Sauvignon Blanc, St Helena Pinot Noir, Lancaster Pinot Noir.
summary Rossendale is the highly successful venture of beef exporter Brent Rawstron, who ventured into viticulture on his farm in 1987. A 120-year-old gatekeeper's lodge on the farm has been converted into a restaurant and sales area, nestling in a century-old forest. All this, and situated only 15 minutes from the centre of Christchurch, making it the closest winery to that city. The skills of winemaker Grant Whelan brought the initial vintages of 1993 and 1994 gold and silver medals.

ruby bay wines ★★☆

Korepo Road, RD1, Upper Moutere, Nelson **region** Nelson
ph (03) 540 2825 **fax** (03) 540 2105 **open** 7 days 11–6
winemaker Anita Croy, Philip Croy **prod.** 2500 **est.** 1976
prod. range ($15–22.50 CD) Chardonnay, Sauvignon Blanc, Riesling, Gewurztraminer, Pinot Noir, Cabernet Sauvignon Merlot.
summary The beautifully sited former Korepo winery, purchased by the Moore family in 1989, is well known for its restaurant. It appears that Ruby Bay has changed hands, and it will be interesting to see whether Anita and Philip Croy can maximise the potential of the site.

sacred hill ★★★★

Dartmoor Road, RD6, Napier **region** Hawke's Bay
ph (06) 844 0138 **fax** (06) 844 3271 **open** 7 days Jan, weekends 11–4 Mar–Apr, winter by appointment
winemaker Tony Bish **prod.** 19 500 **est.** 1986
prod. range ($14.95–35 CD) There are three ranges of wines; at the very top come intermittent releases of special selection wines, including 1995 Rifleman's Chardonnay and Brokenstone Merlot; then comes the Reserve range, although not necessarily carrying that word in the name (just to confuse the unwary, typically being described as barrel-fermented or basket press); and at the bottom the Whitecliff range.
summary The Mason family are pastoralists-turned-grape growers and thereafter winemakers (Mark Mason is a Roseworthy graduate). Sacred Hill has had its ups and downs since it was founded in 1986, but has steadied significantly since 1995. The top-of-the-range Rifleman's Chardonnay, Brokenstone Merlot, and Basket Press Cabernet Sauvignon are high-quality wines. Exports to the UK and US.

Sacred Hill Reserve Barrel Fermented Hawke's Bay Chardonnay

Produced from the small-berried Mendoza clone, which is a far more controllable clone in Hawke's Bay than it is in Marlborough, and which provides greater intensity and density of fruit than Clone 5. It also lends itself to the whole-bunch pressing technique used to make this wine, which is fermented in new French oak.

YYYY **1997** Medium to full yellow-green; the bouquet is rich and complex with ripe nectarine fruit and attractive toasty oak. The palate opens with all the promise of the bouquet, with tangy fruit, but the 14 degrees alcohol starts to make its presence felt towards the slightly heavy finish. Uncompromising food style. **rating:** 86

best drinking 1998–1999 **best vintages** NA **drink with** Pasta carbonara • $19.95

Sacred Hill Rifleman's Chardonnay

The top-of-the-range Chardonnay from Sacred Hill, given the full Monty of winemaker's tricks; hand-picked, whole-bunch pressed, barrel-fermented with some wild yeast, and left on lees for 12 months. A mix of new and one-year-old French oak barriques are used, and the grapes are the small-berried Mendoza clone. Gold medal winner 1999 Liquorland Royal Easter Wine Show.

🍷🍷🍷🍷🍷 **1997** Medium to full yellow-green; complex toasty oak runs through the heart of both the bouquet and palate, surrounded by ripe fig, melon and faintly citrussy fruit. The alcohol in the wine gives some impression of sweetness, which is the only question mark about it. **rating:** 93

best drinking 1999–2003 **best vintages** '97 **drink with** Turkey • $35

saint clair estate ★★★★☆

739 New Renwick Road, RD2, Blenheim **region** Marlborough

ph (03) 578 8695 **fax** (03) 578 8696 **open** At Country Life Craft Shop, Aberharts Road, Blenheim

winemaker Kim Crawford (consultant), Matt Thomson **prod.** 25 000 **est.** 1978

prod. range ($14.95–26.95 R) Marlborough Riesling, Marlborough Noble Riesling, Marlborough Sauvignon Blanc, Awatere Valley Oak Aged Sauvignon Blanc, Marlborough Chardonnay, Single Vineyard Rapaura Road Chardonnay, Marlborough Merlot, Single Vineyard Rapaura Road Merlot.

summary Owners Neal and Judy Ibbotson have followed the tried-and-true path of growing grapes for 15 or so years before venturing into wine production, which they did with spectacular success in 1994. Since that year, ever-increasing quantities of grapes from their 56 hectares have been vinified, with the exceptional skills of Kim Crawford leading to a cascade of show awards. The wines are exported to 16 countries across Europe and North America.

Saint Clair Estate Marlborough Riesling

Since 1997 estate-grown, and made by consultant winemaker Kim Crawford in a typically flamboyant style.

🍷🍷🍷🍷 **1998** Light yellow-green; the bouquet has rich lime fruit with hints of pineapple yet paradoxically seems quite fine. The palate has abundant total flavour, and carries its 16 grams per litre of residual sugar surprisingly well. **rating:** 86

best drinking 1999–2003 **best vintages** NA **drink with** Fresh fruit • $16

Saint Clair Estate Omaka Valley Oak Aged Sauvignon Blanc

Another stylish wine showing the sensitive use of oak.

🍷🍷🍷🍷🍷 **1997** Medium yellow-green; the fruit aromas are quite exotic, guava-like, with the oak well integrated. A very interesting wine, rippling with flavours on the palate, sweet but not heavy. **rating:** 90

best drinking 1999–2001 **best vintages** NA **drink with** Marinated octopus • $19

Saint Clair Omaka Valley Reserve Chardonnay

A beautifully made wine with real elegance, and which deservedly won a gold medal at the 1999 Liquorland Royal Easter Wine Show.

🍷🍷🍷🍷🍷 **1998** Light to medium yellow-green; the complex, creamy, nutty bouquet has a real touch of elegance, elegance which carries through onto the palate where lively melon fruit is woven through gently spicy oak; long finish. **rating:** 94

best drinking 1999–2003 **best vintages** '98 **drink with** Whitebait fritters • $23

Saint Clair Estate Single Vineyard Rapaura Road Merlot

As the name suggests, produced entirely from grapes grown on the Rapaura Road Vineyard. Spends 12 months in oak, and is a more than useful wine.

🍷🍷🍷🍷 **1997** Bright red-purple; fresh red cherry fruit mingles with attractive oak on the bouquet. The medium-weight palate is well balanced, with moderately ripe tannins to provide a frame for the red cherry/cassis fruit flavours. **rating:** 87

best drinking 1999–2002 **best vintages** NA **drink with** Duck confit • $24

st helena estate ★★★☆

Coutts Island Road, Christchurch **region** Canterbury
ph (03) 323 8202 **fax** (03) 323 8252 **open** Mon–Sat 10–4.30, Sun 12–5
winemaker Alan McCorkindale **prod.** 10 000 **est.** 1978
prod. range ($7.50–30 CD) Riesling, Canterbury Plains Müller Thurgau, Southern Alps Dry White, Chardonnay, Reserve Chardonnay, Noble Bacchus, Pinot Gris, Pinot Blanc, Pinot Noir, Reserve Pinot Noir, Port Hills Dry Red, Peers Port.

summary Whether in its moments of success or otherwise, controversy has never been far from St Helena's door. After a spectacular debut for its Pinot Noir in 1982, there has been a roller-coaster ride since, but much work in the vineyard (and also winery) is starting to pay dividends. The arrival of the immensely talented Alan McCorkindale as winemaker has seen the expected lift in quality, and the wines are now exported to Australia, the UK and Canada.

St Helena Estate Pinot Gris

First made in 1984, but has had a chequered career; no wine was made in 1992, 1993 or 1994 because of frost damage, and even the '95 yielded 1 tonne per hectare. One needs to have a considerable sense of humour or a healthy bank balance to sustain such an endeavour. The '98 was tasted twice within a week (as with many of the New Zealand wines) and on both occasions impressed, including the 1999 Liquorland Royal Easter Wine Show at which the wine won a silver medal.

🍷🍷🍷🍷🍷 **1998** Light yellow-green; the bouquet is clean, with quite intense flowery citrus and spice notes. The palate has intriguing touches of honey, spice and flower blossom in a basically dry wine, with good balance and length. **rating:** 90

best drinking 1999–2002 **best vintages** '91, '95, '98 **drink with** Delicate Chinese dishes • $14.50

St Helena Estate Reserve Pinot Gris

Only 200 cases made out of a total of 1000 for the variety. The grapes are hand-picked and selected, and a portion of the wine is fermented in new French oak. For some strange reason, did not show as well as the standard release at the 1999 Liquorland Royal Easter Wine Show.

🍷🍷🍷🍷🍷 **1998** Light yellow-green; the oak influence on the bouquet is subtle to the point of subliminal, but the palate has an extra dimension and length to the mouthfeel, although not necessarily to the flavour. **rating:** 90

best drinking 1999–2002 **best vintages** NA **drink with** Shellfish • $25

St Helena Estate Reserve Pinot Noir

A distinct cut above the varietal Pinot Noir from St Helena, which is as it should be. A special fruit selection which spends significantly longer in oak, and with a higher percentage of new barrels

🍷🍷🍷🍷🍷 **1998** Medium red; a quite sophisticated wine with sappy aromas suggesting the use of whole bunches in the ferment. The oak is quite obvious on the palate, but the wine has considerable length and wholly appropriate lingering, fine tannins. **rating:** 90

best drinking 1999–2003 **best vintages** NA **drink with** Quail • $30

st jerome wines NR

219 Metcalfe Road, Henderson **region** Auckland Area
ph (09) 833 6205 **fax** (09) 833 6205 **open** Mon–Sat 9–6, Sun 12–5
winemaker Davorin Ozich, Miro Ozich **prod.** 7000 **est.** 1968
prod. range ($7.50–35 CD) Riesling, Sauvignon Blanc, Chardonnay, Chablis, Gewurztraminer, Cabernet Merlot, Port.
summary The Cabernet Merlots made by Davorin Ozich between 1987 and 1991 reflect his Master of Science degree and practical training at Chateau Margaux and Chateau Cos d'Estournel in Bordeaux. They were hugely powerful wines, the 1991 in particular. It was rated number two in New Zealand's Top Ten Reds of the Year in the September 1994 edition of *Cuisine* magazine, but did not impress the judges at the 1995 Sydney International Wine Competition, being described as 'harsh and over-extractive'. Herein lies the rub: these are wines which demand cellaring and a certain degree of understanding.

st nesbit ★★★★★

Hingaia Road, RD1, Papakura **region** Auckland Area
ph (09) 379 0808 **fax** (09) 376 6956 **open** Not
winemaker Dr Tony Molloy QC **prod.** 800 **est.** 1980
prod. range ($37 CD) A single Cabernet Merlot (Cabernet Sauvignon, Cabernet Franc, Merlot, Malbec, Petit Verdot) Bordeaux-blend has been supplemented more recently by a Rosé, the result of indifferent vintages in 1992 and 1993 and of the effects of leaf roll virus.
summary Tony Molloy is a leading tax lawyer with a weekend passion; his Bordeaux-blend is revered in New Zealand and very well regarded elsewhere. His Cabernet Merlot is produced in minuscule quantities, much of it exported, leaving a mere 250 cases for the New Zealand market. A major replanting programme of the 4-hectare vineyard enforced a hiatus in production between 1994 and 1996.

sandihurst wines ★★★☆

Main West Coast Road, West Melton, Canterbury **region** Canterbury
ph (03) 347 8289 **fax** (03) 347 8289 **open** 7 days 11–5 mid-Oct–Apr
winemaker Tony Coakley **prod.** 7000 **est.** 1992
prod. range ($11–29 CD) Riesling, Reserve Gewurztraminer, Pinot Gris, Pinot Gris Reserve, Chardonnay, Breidecker, Pinot Noir.
summary Yet another of the ever-expanding number of wineries in the Christchurch region, releasing its first wines in November 1993. It is a substantial operation, with 16 hectares of vineyards in bearing, and production increasing year by year. Its Pinot Gris is one of New Zealand's best.

Sandihurst Wines Premier Pinot Noir

First made in 1995; that was an inauspicious start, but the '96 was distinctly better, the '97 better yet again: a gold medal at the 1999 Liquorland Royal Easter Wine Show.
🍷🍷🍷🍷🍷 **1997** Light to medium red-purple; a light but fragrant bouquet with sappy cherry aromas is followed by an equally lively, stylish and fragrant palate with the fruit of the bouquet enhanced by a touch of charry oak. **rating:** 94
best drinking 1999–2003 **best vintages** NA **drink with** Hare pie • $29

sapich bros NR

152 Forest Hill Road, Henderson, Auckland **region** Auckland Area
ph (09) 814 9655 **fax** (09) 814 9655 **open** NA
winemaker Ivan Sapich **prod.** NA **est.** NA
prod. range Chardonnay, Purple Death (Port-based drink).

summary Until the 1995 Air New Zealand Wine Awards, best known for its 'Purple Death' Port-based drink which the label said was 'rough-as-guts ... but has the distinctive bouquet of horse shit and old tram tickets'. Out of the purple came a 1994 Chardonnay which won a gold medal at the 1995 Air New Zealand Wine Awards, with abundant solid peachy fruit with complex chewy oaky/mealy texture. A full-frontal wine in the older New Zealand style. No recent encounters with death or otherwise.

saxton estate NR

774 Main Road, Stoke, Nelson **region** Nelson
ph (03) 547 7517 **fax** (03) 547 7827 **open** 7 days 10–7
winemaker Wayne Laurie, Graeme Moore **prod.** 700 **est.** 1968
prod. range ($12.95–21.95 CD) Richmond Ranges Blend (Breidecker, Reichensteiner, Chasselas blend), Seibel, Pinot Noir, Cabernet Sauvignon.
summary The wines produced from the 2-hectare estate vineyard attest to the fact that this is indeed Nelson's oldest winery, dating back to 1968 when hybrids were an important part of the industry. Saxton Estate, however, remained largely in the shadows until 1995 when it was purchased by Wayne Laurie and Graeme Moore, who have plans to capitalise on its proximity to Nelson by establishing on-site accommodation and a high-quality restaurant.

seibel wines NR

113 Sturges Road, Henderson **region** Auckland Area
ph (09) 836 6113 **fax** (09) 836 6205 **open** Mon–Fri 11–6 or by appointment
winemaker Norbert Seibel **prod.** 2500 **est.** 1988
prod. range ($11.95–19.95 CD) Limited Edition Hawke's Bay Chardonnay, Hawke's Bay Sauvignon Blanc, Select Noble Late Harvest Chardonnay lead the roster; then come Scheurebe, Barrel Fermented White Riesling, Late Harvest White Riesling, Barrel Fermented Chenin Blanc, Semi-Dry Gewurztraminer, Medium Dry Riesling, Cabernet Franc Merlot Cabernet Sauvignon.
summary Significantly increased production shows that Norbert Seibel is doing well; the wine styles have been described as innovative, and those I have tasted have been outside the mainstream. No recent tastings. Exports to Denmark, Taiwan and US.

seifried estate ★★★★

Cnr State Highway 60, Redwood Road, Appleby, Nelson **region** Nelson
ph (03) 544 5599 **fax** (03) 544 5522 **open** 7 days 11–5
winemaker Hermann Seifried, Daniel Schwarzenbach **prod.** 42 000 **est.** 1973
prod. range ($13–25 R) A white specialist with Riesling, Dry Riesling, Oak Aged Riesling, Chardonnay, Sauvignon Blanc, Chablis, Gewurztraminer, Müller Thurgau, Old Coach Road Classic Dry White, Old Coach Road Chardonnay, Late Harvest Riesling, Ice Wine; the two principal red wines are Pinot and Cabernet Sauvignon.
summary With 56 hectares of vineyards established progressively between 1973 and 1988, and a crush in excess of 900 tonnes, Seifried Estate is by far the largest of the Nelson wineries. The production is heavily biased towards white wines, which are of wholly admirable consistency of style and quality. Just prior to vintage in 1996 Seifried moved to its new winery situated in its picturesque Appleby Vineyard, but some of the 1997 wines showed worrying technical problems. Exports to the UK, US, Canada, Germany, Austria, Thailand and Taiwan.

selaks drylands estate winery ★★★★☆

Hammerichs Road, Rapaura **region** Marlborough
ph (03) 570 5252 **fax** (03) 570 5272 **open** Mon–Fri 9–5, Sat 10–5, Sun 11–4
winemaker Darryl Woolley **prod.** NFP **est.** 1934

prod. range ($9–27 CD) Marlborough Sauvignon Blanc, Founders Reserve Matador Estate Sauvignon Blanc, Chardonnay.

summary A state-of-the-art winery drawing upon its own 14 hectares of estate vineyards and a significant part of the adjoining 80-hectare Matador Vineyard owned by John Webber. The Sauvignon Blancs are as good as they come in classic Marlborough-style. Extensive network of exports.

Selaks Premium Selection Riesling

Produced from grapes grown on Selaks Matador Vineyard. The '98 was picked very ripe, with some botrytis, and left with moderately high residual sugar at 14 grams per litre.

🍷🍷🍷🍷🍷 **1998** Light to medium yellow-green; the bouquet is clean and fresh with zesty lime juice and just a hint of botrytis showing. The palate is well-balanced, lingering and easy-drinking, with sweet lime juice flavours running through to the very finish. **rating:** 90

best drinking 1999–2004 **best vintages** NA **drink with** Fresh fruit • $13

Selaks Drylands Estate Marlborough Sauvignon Blanc

Selaks has a multiplicity of Sauvignon Blanc brands and labels these days, but the Marlborough Sauvignon Blanc that started it all when it made its debut at the 1995 Sydney International Winemakers Competition, being very highly pointed in the Top 100 and duly receiving a gold medal, continues to be its flagship, reaching a new high in the difficult vintage of 1998.

🍷🍷🍷🍷🍷 **1998** Light to medium yellow-green; the bouquet offers an extra dimension of fruit aroma with ripe gooseberry and passionfruit. The palate is elegant, but packed with passionfruit and gooseberry flavour which, against all the odds, shows no heaviness nor hot, high alcohol. Well deserved gold medal 1999 Liquorland Royal Easter Wine Show. **rating:** 94

best drinking 1999–2001 **best vintages** '97, '98 **drink with** Stir-fried vegetables • $16

Selaks Marlborough Sauvignon Blanc

Selaks 'commercial' Sauvignon Blanc, cold-fermented in stainless steel and in considerable quantities. Utterly reliable, and the '98 seems to me to have done better than some higher-priced wines.

🍷🍷🍷🍷 **1998** Light to medium yellow-green; a lively, fresh and crisp bouquet with some lemony/citrus notes. The palate is well balanced, with less hot alcohol than many from the vintage, leaving it with good acidity and liveliness. **rating:** 89

best drinking 1999–2000 **best vintages** NA **drink with** Scallops • $17

selaks wines (auckland) ★★★★☆

15 Old North Road, Kumeu **region** Auckland Area

ph (09) 412 8609 **fax** (09) 412 7524 **open** Mon–Fri 9–5, Sat 10–5, Sun 11–4

winemaker Darryl Woolley **prod.** 100 000 plus **est.** 1934

prod. range ($9–27 CD) Super-premium wines under Founders label including Oak Aged Sauvignon Blanc, Chardonnay, Cabernet Sauvignon and Méthode Traditionelle.

summary With Montana, Selaks first brought Sauvignon Blanc to the attention of overseas markets, especially Australia. Its Sauvignon Blanc and Sauvignon Semillon blends continue to be its forte, always good, frequently outstanding. The opening of its large (2000-tonne) Drylands Estate Winery in Marlborough in March 1996 is compelling evidence of the success of Selaks. Extensive network of exports.

seresin estate ★★★★☆

Bedford Road, Renwick, Marlborough **region** Marlborough

ph (03) 572 9408 **fax** (03) 572 9850 **open** Summer 10–4.30

winemaker Brian Bicknell **prod.** 12 000 **est.** 1992

prod. range ($19.90–29.80 R) Sauvignon Blanc, Pinot Gris, Estate Chardonnay, Reserve Chardonnay, Pinot Noir.

summary Seresin has charged into the Marlborough scene since New Zealand filmmaker Michael Seresin purchased a little over 60 hectares of prime alluvial terrace land adjacent to the Wairau River. Forty hectares of vineyard have been established, and the state-of-the-art winery designed by Ian Athfield (of Te Mata Estate fame) constructed. Brian Bicknell, one of New Zealand's most experienced Flying Winemakers (he has worked in NZ, Hungary, France and for three years in Chile) has been installed as chief winemaker. The first vintage in 1996 immediately established Seresin as one of the star performers in the Marlborough scene.

Seresin Estate Chardonnay Reserve

The best of the estate-grown grapes are selected for the Reserve programme; the wine is barrel-fermented in 100% new French oak barriques, and is all taken through malolactic fermentation, with the usual lees contact and stirring.

🍷🍷🍷🍷 **1997** Full yellow-green; a complex but quite subdued bouquet with a mix of honeyed, nutty, buttery mlf characters is followed by a similarly complex palate which clearly shows the fingerprints of the winemaker and his philosophy – perhaps a little too emphatically for my taste.

rating: 88

best drinking 1999–2001 **best vintages** '96 **drink with** Corn-fed chicken • $27.80

Seresin Estate Pinot Noir

Hand-picked, with a portion placed into the tank to give carbonic maceration characters; the remainder destemmed and pumped directly to the tank without crushing. Pre-fermentation cold-maceration followed, then a warm ferment, with the wine being pressed and taken straight to barrel at the end of fermentation.

🍷🍷🍷🍷 **1997** Medium to full red-purple; stylish and concentrated, with complex plummy fruit and secondary aromas on the bouquet are followed by a concentrated and powerful palate, which toughens slightly on the finish, but there is a great deal of flavour and character running through the wine.

rating: 89

best drinking 1999–2003 **best vintages** '97 **drink with** Game • $29.80

settler vineyards NR

Crownthorpe Settlement Road, RD9, Hastings **region** Hawke's Bay

ph (06) 874 3244 **fax** (06) 874 3244 **open** Not

winemaker Evert Nijzink **prod.** 500 **est.** 1993

prod. range ($14–20 R) Chardonnay, Cabernet Merlot.

summary Evert Nijzink has established a 2.5-hectare biodynamically managed vineyard, and since 1997 has relied entirely upon estate production. A cellar-door and café facility opened in late 1996.

shalimar estate NR

RD2, Ngatapa Road, Gisborne **region** Gisborne

ph (06) 862 7776 **fax** (06) 862 7776 **open** 7 days 10–5

winemaker Alexander Stuart **prod.** 900 **est.** 1994

prod. range ($NA) Sauvignon Blanc, Chardonnay, Pinot Gris, Merlot.

summary Having been grape growers for 25 years, the Stuart family took the plunge into winemaking in 1994, drawing upon a recently established (and in Gisborne, rare) terraced hillside vineyard. Alexander Stuart believes greater flavour and character will follow the lower than normal yields.

sherwood estate NR

Weedons Ross Road, Christchurch **region** Canterbury
ph (03) 347 9060 **fax** (03) 347 8225 **open** 7 days 11–5
winemaker Dayne Sherwood **prod.** 7800 **est.** 1987
prod. range ($12.50–26.50 R) Riesling, Müller Thurgau, Sauvignon Blanc, Chardonnay, Unoaked Chardonnay, Reserve Chardonnay, Estate Pinot Noir, Reserve Pinot Noir, Single Vineyard Selection Rivendell Pinot Noir, Cabernet Franc.
summary Sherwood Estate produced its first wines in 1990; situated close to Christchurch (15 minutes drive) it also offers a garden-setting tasting room with snacks and lunches available in the Vineyard Bar throughout summer. Production has risen significantly since the early days, making Sherwood Estate an important part of the Christchurch landscape. After early success, particularly with Pinot Noir, quality seems to have slipped alarmingly in 1996 and 1997.

sileni estates NR

Maraekakaho Road, Bridge Pa, Hastings **region** Hawke's Bay
ph (06) 879 8768 **fax** (06) 879 7187 **open** 7 days 10–5
winemaker Grant Edmonds, Nigel Davies, Eleanor Dodd **prod.** 3500 **est.** 1997
prod. range Chardonnay, Merlot Cabernets
summary Sileni Estates is poised to become a major force in the Hawke's Bay region. The three partners are chief winemaker Grant Edmonds, Graeme Avery and Chris Cowper (and their families) who between them have winemaking, management and marketing and financial backgrounds and skills. With just under 100 hectares of vines planted, Sileni Estates will be self-sufficient, and production will increase in leaps and bounds from the 1998 level of 3500 cases. It is difficult to imagine that wine quality will be anything other than high.

silverstream vineyard NR

64 Giles Road, Clarkville, Kaiapoi **region** Canterbury
ph (03) 327 5231 **fax** (03) 327 5678 **open** By appointment
winemaker Peter Todd **prod.** NFP **est.** 1990
prod. range ($15–20 CD) Chardonnay, Pinot Noir.
summary One of the newest of the Canterbury wineries, situated on the Waimakari Plains north of Christchurch. Owned by Peter and wife Zeke Todd (of Anglo-Italian and Dutch ancestry, respectively), with 4 hectares of estate plantings. No wine was made under the Silverstream label in 1997 or 1998.

smith of martinborough NR

73 Princess Street, Martinborough **region** Wairarapa
ph (06) 306 9280 **fax** (06) 306 9280 **open** Weekends and public holidays
winemaker Roger Smith **prod.** NA **est.** 1985
prod. range Riesling, Pinot Noir, Cabernet Sauvignon.
summary The venture of Sam Lamb and Roger Smith, who in the early years were content to sell the production from their 3-hectare vineyard to other makers, but are now in the course of restoring turn-of-the-century buildings on the property into a visitors centre, craft and wine shop.

soljans wines ★★★

263 Lincoln Road, Henderson **region** Auckland Area
ph (09) 838 8365 **fax** (09) 838 8366 **open** Mon–Sat 9–6, Sun 11–5
winemaker Simone Parker **prod.** 25 000 **est.** 1937
prod. range ($9.20–23.50 CD) White wines under the Ivory label include Müller Thurgau, Marlborough Riesling, Gisborne Gewurztraminer, Hawke's Bay Sauvignon Blanc, Hawke's Bay

Unoaked Chardonnay, Barrique Reserve Chardonnay; red wines under the Ebony label are Pinotage, Barrique Reserve Merlot, Barrique Reserve Cabernet Merlot; also fortifieds and dessert wines; Legacy Méthode Traditionelle.

summary The traditional but immaculately maintained winery and vineyard constitute a major tourist attraction, and the well-made wines are sold at very modest prices. In 1993 Soljans made a heavy investment importing state-of-the-art sparkling winemaking equipment from France and is now a major contractor for New Zealand Méthode Traditionelle producers. Exports to the UK, US, France, Denmark and Poland – the latter a considerable achievement.

Soljans Hawke's Bay Sauvignon Blanc

Typically a small proportion is barrel-fermented (in 1998 10%), the rest is fermented in stainless steel, and left with a little over 5.5 grams per litre of residual sugar, which is barely noticeable.

YYYY **1998** Light green-yellow; the bouquet is clean and fresh with crisp minerally characters, a touch of lemon and just the barest whiff of oak and mlf. The palate works quite nicely, with a clean, lingering finish. Not particularly intense, though. **rating:** 85

best drinking 1999–2000 **best vintages** NA **drink with** Crab soufflé • $13.95

solstone estate NR

119 Solway Crescent, Masterton **region** Wairarapa

ph (06) 377 5505 **fax** (06) 337 7504 **open** Mon–Fri 8–5, weekends and public holidays 10.30–4.30

winemaker Luc des Bonnets **prod.** 3500 **est.** 1981

prod. range ($12–35 CD) Riesling, Rosé, Merlot, Cabernet Franc, Cabernet Merlot Franc.

summary Tiny quantities of the wines sold to date have been eagerly snapped up by the local clientele, but wines are now being distributed (sparingly) through Kitchener Wines. The wines draw upon 2 hectares of estate cabernet sauvignon and a single hectare of pinot noir, sauvignon blanc, merlot and cabernet franc.

spencer hill estate ★★★★

Best Road, RD1, Upper Moutere, Nelson **region** Nelson

ph (03) 543 2031 **fax** (03) 543 2031 **open** Thur–Sat 12–4 Dec–Feb

winemaker Philip Jones **prod.** 15 000 **est.** 1991

prod. range ($15–31 R) Spencer Hill is the top label from single vineyard sources; Tasman Bay is the second but main label made in larger volume. Chardonnay, Sauvignon Blanc, Pinot Noir, Pinot Gris, Riesling, Gewurztraminer, Merlot.

summary Philip Jones is a graduate in Viticulture from UCLA Davis, California, and also undertook an Oenology degree at Fresno State University. The ornately complex Spencer Hill Chardonnays have won a cascade of trophies and gold medals, joined by the Tasman Bay Chardonnay in 1997, in wine competitions running from New Zealand to London. There can be no doubt that Philip Jones is a highly talented winemaker, and an equally accomplished marketer, with exports to Australia, the UK, Hong Kong, Germany and Japan.

Spencer Hill Estate Tasman Bay Marlborough Chardonnay

This complex and rich wine has done very well in New Zealand tastings. Made in a most interesting, if not downright challenging, style with a mix of European, Californian, Australian and New Zealand influences. One hundred per cent malolactic fermentation. Courageously (but commendably) uses a synthetic cork. Received the Mission Hill Trophy for Best Chardonnay Worldwide at the 1997 Wine & Spirit Competition in Bristol, England. Winner of the Riedel Crystal Trophy for Fuller Bodied Dry White Table Wine at the 1998 Sydney International Wine Competition.

🍷🍷🍷🍷 **1997** Medium yellow-green; the fragrant bouquet has lots of tangy, tropical fruit and pleasantly subdued oak. A nicely balanced and lively wine on the palate, with touches of cashew and nectarine, and, once again, subtle oak. Far gentler handling than the other Spencer Hill white wines from 1997. **rating:** 88

best drinking 1998–1999 **best vintages** NA **drink with** Spiced prawns • $25

springvale estate NR

Dunstan Road, Alexandra **region** Otago

ph (03) 449 2333 **fax** (03) 449 2331 **open** 7 days 11–5 Dec–Apr, weekends and public holidays 11–5 May–Nov

winemaker Rudi Bauer (Contract) **prod.** 580 **est.** 1989

prod. range ($16.50–22 CD) Gewurztraminer, Sauvignon Blanc, Chardonnay, Pinot Noir, Cabernet Sauvignon.

summary Tony and Jo-Anne Brun planted their first hectare of vines in 1989, gradually increasing the area to its present level of just under 5 hectares. The neat restaurant caters for all tastes from nibbles to functions. Contract-winemaking by the ubiquitous Mike Wolter does the rest, and a cellar-door sales and tasting facility was opened in December 1997.

Springvale Estate Pinot Noir

The wine is made by Rudi Bauer, who has established an outstanding track record in the South Island, starting at Rippon, then moving on to Giesen and now to his own business, which includes contract-winemaking for others (including Springvale Estate). This gorgeous wine was a gold medal winner at the 1999 Liquorland Royal Easter Wine Show.

🍷🍷🍷🍷🍷 **1998** Medium purple-red; the bouquet is complex and stylish, with gamey, foresty overtones to the fruit. The palate picks up the pace, with plenty of ripe plum fruit, offering substance and style, and with good length. **rating:** 95

best drinking 1999–2003 **best vintages** '98 **drink with** Game • $22

stonecroft vineyard ★★★★

Mere Road, RD5, Hastings **region** Hawke's Bay

ph (06) 879 9610 **fax** (06) 879 9610 **open** Weekends, public holidays 11–5

winemaker Dr Alan Limmer **prod.** 3000 **est.** 1987

prod. range ($17–38 CD) Gewurztraminer, Gewurztraminer Late Harvest, Sauvignon Blanc, Chardonnay, Ruhani, Crofters II, Syrah.

summary Analytical chemist Dr Alan Limmer produces very full-bodied, rich and ripe wines from his 3-hectare vineyard situated on free-draining, gravelly soils which promote early ripening. Most interesting is the almost unprocurable (mailing list only) Syrah, widely regarded as New Zealand's best. Exports to Australia and UK.

stonyridge vineyard ★★★★★

80 Onetangi Road, Waiheke Island **region** Waiheke Island

ph (09) 372 8822 **fax** (09) 372 8822 **open** 7 days 9–5

winemaker Stephen White **prod.** 800 **est.** 1982

prod. range ($30-85 ML) The top label is Larose Cabernets; the second is Airfield Cabernets. Minuscule quantities of Que Sera Syrah are also grown.

summary The winery that justifies the hype about Waiheke Island. Consistently great wines have been produced, albeit in minuscule quantities; small wonder it has established the highest ex-winery price of $50 en primeur, and $85 upon commercial release, with a limit of one bottle per customer! For many, patronising the winery restaurant will be the only means of tasting these exalted wines. Minuscule exports to UK, US, Australia, Japan and Taipei.

Stonyridge Larose Cabernets

Although Te Mata would vigorously dispute this (and no doubt others, too) I would nominate this wine as New Zealand's best Bordeaux-blend. The blend varies from vintage to vintage in the manner of the great estate wines of Bordeaux, but will typically be not less than 60% Cabernet Sauvignon, with Merlot and Malbec varying between 10% and 18%, and then 5% to 6% of Cabernet Franc and a nominal 1% of Petit Verdot. Prolonged ageing in new French oak allows the wine to be bottled unfiltered.

ΥΥΥΥΥ **1996** Medium to full red-purple; the bouquet is strikingly different from the '95, fully ripe, with mulberry, prune and plum aromas, and an intriguing edge of lavender. An exceptionally powerful and complex wine on the palate with layer upon layer of flavour and texture; cedar, vanilla, blackcurrant and chocolate announce a wine out of the ordinary. **rating:** 97

best drinking 2002–2012 **best vintages** '91, '93, '94, '95, '96 **drink with** Fillet of beef • $85

stratford wines of martinborough NR

115 New York Street, Martinborough **region** Wairarapa
ph (06) 306 9257 **fax** (06) 306 8257 **open** By appointment
winemaker Stratford Canning **prod.** 2000 **est.** 1993
prod. range ($16–28 CD) Riesling, Chardonnay, Pinot Noir.
summary Strat Canning has worked as a winemaker in the Martinborough region for a number of years, and is now also producing his own wines from 1 hectare each of pinot noir, chardonnay and riesling.

tai-ara-rau wines NR

Upper Stout Street, Gisborne **region** Gisborne
ph (06) 867 2010 **fax** (06) 867 2024 **open** 7 days 10–5
winemaker Jeff Sinnott **prod.** 1000 **est.** 1989
prod. range ($NA) Estate Chardonnay and Merlot; Waimata Vineyard Chardonnay, Pinot Noir and Merlot.
summary Tairawhiti Polytechnic has followed in the footsteps of Australia's Charles Sturt University in making commercial quantities of wine as part of the wine industry certificate course which the institute offers. It is the only New Zealand institution to do so, drawing upon 3 hectares of chardonnay and 1.25 hectares each of pinot noir and merlot.

te awa farm winery NR

2375 Roys Hill Road, SH 50 RD5, Hastings **region** Hawke's Bay
ph (06) 879 7602 **fax** (06) 879 7756 **open** Mon–Fri 9–5 weekends 10–6
winemaker Jenny Dobson **prod.** 18 000 **est.** 1992
prod. range ($15–28 R) Longlands Sauvignon Blanc, Chardonnay, Merlot, Cabernet Merlot; Boundary Chardonnay and Merlot.
summary The Lawson family is yet another to venture into winemaking after being contract grape growers for over 15 years. Unusually, however, when Gus and Ian Lawson decided to venture into winemaking on their own account, they decided to start again, purchasing a 173-hectare sheep property on Roys Hill Road in 1992, establishing 32 hectares of sauvignon blanc, chardonnay, merlot, cabernet franc, cabernet sauvignon and syrah. The first vintage from Te Awa Farm was in 1994 and an on-site production facility (incorporating a cellar-door sales area) was constructed over 1997, with the peripatetic Jenny Dobson now in charge of winemaking. Exports to the UK and US.

te awanga estate NR

Parkhill Road, RD2, Te Awanga, Hawke's Bay **region** Hawke's Bay
ph (06) 875 1188 **fax** (06) 875 1188 **open** By appointment

winemaker Kim Crawford (Consultant) **prod.** 2500 **est.** 1995
prod. range ($15–23 R) Chardonnay, Sauvignon Blanc, Pinot Noir, Merlot, Cabernet Merlot, Cabernet Sauvignon.
summary Yet another enterprise with spectacular growth planned in the wake of the conversion from contract grape grower for others to self-production and the expansion of the vineyards from 12 to 32 hectares. An on-site winery is planned, and the promising first wines came onto the market in 1997.

te horo vineyards NR

State Highway 1, Te Horo **region** Wairarapa
ph (06) 364 3392 **fax** (06) 364 3392 **open** Fri–Sun and public holidays 10–5
winemaker Alastair Pain **prod.** 1200 **est.** 1985
prod. range ($9.95–19.95 CD) Chardonnay, Sauvignon Blanc, Riesling, Pinot Gris, Gewurztraminer, Merlot, Cabernet Sauvignon Cremant Méthode Traditionale; a selection of fruit wines.
summary Formerly called Grape Republic, a marketing and promotion tour-de-force using direct mail and wine club techniques, with a vast array of flavoured wines and smaller quantities of more expensive table wines which are distinctly austere. An underground cellar and sales area was opened in 1995. Most of the wine is sold direct ex-winery.

te kairanga wines ★★★★

Martins Road, Martinborough **region** Wairarapa
ph (06) 306 9122 **fax** (06) 306 9322 **open** 7 days 10–5
winemaker Peter Caldwell **prod.** 12 000 **est.** 1984
prod. range ($14.95–25 CD) At the top end is the Reserve range of Chardonnay and Pinot Noir; then the Premium range of Sauvignon Blanc, Chardonnay, Pinot Noir, Cabernet Sauvignon and Cabernet Merlot; at the bottom end Castlepoint range of Müller Thurgau, Dry White, Dry Red and Cabernet Sauvignon.
summary Te Kairanga is an enigma. For a long time its wines were disappointing, but then took a distinct turn for the better, suggesting that all of its problems were behind it. More recent tastings are less conclusive, so close and yet so far from the best of what is undoubtedly a great region. The appointment of Peter Caldwell and the fact that it has over 30 hectares of mature vineyards will help Te Kairanga realise its full potential.

Te Kairanga Pinot Noir

Like the Chardonnay, made entirely from estate-grown Martinborough grapes, and matured for ten months in French oak. Four different pickings at different levels of ripeness were utilised. A gold medal winner and Top 100 finalist in the 1995 Sydney International Wine Competition with the '93 vintage. The wines since '94 have been less convincing, although the '97 marks a partial return to form.
🍷🍷🍷🍷 **1997** Light to medium red-purple; the fruit is somewhat subdued on the bouquet, but there are some briary/foresty notes which add complexity. The palate has a touch more fruit, and good length and persistence. One of the better varietal Pinot Noirs from Te Kairanga. **rating:** 87
best drinking 1999–2002 **best vintages** '91, '93, '94, '97 **drink with** Grilled salmon • $25

te mania estate ★★★☆

c/o The Grape Escape, McShanes Road, Richmond, Nelson **region** Nelson
ph (03) 544 4541 **fax** (03) 544 4541 **open** Mon–Fri 10–5 summer and 11–4 winter
winemaker Jane Cooper **prod.** 3000 **est.** 1990
prod. range ($13.95–24 R) Riesling, Late Harvest Riesling, Sauvignon Blanc, Chardonnay, Merlot.

summary Jon and Cheryl Harrey commenced development of their vineyard in 1990, planting 4.5 hectares in that year (which came into production in 1992), and expanded the plantings in 1994 and 1995, with an additional 4 hectares. They previously sold most of their grapes to other Nelson winemakers, but since 1995 have had Jane Cooper as winemaker. Together with another vineyard owner, the Harreys have purchased a property on which they have erected a cellar-door sales facility, café and arts and crafts centre. Has a rapidly growing reputation, particularly for Sauvignon Blanc. The wines are exported to Australia, the UK and the Netherlands.

Te Mania Estate Home Block Sauvignon Blanc

Produced from 1.5 hectares of sauvignon blanc estate-grown on the Te Mania home block situated in Pughs Road. The wine is cool-fermented for four weeks and bottled in October in the year of vintage. The '98 won a gold medal at the 1999 Liquorland Royal Easter Wine Show.

🍷🍷🍷🍷🍸 **1998** Light to medium yellow-green; the clean, crisp aromas of the bouquet are predominantly of mineral and herb; those minerally characters come through on a particularly well-balanced and long palate. **rating:** 93

best drinking 1999–2000 **best vintages** '98 **drink with** Marinated scallops • $16

te mata estate ★★★★★

Te Mata Road, Havelock North, NZ **region** Hawke's Bay

ph (06) 877 4399 **fax** (06) 877 4397 **open** Mon–Fri 9–5, Sat 10–5, Sun 11–4

winemaker Peter Cowley **prod.** 30 000 **est.** 1896

prod. range ($9.95–39.15 CD) Elston (Chardonnay), Castle Hill Sauvignon Blanc, Cape Crest Sauvignon Blanc, Viognier Rosé, Bullnose (Syrah), Coleraine (Cabernet Franc Merlot), Cabernet Merlot, Awatea (Cabernet Merlot – premium).

summary In the eyes of many, New Zealand's foremost producer of Cabernet Merlot, notwithstanding the consistency of the show success of the Vidal/Villa Maria group. The wines of Te Mata are made in a different style, restrained and elegant but always packed with fine fruit. Nor should the consistently stylish and varietally correct white wines be ignored; these too are of the highest quality. Initially only available via cellar door and mailing list, the wines are now conventionally distributed throughout New Zealand, and are exported to Australia, Belgium, France, Germany, Holland, UK and US.

Te Mata Estate Woodthorpe Terraces Viognier

Made in tiny quantities, and very hard to obtain.

🍷🍷🍷🍷 **1998** Bright straw-green; the bouquet is clean but neutral, the palate very pleasant, with a texture not so far removed from Pinot Gris, as it does have viscosity; there are hints of sweet apple, but not the pastille fruit of Viognier from the Northern Rhône. **rating:** 87

best drinking 2001–2006 **best vintages** NA **drink with** Veal • $20.65

Te Mata Estate Elston Chardonnay

Hand-picked and whole-bunch pressed, fermented and matured in a mix of new and older French oak barriques. The extraordinary power, concentration and complexity of the wine comes primarily from the vineyard, but the role of the winemaker should not be taken for granted.

🍷🍷🍷🍷🍷 **1997** Bright but full yellow-green; the bouquet instantly tells you that this is a wine of exceptional quality, complexity and concentration, truly reminiscent of the very best White Burgundies. All the components coalesce, with intense citrus and nectarine fruit woven through some cashew mlf flavours and spicy oak. Firm but not excessive acidity will hold the wine for a long time. They really don't come better than this. **rating:** 97

best drinking 1999–2007 **best vintages** '93, '94, '95, '96, '97 **drink with** Fresh, slow-cooked Tasmanian salmon • $32

te motu waiheke vineyards ★★★★

76 Onetangi Road, Onetangi, Waiheke Island **region** Waiheke Island
ph (09) 486 3859 **fax** (09) 486 2341 **open** Not
winemaker Paul Dunleavy, John Dunleavy, Mark Roberton **prod.** 1300 **est.** 1990
prod. range ($35 R) Te Motu (Cabernet Merlot), Dunleavy Merlot Franc and a proposed third label for a Cabernet Sauvignon Merlot Franc blend.
summary The venture of the Dunleavy family headed by long-term, but now-retired, Wine Institute of New Zealand chief executive Terry Dunleavy. Produced outstanding wines in 1993 and 1994, providing yet further evidence (if any was needed) of the suitability of Waiheke Island for the production of ripe, full-bodied, Cabernets.

te papa wines NR

Martinborough–Pirinoa Road, Martinborough **region** Wairarapa
ph (06) 306 9899 **open** Not
winemaker John Trethewey **prod.** 500 **est.** NA
prod. range ($17–28 ML) Riesling, Sauvignon Blanc, Pinot Gris, Chardonnay, Pinot Noir.
summary With 7 hectares of sauvignon blanc, chardonnay, pinot noir, pinot gris and riesling under vine, the production of Te Papa Wines will increase significantly in the years ahead. The professed aim of John Trethewey is for minimal intervention in vineyard and winery alike.

te whare ra ★★★

Anglesea Street, Renwick, Marlborough **region** Marlborough
ph (03) 572 8581 **fax** (03) 572 8518 **open** 7 days 10.30–4.30
winemaker To be appointed **prod.** 6000 **est.** 1979
prod. range ($16–25 CD) Duke of Marlborough Semillon, Gewürztraminer and Chardonnay; Riesling, Berry Selection Gewürztraminer Riesling, Sarah Jennings, Arabella Gewürztraminer.
summary Te Whare Ra was purchased from the founding Hogan family by Roger and Christine Smith in October 1997. Christine has taken charge of the vineyard, and Roger the winery; they intend to continue to maintain the winemaking style of the past.

te whau vineyard NR

218 Te Whau Drive, Waiheke Island **region** Waiheke Island
ph (09) 377 3119 **fax** (09) 307 2322 **open** By appointment
winemaker Tony Forsyth, Kim Goldwater (Consultant) **prod.** 2000 **est.** 1993
prod. range ($30–50 CD) Chardonnay, Cabernet Sauvignon Cabernet Franc Merlot.
summary Tony Forsyth has established a little over 2 hectares of close-planted vines on a sheltered and steep (20°) north-facing slope. The first wines were made in 1999 at the on-site gravity-feed winery, complete with fully underground barrel hall.

thainstone wines NR

Giffords Road, RD3, Blenheim **region** Marlborough
ph (03) 572 8823 **fax** (03) 572 8623 **open** Not
winemaker Graeme Paul **prod.** 2000 **est.** 1990
prod. range ($14–18 R) Sauvignon Blanc, Chardonnay.
summary Jim and Viv Murray acquired their 7-hectare vineyard in 1990; already planted to sauvignon blanc, the decision was later taken to graft part over to chardonnay, which came into production in 1997. Until 1995 all the grapes were sold, but with gradually increasing amounts of wine made since that time. There are limited exports to Canada and the UK.

the antipodean NR

PO Box 5, Matakana **region** Northland and Matakana
ph (09) 422 7957 **fax** (09) 422 7656 **open** Not
winemaker Michelle Chignell-Vuletic **prod.** 300 **est.** 1977
prod. range ($90 CD) The Antipodean (a blend of Cabernet Sauvignon, Merlot and Malbec), Pot a Pat (non-vintage blended red), 'A' (Sauvignon Blanc, Semillon blend), The Iconoclast (Shiraz), Obiter (Cabernet Sauvignon), Matakana Day (a non-vintage blend).
summary More words have been written about this tiny winery than almost any other in New Zealand, and almost certainly more words than bottles produced. After a hugely publicised burst of production from 1985 to 1987, production ceased until 1990 with fierce family disputes, and sales recommenced in 1994 with the 1991 vintage of The Antipodean. As the product range testifies, not your run-of-the-mill winery.

the brothers vineyards NR

Brancott Road, RD2, Blenheim **region** Marlborough
ph (04) 386 3873 **fax** (04) 386 3853 **open** By appointment
winemaker Allen Hogan (and Consultants) **prod.** 3000 **est.** 1991
prod. range ($14.50–20 R) Chardonnay, Sauvignon Blanc Semillon, Semillon, Merlot.
summary A new arrival drawing upon 25 hectares of vineyards, with much of the grape production sold to other makers. It has wasted no time in establishing export markets in the UK, Canada and Australia.

the denton winery NR

Awa Awa Road, Ruby Bay **region** Nelson
ph (03) 540 3555 **fax** (03) 540 3555 **open** 7 days 11–6 Oct–Apr
winemaker Richard Denton **prod.** 1700 **est.** 1997
prod. range ($15.50–18 CD) Riesling, Sauvignon Blanc, Chardonnay, Pinot Noir.
summary Richard, and wife Alexandra, Denton discovered Nelson while on a world tour and some years later moved from their native England to Nelson to purchase the property in 1995. The first 2 hectares of vines were planted, and a winery – and the seemingly obligatory café and art gallery – appeared in time for the 1997 vintage. Richard Denton was an amateur brewer for many years, and graduated to amateur winemaking before taking the final plunge into commercial winemaking. A further 3 hectares of vineyard are to be planted, and increased production will follow.

the village winery NR

417 Mount Eden Road, Mount Eden, Auckland **region** Auckland Area
ph (09) 638 8780 **fax** (09) 638 9782 **open** Mon–Sat 10–9
winemaker Ken Sanderson **prod.** 1100 **est.** 1994
prod. range ($10.95–32.95 R) Windmill Road range, notably Syrah.
summary Enterprises such as this are not uncommon in California, but it is the only one I know of in either Australia or New Zealand, with the winery established in the main street of suburban Mount Eden, for good measure, until owner Peter Schinckel found a loophole in the law – a 'dry' area. (New Zealand, like Australia, still has areas in which alcohol may not be sold.) The wines are made under the Windmill Road label, the winery-cum-shop stocking a broad range of wines from other producers as well as Windmill Road.

torlesse ★★★☆

Waipara Village, Waipara, Canterbury **region** Canterbury
ph (03) 377 1595 **fax** (03) 377 1595 **open** By appointment
winemaker Kym Rayner **prod.** 3900 **est.** 1990

prod. range ($9.50–24 R) Müller Thurgau, Riesling (Dry and Medium), Southern Blush, Breidecker Dry, Gewurztraminer, Marlborough Sauvignon Blanc, Marlborough Chardonnay, South Island Chardonnay, Waipara Reserve Chardonnay, Marlborough Cabernet Franc.

summary Torlesse was effectively reborn in 1990 when its existing shareholders purchased the business from a receiver. They include Dr David Jackson, author of several books on viticulture, and winemaker Kym Rayner; all have vineyards in the Canterbury region which supply Torlesse with grapes, supplemented by grapes purchased from the Stonier Vineyard in Marlborough. The plans are for production to increase to about 13 000 cases. Exports to Australia, Canada and the UK.

Torlesse Waipara Riesling

Made, as the name indicates, from grapes grown in the Waipara subdistrict of the Canterbury region, and more proof – if any was needed – that riesling is extremely well suited to the climate and soils of the Waipara Valley.

🍷🍷🍷½ **1997** Light green-yellow; the bouquet is clean, with utterly correct varietal fruit of medium intensity; the palate runs in the same classic path, with lots of fresh, quite dry and delicate fruit; great promise for the future. **rating:** 84

best drinking 2000–2005 **best vintages** NA **drink with** Summer salad • $12

Torlesse Marlborough Gewurztraminer

Yet another example of New Zealand Gewurztraminer which shows excellent varietal flavour without undue phenolics.

🍷🍷🍷🍷 **1998** Light green-yellow; the perfumed bouquet offers spice, lychee and rose petal, and these flavours come through in precisely the same fashion on the palate; the finish shows no toughness whatsoever. **rating:** 88

best drinking 1999–2002 **best vintages** NA **drink with** Gently spiced Thai or Vietnamese food • $13.50

Torlesse Marlborough Sauvignon Blanc

Sauvignon Blanc can be grown satisfactorily in the Canterbury region, but its supplies are limited, while those of Marlborough are immense. It is thus no surprise to find wineries such as Torlesse sourcing their grapes from the Marlborough region. First made in 1994, and an exemplary example of Marlborough Sauvignon Blanc.

🍷🍷🍷🍷½ **1998** Light green-yellow; the bouquet is sparkling fresh and crisp with lively passionfruit aromas. The palate runs through passionfruit into sweet nectarine, and while very rich, is not the least heavy. **rating:** 90

best drinking 1999–2001 **best vintages** '98 **drink with** Steamed Chinese fish • $15.50

totara vineyards ★★★

Main Road, Thames **region** Waikato and Bay of Plenty

ph (07) 868 6798 **fax** (07) 868 8729 **open** Mon–Sat 9–5.30

winemaker Gilbert Chan **prod.** 10 000 **est.** 1950

prod. range ($10–16 R) Müller Thurgau, Chardonnay, Reserve Chardonnay, Sauvignon Blanc, Chenin Blanc, Cabernet Sauvignon.

summary A substantial operation which, however, has had its share of problems, leading to a decision to remove all its vineyards in 1986 under the Vine Pull scheme; it now relies on local growers to provide the grapes for its wines. Had its moment of glory in the 1992 Air New Zealand Wine Awards when the '90 Reserve Chardonnay won the Chardonnay Trophy.

trinity hill ★★★★

2396 State Highway 50, RD5, Hastings **region** Hawke's Bay
ph (06) 879 7778 **fax** (06) 879 7770 **open** 7 days 10–5
winemaker John Hancock, Warren Gibson **prod.** 28 500 **est.** 1996
prod. range ($15–44 R) Wairarapa Riesling, Wairarapa Chardonnay (unoaked), Shepherds Croft Chardonnay, Shepherds Croft Sauvignon Blanc, Shepherds Croft Syrah Merlot Cabernet Franc.

summary A fast-rising star in the New Zealand firmament. A joint venture between former Morton Estate winemaker John Hancock, an Auckland businessman and a pair (a couple actually) of London restaurateurs. The venture began with the establishment of a 20-hectare vineyard on a prime Gimblett Road site, followed by the erection of a state-of-the-art winery in 1996, completed just prior to the first vintage. All of the wines so far released have achieved great critical acclaim. The vineyards now comprise 60 hectares, and the wines are exported to Australia, the UK, US and other destinations.

Trinity Hill Gimblett Road Chardonnay

A super-premium wine coming from the Gimblett Estate and Stockbridge Vineyard, situated adjacent to each other in the Gimblett Road area. The grapes are whole-bunch pressed, and the wine was fermented and aged in French oak barriques, 80% new and 20% one-year-old. The malolactic fermentation is suppressed to enhance fruit character, and the wine is blended and bottled after 11 months in oak.

🍷🍷🍷🍷🍸 **1997** Medium yellow-green; the wine immediately proclaims sophisticated barrel-ferment handling with nutty/spicy overtones to both bouquet and palate. Happily, the fruit does come through on the mid to back palate, and the wine is well balanced. **rating:** 92

best drinking 1999–2001 **best vintages** NA **drink with** Roast chicken • $30

Trinity Hill Shepherds Croft Chardonnay

While Trinity Hill has three distinct areas within Hawke's Bay – Gimblett Road, Ngatarawa and Te Awanga – the majority of the initial releases came from the Shepherds Croft Vineyard at Ngatarawa. Between 50% and 75% of the wine is barrel-fermented depending on the vintage (50% in 1997), and given the usual lees contact, although it is not easy to tell how much (if any) was taken through malolactic fermentation.

🍷🍷🍷🍷🍸 **1997** Light to medium yellow-green; there is attractive melon and citrus fruit on the bouquet supported by subtle oak, leading into a wine which has the finesse and delicacy (and length) which so many New Zealand Chardonnays seem to lack. **rating:** 91

best drinking 1999–2002 **best vintages** NA **drink with** Tuna • $20

Trinity Hill Gimblett Road Cabernet Sauvignon Merlot

A blend of 70% Cabernet Sauvignon, 25% Merlot and 5% Cabernet Franc coming from the Gimblett Estate, Hatton Estate and Stockbridge Vineyard, all in the Gimblett Road area and adjacent to each other. Each variety and vineyard parcel is harvested and vinified separately, and is given approximately three weeks post-fermentation maceration on skins. The wine spends 15 months in barrel in French oak (80% new) undergoing malolactic fermentation and subsequent clarification. It is neither fined nor filtered.

🍷🍷🍷🍷 **1997** Medium to full purple-red; the bouquet is potent and concentrated, with pronounced cool-climate characters – slightly gamey – similar to those of the Yarra Valley. On the palate there is an extra dimension of berry fruit, but still with those ever-so-typical hints of tamarillo. **rating:** 88

best drinking 2001–2005 **best vintages** NA **drink with** Lamb Provençale • $44

twin bays vineyard NR

56 Korora Road, Oneroa, Waiheke Island **region** Waiheke Island
ph (09) 372 2441 **fax** (09) 372 7037 **open** By appointment
winemaker John Hancock **prod.** 600 **est.** 1989
prod. range ($45 CD) Premium label is Fenton Cabernet Merlot; second label is The Red.
summary Despite its tiny size (2 hectares) Twin Bays has already made its contribution to the international reputation enjoyed by Waiheke Island, for the 1994 Stonyridge Airfield Cabernet Merlot which won the trophy for Best New Zealand Cabernet Merlot Blend at the Air New Zealand Wine Awards was made from Twin Bays grapes. Since 1995 the wine has been made and released by Twin Bays owners Barry and Meg Fenton, who have plans to double the vineyards to 4 hectares. Retail distribution in Sydney, Melbourne and Adelaide.

unison vineyard NR

2163 Highway 50, RD5, Hastings **region** Hawke's Bay
ph (06) 879 7913 **fax** (06) 879 7915 **open** By appointment
winemaker Anna-Barbara and Bruce Helliwell **prod.** 2500 **est.** 1993
prod. range ($26–38 R) Two wines only are produced: Unison and Unison Reserve, both being blends of Merlot, Cabernet Sauvignon and Syrah.
summary Bruce Helliwell a New Zealand winemaker with MSc Honours degree met his German-born and trained viticulturist and winemaker wife Anna-Barbara while she was managing a small estate in Chianti Classico hill country. Between them, they have winemaking and viticultural experience in New Zealand, Germany, California, Italy, Switzerland and France. The vines on the 6-hectare estate are close-planted at a density of 5000 vines per hectare, with the yield reduced to only 1 kilo per vine. Small wonder their first vintage (1996) sold out in two months. Exports to Australia, the UK, Taiwan and US.

vavasour wines ★★★★★

Redwood Pass Road, Awatere Valley, Marlborough **region** Marlborough
ph (03) 575 7481 **fax** (03) 575 7240 **open** Mon–Sat 10–5 April–Sept, 7 days 10–5 Oct–Mar
winemaker Glenn Thomas **prod.** 25 000 **est.** 1986
prod. range ($14.95–24.95 R) At the top end come Vavasour Single Vineyard Sauvignon Blanc, Chardonnay; then Awatere Valley Sauvignon Blanc, Chardonnay, Pinot Noir and Cabernet Sauvignon; then Dashwood Chardonnay, Sauvignon Blanc and Pinot Noir; and Stafford Brook Chardonnay and Cabernet.
summary Vavasour has established itself as one of the most reliable and best producers in the Marlborough region. The drier, slightly warmer climate of the Awatere Valley and the unique river-terrace stony soils on which the 12.5-hectare vineyard is established are producing grapes of great intensity of flavour, which are in turn being skilfully handled in the winery by Glenn Thomas. Exports to Australia, UK and Switzerland.

Vavasour Riesling

Yet another relative newcomer to the Riesling scene, but not surprisingly coming up with a very good wine.

🍷🍷🍷🍷 **1998** Light yellow-green; the bouquet is delicate, clean and fresh, with a mix of lime and herb aromas. The palate is quite intense and long, with neatly judged residual sugar giving an overall moderately dry impression. **rating:** 89

best drinking 1999–2002 **best vintages** NA **drink with** Lebanese hors d'oeuvres • $19.95

Vavasour Awatere Valley Sauvignon Blanc

Selected from the three finest estate blocks, some being hand-harvested and whole-bunch pressed, others crushed in the conventional manner. Eight per cent is fermented in older French oak and kept on lees until blending and bottling.

🍷🍷🍷🍷 **1998** Light to medium yellow-green; the bouquet seems to have a definite edge of above-average concentration and tightness. In the mouth, too, the wine has some back palate presence and length, with a mix of mineral, herb and gentle tropical fruit running through the wine. **rating:** 89

best drinking 1999–2001 **best vintages** '97 **drink with** Shellfish • $19.95

Vavasour Pinot Noir

The wine is crushed and cooled, with the object of obtaining pre-fermentation maceration for at least three days. Thereafter fermentation commences, peaking at 31 degrees, and after pressing the wine spends 16 months in French oak (20% new). Very limited quantities made.

🍷🍷🍷🍷🍸 **1997** Strong purple-red; the bouquet shows excellent plum and cherry fruit, clean and firm; like the palate which follows, one can see the development potential immediately. The pristine fruit flavours come through on the palate, and the oak has been held in appropriate restraint. Good now but will be better. **rating:** 90

best drinking 2000–2004 **best vintages** '94, '97 **drink with** Venison • $22.95

vidal estate ★★★★☆

913 St Aubyns Street East, Hastings **region** Hawke's Bay

ph (06) 876 8105 **fax** (06) 876 5312 **open** Mon–Sat 11–6, Sun 10.30–5

winemaker Rod McDonald **prod.** 60 000 **est.** 1905

prod. range ($9.50–39.95 CD) Constant revamping of the product range (new brand managers?) keeps everyone on their toes. The Private Bin range has been replaced by the Estate Range; the Bays range introduced for restaurants; only the East Coast (at the bottom) and the Reserve range (at the top) continue (for the time being). The usual spread of varietals are covered by each range.

summary Together with Te Mata, Villa Maria and Esk Valley, consistently produces New Zealand's finest red wines; they have ripeness, richness and balance, a far cry from the reds of bygone years. Surprisingly, notwithstanding the success of Vidal year after year at the Sydney International Wine Competition, and notwithstanding that its wines are distributed in the UK, Canada, much of Asia and the Pacific, it has still not sought distribution in Australia. Surely it will be only a question of time, for production continues to increase.

Vidal Estate Reserve Chardonnay

One of the most highly regarded of the Hawke's Bay Chardonnays, even if overshadowed in the eyes of some by its sister wine from Villa Maria. The style is moving along with the mainstream of New Zealand Chardonnay, away from overblown, over-extractive making to a more restrained mode, but without sacrificing drinkability and accessibility. Barrel-fermented in French oak (two-thirds new), part of the ferment in a cool room, and part ambient. Fifty per cent of the wine was taken through malolactic fermentation, and it was given the usual lees contact.

🍷🍷🍷🍷🍸 **1997** Medium to full yellow-green; the bouquet has retained freshness and life, the palate even more so, with tight citrus/grapefruit flavours; very good length and nicely balanced and integrated oak. A strong silver medal performer at the 1999 Liquorland Royal Easter Wine Show. **rating:** 90

best drinking 1999–2000 **best vintages** '90, '91, '94, '96, '97 **drink with** Veal fricassee • $29

vilagrad wines NR

Rukuhia Road, RD2, Ohaupo **region** Waikato and Bay of Plenty
ph (07) 825 2893 **open** Tues–Sat 10–6
winemaker Peter Nooyen **prod.** 2000 **est.** 1922
prod. range Recently introduced Nooyen Reserve range of Riesling, Gewurztraminer, Chardonnay, Pinot Noir and Cabernet Merlot Malbec head the range.
summary A low-profile operation making wines of modest but consistently acceptable quality that age surprisingly well, but aspiring to greater things with the Nooyen Reserve wines. A winery restaurant is open on Sundays (and for functions at other times by arrangement).

villa maria ★★★★★

5 Kirkbride Road, Mangere, Auckland **region** Auckland Area
ph (09) 255 0660 **fax** (09) 255 0661 **open** 7 days 9–6
winemaker Michelle Richardson **prod.** 120 000 **est.** 1961
prod. range ($7.50–40 CD) A large range of wines under the Private Bin label, basically varietally identified, stands at the bottom end of the portfolio; next comes the Cellar Selection range of Chardonnay, Sauvignon Blanc, Cabernet Merlot; then Wairau Valley Reserve Sauvignon Blanc; at the top end the Reserve Bin range of Barrique Fermented Chardonnay, Marlborough Chardonnay, Sauvignon Blanc, Gewurztraminer, Noble Riesling, Cabernet Merlot and Cabernet Sauvignon.
summary Whether viewed on the basis of its performances at Sydney International Winemakers Competition, or on any other show result over the last few years, Villa Maria has to be rated one of New Zealand's best large wineries. The quality of the wines, both white and red, is exemplary, the flavours magically full without going over the top. Not surprisingly, the wines are exported to Europe, North America and Asia, but – as is the case with Vidal – not so far to Australia.

Villa Maria Cellar Selection Sauvignon Blanc

Yet another good Villa Maria Sauvignon Blanc, drawn from the Rapaura and Awatere Valley regions, and with just a small percentage given a touch of oak.

🍷🍷🍷🍷 **1998** Light to medium yellow-green; the bouquet is rich and ripe with white peach and tropical fruit, the palate quite well balanced but with that slightly spongy feel of the vintage. Silver medal at the 1999 Liquorland Royal Easter Wine Show. **rating:** 85

best drinking 1999–2000 **best vintages** NA **drink with** Fish soup • $17

Villa Maria Reserve Barrique Fermented Chardonnay

One can legitimately argue about the level of oak in this wine; whether one really likes it or not is a question of personal style preference, but there is no doubting the complexity and power of the wine. A gold medallist and Top 100 finalist in both the 1994 and 1995 Sydney International Wine Competition, and a medal winner at the 1994 Australian National Wine Show in Canberra. A Bob Campbell Classic and five stars from Michael Cooper.

🍷🍷🍷🍷🍷 **1997** Light to medium yellow-green; the bouquet is of medium intensity, with pleasant citrus and melon fruit. The new oak component is a fraction obvious, but should settle down with a year or two in bottle. All of the components are well balanced and well integrated on the palate, featuring citrus, mineral, nectarine and apple fruit. Stylish and elegant. **rating:** 94

best drinking 1999–2001 **best vintages** '90, '91, '94, '96, '97 **drink with** Moroccan chicken stuffed with raisins and pine nuts • $29

Villa Maria Reserve Marlborough Chardonnay

As the label suggests, made entirely from Marlborough region grapes drawn from Villa Maria's two best vineyards in the region, predominantly hand-picked. The wine is 100% barrel-fermented in a mix of new (60%) and one-year-old (40%) oak. Given eight months lees contact, with partial barrel stirring, and 25% malolactic fermentation. Sophisticated winemaking, to say the least.

ŶŶŶŶ **1997** Medium yellow-green; the bouquet is quite rich and complex, with peachy/buttery fruit and spicy oak. The palate opens up well enough, but the burning clove oak on the aftertaste detracts from what should have been a much better wine. **rating:** 85

best drinking 1999–2000 **best vintages** '94, '96, '97 **drink with** Coquilles St Jacques • $26

Villa Maria Reserve Noble Riesling (375 ml)

First made its debut in 1991; a new heavily botrytised Malborough-grown riesling; has garnered innumerable trophies and gold medals for Villa Maria, and is the top-ranked sweet white wine in New Zealand wine shows.

ŶŶŶŶŶ **1998** Medium to full yellow-green; the bouquet is powerful, rich and intense with penetrating lime and cumquat aromas. No less rich and powerful on the palate, extremely luscious and sweet on the mid-palate, finishing with soft acidity. **rating:** 95

best drinking 2000–2005 **best vintages** '91, '92, '93, '94, '96, '97, '98 **drink with** Any rich dessert • $40

Villa Maria Cellar Selection Late Harvest Riesling

Received what in my view was a miserable bronze medal at the 1999 Liquorland Royal Easter Wine Show. As my points indicate, I was well into gold territory. The extraordinary thing about this wine so far as I am concerned is its retention of pristine, pure Riesling varietal character, so often lost under the impact of botrytis.

ŶŶŶŶŶ **1998** Glowing yellow-green; intense, pure and entrancing lime juice riesling aromas lead into a perfectly balanced Auslese-weight palate, lingering and harmonious, and with that continued diamond sparkle of varietal riesling character. **rating:** 97

best drinking 2000–2010 **best vintages** '98 **drink with** Nothing; great aperitif • $29

Villa Maria Reserve Hawke's Bay Merlot

A blend of 94% Merlot and 6% Malbec selected from individual blocks on the Ngakirikiri. Open fermentation with hand plunging, and 10% being run off for barrel fermentation is followed by maturation in new and one-year-old barriques for 18 months.

ŶŶŶŶŶ **1996** Medium red-purple; there is attractive, sweet raspberry and plum fruit on the bouquet matched by well-handled oak. The wine is not a heavyweight, but has very attractive red fruit flavour supported by appropriate oak and subtle tannins. Gold medal 1998 Liquorland Royal Easter Show. **rating:** 94

best drinking 2001–2008 **best vintages** NA **drink with** Venison medallions • $32

voss estate ★★★★

Puruatanga Road, Martinborough **region** Wairarapa

ph (06) 306 9668 **fax** (06) 306 9668 **open** 7 days 11–6 summer

winemaker Gary Voss **prod.** 2000 **est.** 1988

prod. range ($16–26 CD) Reserve Chardonnay, Sauvignon Blanc, Pinot Noir, Waihenga Cabernet Merlot Franc.

summary Voss Estate has been established by Annette Atkins, Gary Voss and Murray Voss, with 4 hectares of vineyards (2 hectares of pinot noir, 1.5 hectares chardonnay and half a hectare of cabernet sauvignon and merlot). Production is largely estate-based, with some grapes purchased from other Martinborough growers. Exports to Australia and Canada.

wai-iti river vineyard NR

PO Box 86, Brightwater, Nelson **region** Nelson

ph (03) 542 3205 **fax** (03) 542 3205 **open** By appointment

winemaker Dave Glover (Contract) **prod.** 500 **est.** 1993
prod. range ($16–24 R) Chardonnay, Pinot Noir, Cabernet Sauvignon.
summary Chan and Philip Woollaston planted their small (6-hectare) vineyard on the Waimea Plains near Nelson in the winter of 1993 on old riverbed gravels. Cellar-door sales and tastings, together with a picnic area are planned.

waimarama estate NR

31 Waimarama Road, Havelock North **region** Hawke's Bay
ph (06) 877 4822 **fax** (06) 877 2980 **open** Not
winemaker Elise Montgomery (Consultant) **prod.** 2500 **est.** 1988
prod. range ($19–26.50 ML) Cabernet Sauvignon, Cabernet Merlot, Dessert Cabernet; Undercliffe is the second label.
summary Waimarama Estate was purchased by a Japanese company in 1999, and all of the wine produced now goes to that company in Japan.

waimea estates NR

148 Main Road Appleby, Nelson **region** Nelson
ph (03) 544 6385 **fax** (03) 544 6385 **open** 7 days 11–5 Oct–Feb
winemaker Jane Cooper, Bruce Collie **prod.** 3500 **est.** 1997
prod. range ($13.95–19.95) Riesling, Sauvignon Blanc, Chardonnay, Cabernet Merlot released under the Waimea Plains label.
summary Former orchardists Trevor and Robyn Bolitho diversified into viticulture in 1993, and by 1999/2000 will have four vineyards in production with a total of 75 hectares under vine. Part of the production is sold to other producers, but since 1997 Waimea Estates has produced wine under its own label. In 1998 a winery was built on site, with an ultimate production capacity of 50 000 cases. The quality of the wines so far released has been consistently good.

Waimea Plains Nelson Riesling

Fresh, lively and well balanced; can be drunk immediately or cellared for up to five years.
YYYY **1998** Light green-yellow; the bouquet is clean, moderately intense, with attractive lime fruit touched with passionfruit. The palate is delicate, but nicely balanced in typically New Zealand fashion with residual sugar to counter the acidity. **rating:** 87
best drinking 1999–2004 **best vintages** NA **drink with** Stuffed capsicum • $13.95

Waimea Plains Nelson Sauvignon Blanc

Warm, dry vintages such as 1998 are particularly welcome in areas such as Nelson, which has a higher average rainfall than neighbour Marlborough. This wine is the outcome of such a year. Both the '97 and '98 vintages have been very successful, receiving critical acclaim and wine show success.
YYYY **1998** Light green-yellow; the bouquet is fresh, light and delicate with a mix of grass, passionfruit and gooseberry aromas; the palate is similarly fresh, delicate and crisp, with pleasantly dry acidity. Neither heavy nor flabby. **rating:** 88
best drinking 1999–2000 **best vintages** NA **drink with** Calamari • $14.95

waipara downs NR

State Highway 7, Bains Road, RD3, Amberley **region** Canterbury
ph (03) 314 6873 **fax** (03) 314 6873 **open** By appointment
winemaker Francois Crochet **prod.** 500 **est.** 1989
prod. range ($11–13 CD) Chardonnay, Pinot Noir, Cabernet Sauvignon, Port.
summary Six hectares of vines on a 320-hectare farm puts Waipara Downs into perspective, but does not diminish the enjoyment Ruth and Keith Berry derive from producing their wines from

the limestone soils of the Waipara Valley. The wines have been consistent bronze and silver medal winners in New Zealand wine shows. The wines are distributed in Australia through Negociants.

waipara springs wines ★★★★

State Highway 1 North, Waipara, North Canterbury **region** Canterbury
ph (03) 314 6777 **fax** (03) 314 6777 **open** 7 days 11–5
winemaker Kym Rayner **prod.** 7000 **est.** 1990
prod. range ($12–24 CD) Sauvignon Blanc, Chardonnay, Riesling, Pinot Noir, Ram Paddock Red, Two Terrace Red.
summary Owned by Bruce and Jill Moore and the Grant family, represented by Andrew Grant. The initial plantings of 4 hectares in 1982 have now risen to 20 hectares, providing all of the grapes for the significantly increased production. The wines have deservedly gained an excellent reputation; exports to Australia, the UK and Canada.

Waipara Springs Sauvignon Blanc

Direct, no-frills winemaking places the primary emphasis on the very good fruit base.
🍷🍷🍷🍷 **1998** Light green-yellow; the bouquet is fresh and crisp, offering intensity in a lemon/citrus, herb/grassy spectrum. The palate is in that direct, no-frills style of the vintage, but does have an extra touch of depth, structure and length. **rating:** 85
best drinking 1999–2000 **best vintages** NA **drink with** Deep-fried calamari • $18

Waipara Springs Reserve Barrique Chardonnay

Barrel-fermented in French oak, and given subsequent lees ageing for eight months; some mlf inputs.
🍷🍷🍷🍷 **1997** Light to medium yellow-green; ripe peachy fruit opens up the bouquet, but there is barrel-ferment oak there as one would expect. The palate is likewise dominated by the very sweet peachy fruit, the oak acting to both provide complexity and balance to that fruit. **rating** 86
best drinking 1999–2000 **best vintages** NA **drink with** Roast pork • $21

waipara west NR

376 Ram Paddock Road, Amberley, RD2, North Canterbury **region** Canterbury
ph (03) 314 8699 **fax** (03) 314 8692 **open** By appointment
winemaker Petter Evans **prod.** 6000 **est.** 1989
prod. range ($14.80–20.50 CD) Riesling, Sauvignon Blanc, Chardonnay, Pinot Noir, Ram Paddock Red (Cabernet blend).
summary Waipara West is situated at the gorge of the Waipara River. The vineyard is surrounded by steep banks and planted on naturally sloping terraces which vary in height and aspect. Seventeen hectares of chardonnay, sauvignon blanc, riesling, merlot, cabernet sauvignon, cabernet franc and pinot noir have been planted, with the very experienced Petter Evans (ex-St Helena) in charge of winemaking. Almost all of the wine is exported, chiefly to the UK.

wairau river wines ★★★★

Cnr Rapaura Road and SH 6, Blenheim **region** Marlborough
ph (03) 572 9800 **fax** (03) 572 9885 **open** 7 days 9–5
winemaker John Belsham (Contract) **prod.** 20 000 **est.** 1978
prod. range ($16.95–40 CD) Riesling, Botrytised Riesling Reserve, Sauvignon Blanc, Sauvignon Blanc Reserve, Chardonnay, Chardonnay Reserve; Philip Rose Estate is the second label introduced in 1995.
summary Phil and Chris Rose have been long-term grape growers in the Marlborough region, having established a 60-hectare vineyard progressively since 1978. The first wines were made under the Wairau River label in 1991 by contract-winemaker John Belsham, and all of the

vintages to date have been of exemplary quality, particularly the tropical-accented Sauvignon Blanc. Exports to Australia (all States), Canada, US and UK.

Wairau River Riesling

You might think that there isn't much that can be done with conventionally fermented Riesling, but there is; or, at least, there is in the vineyard. This wine represents the result of three separate hand pickings through the vineyard, not to deal with disease problems, but simply to give a range of flavour and structure profiles.

🍷🍷🍷🍷 **1998** Light yellow-green; the bouquet has sweet lime juice fruit aromas, the palate being quite rich and full, but a fairly soft profile. Perhaps the last picking was very late. **rating:** 86

best drinking 1999–2001 **best vintages** NA **drink with** Oil-marinated baked aubergine • $16.95

Wairau River Sauvignon Blanc

The Sauvignon Blanc accounts for 8000 cases of the total Wairau River production, and is exported to both the UK and Australia with great success. The consistency of the wine over the '91 to '98 vintages shows why, if you except the very difficult '95 vintage.

🍷🍷🍷🍷🍷 **1998** Medium yellow-green; the bouquet is fresh and crisp and moderately intense giving hints of the very attractive sweet peach and passionfruit flavours of the palate. Not flabby or hot, and will very likely develop well over the next year or so. **rating:** 90

best drinking 2000–2003 **best vintages** '91, '92, '93, '94, '96, '97, '98 **drink with** Deep-fried calamari • $19

Wairau River Sauvignon Blanc Reserve

A barrel-fermented version, also utilising the best and ripest fruit. I am not always convinced by oaked Sauvignon Blancs, but it is hard to deny this wine.

🍷🍷🍷🍷🍷 **1997** Medium to full yellow-green; there are many layers on the bouquet, deriving largely from the clever use of barrel-ferment, but also from the fruit. A particularly rich wine on the palate in which the oak works well, adding particular interest to the finish with a twist of spice. **rating:** 91

best drinking 1999–2002 **best vintages** NA **drink with** Pan-fried Shanghai dumplings • $23

Wairau River Botrytised Riesling Reserve (375 ml)

An extraordinary wine of great intensity and style, irresistible at any price.

🍷🍷🍷🍷🍷 **1997** Medium to full yellow-green; the bouquet is flooded with exceptionally intense lime and cumquat aromas, the palate gorgeously rich, melting and totally seductive, with a mix of cumquat, honey, acidity and lime all coalescing. **rating:** 94

best drinking 2000–2010 **best vintages** '97 **drink with** Fruit tart • $40

walker estate NR

Puruatanga Road, PO Box 124, Martinborough, Wairarapa **region** Wairarapa

ph (06) 306 9615 **fax** (06) 306 9615 **open** Not

winemaker James Walker, Chris Lintz (Contract) **prod.** NA **est.** 1988

prod. range ($14.50–22 CD) Riesling, Rosé, Notre Vigne.

summary The Walker family (Liz, Brendan and son James) established what they believe to be a two-variety vineyard, riesling and shiraz, in 1988. Until 1993 the grapes were sold to other Martinborough winemakers, but since that time have been vinified under the Walker Estate label. The intriguingly named Notre Vigne (our vine) stems from the fact that the vines thought to be shiraz are in fact of an as-yet unidentified variety, with DNA testing failing to reveal the answer. The wines made from the mystery grape are extremely powerful, densely coloured and most unusual.

walnut ridge ★★★★

159 Regent Street, Martinborough **region** Wairarapa
ph (06) 306 9323 **fax** (06) 306 9323 **open** 7 days 11–5 Dec–Jan , weekends 11–5 Feb–Nov
winemaker Bill Brink **prod.** 1000 **est.** 1986
prod. range ($17–42 CD) Sauvignon Blanc, Botrytised Sauvignon Blanc, Pinot Noir, Cabernet Sauvignon.
summary While Bill Brink produces both Pinot Noir and Cabernet Sauvignon, he falls on the Pinot Noir side of the argument so far as Martinborough is concerned. That view, mind you, is not so surprising when you find that the first release from Walnut Ridge was the 1994 Pinot Noir which was awarded a silver medal at the 1995 Air New Zealand Wine Awards. As to the rest, I should quote Bill Brink, who says that he 'came to New Zealand via the somewhat circuitous route of Samoa and the Peace Corps in 1973. After a number of years in public service and doing the Dominion crossword, and a change-of-pace year at Victoria University deliberating the obfuscatory logic of "existential deterrence", I came to Martinborough and began the development of what has become Walnut Ridge'.

Walnut Ridge Pinot Noir

After a few disappointing vintages, started to show the character and quality one expects from Martinborough with both the '96 and '97 vintages.
🍷🍷🍷🍷 **1997** Medium red-purple; very ripe and very sweet plummy fruit floods the bouquet, and the palate follows on in the same lush, opulent, ripe plummy fruit style. However, more is not always best. **rating** 85
best drinking 1999–2002 **best vintages** '96 **drink with** Fresh salmon • $42

west brook winery ★★★☆

34 Awaroa Road, Henderson **region** Auckland Area
ph (09) 838 8746 **fax** (09) 838 5021 **open** Mon–Sat 9–6, Sun 12–5
winemaker Anthony Ivicevich **prod.** 10 000 **est.** 1937
prod. range ($11.90–21.90 CD) Blue Ridge Sauvignon Blanc, Sauvignon Blanc Semillon, Semillon, Barrique Fermented Chardonnay, Chenin Blanc, Traminer Riesling, Cabernet Sauvignon, Cabernet Merlot, Henderson Merlot (with increasing quantities sourced from Hawke's Bay and Marlborough).
summary West Brook is in the process of building a new winery complex in Ararimu Valley Road, Waimauku, near Auckland. Winery operations will be relocated to this site around November 1999, and the adjoining land will be planted to vines. Exports to the UK and Canada.

whitehaven wine company NR

1 Dodson Street, Blenheim, Marlborough **region** Marlborough
ph (03) 577 8861 **fax** (03) 577 8868 **open** 7 days 9–5
winemaker Simon Waghorn **prod.** 25 000 **est.** 1993
prod. range ($13.45–19 R) Riesling, Festival Riesling, Sauvignon Blanc, Barrel Fermented Sauvignon Blanc, Chardonnay, Pinot Noir; also Single Vineyard Reserve Riesling, Gewurztraminer, Pinot Gris, Noble Riesling.
summary Whitehaven is a joint venture between Greg and Sue White, and winemaker Simon Waghorn. Waghorn qualified as a winemaker at Roseworthy in Australia, first becoming assistant winemaker at Cooks Wines and thereafter spending five years as senior winemaker at Corbans Gisborne winery, where he is responsible for production of a string of gold medal winning wines. A 200-tonne state-of-the-art winery has been built, which includes a restaurant and wine shop as part of the complex. As one would expect, the initial releases under the Whitehaven label have had great show success. The business has grown dramatically, and the wines are now exported to Australia, the UK and the Netherlands.

william hill winery NR

Dunstan Road, RD1, Alexandra **region** Otago
ph (03) 448 8436 **fax** (03) 448 8434 **open** Mon–Sun 10–4.30
winemaker Gerry Rowland, David Grant **prod.** 1100 **est.** 1982
prod. range ($16–25 CD) Riesling, Gewurztraminer, Chardonnay, Pinot Noir.
summary Notwithstanding that the William Hill vineyards extend to 9 hectares, production grew painfully slowly in the early years. A new winery was commissioned for 1995 (happily an exceptional vintage for Central Otago) which offers contract-winemaking services for other wineries in the region.

winslow wines NR

Princess Street, Martinborough **region** Wairarapa
ph (06) 306 9648 **fax** (06) 306 9271 **open** 7 days 10–6
winemaker Ross Turner **prod.** 650 **est.** 1987
prod. range ($15–32 CD) Riesling, Sauvignon Blanc, Chardonnay, Cabernet Sauvignon Franc, Reserve Cabernet Sauvignon Franc, Petra Cabernet Sauvignon.
summary The Bio-Gro™ managed estate plantings of 2.2 hectares are devoted to cabernet sauvignon (75%), cabernet franc (15%) and merlot (10%); the riesling and chardonnay are contract-grown. It is with the Cabernet Franc that Winslow's aspirations rest.

wither hills vineyards ★★★★☆

c/o 172 Hepburn Road, Henderson, Auckland **region** Marlborough
ph (09) 836 0129 **fax** (09) 836 3282 **open** Not
winemaker Brent Marris **prod.** 600 **est.** 1992
prod. range ($18–25 ML) Chardonnay, Sauvignon Blanc.
summary The family venture for Delegat's winemaker Brent Marris, who has established a 16-hectare vineyard in Marlborough, selling most of the fruit to Delegat's, and making a small portion under the evocative Wither Hills brand, which takes its name from the range of hills to the south of Marlborough's Wairau Valley. Sumptuous gooseberry Sauvignon Blanc and delicately textured Chardonnay are the outcome.

woodfield estate NR

57 Duncan Road, Hamilton **region** Waikato and Bay of Plenty
ph (07) 827 7170 **fax** (07) 827 7140 **open** Tues–Sun 10–5
winemaker Brian Mahoney **prod.** 750 **est.** 1994
prod. range ($14–20 CD) Chardonnay, Cabernet Merlot.
summary June and Brian Mahoney have established a small winery and cellar-door facility in an architect-designed farmhouse style, and use natural winemaking methods (minimal additives, minimal filtration, no stabilisation) in handling the Waikato-grown grapes they use to make their wines.